Monica Heilbron · Umberto G. Cordani
Fernando F. Alkmim
Editors

São Francisco Craton, Eastern Brazil

Tectonic Genealogy of a Miniature Continent

Springer

Editors
Monica Heilbron
Universidade do Estado do Rio de Janeiro
Maracanã, Rio de Janeiro
Brazil

Umberto G. Cordani
Universidade de São Paulo
São Paulo
Brazil

Fernando F. Alkmim
Escola de Minas
Universidade Federal de Ouro Preto
Ouro Preto
Brazil

ISSN 2364-6438 ISSN 2364-6446 (electronic)
Regional Geology Reviews
ISBN 978-3-319-01714-3 ISBN 978-3-319-01715-0 (eBook)
DOI 10.1007/978-3-319-01715-0

Library of Congress Control Number: 2016953009

 Cover Illustration: Panoramic view of the interior plateau of the northern São Francisco craton known as the Chapada Diamantina. The cliffs and mesas expose the Mesoproterozoic Tombador Sandstone (Photo kindly offered Ricardo Fraga Pereira).

Printed on acid-free paper

This Springer imprint is published by Springer Nature
The registered company is Springer International Publishing AG
The registered company address is: Gewerbestrasse 11, 6330 Cham, Switzerland

Foreword

The São Francisco Craton is one of the five cratons that comprise the Brazilian shield and is also a key component of the Gondwanan supercontinent. Located in eastern South America, it is exposed over more than 1 billion square kilometers. Its geology and that of its margins encompasses more than 3 billion years and has provided us with many insights into the Earth's evolution during that time interval. The craton comprises several blocks of Archean basement, separated by orogenic belts of Paleoproterozoic to Early Paleozoic age. These orogenic belts contain many economically important mineral deposits and are a significant contributor to the mining industry in Brazil.

This book provides a state of knowledge overview and synthesis on the geology of the cratonic basement (i.e., prior to 1.8 Ga, Chaps. 3–5), the intracratonic basins and mafic dyke swarms within the craton (Chaps. 6–9), followed by the Brasiliano orogenic belts that flank the craton on all sides (Chaps. 10–15), and finally by an overview and synthesis (Chaps. 16 and 17). These chapters not only document the geology of the craton itself, they also provide insights into global-scale processes that governed its evolution.

This book begins with an overview of the geology (Chap. 1, Heilbron et al.) and geophysics (Chap. 2, Assumpção et al.) craton and its margins. These overview chapters include some spectacular figures that together provide the foundation for understanding the detailed geology that follows.

Chapter 3 (Teixeira et al.) describes the Archean TTG gneisses, granitoids, and greenstones belts and Chap. 4 (Barbosa et al.) the Eastern Bahia orogenic domain which consists of Archean blocks sutured at ca. 2.0 Ga. Originally thought to reflect a portion of the southern São Francisco Craton affected by the 2.2–1.9 Ga Transamazonian orogeny, Chap. 5 (Alkmim et al.) interprets the geology of the Mineiro belt as a Paleoproterozoic intraoceanic arc system of juvenile granitoids and associated volcano-sedimentary rocks that accreted to the craton at about 2.0 Ga. The documentation of these events contributes to our understanding of global-scale collisional and accretionary orogenic processes that attended the amalgamation of a supercontinent (variously known as Nuna or Columbia) in the late Paleoproterozoic.

Chapters 6 and 7 document the evolution of long-lived (ca. 1.8–0.55 Ga) intracratonic basins that nucleate above the Archean and Paleoproterozoic basement. These basins preserve an important archive of the Mesoproterozoic and Neoproterozoic evolution of the craton. According to most paleocontinental reconstructions, the craton occupied a peripheral position along the Rodinian supercontinent at ca. 1.0 Ga, but an interior position within Gondwana by the Early Paleozoic. The Paramirim Aulacogen (Chap. 6, Cruz et al.) is interpreted as a rift-sag basin, and the age of its inversion is controversial. The São Francisco Basin (Chap. 7, Reis et al.) is, from Mesoproterozoic to Early Neoproterozoic, a rift-sag succession that grades laterally into a passive margin sequence developed along the Western São Francisco plate. Ediacaran sequences are interpreted as foreland basin successions that reflect the developing Brasiliano orogenies and the assembly of West Gondwana. Chapter 8 (Girardi et al.) provides an overview of the geology of various mafic dyke swarms that range from Neoarchean to Neoproterozoic in age and provide a window into various processes affecting the mantle source during that time interval.

The connection between the São Francisco Craton and the Congo Craton of Western Africa was sundered by the opening of the South Atlantic during the fragmentation Pangea. Chapter 9 (Gordon et al.) explores stratigraphic and structural evolution of several Lower Cretaceous basins, emphasizing the cratonic nature of the lithosphere on which they nucleated.

The late Neoproterozoic to Early Cambrian time interval is one of the dramatic changes in Earth's evolution. Paleocontinental reconstructions for this interval are dominated by the amalgamation of the supercontinent Gondwana by the convergence between various cratons, whose own evolution can be traced back to the Archean and Paleoproterozoic. Tectonic events that accompanied Gondwanan amalgamation were accompanied by profound changes in the evolution of the biosphere, dramatic "freeze-fry" swings in climate, and possibly by intervals of significant true polar wander. There is an emerging body of evidence that all these features may be related.

The Brasiliano orogenic belts (Chaps. 10–15) provide a most detailed record of events leading to the amalgamation of West Gondwana, from the development and subsequent foundering of Early Neoproterozoic passive margin sequences, the arc successions that record convergence, to the accretion of these arcs and collision between arcs, microcontinents, and the component cratons of Gondwana during the Ediacaran. The 1200-km-long Brasília Belt along the Western flank (Chaps. 10 and 11, Valeriano et al., Fuck et al.) of the craton reflects the formation and subsequent (ca. 900–540 Ma) accretion of several arcs and microcontinents to São Francisco Craton during convergence and collision between the Amazonian and São Francisco paleocontinents. The Rio Preto and Riacho do Pontal belts (Chap. 12, Caxito et al.) together form a 600-km-long orogenic belt developed along the northwestern and northern margins of the craton, whereas the Sergipano belt (Chap. 13, Oliveira et al.) formed between the São Francisco Craton and the Borborema province in the north. The Araçuaí belt (Chap. 14, Alkmim et al.) along southeastern margin of the São Francisco Craton records the subduction and Ediacaran–Cambrian closure of Adamastor Ocean during the amalgamation of West Gondwana. The Ribeira belt along the southern flank of the São Francisco Craton (Chap. 15, Heilbron et al.) resulted from accretionary collisional episodes during the Ediacaran and Cambrian along the margin of the São Francisco-Paranapanema continent, which amalgamated at ca. 640–620 Ma.

Chapter 16 (D'Agrella-Filho and Cordani) provides an overview of the paleomagnetic evidence for the drift of the São Francisco-Congo craton and its changing relationship with neighboring cratons from its amalgamation in the Paleoproterozoic to its breakup during the Lower Cretaceous opening of the South Atlantic Ocean. Chapter 17 (Cordani et al.) provides a big picture overview of the São Francisco Craton in the context of global-scale tectonics as well as some of the abiding controversies about tectonic processes that affect interpretations of its evolution.

J. Brendan Murphy
Department of Earth Sciences
St. Francis Xavier University Antigonish, NS, Canada

Contents

Part I
Overview

The São Francisco Craton and Its Margins

1

Monica Heilbron, Umberto G. Cordani, and Fernando F. Alkmim

Abstract

The Brazilian shield, land surface expression of the Precambrian nucleus of South America, exposes cratons and a network of Neoproterozoic orogenic belts. The cratons correspond to internal parts of plates that amalgamated to form West Gondwana by the end of the Neoproterozoic and beginning of the Paleozoic. The Neoproterozoic Brasiliano belts, on the other hand, encompass the margin of those plates and accreted terranes. As one among five old and differentiated components of the South American lithosphere, the São Francisco craton (SFC) of southeastern Brazil represents the most intensively studied Precambrian terrain of the continent. This chapter contains introductory information on the SFC, which display attributes typical of the ancient lithosphere and hosts a rock record that spans from the Paleoarchean to the Cenozoic. Together with its bounding orogenic belts the SFC can be viewed as continent within a continent or a continent in miniature. Here, we present a panoramic view of the craton and its marginal belts, briefly discuss the history of its definition and delimitation, and conclude with an outline of the present book.

Keywords

Craton • South American platform • Brasiliano event • São Francisco craton • Neoproterozoic

1.1 Introduction

The South American continent consists of a Precambrian nucleus bounded on the west and south by two Phanerozoic accretionary provinces, the Patagonian platform and the Andean chain, respectively. The old and relatively stable portion of the continent, known as the South American platform (Fig. 1.1), is the portion of the Precambrian nucleus that escaped the effects of the Andean orogenies (Almeida et al. 1981, 2000; Schobbenhaus and Brito Neves 2003). A large part of the South American platform is covered by Phanerozoic sedimentary units, which fill large Paleozoic sags, the equatorial and eastern continental margin basins, as well as sub-Andean depocenters and Paleogene rifts. Exposures of the Precambrian basement characterize the Guianas, Central Brazil and Atlantic shields (Almeida et al. 1981, 2000; Brito-Neves 2002; Schobbenhaus and Brito Neves 2003) (Fig. 1.1).

The shield areas of South America, collectively referred to as the Brazilian shield, comprise two distinct lithospheric types: cratons and Brasiliano orogenic systems (Fig. 1.2). Inherited from the residence of South America in Gondwanaland, cratons and Brasiliano systems were shaped during the amalgamation of the supercontinent, as various

M. Heilbron (✉)
Faculdade de Geologia, Tektos Research Group, Universidade do Estado Rio de Janeiro (UERJ), Rua São Francisco Xavier, 524, Bloco A-4020, Rio de Janeiro, RJ 20.550-900, Brazil
e-mail: monica.heilbron@gmail.com

U.G. Cordani
Instituto de Geociências, Universidade de São Paulo, Rua do Lago, 562, São Paulo, SP 05508-080, Brazil

F.F. Alkmim
Departamento de Geologia, Escola de Minas, Universidade Federal de Ouro Preto, Morro do Cruzeiro, Ouro Preto, MG 35.400-000, Brazil

M. Heilbron et al. (eds.), *São Francisco Craton, Eastern Brazil*, Regional Geology Reviews, DOI 10.1007/978-3-319-01715-0_1

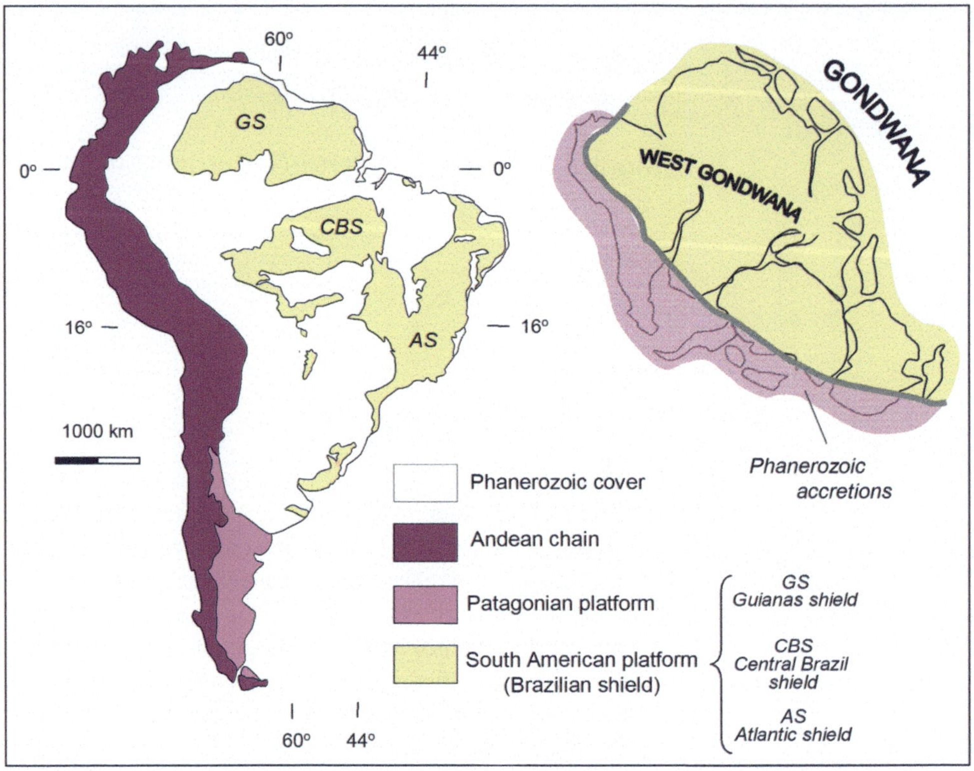

Fig. 1.1 Tectonic subdivision of South America, showing the South America platform and its land surface expression, the Guianas (*GS*), Central Brazil (*CBS*) and Atlantic (*AS*) shields. As a portion of West Gondwana, the Precambrian nucleus of South America received the accretions of Patagonia (Patagonian platform) as well as arcs and terranes that form the Andean chain [based on Almeida et al. (1981)]

plates converged and collided during Ediacaran/Cambrian times (Brito-Neves et al. 1999; Cordani and Sato 1999; Campos Neto 2000; Almeida et al. 2000; Alkmim et al. 2001; Schobbenhaus and Brito Neves 2003) (Fig. 1.2).

According to the traditional definition, cratons are portions of continents, which have attained and maintained tectonic stability at least since the beginning of the Phanerozoic. In other words, pieces of the continental crust not involved in Phanerozoic orogenic processes (Bates and Jackson 1980; Brito-Neves and Alkmim 1993; see Sengör 1999 for a comprehensive review). Following this definition, cratons have been identified and delimited—not necessarily using the same criteria—in all continents, South America included. Geophysical, geochemical and petrological studies conducted in recent years consolidated the modern view of cratons as differentiated components of the continental lithosphere. Their most remarkable attributes are buoyant mantle keels, i.e., anomalously thick (in excess of 370 km) and relatively less dense lithospheric roots detected in many of their Archean nuclei (Jordan 1978, 1988; Forte and Perry 2001; Artemieva and Mooney 2001; Kaban et al. 2003; Eaton et al. 2009; McKenzie et al. 2015). Low heat flow (Pollak and Chapman 1977; Grand 1987; Dawson 1980) and surface expression as low lands are other typical features of the cratons, which as cold and stronger lithospheric segments exhibit differential behavior in tectonic processes.

Five cratons have been recognized in the South America platform: the Amazonian, São Luis, São Francisco, Paranapanema and Rio de la Plata (Fig. 1.2) (Almeida et al. 1981, 2000; Schobbenhaus and Brito Neves 2003). Defined as crustal pieces not significantly affected by the Neoproterozoic Brasiliano collisional deformation and metamorphism (Almeida et al. 1981), the South American cratons correspond to internal portions of the plates that took part in the assembly of West Gondwana (Figs. 1.1 and 1.2) (South America and Africa) (Cordani and Sato 1999; Campos Neto 2000; Alkmim et al. 2001; Brito-Neves 2002). They underlie the low lands of the continent, hosting large Paleozoic sag basins and the main river systems.

The Brasiliano orogenic systems and their counterparts in the African continent—the PanAfrican belts—encompass plate margins, micro-continents and intra-oceanic magmatic arcs also involved in the assembly of West Gondwana between 640 and 500 Ma (Campos Neto 2000; Almeida et al. 2000) (Fig. 1.2). Surrounding the cratons, Brasiliano systems form a network of interfering orogens (Fig. 1.3), whose morphological expressional are the Brazilian highlands (Brito-Neves et al. 1999; Campos Neto 2000; Almeida et al. 2000; Schobbenhaus and Brito Neves 2003; Fuck et al. 2008) (Fig. 1.1).

As relicts of the ancient lithosphere and host of large sedimentary basins, cratons preserve a substantial part of the Earth memory. Their bounding orogenic belts, on the other hand, incorporate pre- to post-collisional rock assemblages that in general record processes occurring over a limited time span. Because of their topographic relief, mountain belts

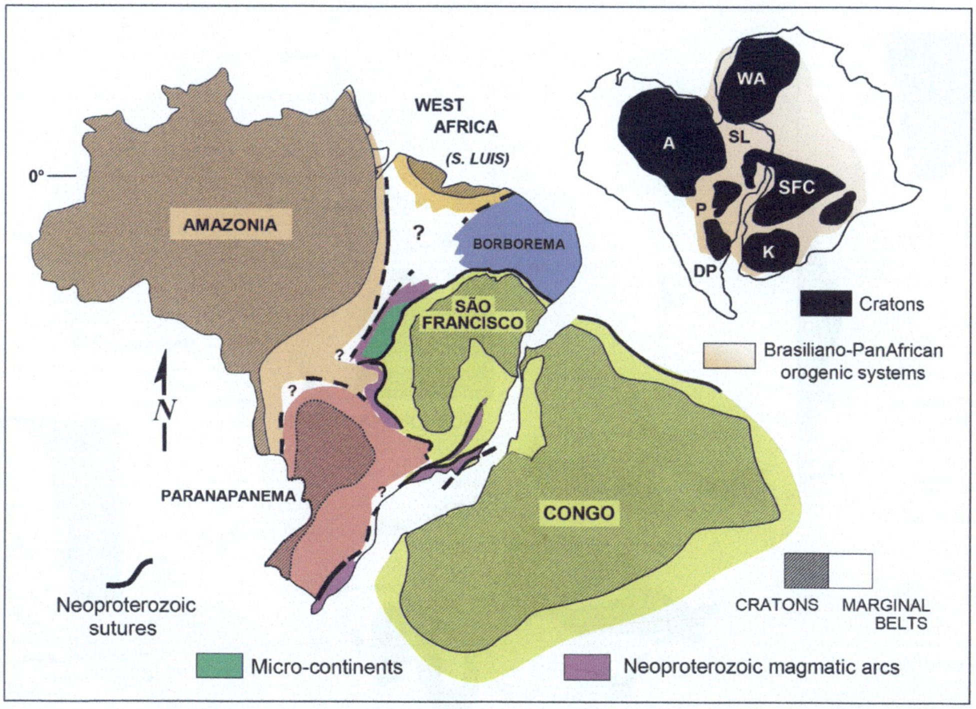

Fig. 1.2 Cratons and Brasiliano/PanAfrican orogenic systems of West Gondwana (South America and Africa) and the geotectonic framework of the Brazilian territory, which encompasses almost the whole Precambrian nucleus of South America. Cratons of South America and Africa (*SFC* São Francisco-Congo, *K* Kalahari, *WA* West Africa, *A* Amazonian, *P* Paranapanema, *DP* Rio de la Plata), correspond to the internal portion of plates that collided during the assembly of West Gondwna by the end of the Neoproterozoic. Mapped and inferred Neoproterzoic sutures bound the Amazonia, West Africa, Borborema, São Francisco-Congo and Paranapanema paleoplates. The margins of these plates, together with magmatic arcs and micro-continents form the Brasiliano orogenic systems. Note that the São Luis craton in northern Brazil represents a fragment of the West Africa left in South America. Tthe São Francisco craton was linked to the Congo craton and the Brasiliano orogenic domain between them corresponds to the Araçuaí-West Congo confined orogen (AWCO). Modified after Alkmim and Martins-Neto (2005)

developed along craton margins provide, however, access to rock successions not exposed in the low lands of the adjacent cratons. The combination of geologic information obtained in cratonic domains and their marginal belts form thus a knowledge basis representative of long periods of the Earth history.

The São Francisco craton (SFC) and its margins exposed in southeastern Brazil (Figs. 1.2 and 1.3) represent the most intensively studied portion of the Precambrian nucleus of the South American plate. Furthermore, the SFC hosts a rock record that spans from the Paleoarchean to the Cenozoic and several Precambrian sedimentary successions that witness Earth processes of global significance. With these and many other attributes, the SFC together with its fringing orogenic belts can be viewed as continent within a continent or a continent in miniature.

This chapter aims to provide the reader with introductory information on the SFC and its margins, as well as on the overall organization of the present book. In the next three sections, we present a panoramic view of the craton and its marginal belts, briefly discuss the history of its definition and delimitation, and conclude with an outline of the book content.

1.2 Geologic Outline of the São Francisco Craton and Its Margins

Surrounded by the southeastern Brazilian highlands and drained by the São Francisco river system, the SFC exhibits the shape of a horse's head, with maximum length and width of ca. 1100 and 900 km, respectively (Fig. 1.4).

The basement of the SFC comprises rock units older than 1.8 Ga (Almeida 1977). Archean TTG-gneisses, granitoids and greenstones belts together with Paleoproterozoic plutons and supracrustal successions are the main lithological assemblages of the basement exposed in its southern tip and northeastern lobe (Teixeira et al. 2000; Barbosa and Sabaté 2004; Alkmim and Martins-Neto 2012). These assemblages form a NS-oriented Archean stable block, bounded on the east by two segments of a Paleoproterozoic orogen (Fig. 1.4b, c). In the southern end of the craton, the

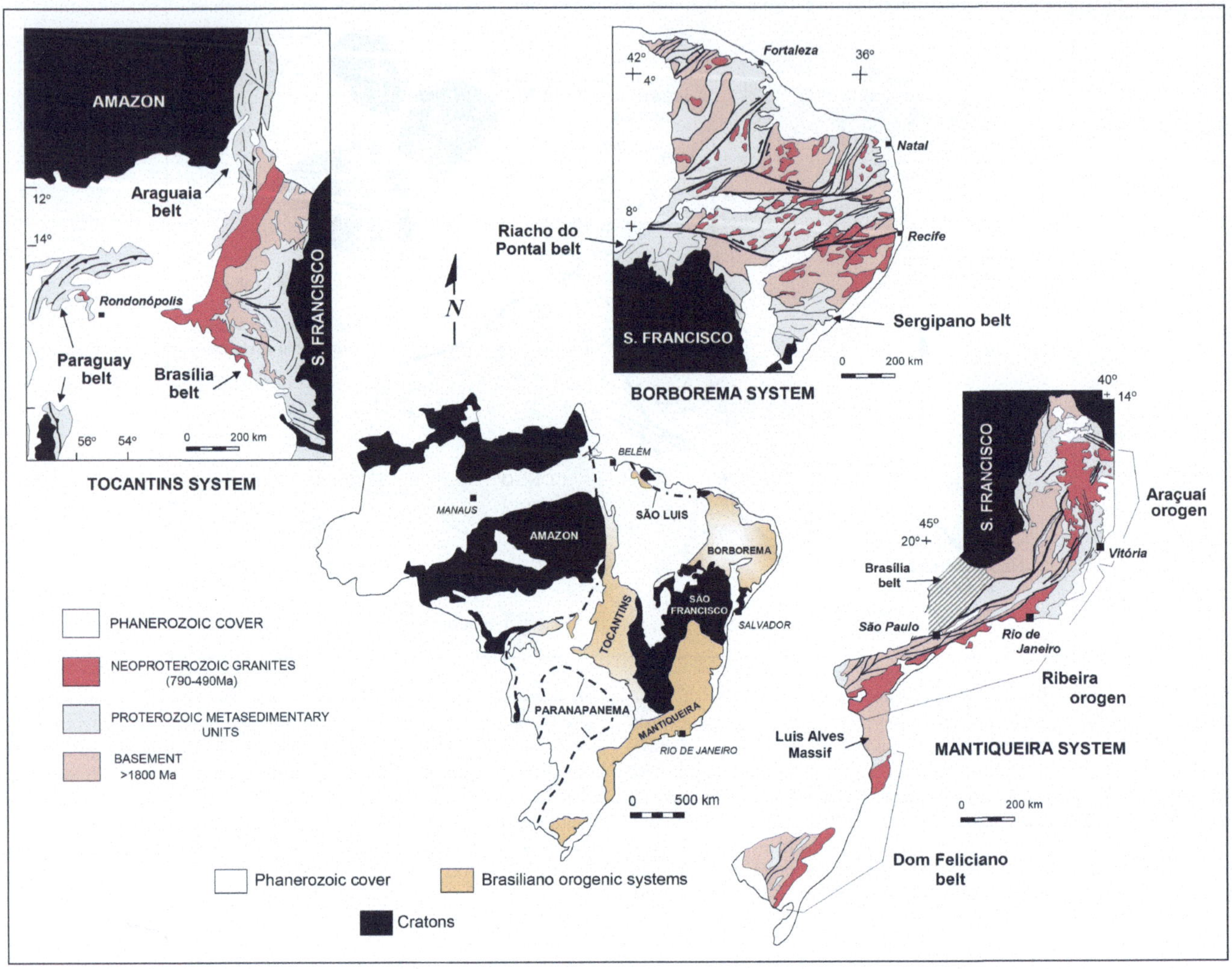

Fig. 1.3 Brasiliano orogenic systems exposed in the Brazilian territory. The Tocantins system, developed between the Amazonian, Paranapanema and São Francisco cratons, comprises the Araguaia, Paraguai and Brasília belts. The Borborema strike-slip province includes the Rio Preto, Riacho do Pontal and Sergipano belts that bound the São Francisco craton to the north. The Mantiqueira system comprises the Araçuaí, Ribeira and Dom Feliciano belts. The northern end of the Mantiqueira system, encompassing the Araçuai orogen in Brazil and the West Congo belt in Africa, is confined to an embayment located between the São Francisco and Congo cratons, as shown on Fig. 1.2. Reproduced from Alkmim (2015) with permission from Springer

Paleoproterozoic orogenic belt encompasses the region of the Quadrilátero Ferrífero (Iron Quadragle) mineral province and the Mineiro belt (Alkmim and Marshak 1998; Teixeira et al. 2015). The Eastern Bahia orogenic domain in the northeastern portion of the craton is made up of a mosaic of various Archean blocks sutured at around 2.0 Ga (Barbosa and Sabaté 2004). Its counterpart in the reconstructions of West Gondwana corresponds to the West Cental African belt exposed in Gabon (Feybesse et al. 1998) (Fig. 1.4b).

The cratonic cover is made up of units younger than 1.8 Ga (Almeida 1977) and occurs in three distinct morphotectonic domains of the craton interior: the São Francisco basin, the Paramirim aulacogen (Cruz and Alkmim 2004; Alkmim and Martins-Neto 2012), and the Recôncavo–Tucano–Jatobá rift (Karner et al. 1992; Milani et al. 1988; Milani and Thomas-Filho 2000) (Fig. 1.4c). Besides Proterozoic sedimentary successions, the São Francisco basin also contains Phanerozoic units (Permo-Carboniferous and Cretaceous rocks) absent in the Paramirin aulacogen and Rio Pardo basin. The Recôncavo–Tucano–Jatobá rift is filled mainly by Upper Jurassic and Lower Cretaceous sediments.

The SFC is fringed on all sides by Brasiliano orogenic belts, except for the Atlantic coast, where the it underlies the Camamu-Almada and Jacuípe basins at the continental margin (Chang et al. 1992) (Figs. 1.4 and 1.5). The Brasiliano belts involve the following lithostratigraphic assemblages in their external sectors: (1) basement complexes, which represent extensions of the SFC crust, bounded by Neoproterozoic sutures (Fig. 1.5); (2) Paleo- to

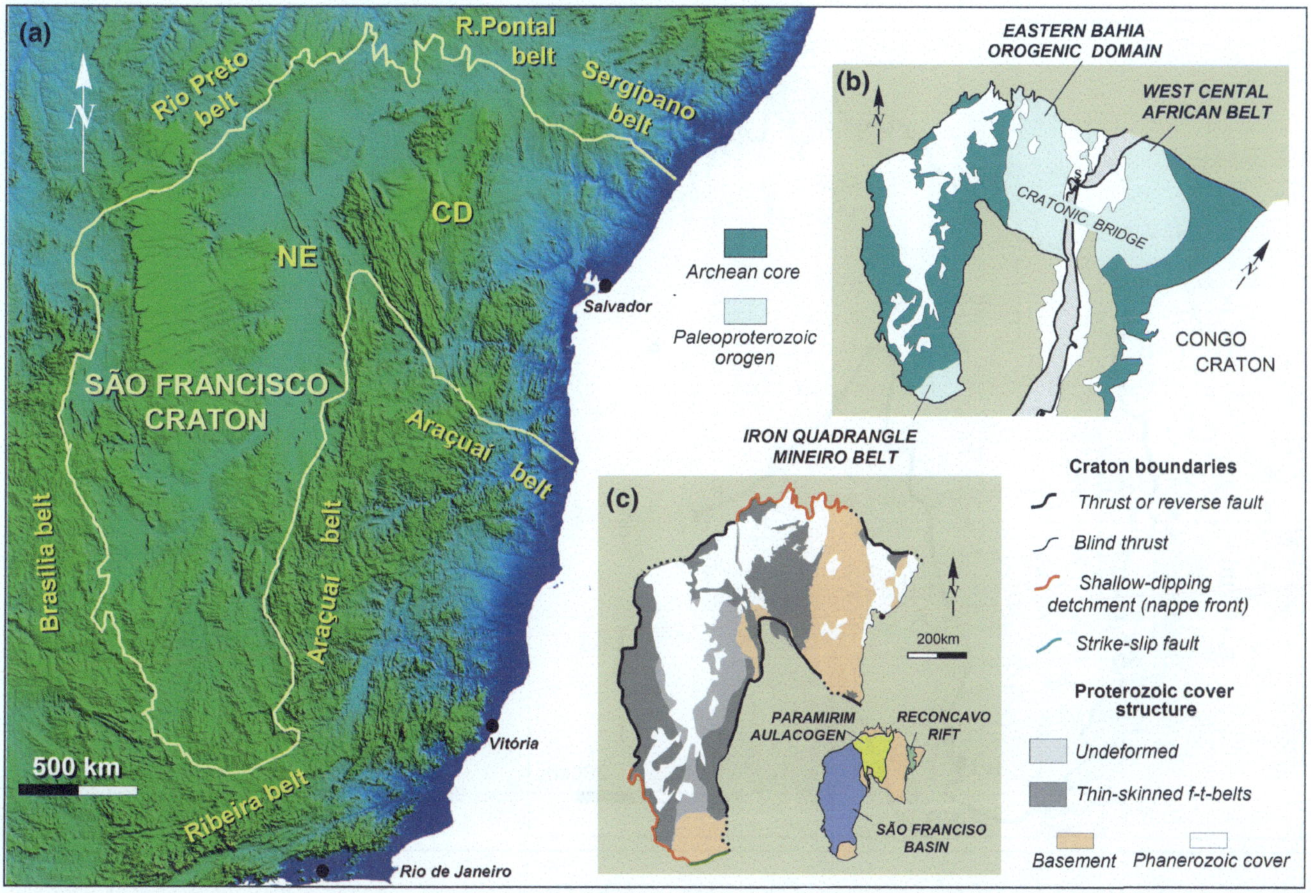

Fig. 1.4 a Digital elevation model of southeastern Brazil, showing the topography associated with the SFC and its marginal belts. Morphotectonic provinces of the craton interior: *NE* northern Espinhaço range, *CD* Chapada Dimantina. **b** The São Francisco and Congo cratons in a reconstruction of West Gondwana, showing their Achean nuclei and Paleoproterozoic orogenic zones [modified from Alkmim and Martins-Neto (2012)]. **c** Nature of the SFC boundaries and structure of the Proterozoic cover units [based on Alkmim et al. (2013)]

Meso-proterozoic (1780–1000 Ma) rift-sag sedimentary sequences; (3) Neoproterozoic (Tonian–Cryogenian) rift to passive margin successions; (4) Neoproterozoic (Cryogenian–Ediacaran) syn-orogenic and subordinate ophiolitic assemblages; (5) syn-collisional granites. These units are metamorphosed under greenschist to granulite facies conditions and involved in fold-thrust or nappe systems, which verge towards the craton. The internal zones comprise a variety of lithological assemblages. The most common are: (1) Neoproterozoic arc-related granitoids and associated volcanosedimentary successions; (2) Archean and Paleoproterozoic complexes representing accreted blocks (micro-continents); (3) syn-, late- and post-collisional granites emplaced in Ediacaran and Cambrian periods.

The long and wide Brasília belt (Pimentel et al. 2000, 2004; Alvarenga et al. 2000; Dardenne 2000; Valeriano et al. 2004), representing the eastern half of the Brasiliano Tocantins orogenic system, occupies the southwestern and western margins of the SFC, whereas the Rio Preto, Riacho do Pontal and Sergipano belts define its northern boundary and root in the giant Borborema strike-slip system of northeastern Brazil (Vauchez et al. 1995; Brito-Neves et al. 2000; Caxito et al. 2014a, b; Oliveira et al. 2010) (Figs. 1.3 and 1.5). Curving along the eastern margin of the craton, the Araçuaí belt corresponds to the external zone of the Araçuaí-West Congo orogen (AWCO) developed as an enclave between the SFC and its African counterpart, the Congo craton (Pedrosa-Soares et al. 2001; Alkmim et al. 2006) (Figs. 1.2 and 1.5). The Araçuaí belt is continuous to the companion Ribeira belt, which borders the southern tip of the craton (Trouw et al. 2000; Heilbron et al. 2004) (Figs. 1.3 and 1.5).

The link between the São Francisco and Congo cratons in West Gondwana was made by a crustal bridge (Figs. 1.2 and 1.4) (Porada 1989), located between the Sergipano-Oubanguides system, on the north and the AWCO, on the south. This bridge, kept intact until the dispersal of West Gondwana in the Lower Cretaceous, preserved a complete segment of the Paleoproterozoic orogen generated at around 2.0 Ga, as the Archean nuclei of the

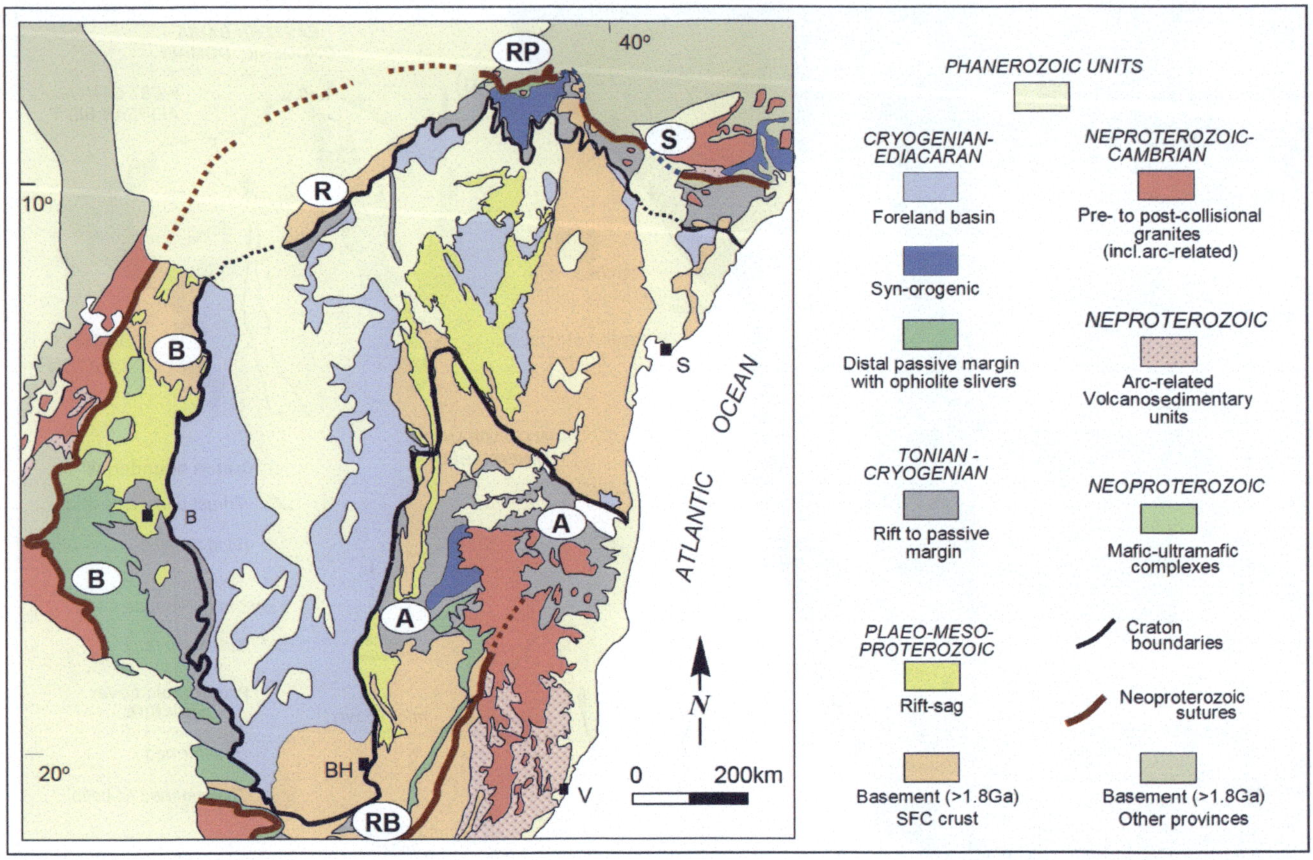

Fig. 1.5 Simplified geologic map of the São Francisco craton and adjacent Brasiliano belts. The craton boundaries in this map were modified from Almeida (1981) by Alkmim (2015) especially in the region where the Araçuaí belt interacts with the intracratonic Paramirim aulacogen (see text for explanation). Maps of the marginal belts compiled from: Valeriano and Fuck et al. (*B* Brasília belt); Caxito et al. (*R* Rio Preto and *RP* Riacho do Pontal belts); Oliveira et al. (*S* Sergipano belt); Alkmim et al. (*A* Araçuaí belt); Heilbron et al. (*RB* Ribeira belt), Chaps. 10–15 of this book. Cities: *B* Brasilia, *S* Salvador, *V* Vitória

cratons stitched together. The Eastern Bahia orogenic domain and the West Central African belt, now exposed, respectively, in northeastern Brazil and Gabon, correspond to the two halves of a sector of the Paleoproterozoic orogen (Fig. 1.4) (Ledru et al. 1994; Feybesse et al. 1998; Barbosa and Sabaté 2004; Thiéblemont et al. 2014).

Neoproterzoic sutures zones were recognized in some sector of the internal zones of the Brasiliano belts (see Chaps. 10–15, this book). In the majority of the cases they can be inferred with basis on geophysical data. Representing the limits of the São Francisco paleo-plate, these zones were tentatively traced in the map of Fig. 1.5.

1.3 Development of the São Francisco Craton Concept

The current image of the SFC shown on Fig. 1.5 was progressively built in the course of a long series of studies, which started with the work by D. Guimarães, who already in 1951 postulated that the Brazilian shield consisted of five relatively stable sialic blocks: the Archean Goiás-Mato Grosso massif and the ancient Guianas, Gondwana, Brazil and Africa (Fig. 1.6a), the latter representing a fragment of proto-Africa left in South America after the opening of the Atlantic. According to Guimarães, oceans ("true geosynclines") separated these blocks by the beginning of the Proterozoic Eon. Ancient Brazil and ancient Africa collided first, during the Huronian orogeny. Later, in the course of the Penokean orogeny, all blocks amalgamated following the consumption of the oceans. Guimarães′ ancient Brazil, whose margins would be marked by Penokean orogenic belts reactivated during the Caledonian orogeny, encompasses the whole NS-trending lobe of the SFC as presently delimited, extending, however, further north up to the Brazilian equatorial margin (Fig. 1.6a).

Barbosa (1966) in his tectonic map of Brazil recognized and traced the boundaries of the "Franciscan craton" (Fig. 1.6b), an Archean stable block, on which various sedimentary units accumulated in the time between the upper Precambrian to the beginning of the Paleozoic. Fringed to the east, south and west by pre-Caledonian orogenic belts,

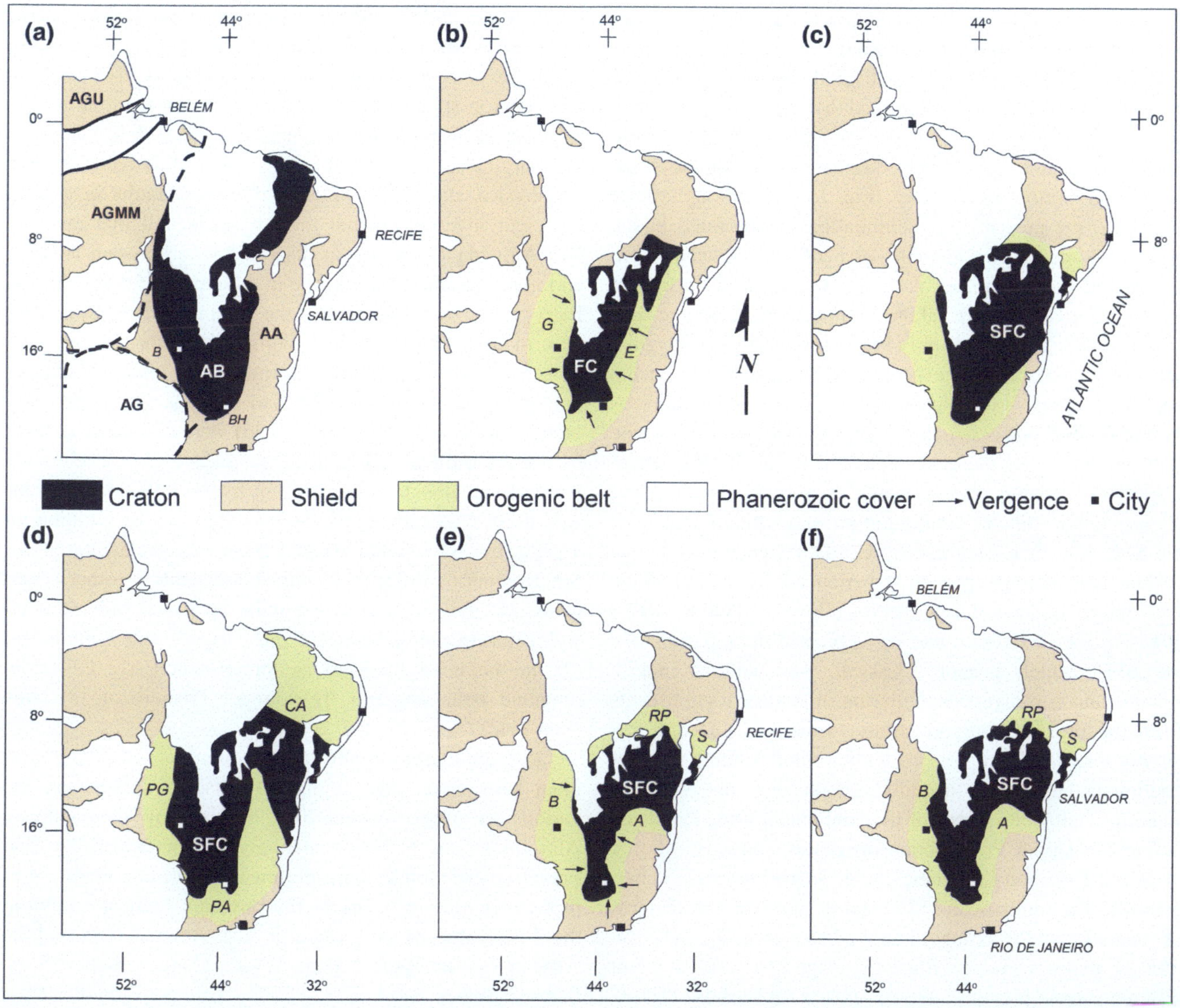

Fig. 1.6 The development history of the São Francisco craton (SFC) concept illustrated by delimitations by different authors. **a** According to Guimarães (1951), the tectonic framework of eastern Brazil included five sialic blocks: the Archean Goiás-Mato Grosso massif (AGMM), as well as ancient Guianas, Brazil, Gondwana and Africa. Ancient Brazil encompasses a substantial part of the São Francisco craton and its margins as present delimited. **b** The Franciscan craton of Barbosa (1966), fringed by the pre-Caledonian orogenic belts Goiano (*G*) and Espinhaço (*E*). **c** Almeida's (1969) tectonic map of eastern Brazil emphasizing the São Francisco craton and its bounding Upper Precambrian belts. **d** The São Francisco craton and its marginal Paraguai (PG), Cariri (CA), and Paraíba (PA) belts, according to Cordani et al. (1968). **e** The SFC and the Brasiliano Sergipano (S), Riacho do Pontal (RP), Brasilia (B), and Araçuaí belts of Almeida (1977). **f** Corrected delimitation of the SFC of Almeida (1981) (see text for explanation)

the Franciscan craton would be locally affected by "germanotype" deformation. Barbosa's delimitation roughly coincides with the covered areas of the craton interior, i.e., the São Francisco basin and Paramirim aulacogen.

After characterizing the "Assintic" orogenic system of southeastern Brazil, Ebert (1968) traced the boundaries of the associated stable foreland, which encompasses a large portion the southern lobe of the craton as presently delimited, as well as the southern segment of the Araçuaí marginal belt.

Almeida (1967, 1969) defined the SFC as one of the stable blocks of the South American platform bounded by the Late Precambrian ("Baikalian") Brasília, Sergipano and Paraíba orogenic belts, respectively, on the west, northeast and east. Interpreted by the author as an older Precambrian feature, the whole Araçuaí belt was incorporated in the cratonic domain (Fig. 1.6c).

Cordani et al. (1968) compiled the available geochronological data and presented for the first time a chronological-tectonic subdivision of the Brazilian shield.

Accordingly, the South American platform was made up of cratonic zones, stabilized at around 1.8 Ga, and metamorphic belts, generated or reactivated between 650 and 450 Ma. The SFC was portrayed by these authors as a stable crustal fragment bounded by the Late Proterozoic Paraíba, Paraguay, and Cariri orogenic belts, on the west, south and north, respectively (Fig. 1.6d). Although quite similar to the present, this delimitation incorporated in the cratonic zone a considerable part of the Araçuaí belt, interpreted by Cordani et al. as a 1.4–1.1 Ga old orogenic feature. Cordani (1973), taking into account K-Ar age data obtained by Távora et al. (1967), changed his original concept, postulating the existence two stable blocks in the region previously designated by him and collaborators as the SFC: the SFC *strictu* sensu, on the west, and the Lençóis craton on the east, separated by a NNW-oriented Upper Proterozoic "mobile belt". A long lasting and intense debate on the configuration and other features of the SFC was then triggered by Cordani's paper.

The considerable growth experienced by the geologic knowledge of Brazil during the 70s allowed Almeida (1977, 1981) to formulate new and more precise delimitations for the SFC, which became classical, not without intense debates and controversies. Differing in some critical aspects from the previous interpretations, Almeida's concept was largely based on comparisons with similar features in other continents and newly acquired geophysical data. Understanding cratons as pieces of the continental crust stabilized before the Upper Precambrian orogenies—already referred to as the Brasiliano orogenic cycle—Almeida (1977) characterized the limits of the SFC taking into account patterns of magnetic and gravimetric anomalies as well as the location of major tectonic structures (Fig. 1.6e). As a major difference from previous authors, except for Barbosa (1966), he defined the Araçuaí belt as another bounding element of the SFC.

Almeida (1981) introduced a significant modification in the western boundary of the craton (Fig. 1.6f) and after reanalyzing its tectonic evolution, postulated the existence of a SFC ancestor, the Paramirim craton, stabilized after the Lower Precambrian Transamazonian orogenic cycle. This older stable block corresponds to the western and NS-trending segment of the SFC in the present configuration.

Starting in 1977, a whole series of meetings focusing the SFC and its margins was organized in Brazil. After many discussions on various conflicting views (e.g., Cordani 1978; Braun and Baptista 1978; Trompette et al. 1992), these events consolidated the SFC concept essentially as formulated by Almeida (1977, 1981).

Alkmim et al. (1993) re-examined many aspects of Almeida's concept in the light of new data. Two main points explored by these authors concerned the tectonic meaning of the craton boundaries and deformation of its Neoproterozoic cover units. In most cases, the craton limits correspond to major reverse or thrust faults, which mark a change in deformation style: thin-skinned tectonics in the craton interior and thick-skinned (basement-involved) deformation in the marginal Brasiliano belts (Fig. 1.4c). Moreover, field and geophysical data also indicated that these faults in general represent inverted normal growth faults. All this revealed that Almeida's criterion to discriminating between cratonic and non-cratonic crust was the involvement of the Archean–Paleoproterozoic basement in the Neoproterozoic Brasiliano deformation. Furthermore, considering that the many bounding faults are inverted normal faults, this also implies that many boundaries of the São Francisco (and possibly other cratons of the Brazilian shield) were ultimately defined by the pre-orogenic extensional phase that generated basins and individualized plates, and not by the collisional tectonism itself (Alkmim et al. 1993). Thus, in two important aspects the São Francisco and other cratons delimited for South America differ from their equivalents in other continents: (1) they include cover units deformed (thin-skinned style) during the same event that shaped their boundaries; (2) are bounded by Late Neoproterozoic/Early Paleozoic orogenic belts and not by younger Phanerozoic systems (Fig. 1.4c).

Geologic evolution and tectonic significance of the marginal belts were other intensively discussed issues in the various meetings on the SFC. Addressing these topics, Alkmim et al. (1993) tentatively characterized the São Francisco paleoplate, considering the few data about Neoproterozoic suture zones in the orogenic belts surrounding the SFC. Studies performed in the Araçuaí belt and its African counterpart, the West Congo belt, (e.g., Pedrosa-Soares et al. 1992, 2001; Trompette et al. 1992; Trompette 1994; Maurin 1993; Tack et al. 2001), as well as on the Paleoproterozoic orogenic belts of the both the SF and Congo cratons (Figueiredo 1989; Ledru et al. 1994; Feybesse et al. 1998) demonstrated the effective link between the São Francisco and Congo cratons in West Gondwana, as emphasized by Porada (1989). As a consequence of that, the SFC and its margins are currently viewed as relatively small part of the São Francisco-Congo paleoplate (Fig. 1.2), individualized and incorporated into Gondwana during the Neoproterozoic.

1.4 Book Overview

This book is organized in five parts subdivided into 17 chapters. The first part discusses the SFC concept. Besides the present chapter, this first section also contains a description of the main attributes of SFC lithosphere by Assumpção and co-authors.

The second part focuses the craton's basement. The tectonic subdivision, composition and significance of units older than 1.8 Ga are addressed in three chapters. Chapter 3 by Teixeira and co-authors discusses the nature and evolution of preserved and reworked Archean nuclei. Chapters 4 and 5, respectively by Barbosa and Barbosa, and Alkmim and Teixeira are devoted to the Paleoproterozoic orogenic domains. The rock record and tectonic evolution of these domains between 2.5 and 2.0 Ga, including the collisional episode that led to the amalgamation São Francisco-Congo cratonic crust are the themes of these chapters.

The third part of the book presents the geological milestones of a ca. 1300 Ma-long period (from the Late Paleoproterozoic up to the Early Neoproterozoic) in which the São Francisco-Congo paleocontinent experienced multiple basin-forming events assisted by acid volcanism and emplacement of mafic intrusions. In addition, this section also focuses the younger Precambrian and Phanerozoic cover units of the craton. Chapters 6–8, by Cruz and Alkmim, Reis et al., and Girardi et al. deals with the Pararamim aulacogen, São Francisco basin and the multiple episodes of mafic magmatism. We also include in this part, Chap. 9 by Gordon and co-authors, which describe the Recôncavo–Tucano–Jatobá rift and the continental margin Camamu-Almada and Jacuípe basins, generated during the final break-up the São Francisco-Congo bridge in the Lower Cretaceous.

The fourth part contains six chapters devoted to the stratigraphy, overall structure and tectonic evolution of the Neoproterozoic Brasiliano that surround the SFC. The evolutionary stages shared by these belts started in the early Neoproterzoic with the development of rift to passive margins basins, coevally with the generation of magmatic arcs in subduction systems outboard of the São Francisco paleocontinent. The Ediacaran period is marked by collision of magmatic arcs, micro-continents and continents, and incorporation of São Francisco–Gongo into Gondwana. Chapters 10 and 11, respectively by Valeriano and Fuck et al., describe the tectonic evolution of the southern and northern portions the Brasília belt. Chapters 12 and 13, by Caxito et al. and Oliveira et al. explore the Rio Preto, Riacho do Pontal and Sergipano belts located along the northern border of the craton. Chapters 14 and 15, by Alkmim et al. and Heilbron et al., deal with the Araçuaí and Ribeira belts developed along the eastern and southern margins of the craton.

The last section of the book contains Chaps. 16 and 17 by D'Agrella and Cordani, and Cordani et al., respectively. These chapters synthesize the trajectory of the SFC in time and space making use of the available paleomagnetic, geotectonic and geologic data.

References

Alkmim, F.F. 2015. Geological background: A tectonic panorama of Brazil. In: Vieira et al. (eds): Landscapes and Landforms of Brazil. World Geomorphological Landscapes, Springer, p. 9–17.

Alkmim, F.F. and Marshak, S., 1998. Transamazonian Orogeny in the Southern São Francisco Craton Region, Brazil: Evidence for Paleoproterozoic collision and collapse in the Quadrilátero Ferrífero. Precambrian Research, 90, 29–58

Alkmim F.F., Martins-Neto M.A. 2005. Brazil. In: Encyclopedia of Geology, vol 1. Elsevier, Amsterdam, p. 306–327

Alkmim, F.F. and Martins-Neto, M.A., 2012. Proterozoic first-order sedimentary sequences of the São Francisco craton, eastern Brazil. Marine and Petroleum Geology, 33, 127–139.

Alkmim, F.F.; Brito Neves, B.B.; Castro Alves, J.A., 1993. Arcabouço tectônico do Cráton do São Francisco - Uma Revisão. In: Dominguez, J.M.L. & Misi, A. (eds) O Cráton do São Francisco. Salvador, SBG/Núcleo BA/SE, SGM/BA. p. 45–62

Alkmim, F.F., Marshak, S. Fonseca, M.A., 2001. Assembling West Gondwana in the Neoproterozoic: Clues from the São Francisco craton region, Brazil. Geology, v. 29, p. 319–322.

Alkmim, F.F., Marshak, S., Pedrosa-Soares, A.C., Peres, G.G., Cruz, S., Whittington, A. 2006. Kinematic evolution of the Araçuaí-West Congo orogen in Brazil and Africa: Nutcracker tectonics during the Neoproterozoic assembly of Gondwana. Precambrian Research 149, 43–64.

Almeida, F.F.M. 1967. Origem e evolução da Plataforma Brasileira. DNPM-DGM, Boletim 241, 36p.

Almeida, F.F.M. 1969. Diferenciação Tectônica da Plataforma Brasileira, Congresso Brasileiro de Geologia, Sociedade Brasileira de Geologia, Salvador, Anais, p. 29–46.

Almeida, F.F.M., 1977. O Cráton do São Francisco. Revista Brasileira de Geociências, 7, 285–295.

Almeida, F.F.M.1981. O Cráton do Paramirim e suas relações com o do São Francisco. Simpósio Sobre o Cráton do São Francisco e Suas Faixas Marginais. Salvador, 1981. SBG-Núcleo Bahia, Salvador, Anais, p. 1–10.

Almeida, F.F. M., Hasui, Y., Brito Neves, B.B., and Fuck, R.A., 1981. Brazilian structural provinces: An introduction. Earth-Sciences Reviews 17, 1–29.

Almeida, F.F.M., Brito Neves, B.B., Carneiro, C.D.R., 2000. Origin and evolution of the South American Platform. Earth-Science Reviews, 50, 77–111.

Alvarenga, C.A.V., C. Moura and P.S.S. Gorayeb 2000. The Araguaia belt. In: Cordani, U.G.; Milani, E.J.; Thomaz Fo, A.; Campos, D.A. (eds) Tectonic Evolution of South America. 31st Intern. Geol. Congr. Rio de Janeiro, p. 335–365

Artemieva, I.M., Mooney, W.D., 2001. Thermal structure and evolution of Precambrian lithosphere: a global study. Journal of Geophysical Research 106, 16387–16414

Barbosa, A.L.M., 1966. Síntese da Evolução Geológica da América do Sul. Boletim do Instituto de Geologia, Escola de Minas de Ouro Preto, 1 (2), 91–111.

Barbosa, J.S.F. and Sabaté, P. 2004. Archean and Paleoproterozoic crust of the São Francisco Craton, Bahia, Brazil: geodynamic features. Precambrian Research, 133, 1–27

Bates, R.L. and Jackson J.A., 1980. Glossary of Geology. American Geological Institute, Falls Church, VA, USA, 751 p.

Braun, O.PG. and Baptista, M.B. 1978. Considerações sobre geologia pré-cambriana da região sudeste e parte da região centro-oeste do Brasil. In: Rocha et al. (eds): Anais da reunião preapratória para Simpósio sobre o Cráton do São Francisco e suas Faixa Marginais. Sociedade Brasileira de Geologia, NBA, Publicação Especial, 3, p. 255–368.

Brito-Neves, B.B., 2002. Main stages of development of sedimentary basins of South America and their relationships of tectonics of supercontinents. Gondwana Reserarch, 5, 175–196.

Brito-Neves, B. B. and Alkmim, F.F., 1993. Cráton: A evolução de um conceito. In: Dominguez, J.M.L. & Misi, A. (Eds). O Cráton do São Francisco.Salvador, Sociedade Brasileira de Geologia, Núcleo BA/SE, p. 1–10.

Brito-Neves, B.B., Campos Neto, M.C., Fuck, R., 1999. From Rodinia to Western Gondwana: An approach to the Brasiliano/Pan-African cycle and orogenic collage. Episodes 22, 155–199

Brito Neves, BB., Santos, EJ., Van Schmus, WR., 2000. Tectonic history of the Borborema Province, Northeastern Brazil. In: Cordani, U., Milani, EJ., Thomaz Filho, A., Campos, DA. (Eds), Tetonic Evolution of South America. Proceedings of the 31 International Geological Congress, Rio de Janeiro, 151–182

Caxito, F.A., Dantas, E.L., Stevenson, R., Uhlein, A., 2014a. Detrital zircon (U-Pb) and Sm-Nd isotope studies of the provenance and tectonic setting of basins related to collisional orogens: The case of the Rio Preto fold belt on the northwest São Francisco Craton margin, NE Brazil. Gondwana Research, 26(2):741–754

Caxito, F.A., Uhlein, A., Stevenson, R., Uhlein, G.J., 2014b. Neoproterozoic oceanic crust remnants in northeast Brazil. Geology, 42 (5):387–390

Campos Neto, M.C. 2000. Orogenic Systems from Southwestern Gondwana: An approach to Brasiliano-PanAfrican Cycle and Orogenic Collage in Southeastern Brazil. In: Cordani, U.G.; Milani, E.J.; Thomaz Filho, A.; Campos, D.A. (eds.) Tectonic Evolution of South América. Rio de Janeiro, 31st International Geological Congress, Rio de Janeiro, p. 335–365.

Chang, H.K.; Kowsmann, R.O.; Figueiredo, A.M.F.; Bender, A.A., 1992. Tectonics and stratigraphy of the East Brazil Rift system: an overview. Tectonophysics, 213: 97–138

Cordani, U.G. 1973. Definição e caracterização do Cráton São Francisco. 27th Congresso Brasileiro de Geologia, Sociedade Brasileira de Geologia, Anais, v. 2, p. 142–145.

Cordani, U.G. 1978. Comentários filosóficos sobre a evolução geológica pré-cambriana. In: Rocha et al. (eds): Anais da reunião preapratória para Simpósio sobre o Cráton do São Francisco e suas Faixa Marginais. Sociedade Brasileira de Geologia, NBA, Publicação Especial, 3, p. 33–65.

Cordani, U.G. and K. Sato, 1999. Crustal evolution of the South American Platform, based on Sm-Nd isotopic systematic on granitoid rocks. *Episodes*, **22**, 167–173

Cordani, U.G., Melcher, G.C., Almeida, F.F.M., 1968. Outline of the Precambrian geochronology of South America. Canadian Journal of Earth Sciences, 5, 629–632.

Cruz, S. & Alkmim, F.F. 2004. The tectonic interaction between the Paramirim Aulacogen and the Araçuaí Belt, São Francisco Craton region, Eastern Brazil. Submetido aos Anais.Acad. Bras. Cienc.

Dawson, J.B. 1980. Kimberlites and their xenoliths. Springer Verlag, Berlin, 252p.

Dardenne, M. A., 2000. The Brasilia Fold belt. In: Cordani, U.G.; Milani, E.J.; Thomaz F°, A.; Campos, D.A. (eds) *Tectonic Evolution of South America*. 31st Intern. Geol. Congr. Rio de Janeiro, p. 231–264

Eaton, D.W., Darbyshire, F., Evans, R.L., Grütter, H., Jones, A.G., Yuan, X., 2009. The elusive lithosphere–asthenosphere boundary (LAB) beneath cratons. Lithos, 109, 1–22.

Ebert, H. 1968. Ocorrências da Fácies Granulito no sul de Minas Gerais e em áreas adjacentes, em dependência da estrutura orogênica. Anais da Acad. Brás. Cien., 40 (Suplemento): 215–229

Figueiredo, M.C.H., 1989. Geochemical evolution of eastern Bahia, Brazil: A probable Early Proterozoic subduction-related magmatic arc. Journal of South American Earth Sciences, 2(2), 131–145.

Feybesse, J.L., Johan, V., Triboulet, C., Guerrot, C., Mayaga-Mikolo, F., Bouchot, V., Eko N'dong, J., 1998. The West Central African belt: a model of 2.5–2.0 Ga accretion and two-phase orogenic evolution. Precambrian Research, 87, 161–216.

Forte, A.M., Perry, H.C.K., 2001. Geodynamic evidence for a chemically depleted continental tectosphere. Science 290, 1940–1944

Fuck, R.A.; Brito Neves, B.B.; Schobbenhaus, C., 2008. Rodinia descendants in South America. Precambrian Research, 160, 108–126

Grand, S. 1987. Tomographic inversion for shear velocity beneath the North American plate. J.Geophys.Res., 92: 14065–14090.

Guimarães, D. 1951. Arqui-Brasil e sua Evolução Geológica. DNPM-DFPM, Boletim 88, 314p.

Heilbron, M.; A.C. Pedrosa-Soares, M.C. Campos Neto, R.A.J. Trouw and V.A. Janasi, 2004. Província Mantiqueira. In: Mantesso-Neto et al. (eds) *Geologia do Continente Sul-Americano.Evolução da obra de Fernando Flávio Marques de Almeida*. Becca, São Paulo, p. 203–234

Jordan, T.H. 1978. Composition and development of the continental tectosphere. Nature, 274: 544–548

Jordan, T.H. 1988. Structure and formation of the continental tectosphere. Journal of Petrology, Spec. Lithosphere Issue, p. 11–37.

Kaban, M.; Schwintzer, P.; Artemeiva, I.M.; Mooney, W.D. 2003. Density of the continental roots: compositional and thermal contributions. Earth Planet. Sci. Lett., 209: 53–69.

Karner, G.D.; Egan, S.S; Weissel, J.K. 1992. Modeling the tectonic development of the Tucano and Sergipe-Alagoas rift basins, Brazil. Tectonophysics, 215: 133–160.

Ledru, P., Johan, V., Milesi, J.P., Tegyey, M. 1994. Evidence for a 2 Ga continental accretion in the circum-south Atlantic provinces. Precambrian Research, 69, 169–191.

McKenzie, D., Daly M.C., Priestley, K. 2015. The lithosphere structure of Pangea. Geology, 43, 783–786

Milani, E.J. and Thomaz-Filho 2000. Sedimentary Basins of South America. In: Cordani, U.G.; Milani, E.J.; Thomaz Fo, A.; Campos, D.A. (eds.) Tectonic Evolution of South América. Rio de Janeiro, 31st International Geological Congress, Rio de Janeiro, p. 389–449

Milani, E.J., Lana, M.C., Szatmari, P., 1988. Mesozoic rift basins around the northeast Brazilian microplate (Recôncavo-Tucano-Jatobá, Sergipe-Alagoas). In: W. Manzpeizer (ed.) Triassic-Jurassic rifting: Continental breakup and the origin of the Atlantic Ocean and passive margins. Amsterdam, Elsevier, 833–858.

Maurin, J.-C., 1993. La chaîne panafricaine ouest-congolienne: correlation avec le domain est-brésilien et hypothese géodynamique. Bulletin de la Société Géologique de France 164, 51–60.

Oliveira E.P., Windley B.F., Araújo M.N.C. 2010. The Neoproterozoic Sergipano orogenic belt, NE Brazil: a complete plate tectonic cycle in western Gondwana. Precambrian Research 181, 64–84.

Pedrosa Soares, A.C.; Noce, C.M.; Vidal, P.; Monteiro, R.L.B.P.; Leonardos, O.H. 1992. Towards a new tectonic model for the upper Proterozoic Araçuai (SE Brazil)-West Congolian(SW Afric) Belt. Journal of South American Earth Sciences, 6, 33–47

Pedrosa Soares, A.C.; Noce, C.M.; Wiedemann, C.M.; Pinto, C. P. 2001. The Araçuaí-West-Congo Orogen in Brazil: an overview of a confined orogen formed during Gondwanaland assembly. Precambrian Research, 110(1–4), 307–323.

Pedrosa-Soares, A.C., Noce, C.M., Wiedemann, C.M., Pinto, C.P., 2001. The Araçuaí-West Congo orogen in Brazil: An overview of a confined orogen formed during Gondwanland assembly. Precambrian Research 110, 307–323.

Pimentel, M.M., R.A. Fuck, H. Jost, C.F. Ferreira Filho and S.M. Araújo 2000. The basement of Brasília fold belt and the Goiás magmatic arc. In: Cordani, U.G., Milani, E.J., Thomaz Filho, A.,

Campos, D.A. (Eds.) *Tectonic Evolution of South America*. International Geological Congress, Rio de Janeiro, p. 195–232
Pimentel, M.M.; H. Jost, and R.A. Fuck 2004. O Embasamento da Faixa Brasilia e o arco magmático de Goiás. In: Mantesso-Neto et al. (eds) *Geologia do Continente Sul-Americano.Evolução da obra de Fernando Flávio Marques de Almeida*. Becca, São Paulo, p. 356–381
Pollak, H.N. & Chapman, D.S. 1977. On the regional variation of heat flow, geotherms and lithospheric thickness. Tectonophysics, 38: 279–296.
Porada, H. 1989. Pan-African Rifting and Orogenesis in Southern to Equatorial África and Eastern Brazil. Precam. Res., 44: 103–136
Schobbenhaus, C. and Brito Neves, B.B., 2003.Geology of Brazil in the context of the South American Platform. In: Bizzi, L.A.; Schobbenhaus, C.; Vidotti, R.M. and Gonçalves, J.H. (eds). Geology, Tectonics and Mineral Resources of Brazil. Serviço Geológico – CPRM, Brasília, p. 5–54
Sengör, A.M.C. 1999. Continental interiors and cratons: any relation? Tectonophysics, 305: 1–42.
Tack, L.; Wingate, M.T.D; Liégeois, J.-P; Fernandez-Alonso, M.; Deblond, A. 2001. Early Neoproterozoic magmatism (1000–910 Ma) of the Zadinian and Mayumbian Groups (Bas-Congo): onset of Rodinia rifting at the western edge of the Congo craton. Precambrian Res., 110(1–4): 277–306
Tavora, FJ., Cordani, UG., Kawashita, K. 1967. Determinações de idade K-Ar em rochas da região central da Bahia. In: Congresso Brasileiro de Geologia, 21. Curitiba, 1967. Anais... Curitiba. SBG, 234–244
Teixeira, W., Sabaté, P., Barbosa, J., Noce, C.M., Carneiro, M.A., 2000. Archean and Paleoproterzoic tectonic evolution of the São Francisco craton, Brazil. In: Cordani, U.G.; Milani, E.J.; Thomaz Filho, A.; Campos, D.A. (eds.) Tectonic Evolution of South América. Rio de Janeiro, 31st International Geological Congress, Rio de Janeiro, 101–137.
Teixeira, W., Ávila, C.A., Dussin, I.A., Neto, A.C., Bongiolo, E.M., Santos, J.O., Barbosa, N.S., 2015. A juvenile accretion episode (2.35–2.32 Ga) in the Mineiro belt and its role to the Minas accretionary orogeny: Zircon U–Pb–Hf and geochemical evidences. Precambrian Research, 256, 148–169.
Thiéblemont, D., Bouton, P., Préat, A., Goujou, J.C., Tegyey, M., Weber F., Obiang. M.E., Joron, J.L., Treuil, M., 2014. Transition from alkaline to calc-alkaline volcanism during evolution of the Paleoproterozoic Francevillian basin of eastern Gabon (Western Central Africa). Journal of African Earth Sciences, 99, 215–227.
Trompette, R. 1994. Geology of Western Gondwana (2000–500 Ma). Pan-African-Brasiliano Aggregation of South América and Africa. Balkema, Rotterdam, 350 p.
Trompette, R.; Uhlein, A.; Egydio Da Silva, M.; Karmann, I. 1992. The Brasiliano São Francisco craton revisited (central Brazil). Journal of South American Earth Sciences, 6, 49–57.
Trouw, R.J.A; M. Heilbron, W.Ribeiro, F. Pacciulo, C.M. Valeriano, J. C.H. Almeida, M.Tupinambá and R.R. Adreis, 2000. The central segment of the Ribeira belt. In: Cordani, U.G.; Milani, E.J.; Thomaz Fo, A.; Campos, D.A. (eds.) Tectonic Evolution of South América. Rio de Janeiro, 31st International Geological Congress, Rio de Janeiro, p. 287–310.
Valeriano, C.M.; Machado, N.; Simonetti, A.; Valladares, C.S.; Seer, H.J.; Simões, L.S.A. 2004. U-Pb Geochronology of the Southern Brasília Belt (SE-BRAZIL): Sedimentary Provenance, Neoproterozoic Orogeny and Assembly of West Gondwana. Precambrian Research, 130, 27–55.
Vauchez, A., Neves, S., Caby, R., Corsini, M. Egydio-Silva, M., Arthau, M., Amaro, V., 1995. The Borborema shear zone system, NE Brazil. Journal of South America Earth Sciences, 8, 247–266

Lithospheric Features of the São Francisco Craton

2

Marcelo Assumpção, Paulo A. Azevedo, Marcelo P. Rocha, and Marcelo B. Bianchi

Abstract

Studying the thick lithosphere of cratons is important to help understand their formation and the mechanisms for their preservation. We present a synthesis of the information available for the deep structure in Eastern Brazil, from seismological and gravity data, to characterize the São Francisco Craton (SFC) and help better define its lateral boundaries at depth. Crustal thicknesses of the SFC, known mainly from receiver function studies, range from 38 to 42 km, except for a localized thickening (up to 44 km) in the northern part, and crustal thinning towards the Atlantic continental margin in Bahia state. Overall, the crust is slightly thicker near the geologically-defined surface boundaries (40–42 km) and slightly thinner in the center (38–40 km), which is consistent with generally low Bouguer anomalies and high topography to the East and to the West of the craton probably defining the suture zones during the Gondwana amalgamation. Modeling of gravity anomalies with some seismic constraints indicates a relatively low-density lithospheric mantle for the SFC, despite higher Pn velocity, which is consistent with a Fe-depleted, buoyant lithosphere, which helps preserve the cratons's root. Surface-wave continental-scale tomography suggested the thickest lithosphere, around 200 km, to be in the Archean southern part of the SFC, consistent with regional P- and S-wave tomography. Both the surface-wave and the body-wave tomographies show high upper mantle velocities beneath the Brasilia fold belt, next to the SFC's surface limits, which is interpreted as a continuation at depth of the craton's lithosphere, beneath the low-grade external metamorphic domain of the Brasilia fold belt. Analysis of the SFC seismicity shows that most earthquakes now occur on shallow (<2 km) normal faults formed during the formation of the Brasiliano continental margin, now reactivated under the present E–W compressional stresses.

Keywords

Tomography • Receiver functions • Cratonic lithosphere • Gravity anomaly

M. Assumpção (✉) · M.B. Bianchi
Geophysics Department, Institute of Astronomy, Geophysics and Atmospheric Sciences, University of São Paulo, Rua Do Matão 1226, São Paulo, SP 05508-090, Brazil
e-mail: marcelo.assumpcao@iag.usp.br

P.A. Azevedo · M.P. Rocha
Seismologial Observatory, Institute of Geosciences, University of Brasília, Brasilia, DF 70910-000, Brazil

M. Heilbron et al. (eds.), *São Francisco Craton, Eastern Brazil*, Regional Geology Reviews, DOI 10.1007/978-3-319-01715-0_2

2.1 Introduction

Studying the thick lithosphere of cratons is important to help understand their formation and the mechanisms for their preservation. The study of the deep crustal structure, which resulted from a long evolution of the lithosphere, is the key to understand the geological processes that shaped the lithosphere (e.g., Artemieva and Meissner 2012). Early compilations of results from deep seismic refraction lines worldwide (Durrheim and Mooney 1991, 1994) suggested that Archean cratons, when compared with Proterozoic provinces, would have thinner and slightly more felsic crust, coupled with a lower density and Fe-depleted upper mantle (Durrheim and Mooney 1994; Hawkesworth et al. 1990; Artemieva 2009). However, more recent global compilations seem to indicate a large variability in crustal properties of Archean provinces (Artemieva 2009; Artemieva and Meissner 2012), with apparently no significant systematic difference from Proterozoic crust.

Here we summarize and discuss the available information regarding crustal thickness (mainly from receiver function studies) and upper mantle seismic velocities from regional tomography studies to characterize the lithosphere of the São Francisco craton and its surrounding belts.

2.2 Crustal Thickness

The first attempt to study the deep crustal structure in the São Francisco craton (SFC) was the recording of an unreversed seismic profile using quarry blasts from Itabira mine (Quadrilátero Ferrífero mining district, see Alkmim and Teixeira, this book) in the early 1970s by Giese and Schutte (1975). No refraction from the upper mantle (Pn phase) was recorded, but the reflections from the Moho indicated a normal crustal thickness of about 40 km. Other unreversed seismic refraction studies in the 1980s, using quarry blasts recorded near the northern limit of the SFC (e.g., Knize et al. 1984), indicated crustal thicknesses greater than 40–42 km (Fig. 2.1).

To compensate for the lack of other active refraction lines, passive seismological methods have been used to estimate crustal thicknesses in SE and Central Brazil using teleseismic receiver functions (RF). Initial studies of RF in SE Brazil (Assumpção et al. 2002) indicated a generally thinner crust in the southern part of the SFC in comparison with the Paraná Basin of southern Brazil (Fig. 2.1). Figure 2.1 shows the crustal thickness map obtained from a compilation by Assumpção et al. (2013a, b), with some additional points from RF analyses of recently deployed stations of the Brazilian Seismographic Network. As shown on Fig. 2.1, the SFC crust exhibits an average thickness of 40 km, a value that approaches 41 km, the average thickness estimated for the Brazilian portion of the South American platform (Assumpção et al. 2013a, b). Thicker crust (exceeding 45 km) occurs in the northern part of the Paraná Basin. A thinner crust is observed in the central part of the Tocantins Province, especially along the Goiás Magmatic Arc, where it is accompanied by high gravity anomalies (Fig. 2.1).

The sparse distribution of data points prevents a more detailed discussion of the observed patterns, but a domain of larger thicknesses can be recognized near the surface limits of the SFC, more or less following the higher topography (Fig. 2.1a) and lower Bouguer anomaly (Fig. 2.2a).

The first detailed study of seismic structure in the SFC was obtained by a 530 km long, E–W deep seismic refraction line, shot in 1998 across the Tocantins Province and the western part of the craton, at 13°–14°S (Berrocal et al. 2004; Soares et al. 2006). The model obtained by Soares et al. (2006) has four crustal layers in the western part of the SFC with Moho depth at about 42 km. Three suture zones (separating blocks with different structures) were identified: one at Minaçu (between the Goiás massif and the external metamorphic zone of the Brasilia belt), one near Cavalcante (in the Brasilia belt external zone) (see Fuck et al. this book), and another near Posse at the edge of the SFC surface limit. All these "collision zones" were modeled as dipping to the west. Another interesting finding was that the upper mantle velocity (Pn) was high (8.2–8.3 km/s) beneath the SFC as well as beneath the external zone, compared to the 8.0–8.1 km/s beneath the Goiás Massif and Magmatic Arc (Berrocal et al. 2004; Soares et al. 2006). All these characteristics led Soares et al. to place the western limit of the craton's deep crust, i.e., the edge of the São Francisco plate, near the boundary between the Goiás massif and the Brasilia belt (near coordinate 230 km in the profile of Fig. 2.2b).

2.3 Gravity Modeling

Modeling of crustal layers from interpretation of deep seismic refraction lines can be non-unique because of different interpretations of weak arrivals. For example, the transition between a thick crust beneath the Araguaia belt to a thin crust beneath the Brasilia belt magmatic arc had been placed between shot points 2 and 3 (coordinate 90 km, as in Fig. 2.2b). Based on gravity modeling and a re-interpretation of Moho reflected phases, Ventura et al. (2011) placed this transition $\sim$40 km to the east, beneath shot point 3. The Moho was modeled about 5 km deeper ($\sim$50 km instead of 45 km) based on receiver function analysis at a station near Porangatu. Towards the east, (beneath shot point 7, Fig. 2.2b), the differences in crustal thicknesses are smaller, less than 4 km. Ventura et al. (2011) did not interpret the eastern section of the seismic experiment, so here we used the model of Soares et al. (2006), which covers part of the SFC, as a basis to extrapolate the crustal structure beneath the central and eastern part of the craton.

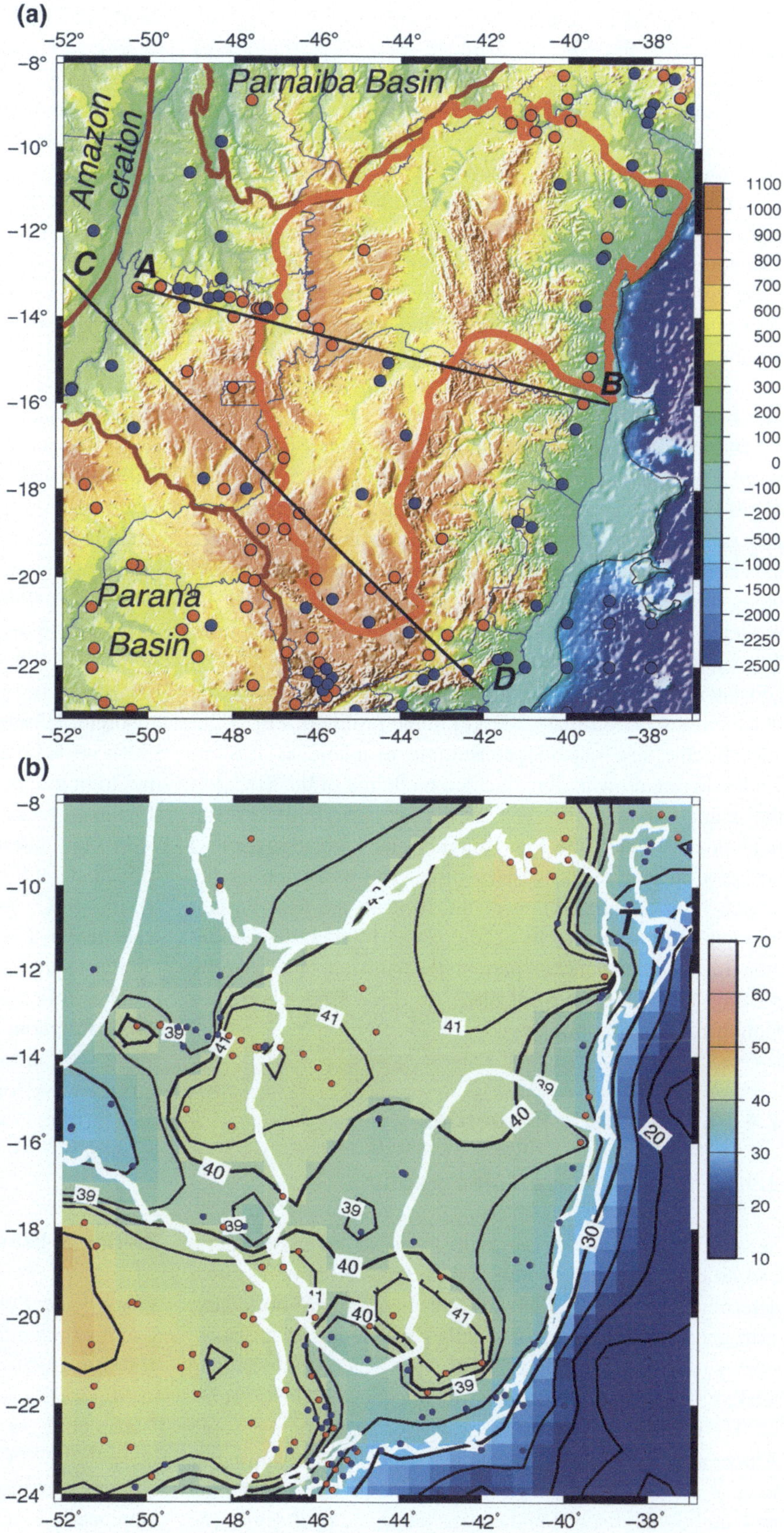

Fig. 2.1 **a** Topography in SE Brazil. *Red/Blue* dots denote crustal thickness more/less than 40 km. Profiles AB and CD refer to Figs. 2.2c and 2.3c, respectively. **b** Contours of crustal thickness from seismic data only. *Red and Blue dots* in the continent are locations where crustal thickness (including topography) were estimated from receiver functions at seismographic stations or from active experiments such as the deep refraction line across the western border of the São Francisco Craton (roughly along 13°–14°S). The data comes from the compilation of Assumpção et al. (2013a, b) complemented with some additional receiver function analyses. "*T*" indicates the Recôncavo–Tucano Basin

We extended the crustal seismic model of Soares et al. (2006) across the SFC to estimate possible variations of crustal thickness as well as to see if the gravity data indicated any changes in upper mantle density. We used the GOCE satellite Bouguer anomalies, which are smoothed enough for the modeling of major regional variations. The P-wave velocities of the seismic model were converted to densities using the linear relation of Christensen and Mooney (1995). The upper mantle density was fixed throughout the profile at 3360 kg/m^3 (corresponding to an average Pn velocity between the 8.05 and 8.23 km/s in the seismic model). The numbers shown in the crustal section of Fig. 2.2b are the density contrasts of the crustal layers relative to the uniform-density upper mantle.

The few Moho depths available in the central part of the profile, obtained from RF studies, indicate an average value of 40 km, slightly less than the 42 km at the easternmost part of Soares et al. (2006) model, but probably within the uncertainty of each method. The low Bouguer anomaly along the eastern limit of the SFC (Fig. 2.2a) could be modeled by a crustal thickening (perhaps reaching 44 km). The high topography along the eastern side of the SFC is consistent with a major boundary, whose limits could be placed around coordinate 900 km in the profile of Fig. 2.2b. Clearly, the few seismic constraints available in Eastern Brazil do not allow a more detailed modeling of the SFC, but the main features observed at the western limit (crustal thickening beneath the low Bouguer and high topography) are also present in the eastern limit. The extension of the crustal limit of the SFC, at depth, beneath the Brasilia belt internal zone, seems to be accompanied by a similar extension in the East beneath part of the Araçuaí belt, thereby placing the boundaries of the São Francisco plate (see Heilbron et al. this book).

2.4 Lithospheric Properties

2.4.1 Surface-Wave Tomography

Seismic tomography is normally used to estimate lateral variations of upper mantle properties. Although direct determination of the lithosphere/asthenosphere boundary, with good resolution, is not usually possible in surface-wave studies, the lateral variations of S-wave velocities help map relative variations of lithospheric thicknesses. Feng et al. (2007) carried out a surface-wave tomography of South America with a method combining both waveform fitting as well as group-velocity measurements. An updated model with additional dispersion data was presented by Assumpção et al. (2013b) and shown on Fig. 2.3. The maximum period of the surface-waves was 150 s, which limits the resolution of the surface-wave tomography down to about 250 km depth. In addition, only features larger than about ~300 km or so can be mapped in such continental-scale surface-wave tomography. At 100 km depth (Fig. 2.3a), the regional pattern shows two main areas with high S-wave velocities: one beneath the southern part of the SFC (where Archean rocks are exposed) and another block beneath the northern part of the Paraná basin, usually attributed to a cratonic nucleus beneath the basin (Cordani et al. 1984; Julià et al. 2008).

At 150 km depth (Fig. 2.3b), the region of high S-wave velocity is concentrated in the southern half of the SFC, which indicates the region of thickest lithosphere. Studies of S-wave receiver functions at station BDFB (Brasília) by Heit (2007) indicated the lithosphere/asthenosphere boundary at about 160 km, shown in the map and profile of Fig. 2.3b. The lithosphere of the SFC is thinner in eastern Bahia, which could be an effect of lithospheric thinning during the South Atlantic opening in the Lower Cretaceous. However, resolution of the surface-wave tomography also decreases considerably towards the coast, where fewer ray paths are observed (Feng et al. 2007; Assumpção et al. 2013b). A thicker lithosphere around the craton's surface limits beneath parts of the Brasilia and Araçuaí belts is also apparent. However, part of this "extended" lithosphere could be an artificial effect due to smoothness constraints used in the tomography inversion. A global map of lithospheric thickness, based on long-period (>50 s) surface waves with ~250 km lateral resolution (McKenzie et al. 2015) also indicated a lithosphere about 200 km thick in the central part of the SFC, and a thinning towards the coast and the Recôncavo–Tucano basin in northern SFC (Fig. 2.1b), similar to the image of Fig. 2.3d.

The limit of the SFC high-velocity region towards the Amazon craton (across the so-called Transbrasiliano Lineament, see Cordani et al. this book) is not well resolved in the surface-wave tomography, but is a relatively sharp transition seen in the P-wave tomography, as described in the next section.

2.4.2 P-Wave Tomography

Teleseismic P-wave tomography (using periods of around 1 s) has potentially better lateral resolution at depths below 100 km, as shown on Fig. 2.4. Previous P-wave tomography in SE Brazil (Schimmel et al. 2003; Assumpção et al. 2004; Rocha et al. 2011) had all shown generally higher velocities beneath the SFC and lower velocities beneath the Cretaceous igneous provinces (Iporá, Alto Paranaíba, and Serra do Mar) emplaced along the orogenic belts that fringe the craton to the south. High velocities have also been observed beneath the northern part of the Paraná basin, not in a single block, but in separate patches.

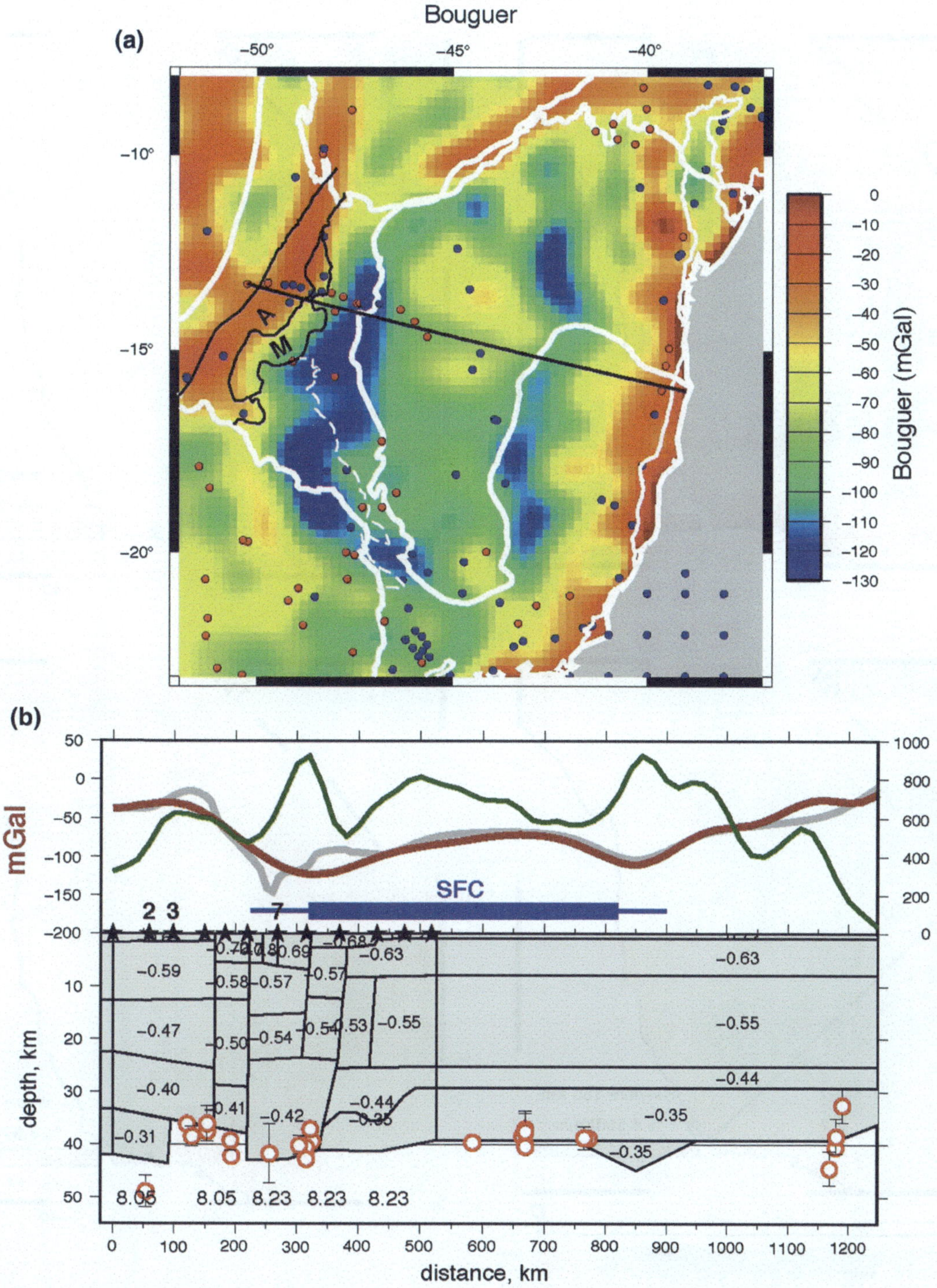

Fig. 2.2 Crustal thickness model across the central part of the SFC. **a** Bouguer map (GOCE) showing low gravity anomalies along the western and eastern borders of the SFC. *Red/blue* dots are Moho depths deeper/shallower than 40 km. *White dashed line* west of the SFC is the separation between external and internal metamorphic zones of the Brasilia Belt. *Black solid lines* indicate limits of the Magmatic Arc ("A") and the Goiás Massif ("M"). **b** Crustal model along the WNW-ESE profile shown in **a**. The *lines* are topography (*green*), observed gravity (*brown*) calculated gravity (*gray*). The *bottom* section from 0 to 530 km is the seismic model of Soares et al. (2006) obtained from 12 explosions (*stars on the left side* of the profile), some of which are numbered as referred to in the text. *Numbers* in the crustal section (*gray area*) are density constrasts relative to a constant density of the upper mantle taken at 3360 kg/m^3. *Numbers* below the crust are the P-wave velocities from Soares et al. (2006). The seismic refraction model was extrapolated across the whole profile—only the Moho depth was changed to fit the low Bouguer anomaly to the east of the craton. *Open red circles* are Moho data points from receiver function studies (as in the map) up to ±200 km from the profile line. The *blue line* indicates the limits of the SFC: *thick line* for the surface limit, and *thin line* for the extrapolated extension at depth

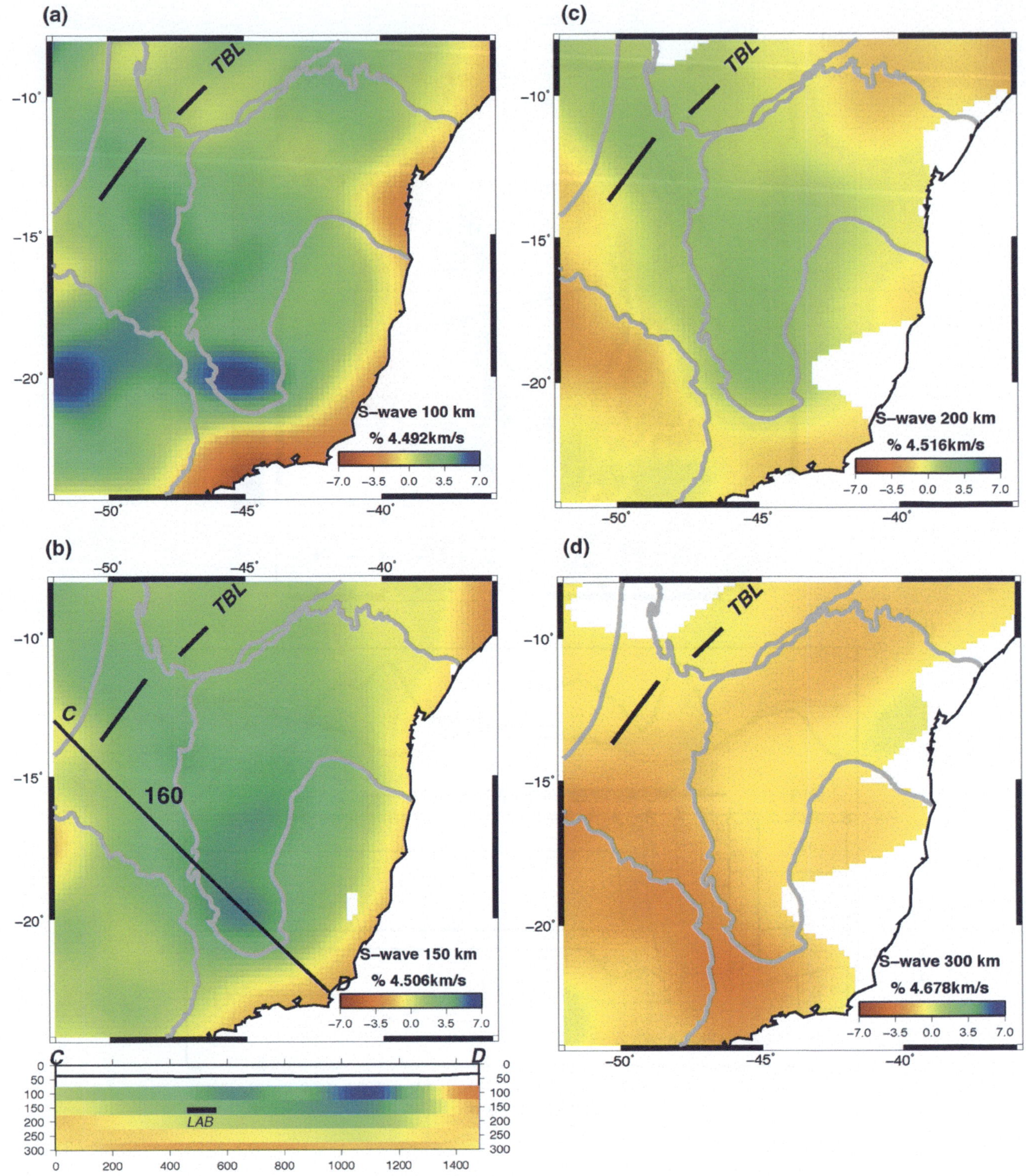

Fig. 2.3 S-wave anomalies from surface-wave tomography using the joint inversion method (waveform + group dispersion) of Feng et al. (2007) presented by Assumpção et al. (2013b) with additional dispersion data. **a–d** Anomalies at depths of 100, 150, 200 and 300 km, respectively. Resolution is poor below 250 km depth. For the 150 km depth (**c**), a vertical NW-SE profile is also shown. Note higher S-wave velocities in the southern part of the SFC, and lower velocities in the northern part, indicating thicker lithosphere in the central and southern parts of the SF craton. The high velocities outside the surface limits of the SFC, especially at 150 km depth, indicate possible extension of the craton beneath the low-grade metamorphic margins of the Brasilia and Araçuaí belts. In the profile "*C-D*" the horizontal bar *LAB* denotes the Lithosphere/Asthenosphere boundary at 160 km depth (Heit 2007)

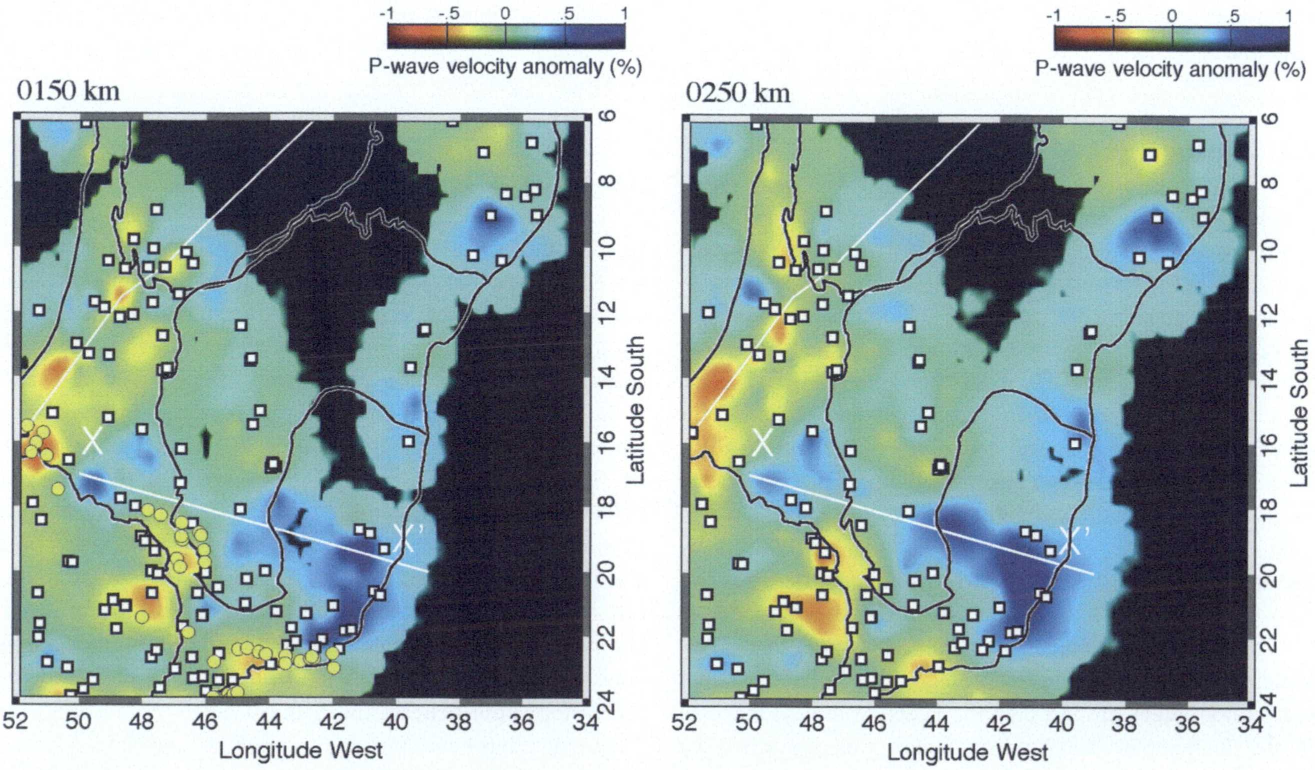

Fig. 2.4 P-wave anomaly at 150 km (*left*) and 250 km (*right*). *White squares* are seismic stations used in the tomography. *Yellow circles* are Cretaceous pipes of the Iporá, Alto Paranaíba (APIP) and Serra do Mar igneous provinces

Here we show an updated P-wave tomography inversion adding more data from the Tocantins Province (Azevedo et al. 2015), as well as from recently deployed stations of the Brazilian Seismographic Network in SE Brazil (Fig. 2.4). The resolution varies considerably because of highly non-even distribution of stations and ray-paths. Where station density is higher, such as in the southern part of the SFC, high velocities can be seen in the southern extreme of the SFC as well as in the Brasilia belt, generally consistent with the larger-scale surface-wave tomography. The Late Cretaceous alkaline intrusions of the Alto Paranaíba Igneous Province (APIP) (Fig. 2.4) correlate with low upper mantle P-wave velocity—consistent with the SFC being "eroded" by metasomatism and losing part of its deep root (Artemieva 2009).

The eastern limit of the high-velocity lithosphere is not well resolved because of few seismic stations in the Araçuaí belt. However, high velocities are suggested all along the coast from Rio de Janeiro northwards. The SFC northern boundary is not resolved with the few stations presently available. The type of tomography method employed here, using relative arrival times, is more suited to map lateral variations in the upper mantle but does not provide good absolute constraints on depths to the bottom of the lithosphere (e.g., Schimmel et al. 2003; Rocha et al. 2011). This is seen in the profile of Fig. 2.5.

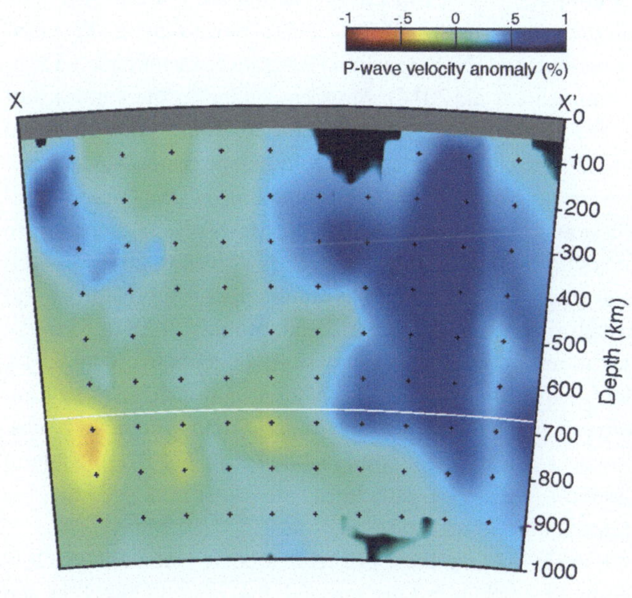

Fig. 2.5 XX′ profile suggesting ~200–300 km lithospheric depth for the Brasilia belt, and maybe 300–400 km for the southern part of the SFC. Depth resolution is very poor in this type of tomography. High velocities beneath the Ribeira belt are poorly resolved because of few stations

The high velocities beneath the Ribeira belt do not seem to be consistent with the surface-wave tomography results. However, the P-wave tomography has poor resolution near the coast north of 22°S due to the few available stations.

2.5 Seismicity and Stresses

Although cratonic areas in Brazil tend to be on average relatively less seismic, when compared to Neoproterozoic Brasiliano belts (Assumpção et al. 2014; Agurto-Detzel et al. 2015a), the southern part of the São Francisco craton has significant seismicity with magnitudes up to about 5 mb (Chimpliganond et al. 2010; Assumpção et al. 2014), as seen on Fig. 2.6a. An interesting feature is that seismicity is concentrated near the central and southern part of the SFC, where the lithosphere is thicker. Although intraplate seismicity occurs at shallow depths in the brittle upper crust, it results from stress concentration that can be caused by deeper structures, especially lateral variations of crustal and lithospheric thicknesses. An analysis of worldwide moderate to large intraplate earthquakes (magnitudes larger than 5) by Mooney et al. (2012) indicated a trend of seismicity to occur near craton edges around cratonic keels. Some evidence of similar features in the Amazon craton was suggested by Assumpção et al. (2014) and can be seen in Fig. 2.6a with epicenters roughly around the tomographic keel in the southern part of the SFC.

Intraplate stresses in Eastern Brazil are not uniform. The southern part of the SFC is characterized by E–W compressional and N–S tensional stresses (Assumpção, 1998; Assumpção et al. 2014). Stresses change in the central part of the craton towards E–W compression (Agurto-Detzel et al. 2015b). In the Tocantins fold belt province, stresses are also compressional but the orientation of the major compressional axis rotates towards SE–NW, as seen from inversion of focal mechanisms (Carvalho et al. 2016) as well as in situ measurements of hydraulic fracturing (Caproni and Armelin 1990). The regional trend, roughly E–W to SE–NW maximum horizontal compression, is probably caused by plate-wide forces such as Mid-Atlantic ridge-push and convergence between the Nazca and South American plates. However, local stress components (probably due to lateral variations of deep structure) are superimposed on the regional component.

Earthquakes in the SFC tend to occur at very shallow depths (less than 5 km), around the cratonic keel in the southern and central parts of the craton (Fig. 2.6a). The few earthquakes with focal mechanisms and hypocentral depths well determined by local stations occur in very shallow (less than 2 km depth) east-dipping faults (Fig. 2.6b). This indicates reactivation of Proterozoic normal faults under the present intraplate compressional stresses. It is not clear yet how exactly the high stresses in the upper crust of the SFC are generated by lateral variations of the craton's deep structure, but models of weak lithospheric mantle at craton's edge and flexural stresses (Assumpção et al. 2004, 2014) are possible candidates.

2.6 Discussion

Despite the few available data on deep crustal structure of the SFC some patterns can be recognized. On the western (Brasilia belt) and eastern border (Araçuaí belt) low Bouguer anomalies, high elevation and receiver function analyses (Figs. 2.1 and 2.2) indicate generally thick crust (down to ~44 km). The central part of the SFC seems to be characterized by slightly thinner crust (~39 km). High velocities in the lithospheric mantle (observed with P- and surface-wave tomographies, Figs. 2.3 and 2.4) are seen extending outside the SFC surface limits. All these features confirm the interpretation of the Tocantins seismic refraction line by Soares et al. (2006), who placed the main suture zone about 50–100 km outside the surface limit of the craton beneath the Brasília belt. A similar feature is proposed for the eastern limit beneath the Araçuaí belt.

On the other hand, the southern limit of the SFC, near the Ribeira belt, despite the high elevations, have high Bouguer anomalies and relatively thin crust. This could be due to the escape tectonics involved with the Ribeira belt (Vauchez et al. 1994), as opposed to collisional tectonics as in the western and eastern borders. Also, the relatively thin crust of the southern part of the SFC could be less dense (more felsic average crustal composition) allowing the crust to be balanced at higher elevations (Assumpção et al. 2002).

Although no direct determination of the depth to the lithosphere-asthenosphere boundary (LAB) has been made, S-wave velocities determined by surface-wave tomography (Fig. 2.3) suggest that the deepest part of the craton, about 200 km deep, lies in its southern part (Minas Gerais state). This is consistent with the expected range of 150–300 km for Archean lithosphere observed worldwide (Artemieva 2006).

In contrast, the northeastern part of the craton (towards the coast of Bahia) seems to have a thinner lithosphere. This is consistent with higher temperatures extrapolated from the geothermal gradient and also with the depths to the Curie temperature derived from spectral analyses of magnetic anomalies (Guimarães et al. 2014). The depths to the Curie temperatures range from 30 to 40 km in the Tocantins province, and from 38 to 44 km in the central part of the craton. At the NE part of the craton the Curie depth becomes shallow again (~30–32 km), consistent with an increase of the geothermal gradient and lithospheric thinning towards the coast of Bahia.

Near the Late Cretaceous Alto Paranaiba igneous province (APIP), the seismic velocities are lower than the

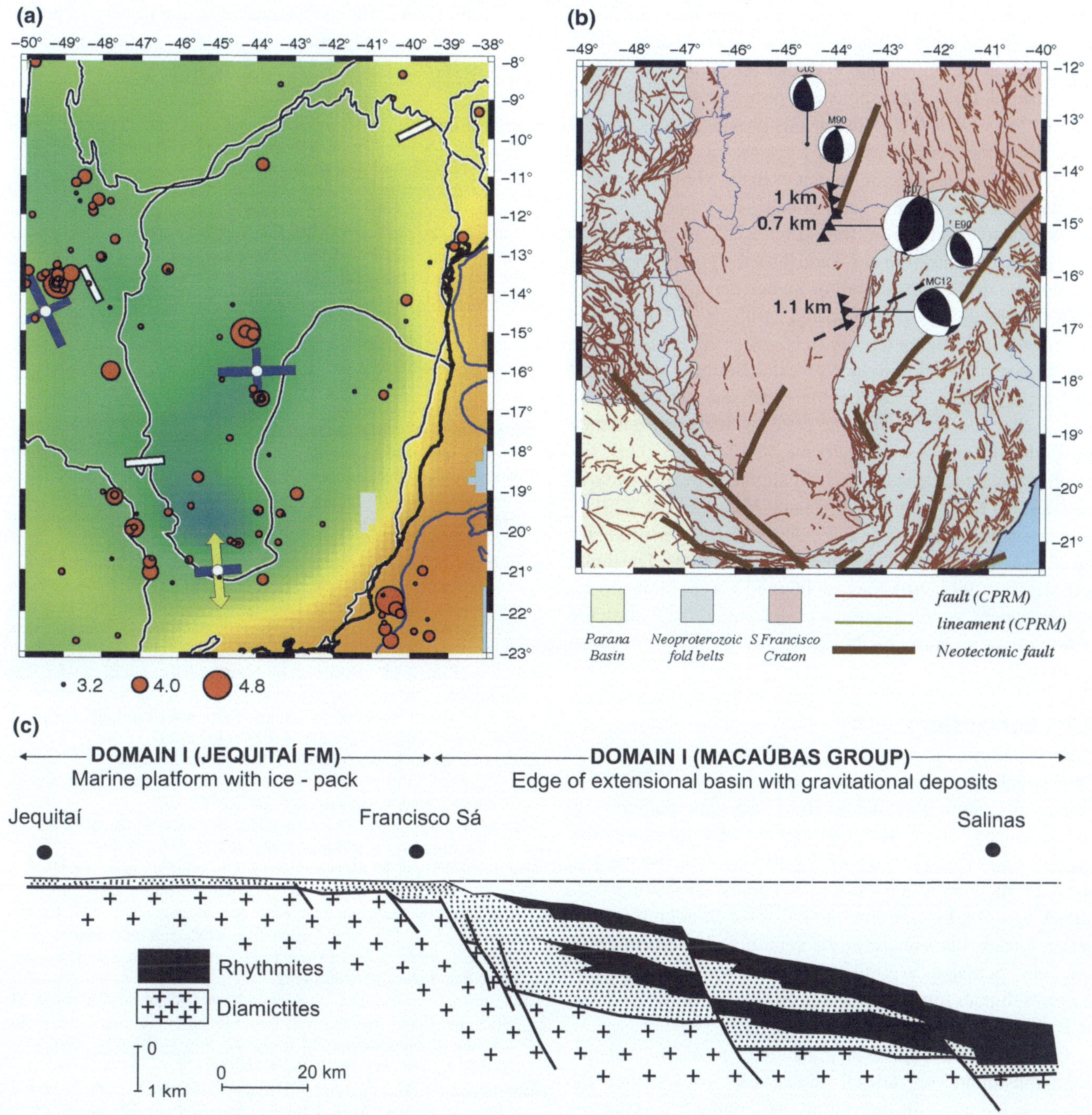

Fig. 2.6 **a** Epicenters and stresses in the SFC region. *Red circles* are epicenters from the uniform catalog of Assumpção et al. (2014). Stress tensors from earthquake focal mechanisms are indicated by *blue bars* (maximum and intermediate compressional stresses) and *yellow arrows* (horizontal tensional deviatoric stress). *Open white bars* are SHmax orientation from in situ measurements (Caproni and Armelin 1990; Magalhães 1999). *Background colors* are S-wave anomalies at 150 km depth from the surface-wave tomography as shown in Fig. 2.3b. *Blue line* offshore is the 200 m bathimetry. **b** Earthquake focal mechanisms and depths in the central part of the São Francisco craton. Events occur in E dipping reverse faults, as reactivation of ancient normal faults formed during the evolution of the craton's margin. *Thin brown lines* are geological faults from CPRM; *thick lines* are lineaments/faults interpreted as neotectonic features (Saadi et al. 2002). **c** Geological profile across the eastern border of the craton between Jequitaí and Salinas, indicated as *dashed line* in the map (Uhlein et al. 1998)

average lithospheric mantle (Fig. 2.4). Several interpretations are possible, such as (1) the original cratonic lithosphere was already thin (and consequently hotter), which facilitated pounding of partially melt material from a deep plume in the Cretaceous (e.g., Gibson et al. 1997), or (2) metasomatism in the Cretaceous, which decreased the seismic velocities, facilitated partial melting and pipe emplacement. Cratonic lithosphere tends to have lower upper-mantle velocities in areas with alkaline intrusions (Artemieva 2009). Both these models are consistent with a weak lithosphere (either due to higher temperature or due to a wet mantle lid) being responsible for higher seismicity near the APIP as well as the Iporá igneous provinces (Assumpção et al. 2004; Azevedo et al. 2015).

Although higher seismic velocities in the SFC lithospheric mantle—compared to the Brasilia belt—have been observed in the seismic refraction line (Fig. 2.2b), as well as in tomographic studies (Figs. 2.3 and 2.4), the gravity modeling did not require a corresponding higher density. If the 2 % higher velocity (8.25 compared with 8.05 km/s) was accompanied by higher density (0.05 g/cm^3 higher), a 50 km lithospheric lid would produce an extra Bouguer anomaly of 100 mGal, clearly incompatible with the observed data (Fig. 2.2b). This means that the craton's lithospheric mantle has higher P- and S-wave velocities ($\sim$2–3 % higher), but roughly the same density, or even slightly lower density as modeled by Koosah et al. (2007) and Ventura et al. (2011). This is consistent with a Fe-depleted composition for Archean cratons (e.g., Hawkesworth et al. 1990; Durrheim and Mooney 1991; Artemieva 2009) and lower temperatures, which account for the buoyancy of the craton's root allowing stability over its entire evolution.

2.7 Conclusions

Few geophysical data on the deep structure of the São Francisco Craton is available. However, some patterns can be recognized. The central part of the craton has normal to slightly thin crust (average of 39–40 km), and the western and eastern borders are thicker, generally correlating with higher topography. Surface- and P-wave tomography suggest a thicker lithosphere in the central and southern parts, reaching perhaps $\sim$200 km. Surface-wave tomography suggests thinner lithosphere in the NE part, towards the coast of Bahia and Tucano basin.

Gravity modelling using the few available seismic point-constraints on crustal thicknesses shows that the lithospheric mantle in the central and southern parts of the craton has about the same density as the surrounding Neoproterozoic belts. This is consistent with an Fe depleted Archean lithospheric mantle, which, despite its lower temperature, is buoyant enough to remain stable throughout its evolution.

Seismicity in the SFC shows that the present E–W compressional stresses are reactivating old, shallow faults, which were formed during the evolution of the Brasiliano marginal basins. Variations of the crustal stresses indicate that strong local components, probably due to deep structure of the SFC, are superimposed on the plate-wide stress components generated at plate boundaries.

Acknowledgments We thank Petrobras Geotectonic Program for supporting the Brazilian Seismographic Network and the Transbrasiliano Project, which allowed new crustal thickness estimates and better tomographic coverage at recently deployed stations. We used results from several temporary experiments carried out with instruments from the Pool of Geophysical Equipment (PEG-BR, National Observatory, Rio de Janeiro). We also thank FAPDF and INCT (Estudos Tectônicos) for supporting geophysical projects in central Brazil. Work carried out with CNPq Grant 30.6547/2013-9 (M.A.) and CAPES PhD scholarship (P.A.A).

References

Agurto-Detzel, H., M, Assumpção, M. Bianchi, and M. Pirchiner, 2015a. Intraplate seismicity in mid-plate South America: correlations with geophysical lithospheric parameters. Geol. Soc. London, Special Publication in "Seismicity, Fault Rupture and Earthquake Hazards in Slowly Deforming Regions". 432, first published on November 2, 2015, doi:10.1144/SP432.5.

Agurto-Detzel, H., M. Assumpção, C. Ciardelli, D.F. Albuquerque, L. V. Barros, G.S.L., França, 2015b. The 2012–2013 Montes Claros earthquake series in the São Francisco Craton, Brazil: new evidence for non-uniform intraplate stresses in mid-plate South America. *Geophys. J. Int.*, 200, 216–226. Doi:10.1093/gji/ggu333.

Azevedo, P.A., M.P. Rocha, J.E. Soares, R.A. Fuck, 2015. Thin lithosphere between the Amazon and São Francisco Cratons, in Central Brazil, revealed by seismic P-wave tomography. *Geophysical J. Int.*, 201, 61–69. doi:10.1093/gji/ggv003

Artemieva, I.M., 2006. Global 1°x1° thermal model TC1 for the continental lithosphere: Implications for the lithosphere secualr evolution. *Tectonophysics*, 416, 245–277.

Artemieva, I.M., 2009. The continental lithosphere: reconciling thermal, seismic and petrological data. *Lithos*, 109, 23–46.

Artemieva, I.M., & R. Meissner, 2012. Crustal thickness controlled by plate tectonics: a review of crust-mantle interactions processes illustrated by European examples. *Tectonophys.*, 530–531, 18–49.

Assumpção, M., 1998. Focal mechanisms of small earthquakes in SE Brazilian shield: a test of stress models of the South American plate. *Geophys. J. Int.*, 133, 490–498.

Assumpção, M., D. James, A. Snoke, 2002. Crustal thicknesses in SE Brazilian shield by receiver function analysis: implications for isostatic compensation. J. *Geophys. Res.*, 107(B1), ESE2-1—ESE2-14, 2006, doi:10.1029/2001JB000422.

Assumpção, M., M. Schimmel, C. Escalante, M. Rocha, J.R. Barbosa & Lucas V. Barros, 2004. Intraplate seismicity in SE Brazil: Stress concentration in lithospheric thin spots. *Geophysical J. Int.* , 159, 390–399. doi:10.1111/j.1365-246X.2004.02357.x

Assumpção, M., M.B. Bianchi, J. Julià, F.L Dias; G.S. França, R.M. Nascimento, S. Drouet, C.G. Pavão, D.F. Albuquerque, A.V. Lopes, 2013a. Crustal thickness map of Brazil: Data compilation and main features. *J. South Am. Earth Sci.*, 43, 74–85. doi:10.1016/j.jsames.2012.12.009.

Assumpção, M., M. Feng, A. Tassara, J. Julià, 2013b. Models of crustal thickness for South America from seismic refraction, receiver functions and surface wave dispersion. *Tectonophys.*, 609, 82–96, doi:10.1016/j.tecto.2012.11.014.

Assumpção, M., J. Ferreira, L. Barros, F.H. Bezerra, G.S. França, J.R. Barbosa, E. Menezes, L.C. Ribotta, M. Pirchiner, A. Nascimento, J. C. Dourado, 2014. Intraplate Seismicity in Brazil. In *Intraplate Earthquakes*, chapter 3, ed. P. Talwani, Cambridge U.P., ISBN 978-1-107-04038-0.

Berrocal, J., Y. Marangoni, N.C. Sá, R. Fuck, J.E.P. Soares, E. Dantas, F. Perosia, and C. Fernandes, 2004. Deep seismic refraction and gravity crustal model and tectonic deformation in Tocantins Province, Central Brazil, *Tectonophysics*, *388*, 187–199.

Caproni, N., J. Armelin, 1990. Instrumentação das escavações subterrâneas da UHE Serra da Mesa. In *Simpósio sobre Instrumentação Geotécnica de Campo - SINGEO-90*, Assoc.Bras.Geol. Eng., São Paulo, vol 1, p 249–257.

Carvalho, J.M., L.V. Barros, and J. Zahradnik, 2016. Focal mechanisms and moment magnitudes of micro-earthquakes in Central Brazil by waveform inversion with quality assessment and inference of the local stress field. *J. South Am. Earth Sci.*, 71, 333–343, doi:10.1016/j.jsames.2015.07.020

Chimpliganond, C., M. Assumpção, M. von Huelsen & G.S. França, 2010. The intracratonic Caraíbas-Itacarambi earthquake of December 09, 2007 (4.9 mb), Minas Gerais State, Brazil. *Tectonophysics*, 480, 48–56.

Christensen, N.I., Mooney, W.D., 1995. Seismic velocity structure and composition of the continental crust: a global view. *J. Geophys. Res.* 100 (B6), 9761–9788.

Cordani, U.G., B.B. Brito Neves, R.A. Fuck, R. Porto, A.T. Filho, and F.M.B. Cunha, 1984. Estudo preliminar de integração do pré-Cambriano com os eventos tectônicos das bacias sedimentares brasileiras. *Rev. Ciência Técnica Petróleo*, 15, Petrobrás, CENPES, Rio de Janeiro.

Durrheim, R.J., Mooney, W.D., 1991. Archean and Proterozoic crustal evolution: Evidence from crustal seismology. *Geology*, 19, 606–609.

Durrheim, R.J., Mooney, W.D., 1994. Evolution of the Precambrian lithosphere: seismological and geophysical constraints. *J. Geophys. Res.* 99 (B8), 15359–15374.

Feng, M., S. Van der Lee and M. Assumpção, 2007. Upper mantle structure of South America from joint inversion of waveforms and fundamental-mode group velocities of Rayleigh waves. *J. Geophys. Res.*, 112, B04312, doi:10.1029/2006JB004449.

Gibson, S.A., Thompson, R.N., Weska, R.K., Dickin, A.P. & Leonardos, O.H., 1997. Late Cretaceous rift-related upwelling and melting of the Trindade starting mantle plume head beneath western Brazil, *Contrib. Mineral Petrol.*, 126, 303–314.

Giese, P., J. Schutte, 1975. Preliminary report on the results of seismic measurements in the Brazilian coastal mountains. Unpublished Report, Free Univ. of Berlin, Berlin, Germany.

Guimarães, S.N.P., D. Ravat, & V.M. Hamza, 2014. Combined use of the centroid and matched filtering spectral magnetic methods in determining thermomagnetic characteristics of the crust in the structural provinces of Central Brazil. *Tectonophys.*, 624–625, 87–99 , doi:10.1016/j.tecto.2014.01.025

Hawkesworth, C.J., Kempton, P.D., Rogers, N.W., Ellam, R.M., van Calsteren, P.W., 1990. Continental mantle lithosphere, and shallow level enrichment processes in the earth's mantle. *Earth Planet. Sci. Lett.* 96, 256–268.

Heit, B., F. Sodoudi, X. Yuan, M. Bianchi, and R. Kind, 2007. An S receiver function analysis of the lithospheric structure in South America. *Geophys.Res.Lett., 34, L14307, doi:*10.1029/2007GL030317.

Julià, J., M. Assumpção & M.P. Rocha, 2008. Deep crustal structure of the Paraná Basin from receiver functions and Rayleigh-wave dispersion: Evidence for a fragmented cratonic root. *J. Geophys. Res.*, 113, B08318, doi:10.1029/2007JB005374.

Knize, S., Berrocal, J., Martins, D., 1984. Modelo Preliminar de Velocidades Sismicas da Crosta Atraves de Explosões Locais Registradas Pela Rede Sismográfica de Sobradinho,BA. *Rev. Bras. Geoc.*, 2, 95–104.

Koosah M., Vidotti R., Soares J.E.P., Fuck R.A. 2007. Gravimetric and seismic data integration in a 2D forward gravimetric modeling for the crust and lid mantle beneath northern Brasilia Belt. In: SBGf, Internat. Cong. of the Brazilian Geophys. Soc., 10th, Rio de Janeiro, Expanded Abstract Volume, CD-ROM.

Magalhães, F.S., 1999. Tensões regionais e locais: Casos no território brasileiro e padrão geral. *PhD Thesis*, Escola de Engenharia de São Carlos, USP, 225 pp.

McKenzie, D., M.C. Daly, and K. Priestley, 2015. The lithospheric structure of Pangea. *Geology*, doi : 10.1130/G36819.1

Mooney, W.D., J. Ritsema, and Y. Hwang (2012), Crustal seismicity and maximum earthquake magnitudes (Mmax) in stable continental regions (SCRs): correlation with the seismic velocity of the lithosphere, *Earth Planet. Sci. Lett.*, 357–358, 78–83, doi:10.1016/j.epsl.2012.08.032.

Rocha, M.P., Schimmel, M., and Assumpção, M., 2011. Upper-mantle seismic structure beneath SE and Central Brazil from P- and S-wave regional traveltime tomography. *Geophysical Journal International*, 184, 268–286, doi:10.1111/j.1365-246X.2010.04831.x.

Saadi, A., M.N. Machette, K.M. Haller, R.L. Dart, L.-A. Bradley, and A.M.P.D. Souza, 2002. Map and Database of Quaternary Faults and Lineaments in Brazil. USGS Open-File Report 02-230 (2002).

Schimmel, M., M. Assumpção & J. VanDecar, 2003. Upper mantle seismic velocity structure beneath SE Brazil from P- and S-wave travel time inversions. *J. Geophys. Res.*, 108(B4), 2191, doi:10.129/2001JB000187.

Soares, J.E., J. Berrocal, R.A. Fuck, W.D. Mooney, and D.B.R. Ventura (2006), Seismic characteristics of central Brazil crust and upper mantle: A deep seismic refraction study, *J. Geophys. Res.*, *111*, B12302, doi:10.1029/2005JB003769.

Uhlein, A., R.R. Trompette and M. Egydio-Silva, 1998. Proterozoic rifting and closure, SE border of the São Francisco Craton, Brazil. *J. South Am. Earth Sci.*, 11(2), 191–203.

Vauchez, A., A. Tommasi, & M. Egydio-Silva, 1994. Self-indentation of a heterogeneous continental lithosphere. *Geology*, 22, 967–970.

Ventura, D.B.R., J.E.P. Soares, R.A. Fuck & L.C.C. Caridade, 2011. Caracterização sísmica e gravimétrica da litosfera sob a linha de refração sísmica profunda de Porangatu, Província Tocantins, Brasil Central. *Rev. Bras. Geoc.*, 41(1), 130–140.

Part II

The Craton Basement

Nature and Evolution of the Archean Crust of the São Francisco Craton

3

Wilson Teixeira, Elson Paiva Oliveira, and Leila Soares Marques

Abstract

We overview the Archean tectonic framework the São Francisco craton based on geologic constraints, integrated geochronologic interpretation and isotopic-geochemical evidence of basement rocks. U–Pb provenance studies of Archean and Paleoproterozoic supracrustal sequences are also used to provide additional inferences about the geodynamic scenario. The Archean rocks crop out mainly in two large areas in the southern and northern portions of the craton, surrounded and/or in tectonic contact with Paleoproterozoic orogenic belts. The ancient substratum is essentially composed of medium- to high-grade gneissic-migmatitic rocks including TTG suites and coeval granite-greenstone associations that collectively provide an isotopic record as old as 4.1 Ga. The combined U–Pb and Sm–Nd T_{DM} age peaks coupled with U–Pb inherited ages in detrital zircons from the supracrustal sequences indicate that very ancient continental crust (>3.5 Ga) exist, particularly in the northern portion of the craton. Mesoarchean events are episodic between 3.6–3.3 and 3.2–2.9 Ga, as for the Neoarchean (2.8–2.6 Ga) in both cratonic portions. This isotopic record indicates a protracted Archean history for the São Francisco craton, highlighted by peculiar tectonic-metamorphic histories of the basement rocks. From a tectonic point of view the compiled data concur with a diachronic evolution from Paleo- to Neoarchean times by means of juvenile accretion/differentiation events characterized by multiple TTG plutonism in genetic association with greenstone belts, coupled with partial melting events of earlier-formed material. All ancient basement complexes and/or continental blocks assembled diachronically during the Late Neoarchean by convergence-related processes akin to plate dynamics. Late-tectonic K-rich granitoids, mafic-ultramafic complexes and mafic dikes collectively mark the Neoarchean thickening and final cratonization of the continental crust.

Keywords

Archean • TTG crust • Greenstone belts • K-rich granitoids

3.1 Introduction

The São Francisco craton (SFC) offers an ideal scenario to study aspects of the Archean geology, leading to the understanding of the early geodynamics of Earth, when significant portions of the continental crust formed and stabilized. The oldest components so far detected in the craton include tonalite-trondhjemite-granodiorite (TTG) rocks and

W. Teixeira (✉)
Instituto de Geociências, Universidade de São Paulo, Rua do Lago, 562, São Paulo, SP 05508-080, Brazil
e-mail: wteixeir@usp.br

E.P. Oliveira
Instituto de Geociências, Universidade de Campinas, Campinas, SP 13083-970, Brazil

L.S. Marques
Instituto de Astronomia, Geofísica e Ciências Atmosféricas, Universidade de São Paulo, Rua do Matão, 1226, São Paulo, SP 05508-090, Brazil

M. Heilbron et al. (eds.), *São Francisco Craton, Eastern Brazil*, Regional Geology Reviews, DOI 10.1007/978-3-319-01715-0_3

granite-greenstone associations along with high-grade rocks, all of them providing an isotopic record that extends from ca. 4.1 to 2.5 Ga.

In this chapter, we overview the geologic-tectonic framework of the Archean basement based on the integrated interpretation of U–Pb ages for gneissic and granitic rocks in conjunction with detrital zircon geochronology of the Archean and Paleoproterozoic supracrustal sequences of SFC. An integrated interpretation of U–Pb ages and isotope data with major and trace element geochemistry from Archean igneous rocks provides additional clues about the geodynamic evolution. However, we are aware that this knowledge is limited, owing to the lack of preservation of the ancient crust and/or reworking by successive orogenic events (Cordani et al., this volume). The text follows the classical systematization of Archean basement complexes (e.g., gneissic-granitic, granite-greenstone, high-grade terrains) and Archean/blocks/fragments to address the most relevant components of the crystalline basement in the southern and northern portions of the SFC in extent and ages (Fig. 3.1). As a whole, our tectonic interpretation was based on the evaluation of published and new isotopic and geochemical data, as follows: (i) U–Pb ages on zircons (SHRIMP, TIMS, LA-ICP-MS) are considered as recording the timing of the major accretionary episodes; (ii) U–Pb ages of detrital zircons (supracrustal sequences) are also merged to infer the ages of the Archean sources and related crust-formation episodes; (iii) depleted mantle Sm–Nd model ages (T_{DM}) and their coherence with U–Pb dating are used for inferences about the time of mantle/differentiation episodes; (iv) $\varepsilon_{Nd(t)}$ and $\varepsilon_{Hf(t)}$ isotopic constraints are used for identifying the major petrogenetic components of a given magmatic event, whether juvenile or reworked ones; (v) geochemical data (e.g., REE) of igneous and metaigneous rocks are combined with the bulk isotopic signature of a given block or regional metamorphic complex to evaluate the nature of the accretion events and to infer the likely tectonic setting.

3.2 The Archean Record of the São Francisco Craton

The Archean basement assemblages crop out in the southern and northern lobes of the craton. Smaller exposures include a fault bounded block in the central portion of the Paramirim aulacogen in the northern portion of the craton, and stratigraphic windows in the interior and margins of the São Francisco basin (see Reis et al.; Cruz and Alkmim, in this book) (Fig. 3.1b, c). Different approaches, and consequently, different subdivision criteria have been used by authors working in the Archean terrains of the southern and northern cratonic portions. The southern exposure encompasses the Quadrilátero Ferrífero mining district (see Alkmim and Teixeira, this book) and adjoining areas, where the Archean substratum has been traditionally subdivided into various metamorphic complexes and greenstone belts (e.g., Rio das Velhas Supergroup), largely overprinted by Paleoproterozoic episodes. The Archean assemblages exposed in the northern portion comprise, on the other hand, a mosaic of individual Archean blocks bounded by the Paleoproterozoic orogenic domain of eastern Bahia (see Barbosa and Barbosa, this book). Correlations between the southern and northern portions regarding the Archean evolution are yet imprecise, which led us to present separated descriptions for them.

The Archean/Paleoproterozoic basement of the SFC is an extension of the much larger Congo craton of western-central Africa (e.g., Trompette 1994) (Fig. 3.1a). It consists of distinct Archean blocks, dated between 3.5 and 2.5 Ga. They are composed of polyphase medium- to high-grade metamorphic rocks and granite-greenstone associations (e.g., Chaillu-Gabon, Angola, Kasai, Uganda, Tanzania, E-Zaire) intruded in places by granitoid rocks, gabbro-anorthosite complexes and mafic dikes. In a similar way as the SFC, these blocks were extensively affected by Paleoproterozoic orogenic episodes (the Eburnean orogeny; 2.3–1.9 Ga), during which amalgamation of the proto-West Congo craton occurred (Cahen et al. 1984; Teixeira and Figueiredo 1991; Borg and Shackleton 1997; Ernst et al. 2013). In particular, Iizuka et al. (2010) used U–Pb dating of detrital zircons from the Congo river sands to evaluate the relevant age peaks in relation with the timing of supercontinent assembly. Coupled Hf isotopic constraints provided a further clue for the nature of the crustal components that participated in the sedimentary system through time. For instance, the slightly positive to moderately negative $\varepsilon_{Hf(t)}$ values suggest that episodes of crustal derived granitoid magmatism (Hf T_{DM} ages in the range of 3.3–3.0 and 2.9–2.6 Ga) have been the primary agent of differentiation of the continental crust (plus sedimentary recycling) since the Archean Eon. As such, this evolutionary history is analogous to that of the SFC as presented below.

3.3 Archean Assemblages of the Southern Portion of the Craton

Distinct Archean gneissic-granitic complexes characterize the southern portion of the SFC (e.g., Campo Belo, Santa Bárbara, Belo Horizonte, Bonfim, and Passa Tempo) (Fig. 3.2). They constitute a medium- to high-grade metamorphic terrain that crops out from the Quadrilátero Ferrífero towards the west, and mainly comprises TTG rocks, migmatites and K-rich granitic plutons. Remnants of supracrustal rocks, as well as mafic-ultramafic layered bodies and mafic dikes are also present (e.g., Machado et al.

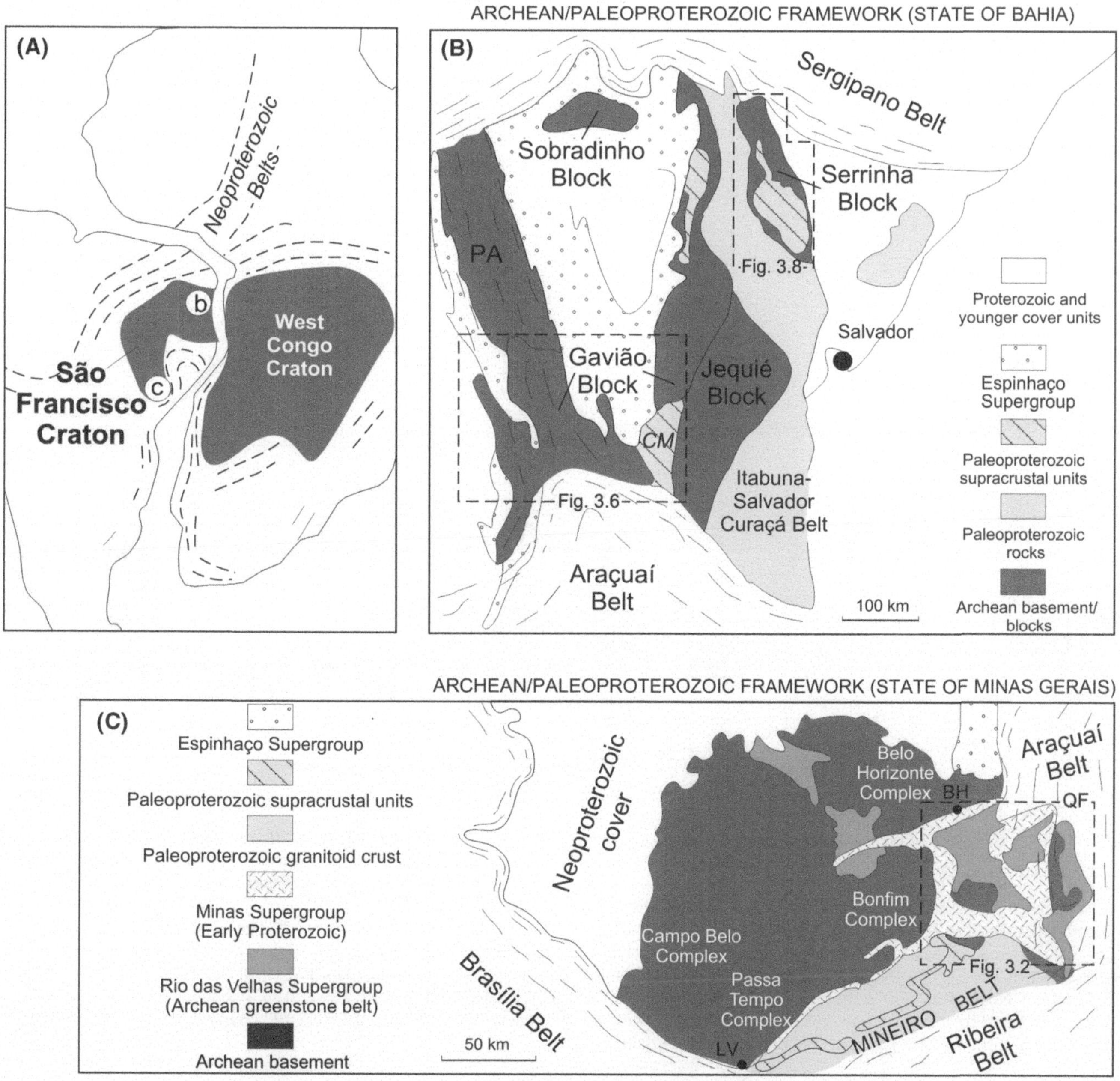

Fig. 3.1 A Tectonic sketch of the São Francisco craton and the adjoining West Congo craton, outlined by the Neoproterozoic marginal belts. *Insets* **B** and **C** show the geologic framework on the northern and southern portions of the craton, emphasizing the extent of the Archean basement and the adjoining Paleoproterozoic belts. Keys (**b**): *PA* Paramirim aulacogen, *CM* Contendas-Mirante supracrustal belt. See text for details

1992; Teixeira et al. 2000; Pinese et al. 1995; Alkmim and Noce 2006; Romano et al. 2013; Lana et al. 2013; Goulart et al. 2013). These metamorphic complexes show contrasting orientations of both mylonitic foliation and transpressional structures, and record three deformational and metamorphic Archean episodes, namely the Rio das Velhas I, II, III events (see Table 3.1 in Campos and Carneiro 2008 and references therein).

Notably, Lana et al. (2013) who investigated the polydeformed TTG rocks in the Quadrilátero Ferrífero by means of detailed U–Pb geochronology deciphered an episodic accretion of the crust in the time intervals of 3.21–3.10, 2.93–2.90, and 2.80–2.77 Ga. The youngest episode includes widespread K-rich plutonism, recently reassessed by Romano et al. (2013) based on extensive zircon U–Pb work and previously published information. The Neoarchean

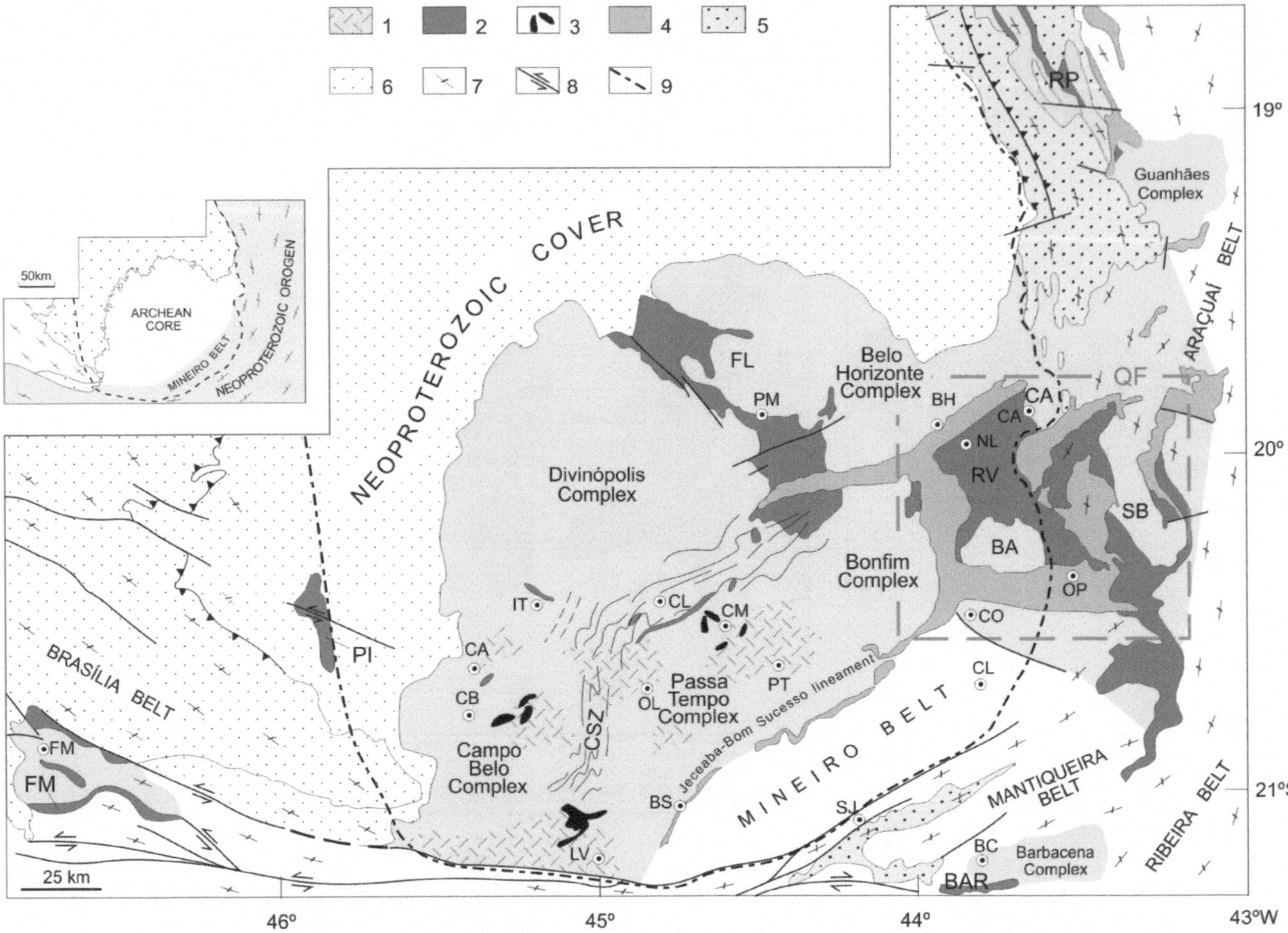

Fig. 3.2 Geologic sketch of the southern segment of the SFC showing the Archean landmass, distinguished by the Campo Belo, Passa Tempo, Bonfim and Belo Horizonte metamorphic complexes, as well as by gneissic-migmatitic domes: SB (Santa Bárbara), CA (Caeté), FL (Florestal), BA (Bação). The Paleoproterozoic Mineiro and Mantiqueira belts and scattered Archean sialic remnants (Barbacena, Guanhães, Gouveia) within the Neoproterozoic marginal belts are also shown (adapted from Noce et al. 2007a, b; Seixas et al. 2012; Lana et al. 2013). *1* Archean crust; *2* Archean greenstone belts: RV (Rio das Velhas Supergroup), RP (Rio Paraúna), PI (Piumhi), FM (Fortaleza de Minas), BAR (Barbacena), *3* Mafic-ultramafic complexes. *4* Minas Supergroup, *5* Espinhaço Supergroup, *6* Bambuí Group, *7* Neoproterozoic reworking/overprint, *8* faults, *9* limit of the SFC. Keys: CSZ = Cláudio shear zone. Towns: FM (Fortaleza de Minas), CB (Campo Belo), CA (Candeias), IT (Itapecerica), CL (Cláudio), CM (Carmópolis de Minas), OL (Oliveira), PT (Passa Tempo), LV (Lavras), BS (Bom Sucesso), PM (Pará de Minas), NL (Nova Lima), CA (Caeté), OP (Ouro Preto), CO (Congonhas), CL (Conselheiro Lafaiete), SJ (São João del Rei), BC (Barbacena). See text for details

magmatic event, regionally dated between 2.79 and 2.70 Ga by several authors in the southern SFC, has been traditionally attributed to the so-called Rio das Velhas orogeny, during which the Rio das Velhas greenstone belt and coeval gneissic-granitic complexes were formed in an active continental margin setting (e.g., Noce et al. 2005, 2007b; Machado and Carneiro 1992). Subordinate calc-alkaline and tholeiitic magmatism has been ascribed to the syn- to late orogenic stages (e.g., Engler et al. 2002; Goulart et al. 2013). A regional NE–SW trending shear zone (Claudio Shear Zone) transects the central portion of the ancient substratum as evidenced by aeromagnetic data. Along this shear zone the country rocks show dextral kinematics and mylonitic character (Oliveira and Carneiro 2001; Campos et al. 2003; Campos and Carneiro 2008) (Fig. 3.2).

A precursor Paleoarchean continental crust can be envisaged by few isotopic data from basement and supracrustal rocks of the southern SFC. The evidences are: (i) subordinate distribution of whole rock Sm–Nd T_{DM} ages in orthogneisses varying from 3.5 to 3.3 Ga (Carneiro et al. 1998; Teixeira et al. 2010; W. Teixeira, unpublished data); (ii) inherited zircon population dated at 3374 Ma (concordant $^{207}Pb/^{206}Pb$ age) from the Neoarchean Carmópolis de Minas mafic-ultramafic suite (see Goulart et al. 2013); (iii) U–Pb ages of detrital zircons from Archean (Rio das Velhas Supergroup) and Paleoproterozoic metasedimentary

Table 3.1 Characteristics of the Archean evolution of the southern SFC based on new data and compiled information

Events	U–Pb age range (Ga)	Geologic-tectonic characteristics	Isotopic constraints
Neoarchean/Early proterozoic	–	Crustal exhumation and tectonic stability of the continental mass	–
Neoarchean 2: Rio das Velhas orogeny (late episode)	2.72–2.70$^{(ca)}$ 2.61–2.55$^{(gp)}$	Passa Tempo Complex: crustal reworking under granulite facies conditions and collisional tectonics. Mafic-ultramafic layered intrusions (Carmópolis de Minas$^{(cms)}$; Ribeirão dos Motas). Basin tectonics. High grade metasedimentary rocks. Claudio shear zone. Post-tectonic to anorogenic plutons and basic dikes	Nd crustal derived isotopic signatures in granulitic rocks and migmatites. Chemical affinity of tholeiitic to calc-alkaline magmatism$^{(cms)}$ coupled with juvenile Nd isotopic signatures U–Pb ages (2.75 and 2.73 Ga) in detrital zircons from the Rio das Velhas greenstone belt. Local quartzites with restricted detrital age components (2.75 and 2.65 Ga)
Neoarchean 1: Rio das Velhas orogeny (early episode):	2.80–2.75$^{(cv)}$ (may include inherited ages up to 3.0)	Granite-greenstone terrain. Accretion the Bonfim and Belo Horizonte Complexes, and domic gneisses and migmatites in the Quadrilátero Ferrífero. Regional amphibolite facies conditions. Rio das Velhas greenstone belt volcanics. Island arc accreted to the Mesoarchean core	Gneissic rocks with juvenile-like Nd signatures (Bonfim and Belo Horizonte Complexes) with geochemical affinity with convergent margin setting of island arcs
Mesoarchean 2: Campo Belo orogeny	3.05–2.92 (including inherited ages). 2.84 Ga$^{(mg)}$	Progressive magmatic accretion (granite-greenstone terrain). Juvenile TTG gneisses and migmatites Campo Belo Complex and Santa Bárbara and Bação domes (Quadrilátero Ferrífero). Regional medium-grade metamorphism. Greenstone belts (dismembered): Piumhi, Rio Paraúna, Barbacena, Fortaleza de Minas	Juvenile-like Nd isotopic signatures. Pb inherited componentes in the Bonfim and Belo Horizonte Complexes. Regional distribution of Sm–Nd T_{DM} ages (3.2–3.0 Ga)
Mesoarchean 1: Campo Belo orogeny (includes the Santa Bárbara event)	3.22–3.20 (including inherited ages)	Juvenile generation of TTG gneisses (medium-to high grade terrain): Campo Belo Complex and Santa Bárbara dome	Protholiths of Paleoproterozoic gneisses (outboard the Quadrilátero Ferrífero)
Paleoarchean (Archean core)	>3.3	–	Pb-inherited age component in country rocks. U–Pb detrital zircon ages in supracrustal strata. Minor cluster of Sm–Nd T_{DM} ages (3.5–3.3 Ga)

Keys $^{(ih)}$inherited; $^{(zc)}$zircon core; $^{(pl)}$felsic plutons; $^{(mg)}$migmatization; $^{(cv)}$coeval felsic volcanism in the Rio das Velhas greenstone belt; $^{(cms)}$Carmópolis de Minas suite; $^{(gp)}$late-tectonic to anorogenic plutonism; $^{(ca)}$calc-alkaline(See Fig. 3.2 and text for details).

rocks (Minas Supergroup) ranging between 3.3 and 3.5 Ga, and one zircon core is as old as 3809 Ma, among other younger age peaks (Machado et al. 1996; Hartmann et al. 2006; Rosière et al. 2008; Moreira et al. 2016). Collectively these results offer a further clue for very ancient material as source for these units.

Detailed geological mapping with the support of zircon U–Pb geochronology, whole rock Nd–Sr isotopic and geochemical data is available for most of the Archean metamorphic complexes, as well as for two classical supracrustal units exposed in the Quadrilátero Ferrífero: (i) the Rio das Velhas Supergroup, an Archean greenstone belt that hosts world-class gold deposits (e.g., Ribeiro-Rodrigues 2001; Baltazar and Zucchetti 2007; Lobato et al. 2007 and references therein); and (ii) the Minas Supergroup (<2.58 Ga; Hartmann et al. 2006), an Early Paleoproterozoic platformal to syn-orogenic succession containing high-grade iron ore deposits hosted by a Lake Superior type banded iron formation (see Alkmim and Teixeira, this volume). Recent combined U–Pb and Lu–Hf isotope studies on detrital zircon and xenotime grains of the Moeda Group metaconglomerates and quartzites indicated that the sediments from the Minas Supergroup derived mainly from an evolved Archean continental crust (up to 3.3 Ga old) that was subjected to episodic crustal recycling and limited juvenile accretion (Koglin et al. 2014), i.e., the same sources of the underlying Archean Rio das Velhas Supergoup (see below). Moreover, the Archean and Paleoproterozoic supracrustal sequences and the TTG gneisses were largely affected by Paleoproterozoic metamorphism and extension tectonics dated between ca. 2.1 and 2.0 Ga. This event, usually known as the "Minas diastrophism" or "Transamazonian orogeny" (e.g., Alkmim and Noce 2006 and references therein), is now attributed to the Paleoproterozoic Minas accretionary orogen (e.g., Teixeira et al. 2015).

The protracted Archean history of the basement rocks agrees well with the idea of the existence of distinct crustal segments, originated through successive accretion episodes. In the next sections, we review the main geological and tectonic features of the Archean crust exposed in the southern portion of the SFC, bounded by the Paleoproterozoic Mineiro belt (Fig. 3.2). Therefore, a tectonic model is addressed, supported by the geochronological, isotopic and geochemical information. Two crustal segments can be distinguished in the region: (i) the Campo Belo metamorphic complex which includes a Mesoarchean inherited U–Pb age component first time identified in the southern SFC (Teixeira et al. 1998 and references therein) and a coeval remnant in the Quadrilátero Ferrífero known as the Santa Bárbara dome (Fig. 3.2); and (ii) the Belo Horizonte, Divinópolis, Bonfim, and Passa Tempo Complexes that constitute much of the Neoarchean continental crust.

3.3.1 The Mesoarchean Campo Belo Metamorphic Complex

The Campo Belo Complex (Fig. 3.3) is a medium to high-grade terrain, predominantly composed of amphibolite facies gray-greenish TTG orthogneisses, diorites, enderbites and pink granitic plutons. Migmatites are widespread, containing lenticular bodies of metabasites, often cut by aplites and pegmatites, and occasionally by Archean and Proterozoic mafic dykes (Teixeira et al. 2000). These country rocks typically exhibit a NS to NW-trending compositional banding. Within the Campo Belo Complex, relicts of clastic and chemical sedimentary units, which can host large graphite deposits, are locally exposed (e.g., Teixeira et al. 1996, 1998; Fernandes and Carneiro 2000; Oliveira and Carneiro 2001; Goulart et al. 2013) (Fig. 3.2).

Rb–Sr whole-rock isochron ages between 2904 ± 56 and 2881 ± 54 Ma with very low initial $^{87}Sr/^{86}Sr$ ratios ($\leq$ 0.7018) in the nearby gneissic rocks of the Campo Belo Complex firstly indicated the occurrence of Mesoarchean crust in this portion of the SFC, as also suggested by compatible Pb evolution in a mantle-like reservoir with Pb-single stage (μ_1) = 8.18 (Teixeira et al. 1996 and references therein). Further U–Pb SHRIMP zircon dating by Teixeira et al. (1998) on a migmatite of the complex revealed three melting events. The oldest zircons yielded a concordant age of 3205 ± 25 Ma, interpreted as inherited age, reflecting the contribution of the primeval continental crust. A second concordant age cluster around 3047 ± 25 Ma was interpreted as recording the main accretion episode in the Campo Belo Complex. The youngest zircon population in the Campo Belo migmatite yielded 2840 ± 17 Ma, considered to be the crystallization age of the neosome. The observed variation of the $\varepsilon_{Nd(2.84\ Ga)}$ values (+0.9 and −3.0) is consistent with the identified melt generations in the migmatite. Therefore, the 3.2–2.9 Ga Campo Belo Complex originated from a juvenile accretionary process leading to a primeval continental core, which is here referred to as "the Campo Belo orogeny" for the first time. This tectonomagmatic episode matches in age with both the Santa Bárbara (ca. 3.21 Ga) and Rio das Velhas I (2.93–2.90 Ga) tectonic-magmatic events according to previous published U–Pb zircon ages in the Santa Bárbara dome (eastern portion of the Quadrilátero Ferrífero), as well as in the gneissic basement of the adjoining Proterozoic belt (Corrêa Neto et al. 2012; Noce et al. 2007c; Lana et al. 2013) (Fig. 3.2 and Table 3.1). In particular, the crystallization of the Santa Bárbara TTG gneisses is well constrained by U–Pb ages between 3210 ± 8 and 3212 ± 9 Ma (Lana et al. 2013). This provides the first direct evidence of the existence of an undisturbed ancient sialic crust in the southern SFC, previously inferred by the inherited U–Pb age component obtained in the Campo Belo migmatite (see above). In a

Fig. 3.3 **a** TTG-Orthogneisses of the Campo Belo Complex, exposed in quarry. **b** Key exposure of the Alberto Flores orthogneiss (2772 ± 6 Ma), the oldest lithostratigraphic unit of the Bonfim complex. Zircon cores from the orthogneiss are as old as 2.92 Ga. The Alberto Flores orthogneiss is cut by folded and weakly foliated dikes termed as Brumadinho granite (2702 ± 24 Ma) (Photo: E. P. Oliveira). **c** Field aspect of the Alberto Flores orthogneiss exhibiting a well developed N–S foliation, Neoarchean in age (Photo: E. P. Oliveira). **d** Field aspect of the Fontex stone quarry where the Neoarchean migmatite of the Divinópolis metamorphic Complex is locally rich in amphibolite boudins (Photo W. Teixeira). **e** Field aspect of the Guanhães TTG banded migmatitic gneiss exhibiting complex folding. **f** Banded trondhjemitic gneiss (2.90 Ga) exposed in the Serrinha quarry, southwestern Bação dome (Quadrilátero Ferrífero) (Photo: F. Alkmim)

similar manner, Lana et al. (2013) reported an identical U–Pb age on zircon cores (3213 ± 13) from a gneissic rock within the Quadrilátero Ferrífero. These crystallization ages, on the other hand, compare well with T_{DM} whole rock ages (3.2–3.0 Ga) and coupled juvenile-like $\varepsilon_{Nd(t)}$ values (see Teixeira et al. 2000 for review) for the gneissic-granitic rocks in the entire region, and point to the role of continental growth in Mesoarchean times (Teixeira et al. 2000, 2010; Silva et al. 2002, 2012b).

Scattered relicts of Mesoarchean greenstone belts (e.g., Piumhí, Fortaleza de Minas-Morro do Ferro, Barbacena and Rio Paraúna) crop out within the Proterozoic framework outside the craton boundaries. These units comprise strongly deformed and metamorphosed Mg-rich mafic and ultramafic volcanics, chemical and clastic sedimentary rocks. Some of the meta-igneous rocks have been dated by various methods, yielding ages between 3.2 and 2.9 Ga, as reviewed by Baars (1997). This suggests at first glance a tectonic relationship with the Campo Belo orogeny (see above). Notably, due to the scattered geographic distribution of these greenstone belt remnants, a larger extent for the Paleoarchean crust of the proto-SFC should be considered (Fig. 3.2). Table 3.2 presents the geologic-isotopic characteristics of the greenstone belt occurrences.

3.3.2 The Neoarchean Metamorphic Complexes

The Neoarchean gneisses and migmatites form much of the Archean crust. They include Mesoarchean dismembered amphibolites are intruded in places by tonalites, granodiorites and granites. They are usually deformed and metamorphosed under amphibolite facies conditions with greenschist facies

Table 3.2 Geologic-geochronologic characteristics of Mesoarchean greenstone belts in the southern SFC (see Fig. 3.2)

Greenstone belts	Distinguished geologic features	Age dating (Ma)	Occurrence/tectonics	References
Piumhí	Low grade metamorphic mafic-ultramafic rocks, volcaniclastics, intermediate lavas, chemical sedimentary rocks, intruded by felsic subvolcanic rocks. Interlayered sills of anorthosite-gabbro-piroxenite-peridotite. Late tectonic felsic intrusions	Anorthosite sill (U–Pb zircon age: 3116 ± 10); riodacitic intrusive dome ($^{207}Pb/^{206}Pb$ minimum ages: 3000 and 2965)	Uplift, allochthonous to paraauthocthonous volcanic-sedimentary sequence onto the Neoproterozoic cover. Crustal compression tectonically related with the Brasilia marginal belt	Alkmim (2004), Alkmim and Danderfer (1998)
Fortaleza de Minas-Morro do Ferro	Remnants of mafic and ultramafic rocks, chemical and clastic sedimentary rocks and tuffs, metamorphosed up to amphibolite facies	Basal komatiites (Sm–Nd whole rock isochron age: 2863 ± 65) Coeval migmatites (host rocks): Rb–Sr isochron age: 2918 ± 105	Dismembered tectonic slivers within an Archean basement inlier deeply reworked in Proterozoic times (basement of the Brasilia marginal belt). Greenstone rocks are in thermal equilibrium with nearby migmatites, both ductile deformed	Alkmim and Marshak (1998), Alkmim and Martins-Neto (2012)
Barbacena	Barbacena Group: differentiated mafic-ultramafic association (e.g., komatiites) and metasedimentary rocks, truncated by gabbroic dikes	Coeval, differentiated metakomatiites (Sm–Nd and Rb–Sr errorchrons between 3190 and 3220). Syntectonic intrusive trondhjemite (U–Pb inheritance of 3218 ± 16). Gabbroic dike (2706 ± 9 U–Pb age)	NE-trending greenstone association intermingled with basement rocks, syntectonically intruded by trondhjemite-tonalite rocks. Basement remnant within the Neoproterozoic marginal Ribeira belt	Alkmim and Noce (2006), Angelim (1997), Baars (1997)
Rio Paraúna	Rio Paraúna Supergroup: low- to medium grade metamorphic sequence: ultramafic to felsic metavolcanics, chemical (BIF) and clastic sediments, schist, phyllite, quartzite, metaconglomerate	Acid volcanic rock (U–Pb age of 2971 ± 8 Ma)	Tectonic slices of encompassing basement rocks, overlying the southern portion of the Espinhaço Supergroup. The supracrustal sequence is largely obliterated by Neoproterozoic compression (Araçuaí belt)	Baltazar and Zucchetti (2007), Barbosa et al. (2008, 2012)

overprint (Machado et al. 1992; Teixeira et al. 2000; Romano et al. 2013; Lana et al. 2013) (Fig. 3.3). In conjunction with the Rio das Velhas greenstone belt in the Quadrilátero Ferrífero the basement rocks make up an extensive granite-greenstone terrain in the southern portion of the SFC (e.g., Baars 1997; Noce et al. 2007c; Lobato et al. 2007). From a tectonic perspective, the Quadrilátero Ferrí-fero shows a characteristic dome and keel geometry (Marshak et al. 1992 and references therein), highlighted by gneiss domes of granitic, granodioritic and tonalitic composition, and subordinate migmatites encircled by supracrustal sequences (e.g., Rio das Velhas and Minas Supergroups. The domes are locally known as Florestal, Caeté and Bação domes and composed essentially of banded gneisses and migmatites (Fig. 3.2). To the southwest of the Quadrilátero Ferrífero, the Neoarchean crust is composed of medium to high-grade TTG-gneisses locally referred to as Bonfim, Belo Horizonte, Divinópolis and Passa Tempo Complexes with coeval metasedimentary remnants. The geologic relationships between these metamorphic complexes are still tentative, due to polyphase deformation, metamorphism in Archean and Proterozoic times, as well as intensive weathering. These complexes are, on the other hand, crosscut to the west by the Claudio Shear Zone (see above and Fig. 3.2).

The Bonfim Complex (Fig. 3.2) includes two types of amphibolite facies, banded gneisses with strongly deformed, dismembered amphibolites and locally sheared tonalites. Weakly foliated and folded K-rich aplites crosscut the regional foliation of the country rocks (Fig. 3.3a–c). The oldest orthogneiss (Alberto Flores) shows geochemical composition akin to the high Al_2O_3 TTG rock suite, whereas the younger Souza Noschese granitic gneiss, which is intrusive into the Alberto Flores orthogneiss, shows geochemical signature suggestive of derivation from partial melting of trondhjemitic crust (Carneiro et al. 1998; Machado and Carneiro 1992 and references therein).

The lithostratigraphy of the Belo Horizonte Complex (Fig. 3.3) is lesser known than that of the Bonfim Complex (Machado and Carneiro 1992; Noce et al. 1997, 1998; Silva et al. 2012a). The Belo Horizonte Complex consists essentially of banded orthogneisses with amphibolite xenoliths and deformed aplites and felsic veins. Migmatites are very common. These rocks locally show a well-developed, NS-striking, low angle mylonitic foliation, which is also characteristic of the adjoining Guanhães gneissic-migmatitic complex and the Bação gneissic dome (Fig. 3.3a, d, e). The less migmatized orthogneisses of the complex shows a predominantly trondhjemitic composition and a REE pattern comparable with that of the nearby Bonfim and Bação complexes.

The Belo Horizonte Complex was originally dated at 2860 ± 17 Ma (ID-TIMS U–Pb) in zircon from a leucossome of a migmatitic gneiss. An inherited $^{207}Pb/^{206}Pb$ age component yielded 2.92 Ga (Noce et al. 1998). New U–Pb SHRIMP ages on both igneous zircon cores and overgrowths recovered from the same outcrop (Silva et al. 2012b) indicated an upper intercept of a concordia diagram with an age of 2787 ± 14 Ma that defines the crystallization of the magmatic precursor of this gneiss. The previous published U–Pb age is therefore probably a mixed result.

The Passa Tempo Complex comprises granulite facies rocks showing regional high-grade ductile deformation. The granulitic complex, limited to the south by the Jeceaba-Bom Sucesso lineament Fig. 3.2), occurs along the southern end of the Claudio shear zone (Campos et al. 2003 and references therein). The Passa Tempo complex mainly comprises hypersthene-bearing gneissic rocks (mainly charnockite and enderbite), showing characteristic NNW layering. Subordinate lens-shaped serpertinites and gabbros are remarkable, where local boudinage and small shear bands affect the mafic and mafic-ultramafic rocks. Charnockites and enderbites (Engler et al. 2002) that are widespread in the neighborhoods of Lavras and Perdões, close to the inferred limit with the Campo Belo Complex, can be also assigned to the Passa Tempo Complex. In this area migmatites are subordinate, whereas granodioritic to alkali-granitic plutons are locally present. These plutons are products of crustal anatexis, which is coeval with the high-grade metamorphism (Engler et al. 2002). According to these authors, the Passa Tempo rocks usually exhibit a regional retrograde amphibolite facies metamorphism (PT conditions of 600–700 °C and 5–6 kb) with a later weak greenschist facies overprint. The latter episode is considered here to signalize the time of the last uplift of the region. The charnockites exhibit REE patterns characteristic of K-rich granites, and were produced by partial melting of TTG crust. This process must have taken place at less than 40 km depth, where plagioclase is a stable residual phase. By contrast, the enderbites, which are interpreted as part of a bimodal suite, were formed through a dual phase partial melting process at mantle or lower crustal depths, where garnet occurs as a stable residual phase.

The main Neoarchean metavolcanic-sedimentary association (Rio das Velhas Supergroup) occurs in the Quadrilátero Ferrífero and in the region to the west of it, in the vicinity of Pará de Minas and Florestal. This unit constitutes a tectonically dismembered, strongly hydrothermally altered greenstone belt that hosts world-class gold deposits (e.g., Schorscher et al. 1982; Ribeiro-Rodrigues 2001; Baltazar and Zucchetti 2007; Lobato et al. 2007; Noce et al. 2007c). The Rio das Velhas Supergroup is subdivided from base to top into the Nova Lima and Maquiné groups. The lower unit is composed of ultramafic and mafic volcanic rocks, minor felsic volcanics, as well as chemical and clastic sedimentary rocks (Fig. 3.4 a, b). From the tectonic point of view, the Rio das Velhas Supergroup is coeval with the Bonfim and Belo

Fig. 3.4 Rocks of the Rio das Velhas greenstone belt succession. **a** Komatiite of the basal portion of the Rio das Velhas Supergroup showing spinifex texture (Photo: F. Alkmim). **b** Deformed pillow lavas of the basal Rio das Velhas greenstone belt succession. **c** Cross-bedded quartzite of the clastic upper portion of the Nova Lima Group (Photo: C.M. Noce). **d** Ribeirão dos Motas stratiform sequence (ca. 2.70 Ga old): the composition of the subhorizontal ultramafic layers ranges from peridotite to pyroxenite (Photo: M.A. Carneiro)

Horizonte Complexes (Fig. 3.2). Collectively this Neoarchean granite-greenstone terrain was affected by the so-called Rio das Velhas orogeny (e.g., Machado et al. 1992; Machado and Carneiro 1992; Teixeira et al. 2000 and references therein).

Scattered mafic-ultramafic layered bodies occur in the southern SFC, such as the Carmópolis de Minas Suite and the Ribeirão dos Motas stratiform sequence. The Carmópolis de Minas Suite, intrusive into the Passa Tempo metamorphic complex, occurs not far from the southernmost end of the Claudio Shear Zone. According to Goulart et al. (2013), it consists of a layered and massive plutonic or sub-volcanic, mafic-ultramafic to felsic association, complexly deformed and metamorphosed under upper amphibolite-granulite facies conditions (like the Passa Tempo metamorphic complex), and includes rocks with tholeiitic and calc-alkaline affinity. U–Pb zircon ages for amphibolite and meta-rhyolite from this suite indicated crystallization ages of 2752 ± 18 and 2713 ± 10 Ma respectively, with a number of inherited grains of Mesoarchean age (Table 3.1). Very subordinate metasedimentary rocks crop out close to the Carmópolis de Minas Suite, occasionally intercalated with the amphibolite lenses. The nearby Ribeirão dos Motas stratiform sequence comprises slightly deformed and metamorphosed layers of peridotite and pyroxenite, and subordinate amphibolite and gabbronorite (Fig. 3.4c). This sequence was emplaced at ca. 2.79 ± 0.3 Ga ($\varepsilon_{Nd(t)} = +0.5$), as indicated by a Sm–Nd whole rock isochron using several of the lithotypes (Carneiro et al. 2004).

Granulite facies metamorphism predominates over the ultramafic rocks, overprinted by amphibolite and greenschist facies paragenesis (Fernandes and Carneiro 2000). Locally, the Ribeirão dos Motas rocks were affected by dextral mylonitic foliation related to the Claudio Shear Zone (Carneiro et al. 1997).

3.3.3 Crustal Evolution

Available U–Pb ages and geochemical data for medium- to high-K granitoids and orthogneisses (2920–2850, 2800–2760 Ma, 2750–2720 Ma) in the Quadrilátero Ferrífero and surroundings characterize not only the polycyclic nature of the Meso- and Neoarchean continental crust of the southern SFC (as old as 3.3 Ga), as already demonstrated by Lana et al. (2013), but also a major change in the composition that reflect melting of different sources through time (Farina et al. 2015). Considering the Neoarchean era, the isotopic and geochemical data on basement rocks, and granitic and mafic-ultramafic intrusions (e.g., Teixeira et al. 2000; Goulart et al. 2013; Romano et al. 2013; W. Teixeira, unpublished data) suggest that the Rio das Velhas orogeny (after Machado and Carneiro 1992) would include two distinct crustal growth episodes dated at ca. 2790–2750 and 2730–2700 Ma (Tables 3.1, 3.2 and Fig. 3.2).

The early Neoarchean episode originated the extensive granite-greenstone terrain, represented by the Bonfim and Belo Horizonte complexes and nearby (domed) gneisses (e.g., Florestal, Caeté, Bação), to which the Rio das Velhas Supergroup is genetically related (e.g., Machado et al. 1992; Teixeira et al. 1996; Campos et al. 2003; Romano et al. 2013; Lana et al. 2013). For instance, the Alberto Flores orthogneiss (Bonfim complex) yields U–Pb crystallization age of 2772 ± 6 Ma (zircon overgrowth), whereas the zircon core gives 2.92 Ga, which is considered a minimum age for the protolith (Machado and Carneiro 1992; Teixeira et al. 2000). This hypothesis agrees well with a group of whole rock Sm–Nd T_{DM} ages (2.80–2.94 Ga) in gneissic rocks of the southern SFC, as well as with recently published U–Pb zircon ages in the Quadrilatero Ferrífero rocks (Farina et al. 2015). The bulk REE pattern and $\varepsilon_{Nd(t)}$ isotopic characteristics of these rocks are consistent with juvenile accretion in a convergent margin setting (e.g., Machado and Carneiro 1992). Specifically the Samambaia intrusive tonalitic gneiss (Bação complex), which encloses mafic volcanics (now amphibolites), yields U–Pb zircon and titanite ages of 2780 +3/−2 Ma and 2774 ± 6 Ma respectively. The slightly negative $\varepsilon_{Nd(t)}$ values calculated for the Bonfim gneisses and amphibolites suggest that much of the crust formed during the 2780–2700 Ma orogenic episode (Teixeira et al. 1996 and references therein). Granodioritic to granitic plutons, such as the Mamona Granitoid and the Brumadinho granitic aplite yield U–Pb ages of 2721 ± 3 and 2703 ± 24/−20 Ma, respectively, and are related to the late-tectonic evolution of the Bonfim complex (e.g., Machado and Carneiro 1992; Alkmim and Noce 2006; Farina et al. 2015)—see below. The Belo Horizonte Complex (2787 ± 14 Ma; Silva et al. 2012a) can be similarly attributed to the early episode of the Rio das Velhas orogeny (Table 3.1).

The generation of such a continental crust (e.g., Belo Horizonte and Bonfim complexes) is roughly concomitant with various granitic plutons in the Quadrilátero Ferrífero region, as well as with felsic volcanics and deposition of siliciclastic rocks in the Rio das Velhas greenstone belt (Noce et al. 1997; Hartmann et al. 2006). For instance, Campos and Carneiro (2008) first demonstrated that the oldest production of TTG crust (ca 2780–2770 Ma) is a regional scale phenomenon (see Table 3.1). According to the same authors, these plutons mainly derived by juvenile accretion at the time, although a few inherited zircon ages (up to 3.0 Ga) indicate that Mesoarchean crust was assimilated during magma genesis, as also suggested by low negative $\varepsilon_{Nd(t)}$ values (see Table 3.1). Some migmatitic gneisses outcropping to the west of the Belo Horizonte and Bonfim Complexes, locally affected by the Claudio Shear Zone (Fig. 3.2), can be tentatively attributed to the early episode of the Rio das Velhas orogeny due to their U–Pb zircon ages (upper intercept with the Concordia) between 2765 and 2750 Ma. These migmatitic gneisses contain again a number of 2.90 Ga inherited zircon grains, in a similar way to the Bonfim and Belo Horizonte rocks (Oliveira 2004).

Lana et al. (2013) reported a detailed U–Pb survey on Neoarchean TTG gneisses and granitoids in the Quadrilátero Ferrífero and surroundings. They confirmed that the main magmatism, deformation and amphibolite-facies metamorphism took place between 2800 and 2770 Ma—attributed to their Rio das Velhas II event. According to these authors, the available geologic and isotopic evidences support a model in which both the TTG rocks and the Rio das Velhas volcanics are genetically related with an island arc accreted to the Mesoarchean continental margin. This model was recently demonstrated by detailed U–Pb work on distint lithoestratigraphic units of the Rio das Velhas Supergroup (e.g., Moreira et al. 2016), indicating the Meso- and Neoarchean basement rocks as the main sources (Fig. 3.5). Few significantly older detrital zircons from the siliciclastic sequences were also reported by Hartmann et al. (2006).

The youngest detrital zircon grains of the Maquiné Group (upper unit of the Rio das Velhas Supergroup) yield ages between 2718 ± 17 Ma (concordant) and 2746 ± 6 (−2 % discordance) and 2749 ± 6 Ma (1 % discordance), determining the maximum depositional age and the terminal evolutional stage of the Rio das Velhas greenstone belt (Hartmann et al. 2006; Moreira et al. 2016). U–Pb zircon ages of felsic volcanics concomitant with the deposition of the Nova Lima Group sediments (Rio das Velhas greenstone belt) defined three eruption episodes at ca. 2792 ± 11, 2773 ± 7 and 2751 ± 9 Ma (Fig. 3.5), suggesting a range of about 40 Ma for the eruptives (Machado et al. 1992; Noce et al. 1997 and references therein). This time interval is therefore consistent with the first episode of the Rio das Velhas orogeny suggested

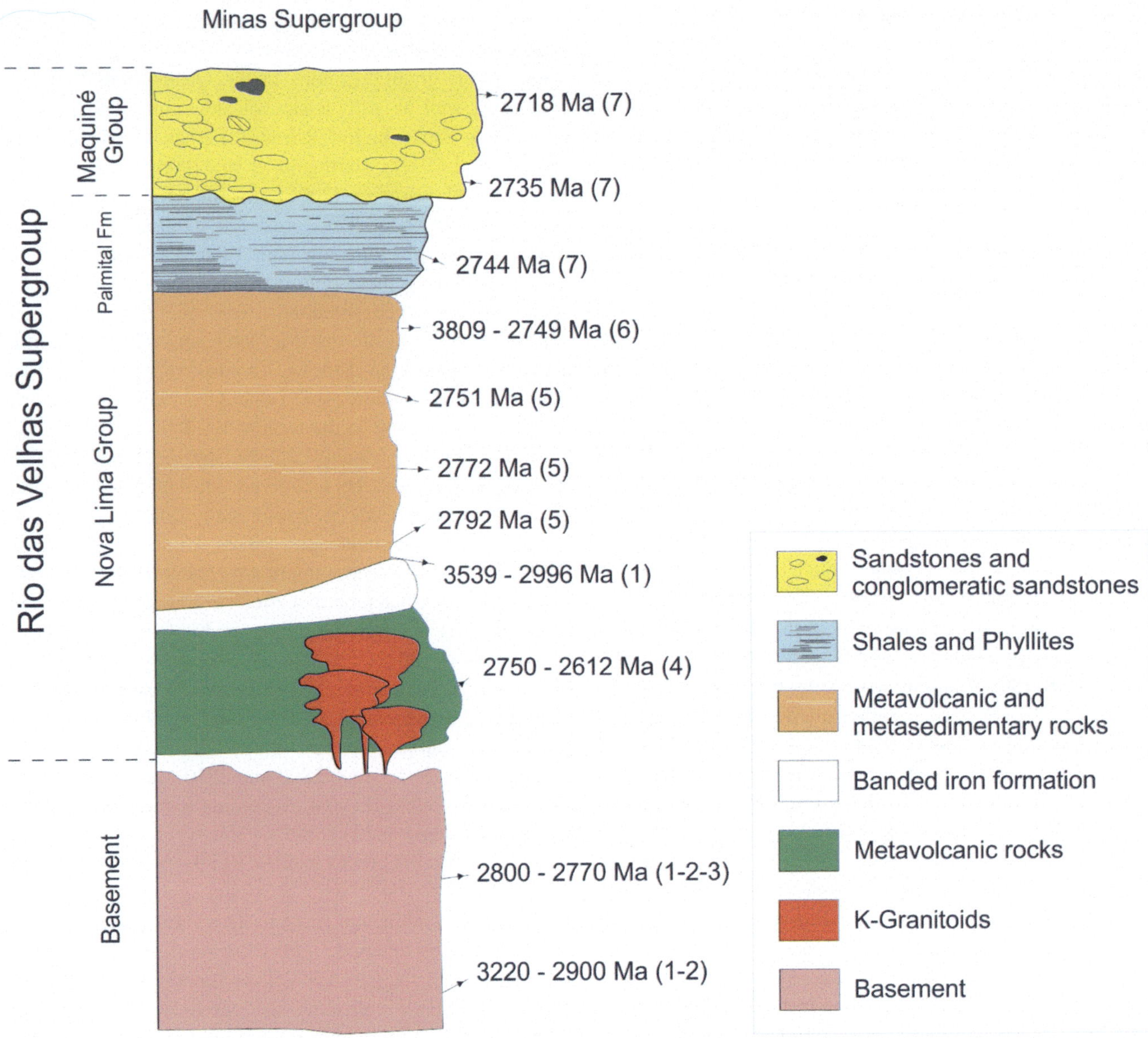

Fig. 3.5 Stratigraphic column of the Rio das Velhas Supergroup (after Moreira et al. 2016 and references therein). Numbers in parentheses after the age data are referred to: (1) Machado et al. (1996); (2) Lana et al. (2013); (3) Machado et al. (1992); (4) Romano et al. (2013); (5) Noce et al. (2005); (6) Hartmann et al. (2006); (7) Moreira et al. (2016)

in this work (Table 3.1). The two oldest volcanic pulses match in age the nearby tonalitic-granodioritic intrusions (e.g. Caeté and Bonfim Complexes), which yielded U–Pb crystallization ages of 2776 +7/−6 Ma and 2780 +3/−2 Ma respectively (Machado et al. 1992; Machado and Carneiro 1992). These plutons are partly or totally emplaced into the domed Archean gneissic-granitic complexes encircled by the Rio das Velhas greenstone belt (Noce et al. 1998). In the Caeté dome (Fig. 3.2), a deformed felsic volcanic rock of the Rio das Velhas Supergroup has inherited zircons that yielded a slightly discordant $^{207}Pb/^{206}Pb$ age of 3029 ± 6 Ma. Therefore, this is again an indirect evidence of a Mesoarchean precursor continental source (e.g., Machado and Noce 1993) in accordance with previous published U–Pb dating for the Bonfim and Belo Horizonte Complexes, including Hf zircon isotopic evidence (Koglin et al. 2014)—see Sect. 3.3.

Ore samples of the Morro Velho and Cuiabá gold deposits yield a U–Pb SHRIMP Concordia age of 2672 ± 14 Ma on hydrothermal monazite grains, which defines the timing of mineralization associated with the Rio das Velhas Supergroup (Lobato et al. 2007). However, Tassinari et al. (2013) report a much younger gold remobilization event under medium grade metamorphism (2.2–2.1 Ga) at the northwestern occurrence of the greenstone belt.

Noce et al. (1997) first evaluated the role of granitic plutons in the Late Archean evolution in southern SFC based on geologic information and U–Pb isotopic evidence, distinguishing three generations of plutons dated at ca. 2780–2760; 2720–2700 and 2600 Ma. Among these events, the younger pulses are considered of crustal origin (e.g., Campos and Carneiro 2008). More recently, Romano et al. (2013) defined two consecutive pulses of K-rich granitic rocks in neighborhoods of the Quadrilátero Ferrífero, dated at 2760–2750 and 2730–2700 Ma. Specifically, the Bação dome (Fig. 3.2), which consists of 2925 ± 5 to 2715 ± 5 Ma TTG gneisses and potassic granitoids, could indicate the ultimate accretion in an island arc setting close to the Mesoarchean landmass, as envisaged by Lana et al. (2013). According to this model the peculiar high-K rocks of the southern SFC should record the final cratonization and consolidation of the continental crust between 2750 and 2720 Ma (Farina et al. 2015). From the above, we suggest the youngest and widespread Neoarchean plutonism of the latest episode of the Rio das Velhas orogeny postulated here (see Table 3.1), as markers of progressive nucleation of the continental mass. This episode probably includes calc-alkaline magmatism (e.g., Carmópolis de Minas layered intrusion) and subsequent high-grade metamorphism having the imprint of polyphase deformation and crustal anatexis (e.g., Passa Tempo granulitic complex) (see above and Fig. 3.2).

The Carmópolis de Minas Layered Suite (2752–2713 Ma) shows predominant low Mg# values that coupled with the presence of low-K_2O rocks in close association with metasedimentary remnants suggest that the protolith may have been derived from evolved magmas in an oceanic arc setting. Specifically, the amphibolite and metarhyolite show slightly positive to slightly negative $\varepsilon_{Nd(t)}$ values respectively, pointing to the involvement of two magmatic sources (restricted to ca. 40 Ma) such as depleted mantle reservoirs and/or enriched mantle sources (Goulart et al. 2013).

We suggest that the Carmópolis de Minas Suite could be related with the Neoarchean arc accreted to the Mesoarchean continental core (i.e., Campo Belo and Santa Barbara complexes), in roughly agreement with the tectonic model of Moreira et al. (2016). During this event a syn-orogenic, tholeiitic phase was shortly followed by sub-arc crustal anatexis to allow for calc-alkaline melts and crustal reworking of short-lived material (Goulart and Carneiro 2013). In other words, there is again evidence for the latest accretionary episode of the Rio das Velhas orogeny postulated here (see Table 3.1). In consequence, the evolving magmatic arc contributed to the ultimate crustal growth of the Neoarchean landmass, as also reflected by a widespread granite-genesis at that time (Noce et al. 1997; Romano et al. 2013).

The ca. 2.7 ± 0.3 Ga Ribeirão dos Motas stratiform sequence that crops out near the Carmópolis de Minas Suite shows a positive $\varepsilon_{Nd(t)}$ value (+0.5) that suggests a derivation from a slightly enriched magma source. The isotopic evidence allows a genetic relationship with the slab subduction process that formed the Carmópolis de Minas Suite (see above). However, the Ribeirão dos Motas mafic-ultramafic rocks show high MgO, Cr, and Ni contents and relatively low SiO_2, TiO_2, K_2O and REE elements akin to komatiite magmas or high magnesian basalts (Carneiro et al. 1997). This geochemical signature is consistent with our hypothesis of partial melting of a peculiar mantle source pertinent to the late episode of the Rio das Velhas orogeny (Table 3.1).

The minimum age for the granulitic facies metamorphism and migmatization of the Passa Tempo complex was first estimated at 2661 ± 36 Ma (Rb–Sr whole rock isochron; Teixeira et al. 1998) later supported by U–Pb dating with an upper intercept of a concordia diagram at 2622 ± 18 Ma (Campos et al. 2003). New U–Pb concordant or nearly concordant zircon U–Pb ages for the Passa Tempo migmatites and gneisses predominating nearby Lavras indicate identical, within error, ages of 2715 ± 12, 2701 ± 5 and 2682 ± 15 Ma (W. Teixeira, unpublished data). These data also determine a reliable age for the anatexis of the Passa Tempo Complex. The rocks yield variable negative $\varepsilon_{Nd(2.70\ Ga)}$ values between −5.1 and −1.5 indicating the important role of crustal reworking at that time (Oliveira 2004; W. Teixeira, unpublished data).

From a geodynamic point of view, development of partial crustal melting under granulite facies concomitant with horizontal shortening is an unequivocal evidence of a compressional setting, syn-kinematic with the 2730–2700 Ma episode of the Rio das Velhas orogeny (see above and Table 3.1). This is also consistent with the nearby occurrence of a large layered mafic-ultramafic intrusions (Carmópolis de Minas, Ribeirão dos Motas) that requires emplacement conditions into a thickened continental crust, now exposed as a deeply eroded root zone (i.e., Passa Tempo granulitic rocks). As a corollary, the geologic boundary between the Neoarchean sialic fragment and the Mesoarchean one (i.e., Campo Belo Complex) is probably obliterated.

Between 2615 and 2550 Ma crustal derived granite plutons (Table 3.1) intruded the Archean substratum of the southern SFC (Noce et al. 1997; Campos and Carneiro 2008; Romano et al. 2013). They were emplaced under brittle-ductile, dextral transpressional conditions with tectonic transport from NE to SW (Rio das Velhas III event; Table 3.1 in Campos et al. 2003). Such a tectonic framework, according to these authors, indicates that a potential continent-continent collision event occurred at the late Archean, as also suggested by the structural/metamorphic framework of the Passa Tempo Complex. In this regard, we suggest that the Claudio Shear Zone with dextral transpression kinematics overprinting coeval migmatitic rocks along the western edge of the Bonfim Complex could be tentatively considered as the collisional front. In other words, the crustal derived granites would indicate the last remobilization and thickening of the continental crust, leading to a coherent, tectonically stable Neoarchean landmass (e.g., Noce et al. 1997, 1998; Romano et al. 2013) (Table 3.1; Fig. 3.2). This coherent lithosphere eventually allowed deposition of the Minas Supergroup in the passive margin setting (e.g. Alkmim and Noce 2006 and references therein). The evolutionary stage was accompanied by extensional tectonics with emplacement of NW-trending, noritic-gabbroic dikes dated at 2658 ± 44 Ma (Sm–Nd isochron; Pinese et al. 1995). This peculiar swarm transects both the Campo Belo metamorphic complex and the Ribeirão dos Motas mafic-ultramafic layered body (Table 3.1).

Final regional cooling of the gneissic-granitic complexes occurred between ca. 2.1–1.9 Ga, as supported by K–Ar and ^{40}Ar-^{39}Ar geochronology over the metamorphic complexes to the west of the Quadrilátero Ferrífero (e.g., Teixeira 1982; Teixeira et al. 1997, 2000; Oliveira 2004). Crustal exhumation occurred concomitantly with Paleoproterozoic retrograde metamorphism overprinting the basement rocks, as well as fold-thrust tectonics as a response from an outboard accretionary orogenic event (Machado et al. 1992; Alkmim and Marshak 1998; Teixeira et al. 2000; Oliveira 2004) (for details see Alkmim and Teixeira, this book).

3.4 Archean Assemblages of the Northern Portion of the Craton

A large number of Archean gneissic migmatitic rocks, granulites, greenstone belts and granitoids makes up the crystalline basement of the northern SFC, which consists of various sialic blocks (Gavião, Sobradinho, Serrinha and Jequié blocks, see Fig. 3.1). This portion of crust was partially or intensively affected by Paleoproterozoic metamorphism and deformation (Barbosa et al. 2012, and references therein) in a similar manner as the southern lobe of the craton.

Two distinct Archean domains separated by the Paleoproterozoic Contendas-Jacobina Lineament (or shear zone) (Fig. 3.6) can be distinguished on tectonic grounds (Barbosa and Sabaté 2004 and references therein): (i) a western domain corresponding to the Gavião Block and its potential northern correlative, the Sobradinho Block; and (ii) an eastern domain that includes the Serrinha and Jequié Blocks, intensively affected by a Paleoproterozoic collisional event responsible for the development of the Itabuna–Salvador–Curaçá belt (see Barbosa and Barbosa, this book).

3.4.1 The Gavião and Sobradinho Blocks

The Gavião Block is one of the most important areas to understand the early crustal evolution in South America, given the extensive exposures of the oldest rocks within the SFC. It is composed of a variety of TTG gray gneisses and migmatites, granites, and dismembered greenstone belt associations (Fig. 3.6). The block is bounded to the east by the Paleoproterozoic Contendas-Jacobina Lineament and to the south by the Neoproterozoic Araçuaí belt. Relicts of the Gavião Block may also form the crystalline basement of volcanic-metasedimentary sequences, such as the Sete Voltas and Boa Vista gneissic domes (Fig. 3.7a) within the Contendas-Mirante supracrustal belt.

The Gavião Block is overlain by metasedimentary sequences of the Paleo/Mesoproterozoic Espinhaço Supergroup and the Neoproterozoic platformal Bambuí Group and correlative units (e.g., Barbosa et al. 2012 and references therein). Of note, basement rocks along the Paramirim aulacogen (Alkmim and Danderfer 1998) at the northwestern portion of the Gavião Block (Fig. 3.1) were thrust over the Espinhaço sequence due to far-field stresses developed during the evolution of the Neoproterozoic Araçuaí belt marginally to the SFC.

3.4.1.1 Granite-Gneissic Terrains

Detailed geologic-geochronologic studies on the Gavião Block have established a protracted evolution from the Paleoarchean to the Neoarchean (Table 3.3), similar to the evolution of the southern SFC. The continental crust formed in the Paleo- and Mesoarchean underwent metamorphism and partial recycling into migmatites and granites between 2.7 and 2.6 Ga (Santos-Pinto et al. 2012). In addition, the Gavião Block includes some of the oldest juvenile granitoid rocks, TTG orthogneisses of the SFC, dated between 3650–3260 and 3180–3000 Ma (Nutman and Cordani 1993; Martin et al. 1997; Peucat et al. 2002; Dantas et al. 2013). Indirect evidence for a Hadean primeval crust (4.1 Ga) is also noticeable by the detrital zircon age in one supracrustal sequence of the Gavião Block. This is also consistent with previously published Sm–Nd T_{DM} model ages on orthogneisses from the same domain (Paquette et al. 2015).

Most of these ancient granitoids intrude or encompass ancient metavolcanic-sedimentary sequences that are scattered over the Gavião Block such as the Ibitira-Ubiraçaba, Mundo Novo and Guajeru greenstone belts (Leahy et al. 1999; Nutman and Cordani 1993; Leal et al. 1996) (Fig. 3.6). Moreover, they constitute the crystalline basement (i.e., the Santa Isabel and Riacho de Santana Complexes) of the Paleoproterozoic Riacho de Santana

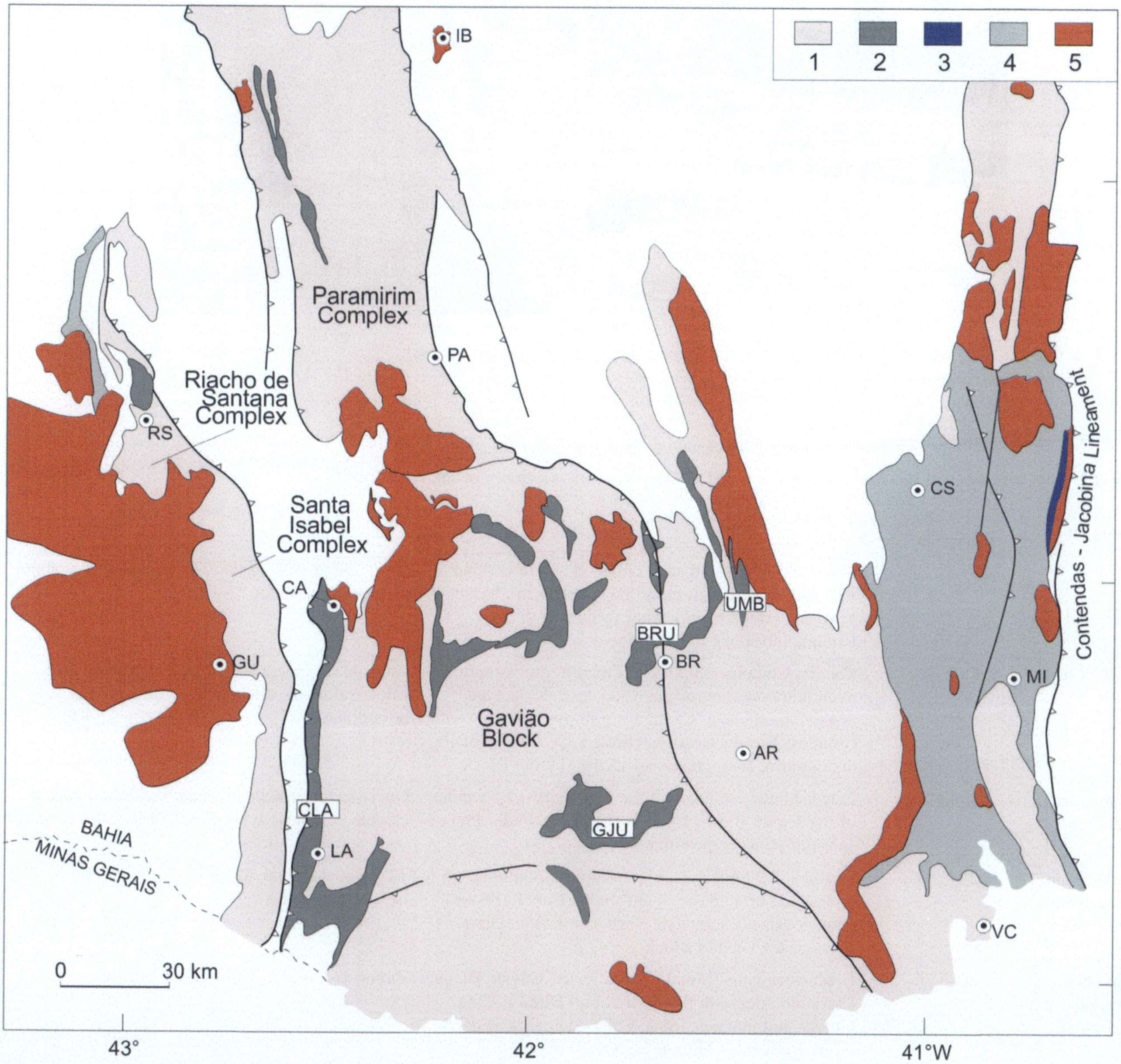

Fig. 3.6 Geologic sketch of the northern portion of the SFC showing the Archean Gavião Block and adjacent terrains, articulated by the Contendas-Jacobina Lineament—CJL (adapted from Barbosa et al. 2012). *1* Archean basement including the Santa Isabel (CSI), Riacho de Santana (RC), and Paramirim (CPA) Complexes and unnamed TTG gneisses and migmatites (CGV), and granitoid rocks. *2* Volcanic-sedimentary sequences including greenstone belts [e.g., Caetité-Licínio de Almeida (*CLA*), Brumado (*BRU*), Guajerú (*GJU*), Umburanas (*UMB*)]; *3* Rio Jacaré sill; *4* Archean-Paleoproterozoic volcanic-sedimentary sequences (e.g. Contendas Mirante); *5* Paleoproterozoic granitoid rocks; *6* cover. Towns: RS (Riacho de Santana), GU (Guanambi), LA (Licínio de Almeida), CA (Caetité), IB (Ibitiara), PA (Paramirim), BR (Brumado), AR (Aracatu), CS (Contendas do Sincorá), MI (Mirante), VC (Vitória da Conquista). See text for details

greenstone belt occurring in the western portion of the Block (Barbosa et al. 2013 and references therein).

Rocks of the oldest age group (3650–3260 Ma) are common in the western, central and eastern border of the Gavião Block, such as the the gneissic-migmatites of the Riacho de Santana Complex, the Sete Voltas and Boa Vista gneissic domes and associated felsic volcanics. The Riacho de Santana Complex (Fig. 3.6; Table 3.3) is the oldest fragment so far recognized in the Block. It yields a ID-TIMS U–Pb zircon upper intercept crystallization age of 3648 $\pm$ 69 Ma and Sm–Nd T_{DM} model age of 3.9 Ga; its negative $\varepsilon_{Nd(t)}$ value is compatible with a protracted history which included a migmatization episode dated at 3.25 Ga (Barbosa et al. 2013). In a similar manner, the Sete Voltas gneissic dome shows a polycyclic evolution, highlighted by TTG gneiss xenoliths as old as 3403 Ma (Sm–Nd T_{DM} age = 3.6–3.7 Ga) within

Fig. 3.7 a Gneissic rocks of the eastern margin of the Paleoarchean Sete Voltas basement dome, where the tonalitic orthogneiss shows E-dipping foliation (Photo: E.P. Oliveira). **b** Pillow basalt of the Paleoarchean Mundo Novo greenstone belt, Cachoeira da Fumaça (Pindobaçu, Bahia) (Photo E.P. Oliveira)

Table 3.3 Characteristics of the Archean evolution and Paleoproterozoic assembly of the northern SFC

Events	U–Pb age range (Ga)	Geologic-tectonic features	Geochemical and isotopic constraints
Early Proterozoic; continental and oceanic arcs	2.17–2.04	Ocean closure, island arcs, collision and reworking of Archean blocks, high-grade metamorphism, terrane extrusion (Uauá block), final formation of the Itabuna-Salvador-Curaçá orogeny	Juvenile Nd isotope data, calc-alkaline signature, metamorphic zircon U–Pb ages
Neoarchean 2: Caraíba	2.65–2.56	Caraíba Complex: continental margin arc batholiths, mafic-ultramafic intrusions (São José do Jacuipe layered anorthosite, Caraiba norite-pyroxenite complex, Rio do Jacaré gabbroic sill). Uauá tholeiite dike swarm. Late granitoids in the Gavião Block	Juvenile to crustal Nd isotope signatures and calc-alkaline geochemistry of Caraíba orthogranulites
Neoarchean 1: Jequié	2.80–2.70	Late tectonic granitoids in the Gavião Block, granites of the Jequié Block, Uauá norite dikes, Pedra Preta chromitiferous serpentinite body	Gneissic rocks with juvenile-like Nd signatures (Jequié complex) with geochemical affinity with continental and oceanic arcs
Mesoarchean: Serrinha-Uauá	3.20–2.98	Serrinha TTG gneisses, Uauá-Capim granitoids, high-grade orthogneisses and Santa Isabel gneisses (Gavião Block). Lagoa da Vaca anorthosite. Granitoid rocks in the Gavião Block	TTG geochemical signatures and juveline Nd isotope signatures
Paleoarchean: Sobradinho and Sete Voltas	3.40–3.26	Greenstone belts (Mundo Novo, in the Gavião Block, Lagoa do Alegre in the Sobradinho Block), TTG plutons and gneisses (Sete Voltas dome), and dacite in the Contendas-Mirante supracrustal belt	Zircon U–Pb ages and Nd–Hf signatures
Paleoarchean to Eoarchean: Enclaves	3.5–3.7	Ancient gneiss-migmatite complexes (e.g., Riacho de Santana gneissic-migmatitic Complex and gabbro-diorite enclaves in Sobradinho migmatitic gneiss)	Zircon U–Pb and Hf ages of diorite enclaves in migmatite (Sobradinho Block), and inherited zircon in the Paleoproterozoic Quijinge granite (Rio Itapicuru greenstone belt)

See text for details

3243 Ma porphyritic granodiorite (T_{DM} age = 3.5–3.7 Ga), or in 3158 Ma gray gneiss (T_{DM} age = 3.5–3.6 Ga). Both country rocks also yield slightly negative $\varepsilon_{Nd(t)}$ values (Nutman and Cordani 1993; Martin et al. 1997). By contrast, the Boa Vista granitic gneiss dome is more homogenous and is 3353 Ma old (Nutman and Cordani 1993). Another roughly contemporary bodies such as the Bernarda tonalitic (3386 Ma) and Aracatu trondhjemitic (3325 Ma) massifs (Santos-Pinto et al. 2012) and the Mariana TTG (3259 Ma) suite (Bastos Leal et al. 2003) occur in the central part of the Gavião Block.

According to Martin et al. (1997), the protoliths of the TTG gneiss xenoliths in the Sete Voltas dome were produced by partial melting of an Archean tholeiite, leaving a hornblende garnet residue, whereas their host granodiorite and gray gneiss have geochemical composition different from that of typical TTG and were probably produced by partial melting of an older continental crust. The authors also suggested that partial melting of the older crust took place at depths of 30–45 km. This information, combined with the rocks deformation characteristics and migmatization, was related with an Archean collisional thickening akin to modern type plate tectonics.

Examples of the youngest age group of rocks (3180–3000 Ma, see Table 3.3) occur in the eastern portion of the Gavião Block, such as a granite gneiss (3158 ± 2 Ma) emplaced into the Sete Voltas gneissic dome (Martin et al. 1997), the Lagoa da Macambira granite (3146 ± 24 Ma, Leal et al. 1998), the Lagoa do Morro granodiorite (3184 ± 6 Ma, Nutman and Cordani 1993), and a gneiss of the Mairi Complex (3040 ± 15 Ma, Peucat et al. 2002). Similarly, in the western part of the Gavião Block, the gneissic rocks of the Santa Isabel Complex (Fig. 3.6) may be contemporary with this orogenic stage, as suggested by reconnaissance U–Pb work that yields an upper intercept at ca. 2.95 Ga (with migmatization at ca. 2.75 Ga). However, Rosa (1999) reports U–Pb crystallization age of 3.35 Ga for a gneiss xenolith in the Paleoproterozoic Cara Suja pluton emplaced into the Santa Isabel Complex—in the western edge of the block. Consequently the latter unit may be significantly older, and related to an early accretionary stage of the Gavião Block. The Santa Isabel Complex displays T_{DM} ages of 3.3–3.1 Ga and variable $\varepsilon_{Nd(t)}$ values (−4.7 to +0.3), suggesting juvenile and reworking processes from the Mesoarchean protholiths (Barbosa et al. 2013). Subsequently, episodes of new magma intrusion took place between 2850 and 2520 Ma all over the Gavião Block. The Serra dos Pombos alkaline granitoid (2845 ± 45 Ma; Marinho et al. 1994a, b), the Caraguataí syenitic augen gneiss (2693 ± 5 Ma; Cruz et al. 2012), and the Serra do Eixo alkaline granite gneiss (2693 ± 5 Ma, Santos-Pinto et al. 2012; or 2524 ± 14 to 2656 ± 10 Ma; Bastos Leal et al. 2003) are examples of this assemblage (Table 3.3).

Between Sobradinho and Campo Alegre de Lourdes (northernmost SFC), gneisses and migmatites that are collectively referred to as the Sobradinho Block (Fig. 3.1; Table 3.3) contain rare gabbro-diorite enclaves, on which Dantas et al. (2010) found an old zircon population dated at 3537 ± 8 Ma. The Sm–Nd T_{DM} model ages for the samples yielded ages up to 3.7 Ga, while zircon Hf T_{DM} model ages are between 3.7 and 3.9 Ga (Dantas et al. 2010, 2013). These data provide robust evidence of Eoarchean juvenile components in the block, like the Riacho de Santana gneissic-migmatitic rocks (see above). According to Dantas et al. (2013), the Sobradinho Block may be an independent micro-continent that accreted to the São Francisco paleocontinent during the Paleoproterozoic orogeny. Alternatively, Barbosa et al. (2012) reviewing the geologic-tectonic framework of the northern SFC interpret the Sobradinho Block as the northern portion of the Gavião Block.

Both the Gavião and Serrinha Blocks include a group of rocks dated between 3180 and 3000 Ma (Table 3.3). These ages agree well with U–Pb crystallization ages from orthogneisses of the Campo Belo—Santa Barbara metamorphic Complexes in the southern portion of the SFC (see previous section). However, a similar crustal evolution does not imply initial proximity of the sialic fragments.

3.4.1.2 Greenstone Belts

Relicts of Archean greenstone belt sequences (e.g., Lagoa do Alegre, Mundo Novo, Umburanas, Guajeru, Ibitira-Ubiraçaba, Brumado, including the lower unit of the Contendas-Mirante belt; Table 3.4) are present in both the Gavião and Sobradinho Blocks (Fig. 3.6). Peucat et al. (2002) obtained a zircon U–Pb SHRIMP crystallization age of 3305 ± 9 Ma for a metadacite of the Mundo Novo greenstone belt, showing that this is also one of the most ancient supracrustal sequences in the craton, although few outcrops exhibit well preserved primary features (Fig. 3.7b). Sm–Nd T_{DM} ages for the metadacites vary from 3.6 to 3.4 Ga, with $\varepsilon_{Nd(3.30)}$ values in the range +0.5 to −1.9 (Oliveira et al. 2010). Although few whole rock major and trace element data are available for the volcanic rocks for petrogenetic inferences (e.g., Mascarenhas et al. 1998), the variation of the $\varepsilon_{Nd(t)}$ values suggests that the Mundo Novo rocks may belong to an arc assemblage accreted onto the Gavião Block (Oliveira et al. 2010).

The lower unit of the Contendas-Mirante supracrustal sequence (Table 3.3) contains metabasalt, dacite, BIF, metacherts and marble (Marinho et al. 1994a). TIMS U–Pb zircon dating of a dacite yielded an upper intercept age of 3304 +31/−24 Ma, whereas the Sm–Nd T_{DM} ages are 3.3-3.4 Ga). In addition, BIF samples gave a whole rock Pb–Pb isochron with 3265 ± 51 Ma (Marinho et al. 1994a). This suggests a time correlation between the Contendas-Mirante lower unit and the Mundo Novo greenstone belt, although additional U–Pb provenance studies on the siliciclastic sequences are crucial for testing this hypothesis. However, the upper units of the Contendas-Mirante greenstone belt are Paleoproterozoic in age according to SHRIMP U–Pb work (Nutman et al. 1994).

The Umburanas, Brumado and Ibitira-Ubiraçaba greenstone belts (Fig. 3.6; Table 3.4) are similar and contain metakomatiites and metabasalts at the base (Fig. 3.7b), followed up section by felsic metavolcanics and metasedimentary rocks. Detrital zircons recovered from one Ibitira-Ubiraçaba schist collected close to Caetité town indicated ages ranging between 3.5–2.5 and 2.0–2.1 Ga (prevailing metamorphic event), except one grain (4.1 Ga)—the oldest zircon so far found in the SFC (Paquette et al. 2015). According to these authors this indicates that Hadean-aged materials incorporated to the SFC, in agreement with published Sm–Nd T_{DM} ages (see below). The

Table 3.4 Geologic and geochronology characteristics of greenstone belts in the northern SFC

Greenstone belts	Distinguished geologic features	Age (Ma)	Occurrence/tectonic distribution	References
Mundo Novo	Low grade metamorphic pillow basalts, intermediate lavas, chemical and clastic sedimentary rocks	Metadacite (zircon SHRIMP U–Pb age: 3305 ± 9)	Allochthonous supracrustal sequence accreted onto grey gneisses. Crustal compression tectonically related with the Paleoproterozoic Itabuna–Salvador–Curaçá belt	1
Lagoa do Alegre	Low- to medium grade metamorphic mafic and ultramafic (komatiite) rocks, chemical and clastic sedimentary rocks	Intrusive granite minimum age: 3300	Dismembered tectonic slivers within Archean gneiss-migmatite basement	2, 3
Guajeru	Low- to medium grade metamorphic mafic and ultramafic rocks, chemical and clastic sedimentary rocks	Intrusive alkaline granites (zircon Pb evaporation minimum age: 2649 ± 15 to 2670 ± 15)	Slievers of greenstone associations intermingled with basement rocks, syntectonically intruded by 3.1 Ga TTG rocks	3, 4
Umburanas, Brumado, Ibitira-Ubiraçaba	Low- to medium grade metamorphic supracrustal sequences: ultramafic to felsic metavolcanics, chemical (BIF) and clastic sediments	Meta-andesite (Umburanas greenstone belt): U–Pb age of 2744 ± 15. Minimum age of 2693 ± 5	Greenstone associations intermingled with basement rocks, intruded by TTG suites	3, 5
Boquira	Meta-ultramafics, metabasalts, chemical and clastic sediments	Galena (Pb–Pb age of 2.5–2.7 Ga)	Greenstone association with basement gneisses and migmatites	3

References: (1) Peucat et al. (2002); (2) Angelim (1997); (3) Cunha et al. (2012); (4) Lopes (2002); (5) Bastos Leal et al. (2003). See also Table 3.3.

Umburanas greenstone belt was partially studied by means of radiometric ages. A meta-andesite from the intermediate unit yielded zircon Pb evaporation age of 2744 ± 15 Ma (Bastos Leal et al. 2003). No age determinations are available for the basaltic-komatiitic lower unit or for the carbonate upper unit of the Umburanas greenstone belt. It is, however, definitely older than 2693 ± 5 Ma, which is the age of the Serra do Eixo intrusive granite (Santos-Pinto et al. 2012). Similar field relationships were found in the Guajeru greenstone belt, where alkaline granitoids intrusive into supracrustal rocks of the belt yielded zircon Pb evaporation ages between 2649 ± 15 and 2670 ± 15 Ma (Barbosa et al. 2012).

Other relicts of greenstone belts in the Gavião Block have ages poorly constrained or no age data. For instance, the Boquira greenstone belt (Table 3.4) is composed of ultrabasic rocks (metakomatiites?), metabasalts, and clastic and chemical metasedimentary rocks. Pb–Pb analyses for galena samples from the Boquira mine yielded model ages in the time interval 2.7–2.5 Ga, interpreted as the minimum age of the Pb and Zn sulfides (Cunha et al. 2012). These authors report a list of the main exposures of greenstone belts in the Gavião Block. Finally, in the Sobradinho Block, the Lagoa do Alegre greenstone belt comprises a lower volcanic-sedimentary sequence and an upper metasedimentary sequence. According to Angelim (1997) granitoid intrusions dated at 3300 Ma constrain a Mesoarchean age for the metavolcanic-sedimentary association.

3.4.2 The Serrinha Block

The Serrinha is the smallest Mesoarchean block in the northern SFC (Fig. 3.8), but the wealth of field relationships coupled with geochronological and geochemical data makes it one of the best-known areas of the craton. The Serrinha Block occurs as an elongated N–S segment up to 100 km wide, limited to the west and to the south by tectonic contacts with the Paleoproterozoic Itabuna–Salvador–Curaçá belt. Two migmatite-gneiss complexes (Santa Luz and Uauá) can be distinguished in this block, intruded by mafic dikes and mafic–ultramafic complexes (see below). The country rocks were metamorphosed under amphibolite to granulite facies conditions and lie in tectonic contact with Paleoproterozoic rocks of the Rio Capim and Rio Itapicuru greenstone belts (Schrank and Silva 1993; Oliveira et al. 2010 and references therein). Syn- to late-tectonic granitoid plutons are emplaced into these greenstone belts, where the domic structures are related with the Paleoproterozoic orogeny (Mello et al. 2006).

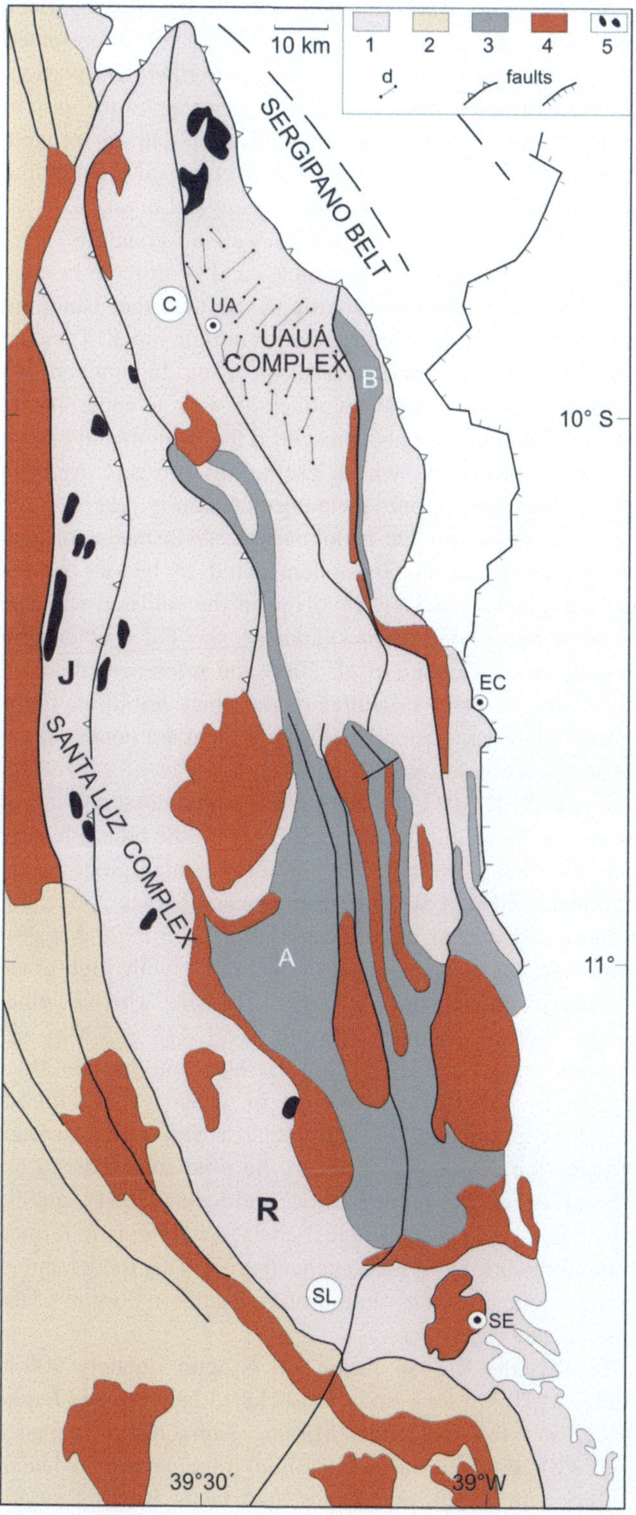

Fig. 3.8 Geologic sketch of the Mesoarchean Serrinha Block (adapted from Oliveira et al. 2010; Barbosa et al. 2012). *1* Archean substratum partly reworked in Paleoproterozoic times: C = Caldeirão belt; SL = Santa Luz complex with the Jacurici (J) and Retirolândia (R) domains; Uauá Complex intruded by Neoarchean mafic dikes. *2* Caraíba complex. *3* Paleoproterozoic greenstone belts: Itapicuru (A) and Rio Capim (B). *4* Paleoproterozoic granites. *5* Mafic-ultramafic complexes. Towns: EC = Euclides da Cunha; SE = Serrinha; UA = Uauá. See Table 3.3 and text for details

The Santa Luz Complex (Fig. 3.8; Table 3.3) contains Mesoarchean (3085–2983 Ma) migmatites, banded gneisses, TTG orthogneisses, mafic–ultramafic complexes, and mafic dikes (Oliveira et al. 2010). The banded gneiss unit is the end product of deformation of migmatites and mafic dikes during the accretion of the Paleoproterozoic Rio Itapicuru greenstone belt onto the Archean basement (Oliveira 2011). Paleoproterozoic plutons related to the Itabuna–Salvador–Curaçá belt intrude both the Archean substratum and the Paleoproterozoic supracrustal units. One of those (Quijinge trondhjemite pluton) yielded U–Pb age determinations in inherited zircons of 3314 and 3620 Ma (Rios et al. 2008). Other plutons like the Ambrósio granitic dome inside the Rio Itapicuru greenstone belt hosts migmatitic gneiss enclaves as old as 3.1 Ga (Mello et al. 2006). This suggests that Paleoarchean to Mesoarchean material formed the precursor crust of the Serrinha Block. From a geochronological point of view the Santa Luz Complex can be divided into two domains of grey gneisses, i.e. the Retirolândia (ca. 3085 Ma) and the younger Jacurici domain (ca. 2980 Ma) (Oliveira et al. 2010) (Fig. 3.8). Both domains are composed of TTG suites respectively with Sm–Nd T_{DM} model ages in the time intervals 3.1–3.2 Ga and 2.9–3.1 Ga. The $\varepsilon_{Nd(t)}$ values vary from +1.1 to zero and +2.7 to −0.8, respectively (Oliveira et al. 2010), indicating a predominant juvenile origin.

Conversely, the Uauá Complex (Fig. 3.8; Table 3.3), which may be an allochthonous block (Oliveira 2011), is bordered to the west by the Archean-Paleoproterozoic Caldeirão belt and to the east by the Paleoproterozoic Rio Capim greenstone belt (Oliveira et al. 2011, 2013). The Uauá basement consists mostly of NW-trending, ancient banded gneisses (of unknown age) intruded by Mesoarchean layered anorthosite, peridotite and diorite complexes, metamorphic mafic dikes, and tonalite-granodiorite bodies. Neoarchean non-metamorphic norite-tholeiite dikes crosscut in places the country rocks (Oliveira 2011; Oliveira et al. 2013, and references therein). Except the Neoarchean dikes (norite 2726 Ma and tholeiite 2623 Ma; Oliveira et al. 2013), much of the basement rocks underwent granulite facies metamorphism and were later retrogressed to amphibolite grade.

The granulite-facies felsic igneous bodies with intercalated garnet-bearing mafic granulites are strongly foliated, often with horizontal, or south-dipping low-angle foliation. They show stretching lineations and asymmetric folds indicative of having been thrust northward (Oliveira et al. 2013 and references therein). The banded gneisses, 3.0 Ga old, show slightly negative to positive $\varepsilon_{Nd(3.0Ga)}$ values and major and trace element geochemical characteristics akin to a subduction zone setting (Oliveira et al. 2010). As such, these rocks potentially represent exhumed tectonic slices of the roots of a Mesoarchaean continental arc (Oliveira et al. 2010), as suggested by the widespread U–Pb ages. Cordani et al. (1999) obtained ages between 3120 and 3130 Ma for the Capim tonalite, whereas Oliveira et al. (2010) reported on felsic igneous bodies with ages between 3072 and 2991 Ma. The younger tholeiite dike swarm (2623 Ma) that crosscuts the Uauá crust may represent a failed arm of a Neoarchean rift system (Oliveira 2011; Oliveira et al. 2013).

The Caldeirão belt (Caraíba Complex), bounded to the east by the Uauá Complex (Fig. 3.8), comprises a 10-km wide sheared sequence of steeply dipping quartzites, sillimanite-cordierite-garnet gneisses, granodioritic orthogneisses, mafic rocks and migmatites, all metamorphosed under amphibolite to granulite facies conditions. SHRIMP U–Pb age indicates a 3150 Ma age for one orthogneiss and 2076–2040 Ma dating for the regional deformation and metamorphism (Oliveira et al. 2002). Deformation of the Caldeirão belt took place during northward lateral extrusion of the Uauá Block in the Paleoproterozoic (Oliveira et al. 2013)—see Table 3.3.

Two major Archean mafic-ultramafic intrusions occur in the Serrinha Block: the Lagoa da Vaca anorthosite complex and the Pedras Pretas mafic-ultramafic complex. The first one (Paixão and Oliveira 1998) crops out close to a major ductile shear zone that separates the Uauá banded gneisses from the Caldeirão migmatite-quartzite-orthogneiss belt (Fig. 3.8). The complex makes up a metamorphosed layered igneous body composed of plagioclase and amphibole-rich bands. A whole-rock Pb–Pb isochron yielded the age of 3160 $\pm$ 65 Ma, suggesting that the anorthosite is the oldest ever found in the SFC. This isochron presents a model μ_1 Pb value of 8.8 $\pm$ 0.6, compatible with its derivation from the sub-continental lithospheric mantle or from the depleted mantle, in both cases followed by crustal contamination. Therefore the anorthosite emplacement is likely to have occurred in a continental setting, most likely similar to a present-day passive continental margin (Paixão and Oliveira 1998). The Pedras Pretas mafic-ultramafic complex is crosscut by mafic dikes with a minimum Neoarchean age (Oliveira et al. 2012). It comprises a small body of chromitite-rich serpentinite interleaved with banded gneisses of the Santa Luz complex. Further geochronological and isotopic studies are needed to constrain the emplacement age and genesis of the mafic-ultramafic complex.

3.4.3 The Jequié Block

This block (Fig. 3.1b; Table 3.3) occurs to the SE of the Contendas-Jacobina Lineament, stretching eastwards almost to the Atlantic coast. It contains several north-south trending belts composed of Archean high-grade rocks, separated by faults, and showing strong deformation and emplacement of granitic material tectonically related to the Paleoproterozoic orogeny (e.g., Nutman and Cordani 1993).

According to Barbosa et al. (2012) the Jequié Block consists of calc-alkaline plutonic rocks mainly represented by gneisses and acid granulites. On the northeastern portion, gabbro-anorthosite bodies dominate, whereas on the eastern portion, close to the Itabuna–Salvador–Curaça orogen, Barbosa (1990 and references therein) mapped amphibolite facies banded gneisses with amphibolite and quartz-feldspar bands. Charnockitic granulites as well as granulite facies supracrustal rocks (i.e., garnet- and cordierite-bearing charnockites of sedimentary origin), on the other hand, are conspicuous over the northern portion of the block. They are believed to be intracratonic basin deposits. Inhomogeneous ortho-, and para-derived granulites as well as enderbitic to charnockitic rocks are also present in the western and central portion of the block, whose protholiths preserve Archean ages, though the tectonic-metamorphic history relates to the nearby Paleoproterozoic geodynamics. As a matter of fact, the Jequié Block has been considered to be one of the Archean blocks that were involved in the collision with the Serrinha and Gavião Blocks during the Paleoproterozoic orogeny (e.g., Teixeira et al. 2000 and references therein). Therefore, the main structures of the block are thrust faults that set the block's boundaries; all of them are related to the Paleoproterozoic orogeny.

Zircon SHRIMP U–Pb ages for charnockitic granulites of the Jequié Block cluster around 2.7 Ga for the protolith ages, and between 2.06 and 2.05 Ga for the high-grade metamorphism (Silva et al. 2002; Barbosa and Sabaté 2004). For instance, Silva et al. (2002) reported a 2715 $\pm$ 29 Ma protolith age for one charnockitic granulite with high-grade metamorphic overprint at 2047 $\pm$ 14 Ma, whereas other sample yielded younger protolith age (2473 $\pm$ 5 Ma) and metamorphism at 2061 $\pm$ 6 Ma. Barbosa and Sabaté (2004) reported a SHRIMP zircon age of 2689 $\pm$ 7 Ma for an enderbitic granulite in the Mutuipe area. Sm–Nd T_{DM} model ages for the Jequié rocks fall in the time span 3.3–2.9 Ga (Marinho et al. 1994a, b; Barbosa and Sabaté 2004), and the negative $\varepsilon_{Nd(t)}$ values suggest that the plutonism formed through reworking of Mesoarchean crust. This model agrees well with the crustal evolution of the southern portion of the SFC, as envisaged here.

Finally, the Rio do Jacaré sill is approximately 900 m thick and 70 km long body sandwiched between the Jequié Block and the Contendas-Mirante supracrustal sequence (Fig. 3.8). It varies in composition from gabbro to leucogabbro with minor pyroxenite, and hosts Brazil's largest vanadium deposit (Brito 2000). Marinho et al. (1994a, b) dated the sill at 2474 $\pm$ 72 Ma by means of a Pb–Pb whole rock isochron, but citation of unpublished data in Leite and Marinho (2012) indicate zircon LA-ICP-MS and Pb evaporation ages respectively of ca. 2630 and 2650 Ma.

3.4.4 The Basement of the Itabuna–Salvador–Curaçá Belt

Remnants of Archean rocks are also found within high-grade orthogneisses of the Paleoproterozoic Itabuna–Salvador–Curaçá belt (e.g., Oliveira et al. 2010 and references therein) (Fig. 3.1b; Table 3.3). The orogen formed by accretion of ca. 2.19 Ga arc-like rocks onto the Jequié Block located to the south (Peucat et al. 2011 and references therein). Crustal evolution included Paleoproterozoic reworking of the arc rocks and their Jequié Archean substratum in the south, as well as the Neoarchean continental arc crust in the north (Caraiba Complex; see Fig. 3.8), and final intrusion of syn- to post-tectonic igneous bodies (Barbosa et al. 2008; Oliveira et al. 2010). Barbosa and Barbosa (this book) presents a more detailed description of the Paleoproterozoic geologic-tectonic setting.

In the southern portion of the Paleoproterozoic Itabuna–Salvador–Curaçá belt, Silva et al. (2002) report Neoarchean ages on scattered rocks such as on charnockitic granulites with zircon SHRIMP U–Pb ages of 2719 ± 10 Ma at the Pernambuco hill in Ilhéus, 2799 ± 18 Ma on roadside to Ipiaú, 2847 ± 7 Ma at the Coaraci quarry, and 2715 ± 29 Ma at Jitauna, road Ipiaú to Jequié. Most of the zircon grains dated by Silva et al. (2002) show metamorphic rims with ages between 2078 and 2047 Ma. Peucat et al. (2011) found a slightly younger age of 2675 ± 11 Ma on zircon cores from an enderbitic granulite west of Itapitanga with metamorphic rims at 2080 ± 21 Ma, as well as arc-related intrusions ca. 2.19 Ga old.

In the northern portion of the orogen (Fig. 3.8), the 2634–2695 Ma Caraíba charnockitic granulites close to São José do Jacuípe (see previous section) also indicated Paleoproterozoic reworking on zircon metamorphic rims, which dated mostly between 2082 and 2072 Ma (Silva et al. 1997). The latter ages match with the time interval of high-grade metamorphism and extremely reworking of Archean protoliths along the orogen, as well as the intrusion of syn- to late tectonic granitic bodies (Oliveira et al. 2010; Barbosa et al. 2008). Neoarchean remnants younger than those at south are also present, as indicated by SHRIMP ages on zircon cores from 2574 Ma Caraíba charnockitic granulite (Oliveira et al. 2010). In a similar way, the Neoarchean protholiths of the Jequié Block exhibit extensive recrystallization of zircon rims due to Paleoproterozoic high-grade metamorphism (see previous section).

Finally, two major Archean mafic-ultramafic complexes occur in the northern extension of the Itabuna–Salvador–Curaçá belt, namely the Caraíba norite-pyroxenite and the São José do Jacuípe leucogabbro complexes. Both complexes are similar in age, i.e. 2580 ± 10 Ma (zircon U–Pb SHRIMP) for Caraíba and 2583 + 8 Ma (zircon U–Pb SHRIMP) for São José do Jacuípe (Oliveira et al. 2010). The former hosts a copper sulphide deposit and occurs as deformed dikes and large bodies intruded into migmatite gneisses (Oliveira et al. 2004). Conversely, the São José do Jacuípe complex is non-mineralized and occurs as dismembered small bodies of layered anorthosite-gabbro-pyroxenite in syn-, to late-tectonic granites (Piaia 2011). According to Oliveira et al. (2010) and Piaia (2011), the two mafic-ultramafic complexes may be part of the roots of a Neoarchean continental arc represented by the ca. 2.65–2.56 Ga tonalite-granodiorite protoliths of the Paleoproterozoic Caraiba Complex. As a whole, this scenario allows a correlation with the Neoarchean collisional event hyphotesized here for the southern portion of the SFC.

3.5 Concluding Remarks

The SFC, like its African counterpart (the West Congo Craton), results from accretion of distinct Archean fragments essentially composed of medium- to high-grade gneisses including TTG suites, and granite-greenstone associations. This polycyclic substratum was assembled during late Neoarchean times under high-grade metamorphic conditions, and intruded by late tectonic K-rich granites, mafic-ultramafic suites, and mafic dikes.

The new and compiled data indicate that the Sobradinho, Gavião, Serrinha and Jequié Blocks in conjunction with the southernmost basement segment of the SFC (Fig. 3.1) show diachronic evolution from Paleo- to Neoarchean times by means of juvenile accretion/differentiation of continental crust, coupled with partial melting events of earlier-formed material. This tectonic evolution is coherent with the geochemical and U–Pb isotopic data on both igneous zircon cores and their metamorphic rims of igneous and metaigneous rocks, as well as with detrital zircon ages of supracrustal sequences (e.g., Ibitira-Ibiraçaba sequence; see previous section). We interpret these results as a robust evidence of multiple pulses of TTG plutonism in genetic association with greenstone belts, as in modern oceanic arcs and active continental margins. In particular, both the TTG suites and granitoid rocks, varying in time and extent, are time markers of the progressive thickening and differentiation of the continental crust through time, leading to tectonic stabilization of the landmasses at the end of the Archean.

Figure 3.9 presents the U–Pb zircon crystallization ages and Sm–Nd T_{DM} whole rock model ages for Archean country rocks of the SFC, based on published data. Considering that Sm–Nd T_{DM} model ages are usually ca. 200 m.y. older than the corresponding U–Pb age for a given rock, the observed peaks (coupling both methods) can be taken as reasonable estimates of the bulk Archean accretion/crustal differentiation

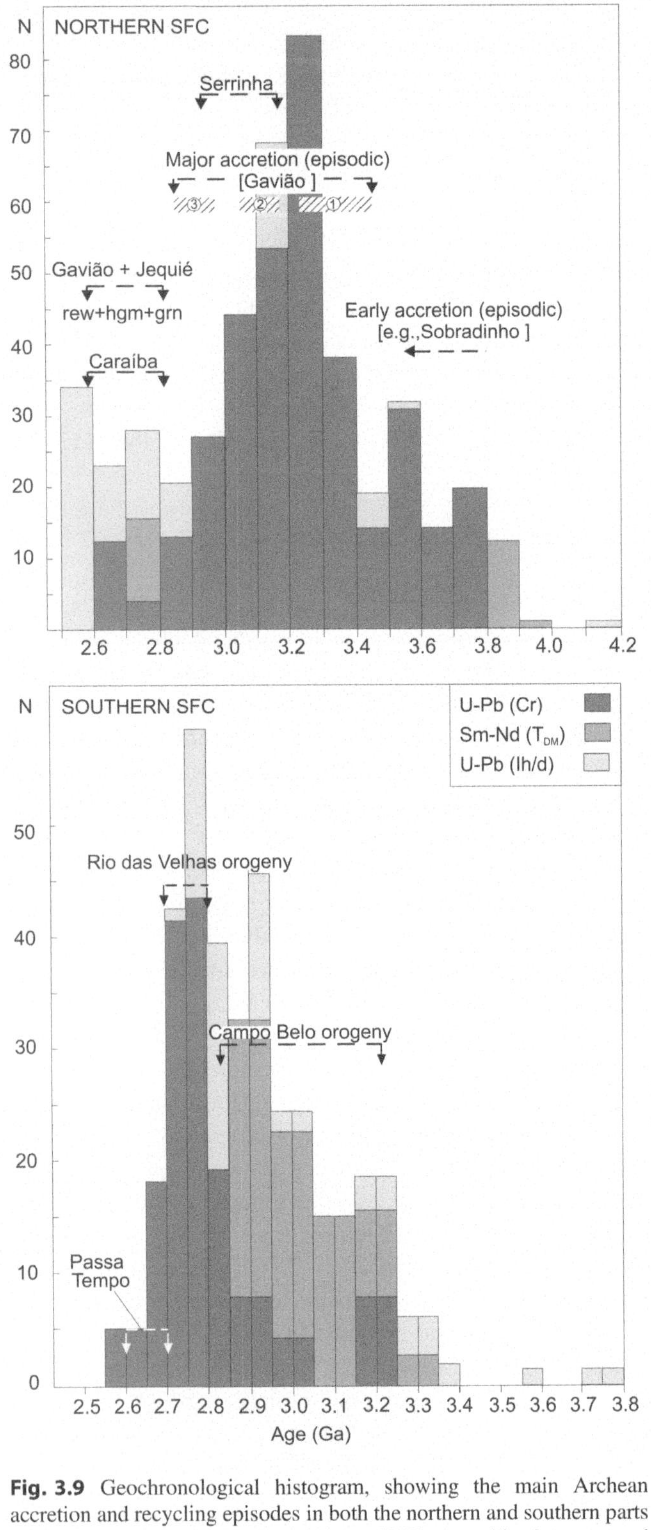

Fig. 3.9 Geochronological histogram, showing the main Archean accretion and recycling episodes in both the northern and southern parts of the Craton, given by peaks of zircon U–Pb crystallization ages and Sm–Nd T_{DM} model ages (whole rocks). Keys: rew (crustal reworking), hgm (high-grade metamorphism), grn (granite-genesis); Cr (crystallization age), Ih/d (inherited or detrital grains ages). Note that the histograms have different scale. See text for details

events. However, we are aware that there might be a bias due to the heterogenous number of U–Pb ages available for the southern and northern parts of the SFC. Nevertheless, the resulting geochronological scenario, which is roughly similar for both portions of the craton, is consistent with a protracted Archean history for the proto-SFC, starting in the Hadean time, and highlighted by the peculiar tectonic-metamorphic histories of the basement rocks and coeval greenstone belts. The possible physical link among the primitive fragments was obliterated by crustal remobilization, high-grade metamorphism, shortening processes, and granite emplacement during subsequent collisions of the intervening Paleoproterozoic belts (Teixeira et al. 2015). This complexity hampers a correlation among the Archean units.

Scattered zircon U–Pb crystallization and inherited ages > 3.6 Ga on Fig. 3.9 refer to the oldest Paleoarchean components in South America, such as a Hadean xenocryst (4.1 Ga) in the Ibitira-Ibiraçaba schist, and igneous zircons from Riacho de Santana gneiss (3.6 Ga) in the Gavião block (Sm–Nd T_{DM} age of 3.9 Ga). A gabbro-diorite enclave in gneiss (3.5 Ga old) of the Sobradinho block yields Sm–Nd T_{DM} model age of 3.7 Ga and zircon Hf T_{DM} model ages between 3.7 and 3.9 Ga. Moreover, the Neoarchean Rio das Velhas greenstone belt in the southern SFC similarly contains detrital zircons as old as 3.8 Ga. Therefore there is isotopic evidence for Hadean differentiated crust incorporated to the SFC sialic core.

The ages older than 3.6 Ga on Fig. 3.9 suggest that Paleoarchean crustal growth was episodic. For instance, inherited age clusters in detrital zircons in the Proterozoic Tombador cover (Guadagnin et al. 2015) and in the northern SFC suggest that sediment accumulation accompanied by sediment recycling has operated since Paleoarchean times. This concurs with models of accretion and recycling of the early continental crust, via subduction due to vigorous and hotter thermodynamic mantle conditions (e.g., Rey and Coltice 2008 and references therein).

Major Mesoarchean accretion is apparent between 3.5–3.3 and 3.2–2.9 Ga (Fig. 3.9), according to the age peaks (U–Pb + Sm–Nd T_{DM}) for both the northern (episodes 2 and 3; Gavião Block) and southern portions of the SFC (Campo Belo orogeny). Specifically, the orthogneisses and migmatites of the Serrinha Block (northern portion) and the Campo Belo and Santa Bárbara Complexes (southern portion; Campo Belo orogeny) accreted between 3.2 and 2.9 Ga (Fig. 3.9). Taking into account the entire ancient core of the SFC, these data indicate again vigorous continental growth at this time.

The Rio das Velhas orogeny (2790–2750 and 2730–2700 Ma; Fig. 3.9) produced gneissic-granitic complexes and coeval granite-greenstone associations in an active continental margin setting. The final phase of this orogeny also led to pervasive crustal remobilization of older rocks and high-grade metamorphism representing a late collision stage (e.g., Passa Tempo Complex; Fig. 3.9) of the Archean fragments. Crustal derived granite plutons dated from ca 2.65 to 2.55 Ga led to tectonic stability of the continental crust. Such a model compares well with the geochronological evolution (2.70–2.55 Ga) of the Jequié Block as well as with that of the Caraíba Complex (Fig. 3.9), which are the reflection of similar high-grade metamorphism, granite intrusion and crustal reworking of Archean protoliths. Collectively, this evidence points to the important role of Neoarchean assembly of older polycyclic rocks. Whether the Neoarchean Caraíba Complex (basement of the Paleoproterozoic orogen) represents or not an extension of the Jequié block deeply reworked by the Paleoproterozoic orogeny is still uncertain.

Limited structural evidence (e.g., flat foliation with sub-horizontal sheath folds) in local exposures of the Archean crust, such as in the Aracatu region (Gavião block), Ipiaú band (Jequié Block) and Passa Tempo Complex, favors the model of Neoarchean contractional tectonics, as also suggested by the isotopic evidence. However, vertical tectonics has been reported in places of the Jequié Block (Teixeira et al. 2000 and references therein). In any case, these models are dependent on further robust geochronology and isotopic-geochemical evidence.

It has been of widely acceptance that some kind of plate tectonics was in operation since early Archean times (e.g., Condie 2007; Martin et al. 2010; Shirey and Richardson 2011; Dhuime et al. 2012). From the geochemical point of view, the chondrite-normalised La/Yb versus Yb plot for Archean rocks, among other correlative diagrams, helps to constrain the operating tectonic processes, as shown on Fig. 3.10. In this diagram, the trace element data for selected rock units in both the northern and southern cratonic areas are plotted. Most gray gneisses older than 3.0 Ga of the Gavião Block, the ca. 3.0 Ga Santa Luz Complex in the Serrinha Block, and the ca. 2.7 Ga Passa Tempo Complex (southern SFC) show chemistry compatible with TTG suites, enhanced by strong fractionation of light from heavy REE patterns and minor or absent Eu anomalies (Teixeira et al. 2010). Therefore, these rocks may have been originated by partial melting of a garnet-bearing mafic source at great depths (e.g., Rapp et al. 1991). Moreover, some REE patterns for the 3.4 Ga Mariana dome (northern SFC) display little fractionation of light from heavy REEs and pronounced negative Eu anomalies (Teixeira et al. 2010). This indicates that plagioclase fractionation played a significant role. In other words, this may be alternatively explained by partial melting at the base of an oceanic plateau, such as what was recently postulated for the Earth's earliest evolved crust (e.g., Reimink et al. 2014). However, in the case of the SFC rocks, such interpretation still depends on further detailed geochemical and isotopic studies plus re-evaluation of previous published data. On the other hand, the 3.0 Ga rocks of the Uauá Complex (Serrinha Block), the 2.7 Ga gneisses of the Jequié Block, and the granite gneisses of the Meso- and Neoarchean Campo Belo, Bonfim, and Belo Horizonte Complexes (southern SFC) show geochemical characteristics very much like the calc-alkaline series. Therefore, there is robust evidence for origin of these units in tectonic settings similar to modern subduction zones. For instance, the younger Neoarchean alkaline granites in the Gavião Block may have had the TTG terrains as their crustal sources. After Neoarchean tectonic stability and ultimate thickness of the continental crust, an emplacement of mafic ultramafic suites and mafic dikes occurred, such as those emplaced into the Uauá Block and in the Passa Tempo complex.

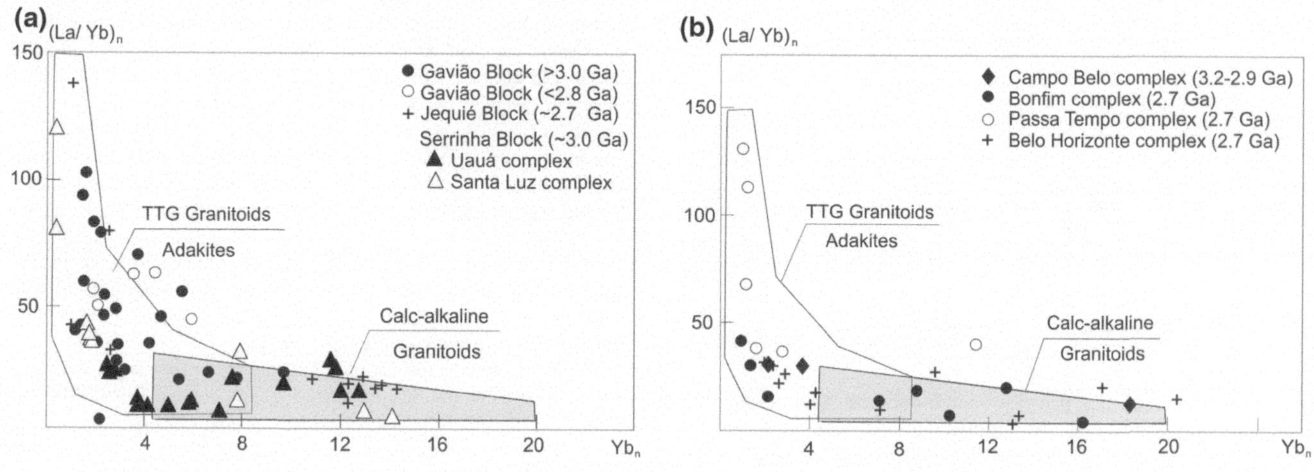

Fig. 3.10 Chondrite-normalised La/Yb versus Yb, showing the geochemical characteristics of Archean granitoids and *gray* gneisses of the northern (**a**) and southern (**b**) parts of the craton (see Fig. 3.1)

Figure 3.11 presents a paleotectonic cartoon for the SFC based on geologic and geochronologic information that highlights a mosaic of individual Archean segments encircled by Paleoproterozoic orogenic domains. For instance, previously published gravimetric and aeromagnetic interpretations dealing with garnet bearing rocks over the Serrinha Block indicated variations of the pressure parameters. They are consistent with the idea of a thicker lithospheric keel with thinner edges, due to Paleoproterozoic reworking during the Itabuna–Salvador–Curaçá orogen (Pereira and Fuck 2005). As a corollary, we can similarly think about a thick, mantle keel in the Gavião Block, given the long-lived production of continental crust since Paleoarchean times, thus providing the proto-craton's stability and consequently the rheological conditions for emplacement of huge amount of mafic-ultramafic magmas in the late Neoarchean.

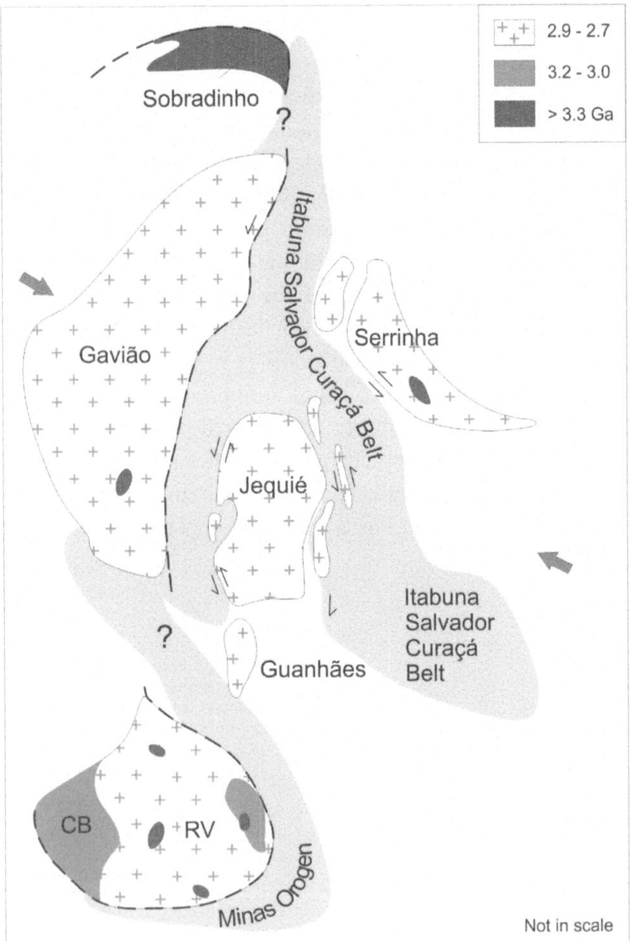

Fig. 3.11 Cartoon of the paleotectonic framework (São Francisco proto-craton) distinguished by individual Archean continental fragments/blocks, and the adjoining Paleoproterozoic collage, given by the Minas orogen and the coeval Itabuna–Salvador–Curaçá belt. Keys: CB (Campo Belo orogeny), RV (Rio das Velhas orogeny). (Adapted from Alkmim and Noce, 2006). See text for details

Additional geophysical data (Pereira and Fuck 2005) in the southern SFC similarly indicate the transition from thicker lithosphere (typical of Archean terrains, as supported by the radiometric ages) to thinned lithosphere of the Neoproterozoic marginal belt nearby the Piumhi greenstone belt (Table 3.2; Fig. 3.2). In the African counterpart, Tadjou et al. (2009) report a similar picture, based on interpretation of gravity-based geophysical study involving the Congo Craton and the allochtonous Pan-African belt in Cameroon; whereas the ancient crust of the Congo craton is relatively thick, consisting mainly of low-density rocks, most Pan-African belt rocks are relatively denser.

Finally, from a geodynamic perspective, the integrated data suggest that juvenile accretion played a fundamental role from Paleoarchean to Neoarchean times. The eventual Neoarchean assembly of distinct blocks (e.g., Gavião, Jequié) and/or sialic fragments (e.g., Bonfim and Belo Horizonte Complexes; Quadrilátero Ferrífero) took place diachronically between ca. 2.80 and 2.70 Ga, including high-K plutonism, essentially by means of plate tectonics processes. The Sobradinho and the Serrinha Blocks may represent allochthonous blocks due to their peculiar age records without Neoarchean imprint, although this hypothesis must be tested by robust geochronology and coupled isotopic and geochemical studies.

Acknowledgments The authors are grateful to the support of FAPESP (Research Foundation of the São Paulo state) and to CNPq (Brazilian Research Council): grants 2009/53818-5, 2009-8/302917 and 2011-5/308424. This work is a contribution to FAPESP thematic project 2012//15824-6. The authors also acknowledge the criticism and suggestions from F.F. Alkmim and U.G. Cordani.

References

Alkmim FF (2004) O que faz de um cráton um cráton? O cráton do São Francisco e as revelações Almeidianas ao delimitá-lo. In: Mantesso-Neto V, Bartorelli A, Dal Ré Carneiro C (Eds.) Geologia do Continente Sul-Americano: Evolução da Obra de Fernando Flávio Marques de Almeida, Ed Beca, 17–34.

Alkmim FF, Danderfer A (1998) Basement-involved and detached phases of deformation in the evolution of fold-thrust belts: Examples from the Southern Cordilheira do Espinhaço and the Paramirim deformation corridor, Brazil. Basement Tectonics 12 - Proceedings of the International Conferences on Basement Tectonics, pp 211–212

Alkmim FF, Marshak S (1998) Tranzamazonian Orogeny in the southern São Francisco Craton region, Minas Gerais, Brazil: evidence of Paleoproterozoic collision and colapse in the Quadrilátero Ferrífero. Precam Res 90: 29–58.

Alkmim FF, Martins-Neto MA (2012) Proterozoic first-order sedimentary sequences of the São Francisco craton, eastern Brazil. Mar Pet Geol 3: 127–139.

Alkmim FF, Noce CM (2006) Outline of the geology of Quadrilátero Ferrífero. In: Alkmim FF, Noce CM (eds) The Paleoproterozoic Record of São Francisco Craton. IGCP 509 Field Workshop, Bahia and Minas Gerais. Field Guide and Abstracts, 37–73.

Angelim LAA (1997) Petrolina, folha SC24-V-C estados da Bahia, Pernambuco e Piauí: escala 1: 250.000, Brasilia, DF: CPRM. Programa de Levantamentos Geológicos Básicos do Brasil.

Baars FJ (1997) The São Francisco Craton. In: De Witt M, Ashwall LD (eds) Greenstone belts. Oxford Monographs of Geology and Geophysics 35: 529–557.

Baltazar OF, Zucchetti M (2007) Lithofacies associations and structural evolution of the Archean Rio das Velhas greenstone belt, Quadrilátero Ferrífero, Brazil: A review of the setting of gold deposits. Ore Geology Reviews 32: 471–499.

Barbosa JSF (1990) The granulites of the Jequié Complex and Atlantic Mobile Belt, Southern Bahia, Brazil – An expression of Archean Paleoproterozoic Plate Convergence. In: Vielzeuf D, Vidal P (eds) Granulites and Crustal Evolution. Springer-Verlag, Clermont Ferrand, France, 195–221.

Barbosa JSF, Sabaté P (2004) Archean and Paleoproterozoic crust of the São Francisco Craton, Bahia, Brazil: geodynamic features. Precam Res 133: 1–27.

Barbosa JFS, Peucat JJ, Martin H, da Silva FA, de Moraes AM, Corrêa-Gomes LC, Sabaté P, Marinho MM, Fanning CMF (2008) Petrogenesis of the late-orogenic bravo granite and surrounding high-grade country rocks in the paleoproterozoic orogen of Itabuna–Salvador–Curaçá block, Bahia, Brazil. Precam Res 167: 35–52.

Barbosa JFS, Cruz SP, Souza JS (2012) Terrenos metamórficos do embasamento. In: Barbosa JFS (ed) Geologia da Bahia: pesquisa e atualização, CBPM 1: 101–201.

Barbosa N, Teixeira W, Leal LRB, Leal ABM (2013). Evolução crustal do setor ocidental do Bloco Arqueano Gavião, Cráton do São Francisco, com base em evidências U–Pb, Sm–Nd e Rb–Sr. Geologia USP. Série Cient. 13: 63–88.

Bastos Leal LR, Cunha JC, Cordani UG, Teixeira W, Nutman A, Menezes Leal AB, Macambira MJB (2003) SHRIMP U–Pb, $^{207}Pb/^{206}Pb$ zircon dating and Nd isotopic signature of the Umburanas greenstone belt, Northern São Francisco Craton, Brazil. J S Am Earth Sci 15: 775–785.

Borg G, Shackleton RM (1997) The Tanzania and NE-Zaire Cratons. In: de Wit, M.J., Ashwal, L.D. (Eds.), Greenstone Belts. Oxford University Press, Oxford, pp. 608–619.

Brito RSC (2000) Geologia e petrologia do sill máfico-ultramáfico do Rio Jacaré - Bahia e estudos das mineralizações de Fe-Ti-V e platinóides associadas. Doctor thesis. Geosciences Institute, University of Brasília, Brazil.

Cahen I, Snelling NJ, Dehal J, Vail JR (1984) The Geochronology and Evolution of Africa. Clarendon Press, Oxford. 512 pp.

Campos JCS, Carneiro MA (2008) Neoarchean and Paleoproterozoic granitoids marginal to the Jeceaba-BomSucesso lineament (SE border of the southern São Francisco craton): Genesis and tectonic evolution. J S Am Earth Sci 26: 463–484.

Campos JCS, Carneiro M, Basei MAS (2003) U–Pb evidence for late Neoarchean crustal reworking in the Southern São Francisco craton (Minas Gerais, Brazil) An Acad Bras Ciênc 75: (4) 2–17.

Carneiro MA, Jordt-Evangelista H, Teixeira, W (1997) Eventos magmáticos arqueanos de natureza cálcio-alcalina e tholeiítica no Quadrilátero Ferrífero e suas implicações tectônicas. R Bras Geoc 27(1): 121–128.

Carneiro MA, Teixeira W, Carvalho Junior IM, Fernandes RA (1998) Ensialic tectonic setting of the Archean Rio das Velhas Greenstone Belt: Nb and Pb Isotopic evidence from the Bonfim Metamorphic Complex, Quadrilátero Ferrífero, Brazil. R Bras Geoc 28(1): 71–82.

Carneiro MS, Teixeira W, Carvalho-Júnior IM, Pimentel MM, Oliveira AH (2004) O comportamento dos sistemas Sm–Nd e Rb–Sr da Sequência Acamadada Máfico-Ultramafica Ribeirão dos Motas (Arqueano) Craton São Francisco Meridional: Evidências de Enriquecimento Mantélico e Fracionamento Isotópico. Geol USP, Sér Cient 4(2):13–26.

Condie K (2007) Accretionary orogens in space and time. In: Hatcher Jr., R.D.,Carlson, M.P., McBride, J.H., Martínez Catalan, J.R. (Eds.), 4-D Framework of Continental Crust, Geological Society of America, Memoir 200, pp. 145–158.

Cordani UG, Sato K, Nutman A (1999) Single zircon SHRIMP determination from Archean tonalitic rocks near Uauá, Brazil. II SSAGI South Am Symp Isotope Geology, Actas, 27–30.

Corrêa Neto AV, Modesto AM, Caputo Neto V, Guerrero JC (2012) Alteração hidrotermal em zona de cisalhamento associada ao Lineamento Congonhas, sul do Quadrilátero Ferrífero, Minas Gerais. Anu Inst Geoc 35 (2): 55–64.

Cruz, SCP, Peuca, JJ, Carneiro MA, Teixeira L, Santos de Souza J, Barbosa JSF, Leal ABM, Martins AAM (2012) The Caraguataí syenitic suite, a ca 2.7 Ga alkaline magmatism (petrology, geochemistry and U–Pb zircon ages): a Neoarchean crustal melting event in the southern part of the Gavião block (São Francisco Craton), Brazil. J. S Am Earth Sci 37: 95–112.

Cunha JC, Barbosa JSF, Mascarenhas JF (2012) Greenstone belts e sequências similares. In: Barbosa JSF (Coord) Geologia da Bahia: pesquisa e atualização, CBPM. Chapter IV, pp. 203-326. Bahia, BA.

Dantas EL, Brito Neves BB, Fuck RA (2010) Looking for the oldest rocks of South America: Paleoarchean orthogneiss of the Sobradinho Block, northernmost foreland of the São Francisco Craton, Petrolina, Pernambuco, Brazil. VII SSAGI South Am Symp Isotope Geology, S0136.

Dantas EL, Brito Neves BB, Fuck RA (2013) Looking for the Early Archean rocks in South America: U–Pb dating and Hf isotopes in zircons from the north São Francisco Craton, Brazil. Geol Soc Amer, Annual Meeting 2013, Paper 269-5.

Dhuime B, Hawkesworth CJ, Cawood PA, Storey CD (2012) A change in the geodynamics of continental growth 3 billion years ago. Science 335: 1334–1336.

Engler A, Koller F, Meisel T, Quéméneur J (2002) Evolution of the Archean/Proterozoic crust in the southern São Francisco craton near Perdões, Minas Gerais, Brazil: petrological and geochemical constraints. J S Am Earth Sci 15: 709–723.

Ernst RE, Pereira E, Hamilton MA, Pisarevski AS, Rodrigues J, Tassinari CCG, Teixeira W, Van-Dunem V (2013) Mesoproterozoic intraplate magmatic "barcode" record of the Angola portion of the Congo craton: newly dated magmatic events at 1500 and 1110 Ma and implications for Nuna (Columbia) supercontinent reconstructions. Precam Res 230: 103–118.

Farina F, Albert C, Lana C (2015) The Neoarchean transition between medium- and high-K granitoids: Clues from the Southern São Francisco Craton (Brazil). Precam Res 266: 375–394.

Fernandes RA, Carneiro MA (2000) O Complexo Metamórfico Campo Belo (Craton São Francisco Meridional): unidades litodêmicas e evolução tectônica. R Bras Geoc 30 (4): 671–678.

Goulart LEA, Carneiro MA (2013) Evolution of arc magmatism in the Carmópolis de Minas Layered Suite, Minas Gerais, Brazil: Sm–Nd and Rb–Sr isotope geochemistry. R Esc Minas, Ouro Preto, 66(4): 447–454. doi: 10.1590/S0370-44672013000400007

Goulart LEA, Carneiro MA, Endo I, Suita MTF (2013) New evidence of Neoarchean crustal growth in southern São Francisco Craton: the Carmópolis de Minas Layered Suite, Minas Gerais, Brazil. Braz. J. Geol., São Paulo, 43(3): 445–459. doi: 10.5327/Z2317-48892013000300003.

Guadagnin F, Chemale Jr F, Magalhães AJC, Santana A, Dussin I, Takehara L (2015) Age constraints on crystal-tuff from the Espinhaço Supergroup - Insight into the Paleoproterozoic to Mesoproterozoic intracratonic basin cycles of the Congo – São Francisco Craton. Gondwana Res 27: 363–376.

Hartmann LA, Endo I, Suita MTF, Santos JOS, Frantz JC, Carneiro MA, McNaughton NJ, Barley ME (2006) Provenance and age delimitation of Quadrilátero Ferrífero sandstone based on zircon U–Pb isotopes. J S Am Earth Sci 20: 273–285.

Iizuka T, Komiya T, Rino S, Maruyama S, Hirata T (2010). Detrital zircon evidence for Hf isotopic evolution of granitoid crust and continental growth. Geoch Cosmoch Acta 74: 2450–2472.

Koglin N, Zeh A, Cabral AR, Gomes AAS, Neto AVC, Brunetto WJ, Galbiatti H (2014). Depositional age and sediment source of the auriferous Moeda Formation, Quadrilátero Ferrífero of Minas Gerais, Brazil: New constraints from U–Pb–Hf isotopes in zircon and xenotime. Precam Res 255: 96–108.

Lana C, Alkmim FF, Armstrong R, Scholz R, Romano R, Nalini HA (2013) The ancestry and magmatic evolution of Archaean TTG rocks of the Quadrilátero Ferrífero province, southeast Brazil. Precam Res 230: 1–30.

Leahy GAS, Rosa MLS, Conceição H, Scheller T, Macambira MJB, Oberli F, Meier M, Martin H, Barreto Santos E, Paim MM (1999) Geochronology and isotopic signature of the syenitic intrusion of Ceraíma, southwest Bahia, Brazil. In: S. Sudamericano Geol. Isot., Abstracts, 2: 234–237.

Leal LRB, Teixeira W, Cunha JC, Macambira MJB, Cordani U (1996). Evolução Crustal dos Terrenos TTGs Arqueanos do Bloco do Gavião, Craton do São Francisco: Geocronologia U–Pb (Shrimp) e Pb–Pb Em Zircões. In: Anais do XXXIX Congresso Brasileiro de Geologia, Salvador. SBG 6, 539–541.

Leal LRB, Teixeira W, Cunha JC, Macambira MJB (1998). Archean Tonalitic-Trondhjemitic and Granitic Plutonism. In: The Gavião Block, São Francisco Craton, Bahia, Brasil: Geochemical And Geochronological Characteristics. Rev Bras Geoc 28 (2): 209–220.

Leite CMM, Marinho MM (2012) Serra de Jacobina e Contendas-Mirante. In: Barbosa JSF, Mascarenhas JF, Correa-Gomes LC, Dominguez JML, Souza JS (eds) Geologia da Bahia Pesquisa e Atualização. Salvador: CBPM - Companhia Baiana de Pesquisa Mineral, p. 397–441.

Lopes GAC (2002) Projeto Guajeru-CBPM. Salvador, Bahia 1, 408 p.

Lobato LM, Santos JOS, McNaughton NJ, Fletcher, IR, Noce, CM (2007) U–Pb SHRIMP monazite ages of the giant Morro Velho and Cuiabá gold deposits, Rio das Velhas greenstone belt, Quadrilátero Ferrífero, Minas Gerais, Brazil. Ore Geol Rev 32: 674–680.

Machado N, Carneiro MA (1992) U − Pb evidence of late Archean tectono-thermal activity in the southern São Francisco shield, Brazil. Can J Earth Sci 29: 2341–2346.

Machado N, Noce CM (1993) A evolução do setor sul do Cráton do São Francisco entre 3.1 e 0.5 Ga baseada em geocronologia U–Pb. Anais do II Simp. Cráton do São Francisco, 100–102.

Machado N, Noce CM, Ladeira EA, Belo de Oliveira OA (1992) U–Pb geochronology of Archean magmatism and Proterozoic metamorphism in the Quadrilátero Ferrífero, southern São Francisco craton, Brazil. Geol Soc Am Bull 104 (9): 1221–1227.

Machado N, Schrank A, Noce CM, Gauthier G (1996) Ages of detrital zircon from Archean-Paleproterozoic sequences: Implications for greenstone belt setting and evolution of a Transamazonian foreland basin in Quadrilátero Ferrífero, southeast Brazil: evidence from zircon ages by laser ablation ICP-MS. Earth Plan Sci Letters, 141, 259–276. Geol Soc Am Bull 104: 1221-1227.

Marinho MM, Sabaté P, Barbosa JSG (1994a) The Contendas Mirante volcano-sedimentary belt. In: MCH Figueiredo and AJ Pedreira (eds.) Petrologic and Geochronologic Evolution of the oldest segments of the São Francisco Craton, Brazil. Bol. IG-USP 17: 37–72.

Marinho MM, Vidal Ph, Alibert C, Barbosa JSF, Sabaté P (1994b) Geochronology of the Jequié-Itabuna granulitic belt and the Contendas Mirante volcano-sedimentary belt. In: Figueiredo MCH, Pedreira AJ (eds) Petrological and geocronologic evolution of the oldest segments of the São Francisco Craton, Brazil. Bol IG-USP 17:73–96.

Marshak S, Alkmim FF, Jordt-Evangelista, H (1992) Proterozoic crustal extension and the generation of dome-and-keel structure in an Archean granite-greenstone terrane. Nature 357: 491–493.

Martin H, Peucat JJ, Sabaté P, Cunha JC (1997) Crustal evolution in the early Archaean of South America: example of the Sete Voltas Massif, Bahia State, Brazil. Precam Res 82:35–62.

Martin H, Moyen JF, Rapp R (2010) The sanukitoid series: magmatism at the Archaean–Proterozoic transition. Earth and Environmental Science Transactions of the Royal Society of Edinburgh 100: 15–33.

Mascarenhas JF, Ledru P, Souza SL, Conceição Filho VM, Melo LFA, Lorenzo CL, Milési JP 1998. Geologia e recursos minerais do Grupo Jacobina e da parte sul do Greenstone Belt de Mundo Novo. Série Arquivos Abertos. CBPM, Salvador, 58p.

Mello EF, Xavier RP, McNaughton NJ, Hagemann SG, Fletcher I, Snee L (2006) Age constraints on felsic intrusions, metamorphism and gold mineralisation in the Paleoproterozoic Rio Itapicuru greenstone belt, NE Bahia State, Brazil. Mineral Deposita 40: 849–866.

Moreira, H.; Lana, C.; Nalini Jr., H.N. 2016. The detrital zircon record of an Archaean convergent basin in the Southern São Francisco Craton, Brazil. Precam Res 275: 84–99.

Noce CM, Teixeira W, Machado N (1997) Geoquímica dos gnaisses TTG e granitóides neoarqueanos do Complexo Belo Horizonte, Quadrilátero Ferrífero, Minas Gerais. R Bras Geoc 27: 25–32.

Noce CM, Machado N, Teixeira W (1998) U–Pb geochronology of gneisses and granitoids in the Quadrilátero Ferrífero (Southern São Francisco Craton): age constraints for Archean and Paleoproterozoic magmatism and metamorphism. R Bras Geoc 28(1): 95–102.

Noce CM, Zucchetti M, Baltazar OF, Armstrong R, Dantas EL, Renger FE, Lobato LM (2005) Age of felsic volcanism and the role of ancient continental crust in the evolution of the Neoarchean Rio das Velhas greenstone belt (Quadrilatero Ferrifero, Brazil): U − Pb zircon dating of volcaniclastic graywackes. Precam Res 141: 67–82.

Noce CM, Pedrosa-Soares AC, Silva LC, Alkmim FF (2007a) O embasamento arqueano e paleoproterozoico da faixa Araçuaí. Geonomos 15(1): 17–23.

Noce CM, Pedrosa-Soares AC, Silva LC, Armstrong R, Piuzana D (2007b) Evolution of polycyclic basement complexes in the Araçuaí Orogen, based on U–Pb SHRIMP data: implication for Brazil-Africa links in Paleoproterozoic time. Precam Res 159: 60–78.

Noce CM, Tassinari CCG, Lobato LM (2007c) Geochronological framework of the Quadrilátero Ferrífero, with emphasis on the age of gold mineralization hosted in Archean greenstone belts. Ore Geol Rev 32: 500–510.

Nutman AP, Cordani UG (1993) SHRIMP U–Pb zircon geochronology of Archaean granitoids from the Contendas-Mirante area of the São Francisco craton, Bahia, Brazil. Precam Res 63: 179–188.

Nutman AP, Cordani UG, Sabatè P (1994). SHRIMP U–Pb ages of detrital zircons from the early Proterozoic Contendas-Mirante Supracrustal Belt, São Francisco Craton, Bahia, Brazil. J S Am Earth Sci 7 (2): 109–114, 1994.

Oliveira AH (2004) Evolução tectônica de um fragmento do Cráton São Francisco Meridional com base em aspectos estruturais, geoquímicos (rocha total) e geocronológicos (Rb–Sr, Sm–Nd, Ar–Ar, U–Pb) Tese de Doutorado, Universidade Federal de Ouro Preto.

Oliveira EP (2011) The Late Archaean Uauá Mafic Dyke Swarm, São Francisco Craton, Brazil, and Implications for Palaeoproterozoic extrusion tectonics and Orogen Reconstruction. In: Srivastava RK (Ed) "Dyke Swarms: Keys for Geodynamic Interpretation". Springer-Verlag Berlin Heidelberg, p. 19–31. DOI:10.1007/978-3-642-12496-9-2.

Oliveira AH, Carneiro MA (2001) Campo Belo Metamorphic Complex: Tectonic Evolution of an Archaean sialic crust of the southern

São Francisco Craton in Minas Gerais (Brazil). An Acad Bras Ciênc 73 (3): 397–415.

Oliveira E.P., Mello E.F., McNaughton N., 2002. Reconnaissance U–Pb geochronology of early Precambrian quartzites from the Caldeirão belt and their basement, NE São Francisco Craton, Bahia, Brazil: Implications for the early evolution of the Palaeoproterozoic Salvador-Curaçá Orogen. J. South Am Earth Sci 15: 349–362.

Oliveira EP, Windley BF, McNaughton N, Pimentel M, Fletcher IR (2004) Contrasting copper and chromium metallogenic evolution of terranes in the Palaeoproterozoic Itabuna–Salvador–Curaçá Orogen, São Francisco Craton, Brazil: new zircon (SHRIMP) and Sm–Nd (model) ages and their significance for orogen-parallel escape tectonics. Precam Res 128: 143–165.

Oliveira EP, McNaughton N, Armstrong R (2010) Mesoarchaean to Palaeoproterozoic growth of the northern segment of the Itabuna–Salvador–Curaçá Orogen, São Francisco Craton, Brazil. In: Kusky T, Mingguo Z, Xiao W (eds.) The Evolving Continents: Understanding Processes of Continental Growth. Geol Soc London Spec Publ 338: 263–286.

Oliveira EP, Souza ZS, McNaughton NJ, Lafon J-M, Costa FG, Figueiredo AM (2011) The Rio Capim volcanic-plutonic-sedimentary Belt, São Francisco Craton, Brazil: Geological, geochemical and isotopic evidence for oceanic arc accretion during Palaeoproterozoic continental collision. Gondwana Res 19: 735–750.

Oliveira E, Silveira E, Sõderlund U, Ernst R, Evans DAD (2012) New U-Pb zircon-baddeleyite ages on Archaean to Neoproterozoic LIPs (mafic dykes) of the São Francisco Craton, Brazil, and their potential use for Palaeocontinent reconstruction. Symposium Supercontinent 2012. September 25-28, 2012, University of Helsinki, Finland. Programme and Abstracts, pp. 94-95.

Oliveira EP, Silveira EM, Söderlund U, Ernst RE (2013) U–Pb ages and geochemistry of mafic dyke swarms from the Uauá Block, São Francisco Craton, Brazil: LIPs remnants relevant for Late Archaean break-up of a supercraton. Lithos 174: 308–322.

Paixão MAP, Oliveira EP (1998) The Lagoa da Vaca complex: An Archaean layered anorthosite body on the western edge of the Uauá Block, Bahia, Brazil. Rev Bras Geoc 28: 201–208.

Paquette JL, Barbosa JSF, Rohais S, Cruz SCP, Goncalves P, Peucat JJ, Leal ABM, Santos-Pinto M, Martin H (2015) The geological roots of South America: 4.1 Ga and 3.7 Ga zircon crystals discovered in N.E. Brazil and N.W. Argentina. Precam Res 271: 49–55.

Pereira RS, Fuck RA (2005) Núcleos arqueanos no Craton do São Francisco. III Simpósio sobre o Craton do São Francisco, Salvador, 2005, pp. 247–250.

Peucat JJ, Mascarenhas JF, Barbosa JSF, Souza SL, Marinho MM, Fanning CM, Leite CMM (2002) 3.3 Ga SHRIMP U–Pb zircon age of a felsic metavolcanic rock from theMundo Novo greenstone belt in the São Francisco craton, Bahia (NE Brazil). J S Am Earth Sci 15: 363–373.

Peucat JJ, Barbosa JSF, Pinho ICA, Paquette JL, Martin H, Fanning CM, Leal ABM, Cruz S (2011) Geochronology of granulites from the south Itabuna–Salvador–Curaçá Block, São Francisco Craton (Brazil): Nd isotopes and U–Pb zircon ages. J S Am Earth Sci 31: 397–413.

Piaia P (2011) Geochemistry of mafic rocks from the São José do Jacuípe area, northern segment of the Itabuna–Salvador–Curaçá Orogen, São Francisco Craton, Bahia. Dissertação de Mestrado. Instituto de Geociências , Universidade Estadual de Campinas Campinas, Brazil.

Pinese JPP, Teixeira W, Piccirillo EM, Quéméneur JJG, Bellieni G (1995) The Precambrian Lavras mafic dykes, southern São Francisco Craton, Brazil: preliminary geochemical and geochronological results. In: Baer G, Heimann A (eds.) Physics and Chemistry of Dykes, Rotterdam – Netherlands; 3rd Int Dyke Conf, pp. 205–219.

Rapp RP, Watson EB, Miller CF (1991) Partial melting of amphibolites/eclogite and the origin of Archean trondhjemites and tonalites. Precam Res 51: 1–15.

Reimink JR, Chacko T, Stern RA, Heaman, LM (2014) Earth's earliest evolved crust generated in an Iceland-like setting. Nature Geoscience 7: 529–533.

Rey PF, Coltice N (2008) Neoarchean lithospheric strengthening and the coupling of Earth's geochemical reservoirs. Geology 36: 635–638.

Ribeiro-Rodrigues LC (2001) Brazil's premier gold province. Part I: the tectonic, magmatic and structural setting of the Archaean Rio das Velhas greenstone belt, Quadrilátero Ferrífero. Mineral Deposita 36: 228–248.

Rios DC, Davis DW, Conceição H, Rosa MLS, Davis WJ, Dickin AP, Marinho MM, Stern R (2008) 3.65–2.10 Ga history of crust formation from zircon geochronology and isotope geochemistry of the Quijingue and Euclides plutons, Serrinha nucleous, Brazil. Precam Res 167: 53-70.

Romano R, Lana C, Alkmim FF, Stevens G, Armstrong R (2013) Stabilization of the Southern São Francisco Craton, SE Brazil, through a long-lived and episodic period of potassic magmatism. Precam Res 224: 1–20.

Rosa MLS (1999) Geologia, Geocronologia, Mineralogia, Litogeoquímica e Petrologia do Batólito Monzo-Sienítico Guanambi-Urandi (SW-Bahia). Tese de Doutorado. Salvador. Instituto de Geociências, Universidade Federal da Bahia, Bahia, Brazil.

Rosière CA, Spier CA, Rios FJ, Suckau VE (2008) The Itabirites of the Quadrilátero Ferrífero and Related High-Grade Iron Ore Deposits: An Overview. Rev Econ Geol 15: 223–254.

Santos-Pinto M, Peucat JJ, Martin H, Barbosa JSF, Fanning CM, Cocherie A, Paquette JL (2012) Crustal evolution between 2.0 and 3.5 Ga in the southern Gavião block (Umburanas-Brumado-Aracatu region) São Francisco Craton, Brazil. A 3.5–3.8 Ga proto-crust in the Gavião block? J S Am Earth Sci 40: 129-142.

Schorscher, HD, Santana, FC, Polonia, JC, Moreira, JMP (1982) Quadrilátero Ferrífero-Minas Gerais State: Rio das Velhas greenstone Belt and Proterozoic Rocks. Int. Symp. On Archean and Early Proterozoic Geologic Evolution and Metallogenesis, Excursion Annex. Salvador, 43p.

Schrank A, Silva MG (1993). Greenstone belt do cráton do São Francisco, Brasil. In: JML Dominguez and A Misi (eds.) O craton do São Francisco. SBG/SGM/CNPq, Salvador, pp. 85–118.

Seixas LAR, David J, Stevenson R (2012) Geochemistry, Nd isotopes and U–Pb geochronology of a 2350 Ma TTG suite, Minas Gerais, Brazil: Implications for the crustal evolution of the southern São Francisco craton. Precam Res (196–197): 61–80. DOI:10.1016/j.precamres.2011.11.002.

Shirey SB, Richardson SH (2011) Start of the Wilson Cycle at 3 Ga shown by diamonds from subcontinental mantle. Science 333: 434–436.

Silva LC, McNaughton NJ, Melo RC, Fletcher IR (1997). U–Pb SHRIMP ages in the Itabuna-Caraíba TTG hight-grade Complex: the first window beyond the Paleoproterozoic overprint of the eastern Jequié Craton, NE Brazil. In: ISGAM – Inter. Symp. Granites and Associated Mineralization. Salvador, Abstracts, SBG 1: 282–283.

Silva LC, Armstrong R, Delgado IM, Pimentel MM, Arcanjo JB, Mello RC, Teixeira LR, Jost H, Cardoso Filho JM, Pereira LHM (2002) Reavaliação da evolução geológica em terrenos pré-cambrianos brasileiros com base em novos dados U–Pb SHRIMP, Parte I: Limite centro-oriental do Cráton São Francisco na Bahia. R Bras Geoc 32(4): 501–512.

Silva LC, Pedrosa-Soares AC, Dussin I, Armstrong R, Noce CM (2012a) O complexo Belo Horizonte de Carlos Noce revisitado. Open file (ppt) report, http://www.46cbg.com.br/0110/sala3/10h10_luiz_carlos_01-10_sl03.ppt.pdf.

Silva LC, Pedrosa-Soares AC, Pimentel MM, Pinto CP, Armstrong R (2012b). Paleoarchean continental crust in central Brazil: first record of a ca. 3.4 Ga old TTG sliver within the southern domain of the São Francisco Craton, Minas Gerais. VIII South Am Symp Isotope Geol, Medelin. 2012 (CD-ROM).

Tadjou JM, Nouayou R, Kambuia J, Kande HJ, Manguelle-Dicoum E (2009) Gravity analysis of the boundary between the Congo craton and the Pan-African belt of Cameroon. Austrian J. Earth Sci. 102, 71–79.

Tassinari CCG, Velàsquez ME, Munhá JMU, Lobato LM, Mateus AM, Campos WF (2013) Evidences of Paleoproterozoic metamorphism in the NW region of the Quadrilátero Ferrífero area, Minas Gerais, Brazil: Implications for gold mineralizations. Goldschmidt 2013 Conference Abstracts, DOI:10.1180/minmag.2013.077.5.20.

Teixeira W (1982). Geochronology of the southern portion of the São Francisco Craton. Rev Bras Geoc 12 (1–3): 268-277

Teixeira and Figueiredo (1991).Teixeira W and Figueiredo MCH (1991) An outline of Early Proterozoic crustal evolution in the Sao Francisco craton, Brazil: a review. Precam Res 53: 1-22.

Teixeira W, Carneiro MA, Noce CM, Machado N, Sato K, Taylor PN (1996) Pb, Sr and Nd isotopic constraints on the Archean evolution of gneissic-granitoid complexes in the southern São Francisco Craton, Brazil. Precam Res 78: 151–164.

Teixeira W, Noce CM, Quemeneur JJG, Martins VTS (1997) Sr, Nd and Pb isotopic signatures of intrusive granitoids of the Paleoproterozoic Mineiro magmatic arc, southern São Francisco Craton, Brazil. In: Int Symp on Granites and Associated Mineralizations, Salvador, Brazil. Extended Abstracts, SBG, pp. 288–289.

Teixeira W, Cordani UG, Nutman AP, Sato K (1998) Polyphase Archean evolution in the Campo Belo metamorphic complex, Southern São Francisco Craton, Brazil: SHRIMP U–Pb zircon evidence. J S Am Earth Sci 11 (3): 279–289.

Teixeira W, Sabaté P, Barbosa J, Noce CM, Carneiro MA (2000) Archean and Paleoproterozoic tectonic evolution of the São Francisco Craton. In: Cordani UG, Milani EJ, Thomaz-Filho A, Campos DA (eds) Tectonic Evolution of South America. 31st IGC, p. 101–137.

Teixeira W, Marques LS, Petroni, CR (2010). Nature and evolution of the Archean crust of the São Francisco Craton. Fifth International Archean Symposium abstracts (edited by I.M. Tyler and C.M. Knox-Robinson). Record 2010/18. Geological Survey of Western Australia, p. 229.

Teixeira W, Geraldes MC, Matos R, Ruiz AS, Saes G, Vargas-Mattos G (2010) A review of the tectonic evolution of the Sunsás belt, SW Amazonian Craton. J S Am Earth Sci 29: 47–60.

Teixeira W, Ávila, CA, Dussin, IA, Corrêa Neto, AV, Bongiolo, EM, Santos, JOS, Barbosa, N (2015) Zircon U–Pb–Hf, Nd–Sr constraints and geochemistry of the Resende Costa Orthogneiss and coeval rocks: new clues for a juvenile accretion episode (2.36–2.33 Ga) in the Mineiro belt and its role to the long-lived Minas accretionary orogeny. Precam Res: 256: 148-169.

Trompette R (1994) Geology of Western Gondwana (2000-500 Ma) Balkema, Netherlands.

The Paleoproterozoic Eastern Bahia Orogenic Domain

4

Johildo Salomão Figueirêdo Barbosa and Rafael Gordilho Barbosa

Abstract

The Paleoproterozoic Eastern Bahia orogenic domain occupies an approximately 200 km-wide area in the northern São Francisco craton, located between the Atlantic coast and the interior plateau of Bahia state known as the Chapada Diamantina. Together with its African counterpart, the West Central African belt of Gabon, the Eastern Bahia orogenic domain form a complete segment of a collisional orogen developed during the transition between the Rhyacian and Orosirian periods. Exposed in the level of its roots, the Eastern Bahia orogenic domain involves mainly amphibolite to granulite facies rocks that characterize three distinct Archean basement blocks (the Gavião, Jequié and Serrinha blocks) and a Neoarchean magmatic arc (the Itabuna-Salvador-Curaça belt, ISAC belt). The Paleoproterozoic rock record of the domain includes volcanic, sedimentary and granitic assemblages deposited and emplaced during its pre-, syn- and post-collisional evolutionary stages. The supracrustal rocks are represented by continental to marine metasedimentary successions of presumed Siderian to Rhyacian ages, as well as volcano-sedimentary accumulated in intra- and back-arc basins settings. The granitic rocks comprise five distinct groups of relatively small plutons emplaced between 2320 and 1960 Ma, mainly in the Gavião and Serrinha blocks. The structural framework of the domain involves sinuous NS-trending and E-dipping reverse, reverse-sinistral to sinistral strike-slip faults and ductile shear zones. These structures nucleated during a second and syn-metamorphic deformation phase dated at ca. 2080 Ma, which was preceded by the convergence and collision of the various basement blocks of the domain at around 2100 Ma. This chapter contains a synthesis on the stratigraphy, overall structure and Paleoproterozoic evolutionary history of the domain.

Keywords

Eastern bahia orogenic domain • Paleoproterozoic • Granites • São Francisco craton

J.S.F. Barbosa (✉) · R.G. Barbosa
NGB-Nucleo de Geologia Básica – Mapeamento Geológico e Metalogenese, Instituto de Geociências, Universidade Federal da Bahia, Rua Barão de Jeremoabo, s/n, Campus Universitário de Ondina, Salvador, BA CEP 40.170-020, Brazil
e-mail: johildo.barbosa@gmail.com

M. Heilbron et al. (eds.), *São Francisco Craton, Eastern Brazil*, Regional Geology Reviews, DOI 10.1007/978-3-319-01715-0_4

4.1 Introduction

The northern lobe of the São Francisco craton in state Bahia exposes, between the Atlantic coast and the interior plateau known as the Chapada Diamantina, a roughly NS-trending orogenic zone developed during the transition between the Rhyacian and Orosirian periods of the Paleoproterozoic Era (Fig. 4.1). Involving the convergence and collision of four major Archean blocks, this orogenic domain is currently viewed as part of a Paleoproterozoic orogen that encompasses: (1) the Mineiro belt and Quadrilátero Ferrífero region in the southern portion of the craton (Alkmim and Teixeira, this book); (2) the substratum of the Neoproterozoic Araçuaí-West Congo orogen (Pedrosa-Soares et al. 2001; Alkmim et al. 2006), developed between the São Francisco and Congo cratons; and (3) its African counterpart, the West Central African belt preserved in the Congo craton in Gabon (Figueiredo 1989; Ledru et al. 1993; Feybesse et al. 1998; Barbosa and Sabaté 2002, 2004) (Fig. 4.1). In the West-Gondwana scenario that preceded the opening of the South Atlantic in the Eocretaceous, the Eastern Bahia orogenic domain together with the West Central African belt formed a crustal bridge connecting the São Francisco and Congo Cratons (Porada 1989).

This chapter contains an outline of the geology of Eastern Bahia orogenic domain. The lithostratigraphic assemblages, overall structure and metamorphic zoning of the domain are discussed in the first sections, which are followed by a summary of the Paleoproterozoic evolutionary history, focusing mainly the Rhyacian–Orosirian collisional event (referred to as the Transamazonian event in the Brazilian literature). Detailed geological mapping along with petrological, geochemical and geochronological studies carried out in the region over the last 30 years form the basis of the present synthesis.

4.2 Stratigraphy

Differently from its African counterpart, the Rhyacian–Orosirian Eastern Bahia orogenic domain is exposed in the level of its roots, so that high grade Archean rocks frankly predominate throughout the whole region. Paleoproterozoic rocks, including various generations of granitic plutons and supracrustal assemblages are relatively scarce, occupying small areas mainly in the western portion of the domain (Barbosa and Sabaté 2004; Barbosa et al. 2012).

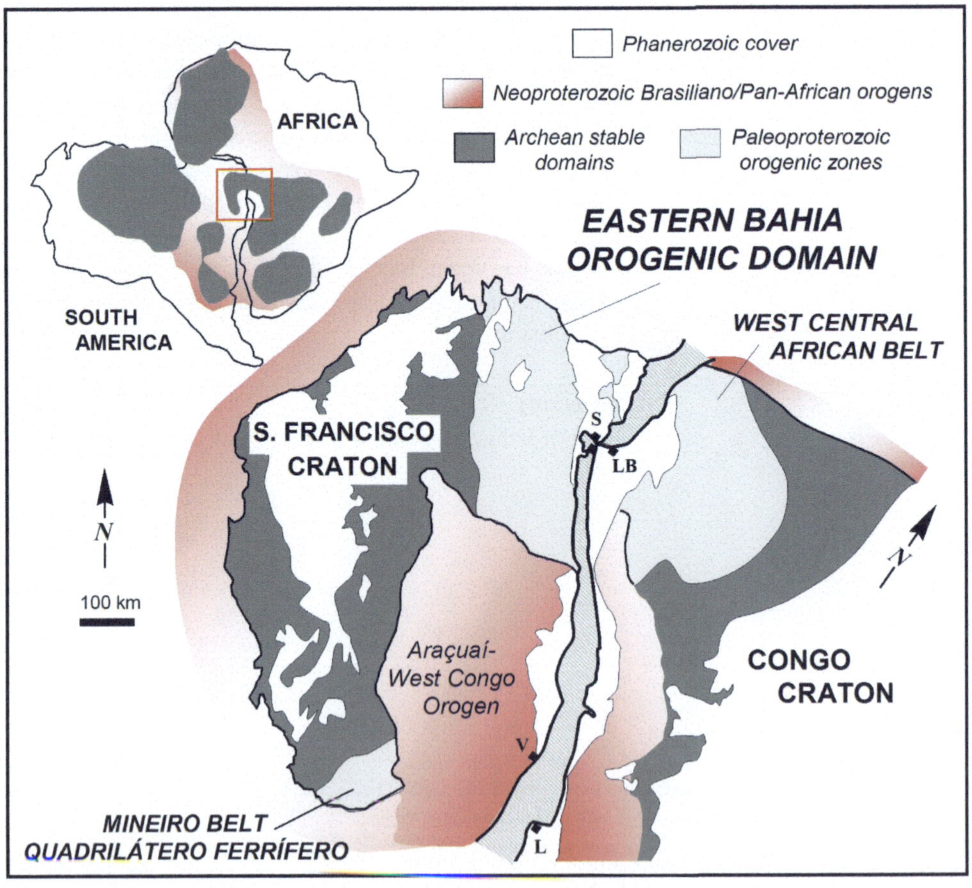

Fig. 4.1 The Paleoproterozoic Eastern Bahia orogenic domain in a reconstruction of West Gondwana. The Eastern Bahia orogenic domain represents the western half (present coordinates) of the Rhyacian–Orosirian orogen that formed the São Francisco-Congo cratonic bridge. The eastern half of the orogen corresponds to the West Central African belt of Gabon (Modified after Alkmim and Martins-Neto 2012)

4.2.1 Basement

The basement of the domain is made up of three major Archean nuclei, the Gavião, Jequié, and Serrinha blocks (see Teixeira et al. in this book for further details), as well as a belt involving Meso/Neoarchean orthogneisses and high grade metasedimentary rocks, known as the Itabuna-Salvador-Curaçá orogen in the Brazilian literature (Barbosa and Sabaté 2002, 2004) (Fig. 4.2a, b) and hereafter referred to as the ISAC belt. Bounded by major shear zones, these blocks show distinct geophysical (Fig. 4.2a), petrological, geochronological and geochemical attributes, especially Sm–Nd model age patterns, as well as ε_{Nd} and ε_{Sr} signatures (Fig. 4.3). Furthermore, Sm–Nd T_{DM} values show a progressive decrease towards east, i.e., from the Gavião Block, the oldest, to the ISAC belt, the youngest (Fig. 4.3a) (Barbosa and Sabaté 2004; Barbosa et al. 2012).

The Gavião block, as the largest component of the basement, probably continues further west and south beneath the Proterozoic cover forming the substratum of the NS-trending lobe of the craton (Figs. 4.1 and 4.2). Located in the tectonically most distal position of the domain, the Gavião block consists essentially of Paleo, Meso and Neoarchean orthogneiss complexes, including various TTG assemblages metamorphosed under amphibolites facies conditions (Barbosa and Sabaté 2004, Teixeira et al. this book). Some TTG complexes show U-Pb crystallization ages from 3.4 to 3.1 Ga and Sm–Nd model ages around 3.6 Ga, which place them among the oldest crustal pieces so far recognized in South America (Martin et al. 1997; Santos Pinto et al. 1998; Martins 2014). The gneissic complexes are locally covered or associated with volcanosedimentary successions, which include: (1) komatiites, mafic lavas and metachert at the base; (2) felsic volcanic and piroclastic rocks in the middle portion; and (3) terrigenous sediments in the upper section (Mascarenhas and Silva 1994; Cunha et al. 1996; Angelim 1997; Bastos Leal et al. 1998; Silva and Cunha 1999; Arcanjo 2000; Guimarães 2005; Barbosa et al. 2012). With deposition ages between 3.3 and 2.8 Ga, these assemblages underwent metamorphism and deformation in the time interval of 2.8 and 2.7 Ga, leading to the full development of greenstone belts and partial melting of the TTG substratum (Santos Pinto et al. 1998).

The Serrinha block, located in the northeastern sector of the orogenic domain (Fig. 4.2), is made up of Mesoarchean (3.0–2.9 Ga) orthogneisses, migmatites and tonalites with gabbroic enclaves metamorphosed under amphibolite facies conditions (Barbosa and Sabaté 2004; Barbosa et al. 2012). According to Oliveira et al. (2004a, b) and Oliveira (2010) (see also Teixeira et al. this book), the eastern portion of the

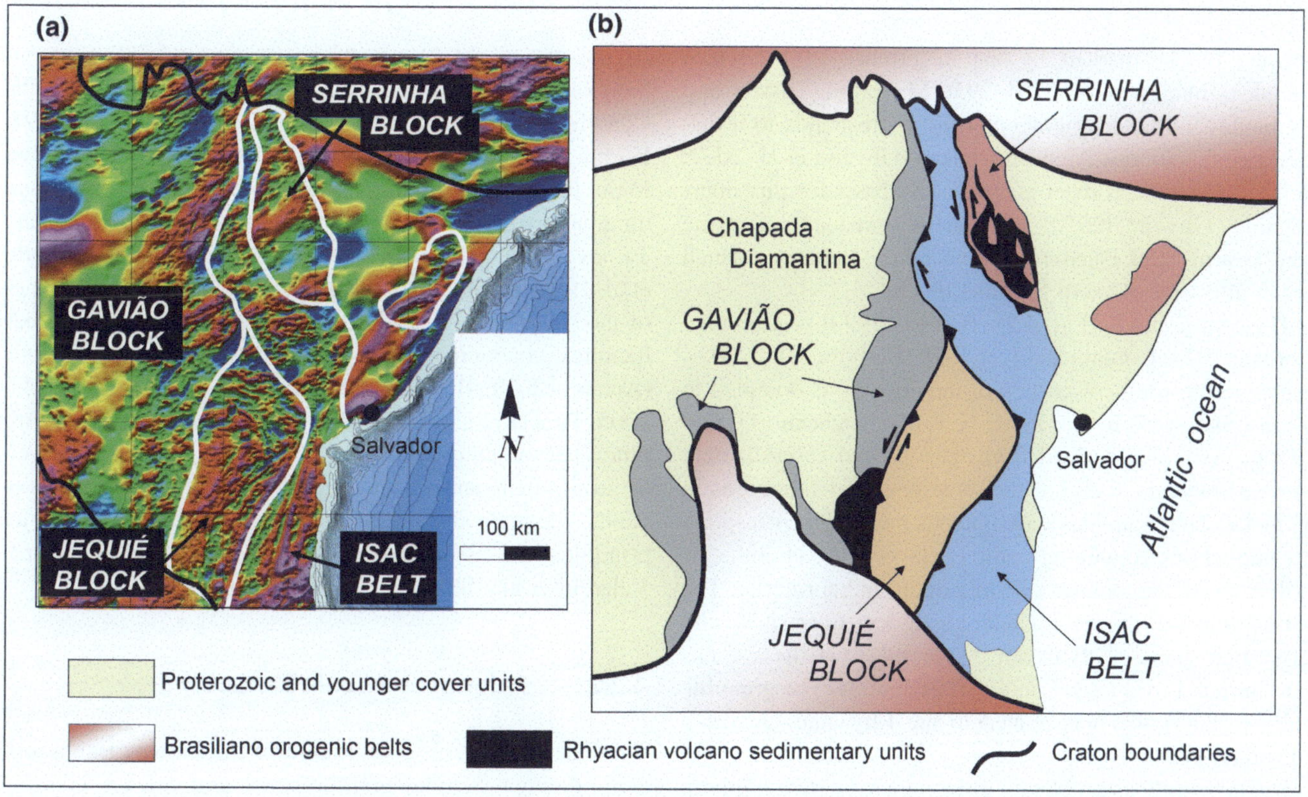

Fig. 4.2 **a** Expression of the various basement blocks of the Eastern Bahia orogenic domain in a magnetic anomaly map (Based on Bizzi et al. 2004). **b** The Archean basement blocks of the domain as delimited by Barbosa and Sabaté (2004)

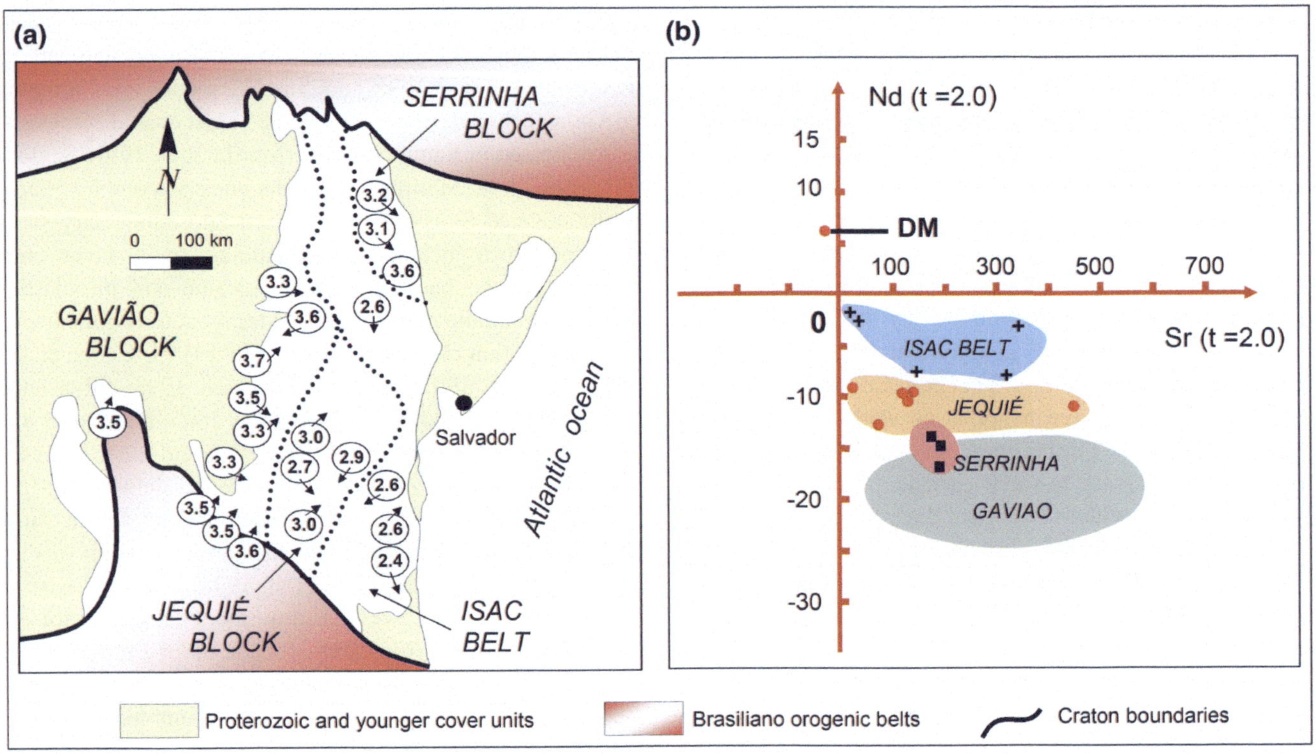

Fig. 4.3 **a** Distribution of the Sm–Nd (T_{DM}) ages along the basement blocks of the Eastern Bahia orogenic domain. *Note* that the ages become progressively older towards west (Based on Barbosa and Sabaté 2002, 2004). **b** εNd × εSr (t = 2.0 Ga) diagram showing the distinct fields characterized by data from the ISAC belt (*ISAC*) and the Gavião (*GB*), Serrinha (*SB*), and Jequié (*JB*) blocks (Reproduced from Barbosa et al. 2012)

Serrinha block, marked by the occurrence of the Uauá Complex, made up of 3204–2956 Ma orthogneisses and migmatites intruded by granodiorites, corresponds to a distinct and alloctonous Archean nucleus (Oliveira et al. 2002; Mello et al. 2006; Rios et al. 2009). As basis for this interpretation, Oliveira (2010) invoked the existence of mafic dyke swarms and Paleoproterozoic fabric elements, which do not affect the western Serrinha block.

The Jequié block (Fig. 4.2) (Barbosa 1990) or Complex (Cordani 1973; Cordani and Iyer 1979; Barbosa et al. 2003), located in the southern portion of the domain, is bounded to the east and north by the ISAC belt. Mesoarchean (3.0 and 2.9 Ga) granulites derived from igneous and subordinately from sedimentary rock are the most common lithotypes in the block. The granulites with igneous protoliths correspond to charnockites containing mafic enclaves and xenoliths, as well as leucocharnockites rich in garnets and cordierite. The metasedimentary rocks include kinzigites, quartzites, and garnet-rich gneisses. Between 2.7 and 2.8 Ga these rocks were intruded by a large volume of granitoids. The granulite facies metamorphism is related to the Rhyacian–Orosirian collisional event.

Distinct rock assemblages predominate in the northern and southern segments of the ISAC belt (Fig. 4.2), which represents a Meso to Neoarchean island arc complex (Barbosa et al. 2012). The northern portion consists essentially of granulite facies Mesoarchean orthogneisses (Caraíba Complex, Figueiredo 1980; Loureiro 1991; Melo 1991; Pereira 1992; Sampaio 1992; Melo et al. 1995; Nunes and Melo 2007; Oliveira 2010), Meso to Neoarchean mafic and ultramafic rocks (São José do Jacuípe Suite, Oliveira and Tarney 1995) and Neoarchean metasedimentary units (Melo et al. 1995; Kosin et al. 1999). The most common lithotypes of the southern segment of the belt are tonalites and trondhjemites containing wide variety of mafic intercalations (Arcanjo 2000; Barbosa 1990; Barbosa et al. 2008; Pinho 2005; Peucat et al. 2011), also metamorphosed under granulite facies conditions. Quartzites and paragneisses, as well as iron-, manganese- and graphite-rich rocks form large bands tectonically intercalated with the tonalitic/trondhjemitic assemblage (Toniatti and Barbosa 1973; Valarelli et al. 1982; Barbosa et al. 2012).

4.2.2 Paleoproterozoic Units

The Paleoproterozoic rock record of the Eastern Bahia orogenic domain includes granitic suites and volcanosedimentary assemblages emplaced and accumulated during the pre-, syn- and post-collisional evolutionary stages. Tectonic setting and age of these units vary along the various blocks of the domain.

4.2.2.1 Supracrustal Assemblages

Metasedimentary successions composed of pelitic schists, Lake Superior type banded iron-formations, quartzites, dolomitic marbles and manganese-rich schists occur in the northern part and especially in the southeastern border of the Gavião block. Grouped into the Colomi, Urandi, Licínio de Almeida, and Caetité sequences (Barbosa et al. 2012), these units are traditionally correlated to the passive margin to syn-orogenic sequence of the Siderian–Rhyacian Minas Supergoup exposed in southern portion of the craton (Quadrilátero Ferrífero region, see Alkmim and Teixeira, this book). In various aspects similar to the Francevillian formations of the West Central African belt of Gabon (Ledru et al. 1993; Feybesse et al. 1998; Thiéblemont et al. 2014), these units are not properly dated yet.

The best preserved and geographically most expressive Paleoproterozoic supracrustal assemblage of the Eastern Bahia orogenic domain corresponds to the Rio Itapicuru greenstone belt (Silva et al. 2001) succession (Fig. 4.4). Covering an approximately 150 km long and 50 km wide area in the western Serrinha block, this sequence comprises three units. The basal unit consists of a ca. 5000 m-thick succession of basalts and pillowed basalts flows containing minor lenses and layers of flow breccias, cherts, and banded iron formations. The middle unit is composed of dacitic and andesitic lavas and volcanoclastic rocks also showing intercalations of cherts and banded iron formations. The upper unit is essentially epiclastic, made up of conglomerates, sandstones and pelites derived from the older volcanic members (Mascarenhas 1979; Kishida 1979; Silva 1983; Silva et al. 2002; Oliveira et al. 2007). Metamorphosed under amphibolite to greenschist facies conditions the Rio Itapicuru sequence was deposited between 2145 and 2080 Ma, as indicated by the U-Pb age determinations on zircons from all members performed by Silva et al. (2002), Oliveira (2010) and Ruggiero and Oliveira (2010).

The Rio Capim succession is another greenstone belt sequence of similar lithological content and age that occupies a small area in the northeastern portion of the Serrinha block. According to Silva and Cunha (1999), the Rio Itapicuru and Rio Capim sequences represent volcanic and sedimentary deposits accumulated in intra- and back-arc basins developed in the Serrinha block between 2145 and 2080 Ma.

The Areião Formation that unconformably overlays the Archean Contendas Mirante greenstone belt (Marinho 1991; Teixeira et al. this book) in the southern portion of the Gavião block is a terrigenous unit deposited between 2160 (its maximum depositional age) and 2080 Ma (Nutman and Cordani 1994). The Formation consists of feldspar-rich metasandstones and metaconglomerates (Marinho 1991), interpreted by Ledru et al. (1997) as foreland basin deposits, generated during the collision between the Gavião and Jequié blocks in the beginning of the Orosirian period.

4.2.2.2 Granitic Rocks

Despite their relatively little expression on maps, numerous granitoid plutons emplaced between 2.3 and 1.9 Ga cut the basement and Paleoproterozoic supracrustal units, especially in the Gavião and Serrinha blocks. Assuming that the collisional event that led to the development of the domain was synchronous to the peak of the regional metamorphism dated at 2083 $\pm$ 4 Ma (Peucat et al. 2011), the Siderian–Orosirian granite plutons can be grouped into five distinct categories according to their geochemical signatures and ages. These groups are (Table 4.1):

- Pre- to syn-collisional, calc-alkaline to shoshonitic, K-rich, metaluminous to peraluminous granites and monzogranites dated between 2300 and 2065 Ma that occur mainly in the Gavião block (Barbosa et al. 2012; Cruz et al. 2015). Examples are Rio do Paulo and Ibitiara granites (Santos Pinto et al. 1998; Teixeira 2005)
- Pre- to syn-collisonal, calc-alkaline, peraluminous trondhjemites, tonalites, granodiorites, and granites dated between 2164 and 2109 Ma, which intrude the basement and the Rio Itapicuru volcanosedimentary sequence in the southern Serrinha block (Silva 1983; Chauvet et al. 1997; Oliveira et al. 2007; Barbosa et al. 2012). Examples for this category are Teofilandia and Itareru tonalities (Mello et al. 2006; Carvalho and Oliveira 2003).
- Syn- to late-collisional alkaline plutons (mainly syenites) emplaced between 2098 and 2060 Ma in the ISAC belt. The ca. 150 km-long to 10 km-wide Itiuba syenite body (Conceição 1990; Conceição et al. 2003) dated at 2084 Ma (Oliveira et al. 2004a, b), is the best representative of this category.
- Post-collisional syenites dated between 2054 and 2049 Ma, which form the Urandi-Guanambi batolith in the southern Gavião block (Rosa et al. 1996).
- Post-collisional peraluminous, muscovite- and biotite-bearing leucogranites emplaced along the eastern border of the Gavião Block (Sabaté et al. 1990; Teixeira et al. 2000), probably around 1.970 Ma (Mougeot 1996).

According to Cruz et al. (2015) the granitoids of first group together with metavolcanic rocks and amphibolites intercalated with Rhyacian metasedimentary units characterize the development and evolution of a magmatic arc along the margin of the Gavião block between 2300 and 2000 Ma. The second group of granitoids also seems to

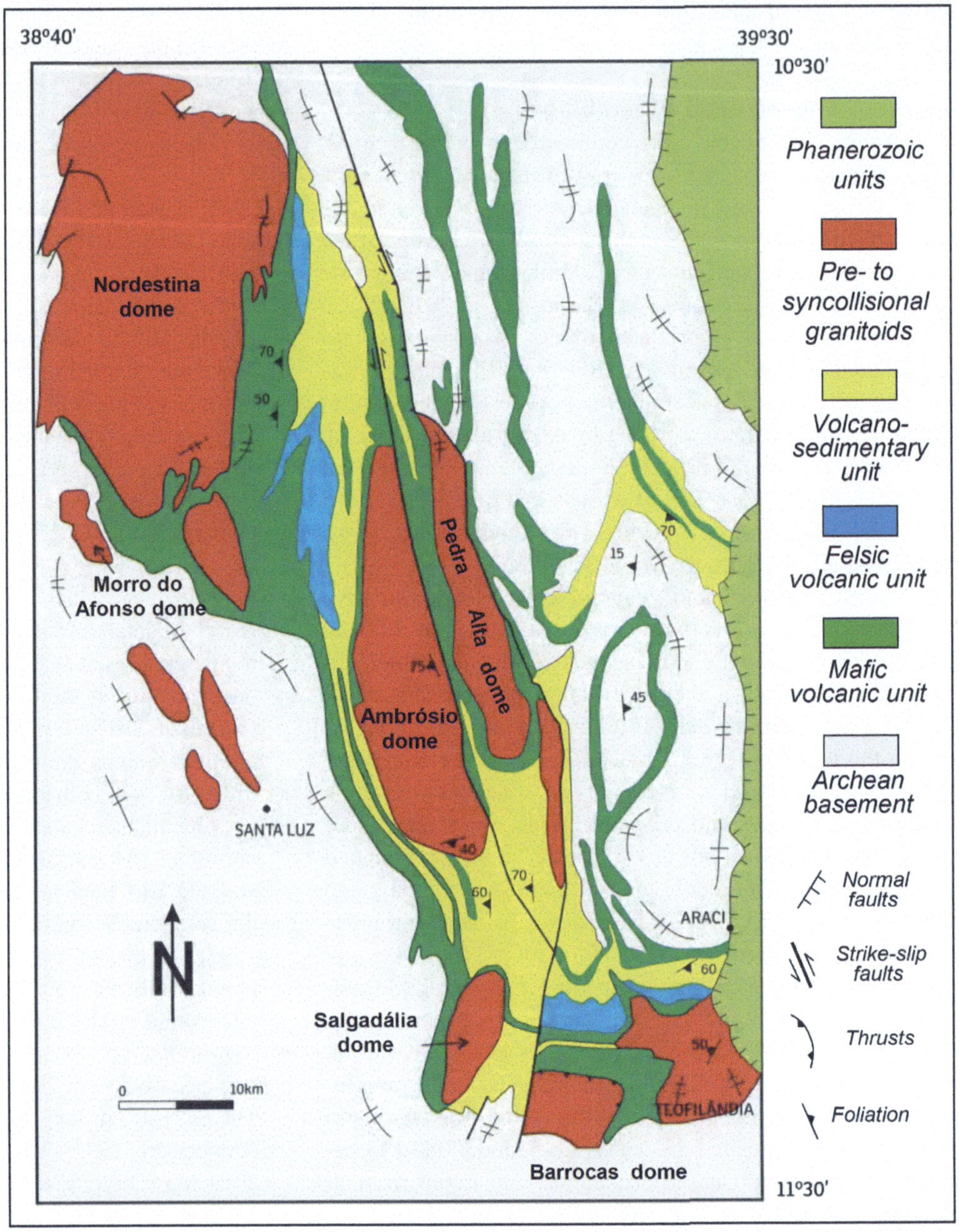

Fig. 4.4 Simplified geologic map of the Itapicuru greenstone belt, emphasizing the distribution of the supracrustal units. The elongated domes of the belt are the structural expression of pre- to syn-collisional granite plutons (Extrated from Rocha-Neto 1994)

record the development of a magmatic arc, in this case along the southwestern margin of the Serrinha block and slightly later, between 2100 and 2080 Ma (Barbosa et al. 2012; Oliveira et al. 2007).

4.3 Tectonic Framework and Metamorphism

Reverse, reverse-sinistral to sinistral strike-slip faults and ductile shear zones, thrusts, folds of variable geometry, and granitoid-cored domes are the large-scale structures of the Eastern Bahia orogenic domain (Barbosa and Sabaté 2004; Barbosa et al. 2004, 2012). Showing quite sinuous map traces around the NS direction and associated with a dominant WNW-directed tectonic motion, these structures compose different regional geometries in the northern and southern sectors of the domain (Figs. 4.5 and 4.6). The architecture of northern portion of domain is marked by a slightly asymmetrical flower structure (Fig. 4.5), whose central and high strain zone coincides with the northern and narrower branch of the ISAC belt. Along the southern portion of the domain, the fabric elements display a similar

Table 4.1 Paleoproterozoic granitoids of the Eastern Bahia orogenic domain (ages in Ma)

Pre- to syncollisional granitoids, Gavião block		
Rio do Paulo granite	2324 ± 6 (Zr Pb–Pb)	Santos Pinto et al. (1998)
Jussiape granite	2125 ± 2 (U–Pb LA–ICPMS)	Guimarães (2005)
Aracatu granite	2091 ± 6,6(U–Pb LA–ICPMS)	Campos (2013)
Ibitiara granite	2091 ± 7 (U–Pb LA–ICPMS)	Teixeira (2005)
Pre- to syncollisional granitoids, Serrinha block		
Cipó granite	2164 ± 2 (U–Pb SHRIMP)	Rios et al. (1998)
Eficeas granodiorite	2163 ± 5 (U–Pb TIMS)	Rios et al. (2009)
Trilhado granodiorite	2155 ± 9 (U–Pb SHRIMP)	Rios (2002)
Quijingue tonalite	2155 ± 3 (U–Pb SHRIMP)	Rios et al. (2008)
Nordestina trondhjemite	2152 ± 6 (U–Pb SHRIMP)	Cruz Filho et al. (2005)
Teofilandia granodiorite	2127 ± 7 (U–Pb SHRIMP)	Mello et al. (2006)
Barrocas tonalite	2127 ± 4 (Zr Pb–Pb)	Rios et al. (2000)
Lagoa dos bois granite	2107 ± 23 (U–Pb SHRIMP)	Rios et al. (2009)
Itareru tonalite	2109 ± 5 (U–Pb SHRIMP)	Carvalho and Oliveira (2003)
Syn- to late collisional granitoids, ISAC belt		
São felix syenite	2098 ± 1 (Zr Pb–Pb)	Rosa et al. (2001)
Pedra solta granite	2088 ± 9 (Zr Pb–Pb)	Otero (1997)
Itiuba syenite	2084 ± 9 (U–Pb SHRIMP)	Oliveira et al. (2004a, b)
Morro do Juá	2082 ± 2 (Zr Pb–Pb)	Otero (2006)
Pé Serra-Camará granite	2078 ± 4 (Zr Pb–Pb)	Otero (1997)
Bravo granite	2060 ± 6 (U–Pb SHRIMP)	Barbosa et al. (2008)
Post-collisional syenites, Gavião block		
Guanambi syenite	2054 ± 8 (U–Pb TIMS)	Rosa (1999)
Ceraima syenite	2054 ± 3 (Zr Pb–Pb)	Leahy (1998)
Cara Suja syenite	2053 ± 3 (U–Pb TIMS)	Paim (1998)
Estreito syenite	2041 ± 2 (Zr Pb–Pb)	Santos et al. (2000)
Post-collisional leucogranites, Gavião block		
Campo Formoso granite	1969 ± 29 (Rb–Sr)	Rudowski (1989)
Carnaiba granite	1960 ± 16 (Rb–Sr)	Celino (1991)
Jaguarari granite	1883 ± 87 (Rb–Sr)	Cuney et al. (1990)

geometry, becoming progressively steeper towards east (Fig. 4.6). The sector of highest strains also coincides with the ISAC belt (Barbosa et al. 2012).

In outcrops, the basement rock assemblages exhibit in general very complex structural patterns, which reflect the superposition of Paleoproterozoic on Archean fabric elements. Furthermore, the Paleoproterozoic deformation took place in two main phases, as recorded by the younger supracrustal units of the domain (e.g., the Rio Itapicuru greenstone belt, Fig. 4.4). The first phase, essentially contractional, led to development of NW-verging F_1-folds and thrusts in association of a variety of small-scale structures. The second phase reflects an overall sinistral transpression that causes renewed folding (F_2), reactivation of preexistent elements and nucleation of the regionally dominant sinistral-reverse to sinistral shear zones (Alves da Silva 1994; Chauvet et al. 1997; Barbosa and Sabaté 2004; Barbosa et al. 2012; Fig. 4.7).

Granulite facies metamorphic conditions predominate in the ISAC belt and Jequié block, whereas amphibolite to greenschist facies mineral assemblages characterize the Gavião and Serrinha blocks. P-T estimations for the granulite facies rocks have yielded background values of 830–850 °C and 5–7 kbar, respectively (Barbosa. 1994; Barbosa et al. 2012). P-T values in the order of 1000 °C and 8–10 kbar have been locally documented in Mg-rich granulites of the northern portion of ISAC belt by Leite et al. (2009).

Barbosa et al. (2012) estimated that the syn-kinematic regional metamorphism lasted from 2080 to 2050 Ma. Summarizing results of various age determinations, Peucat et al. (2011) concluded that the peak conditions of the regional metamorphism were reached at 2083 ± 4 Ma. The

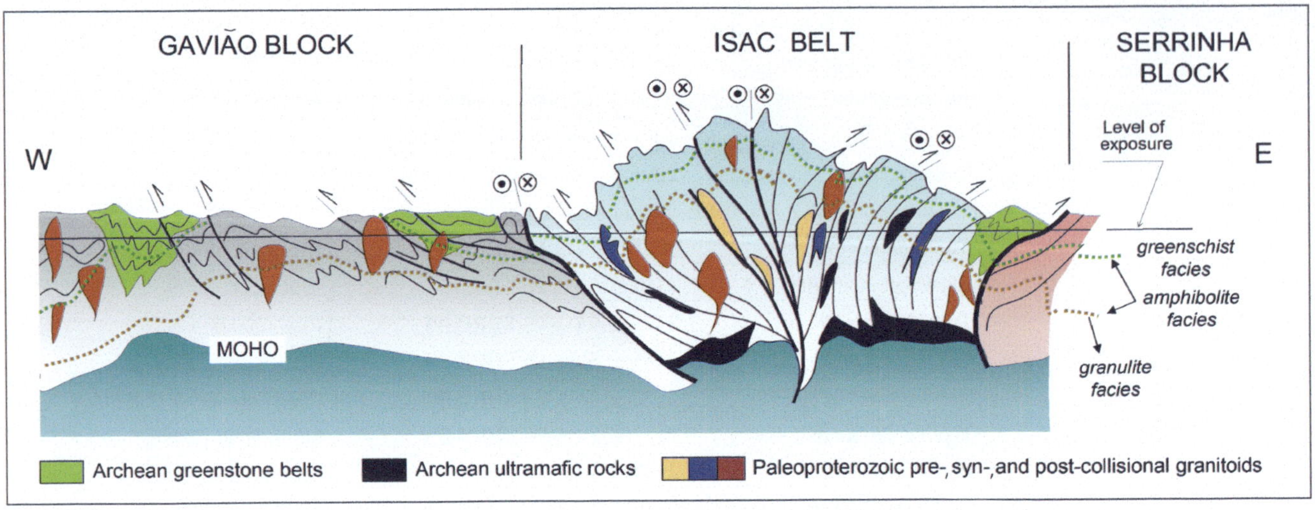

Fig. 4.5 Schematic cross-section representative of the northern portion of the Eastern Bahia orogenic domain (not to scale). Modified from Barbosa et al. (2012)

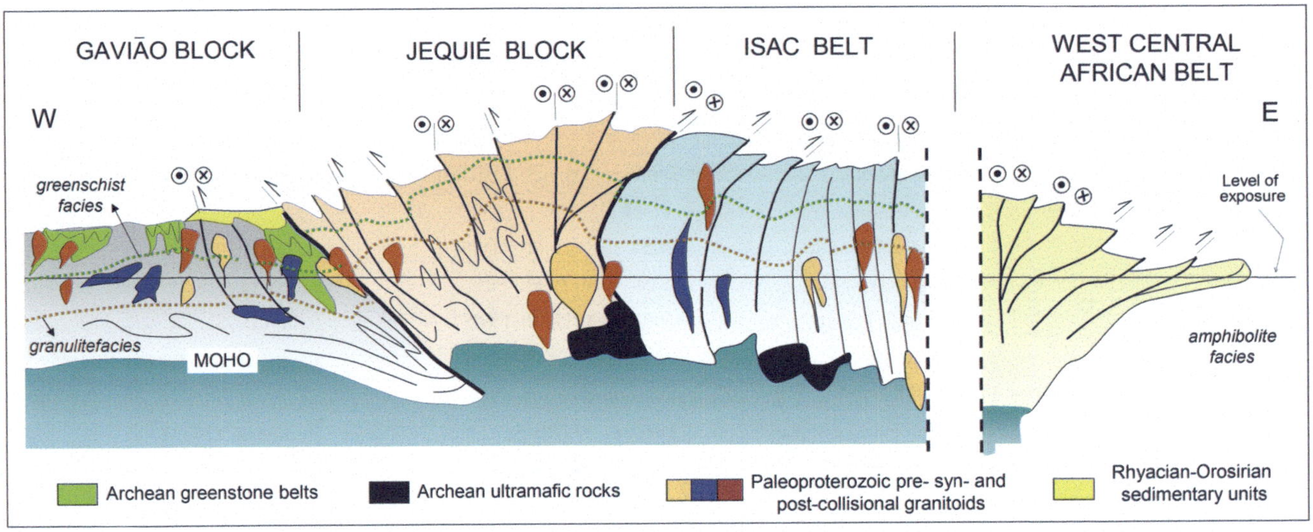

Fig. 4.6 Schematic cross-section representative of the southern portion of the domain (not to scale). Modified from Barbosa et al. (2012)

available data also indicates that this age reflects the sinistral transpressional deformation that took place during the climax of the collisional event responsible for the development of the Eastern Bahia orogenic zone (Barbosa and Sabaté 2004; Barbosa et al. 2012; Peucat et al. 2011).

4.4 Evolutionary History

The synthesis presented in the previous sections along with literature data indicate that the Eastern Bahia orogenic domain together with its African counterpart, the West Central African belt, form a segment of an orogen developed in the course of a major collisonal event that took place between 2125 and 2050 Ma (Barbosa and Sabaté 2002, 2004). This event involves the convergence and collision of the previously described Archean blocks, which represent distinct terranes, as illustrated by the Figs. 4.8 and 4.9.

In pre-collisional scenario, the Archean blocks of the domain were represented by micro-continents (the Gavião, Jequié and Serrinha blocks) and a Neoarchen magmatic arc (the ISAC belt) separated by oceanic basins, which underwent progressively consumption along convergent systems installed along the margins of the Gavião and Serrinha blocks (Fig. 4.9). This phase is recorded by the emplacement of subduction-related granites and deposition of volcanosedimentary sequences in intra- and back-arc basins (e.g., the Rio Itapecuru sequence) developed on both blocks.

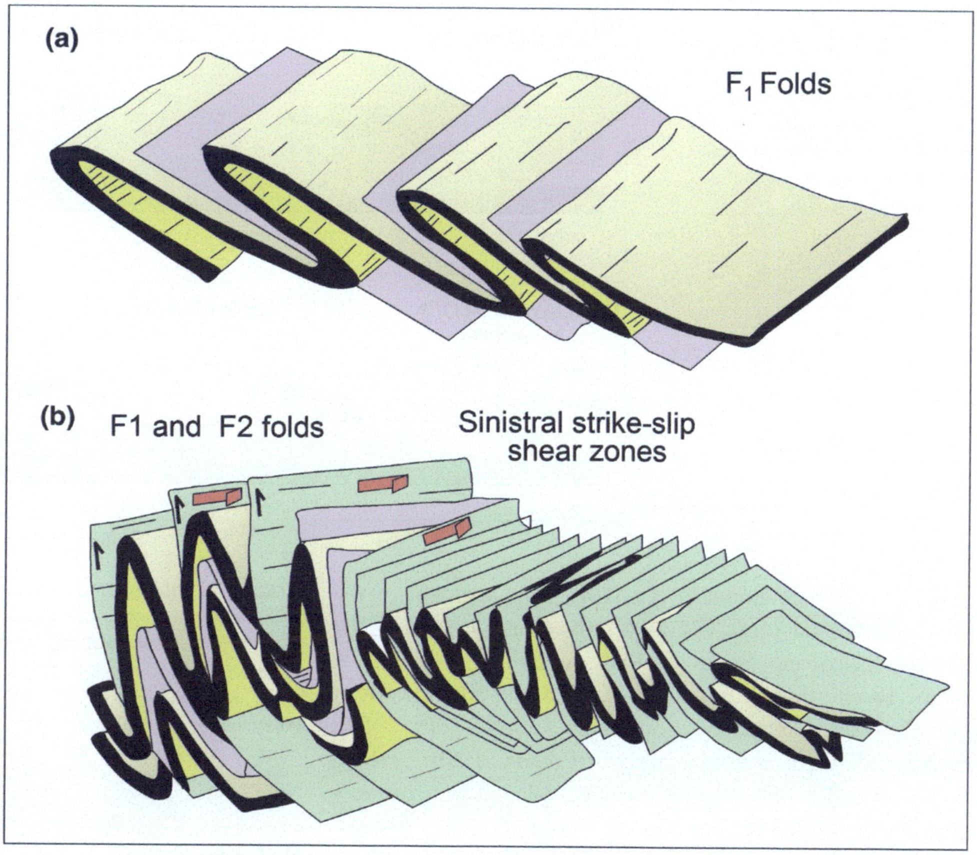

Fig. 4.7 Schematic representation of the structures related to the main deformation phases recorded in the Eastern Bahia domain and the geometries resulting from their superposition. **a** W-verging F1 folds and associated axial plane S1 foliation that characterize the first phase of deformation. **b** F2 folds associated with NNW-trending sinistral strike-slip shear zones overprinting the fabric elements of the first deformation phase. (Reproduced from Barbosa et al. 2012)

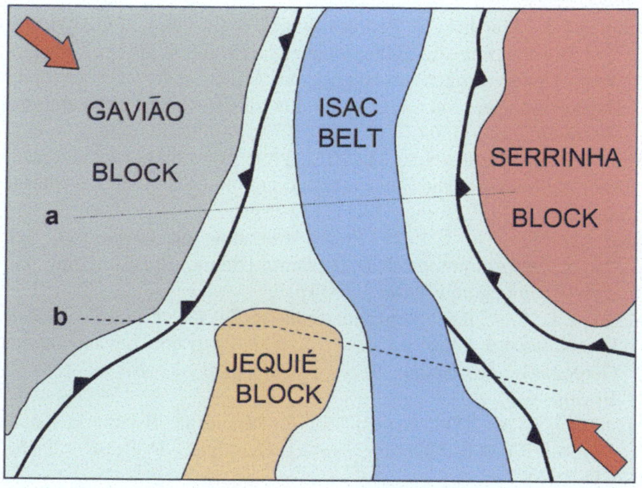

Fig. 4.8 Cartoon illustrating the relative positions of the various Archean blocks prior to the 2080 Ma collision that led to development of the Eastern Bahia orogenic domain (Modified from Barbosa et al. 2012). The *lines a* and *b* indicate the location of the schematic cross sections shown on Fig. 4.9

The progressive convergence of the Gavião block, representing the largest Archean nucleus of the São Francisco craton, with the Gabonian block (Thiéblemont et al. 2014), which corresponds to one of the Archean nuclei of the Congo craton, led to the amalgamation of the various intervening blocks and ultimately to the development of a collisional orogen at around 2100 Ma. The ISAC belt and the Serrinha block were intensively affected by NW-directed thrusting in the northern portion of the domain. An analogous structural picture was generated by the collision involving the Gavião and Jequié blocks with the ISAC belt in the southern portion of the domain. As indicated by metamorphic age determinations, immediately after the collision, the sinistral transpression affected the whole orogen, thereby generating the strike-slip system that characterizes the present-day architecture of the domain (Figs. 4.5 and 4.6).

4.5 Final Remarks

The Rhyacian–Orosirian Eastern Bahia orogenic domain, together with the West Central African belt exposed mainly in Gabon forms a complete segment of the orogen that resulted from the collision of Archean landmasses, represented by the Gavião block and the Gabonese massif, along with the intervening Jequié and Serrinha microcontinents, as well as the Neoarchean ISAC magmatic arc (Figueiredo 1980; Ledru et al. 1997; Barbosa and Sabaté 2004; Feybesse

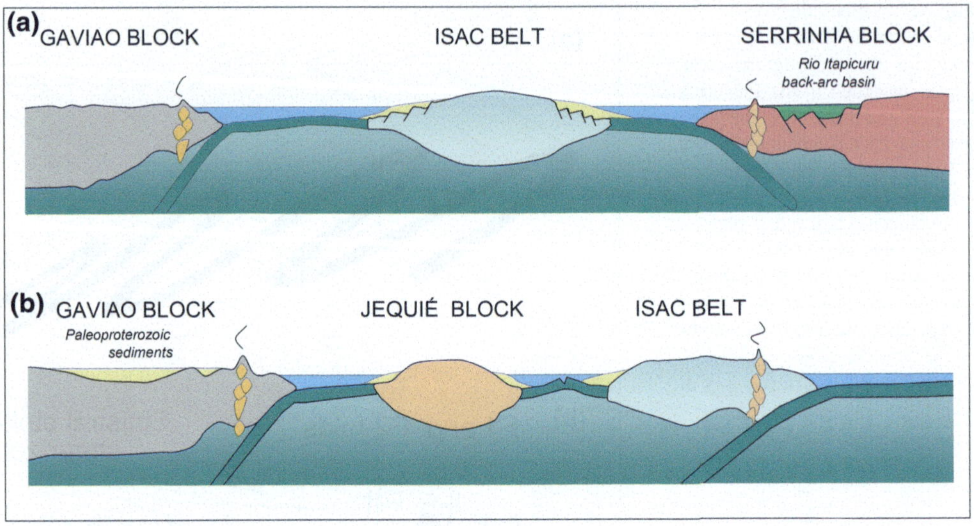

Fig. 4.9 Crossectional illustrations of the pre-collisonal scenario of the northern (**a**) and southern (**b**) portions of the Eastern Bahia orogenic domain. Modified from Barbosa et al. (2012). For location of the sections see Fig. 4.8

et al. 1998; Thiéblemont et al. 2014). This segment of the orogen remained preserved as a crustal bridge linking the São Francisco and Congo cratons until the Lower Cretaceous. The northward migration of the South Atlantic rift in the Lower Cretaceous eventually broke the São Francisco-Congo cratonic bridge (Porada 1989), which exerted a major influence on the post-Paleoproterozoic tectonic history of the cratons and the Neoproterozoic Araçuaí-West Congo orogen developed between them (Pedrosa-Soares et al. 2001; Alkmim et al. 2006). For instance, as demonstrated by Gordon et al. (this book), the differentiated nature of lithosphere along the cratonic bridge has a major impact on the architecture and evolution of the Brazilian continental margin basins generated through its extension and break-up.

Remarkably, major differences can be observed between the lithological assemblages and overall architecture of the Brazilian and African counterparts of the São Francisco-Congo cratonic bridge. These differences result basically from their level of exposure. As previously mentioned, the Brazilian side is deeply eroded and exposed in the level of the mid- to lower-crust. Studies performed in the Cretaceous Recôncavo rift basin (see Gordon et al., this book) show that the clasts of the conglomerates accumulated along its border faults derive essentially from granulite facies rocks. Therefore, the orogenic domain was already exposed in the level of its roots by the beginning of the Mesozoic Era (Barbosa et al. 2007). Further investigations are however required to better constrain the contrasting exhumation history the Brazilian and African counterparts of the São Francisco-Congo cratonic bridge.

Acknowledgments The authors thank F. Alkmim for his suggestions, comments, and editorial reviews, which improved the original text, making our ideas and interpretations sound simple and clear. For financial and logistic support, we thank the National Council for Scientific and Technological Development-CNPq and Companhia Baiana de Pesquisa Mineral- CBPM.

References

Alkmim F.F., Marshak, S., Pedrosa-Soares, A.C.; Peres, G.G., Cruz, S. C.P.,Whittington, A. 2006. Kinematic evolution of the Araçuaí-West Congo orogen in Brazil and Africa: nutcracker tectonics during the Neoproterozoic assembly of Gondwana. Precambrian Research, 149, 43–64.

Alkmim F.F., Martins-Neto M., 2012. Proterozoic first-order sedimentary sequences of the São Francisco craton, eastern Brazil. Marine and Petroleum Geology, 33(1), 127–139.

Alves da Silva, F. C.1994. Étude structurale du Greenstone Belt Paleoproterozoique du Rio Itapicuru (Bahia, Brésil). Unpl. Dr. Thesis, Université d'Orléans, 311p.

Angelim, L. A. A. 1997. Petrolina, folha SC.24-V-C: estados da Bahia, Pernambuco e Piauí: escala 1:250.000. Programa Levantamentos Geológicos do Brasil. Serviço Geológico do Brasil, CPRM. Brasília, DF.

Arcanjo, J. B. A. 2000. Projeto Vale do Paramirim: Estado da Bahia, escala 1:100.000. Salvador, Serviço Geologico do Brasil, CPRM, CD-ROM.

Barbosa, J.S.F., Correa-Gomes, L.C., Marinho, M.M. 2008. Geologia do segmento sul do Orógeno Itabuna-Salvador-Curaçá. In: BAHIA. Secretaria da Indústria, Comércio e Mineração. Geologia da Bahia: 50 anos de contribuição. p. 7–29.

Barbosa, J. S. F. 1994. Projeto petrologia e implicações metalogenéticas das rochas granulíticas do segmento Boa Nova – Itajibá, sudeste da Bahia: relatório final resumido. Salvador: UFBA. Convênio CBPM - IGEO/Curso de Pós-Graduação em Geologia.

Barbosa, J. S. F. 1990. The granulites of the Jequié Complex and Atlantic Mobile Belt, Southern Bahia, Brazil - An expression of Archean Proterozoic Plate Convergence. In: Vielzeuf, D.; Vidal, P. H. (eds.). Granulitization and crustal evolution. Clermont Ferrand, FR: Springer-Verlag, (Nato Science Series C, v. 311).

Barbosa, J. S. F., Dominguez, J. M. L.; Correa-Gomes, L. C.; Marinho, M. M. 2007. O conglomerado de Mont Serrat e suas relações com o alto e a falha de Salvador, Bahia. Revista Brasileira de Geociências, 37, p. 324–332.

Barbosa, J. S. F.; Correa-Gomes, L. C.; Marinho, M.M.; Alves da Silva, F.C. 2003.Geologia do segmento sul do orógeno Itabuna-Salvador-Curaçá. Revista Brasileira de Geociências, 33, p. 33–48.

Barbosa, J. S. F.; Martin, H.; Peucat, J. J. Archean/Paleoproterozoic 2004.Crustal domic evolution of the Lage, Mutuipe, Brejões and Santa Inês région. Jequié Block, Bahia, Brazil. Precambrian Research, Amsterdam, v. 135, p. 105–131.

Barbosa, J. S. F. and Sabaté, P. 2002. Geological features and the Paleoproterozoic collision of four Archean crustal segments of the São Francisco Craton, Bahia, Brazil. A synthesis. Anais da Academia Brasileira de Ciências, 74, 343–359.

Barbosa, J. S. F. and Sabaté, P. 2004.Archean and Paleoproterozoic crust of the São Francisco Craton, Bahia, Brazil: Geodynamic features. Precambrian Research, 133, 1–27.

Barbosa, J.S.F., Mascarenhas, J.F., Correa-Gomes, L.C., Dominguez, L.M., Santos de Souza, J. 2012. Geologia da Bahia, Pesquisa e Atualização. Salavador, CBPM, Série publicações especiais, v.1, 559p.

Bastos Leal, L. R., Teixeira, W., Cunha, J.C., Macambira, M. J. B. 1998. Archean tonalitic-trondhjemitic and granitic plutonism in the Gavião Block, São Francisco Craton, Bahia, Brasil: geochemical and geochronological characteristics. Revista Brasileira de Geociências, 28, 209–220.

Bizzi, L. A.; Schobbenhaus, C.; Gonçalves, J. H.; Baars, F. J.; Delgado, I.de M.; Abram, M. B.; Leão Neto, R.; Matos, G. M. M. De; Santos, J.O.S. 2004. Mapa geológico do Brasil. Escala 1:2.500.000 In: Bizzi, L. A. et al. Geologia, tectônica e recursos minerais do Brasil: Sistema de Informações Geográficas-SIG. Brasília, DF: CPRM,. v. 3. 3/4 CD-ROM.

Campos, L. D. 2013. O depósito de Au-Cu Lavra Velha, Chapada Diamantina Ocidental: um exemplo de depósito da classe IOCG associado aos terrenos paleoproterozoicos do Bloco Gavião. Unpl. MSc Thesis, Instituto de Geociências, Universidade Federal da Bahia, 113p.

Carvalho, M. J. and Oliveira, E. P. 2003 Geologia do tonalito Itareru, bloco Serrinha, Bahia: uma intrusão sin-tectônica do início da colisão continental no segmento norte do orógeno Itabuna - Salvador - Curaçá. Revista Brasileira de Geociências, 33 (Suplemento), 55–68.

Celino, J. J. 1991.Maciço de Jaguarari: arquitetura, impressões petrológicas e a geoquímica de acreçoes leucograníticas no cinturão transamazônico da Serra de Jacobina – (Bahia, Brasil). Unpl. MSc. Thesis, Instituto de Geociências, Universidade Federal da Bahia, Salvador, 163p.

Chauvet, A., Alves da Silva, F.C., Faure, M., Guerrot, C. 1997. Structural evolution of the Paleoproterozoic Rio Itapicuru granite-greenstone belt (Bahia, Brazil): the role of synkinematic plutons in the regional tectonics. Precambrian Research, 84, 139–162.

Conceição, H. 1990. Petrologie du massif syenitique d'Itiuba: contribuition à l'etude mineralogique des roches alcalines dans l'Etat de Bahia (Brésil). Unpl. Dr. Thesis,Centre D'Orsey, Université Paris-Sud, Paris, 395p.

Conceição, H., Silva, R. M. L., Macambira, M. J. B., Scheller, T., Marinho, M. M., Correia, R. D. 2003. 2,09 Ga idade mímina da cristalização do batólito sienítico Itiúba: um problema para o posicionamento do clímax do metamorfismo granulítico (2,05–2,08 Ga) no cinturão móvel Salvador-Curaçá, Bahia. Revista Brasileira de Geociências, 33, 395–398.

Cordani, U. G. 1973. Evolução geológica pré-cambriana da faixa costeira do Brasil, entre Salvador e Vitória. Unpl. Thesis (Livre Docência), Instituto de Geociências, Universidade de São Paulo, São Paulo, 107p.

Cordani, U. G.and Iyer, S. S. 1979. Geochronological investigation on the Precambrian granulitic terrains of Bahia, Brazil. Precambrian Research, 9, 255–274.

Cruz Filho, B. E., Conceição, H, Silva, M. L. R., Rios, D. C., Marinho, M. M., Macambira, J. B.2005. Geocronologia e assinatura isotópica (Rb-Sr e Sm-Nd) do batólito trondhjemítico Nordestina, Núcleo Serrinha, nordeste do Estado da Bahia. Revista Brasileira de Geociências, v. 35, (Suplemento), 1–8.

Cruz, S.C.P., Barbosa, J.S.F., Teixeira, L., Alkmim, F.F., Paquette, J. L., Peucat, J.J., 2015. O arco magmático sideriano-Riaciano (2,324–2,050 Ma) desenvolvido na margem continental Gavião, Bahia, Brasil. 15th Simpósio Naciaonal de Estudos Tectônicos, SNET, Vitória, Anais, p. 205–208.

Cuney M.; Sabaté, P.; Vidal, P.; Marinho, M.M.; Conceição, H. 1990. The 2 Ga peraluminous magmatism of the Jacobina-Contendas Mirante belt (Bahia) (Brazil): major and trace element geochemistry and metallogenic potential. Journal of Volcanology and Geothermal Research, 44, 123–141.

Cunha, J.C., Froes, R. J., Bastos Leal, L. R., Macambira, M. J. B., Teixeira, W. 1996. Idade dos greenstone belts e dos terrenos TTG's associados da região de Brumado, Centro-Oeste do Cráton do São Francisco (Bahia, Brasil). In: 39 Congresso Brasileiro de Geologia, Salvador. Sociedade Brasileira de Geologia, Anais, 1, p. 62–65.

Feybesse J.L, Johan V, Triboulet C, Guerrot C, Mayaga-Mikolo F, Bouchot V, Eko N'dong J. 1998. The West Central African belt: a model of 2.5–2.0 Ga accretion and two-phase orogenic evolution. Precambrian Research, 87, 161–216.

Figueiredo, M. C. H. 1989, Geoquímica das rochas metamórficas de alto grau no nordeste da Bahia, Brasil.. In: Inda, H. A. V.; Marinho, M. M.; Duarte, F. B. (Org.). Geologia e recursos minerais do Estado da Bahia: textos básicos. Salvador, Secretaria das Minas e Energia, p. 1–71.

Figueiredo, M.C.H. 1980. Geochemistry of High-Grade Metamorphic Rocks, Northeastern Bahia, Brazil.Unpl. PhD Thesis.The University of Western Ontario, 169p.

Guimarães, J. T. 2005. Projeto Ibitiara-Rio de Contas, Estado da Bahia, Escala 1:200.000. Serviço Geologico do Brasil, CPRM, 217p.

Kishida, A. 1979. Caracterização geológica e geoquímica das seqüências vulcanossedimentares no Médio Rio Itapicuru, Bahia. Unpl. MSc Thesis, Instituto de Geociências, Universidade Federal da Bahia. Salvador, 98p.

Kosin, M.; Guimarães, J. T.; Abram, M. B. (Org.). 1999. Folha Aracaju-SW, folha SC.24-Y. Serviço Geologico do Brasil, CPRM, 115p.

Leahy, G. A. S.; Silva, M. L. R.; Macambira, M. J. B.; Herve, M.; Paim, M. M.; Santos, E. B 1998. Maciço Sienítico de Ceraíma (sudoeste da Bahia): idade, petrografia e geoquímica do magmatismo pós-orogênico alcalino-potássico com afinidade lamprofírica. In: Conceição, H. (ed.). Contribuição ao estudo dos granitos e rochas correlatas. Sociedade Brasileira de Geologia, Salvador, Publicação Especial, 5, p. 61–77.

Ledru, P., Cocherie, A.,Barbosa, J., Johan, V., Onstott, T. 1993. Âge du métamorphisme granulitique dans le Craton du São Francisco (Brésil). Implications sur la nature de l'orogène transmazonien. Comptes rendus de l' Academie des Sciences, 211, 120–125.

Ledru, P.; Milesi, J.P.; Johan, V.; Sabate, P.; Maluski, H. 1997. Foreland basins and gold- bearing conglomerates: a new model for the Jacobina Basin (São Francisco Province, Brazil). Precambrian Research, 86, 155–176.

Leite, C. M. M., Barbosa, J.S.F., Gonçalves, P., Nicollet, C., Sabaté, P. 2009 Petrological evolution of silica-undersaturated sapphirine-bearing granulite in the Paleoproterozoic Salvador-Curaçá Belt, Bahia, Brazil. Gondwana Research, 15, p. 49–70.

Loureiro, H. S. C. (Org.) 1991. Mundo Novo, folha SC.24-Y-D-IV: Estado da Bahia: escala 1:100.000. Serviço Geológico do Brasil, CPRM, 47p.

Marinho, M. M. 1991. La sequence volcano - sedimentaire de Contendas - Mirante et la bordure occidentale du bloc Jequié (Craton du São Francisco, Brésil): un exemple de transition Archeen - Proterozoique. Unpl. Dr. Thesis, Université Blaise Pascal, Clermont Ferrand II, 388p.

Martin, H.; Peucat, J.J.; Sabate, P.; Cunha, J.C. Crustal evolution in the early Archaean of South America: example of the Sete Voltas Massif, Bahia State, Brazil. 1997. Precambrian Research, 82, p. 35–62.

Martins, A.A.M. 2014 (Org.). Projeto Brumado–Condeúba. Salvador: Serviço Geológico do Brasil, CPRM, 67p.

Mascarenhas, J. F. 1979. Estruturas do tipo "greenstone belt" no leste da Bahia. In: Inda, H. A. V. (Org.). Geologia e recursos minerais do Estado da Bahia: textos básicos. Salvador: CBPM,. v. 2, p. 24–53.

Mascarenhas, J. F. and Silva, E. F. A. 1994. Greenstone belt de Mundo Novo: caracterização e implicações metalogenéticas no Cráton do São Francisco. Salvador: CBPM, Série Arquivos Abertos, 5, 69p.

Melo, R. C. (Org.). 1991. Pintadas, folha SC.24-Y-D-V: Estado da Bahia: texto explicativo. Salvador: Serviço Geologico do Brasil, CPRM, 74p.

Melo, R. C.; Loureiro, H. S. C.; Pereira, L. H. M. 1995, Serrinha, folha SC.24-Y-D1, Estado da Bahia, escala 1:250.000. Serviço Geológico do Brasil, CPRM, 83p.

Mello E. F., Xavier, R.P., McNaughton, N.J. Hagemann, S.G., Fletcher I., Snee L. 2006 Age constraints of felsic intrusions, metamorphism, deformation and gold mineralization in the Paleoproterozoic Rio Itapicuru greenstone belt, NE Bahia State, Brazil. Mineralium Deposita, 40, 849–866.

Mougeot, R. 1996. Étude de la limite Archéen–Protérozoique et des mineralisations Au, ± U associées: exemples des régions de Jacobina (Etat de Bahia, Brésil) et de Carajás (Etat de Pará, Brésil). Doctoral Thesis,Université Montepellier II, Montpellier, France, 301p.

Nunes, N. S. V. and Melo, R. C. 2007. Região central do Cinturão Bahia Oriental: geologia e recursos minerais. Salvador, CBPM, Série Arquivos Abertos, 26, 78p.

Nutman, A. P. and Cordani, U. C. 1994. SHRIMP U-Pb zircon geochronology of Archean gneisses and Contendas-Mirante conglomerates, São Francisco Cráton. Boletim IG-USP. Publicação Especial, São Paulo, v. 17, p. 97–107.

Oliveira, E. P. 2010. The Palaeoproterozoic Uauá mafic dyke swarm, São Francisco Craton, Brazil and implications for orogen reconstruction. In: 6th International Dyke Conference, Abstracts, p. 104–104.

Oliveira, E. P., Carvalho M. J., McNaughton N. J.. 2004a. Evolução do segmento norte do orógeno Itabuna-Salvador-Curaçá: cronologia da acresção de arcos, colisão continental e escape de terrenos. Geologia USP, Série científica, 4 (1), 24–39.

Oliveira, E. P., Windley B. F, McNaughton N. J., Pimentel M., Fletcher I. R. 2004b. Contrasting copper and chromium metallogenic evolution of terranes in the Palaeoproterozoic Itabuna-Salvador-Curaçá Orogen, São Francisco Craton, Brazil: new zircon (SHRIMP) and Sm-Nd (model) ages and their significance for orogen-parallel escape tectonics. Precambrian Research, 128, 143–165.

Oliveira, E. P., Mello, E. F., McNaughton, N. T. 2002. Reconnaissance U-Pb geochronology of early Precambrian quartzites from the Caldeirão belt and their basement, NE São Francisco Craton, Bahia, Brazil: implications for the early evolution of the Palaeoproterozoic Salvador-Curaçá Orogen. Journal of South American Earth Sciences, 15, 349–362.

Oliveira, E. P. and Tarney, J. 1995. Genesis of the copper-rich Caraiba norite-hypersthenite complex, Brazil. Mineralium Deposita, 30, 351–373.

Oliveira, E.P., Donatti Filho, J.P., Ruggiero, A., Costa, F.G. 2007. The birth of the Rio Itapicuru greenstone belt, Bahia-Brazil, at a Palaeoproterozoic magma-poor rifted continental margin - a working hypothesis. 11th Simpósio Nacional de Estudos Tectonicos, Anais, p. 122–124.

Otero, O. M. F. 2006. Granitogênese paleoproterozóica e pós-tectônica no cinturão móvel Salvador-Curaçá: aspectos geológicos, petrográficos, geoquímicos e geocronológicos de corpos da Zona Axial. Unpl. Dr. Thesis, Instituto de Geociências, Universidade Federal da Bahia, 160p.

Otero, O. M. F. 1997. Petrografia, mineralogia e geoquímica do Plutão de Pedra Solta-Bahia. Unpl. Msc Thesis, Instituto de Geociências, Universidade Federal da Bahia, 196p.

Paim, M. M. 1998. Petrologia da intrusão sienítica potássica de Cara Suja (Sudoeste da Bahia). Unpl. MSc Thesis. Instituto de Geociências, Universidade Federal da Bahia, Salvador, 116p.

Pedrosa-Soares, A.C, Noce, C.M, Wiedemann, C.M., Pinto, C.P. 2001. The Araçuaí-West-Congo orogen in Brazil: an overview of a confined orogen formed during Gondwanaland assembly. Precambrian Research 110, 307–323.

Pereira, L. H. (Org.).1992. Serrinha, folha SC.24-Y-D-V: Estado da Bahia. Serviço Geologico do Brasil, CPRM, 57p.

Peucat, J. J., Barbosa, J. S. F., Araújo Pinho I. C., Paquette J. L., Martin H., Fanning, C. M., Menezes Leal A. B.,Cruz C. P.S., 2011. Geochronology of granulites from the south Itabuna-Salvador-Curaçá Block, São Francisco Craton (Brazil): Nd isotopes and U-Pb zircon ages. Journal of South American Earth Sciences, 31, 397–413.

Pinho, I. C. A. 2005. Geologia dos metatonalitos/metatrondhjemitos e granulitos básicos das regiões de Camamu-Ubaitaba-Itabuna, Bahia. Unpl. Dr. Thesis. Instituto de Geociências, Universidade Federal da Bahia, Salvador, 156p.

Porada, H. 1989. Pan - African rifting and orogenesis in southern to equatorial Africa and eastern Brazil. Precambrian Research, 44, 103–136.

Rios, C. D. Conceição, H.; Macambira, M. J. B.; Burgos, C. M. G. de; Peixoto, A. A. de; Cruz Filho, B. E. da; Oliveira, L. L.; Lisboa, M. P. 1998. Granitogênese da parte meridional-oriental do núcleo Serrinha: idade, petrografia e geoquímica. In: Conceição, H. et al. (ed.). Contribuição ao estudo dos granitos e rochas correlatas. Publicação Especial da Sociedade Brasileira de Geologa, Núcleo Bahia-SE, 5, p. 91–113.

Rios, C. D. 2002. Granitogênese no Núcleo Serrinha, Bahia, Brasil: geocronologia e litogeoquimica. 2002. Unpl. Dr. Thesis. Instituto de Geociências, Universidade Federal da Bahia, Salvador, 233p.

Rios, C. D.; Davis D. W.; Conceição, H.; Macambira, M. J. B.; Peixoto, A. A.; Cruz Filho, B. E. da, Oliveira L. L. 2000. Ages of granites of the Serrinha nucleus, Bahia (Brazil): an overview. Revista Brasileira de Geociências, 30, 74–77.

Rios, C. D., Davis D.W., Conceição, H., Rosa M.L.S., W.J.; Dickin D. A.P., Marinho M.M., Stern R., 2008. 3.65–2.10 Ga history of crust formation from zircon geochronology and isotope geochemistry of the Quijingue and Euclides plutons, Serrinha nucleus, Brazil. Precambrian Research, 167, 53–70.

Rios, C. D., Davis, D.W., Conceição, H., Davis W.J., Rosa M.L.S., Dickin A.P. 2009. Geologic evolution of the Serrinha nucleus granite-greenstone terrane (NE Bahia, Brazil) constrained by U-Pb single zircon geochronology. Precambrian Research, 170, 175–201.

Rocha-Neto, M.B. 1994. Geologia e recursos minerais do greenstone belt do Rio Itapicuru, Bahia. CBPM, Salvador, Série Arquivos Abertos, v.4, 47p.

Rosa, M. L. S. 1999. Geologia, geocronologia, mineralogia, litoquímica e petrologia do batólito monzo - sienítico Guanambi - Urandi (SW - Bahia). Unpl. Dr. Thesis. Instituto de Geociências, Universidade Federal da Bahia, Salvador, 186p.

Rosa, M. L. S., Conceição, H., Cruz Filho, B. E., Leahy, G. A. S., Martin, H., Paim, M. M., Peixoto, A. A., Santos, E. B. 1996. Batólito monzo - sienítico Guanambi - Urandi: mais um exemplo de magmatismo potássico/ultrapotássico na Bahia? In: 39th Congresso Brasileiro de Geologia, Anais, p. 406–407.

Rosa, M. L. S., Conceição, H., Macambira, M. J. B., Scheller, T., Martin, H., Bastos Leal, L. R. 2001. Idade Pb-Pb e assinatura isotópica Rb-Sr e Sm-Nd do magmatismo sienítico paleoproterozóico no sul do cinturão móvel Salvador-Curaçá: maciço sienítico de São Félix, Bahia. Revista Brasileira de Geociências, 31, 397–400.

Rudowski, L. Petrologie et geochimie des granites transamazoniens de Campo Formoso et Carnaíba (Bahia, Brésil), et des phlogopitites a emeraudes associées. 1989. 290 f. These (Doctorat en Pétrologie, Géochimie et Métallogénie)-Laboratoire de Géologie Appliquée, Université Paris, Paris, 1989.

Ruggiero, A. and Oliveira, E. P. 2010. Caracterização de vulcânicas adakíticas e cálcio - alcalinas no greesntone belt do Rio Itapicuru: petrogênese e implicações geodinâmicas. Revista Brasileira de Geociências, 40, 1–18.

Sabaté, P., Cuney, M., Vidal, P. 1990. Expressão estrutural e plutônica de uma colisão transamazônica NS no Cráton do São Francisco (Bahia - Brasil). 36th Congresso Brasileiro de Geologia, Boletim de resumos, p. 323.

Sampaio, A. R. (Org.). 1992. Gavião, folha SC.24-Y-D-II: Estado da Bahia, escala 1:100.000. Brasília, Serviço Geologico do Brasil, CPRM, 57p.

Santos, E.B., Rosa, M. L. S., Conceição, H., Macambira, M. J. B., Scheller, T., Paim, M. M., Leahy, G.A.S. 2000. Magmatismo alcalino-potássico paleoproterozóico no SW da Bahia e NE de Minas Gerais, maciço do Estreito: geologia, idade, petrografia e litogeoquímica. Geochimica Brasiliensis, 14, 249–267.

Santos Pinto, M. A., Peucat, J.J., Matin, H., Sabaté, P. 1998. Recycling of the Archaean continental crust: the case study of the Gavião Block, Bahia, Brazil. Journal of South American Earth Science, 11, 487–498.

Silva, L. C.; Armstrong, R.; Noce, C. M.; Carneiro, M. A.; Pimentel, M.; Pedrosa Soares, A. C.; Leite, C. A.;Vieira, V. S.; Silva, M. A. da; Paes, J. de. C.; Cardoso Filho, J. M. 2002. Reavaliação da evolução geológica em terrenos pré - cambrianos brasileiros com base em novos dados U-Pb SHRIMP, Parte II: orógeno Araçuaí, cinturão mineiro e Cráton do São Francisco meridional. Revista Brasileira de Geociências, 32, 513–528.

Silva, M. G. 1983. A sequência vulcanossedimentar do médio Rio Itapicuru, Bahia: caracterização petrográfica, considerações petrogenéticas preliminares e zoneografia metamórfica. Unpl. MSc Thesis, Institudo de Geociências, Universidade Federal da Bahia, Salvador, 88p.

Silva, M. G., Coelho, C. E. S., Teixeira, J. B. G., Alves da Silva, F. Ç., Silva, R. A. 2001. The Rio Itapicuru greestone belt, Bahia, Brazil: geologic evolution and review of gold mineralization. Mineralium Deposita, 36, 345–357.

Silva, M. G. and Cunha, J. C. 1999. Greenstone belts and equivalent volcano - sedimentary sequences of the São Francisco Craton, Bahia, Brazil: geology and mineral petential. In: Silva, M. G. and Misi, A. (eds.). Base metal deposits of Brazil. Serviço Geológico do Brasil, Belo Horizonte, p. 92–99.

Teixeira, L. R. (Org.) 2000. Projeto Vale do Paramirim: relatório temático de litogeoquímica. Salvador: Serviço Geologico do Brasil, CPRM, 33p.

Teixeira, L. R. 2005 (Org.). Projeto Projeto Ibitiara-Rio de Contas: relatório temático de litogeoquímica. Salvador: Serviço Geologico do Brasil, CPRM, 43p.

Thiéblemont, D., Bouton, P., Préat, A., Goujou, J.C., Tegyey, M., Weber F., Obiang. M.E., Joron, J.L., Treuil, M., 2014. Transition from alkaline to calc-alkaline volcanism during evolution of the Paleoproterozoic Francevillian basin of eastern Gabon (Western Central Africa). Journal of African Earth Sciences, 99, 215–227.

Toniatti, G. and Barbosa, J. S. F. 1973. O manganês de Maraú, Bahia. 27th Congresso Brasileiro de Geologia, Anais, v. 2, p. 421–430.

Valarelli, J. V., Barbosa, J.S.F., Mello, R. M. da S., Hypolito, R., 1982 Paragênese do protominério metamórfico de manganês de Maraú, Bahia. 32th Congresso Brasileiro de Geologia, Anais, v. 3, p. 819–825.

The Paleoproterozoic Mineiro Belt and the Quadrilátero Ferrífero

5

Fernando F. Alkmim and Wilson Teixeira

Abstract

Located in central Minas Gerais state, the Mineiro belt and the adjacent mining district known as the Quadrilátero Ferrífero (QF) represent segments of a Paleoproterozoic orogen preserved in the southern São Francisco craton. Siderian to Rhyacian juvenile granitoids associated with volcanic and sedimentary rocks are the main rock assemblages involved in the evolution of the Mineiro belt. The plutonic rocks characterize four distinct groups of subduction-related granitoids generated between 2.47 and 2.10 Ga. The supracrustal units comprise four volcano-sedimentary successions deposited in island arc settings during the same age interval of the granitoids. The structural picture of the Mineiro belt is dominated by NE-trending oblique to strike-slip shear zones associated to a regionally penetrative SE-dipping foliation. The Paleoproterozoic rock record of the QF includes greenschist to amphibolite facies metasedimentary successions, which unconformably overlie a Meso- to Neo-archean granite-greenstone basement assemblage. The Minas Supergroup comprises passive margin and syn-orogenic sediments, deposited during the 2.58–2.42 Ga interval and around 2.10 Ga, respectively. The youngest Paleoproterozoic unit of the QF, the Itacolomi Group, comprises post-orogenic intermontane deposits, accumulated between 2.05 and 1.74 Ga. The large-scale structures of the QF, an association of basement domes and synclines containing the Paleoproterozoic units, characterize a typical dome-and-keel architecture, whose development at around 2.05 Ga was preceded and followed by contractional deformation events. As components of a Paleoproterozoic orogen that also encompasses the Eastern Bahia orogenic domain of the northern portion of the craton and its African counterpart, the Mineiro belt and the QF are presently viewed respectively as an intra-oceanic arc system and a continental margin basin, which interacted and collided during the transition between the Rhyacian and Orosirian periods.

Keywords

Mineiro belt • Quadrilátero Ferrífero • Paleoproterozoic • São Francisco craton

F.F. Alkmim (✉)
Departamento de Geologia, Escola de Minas, Universidade Federal de Ouro Preto, Morro do Cruzeiro, Ouro Preto, Minas Gerais 35.400.000, Brazil
e-mail: alkmim@degeo.ufop.br; ffalkmim@gmail.com

W. Teixeira
Departamento de Mineralogia e Geotectônica, Instituto de Geociências, Universidade de São Paulo, Rua do Lago, 562, São Paulo, SP 05508-080, Brazil

M. Heilbron et al. (eds.), *São Francisco Craton, Eastern Brazil*, Regional Geology Reviews, DOI 10.1007/978-3-319-01715-0_5

5.1 Introduction

The Mineiro belt and the mining district of central Minas Gerais state known as the Quadrilátero Ferrífero (QF) ("Iron Quadrangle") together with the Paleoproterozoic Eastern Bahia orogenic domain (see Barbosa and Barbosa this book) represent distinct portions of an orogen developed during the transition between the Rhyacian and Orosirian periods. From a much broader perspective, this orogen also includes the units that form the basement of the adjoining Brasiliano/PanAfrican Araçuaí-West Congo orogen, developed between the São Francisco and Congo cratons (Fig. 5.1) (Teixeira and Figueiredo 1991; Heilbron et al. 2010; Alkmim and Noce 2006; Noce et al. 2007 and references therein).

Teixeira and Figueiredo (1991) defined the Mineiro belt (Fig. 5.1) as the extensive segment of the southern São Francisco craton affected by plutonism, metamorphism and deformation in the time interval of 2.2–1.9 Ga, i.e., in the course of the Transamazonian thermo-tectonic event. Petrological, geochemical and geochronological studies performed in this region during the last decade have revealed that the Mineiro belt in reality encompasses a crustal fragment made up essentially of juvenile granitoids in association with volcanic and sedimentary rocks emplaced and accumulated between the Siderian and Rhyacian periods (e.g., Noce et al. 2000; Ávila et al. 2014; Barbosa et al. 2015; Seixas et al. 2012, 2013; Teixeira et al. 2015).

The QF, located immediately to the northeast of the Mineiro belt, exposes, besides a complete section of the Archean basement of the craton (see Teixeira et al. this book), the Paleoproterozoic Minas and Itacolomi sedimentary successions (Dorr 1969; Renger et al. 1995). Together, these sequences record a passive margin to post-orogenic basin evolution between the Siderian and Orosirian periods (Alkmim and Marshak 1998; Alkmim and Martins-Neto 2012). Widely known for its large gold and iron deposits, the QF has been focus of systematic geologic studies since the eighteen century (see Farina et al. 2016 for a recent review). In this regard, the synthesis by Dorr (1969) of a detailed mapping program covering the whole district is considered to be a milestone in the development of geologic knowledge of the region.

In this chapter, we provide first a brief description of the lithological assemblages and structures of both the Mineiro belt and the QF. After discussing an evolutionary model for these regions, we examine them in the light of the Paleoproterozoic framework of the craton and its margins, indicating their geotectonic significance and the main open questions for future research.

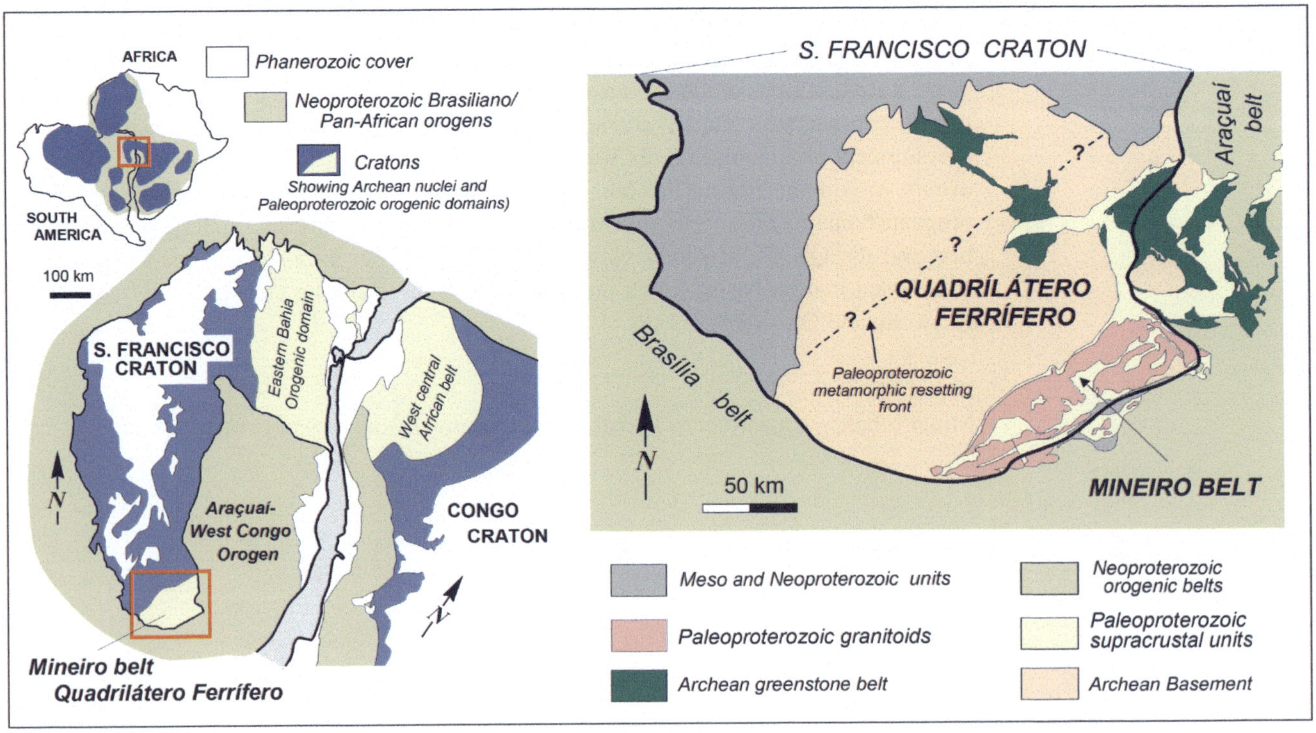

Fig. 5.1 The São Francisco and Congo cratons in a reconstruction of West Gondwana (South America and Africa) showing the location of the Mineiro belt, the Quadrilátero Ferrífero, the Eastern Bahia orogenic domain and the West Central African belt as segments of a Paleoproterozoic orogen preserved in the craton interior. Based on Alkmim and Martins-Neto (2012) and Barbosa et al. (2015)

5.2 The Mineiro Belt

As presently portrayed in the literature, the NE-trending Mineiro belt (Teixeira 1985) occupies a ca. 180 km-long area located to southwest of the QF (Figs. 5.1 and 5.2), bounded to the northwest and northeast by major shear zones referred to as the Jeceaba-Bom Sucesso and Congonhas-Itaverava lineaments, respectively (Campos and Carneiro 2008; Ávila et al. 2010; Corrêa Neto et al. 2012; Teixeira et al. 2015). The southern boundary of the belt (Teixeira and Figueiredo 1991) is probably located beyond the major Lenheiro shear zone, which is currently assumed as the local São Francisco craton boundary. In fact, lithological assemblages and fabric elements characteristic of the belt continue further south, extending over a large area beyond the limit of the craton (Fig. 5.2) (Ávila et al. 2010, 2014; Teixeira et al. 2015).

5.2.1 Stratigraphy

The Mineiro belt comprises different types of Paleoproterozoic orthogneisses (trondhjemitic, granodioritic, granitic), undeformed igneous bodies (gabbro, diorite, granite) and coeval supracrustal sequences (e.g., Noce et al. 2000; Seixas et al. 2012, 2013; Ávila et al. 2014; Teixeira et al. 2015; Barbosa et al. 2015). Occurrences of the Archean basement are very subordinate within the belt, as indicated by current geologic mapping and U–Pb dating (C.A. Ávila, oral communication, 2016; W. Teixeira, unpublished U–Pb dating). Some granitoid assemblages show, however, evidence of the involvement of Archean components in their generation.

5.2.1.1 Paleoproterozoic Granitic Rocks

As its most distinctive geologic trace, the Mineiro belt contains a large number of Siderian and Rhyacian granitic bodies (Fig. 5.2), grouped into plutons, suites or batholiths by the various authors working the region (Table 5.1).

In general, the orthoderived rocks display calc-alkaline affinity (e.g., Noce et al. 2000; Cherman and Valença 2005; Ávila et al. 2006; Teixeira et al. 2015) and formed in association with oceanic slab subduction and magma underplating, coupled with variable amounts of continental crust assimilation, as envisaged by Noce et al. (2000). From the list of granitoid bodies shown on Table 5.1, the most representative in terms of petrogenetic significance and age

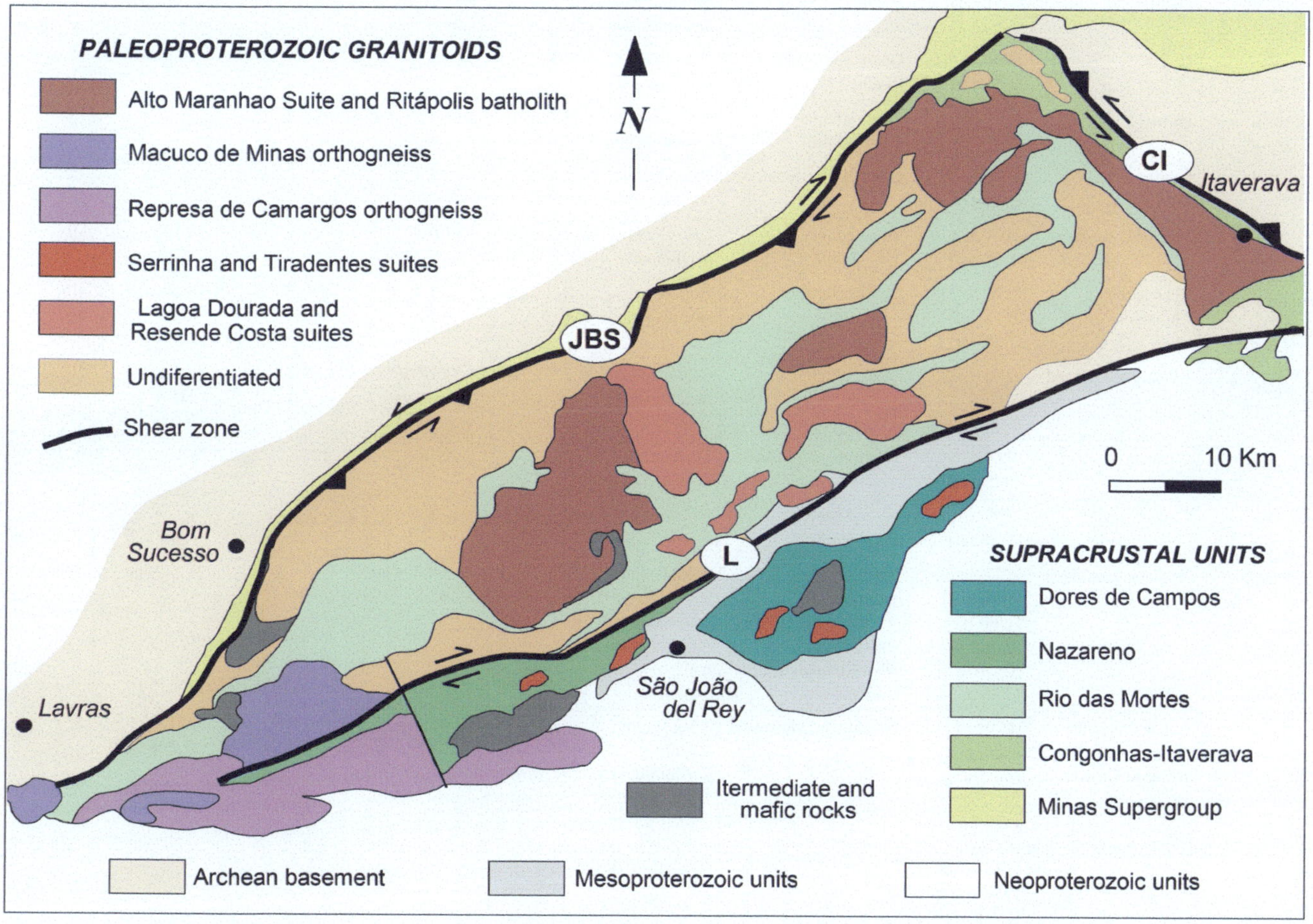

Fig. 5.2 Simplified geologic map of the Mineiro belt emphasizing the distribution of the main granitoid and supracrustal assemblages. Major shear zones: *JBS* Jeceaba Bom Sucesso, *CI* Congonhas Itaverava, *L* Lenheiro. Modified from Barbosa et al. (2015)

Table 5.1 Summary of isotopic and chemical characteristics of main Paleoproterozoic plutons of the Mineiro belt

Age range (Ma)	Plutons, batholiths suites	Age (Ma)	Rock types	Geochemistry	Isotopic constraints				References
					(wr) Nd		(zr) Hf		
					T_{DM} ages (Ga)	$\varepsilon_{Nd(t)}$	T_{DM} ages (Ga)	$\varepsilon_{Nd(t)}$	
2.19–2.10	Tabuões	1962 ± 13 Rb–Sr (wr), (ir) = 0.7024	Trondhjemite	Metaluminous–peraluminous	2.4	−2.8			1
	Represa de Camargos	2086 ± 12 U–Pb (zr)	Tonalite, granite	peraluminous	2.4	-2.8			10
	Rio Grande	2095 ± 12 U–Pb (zr)	Tonalite	N/A	N/A	N/A			10
	Alto Maranhão	2128 ± 10, 2130 ± 2 U–Pb (zr) 212 4 ± 2 U–Pb (ti)	Tonalite with dioritic enclaves	Calc-alkaline, TTG affinity, high- Al_2O_3, metaluminous.	2.3 to 2.4	−1.0 to −1.3			1, 2
	Ritápolis	2123 ± 33 U–Pb (zr) 2149 ± 10 U–Pb (zr)	Tonalite–granite	Calc-alkaline, peraluminous to slightly metaluminous	2.4–3.1	−7.3 to −1.0			1, 7, 8, 9
	Macuco de Minas, Itumirim	2131 ± 5 to 2126 ± 21 U–Pb (zr)	Granodiorite, tonalite	Calc-alkaline, peraluminous and metaluminous	2.8 to 2.5	−6.7 to −3.2			10, 14
	Lavras-Poço de Pedra	2111 ± 4 U–Pb (zr)	Tonalite, granodiorite, granite	Calc-alkaline, peraluminous–metaluminous	2.5–2.3	−0.2 to −6.3	2.7–3.1	−0.4 to −6.7	10
	Glória	2188 ± 29 U–Pb (zr)	Quartz-monzodiorite, tonalite	Calc-alkaline, metaluminous	2.7	−5.5			11
	Fé	2191 ± 9 U–Pb (zr)	Monzogranite–syenogranite	Calc-alkaline, peraluminous	2.7	−5.2			12
	Lajedo	2208 ± 26 U–Pb (zr)	Granodiorite–tonalite	Calc-alkaline, peraluminous	2.6	−2.6			12
	Morro do Resende	2174 ± 4 U–Pb (zr)	Monzogranite–granodiorite	Calc-alkaline, peraluminous	2.5	−0.3	2.3–2.7	+4.1 to CHUR	10
	Represa de Camargos	2170 ± 36, 2171 ± 24 U–Pb (zr)	Granite, tonalite	High-K calc-alkaline	3.0–3.2	−4.4, −7.2	2.4–2.7 2.8–3.2	+4.1 to CHUR −0.5 to −7.0	10
	Ribeirão do Amaral	2170 ± 12, 2166 ± 5 U–Pb (zr)	Quartz-diorite, tonalite	Calc-alkaline, metaluminous	2.5	CHUR	2.5–2.6 2.7 to 2.9	+2.5 to CHUR −0.5 to −3.2	10
	Nazareno	2165 ± 7 U–Pb (zr)	Tonalite, granodiorite	Tholeiitic, metaluminous	N/A	N/A	N/A	N/A	9, 10
	Congonhas	2195 ± 15 U–Pb (zr)	Granodiorite, trondhjemite	N/A	3.0–2.9	−3.0			2
2.23–2.20	Serrinha-Tiradentes	2227 ± 22–2204 ± 11 U–Pb (zr)	Gabbro, quartz-diorite, granodiorite, granophyric subvolcanic and volcanics	Sub-alkaline and calc-alkaline, metaluminous and peraluminous	2.6–2.3	−0.8; +1.1–+2.3			4, 5
2.36–2.32	Lagoa Dourada (LD) Resende Costa (RC)	(LD): 2356 ± 3 and 2350 ± 4 U/Pb (zr) (RC): 2351 ± 48–2317 ± 16 U–Pb (zr)	Tonalite, trondhjemite	TTG affinity, slightly peraluminous. High-Al_2O_3	2.4–2.5	+1.1–+3.2	2.3–2.7 3.0–3.4	+1.0–+6.1 −3.2 to −9.0	6, 7

(continued)

Table 5.1 (continued)

Age range (Ma)	Plutons, batholiths suites	Age (Ma)	Rock types	Geochemistry	Isotopic constraints				References
					(wr) Nd		(zr) Hf		
					T_{DM} ages (Ga)	$\varepsilon_{Nd(t)}$	T_{DM} ages (Ga)	$\varepsilon_{Nd(t)}$	
2.47–2.41	Cassiterita	2472 ± 11 to 2414 ± 29 U–Pb (zr)	Tonalite, trondhjemite	TTG affinity; peraluminous; high-Al_2O_3	2.5	+2.0[a]			7, 10

Data compiled from: (1) Noce et al. (2000), (2) Seixas et al. (2013), (3) Barbosa et al. (2015), (4) Ávila et al. (2010), (5) Ávila et al. (2014), (6) Seixas et al. (2012), (7) Teixeira et al. (2015), (8) Ávila et al. (2003), (9) Vasconcelos (2015), (10) Barbosa (2015), (11) Ávila et al. (2006), (12) Teixeira et al. (2008), (13) Oliveira (2004), Cherman (2004)
[a]Recalculated from (8), *wr* whole rock isochron, *in* initial $^{87}Sr/^{86}Sr$, *zr* zircon crystallization age (TIMS, SHRIMP, ICPMS-LA, Pb evaporation), *ti* titanite age (ID-TIMS), *N/A* not reported

are the Lagoa Dourada-Resende Costa (ca. 2.35 Ga), Serrinha-Tiradentes (2.23–2.21 Ga) and Alto Maranhão (2.13–2.12 Ga) suites, as well as the Ritápolis batholith (2.15–2.12 Ga). Aplite and pegmatitic dykes are common in most of the plutonic rocks.

The Lagoa Dourada Suite occurs as a ca. 18 km-long and 8 km-wide intrusion in the central portion of the belt (Fig. 5.2). This body consists of a series of biotite-hornblende tonalites and biotite trondhjemites with commingled dioritic mafic magmatic enclaves (Fig. 5.3a). These rocks are metaluminous to slight peraluminous with low-Mg# and high-Ca contents (Seixas et al. 2012). Dated at 2356 ± 3–2350 ± 4 Ma, the Lagoa Dourada Suite is peculiar for two reasons. First, as attested by the geochemical signature and $\varepsilon_{Nd(t)}$ values between +2.1 and +1.0, this assemblage originated from melts derived from a juvenile and short-lived tholeiitic basalt source (Seixas et al. 2012). Second, it represents an igneous event that took place in a time interval normally associated to a global magmatic quietude (e.g., Condie et al. 2009).

The Lagoa Dourada Suite is roughly coeval with the Resende Costa peraluminous orthogneiss (Figs. 5.2 and 5.3b), TTG-like assemblages, and amphibolites that yield U–Pb crystallization ages from 2351 ± 48 to 2317 ± 16 Ma and coherent Sm–Nd T_{DM} ages between 2.3 and 2.5 Ga (Teixeira et al. 2015). Coupled $\varepsilon_{Nd(t)}$ and $\varepsilon_{Sr(t)}$ values obtained for the Resende Costa gneissic rocks suggest derivation from a short-lived, slightly depleted Paleoproterozoic source, and the available geochemical data is consistent with a tholeiitic parental magma (Teixeira et al. 2015). The positive to negative zircon $\varepsilon_{Hf(t)}$ values point to the subordinate involvement of crustal components during the hypothesized 2.35 Ga subduction event. Together, the Lagoa Dourada and Resende Costa granitoids form a significant piece (>250 km^2) of ca. 2.35 Ga juvenile crust surrounded by younger units of the belt (Fig. 5.2). Moreover, two U–Pb zircon age determinations recently performed on the nearby Cassiterita orthogneiss yielded the oldest crystallization ages so far found in the Mineiro belt: 2468 ± 8 and 2414 ± 29 Ma (Barbosa 2015).

The Serrinha Suite (Ávila et al. 2010) and the coeval Tiradentes Suite (Ávila et al. 2014) comprise relatively small plutons intruding volcanosedimentary sequences in areas to the south of the Lenheiro shear zone (Fig. 5.2). The Tiradentes Suite includes well-preserved mafic andesites (Fig. 5.3c), dacites, granophyres in association with metasedimentary rocks (Fig. 5.3d, e). With U–Pb zircon crystallization ages between 2227 ± 22 and 2204 ± 11 Ma, these suites are composed mainly of metaluminous to peraluminous tonalites and granites. Sm–Nd T_{DM} whole rock ages between 2.6 and 2.3 Ga and $\varepsilon_{Nd(t)}$ values between −0.9 and +2.3 attest the juvenile character of these granitoids (Ávila et al. 2010, 2014).

The Alto Maranhão Suite consists essentially of strongly foliated tonalites of adakitic affinity, locally rich in mafic enclaves and mingling structures (Seixas et al. 2013). This suite intrudes the volcanosedimentary sequence that crops out in the northeastern portion of the Mineiro belt (Fig. 5.2). Available U–Pb ages obtained in titanite and zircon grains indicate that the Alto Maranhão pluton crystallized between 2130 ± 2 and 2124 ± 1 Ma (Noce et al. 2000; Seixas et al. 2013). Sm–Nd T_{DM} ages of 2.4 and 2.3 Ga, coupled with $\varepsilon_{Nd(t)}$ values in the range of −0.9 to +1.3, are evidence for the juvenile character and a short-lived history for their magmatic precursors. According to Seixas et al. (2013), the Alto Maranhão Suite likely formed from a mantle wedge source metasomatized to different degrees by slab-derived melts. Several plutons intruding the supracrustal assemblages in areas southwest of the Congonhas–Itaverava shear zone exhibit U–Pb crystallization ages between 2131 ± 4 and 2121 ± 2 Ma, i.e., in the same range of the Alto Maranhão Suite (Barbosa et al. 2015).

The Ritápolis Batholith forms the largest intrusion in the central portion of the belt (Fig. 5.2), composed of foliated leucogranites and tonalites that occur in different

Fig. 5.3 **a** Commingled dioritic mafic enclaves in the Lagoa Dourada orthogneiss (photo: W. Teixeira). **b** Felsic dykes cutting the Resende Costa orthogneiss (photo: C.A. Ávila). **c** Andesite of the Tiradentes Suite, intruded by a felsic dike (photo: C.A. Ávila). **d** Diamictite containing volcanic clasts of the coeval of the Tiradentes Suite. **e** Carbonaceous phyllite of the Tiradentes Suite (photo: C.A. Ávila)

Fig. 5.4 **a** Contact between medium-grained equigranular facies (*blue pen*) and coarse-grained facies of the Ritápolis granite, cut by a leucocratic pegmatite dyke (photo: C.A. Ávila). **b** Amphibolite xenolith in a granite coeval with the Ritápolis batholith (photo: C.A. Ávila). **c** Deformed orthogneiss affected by the Lenheiro shear zone (photo: C. A. Ávila)

macroscopic facies (Fig. 5.4a). Abundant xenoliths (Fig. 5.4b) sourced from the country rocks, as well as pegmatite dykes, are common features in the occurrences of the unit. Supracrustal sequences interpreted as roof pendants (C.A. Ávila, written communication, 2014) are also present in some outcrops. The available geochemical data indicate that the Ritápolis rocks are calc-alkaline, peraluminous to light metaluminous, and bearing a crustal signature akin to an active continental margin setting (Barbosa et al. 2015 and references therein). Zircon U–Pb SHRIMP ages of a metagranite yield a crystallization age of 2123 ± 33 Ma (MSWD = 8.5) and a lower intercept at 697 ± 96 Ma, reflecting Pb episodic loss (Teixeira et al. 2014). A few additional spots in inherited zircons from the same sample yielded concordant ages of 2354 ± 9 Ma (MSWD = 1.01). These data highlights the contribution of Siderian crustal components (such as the Lagoa Dourada-Resende Costa rocks) in the melt generation. In addition, the large range of negative $\varepsilon_{Nd(t)}$ values (between −7.3 and −1.8), coupled with Archean Sm–Nd T_{DM} ages (up to 3.1 Ga) and high initial $^{87}Sr/^{86}Sr$ ratios point toward a significant contribution of ancient crustal components during the formation of the Ritápolis Batholith (Barbosa et al. 2015).

Barbosa et al. (2015) conducted a detailed geochemical and geochronological investigation of the U–Pb, Hf and Nd–Sr isotopic systems on zircons extracted from several metaluminous to peraluminous metaigneous rocks occurring along the western portion of the Mineiro belt. Combining their results with literature data, the mentioned authors characterized a major calc-alkaline magmatic arc developed in the 2.17–2.10 Ga interval (see Table 5.1). The new zircon U–Pb ages define a two-stage plutonic process, dated at 2174–2149 Ma (metaquartz-diorite, metagranitoids and orthogneisses), and 2145–2100 Ma (metadiorites, metagranitoids and orthogneisses). The Alto Maranhão and Ritápolis plutons are the main components of the youngest magmatic episode. Moreover, the isotopic analyses on inherited zircons from several plutons yielded $^{207}Pb/^{206}Pb$ ages as old as 2785 Ma, highlighting again the important role of crustal contamination/assimilation in the magma genesis. Some samples displayed marked positive Pb anomalies, which reinforce the important role of crustal derivation. The zircon $\varepsilon_{Hf(t)}$ (from +8 to −10) and $\varepsilon_{Nd(t)}$ (from −7 to +0.3) values, as well as the Lu–Hf T_{DM} ages that largely range between 2.3 and 3.2 Ga, are consistent with a subduction zone, which involved juvenile Paleoproterozoic melts and Archean components.

The information summarized above, along with a robust petrological, geochemical, and geochronological database, indicate that the generation of the granitic assemblages forming the Mineiro belt resulted from a relatively long lasting (>350 Ma) process of magmatic accretion in intra-oceanic subduction systems (Ávila et al. 2010, 2014; Teixeira et al. 2015; Barbosa et al. 2015). This process may have started at ca. 2.47 Ga, given the U–Pb crystallization age of the Cassiterita Batholith (Barbosa 2015).

5.2.1.2 Supracrustal Units

According with several authors (e.g., Toledo 2002; Ávila et al. 2014; Corrêa Neto et al. 2012; Teixeira et al. 2008), the supracrustal assemblages involved in the Mineiro belt characterize four distinct Paleoproterozoic informal units, namely: the Congonhas-Itaverava, Rio das Mortes, Nazareno and Dores de Campos. These assemblages were previously portrayed as an extension of the Neoarchean Rio das Velhas Supergroup exposed in the QF (Seixas 1988; Zucchetti et al. 1996; Baltazar and Zucchetti 2007) or as a distinct Archean sequence (the Barbacena greenstone belt, Barbosa 1985). Dated between 2350 and 2100 Ma by means of zircon provenance studies and crystallization ages of coeval amphibolite rocks, the supracrustal units of the belt share as common traces a volcanosedimentary origin and the manganese-rich character.

The Congonhas-Itaverava volcanosedimentary sequence occurs in a ca. 50 km-long and 3 km-wide band along the northeastern boundary of the belt (Fig. 5.2). Intruded by the 2195 ± 15 Ma Congonhas stock (Seixas et al. 2013), this unit includes a basal package of mafic and ultramafic volcanics (including komatiites), a middle layer containing banded iron formations, marbles and pelites, and a upper succession of black shales and turbiditic graywackes (Corrêa Neto et al. 2012). U–Pb dating of detrital zircons extracted from the upper succession indicates contribution from multiple Archean sources and a maximum depositional age of 2349 ± 14 Ma (Teixeira et al. 2015). Metamorphosed under greenschist facies conditions, this sequence contains metabasalts, which yield positive to slightly negative $\varepsilon_{Nd(2.35Ga)}$ values and variable $\varepsilon_{Sr(2.35Ga)}$ values. The Dy/Dy*and Dy_N/Yb_N variations determined for these rocks are consistent with a LREE enriched source for the parental magma in accordance with an intra-oceanic environment (Teixeira et al. 2015).

The Rio das Mortes sequence occupies a large area along the portion of belt located between the Jeceaba-Bom Sucesso and Lenheiro lineaments (Fig. 5.2). It encircles plutonic rocks of diverse ages, such as the Resende Costa and Cassiterita orthogneisses (Barbosa et al. 2015), and is intruded by the 2.15–2.10 Ga Ritápolis Batholith among other coeval plutons and pegmatite injections (Barbosa et al. 2015). This sequence is composed mainly of phyllites in close association with gondites, as well as quartzites and tholeiitic amphibolites (originally basalts). Minor ultramafic rocks bodies are also present (Ávila 2000; Ávila et al. 2010). Detrital zircons extracted from a phyllite of the succession indicate a major age mode (82 % of the population) between 2300 and 2490 Ma. Neoarchean ages constitute a minor component. The youngest detrital grain in the population, dated at 2165 Ma, constraints the maximum depositional age of the sequence (Teixeira et al. 2012). Two amphibolites ascribed to this unit yielded U–Pb crystallization ages of 2231 ± 5 and 2202 ± 11 Ma (Ávila et al. 2012).

The Nazareno sequence (previously referred to as the Nazareno greenstone belt, Toledo 2002) occupies a ca. 45 km-long area located immediately to the south of the Lenheiro shear zone (Ávila et al. 2010) (Fig. 5.2). Abundant mafic and ultramafic rocks, including komatiites with spinifex textures, characterize the basal portion of the sequence. The middle and upper sections include amphibolites (metabasalts), and subordinate pelitic phyllites, quartzites and gondites (Toledo 2002; Ávila et al. 2010). Detrital zircons recovered from two quartzites display predominant Paleoproterozoic ages, indicating provenance probably from the Serrinha and Resende Costa-Lagoa Dourada rocks (Teixeira et al. 2012). In addition, two Nazareno amphibolites yielded zircon U–Pb crystallization ages of 2223 ± 4 and 2276 ± 14 Ma (Ávila et al. 2010, 2012). The latter age suggests that the deposition of the Nazareno sediments had started significantly earlier than the Rio das Mortes sequence.

The fourth supracrustal informal unit of the Mineiro belt, the Dores de Campos sequence (Ávila et al. 2014), is also made up essentially of amphibolites, phyllites, gondites and quartzites, containing subordinate metaultramafic lenses and pods at the base. The occurrences of this sequence are restricted to the region south of the craton boundary Lenheiro shear zone in the vicinity of the town of São João del Rey (Fig. 5.2). Teixeira et al. (2012) and Ávila et al. (2014) conducted provenance studies on zircons extracted from a ferruginous quartzite, an interbedded carbonaceous phyllite, and a diamictite. The obtained results indicate that the deposition of the Dores de Campos sequence took place after 2088 ± 12 Ma according to the youngest age peak in the zircon population (Ávila et al. 2014). Additional Paleoproterozoic age peaks suggest provenance from the Lagoa Dourada-Resende Costa rocks and Tiradentes suite, although subordinate contribution from Archean sources is also present. Furthermore, Ávila et al. (2014) reported the U–Pb zircon age of 2255 ± 51 Ma for an amphibolite layer in the Dores de Campo sequence.

The geochemical signature and the juvenile character of the mafic rocks included in the metavolcanic-sedimentary sequences of the Mineiro belt point toward an origin within intra-oceanic settings (Ávila et al. 2010, 2014). Furthermore, their ages fall in the same time interval, during which a substantial volume of juvenile granitoids characteristic of the belt was emplaced (Ávila et al. 2010, 2014; Teixeira et al. 2015).

5.2.2 Structure and Metamorphism

According to Endo (1997), the dominant NE-oriented structural traces of the Mineiro belt (Fig. 5.2) correspond to SE-dipping or near vertical shear zones, upright to NW-verging folds in association with a regionally pervasive foliation. Motion along the shear zones is dominantly dextral strike-slip, as attested by kinematic indicators of different scales. However, as pointed out by the same author, alternating dextral and sinistral motions can be observed along some of these structures. Moreover, fabric elements of an older deformation phase occur in relict form within the shear zones and especially in the domains between them (Endo 1997). Toledo (2002), in a detailed study performed in the Nazareno greenstone sequence exposed in the southwestern portion of the belt, ascribed the documented structures to three distinct deformation phases. Accordingly, elements of the first phase are restricted to domains preserved from the effects of the second and regionally dominant deformation episode, which led to the development of NW-directed thrusts, NE-trending folds, strike-slip shear zones, and a regionally pervasive foliation. The third phase was responsible for the nucleation of NW-trending strike-slip faults.

The Jeceaba-Bom Sucesso lineament that bounds the Mineiro belt to the north (Fig. 5.2) represents a NE-striking left-lateral shear zone, which affects the basement and the basal units of the Paleoproterozoic Minas Supergroup (Moeda and Cauê formations) (Endo 1997; Campos and Carneiro 2008; Neri et al. 2013). The Lenheiro lineament on the other hand corresponds to a NE-oriented dextral strike-slip shear zone (Endo 1997; Heilbron et al. this book) (Fig. 5.4c). According to Ávila (2000), Ávila et al. (2014), the Lenheiro shear zone divided the belt into two crust blocks characterized by distinct Paleoproterozoic lithological assemblages. The NW-trending Congonhas-Itaverava lineament (Fig. 5.2) marks the northeastern limit of the belt and corresponds to a NE-dipping, reverse-sinistral shear zone. Along this shear zone, the rocks ivolved exhibit the effects of an intensive hydrothermal alteration associated with a gold mineralization (Corrêa Neto et al. 2012).

Taking into account the juvenile signature and the oceanic setting of both the granitic and supracrustal assemblages involved in the Mineiro belt, in clear contrast to the Archean age and polycyclic nature of the adjacent terrains, the bounding shear zones of the belt were interpreted as Paleoproterozoic sutures by Campos and Carneiro (2008), as well as Ávila et al. (2010, 2014).

The rock units of the Mineiro belt record at least two distinct metamorphic episodes (Ávila et al. 2010 and references therein). The older, dated at ca. 2190 Ma, is an amphibolite facies episode recorded by the mafic rocks of the Nazareno and Rio das Mortes sequences. The younger, also recorded by the adjoining plutons, is marked by retrogressive low amphibolite to greenschist facies parageneses developed between 2130 and 2100 Ma. Local K–Ar resetting of biotite ages (ca. 600; Noce et al. 2000), as well as Neoproterozoic lead loss detected in U–Pb Concordia lower intercepts (see previous section) suggest that the Mineiro belt undergone a significantly younger low grade thermal event.

5.3 The Quadrilátero Ferrífero

The QF iron and gold-mining district extends over an area of approximately 15,000 km^2 located in the vicinity of Belo Horizonte (Fig. 5.5). This region is characterized by four long ridges underlain by quartzite and banded iron formation (Figs. 5.5 and 5.6), whose collective expression in maps and remote sensing images resembles a square, hence the name Quadrilátero Ferrífero, i.e., "Iron Quadrangle" (Fig. 5.5).

From a tectonic point of view, the QF lies in the external zone of two superimposed orogens. The district is located in the foreland zone of the Rhyacian–Orosirian orogen that encompasses the Mineiro belt and the substratum of the eastern reworked margin of the São Francisco craton. At the

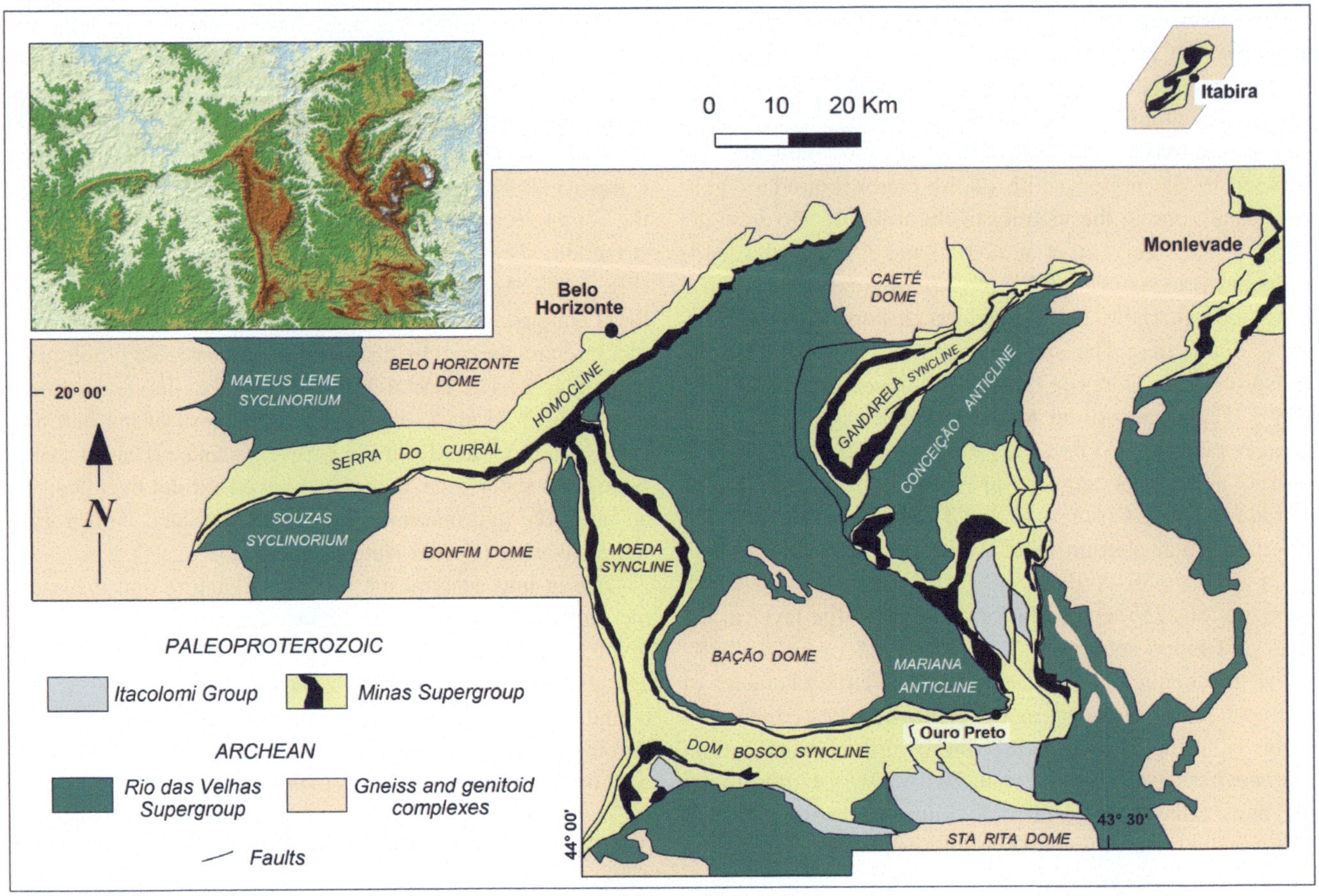

Fig. 5.5 Digital elevation model and simplified geologic map of the Quadrilátero Ferrífero region, emphasizing the main lithological assemblages and the large scale structures. Based on Dorr (1969) and Romano (1989)

same time, more than a half of the QF seems to have been affected by the Ediacaran Brasiliano orogenic front responsible for the development of the Araçuaí-West Congo orogen, which bounds the craton to the east (Fig. 5.1) (Machado et al. 1996a, b; Alkmim and Marshak 1998; Alkmim 2004; Teixeira et al. 2015). According to our interpretation, this peculiar location is responsible for many of the structural complexities that characterize the district.

5.3.1 Stratigraphy

As a Paleoproterozoic foreland, the QF exposes a Meso- to Neo-archean basement unconformably overlain by metasedimentary successions of the Minas Supergroup and Itacolomi Group (Fig. 5.7), deposited between the Siderian and Orosirian periods (Machado et al. 1996a, b; Hartmann et al. 2006). Small Paleoproterozoic granite bodies and pegmatite veins intrude the basement and the Minas units in the southern half the QF (Machado et al. 1992; Noce 1995). Moreover, abundant Proterozoic mafic dykes and sills cut both the Minas and Itacolomi sequences throughout the whole region (Girardi et al. this book). For instance, one of these intrusions, the Ibirité Gabbro, cuts the upper Minas Supergroup near Belo Horizonte. Dated by Silva et al. (1995), this dyke yielded a U–Pb baddeleyite upper intercept age of 1714 ± 5 Ma. Furthermore, three generations of NW-trending dykes that intrude the basement and the overlying Minas Supergroup in the western QF yielded U–Pb baddeleyite ages of 1798 ± 4 and 1793 ± 18 Ma, between 1702 ± 13 and 1717 ± 11, and 766 ± 36 Ma (Cederberg et al. 2016). Small fault bounded basins filled with Paleogene coal, clays, sand and gravel (Florália Formation, Dorr 1969) occur in some plateaus of the QF.

5.3.1.1 Basement

The Archean basement in the QF is made up of two major lithostratigraphic units, discussed in detail by Teixeira et al. (this book) (Fig. 5.7). The first includes Paleo- to Neo-archean (3220–2680 Ma) complexes composed of TTG-gneisses, migmatites and granitoids that form the core of various domal structures exposed in the region (Machado and Carneiro 1992; Lana et al. 2013; Romano et al. 2013; Farina et al. 2015, 2016; Teixeira et al. this book). The

Fig. 5.6 Panoramic view of the western Curral ridge, the most prominent morphological element of the northern QF. Extending for ca. 80 km in the NE direction, the crest of the ridge is underlain by itabirites of the Siderian Cauê Formation (Minas Supergroup). The picture faces NE and shows the Paleoproterozoic Minas strata dipping NNW and curving around the northern border of the Bonfim dome that lies to the south

second is the Neoarchean (2790–2750 Ma) Rio das Velhas Supergoup (Dorr 1969), which comprises a typical Au-mineralized greenstone belt sequence (Baltazar and Zucchetti 2007; Machado et al. 1992, 1996a, b; Schrank and Machado 1996; Noce et al. 2005; Hartmann et al. 2006; Moreira et al. 2016).

5.3.1.2 Minas Supergroup

The Paleoproterozoic Minas Supergroup, subdivided into five groups and eight formations (Fig. 5.7), comprises a ca. 8000 m-thick succession of continental to marine sedimentary rocks (Dorr 1969; Renger et al. 1995; Alkmim and Martins-Neto 2012). The rocks of this unit were metamorphosed under greenschist to amphibolite facies conditions. In spite of this, relict sedimentary features allow inferences on their depositional environments and paleogeographic significance. In the following descriptions, we will often refer to them according to their protoliths.

The basal Tamanduá and Caraça Groups comprise essentially alluvial quartz-rich sandstones and subordinate conglomerates (Moeda Formation, Fig. 5.8), which grade laterally and upwards into marine deposits, represented by pelites containing chert and dolomites lenses of the Batatal Formation (Dorr 1969; Villaça 1981; Renger et al. 1995, 1998; Koglin et al. 2012, 2014).

The marine transgression initiated in the Caraça Group continues further up section, being recorded first by the chemical sediments of the Itabira Group, subdivided into the Cauê and Gandarela formations (Fig. 5.7). The Cauê Formation, the marker bed of the Minas Supergroup, consists of a Lake Superior-type banded iron formation metamorphosed to itabirites (of siliceous, dolomitic, amphibolitic, and Mn-rich varieties), dolomitic marbles and large bodies of high grade, hypogene and supergene iron ores (Dorr 1964; Klein and Ladeira 2000; Rosière et al. 2008; Spier et al. 2003, 2007; Amorim and Alkmim 2011; Alkmim 2014)

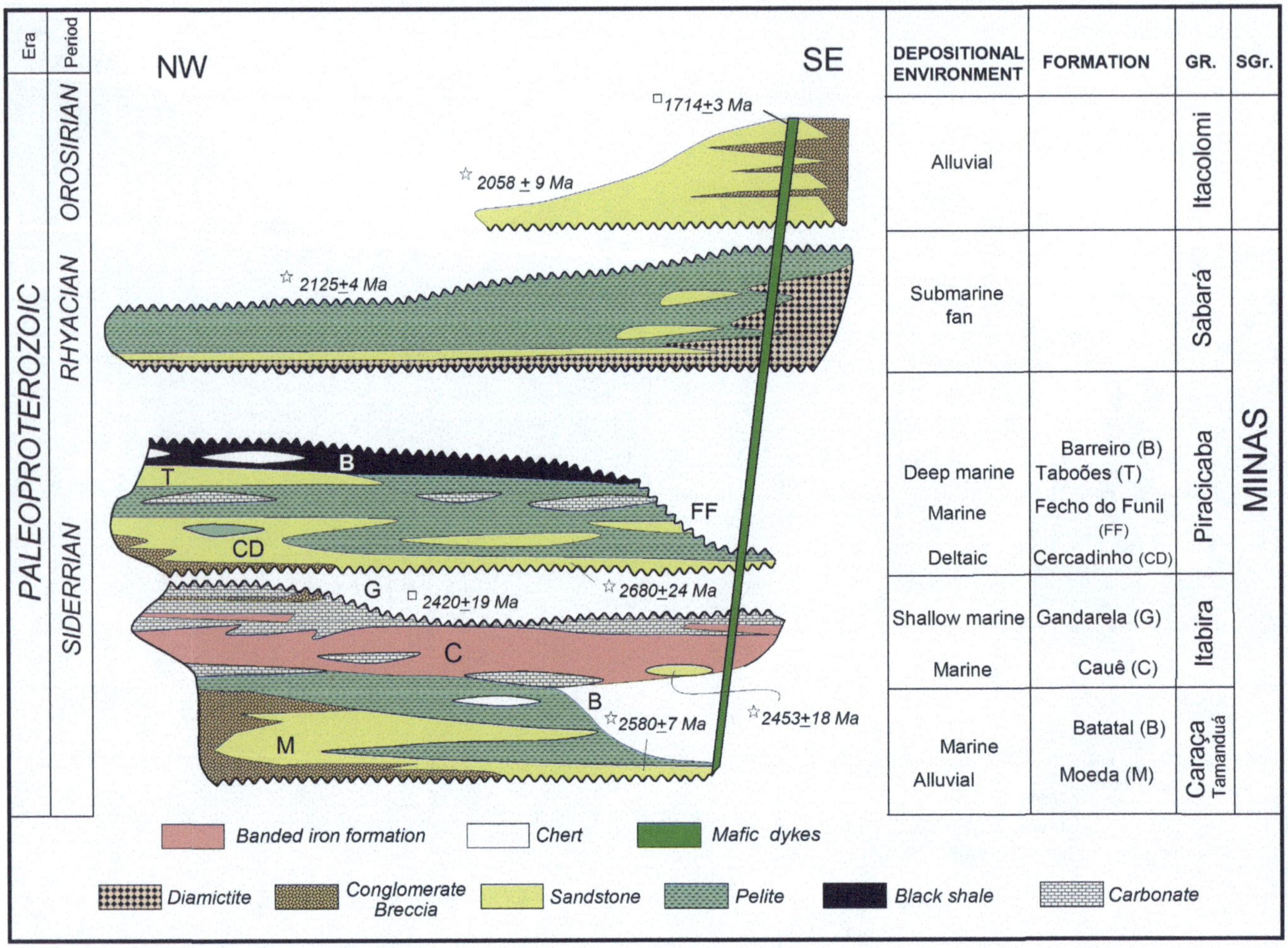

Fig. 5.7 Stratigraphic chart for the Paleoproterozoic units of the Quadrilátero Ferrífero. *Numbers with stars* indicate the youngest detrital zircon age; *numbers with squares* denote deposition or crystallization ages. Geochronological data from Babisnki et al. (1995), Silva et al. (1995), Machado et al. (1996a, b), Hartmann et al. (2006), Mendes et al. (2014), Alkmim et al. (2014), Cassino (2014). Reproduced from Alkmim and Martins-Neto (2012) with permission from Elsevier

(Fig. 5.9). The Gandarela Formation is made up of limestones (locally stromatolitic) and dolomites, with intercalations of dolomitic iron formation and pelites (Dorr 1969; Souza and Müller 1984; Babinski et al. 1995). A layer of coarse-grained breccias, composed of iron formation, carbonate, chert and pelite fragments imbedded in a hematite-rich carbonate matrix, regionally marks the upper portion of this formation (Guild 1957; Moore 1968).

An erosional surface locally separates the Gandarela carbonates from the overlaying Piracicaba Group (Fig. 5.7), which comprises a thick package of deltaic sediments (including hematite-rich sandstones and pelites), covered by marine pelites, sandstones and black shales (Dorr et al. 1957; Pomerene 1964; Renger et al. 1995).

The youngest unit of the Minas Supergroup, the Sabará Group, is separated from the older and younger sequences by regional unconformities (Fig. 5.7). Pelites and greywackes, some of them Mn-rich, with layers and lenses of diamictites, conglomerates and sandstones, predominate in the approximately 3500 m-thick turbiditic sequence that characterizes this group (Dorr 1969; Barbosa 1979; Renger et al. 1995; Reis et al. 2002) (Fig. 5.10).

The maximum depositional ages of the basal Tamanduá and Moeda sandstones can be estimated at 2580 ± 7 Ma, as indicated by the youngest detrital zircon concordant ages (Hartmann et al. 2006). Babinski et al. (1995) using the Pb–Pb whole-rock method obtained an isochron age of 2.419 ± 19 Ma for a few limestone samples from the middle

Fig. 5.8 Gold and uranium-bearing metaconglomerate of the Moeda Formation, Minas Supergroup. Vein-quartz and phyllite pebbles are imbedded in a matrix rich in pyrite nodules and clasts. The *arrow* in the *left side* of the photo points to a large pyrite aggregate

Gandarela Formation. These data indicate that the deposition of the QF guide layer, the Cauê Banded Iron Formation, occurred between 2580 and 2420 Ma, reproducing the same age interval of the large Paleoproterozoic banded iron formation deposits of Africa, Australia and North America (e.g., Klein 2005; Clout and Simonson 2005). Based on the age of zircon grains extracted from a volcanic layer, Cabral et al. (2012) postulated that the Cauê Formation accumulated earlier, at around 2650 Ma. However, more recently, Cassino (2014) dated a large population of detrital zircon grains extracted from a quartzite lens found at the base of the Cauê Formation, obtaining 2453 $\pm$ 18 Ma as the youngest concordant age. This result reinforces the estimation of an Early Paleoproterozoic depositional age for the lower Minas Supergroup by the authors preceding Cabral et al. (2012). Finally, Mendes et al. (2014) investigated the provenance of the Cercadinho Formation that marks the base of the Piracicaba Group using U–Pb age determinations on detrital zircons. The youngest population provided an average age of 2680 $\pm$ 24 Ma. The contribution of Meso- to Neo-archean crustal sources is attested by the age clusters of 2800–2900 and 3210–3270 Ma.

The geochronological background of the Minas Supergroup indicated that the unconformity at the base of the uppermost Sabará Group probably corresponds to a hiatus of ca. 300 Ma. The deposition of the Sabará Turbidites must have occurred soon after 2125 $\pm$ 4 Ma, the age of the youngest detrital zircon so far determined in metagraywackes of this unit (Machado et al. 1996a, b). A Sm–Nd T_{DM} age of 2.3 Ga, coupled with an $\varepsilon_{Nd(t)}$ value of -1.0 obtained by Brueckner et al. (2000) for a garnet-schist occurring in the northern QF, indicate that the Sabará sediments derived essentially from Paleoproterozoic sources with a relatively short residence time in the crust.

From a tectonic perspective, the lower and middle portions of the Minas Supergroup (Tamanduá, Caraça, Itabira and Piracicaba groups) constitute a megasequence deposited in a basin developed along the southern margin (present

Fig. 5.9 Folded and deeply weathered banded iron formation (Cauê Formation, Minas Supergroup), in this case converted into iron ore. The hematite-rich micro and mesobands are separated by large pore spaces left by the supergene leaching of quartz and carbonate

reference frame) of the São Francisco proto-craton (i.e., its Archean core) in the course of the Siderian period (Alkmim and Marshak 1998; Alkmim and Martins-Neto 2012). Fed by Archean sources located to the north of the QF region (Dorr 1969; Renger et al. 1995; Machado et al. 1996a, b), the precursor Minas passive margin basin was converted into an active system much later, as recorded by the onset of the Sabará deposition by the end of the Rhyacian period. As already pointed out by Barbosa (1968, 1979) and Dorr (1969), the Sabará Group represents a syn-orogenic flysch assemblage, reflecting a change on sediment source and basin regime. In fact, the Sabará sediments were shed from Paleoproterozoic sources located to the south and east of the QF (Machado et al. 1996a, b; Brueckner et al. 2000; Reis et al. 2002), i.e., from the region that now encompasses the Mineiro belt and the basement exposures of the Neoproterzoic Araçuaí and Ribeira belts.

5.3.1.3 Itacolomi Group

Restricted to the southern half the QF (Fig. 5.5), the Itacolomi Group (Dorr 1969) comprises an up to 1,800 m-thick sequence of quartz-rich sandstones (Fig. 5.11), conglomerates and minor pelites (Dorr 1969; Glöckner 1981; Tessari and Amorim 1984). The group corresponds to an alluvial complex with local marine transitions (Alkmim 1987), and lies unconformably on top of the Minas succession (Fig. 5.7). It contains clasts of sandstones, pelites and especially banded iron formation propably derived from the Minas Supergroup (Guimarães 1931; Dorr 1969). These fact led Barbosa (1968) and Dorr (1969) to interpret the Itacolomi as a post-orogenic intermontane molasse. Geochronological determinations on detrital zircons extracted from some Itacolomi sandstones point towards a predominance of Paleoproterozoic sources and a maximum depositional age of 2059 ± 58 Ma (Machado et al. 1996a, b; Hartmann et al. 2006). A recent provenance study, carried out on all occurrences of the Itacolomi Group, confirmed the previous results and indicated a maximum deposition age of 2058 ± 9 Ma for this unit (Alkmim et al. 2014).

5.3.2 Structure and Metamorphism

As suggested by the simplified geologic map of Fig. 5.5, the structural panorama of the QF is characterized by an association of basement domes and synclines containing the supracrustal units, thereby defining a dome-and-keel geologic architecture, a large-scale structure commonly exhibited by Archean and some Paleoproterozoic provinces (Marshak et al. 1992, 1996; Alkmim and Marshak 1998). Indeed, as first pointed out by Barbosa (1968), there is a clear geometrical relationship, between the Bação and Bonfim domes and the Moeda/Dom Bosco syncline (Fig. 5.5). A closer examination of Fig. 5.5 reveals, however, two other sets of structures. The most evident is represented by NE-oriented regional structures present in the northern half of the QF; the second, consists of a series of thrusts that cut the entire Precambrian section in the eastern half of the QF.

The various domes of the region share a series of common features. They are cored by the Archean basement gneisses, which show tectonic contacts marked by a metamorphic aureole with the surrounding supracrustal rocks (Dorr 1969; Herz 1978; Marshak et al. 1992; Chemale et al. 1994; Alkmim and Marshak 1998). Moreover, the shear zones along the contacts with the metasedimentary rocks in general record variable senses of displacement (Hippertt et al. 1992; Endo and Nalini 1992; Marshak et al. 1992, 1996; Chemale et al. 1994; Hippertt 1994; Endo 1997).

The Serra do Curral homocline, the Gandarela syncline, the Conceição anticline, as well as the João Monlevade and Itabira synclinoria (Fig. 5.5) trend preferentially NE, verge NW and do not show any geometrical relationship to the domes. As one of the most prominent and intriguing structures of this set, the Serra do Curral homocline extends over more than 80 km in the ENE direction and involves overturned Minas Supergroup strata (Fig. 5.5) (Dorr 1969).

Fig. 5.10 Conglomerate of the Sabará Formation containing clasts of vein-quartz, quartzites, granitoids, gneisses, and iron formation. The matrix is fine-grained and chlorite-rich

The dominant structures of the Mineiro belt described in the previous section apparently belong to this group, for they also trend NE and show local NW vergence (compare Figs. 5.2 and 5.5).

The thrust system of eastern QF comprises a series of arcuate and WNW-directed faults, showing two culminations (Fig. 5.5). The slip lineations on the fault surfaces and the stretching lineations in the rocks involved show a strong preferential orientation around 110/35. The penetrative foliation developed within the thrust sheets follows the same trajectory as the faults and the associated mylonitic foliation, showing quite variable orientations (Chemale et al. 1994). Remarkably, these small scale fabric elements associated with the WNW-directed thrusts are not restricted to the eastern portion of the QF. They also dominate the rock fabric throughout almost the whole area to the east of the Moeda syncline (Fig. 5.5). Hinges of small scale to km-size folds are in general parallel to the stretching lineation, and like the regional foliation, are not geometrically or kinematically related to any of the regional folds that dominate the large scale structural picture of the QF (Alkmim and Noce 2006).

The various sets of tectonic structures affecting the Minas and Itacolomi rocks in the QF have been described and interpreted in different ways by the authors that worked in the region (e.g., Dorr 1969; Drake and Morgan 1980; Ladeira and Viveiros 1984; Endo 1997; Chemale et al. 1994; Chauvet et al. 1994; Alkmim and Marshak 1998; Endo et al. 2005). These authors agree at least on one point: the mentioned units record a polyphase deformation history, which results locally in very complex structural patterns. Based on crosscut relationships among fabric elements and associated features, Alkmim and Marshak (1998) suggested that the post-Minas structures nucleated according to the following succession: (1) development of a system of NE-striking and NW-verging faults and folds; (2) formation of domes and associated synclines (keels); (3) WNW-verging folding and thrusting, coupled with development of the penetrative small-scale structures characteristic of the region, and reactivation of preexisting structures.

The Minas and Itacolomi strata exhibit mineral parageneses typical of the regional greenschist facies metamorphism in most of the QF area. However, an increase of the

Fig. 5.11 Quartz-rich metasandstone of the Itacolomi Group showing through cross-bedding marked by concentration of heavy minerals

conditions of the regional metamorphism is observed in the eastern QF, where the amphibolite facies parageneses predominate (Dorr 1969; Herz 1978; Müller et al. 1982; Pires 1995) (Fig. 5.12). Besides regional metamorphism, the rocks of the Minas Supergroup (together with those of the Archean Rio das Velhas Supergroup) were affected by high temperature—low pressure aureoles in the areas adjacent to the Belo Horizonte, Bonfim, and Bação domes (Herz 1978; Jordt-Evangelista et al. 1992) (Fig. 5.12).

In a detailed study of a 4 km-wide metamorphic aureole developed in pelitic rocks of the Sabará Group along the contact with the Archean Belo Horizonte Complex in northern QF, Jordt-Evangelista et al. (1992) estimated a maximum temperature of 650 °C and pressures between 3 and 4 kbar for the observed parageneses. A Sm–Nd isochron (whole rock, garnet, plagioclase) obtained from a schist from this aureole yielded an age of 2098 ± 33 Ma (Brueckner et al. 2000). Furthermore, the Ibirité Gabbro, dated at 1714 Ga (Silva et al. 1995), cuts the aureole.

The development of the aureole affecting the Archean Rio da Velhas Supergroup around the Bação dome in the central QF was dated by Vlach et al. (2003) at 2060 ± 20 Ma on monazites using an ion microprobe. In addition, metamorphic U–Pb ages in the interval of 2065 and 2035 Ma were obtained by Noce (1995) and Machado et al. (1992) in titanite crystals extracted from gneisses of the Belo Horizonte and Bação domes. More recently, a large number of U–Pb age determinations were carried out on titanite and monazite grains collected in granitoids and amphibolites of the Bação, Belo Horizonte and Bonfim complexes by Aguilar Gil et al. (2015). The ages obtained range from 2080 to 1950 Ma and cluster around 2058 Ma, in aggrement with the results previously obtained by Noce (1995) and Machado et al. (1992) from the same units, as well as by Brueckner et al. (2000) and Vlach et al. (2003) for the metamorphic aureoles. Admitting a genetic relationship between dome structures and aureoles, an age of ca. 2.05 Ga can be assumed for the process of dome formation in region.

The greenschist to amphibolites facies parageneses of the regional metamorphism are apparently syn-kinematic in respect to the third and presumably youngest set of WNW-verging structures that affect at least two thirds of the QF. However, this metamorphic episode and the associated fabric elements have not been properly dated yet. A large

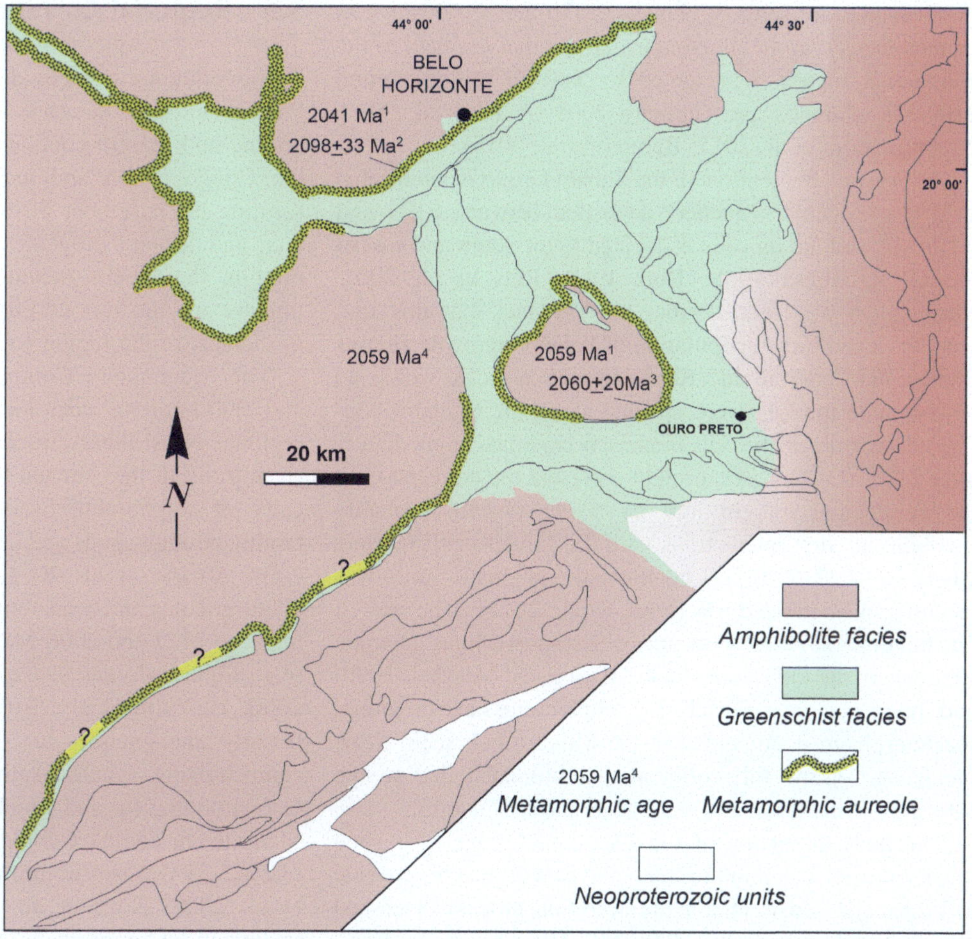

Fig. 5.12 Simplified metamorphic map of the Mineiro belt and Quadrilátero Ferrífero, showing the distribution of the metamorphic facies and contact aureoles developed on metasedimentary rocks around basement domes. The *numbers* denote metamorphic ages obtained in the aureoles and basement complexes by: (1) Noce (1995), (2) Brueckner et al. (2000), (3) Vlach et al. (2003), (4) Machado et al. (1992)

number of K–Ar age determinations performed on both basement and supracrustal rocks by Teixeira (1993) strongly suggests a Neoproterozoic isotopic resetting between 640 and 500 Ma. Brueckner et al. (2000), Silva et al. (2002), and Noce et al. (2007), using the Sm–Nd and U–Pb methods in Archean and Paleoproterozoic rocks exposed to east of the QF (within the Araçuaí belt), dated an amphibolite facies regional metamorphism associated with structures kinematically equivalent to those observed in the QF. The obtained ages between 589 and 538 Ma fall within the interval of the K–Ar apparent ages indicated by Teixeira (1993). In spite of this, the lack of more consistent and accurate geochronological data (e.g., U–Pb monazite, rutile and titanite ages) makes the timing of the regional metamorphism a matter of dispute. For instance, Machado et al. (1992) and Endo (1997) favor the hypothesis of a Paleoproterozoic age, whereas Belo de Oliveira et al. (1987), Chemale et al. (1994), Alkmim and Marshak (1998) and Teixeira et al. (2000) view this metamorphism as a manifestation of the Neoproterozoic Brasiliano collisional event.

No direct age determinations are available for the NE-trending and probably oldest set of post-Minas structures. It is also not clear if the deformation event responsible for their nucleation was assisted by metamorphic reactions in the supracrustal units. However, since the NE trending folds and faults involve the 2125 Ma Sabará Group, which was also caught by the dome emplacement process, their nucleation must have occurred simultaneously or immediately after the deposition of the Sabará turbidites.

5.4 Tectonic Evolution

The information on the stratigraphy, structure and metamorphism of both the Mineiro belt and the QF provided in the previous sections clearly indicate that, despite occupying adjacent areas, they have almost nothing in common. In fact, except for the NE-oriented group of structures and the 2.1 Ga granitic intrusions that also occur in form of relatively small bodies in southern and western QF (Teixeira et al. 1987; Noce 1995; Oliveira 2004), no other rock unit or fabric element are shared by these provinces. The Mineiro belt, underlain essentially by granitic rocks and coeval volcano-sedimentary sequences emplaced and deposited in intra-oceanic arc settings between 2470 and 2100 Ma, can be thus viewed as an alloctonous terrane that accreted to the continental margin represented by the QF region around 2100 Ma (Campos and Carneiro 2008; Ávila et al. 2010, 2014; Teixeira et al. 2015; Barbosa et al. 2015).

As previously mentioned, the Sabará Group is interpreted as a syn-orogenic sequence, deposited between 2125 and 2059 Ma, and fed by sources located to the south and east of the QF (Barbosa 1968; Dorr 1969; Reis et al. 2002; Machado et al. 1992, 1996a, b). The fact that this unit contains a significant population of Paleoproterozoic zircons implies that Siderian and Rhyacian source rocks, which do not occur in the QF, were uplifted along the southern continental margin of the São Francisco landmass immediately after 2125 Ma. This event might represent the conversion of the continental margin into an active system, probably involving its first interactions with the Mineiro belt terrane, where a 2.17–2.10 Ga continental magmatic arc was recently characterized (Barbosa et al. 2015). The set of NE-trending structures, as the oldest post-Minas tectonic elements in the QF (Dorr 1969; Alkmim and Marshak 1998) and the associated greenschist to amphibolite facies regional metamorphism documented in the Mineiro belt seem to be the record of the full collision that ultimately led to the suturing of this terrane to the São Francisco margin.

The detrital zircons of the Itacolomi Group, also shed from sources located to the south and east of the QF, provide a further age constraint for the accretion process discussed above. Paleoproterozoic 2400–2000 Ma zircons predominate in the Itacolomi Group (Machado et al. 1996a, b; Hartmann et al. 2006; Alkmim et al. 2014). Except for small pegmatite veins (Machado et al. 1992), possible sources for the 2400–2000 Ma igneous or metamorphic zircons were not found within the QF yet. Metamorphic ages between 2080 and 1900 Ma do occur in the QF, recorded, however, by titanite and monazite (and not by zircons) in the various basement domes and associated aureoles (Machado et al. 1992; Noce 1995; Brueckner et al. 2000; Vlach et al. 2003; Aguilar Gil et al. 2015). This fact suggests that the metamorphic reactions in the time between 2080 and 1900 processed under P, T conditions not high enough to recrystallize or generate new zircon grains in the QF rocks. Since the Itacolomi Sandstones contain abundant zircons dated in this time interval, they must have been also shed from high-grade rocks uplifted and exposed in the areas to the south and east of the QF. Therefore, the Itacolomi Group must post-date the collision that causes the conversion of the QF into the foreland domain of a mountain belt developed along the São Francisco continental margin. The aftermath of this collision is also marked by the development of domes and synclinal keels between them, a process that more likely preceded or took place simultaneously to the deposition of the Itacolomi Group (Alkmim and Marshak 1998).

5.5 Regional Correlations

The evolutionary stages discussed above took place in a Paleoproterozoic scenario that, besides the Mineiro belt terrane and the QF continental margin basin, also involved the Congo craton and its margins, as well as the units forming the basement of the adjacent Neoproterozoic Araçuaí and Ribeira orogenic belts. Developed on the craton margin, these belts include among other units the Mantiqueira and the Juiz de Fora complexes, presently exposed to the east of the region focused in this chapter (Fig. 5.13).

The Mantiqueira Complex occupies a ca. 30 km-wide and 200 km-long area located between the QF and the regional NS-striking structure known as the Abre Campo shear zone in the Araçuaí belt (Fig. 5.13). It consists basically of amphibolite facies mylonitic orthogneisses containing boudins, pods and layers of amphibolites (Peres et al. 2004; Duarte et al. 2004; Noce et al. 2007). Syenites, alkali-granites and monzonites, often undeformed and dated between 2127 and 2006 Ma, are also present in the complex (e.g., Figueiredo and Teixeira 1996; Jordt-Evangelista et al. 2000; Heilbron et al. 2010; Silva et al. 2015). Compositionally, the gneisses are granitic to tonalitic (TTG-like), characterizing a calc-alkaline series (Duarte et al. 2004). Geochronological and geochemical studies indicate that the Mantiqueira complex comprises a Meso- and Neo-archean (3.20–2.75 Ga) substratum (Teixeira et al. 1987; Brueckner et al. 2000; Noce et al. 2007) that received significant magmatic additions in the time interval of 2200–2044 Ma (Figueiredo and Teixeira 1996; Silva et al. 2002, 2015; Heilbron et al. 2010; Duarte et al. 2005; Noce et al. 2007). Therefore, the Mantiqueira Complex represents a cordilleran magmatic arc developed along the eastern margin of the São Francisco landmass between the Rhyacian an Orosirian periods (Noce et al. 2007).

The Paleoproterozoic Juiz de Fora Complex consists of granulite facies orthogneisses (enderbites, charno-enderbites, charnockites, and basic rocks), often migmatitic, and tectonically intercalated with Neoproterozoic parageneisses and granitoids along the eastern side of the Abre Campo shear zone (Fig. 5.13). These rocks, representing a calcalkaline suite, show U–Pb crystallization ages between 2200 and 2084 Ma (see Noce et al. 2007 and Silva et al. 2015 for a review). Of note, a basic granulite of tholeiitic affinity dated by Heilbron et al. (2010) yielded a significantly older crystallization age of 2427 ± 9 Ma. Most authors interpret the Juiz de Fora Complex as a product of subduction-related magmatism in intra-oceanic settings (Machado et al. 1996a, b; Heilbron et al. 1997, 2010; Silva et al. 2002; Noce et al. 2007).

The ca. 300 km-long Abre Campo shear zone (Fig. 5.13), representing one of the most prominent structures of the Ediacaran Araçuaí-West Congo orogen, is followed along its whole length by a major Bouguer anomaly (Haralyi and

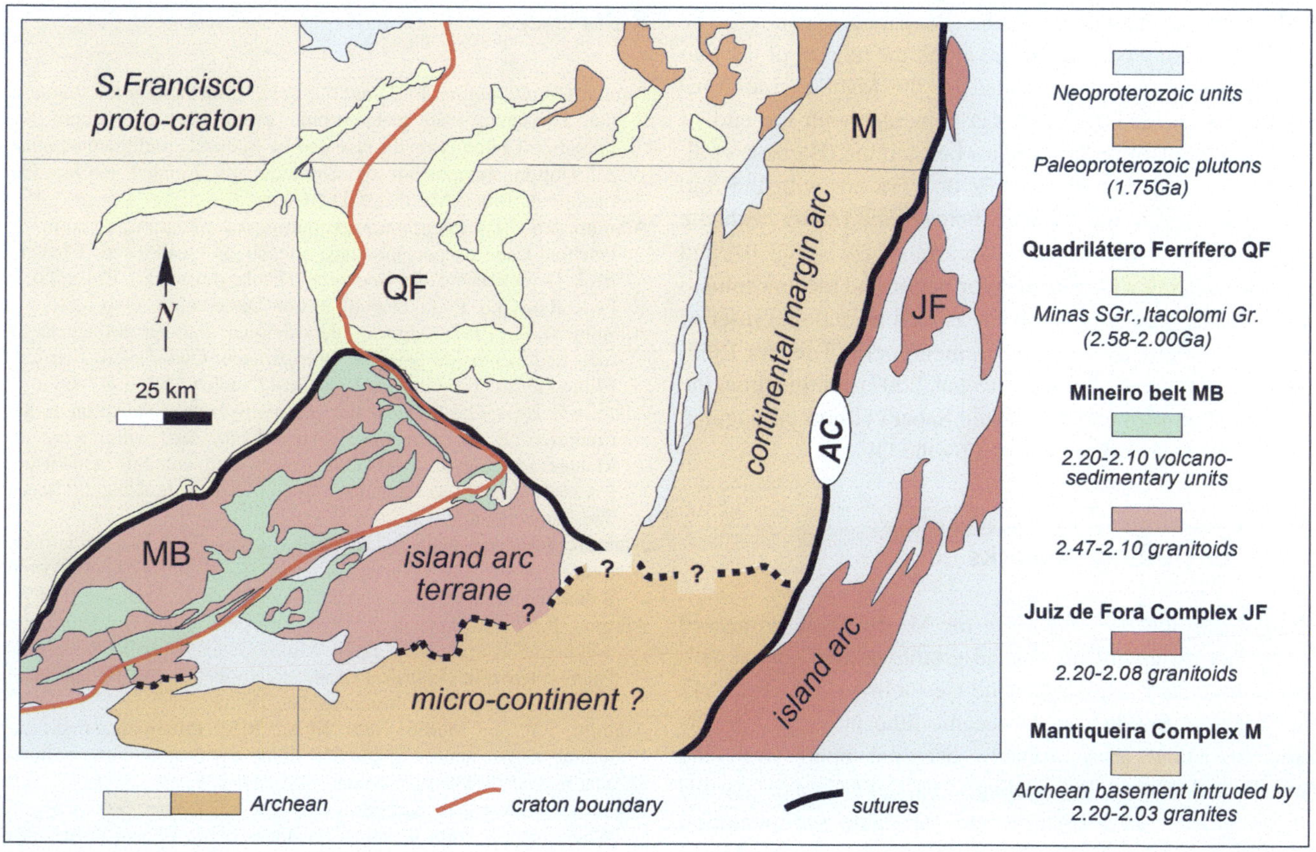

Fig. 5.13 Simplified tectonic map of the southern São Francisco craton and adjacent Neoproterozoic Araçuaí and Ribeira belts, emphasizing the major Paleoproterozoic crustal units present in the region. The QF and Mantiqueira Complex (*M*) represent the proto-São Francisco craton to which the Mineiro belt (*MB*) and Juiz de Fora arc (*JF*) terranes accreted around 2.10 Ga, possibly along with a micro-continent represented by the piece of Archean crust located in the southern portion of the focused region (see text for explanation). *AC* Abre Campo shear zone. Craton boundary according to Alkmim (2004)

Hasui 1982). Juxtaposing the juvenile Juiz de Fora arc on the east with the cordilleran Mantiqueira arc on the west, this shear zone is currently interpreted as a Paleoproterozoic suture that also acted as a plate boundary during the Neoproterozoic (Machado et al. 1996b; Cunningham et al. 1998; Fischel 1998; Alkmim et al. 2006).

In the geotectonic scenario represented by the map of Fig. 5.13, the Mineiro belt appears surrounded by Archean complexes, which are in turn bounded to the east by the Abre Campo suture. As a Paleoproterozoic juvenile terrane, the crustal segment of the Mineiro belt must be separated from the Archean block located to the southeast by a suture as well. Such a structure was not recognized in the region yet. Furthermore, according to Ávila et al. (2014) and Teixeira et al. (2015), the Paleoproterozoic granitoids of the Mineiro belt formed in successive subduction systems within a Siderian-Rhyacian ocean, which also hosted the Juiz de Fora arc and a micro-continent. This micro-continent apparently corresponds to the crustal block exposed in the region to the southeast of the Mineiro belt and south of the Mantiqueira Complex (Fig. 5.13) (see Teixeira et al. this book).

As indicated above, the Mineiro belt (2.47–2.10 Ga), together with the Mantiqueira (2.20–2.00 Ga) and Juiz de Fora (2.20–2.08 Ga) complexes, characterize a long-lived system of oceanic and continental magmatic arcs generated before the assembly of São Francisco-Congo paleocontinent (e.g., Teixeira et al. 2015 and references therein). As postulated by Noce et al. (2007), this amalgamation also involved the Mantiqueira correlative Kimezian Complex that forms the basement of the western margin of the Congo craton (Tack et al. 2001).

How did all the previously mentioned Paleoproteroic crustal units come together? In other words, which was the nature of the structures bounding these pieces of crust during the Paleoproterozoic assembly of the São Francisco Congo paleocontinent? Multiple solutions exist for this problem. In order to understand the pre-collisional paleogeograhy and the role played by the various actors mentioned above, further studies focusing especially the region to the southeast of the Mineiro belt are required.

The available geochronological data allow, however, some inferences on the timing of the interaction between the various geotectonic components of the described scenario. The development of a cordilleran arc along the margin of the São Francisco landmass (represented by the Mantiqueira Complex) between 2.17 and 2.00 Ga coincides with the end of granitic plutonism in the Mineiro belt terrane (Barbosa et al. 2015). The collision of the Juiz de Fora arc with the São Francisco margin and Mineiro belt terrane (along with the intervening micro-continent) occurred between 2.10 and 2.00 Ga, causing widespread deformation and metamorphism in all crustal segments involved, as well as regional reset of the K–Ar mineral ages in the basement rocks (Teixeira 1985, 1993; Teixeira et al. 2000; Noce et al. 2007). At this time, the sedimentation of the syn-orogenic Sabará Group propagated towards the foreland represented by the QF.

5.6 Concluding Remarks

The Paleoproterozoic history of the Mineiro belt, witnessed by distinct groups of arc-related granitoids and coeval volcanosedimentary sequences is now established between 2.47 and 2.10 Ga. Subduction of oceanic lithosphere and generation of island arcs probably occurred apart from the Archean paleocontinent, starting at ca. 2.47 Ga ago. At this time, banded iron formation and carbonate sedimentation (Cauê and Gandarela formations) took place in the Minas passive basin. The existence of an early Siderian oceanic lithosphere is also suggested by the regional distribution of Sm–Nd T_{DM} ages (2.4–2.5 Ga) within the area of Mineiro belt (Seixas et al. 2012; Teixeira et al. 2015). Closure of the Siderian ocean separating the Archean nuclei of the São Francisco and Congo paleocontinents led to their convergence and collision, along with the intervening Mineiro and Juiz de Fora island arcs between 2.10 and 2.00 Ga. The Minas basin, now converted into a retro-arc/foreland depocenter, started to be filled by the Sabará syn-orogenic assemblage. This process culminated with the deposition of the Itacolomi post-orogenic sandstones and conglomerates.

In many aspects, the Paleoproterozoic tectonic evolution of the QF and the Mineiro belt, together with the units forming the basement of São Francisco and Congo craton margins, is analogous to that of the Eastern Bahia orogenic domain (see Barbosa and Barbosa this book).

Acknowledgments This paper summarizes the results obtained in various research projects. F.F. Alkmim projects were supported by the CNPq (Conselho Nacional de Desenvolvimento Científico e Tecnológico), Grant # 308045/2013-0. W. Teixeira acknowledges the CNPq (Grant # 303468/2015-5) and FAPESP (Research Foundation of the São Paulo state)—Grant 2012/15824-6. Review comments provided by U.G. Cordani are also appreciated.

References

Aguilar Gil, C., Farina, F., Lana, C., 2015. Constraining the timing of the Transamazonian metamorphic event in the Southern São Francisco Craton (Brazil): revealed by monazite and titanite dating. 8th Hutton Symposium on Granites and Related Rocks, Florianópolis, Brazil.Abstract, CD-ROM.

Alkmim, A.R. 2014. Investigação geoquímica e estratigráfica da Formação Ferrífera Cauê na porção centro-oriental do Quadrilátero Ferrífero, MG. Departamento de Geologia, Escola de Minas, Universidade Federal de Ouro Preto. Unpubl. Master Dissertation, 178p.

Alkmim, F.F. 1987. Modelo deposicional para a seqüência de metassedimentos da Serra de Ouro Branco, Quadrilátero Ferrífero, Minas Gerais. Boletim Soc.Bras.Geol./Nucleo MG, 6: 47–68

Alkmim F. 2004. O que faz de um cráton um cráton? O Cráton do São Francisco e as revelações Almeidianas ao delimita-lo. In: Mantesso-Neto et al. (eds) Geologia do Continente Sul- Americano. Evolução da obra de Fernando Flávio Marques de Almeida. Becca, São Paulo, 17–35.

Alkmim, F.F. and Noce, C.M. (eds.) 2006. The Paleoproterozoic Record of the São Francisco Craton. IGCP 509 Field workshop, Bahia and Minas Gerais, Brazil. Field Guide & Abstracts, 114 p.

Alkmim, F.F., and Marshak, S. 1998. Transamazonian Orogeny in the Southern São Francisco Craton, Minas Gerais, Brazil: Evidence for Paleoproterozoic collision and collapse in the Quadrilátero Ferrífero. Precambrian Research, 90, 29–58

Alkmim, F.F. and Martins-Neto, M.A., 2012. Proterozoic first-order sedimentarysequences of the São Francisco Craton, eastern Brazil. Marine and Petroleum Geology, 33, 127–139.

Alkmim, F.F., Marshak, S., Pedrosa-Soares, A.C., Peres, G.G., Cruz, S., Whittington, A. 2006. Kinematic evolution of the Araçuaí-West Congo orogen in Brazil and Africa: Nutcracker tectonics during the Neoproterozoic assembly of Gondwana. Precambrian Research 149, 43–64.

Alkmim, FF., Lana, C., Duque, T.R.F., 2014. Zircões detríticos do Grupo Itacolomi e o registro do soerguimento do cinturão Mineiro. 47° Congresso Brasileiro de Geologia. Salvador, Brasil. Anais CD-ROM, p. 1802.

Amorim, L. Q. and Alkmim, F. F., 2011. New ore types from the Cauê Banded Iron Formation, Quadrilátero Ferrífero, Minas Gerais, Brazil: Responses to the growing demand. In: Iron Ore Conference 2011. The Growing Demand. The Australasian Institute of Mining and 1025 Metallurgy, 2011, p. 59–71.

Ávila, C.A., 2000. Geologia, petrografia e geocronologia de corpos plutônicos Paleoproterozóicos da borda meridional do Cráton São Francisco, região de São João del Rei, Minas Gerais. Unpubl. Doctoral Thesis. Universidade Federal do Rio de Janeiro, Rio de Janeiro, 401 pp.

Ávila, C.A.; Valença, J.G.; Moura, C.A.V.; Klein, V.C., Pereira, R.M., 2003. Geoquímica e idade do Tonalito/trondhjemito Cassiterita, borda meridional do Cráton São Francisco, Minas Gerais. Arquivos do Museu Nacional, 61(4): 267–284.

Ávila, C.A., Teixeira, W., Cordani, U.G., Barrueto, H.R., Pereira, R. M., Martins, V.T.S., Dunyi, L., 2006. The Glória quartz-monzodiorite isotopic and chemical evidenceof arc-related magmatism in the central part of the Paleoproterozoic Mineiro belt, Minas Gerais State, Brazil. Anais da Academia Brasileira de Ciências 78, 543–556.

Ávila, C.A., Teixeira, W., Vasques, F.S.G., Dussin, I.A., Mendes, J.C., 2012. Geoquímica e idade U–Pb (LA-ICPMS) da crosta oceânica Riaciana do Cinturão Mineiro, borda meridional do Cráton São Francisco. 46° Congresso Brasileiro Geologia, Boletim de Resumos, Sociedade Brasileira de Geologia, CD-ROM.

Ávila, C.A., Teixeira, W., Bongiolo, E.M., Dussin, I.A., 2014. The Tiradentes suite and its role in the Rhyacian evolution of the Mineiro belt, São Francisco Craton: Geochemical and U-Pb geochronological evidences. Precambrian Research, 243, 221–251.

Ávila, C.A., Teixeira, W., Cordani, U.G., Moura, C.A.V., Pereira, R.M. 2010. Rhyacian (2.23–2.20) juvenile accretion in the souhtern São Francisco craton, Brazil: Geochemical and isotopic evidence from the Serrinha magmatic suite, Mineiro belt. Journal of South American Earth Sciences, 29, 143–159

Babinski, M., Chemale Jr., F., Van Schmus, W.R. 1995. The Pb/Pb age of the Minas Supergoup carbonate rocks, Quadrilátero Ferrífero, Brazil. Precambrian Research, 72, 235–245.

Baltazar, O.F. and Zucchetti, M., 2007. Lithofacies associations and structural evolution of the Archean Rio das Velhas greenstone belt, Quadrilátero Ferrífero, Brazil: a review of the setting of gold deposits. Ore Geology Reviews 32, 1–2.

Barbosa, A.L.M. 1968. Contribuições recentes à geologia do Quadrilátero Ferrífero. Ed. Escola de Minas, Ouro Preto, 47 pp.

Barbosa, A.L.M., 1979. Variacões de fácies na Série Minas. Bol. Soc. Bras. Geol., Núcleo Minas Gerais, 1, 89–100.

Barbosa, M.I.M. 1985. Geoquímica das Faixas Máficas- Ultramáficas, Plutonitos e Magmatitos do "Greenstone Belt" Barbacena, na Região de Conselheiro Lafaiete (MG). Instituto de Geociências, Universidade Federal do Rio de Janeiro. Unpubl. Master Thesis, 227 p.

Barbosa, N.S., Teixeira, W., Ávila, C.A., Montecinos, P.M., Bongiolo, E.M., 2015. 2.17–2.10 Ga plutonic episodes in the Mineiro belt, São Francisco Craton, Brazil: U-Pb ages, geochemical constraints and tectonics. Precambrian Research, 270, 204–225.

Barbosa, N.S., 2015. Evolução Paleoproterozoica do Cinturão Mineiro: Geocronologia U-Pb, isótopos de Nd-Hf-Sr e geoquímica de rochas plutônicas. PhD Thesis. Universidade de São Paulo, 229 pp.

Belo de Oliveira, O.A., Greco, F.M., Vieira, M.B.H., 1987. A relação da tectônica Espinhaço Meridional e Quadrilátero Ferrífero, MG. Anais I Simp. Nacional Estudos Tectônicos, Salvador, Soc.Bras. Geol., Boletim de Resumos, p. 74–76.

Brueckner, H.K., Cunningham, D., Alkmin, F.F., Marshak, S., 2000. Tectonic implications of Precambrian Sm–Nd dates from the southern São Francisco craton and adjacent Araçuai and Ribeira belts, Brazil. Precambrian Research, 99, 3, 255–269.

Cabral, A.R., Zeh, A., Koglin, N., Gomes, A.A.S., Viana, D.J., Lehmann, B. 2012. Dating the Itabira iron formation, Quadrilátero Ferrífero of Minas Gerais, Brazil, at 2.65 Ga: depositional U–Pb age of zircon from a metavolcanic layer. Precambrian Research, 204, 40–45.

Campos, J.C.S. and Carneiro, M.A. 2008 Neoarchean and Paleoproterozoic granitoids marginal to the Jeceaba-Bom Sucesso lineament (SE border of the southern São Francisco craton): Genesis and tectonic evolution. Journal of South American Earth Sciences, 26, 463–484.

Cassino, L.F., 2014. Distribuição de idades U-Pb de zircões detríticos dos Supergrupos Rio das Velhas e Minas na Serra de Ouro Preto, Quadrilátero Ferrífero, MG – Implicações para a evolução sedimentar e tectônica. Departamento de Geologia, Universidade Federal de Ouro Preto. Unpubl. Graduation Thesis, 52p.

Cederberg, J., Söderlund, U., Oliveira, E., Ernst, R., Pisarevsky, S., 2016. U-Pb baddeleyite dating of the Proterozoic Pará de Minas dyke swarm in the São Francisco craton (Brazil) – implications for tectonic correlation with Siberia, Congo and the North China cratons. Journal of the Geological Society of Sweden, GFF-2015, Special Issue, Geologiska Foreningen (Taylor and Francis). doi.org/10.1080/11035897.2015.1093543

Chauvet, A., Faurre, M., Dossin, I., Charvet, J., 1994. Three-stage structural evolution of the Quadrilátero Ferrífero: consequences for Neoproterozoic age and the formation of gold concentrations of the Ouro Preto area, Minas Gerais, Brazil. Precambrian Research, 68, 139–167.

Chemale Jr, F., Rosière, C.A., Endo, I., 1994. The tectonic evolution of the Quadrilátero Ferrífero, Minas Gerais, Brazil. Precambrian Research, 65, 25–54.

Cherman, A.F., 2004. Geologia, petrografia e geocronologia de ortognaisses paleo-proterozoicos da borda meridional do Cráton do São Francisco, na região entre Itumirim e Nazareno, Minas Gerais. Unpubl. PhD Thesis, Universidade Federal do Rio de Janeiro, Rio de Janeiro, 259p.

Cherman, A.F., Valença, J.G., 2005. Geologia e geocronologia dos ortognaisses paleo-proterozóicos da borda meridional do Craton do São Francisco, entre as cidadesde Nazareno e Lavras, sul de Minas Gerais. Anais do Simpósio sobre o craton São Francisco, Salvador. SBG 1, 147–150.

Clout, J.M.F. and Simonson, B.M. 2005. Precambrian Iron Formations and Iron Formation-Hosted Iron Ore Deposits. SEPM 100 Anniversary Volume, p. 203–234.

Condie, K.C., O'Neil, C., Aster, C.R. 2009. Evidence and implications for a widespread magmatic shutdown for 250 My on Earth. Earth and Planetary Science Letters, 282, 294–298.

Corrêa Neto, A.V., Almeida, A.M., Caputo Neto, V., Guerrero, J.C., 2012. Alteração Hidrotermal em Zona de Cisalhamento Associada ao Lineamento Congonhas, Sul do Quadrilátero Ferrífero, Minas Gerais. Anuário do Instituto de Geociências – UFRJ, 35(2), 55–64 (www.anuario.igeo.ufrj.br)

Cunningham, W.D., Alkmim, F.F., Marshak, S., 1998. A strucutural transect across the coastal mobile belt in the Brazilian Highlands (latitude 20°S): the roots of a Precambrian trsnpressional orogen. Prcambrian Research, 92, 251–275

Dorr, J.V.N II, 1964. Supergene iron ores of Minas Gerais, Brazil, Economic Geology, 59:1203–1240.

Dorr, J.V.N II, 1969. Physiographic, stratigraphic and structural development of the Quadrilátero Ferrífero, Minas Gerais, Brazil. U.S. Geological Survey Professional Paper 641-A, p. 1–110

Dorr, J.V.N II, Gair, J.E., Pomerene, J.G., Rynearson, G.A., 1957. Revisão da estratigrafia pré- cambriana do Quadrilátero Ferrífero. DNPM-DFPM, Rio de Janeiro, 31p.

Drake, A.A. and Morgan, B.A., 1980. Precambrian plate tectonics in the Brasilian shield; evidence from the pre-Minas rocks of the Quadrilátero Ferrífero, Minas Gerais. U.S. Geological Survey Professional Paper, 1119, B1-B19.

Duarte, B.P., Valente, S.C., Heilbron, M., Campos Neto, M.C., 2004. Petrogenesis of the orthogneisses of Mantiqueira Complex, central Ribeira belt, SE Brazil: An Archaean to Palaeoproterozoic basement unit reworked during the Pan-African Orogeny. Gondwana Research, 7 (2), 437–450.

Endo, I. and Nalini, H. A. Jr. 1992. Geometria e cinemática das estruturas extensionais e compressionais na borda oeste do Sinclinal Moeda, Quadrilátero Ferrífero. Revista da Escola de Minas, 45(1/2): 15–17.

Endo, I. 1997. Regimes tectônicos do Arqueano e Proterozóico no interior da Placa Sanfranciscana: Quadrilátero Ferrífero e áreas adjacentes, Minas Gerais. São Paulo. Unpubl. Doctoral Thesis, Instituto de Geociências, Universidade de São Paulo, São Paulo, 382p.

Endo, I., Oliveira, A.H., Peres, G.G., Guimarães, M.L.V., Lagoeiro, L. E., Machado R., Zavaglia, G., Rosas, C.F., Melo, R.J.2005 Nappe Curral: Uma megaestrutura alóctone do Quadrilátero Ferrífero e controle da mineralização. X Simposio Nacional de Estudos Tectônicos - SNET, Boletim de Resumos, Sociedade Brasileira de Geologia, Curitiba, p. 279–281.

Farina, F., Albert, C., Lana, C., 2015. The Neoarchean transition between medium-and high-K granitoids: Clues from the Southern São Francisco Craton (Brazil). Precambrian Research, 266, 375–394

Farina, F., Albert, C., Martínez Dopico, C., Aguilar Gil, C., Moreira, H., Hippertt, J.P., Cutts, K., Alkmim, F.F., Lana C. 2016. The Archean-Paleoproterozoic evolution of the Quadrilátero Ferrífero (Brasil): current models and open questions. Journal of South American Earth Sciences, *in press.*

Figueiredo, M.C.H. and Teixeira, W., 1996. The Mantiqueira Metamorphic Complex, Eastern Minas Gerais State: Preliminary geochronological and geochemical results. Anais da Academia Brasileira de Ciências 68, 223–246.

Fischel DP. 1998. Geologia e dados isotópicos Sm-Nd do Complexo Mantiqueira e do Cinturão Ribeira na região de Abre Campo, Minas Gerais. Universidade Nacional de Brasília, Brasilia, Unpubl. Master Dissertation, 98 p.

Glöckner, K.H. 1981. Lithostratigraphie, Sedimentologie, Tektonik und Metamorphose der proterozoischen Itacolomi Serie bei Ouro Preto, Minas Gerais, Brasilien. Clausthaler Geowissenschafte Dissertationen, H.10, 221p.

Guild, P.W. 1957. Geology and mineral resources of the Congonhas district, Minas Gerais, Brazil. U.S. Geological Survey Professional Paper, 290, 90p.

Guimarães, D., 1931. Contribuição à geologia do Estado de Minas Gerais. Bol. Serv. Geol.Min.,55: 1–36.

Haralyi, N.L.E. and Hasui, Y., 1982. The gravimetric information and Archean-Proterozoic structural framework of eastern Brazil. Revista Brasileira de Geociências, 112(1–3): 160–166.

Hartmann L.A., Endo I., Suita M.T.F., Santos J.O.S., Frantz J.C., Carneiro M.A., Naughton N.J., Barley M.E., 2006. Provenance and age delimitation of Quadrilátero Ferrífero sandstones based on zircon U–Pb isotopes. Journal of South American Earth Sciences, 20, 273–285

Heilbron, M., Duarte, B.P., Valeriano, C.M., Simonetti, A., Machado, N., Nogueira, J.R., 2010. Evolution of reworked Paleoproterozoic basement rocks within the Ribeira belt (Neoproterozoic), SE-Brazil, based on U–Pb geochronology: Implications for paleogeographic reconstructions of the São Francisco-Congo paleocontinent. Precambrian Research, 178, 136–148.

Heilbron, M., Machado, R., Figueiredo, M., 1997. Lithogeochemistry of the Paleoproterozoic Granulites of the Bom Jardim de Minas (MG) - Vassouras (RJ) region, Central Segment of Ribeira Belt. Revista Brasileira de Geociências, 27, 83–98.

Herz, N., 1978. Metamorphic rocks of the Quadrilátero Ferrífero, Minas Gerais, Brazil. U.S. Geological.Survey Professional Paper, 641-C, 81p.

Hippertt, J.F., 1994. Structure indicative of helicoidal flow in a migmatitic diapir (Bação Complex, southeastern Brazil). Tectonophysics, 234: 169–996

Hippertt, J.F., Borba, R.P. and Nalini, H.A., 1992. O contato Formação Moeda-Complexo do Bonfim: Uma zona de cislhamento normal na borda oeste do Quadrilátero Ferrífero, MG. Revista da Escola de Minas, 45(1/2): 32–34.

Jordt-Evangelista, H., Alkmim, F.F. and Marshak, S., 1992. Metamorfismo progressivo e a ocorrência dos 3 polimorfos $Al_2O_3SiO_5$ (cianita, andaluzita e sillimanita) na Formação Sabará em Ibirité, Quadrilátero Ferrífero, MG. Revista da Escola de Minas, 45(1/2), 157–160.

Jordt-Evangelista, H., Peres, G.G., Macambira, M.J.B., 2000. Pb/Pb single-zircon dating of Paleoproterozoic calc-alkaline/ alkaline magmatism in the southern São Francisco Craton region, Brazil. Revista Brasileira de Geociências 30 (1), 174–176.

Klein, C. 2005. Some Precambrian banded iron-formations (BIFs) from around the world: Their age, geologic setting, mineralogy, metamorphism, geochemistry, and origins. American Mineralogist, 90, 1473–1499.

Klein, C., Ladeira, E.A., 2000. Geochemistry and petrology of some Proterozoic banded iron formations of the Quadrilatero Ferrifero, Minas Gerais, Brazil. Economic Geology, 95 (2), 405–427.

Koglin, N., Cabral, A.R., Brunetto, W.J., Vymazalová, A., 2012. Gold–tourmaline assemblage in a Witwaterstrand-like gold deposit, Ouro Fino, Quadrilátero Ferrífero of Minas Gerais, Brazil: the composition of gold and metallogenic implications. Neues Jahrbuch für Mineralogie Abhandlungen: Journal of Mineralogy and Geochemistry, 189 (3), 263–273.

Koglin, N., Zeh, A., Cabral, A.R., Gomes, A.A.S., Neto, A.V.C., Brunetto, W.J., Galbiatti, H., 2014. Depositional age and sediment source of the auriferous Moeda Formation, Quadrilátero Ferrífero of Minas Gerais, Brazil: New constraints from U–Pb–Hf isotopes in zircon and xenotime. Precambrian Research, 255, 96–108

Ladeira, E.A. and Viveiros, J.F.M., 1984. Hipótese sobre a estruturação do Quadrilátero Ferrífero com base nos dados disponíveis. Bol.Soc. Bras.Geol./Núcleo MG, 4: 1–14.

Lana, C., Alkmim, F.F., Armstrong, R., Scholz, R., Romano, R., Nalini, H.A., 2013. The ancestry and magmatic evolution of Archaean TTG rocks of the Quadrilátero Ferrífero province, southeast Brazil. Precambrian Research, 231, 157–173.

Machado, N. and Carneiro, M.A., 1992. U-Pb evidence of Late Archean tectonothermal activity in southern São Francisco shield, Brazil. Canadian Journal of Earth Sciences, 29,2341–2346.

Machado, N., Noce, C.M., Ladeira, E.A., Belo de Oliveira, O., 1992. U-Pb Geochronology of Archean magmatism and Proterozoic metamorphism in the Quadrilátero Ferrífero, southern São Francisco craton, Brazil. Geological Society of America Bullettin, 104,1221–1227.

Machado, N., Schrank, A., Noce, C.M., Gauthier, G. 1996a. Ages of detrital zircon from Archean-Paleoproterozoic sequences: Implications for Greenstone Belt setting ans evolution of a Transamazonian foreland basin in Quadrilátero Ferrífero, southeast Brazil. Earth and Planetary Science Letters, 141: 259–276

Machado N, Valladares C, Heilbron M., Valeriano C. 1996b. U-Pb geochronology of the central Ribeira belt (Brazil) and implications for the evolution of the Brazilian Orogeny. Precamb Res 79, 347–361.

Marshak, S., Alkmim, F.F. and Jordt-Evangelista, H., 1992. Proterozoic crustal extension and the generation of dome-and-keel structure in an Archean granite-greenstone terrane. Nature, 357, 491–493.

Marshak, S., Tinkham, D., Alkmim, F.F., Brueckner, H., Bornhorst, T., 1996. Dome-and-keel provinces formed during Paleoproterozoic orogenic collapse - Diapir clusters or core complexes?: Examples from the Quadrilátero Ferrífero (Brazil) and the Penokean Orogen (USA). Geology, 25, 415–418

Mendes, M.D.C.O., Lobato, L.M., Suckau, V., Lana, C., 2014. Datação U-Pb in situ por LA-ICPMS em zircões detríticos da Formação Cercadinho, Supergrupo Minas. Geologia USP. Série Científica, 14 (1), 55–68

Moore, S.L. 1968. Geology and ore deposits of Antônio dos Santos, Congo Soco, and Conceição do Rio Acima quadrangles, Minas Gerais, Brazil. US Geological Survey Professional Paper 341-I, 50p.

Moreira, H., Lana, C., Nalini Jr., H. A., 2016. The detrital zircon record of an Archaean convergent basin in the Southern São Francisco Craton, Brazil. Precambrian Research, 275, 84–99.

Müller, G, Schuster, A., Hoefs, J 1982. Oxygen isotope variations in polymetamorphic iron ores from the Quadrilátero Ferrífero, Brazil, Revista Brasileira de Geociências, 12(1–3), 248–355.

Neri, M.E.N.V., Rosière, C.A., Lana, C.C., 2013. Supergrupo Minas na Serra de Bom Sucesso, extremo sudoeste do Quadrilátero Ferrífero - MG: Petrografia, geoquímica e isótopos de U-Pb. Revista Geologia USP, Série Científica 13 (2), 117–202.

Noce, C.M., 1995. Geocronologia dos eventos magmáticos, sedimentares e metamórficos na região do Quadrilátero Ferrífero, Minas Gerais. Unpubl. PhD Thesis, Instituto de Geociências Universidade de São Paulo, São Paulo, 129 p.

Noce C.M., Teixeira W., Quéméneur J.J.G., Martins V.T.S. Bolzachini E. 2000. Isotopic signatures of Paleoproterozoic granitoids from southern São Francisco Craton, NE Brazil, and implications for the evolution of the Transamazonian Orogeny. Journal of South American Earth Sciences, 13, 225–239.

Noce, C.M., Zucchetti, M., Baltazar, O.F., Armstrong, R., Dantas, E.L., Renger, F.E., Lobato, L.M., 2005. Age of felsic volcanism and the role of ancient continental crust in the evolution of the Neoarchean Rio das Velhas greenstone belt (Quadrilatero Ferrífero, Brazil): U-Pb zircon dating of volcaniclastic graywackes. Precambrian Research. 141, 67–82.

Noce, C.M., Pedrosa-Soares, A.C., Silva, L.C., Armstrong, R., Piuzana, D., 2007. Evolution of polycyclic basement in the Araçuaí Orogen based on U-Pb SHRIMP data: implications for the Brazil-Africa links in the Paleoproterozoic time. Precambrian Research, 159, 60–78.

Oliveira, A.H., 2004. Evolução Tectônica de um fragmento do Cráton São Francisco Meridional com base em aspectos estruturais, geoquímicos (rocha total) e geocronológicos (Rb-Sr, Sm-Nd, Ar-Ar, U-Pb). Universidade Federal de Ouro Preto. Unpubl. Doctoral Thesis. 134p

Peres, G.G., Alkmim, F.F., Jordt-Evangelista, H. 2004. The southern Araçuaí belt and the Dom Silvério Group: Geologic architecture and tectonic significance. Annais da Academia Brasileira de Ciências, 76(4): 771–790.

Pires F. R. M. 1995. Textural and mineralogical variations during metamorphism of the Proterozoic Itabira Iron Formation in the Quadrilátero Ferrífero, Minas Gerais, Brazil. Anais da Academia Brasileira de Ciências, 67(1), 77–105.

Pomerene, J.B., 1964. The geology and ore deposits of the Belo Horizonte, Ibirité and Macacos quadrangles, Minas Gerais, Brazil. US Geol.Surv. Prof Paper 341-D. 57p.

Reis, L.A., Martins-Neto, M.A., Gomes, N.S., Endo, I., 2002. A bacia de antepaís paleoproterzóica Sabará, Quadrilátero Ferrífero, MG. Revista Brasileira de Geociências, 32, 43–58.

Renger, F. E., Noce, C.M., Romano, A.W., Machado, N., 1995. Evolução sedimentar do Supergrupo Minas: 500 Ma de registro geológico no Quadrilátero Ferrífero, Minas Gerais, Brasil. Geonomos, 2(1), 1–11.

Renger, F.E., Silva, R.M.P., Suckau, V.E., 1988. Ouro nos conglomerados da Formação Moeda, Sinclinal de Gandarela, Quadrilátero Ferrífero, Minas Gerais. Congr. Bras. Geol, Soc. Bras. Geol. 35, 44–57.

Romano, A.W., 1989. Evolution Tectonique de la region nord-ouest du Quadrilatère Ferrifere–Minas Gerais–Brésil. Ph.D. Thesis, Université de Nancy I, Nancy, France. 214pp.

Romano, R., Lana, C., Alkmim, F.F., Stevens, G.S., Armstrong, R., 2013. Stabilization of the southern portion of the São Francisco Craton, SE Brazil, through a long-lived period of potassic magmatism. Precambrian Research, 224, 143–159.

Rosière, C. A, Spier, C. A., Rios, F. J., Suckau, V E, 2008. The itabirites from the Quadrilátero Ferrífero and related high-grade ores: An overview, Reviews in Economic Geology, 15:223–254.

Schrank, A. and Machado, N., 1996. Idades U-Pb em monazitas e zircões do distrito aurífero de Caeté, da Mina de Cuiabá e do Depósito de Carrapato, Quadrilátero Ferrífero (MG). Anais 39th Congr.Bras.Geol., Salvador, Soc.Bras.Geol., v. 6, p. 473–475.

Seixas, L.A.R. 1988. Geologia e Metalotectos de Ouro de uma Fração do Lineamento Congonhas, MG. Instituto de Geociências, Universidade Nacional de Brasília. Unpl. Master Dissertation, 116 p.

Seixas, L. A. R., Bardintzeff, J. M., Stevenson, R., Bonin, B., 2013. Petrology of the high-Mg tonalites and dioritic enclaves of the ca. 2130 Ma Alto Maranhão suite: Evidence for a major juvenile crustal addition event during the Rhyacian orogenesis, Mineiro Belt, southeast Brazil. Precambrian Research, 238, 18–41.

Seixas, L.A.R., David, J., Stevenson, R., 2012. Geochemistry, Nd isotopes and U–Pb geochronology of a 2350 Ma TTG suite, Minas Gerais, Brazil: implications for the crustal evolution of the southern São Francisco craton. Precambrian Research, 196, 61–80.

Silva, A.M., Chemale Jr., F., Heaman, L., 1995. The Ibirité gabbro and the Borrachudo granite - The rift-related magmatism of Mesoproteroic age in the Quadrilátero Ferrífero (MG). Anais VIII Simp. Geol. Minas Gerais, Diamantina, Soc. Bras.Geol./Núcleo MG, Bol.13: 89–90.

Silva L.C., Armstrong R., Noce C.M., Carneiro M.A., Pimentel M. M., Pedrosa-Soares A.C., Leite C.A., Vieira V.S., Silva M.A., Paes V.J.C., Cardoso-Filho J.M. 2002. Reavaliação da evolução geológica em terrenos pré-cambrianos brasileiros com base em novos dados U-Pb SHRIMP, parte II: Orógeno Araçuaí, Cinturão Mineiro e Cráton São Francisco Meridional. Revista Brasileira de Geociências, 32, 513–528.

Silva, L.C., Pedrosa-Soares, A.C., Armstrong, R., Pinto, C.S., Magalhães, J.T.R., Pinheiro, M.A.P., Santos, G.G., 2015. Journal of South American Earth Sciences 67, 1–18

Souza, P.C., Müller, G., 1984. Primeiras estruturas algais comprovadas na Formação Gandarela, Quadrilátero Ferrífero. Revista de Escola de Minas, Ouro Preto, 2, 161–198.

Spier, C. A., Oliveira, S. M., Rosière, C. A, 2003. Geology and geochemistry of the Águas Claras and Pico mines, Quadrilátero Ferrífero, Minas Gerais, Brazil, Mineralium Deposita, 38:751–754.

Spier, C A, Oliveira, S M, Sial, A N and Rios, F J, 2007. Geochemistry and genesis of the banded iron formations of the Cauê Formation, Quadrilátero Ferrífero, Minas Gerais, Brazil, Precambrian Research, 152:170–206.

Tack, L., Wingate, M.T.D., Liégeois, J.P., Fernandez-Alonso, M., Deblond, A., 2001. Early Neoproterozoic magmatism (1000–910 Ma) of the Zadinian and Mayumbian Groups (Bas-Congo): onset of Rodinian rifting at the western edge of the Congo Craton. Precambrian Research, 110, 277–306.

Teixeira, W., 1985. A evolução geotectônica da porção meridional do Cráton do São Francisco, com base em interpretações geocronológicas. Unpubl. Doctoral Thesis, Universidade de São Paulo, São Paulo, 207 pp.

Teixeira, W., 1993. Avaliação do acervo de dados geocronológicos e isotópicos do Cráton do São Francisco - Implicações tectônicas. In Dominguez, J.M.L. and Misi, A. (eds.): O Cráton do São Francisco. Soc.Bras.Geol./Núcleo BA/SE, Salvador, p. 11– 3

Teixeira, W. and Figueiredo, M.C.H., 1991. An outline of Early Proterozoic crustal evolution in the São Francisco region, Brazil: A review. Precambrian Research, 53(1–2), 1–22.

Teixeira W, Jordt-Evangelista H, Kawashita K., Taylor P.N. 1987. Complexo granulítico de Acaiaca,MG: idade, petrogênese e implicações tectônicas.In: Simp Geol Minas Gerais, 4., Anais…, Belo Horizonte: SBG Núcleo MG 7, 58–71.

Teixeira, W., Ávila, C.A., Nunes, L.C., 2008. Nd–Sr isotopic geochemistry and U–Pb geochronology of Fé granitic gneiss and Lajedo granodiorite: implications for Paleoproterozoic evolution of the Mineiro belt, southern São Francisco Craton. Geologia USP Série Científica 8, 53–73.

Teixeira W, Sabaté P, Barbosa J, Noce CM, Carneiro MA (2000) Archean and Paleoproterozoic tectonic evolution of the São Francisco Craton. In: Cordani, U.G.;Milani, E.J.; Thomaz Fo, A.; Campos, D.A. (eds.) Tectonic Evolution of South América. Rio de Janeiro, 31st International Geological Congress, Rio de Janeiro, p. 101–137.

Teixeira, W.; Ávila, C.A.; Dussin, I.A.; Vasques, F.S.G.; Hollanda, M. H.M., 2012. Geocronologia U-Pb (LA-ICPMS) em zircão detrítico de rochas metassedimentares paleoproterozoicas da parte sul do Craton do São Francisco: proveniência, delimitação temporal e implicações tectônicas. 12° Simpósio de Geologia do Sudeste/16° Simpósio de Geologia de MG. Nova Friburgo, Sociedade Brasileira de Geologia, Anais, CDROM, p. 12.

Teixeira, W., Ávila, C.A., Bongiolo, E.M., Hollanda, M.H.B., Barbosa, N.S., 2014. Age and tectonic significance of the Ritápolis batholith, Mineiro belt (Southern SãoFrancisco Craton): U-Pb LA-ICPMS, Nd isotopes and geochemical evidence. In: 9th South American Symposium on Isotope Geology, p. 194

Teixeira, W., Ávila, C.A., Dussin, I.A., Neto, A.C., Bongiolo, E.M., Santos, J.O., Barbosa, N.S., 2015. A juvenile accretion episode (2.35–2.32 Ga) in the Mineiro belt and its role to the Minas accretionary orogeny: Zircon U–Pb–Hf and geochemical evidences. Precambrian Research, 256, 148–169.

Tessari, O. and Amorim, L.Q., 1984 Evolução sedimentar e tectônica do Grupo Itacolomi na região de Passagem de Mariana. Revista da Escola de Minas, 37(4), 31–44

Toledo, C.L.B., 2002. Evolução geológica das rochas máficas e ultramáficas no Greenstone Belt Barbacena, na região de Nazareno, MG. Unpubl. Doctoral Thesis, Instituto de Geociências, Universidade Estadual de Campinas, UNICAMP, Campinas, 307 pp.

Vasconcelos, F.F., 2015: Evolução geológica, metamorfismo e metassomatismo da região ao redor da Mina de Volta Grande, Nazareno, Estado de Minas Gerais. Universidade Fedral do Rio de Janeiro, Unpblished Master Dissertation, 129p.

Villaça J.N. 1981. Alguns aspectos sedimentares da Formação Moeda. Sociedade Brasileira de Geologia, Núcleo MG, 2, 92–137.

Vlach, S.R.F.; Campos Neto, M. C.; Caby, R.; Basei, M. A. S. 2003. Contact metamorphism in metapelites from the Nova Lima Group, Rio das Velhas Supergroup, Quadrilátero Ferrífero: a monazite Th-U-PbT dating by the electron-probe microanalyser. In: IV South American Symposium on Isotope Geology, Salvador. Short Papers, v. 1, p. 307–310

Zucchetti, M.; Baltazar, O.F., Raposo, F.O. 1996. Estratigrafia. In: Companhia de Pesquisa de Recursos Minerais. Projeto Rio das Velhas–Texto Explicativo do Mapa Geológico Integrado, escala 1:100.000. Departamento Nacional de Produção Mineral/CPRM–Serviço Geológico do Brasil, Belo Horizonte, p. 13–42.

Part III

Intracratonic Basins and Precambrian Mafic Dyke Swarms

The Paramirim Aulacogen

6

Simone Cerqueira Pereira Cruz and Fernando F. Alkmim

Abstract

The Paramirim aulacogen occupies the region of the northern São Francisco craton that extends from the northern Espinhaço range to a large plateau known as the Chapada Diamantina in the northeastern Brazilian state Bahia. Corresponding to a long-lived and partially inverted rift-sag basin, the aulacogen hosts and exposes the most complete section of the cratonic cover complex accumulated between the Statherian and Edicaran periods. The sedimentary and volcanic rocks of the basin fill successions are grouped into two major lithostratigraphic units, the Espinhaço and São Francisco supergroups. Exhibiting remarkable facies differentiation between their occurrences in the northern Espinhaço range, on the west, and Chapada Diamantina, on the east, these units are associated with Statherian granitic rocks, as well as Statherian, Calymmian and Tonian mafic dykes and sills. The Espinhaço Supergroup comprises five unconformity bounded sequences, which record alternating mechanical and flexural subsidence pulses that took place between 1775 and 900 Ma. Frankly dominated by quartz-rich sandstones, deposited in alluvial, aeolian and shallow marine systems, the supergroup reflects continuous recycling of sediments in the large depocenters of the aulacogen. The Neoproterozoic São Francisco Supergroup is composed of three distinct sequences, deposited between 900 and 500 Ma, among them the glaciogenic Bebedouro Formation and the shallow marine carbonates and pelites of the Salitre Formation. The overall geologic architecture of the aulacogen is dominated by structures generated during its contractional inversion in the Edicaran/Cambrian boundary. The regional structure of the aulacogen corresponds to a NNW-trending basement cored uplift, flanked on both sides by systems of reverse faults, thrusts, folds, and strike-slip shear zones. Structures related to the basin generation processes are preserved only in a small area of the central segment of the aulacogen. The dominant NNW-oriented fabric elements interact or are overprinted by E–W-trending structures in the northern and southern ends of the aulacogen. These structures represent the Neoproterozoic Brasiliano orogenic fronts that propagated from the marginal belts toward the craton interior.

S.C.P. Cruz (✉)
Instituto de Geociências, Universidade Federal da Bahia, Campus da Federação, Salvador, BA 40.000-000, Brazil
e-mail: simonecruzufba@gmail.com

F.F. Alkmim
Departamento de Geologia, Escola de Minas, Universidade Federal de Ouro Preto, Morro Do Cruzeiro s/n, Ouro Preto, MG 35.400-000, Brazil
e-mail: ffalkmim@gmail.com

M. Heilbron et al. (eds.), *São Francisco Craton, Eastern Brazil*,
Regional Geology Reviews, DOI 10.1007/978-3-319-01715-0_6

Keywords
Paramirim aulacogen • Rift-sag basin • Espinhaço Supergroup • São Francisco Supergroup

6.1 Introduction

Originally referred to as the Espinhaço aulacogen (Costa and Inda 1982), the Paramirim Aulacogen (after Pedrosa-Soares et al. 2001) corresponds to a long-lived and partially inverted rift-sag basin nucleated at around 1.78 Ga in the landmass presently represented by the São Francisco craton and its margins (Schobbenhaus 1996; Cruz and Alkmim 2006). Located in the northeastern Brazilian state Bahia, this poly-historical successor basin encompasses the region of the northern Espinhaço range, the Paramirim and São Francisco river valleys, as well as the large plateau known as the Chapada Diamantina (Figs. 6.1 and 6.2).

The Paramirim aulacogen hosts and exposes the whole section of the Precambrian cover complex of the craton, i.e., units younger than 1.8 Ga, which comprise two major lithostratigraphic units, namely, the Espinhaço and São Francisco supergroups (Fig. 6.3). The aulacogen fill succession records a series basin-forming, basin-modification, basin-inversion and climatic events that took place between the Statherian and Ediacaran periods of the Proterozoic Eon, i.e., between ca. 1778 and 500 Ma (Alkmim and Martins-Neto 2012; Pedrosa-Soares and Alkmim 2011; Guimarães et al. 2012; Guadagnin and Chemale Jr. 2015). The geologic evolution of the aulacogen, especially the age of its inversion, was a matter intensively debated, for this

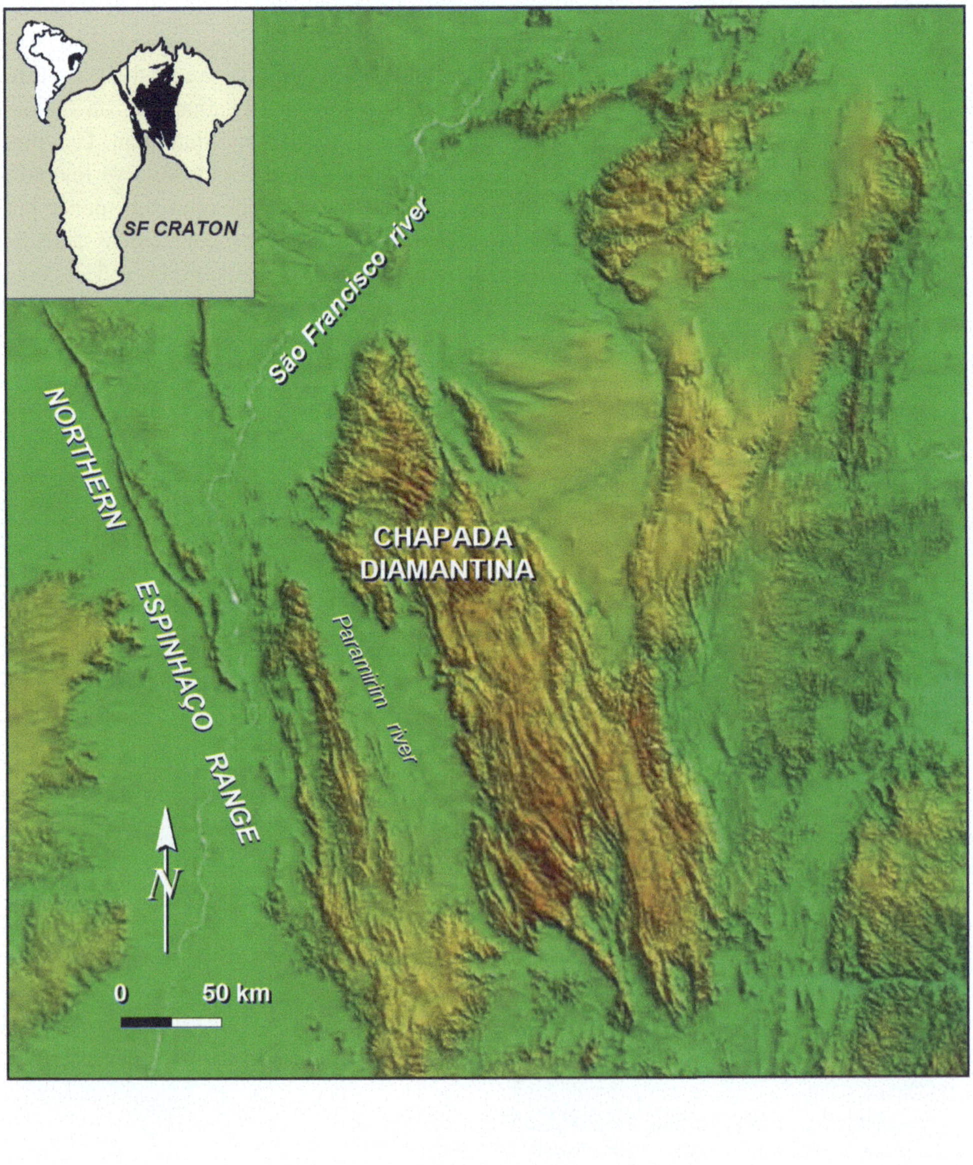

Fig. 6.1 Morphological expression of the Paramirim aulacogen in a digital elevation model of the northern São Francisco craton (*box* on upper left corner), showing the northern Espinhaço range, the Pararamirim and São Francisco river valleys, as well as the plateau known as the Chapada Diamantina

Fig. 6.2 Panoramic view of the eastern Chapada Diamantina (Sincorá ridge), which exposes the Tambador Sandstone, the basal unit of the Mesoproterozoic Chapada Diamantina Group

very topic has major implications for the São Francisco craton concept.

6.2 Stratigraphy

The Paramirim aulacogen developed on the Gavião block, the westernmost component of the basement in the northern portion of the craton (see Teixeira et al. and Barbosa and Barbosa this book). Meso to Neoarchean TTG's, K-rich granitoids and greenstone belts, together with Rhyacian-Orosirian metasedimentary units and granitoids are the dominant rock assemblages in the basement exposures in and around the aulacogen.

Besides the sedimentary and volcanic rocks of the Espinhaço and São Francisco supergroups, the development of the aulacogen was also associated with the emplacement of a substantial volume of Statherian plutonic rocks, as well as Statherian, Calymmian and Tonian mafic dykes and sills (Figs. 6.3 and 6.4).

A remarkable facies differentiation exists between the aulacogen fill units occurring in the northern Espinhaço range, on the west, and Chapada Diamantina, on the east (Figs. 6.3 and 6.4). These regions are separated by the Paramirim river valley that exposes the eponymous basement block (Fig. 6.5). According to various studies carried out in the region, the Paramirim basement inlier acted as a major horst already in the early evolutionary stages of the aulacogen. Thus, two almost independent sub-basins were individualized by the end of the Paleoproterozoic, remaining as such up to end of the Neoproterozoic (Jardim de Sá 1981; Costa and Inda 1982; Dominguez 1993; Danderfer Filho and Dardenne 2002; Schobbenhaus 1996; Cruz and Alkmim 2006; Alkmim and Martins-Neto 2012). The Espinhaço successions in western and eastern portions of the aulacogen were subdivided in distinct ways by different authors, resulting in a plethora of names and stratigraphic classifications over the years. The lithostratigraphic units of both sub-basins are discussed separately in the next sections, following the stratigraphic subdivision proposed by Guimarães et al. (2012) with some modifications in order to incorporate new literature data.

6.2.1 The Espinhaço Supergroup and Associated Plutonic Rocks

Dominated by quartz-rich sandstones and comprising the Paleo to Mesoproterozoic section of the cratonic cover, the Espinhaço Supergroup occupies most of the aulacogen area and reaches a maximum thickness of approximately 8000 m.

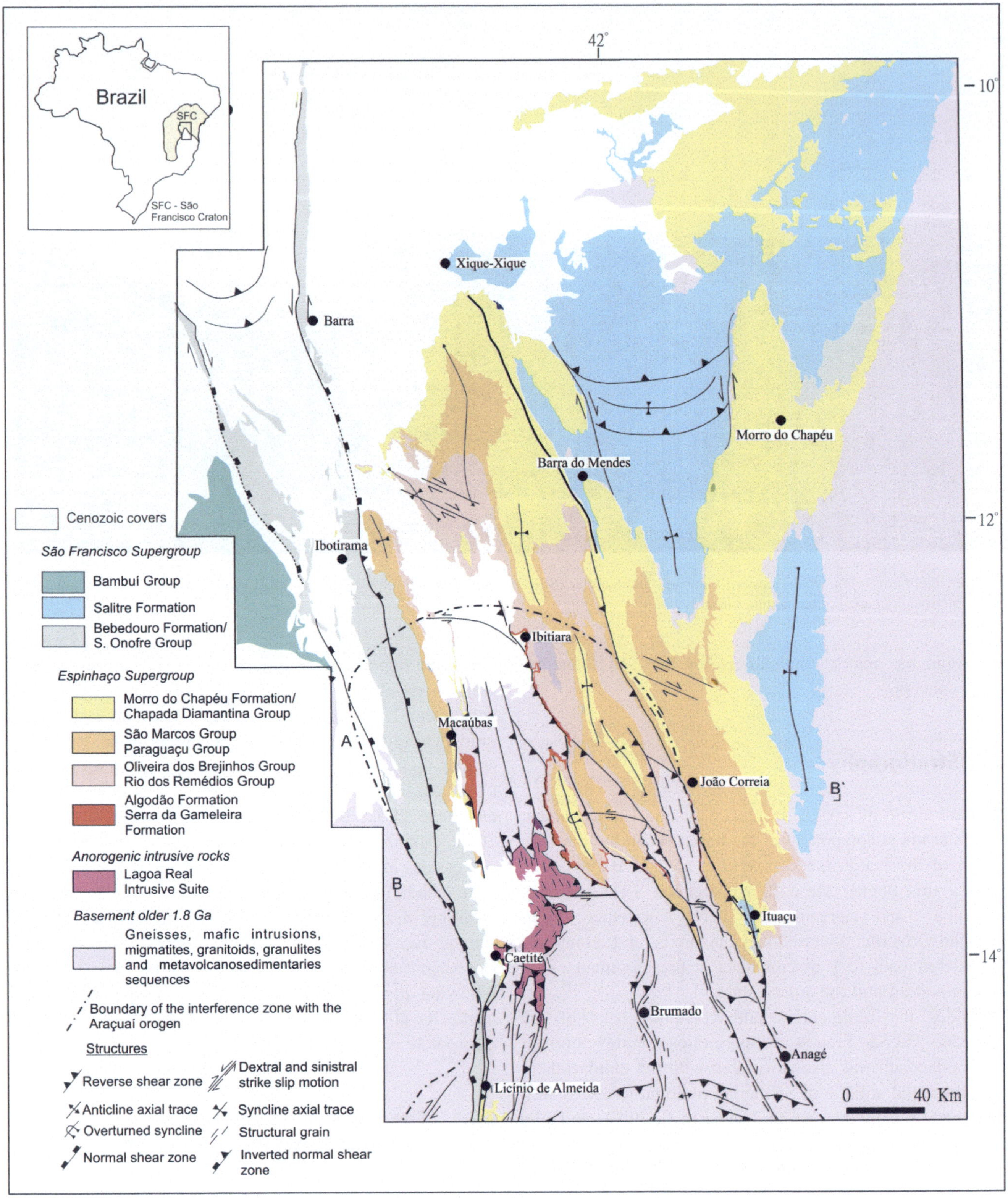

Fig. 6.3 Geologic map of the Paramirim aulacogen emphasizing the lithostratigraphic units and structures. Modified from Cruz and Alkmim (2006)

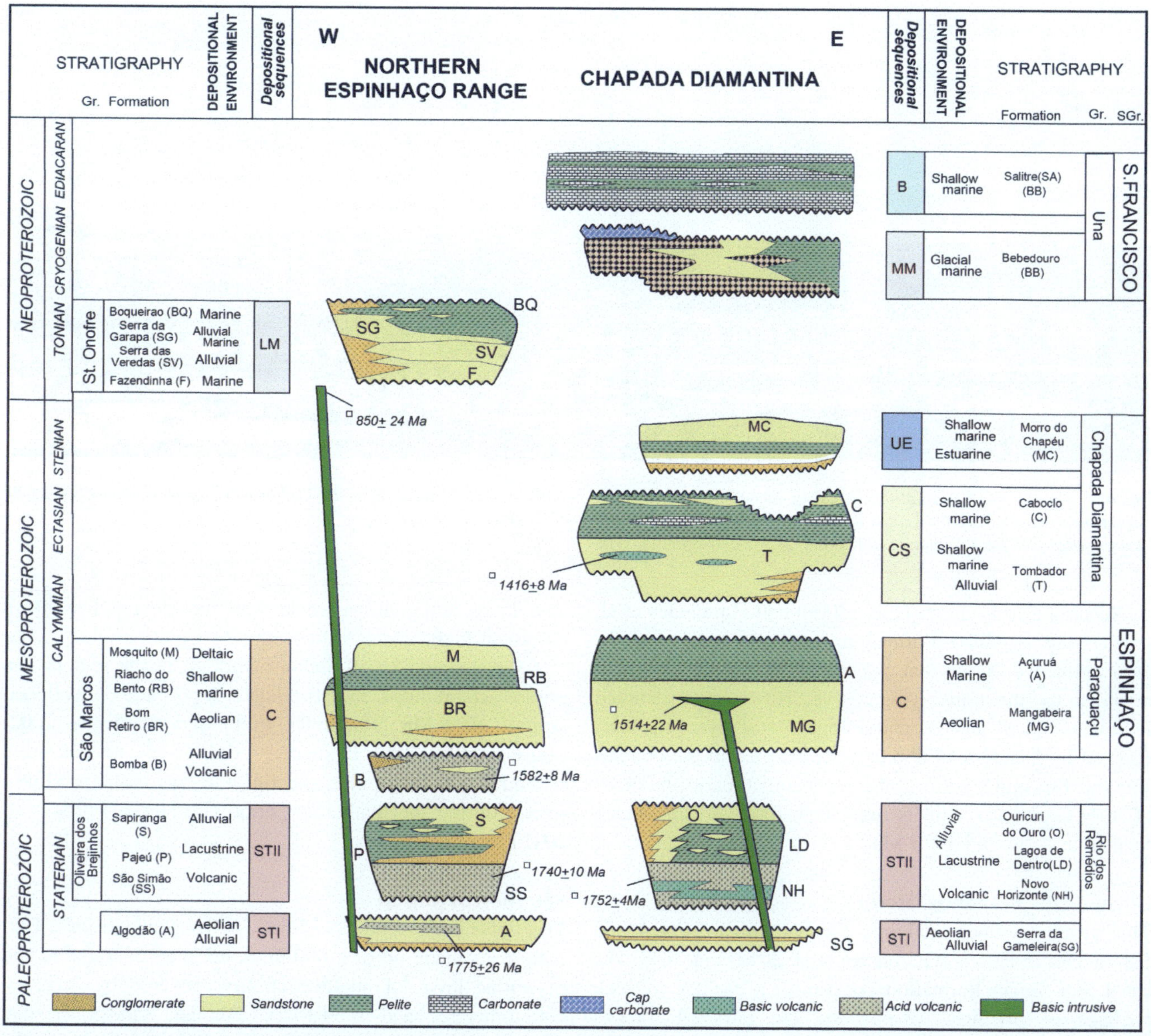

Fig. 6.4 Stratigrphic chart of the northern Espinhaço and Chapada Diamantian sub-basins of the Paramirim aulacogen. Depositional sequences: *STI* Statherian I, *STII* Statherian II; *C Calymmian*; *CE* Calymmian-Stenian; *UE* Upper Espinhaço; *LM* Santo Onofre; *MM* Bebedouro; *B* Salitre. The *numbers* denote crystallization or the youngest detrital zircon ages. See text for explanation and references (Modified from Alkmim and Martins-Neto 2012)

6.2.1.1 Northern Espinhaço Range Sub-Basin

The Espinhaço Supergroup, subdivided into the Algodão Formation and two groups, the Oliveira dos Brejinhos and São Marcos, occurs as a relatively narrow belt that extends for ca. 200 km along the eastern flank of the northern Espinhaço range (Fig. 6.4).

Algodão Formation

With a maximum thickness of approximately 1600 m, the Algodão Formation consists predominantly of feldspar-rich sandstones, containing lenses and layers of conglomerates and breccias representing braided fluvial, alluvial fan and aeolian deposits. The upper portion of the formation contains intercalations of acid lavas flows and pyroclastics rocks (Danderfer Filho and Dardenne 2002; Guimarães et al. 2012; Danderfer Filho et al. 2015).

The detrital zircon age spectra of the lower and middle portions of the Algodão Formation indicate Paleoproterozoic and Archean basement units as their main sources and a maximum deposition age of 1984 ± 21 Ma. The acid lavas of the uppermost section of the formation yielded a crystallization age of 1775 ± 26 Ma (Danderfer Filho et al. 2015).

Fig. 6.5 View of the Paramirim river valley (underlain by basement assemblages intruded by the Lagoa Real anorogenic plutons) and the western edge of the Chapada Diamantina in the background. The 1958 m-high Almas peak in the background is one of the highest points of the Paramirim aulacogen

According to Loureiro et al. (2009) and Guimarães et al. (2012), the Algodão Formation represents relicts of an intracontinental basin that preceded the rift development recorded by the units lying above. However, a detailed sedimentological and geochronological study recently carried out by Danderfer Filho et al. (2015) led to the conclusion that this unit records a rifting event that marks the onset of the Espinhaço Supergroup deposition in the Early Statherian at around 1775 Ma.

Oliveira Dos Brejinhos Group

This group is subdivided into the São Simão, Pajeú and Sapiranga formations (Guimarães et al. 2012; Fig. 6.4). The basal São Simão Formation consists of a ca. 250 m-thick package of rhyolites, rhyodacites and piroclastic rocks (Danderfer Filho and Dardenne 2002; Loureiro et al. 2009; Guimarães et al. 2012). An erosional surface separates the acid lavas from the alluvial to lacustrine conglomerates, arcoses and pelites of the Pajeú Formation, which is in turn is overlain by the ca. 3000 m-thick succession of alluvial, lacustrine and aeolian sandstones and pelites of the Sapiranga Formation (Guimarães et al. 2008).

Rhyolites of the São Simão Formation were dated at 1731 ± 5 and 1740 ± 10 Ma by Danderfer Filho et al. (2009, 2015). The detrital zircon age spectra of the group are characterized by various Archean and Paleoproterozoic age clusters, which peak in the intervals of 2.17–2.07 Ga and 1.81–1.73 Ga (Guadagnin and Chemale Jr. 2015). Data indicate sourcing mainly from rock assemblages involved in the Rhyacian-Orosirian Eastern Bahia orogenic domain that forms the craton basement in areas to east of the aulacogen (see Barbosa and Barbosa this book), as well as Statherian acid lavas and plutonic rocks that intrude the Paramirim basement high.

As postulated by all authors working in the region, this group accumulated in rift-related depocenters developed around 1740 Ma (Danderfer Filho and Dardenne 2002; Loureiro et al. 2009; Guimarães et al. 2012), i.e., during a second extensional pulse affecting the São Francisco lithosphere during the Statherian period (Danderfer Filho et al. 2015).

São Marcos Group

The lower unit of the São Marcos Group, the Bomba Formation, is made up of a 600-thick package of traquitic and rhyolithic lavas of alkaline affinity, covered by breccias, conglomerates and sandstones. Aeolian sandstones that grade up into marine sandstones and pelites characterize the Bom Retiro (maximum thickness 450 m) and Riacho do Bento (500 m) formations, respectively. The upper unit of the group, the Mosquito Formation, comprises deltaic sandstones and pelites (Danderfer Filho 2000; Loureiro et al. 2009; Danderfer Filho et al. 2009).

Danderfer Filho et al. (2009) obtained the U–Pb SHRIMP crystallization ages of 1582 ± 8 and 1569 ± 14 Ma on zircons extracted from the rhyolites of the Bomba Formation. The facies framework and occurrence form of this group led Loureiro et al. (2009) and Guimarães et al. (2012) to suggest its deposition during a sagging episode that affected the whole aulacogen area.

6.2.1.2 Chapada Diamantina Sub-Basin

The exposures over the 450 km-long and 240 km-wide plateau of the Chapada Diamantina are frankly dominated by the Espinhaço Supergroup, which comprises a ca. 5000 m-thick quartz sandstone-dominated succession, subdivided into the Serra da Gameleira Formation and three groups, the Rio dos Remédios, Paraguaçu and Chapada Diamantina (Fig. 6.4).

Serra da Gameleira Formation

This unit occurs as a continuous, 70–200 m-thick layer fringing the basement exposures along the southwestern edge of the Chapada Diamantina. It is composed predominately of aeolian sandstones, which laterally grade to alluvial conglomerates and lacustrine fine-grained sanstones and pelites (Guimarães et al. 2008).

The minimum age of the formation is constrained by the age of the overlying acid volcanics (Rio dos Remédios Group), dated at 1.75 Ga (Danderfer Filho et al. 2009; Schobbenhaus et al. 1994; Babinski et al. 1994). Its maximum depositional age can be estimated in ca. 2040 Ma, the youngest metamorphic age documented in the basement units. Like the basal unit of the supergroup in the northern Espinhaço range (Algodão Formation), this formation was interpreted by Guimarães et al. (2008) as deposited in a sag basin preceding the development of the rifting episode recorded by the overlying volcanic and sedimentary rocks.

Rio dos Remédios Group and Associated Lagoa Real Plutonic Suite

The 2000 m-thick succession of acid lavas and lacustrine to alluvial sediments of the Rio dos Remédios Group (Schobbenhaus and Kaul 1971; Teixeira 2005; Loureiro et al. 2009; Guimarães et al. 2012) is further subdivided into three formations. The Novo Horizonte Formation, at the base of the group, comprises a 600 m-thick succession of felsic lava flows, crystal tuffs (Fig. 6.6a, b) and epiclastic rocks (McReath et al. 1981; Guimarães et al. 2008) of dacitic, rhyolitic and andesitic compositions. The 850 m-thick Lagoa de Dentro Formation, made up of interbeded sandstones and pelites containing conglomerate lenses, grades up into the 550 m-thick package of lithic and arcosic conglomerates and sandstones of the Ouricuri do Ouro Formation (Guimarães et al. 2012).

Geochronological studies performed by Schobbenhaus et al. (1994) and Babinski et al. (1994) in the acid lavas of the Novo Horizonte Formation exposed in the southwestern edge of the Chapada Diamantina yielded crystallization ages of 1752 ± 4 and 1748 ± 4 Ma.

Large bodies of porphyritic and coarse-grained, hastingsite-bearing syenites, syenogranites, and alkali granites (Fig. 6.6c) intrude the Paramirim basement high (Fig. 6.3). Grouped in the Lagoa Real Intrusive Suite (Arcanjo et al. 2005), these plutons characterize a metaluminous, high-K and iron-rich calcalkaline series of anorogenic origin (Maruèjol et al. 1987; Teixeira 2000). Affected by Edicaran shear zones in many places, rocks of the Lagoa Real Suite are converted into gneisses and mylonites containing albitites mineralized in uranium (Pimentel et al. 1994; Cruz et al. 2007c). U–Pb data in zircons and titanites extracted from these rocks have yielded crystallization ages around 1750 Ma, i.e., the same age of the acid lavas of the Novo Horizonte Formation. The albitization associated with the uranium mineralization took place mainly around 950 Ma (Pimentel et al. 1994; Lobato et al. 2015).

The rock assemblage of this group together with the alkaline plutons of the Lagoa Real Suite also record the second 1.74 Ga rifting event represented by the previously described Oliveira dos Brejinhos Group of the northern Espinhaço range (Danderfer Filho et al. 2015; Guadagnin and Chemale Jr. 2015).

Paraguaçu Group

This group is subdivided into the Mangabeira and Açuruá formations. With a maximum thickness of ca. 1500 m, the lower Mangabeira Formation is composed of marine sandstones and pelites that grade upward into aeolian and fluvial sandstones. The ca. 900 m-thick upper formation comprises shallow marine sandstones and pelites (Figs. 6.4 and 6.6d).

Sandstones of the Mangabeira Formation are cut by mafic dykes and sills dated at 1514 ± 22 and 1501 ± 9 Ma (Babinski et al. 1999; Silveira et al. 2013), which constrain its minimum depositional age.

Onlapping the basement over a large area of the central Chapada Diamantina, the Paraguaçu Group apparently corresponds to a sag sequence, accumulated during a considerable expansion of the Espinhaço basin (Loureiro et al. 2009; Guimarães et al. 2012).

Chapada Diamantina Group

This group, subdivided into three formations, occupies the largest area in the aulacogen (Figs. 6.3 and 6.4). The oldest Tombador Formation consists of a ca. 400 m-thick package of braided fluvial to shallow marine and tide-influence sandstones with local lenses and layers of diamond-bearing conglomerates (Magalhães et al. 2014, 2016; Fig. 6.7a, b). The Tombador Sandstones grade upward into the shallow marine pelites with subordinate carbonate lenses of the Caboclo Formation (Figs. 6.4 and 6.7d).

The Morro do Chapéu Formation (Fig. 6.4), separated from the underlying units by a regional unconformity, is made up of a diamond-bearing conglomerate and a ca. 300 m-thick package of fluvial to estuarine sandstones and pelites with intercalations of aeolian deposits (Dominguez 1993; Loureiro et al. 2009). According to Schobbenhaus (1996) and Guimarães et al. (2008), this sequence defines a

Fig. 6.6 **a** A sandstone bed covered by a rhyolite layer of Novo Horizonte Formation, Rio dos Remédios Group (1.73 Ga Statherian II sequence) cut by a thrust fault in the southwestern Chapada Diamantina. **b** Crystal tuff of Novo Horizonte Formation. **c** Deformed porprhyritic granite of the Lagoa Real Suite, exposed in the Pararimirim valley. Rotated and stretched feldspar porphyroclasts are imbedded in a feldspar, quartz and amphibole-rich matrix. **d** Fine-grained marine sandstones (redish) with wave ripple cross-lamination interbeded with pelites exhibiting mud-cracks of the Açuruá Formation, Upper Paraguaçu Group, Calymmian sequence

separated unit, probably of Neoproterozoic age and, thus, not belonging to the Chapada Diamantina Group. In favor of this proposal these authors emphasize the unconformity bounded character of the sequence and the fact that 860 Ma dykes do not cut this unit.

Battilani et al. (2007), using the Ar–Ar method in metasomatic muscovites extracted from an alterated mafic dyke that cuts the Tombador Formation (Fig. 6.7c), estimated the minimum age of the intrusion at 1513 ± 3 Ma. However, as demonstrated by Guadagnin et al. (2015), the upper portion of the Tombador Formation contains crystal-tuffs dated at 1420 Ma. Moreover, the youngest detrital zircons population constrains its maximum depositional age at 1450 Ma (Guadagnin et al. 2015; Guadagnin and Chemale Jr. 2015). Carbonates of the overlaying Caboclo Formation were dated by the whole rock Pb–Pb method at 1140 ± 140 Ma (Babinski et al. 1993).

The deposition of the Chapada Diamantina Group is associated with the maximum increase of the basin area. The spatial distribution of the Tombador Formation defines a regional onlap on the basement over much of the eastern Chapada Diamantina region. The facies framework of the group suggests that the expansion of basin took place in response of a major sagging event (Guimarães et al. 2012; Magalhães et al. 2014, 2016), probably associated to local fault-bounded uplifts, as indicated by the occurrence of conglomerates and breccias in the Tombador Formation.

6.2.2 Santo Onofre Group and São Francisco Supergroup

The Santo Onofre Group in the northern Espinhaço range together with the São Francisco Supergroup in the Chapada Diamantina form the Neoproterozoic fill succession of the aulacogen (Figs. 6.3 and 6.4).

6.2.2.1 Santo Onofre Group

Subdivided into four formations, the ca. 2000 m-thick Santo Onofre Group (Fig. 6.4; Schobbenhaus 1996; Guimarães

Fig. 6.7 **a** Pebble to cobble diamond-bearing conglomerate of the Tombador Formation (Chapada Dimantina Group, CS sequence), showing imbricate clasts (*lower part* of the photo). Sandstones, vein-quartz and quartzites are the predominant compositions of the clasts. **b** Tombador alluvial sandstones showing large-scale cross-bedding. **c** A 40 cm-wide mafic dyke cutting sandstones of the Tombador Formation. Note the sandstone xenoliths in the *middle* of the dyke. **d** Road cut exposure of the marine pelites of the Caboclo Formation, Chapada Diamatina Group (CS sequence)

et al. 2012) occupy a large area in the northern Espinhaço range (Fig. 6.3). Due to the lack on geochronological data, the Santo Onofre Group is here tentatively correlated to the Neoproterozoic fill succession of the aulacogen, following the proposal by Schobbenhaus (1996).

The Fazendinha Formation (Danderfer Filho 2000; Loureiro et al. 2009) comprises a ca. 470 m-thick package of feldspar-rich sandstones, quartz-arenites and subordinate fine-grained conglomerates, and pelites. The sandstones exhibit hummocky cross-stratification, tabular and trough cross-beds, and asymmetrical ripples (Guimarães et al. 2012). A shallow marine environment under influence of winds and tides was suggested for the deposition of the Fazendinha Sandstones (Danderfer Filho 2000; Loureiro et al. 2009; Guimarães et al. 2012).

The Serra das Veredas Formation consists of matrix-supported conglomerates, clast supported and fine-grained conglomerates, lithic sandstones containing scattered pebles, quartz-arenites, and pelites. Characteristic for this formation is a lazulite, dumortierite, and kyanite-bearing quartz-arenite layer of the upper portion of the formation, quarried in the region as dimension stone (Fleischer 1971; Schobbenhaus 1972; Jordt-Evangelista and Danderfer-Filho 2012; Fig. 6.8a). Danderfer Filho (2000) suggests alluvial fans to braided plain in transition to a shallow marine and storm-dominated platform as depositional environment for this unit. The phosphate-bearing blue quartzites are interpreted products of eo-diagenetic alteration of volcanic ashes under arid climate conditions (Jordt-Evangelista and Danderfer-Filho 2012). An U–Pb geochronological investigation via LA-ICMPS carried out in detrital zircons extracted from the blue quartzites by Franz et al. (2014) indicates a maximum depositional age of ca. 1.5 Ga and a syn-kinematic metamorphic event at around 634 Ma.

Shallow marine arcoses, quartz-arenites, ferruginous sandstones and conglomerates that grade upward into hematitic, carbonaceous, and manganesiferous pelites containing lenses of stromatolithic dolarenites characterizes the 200-thick Serra da Garapa Formation (Guimarães et al. 2012).

Fig. 6.8 **a** The lazulite, dumortierite, and kyanite-bearing *blue* quartzite of the Serra das Veredas Formation, St. Onofre Group. **b** *Folded black* shales of the Boqueirão Formation, exposed in the Northern Espinhaço range. **c** Glaciogenic diamictite of the Bebedouro Formation, containing gneiss, granitoid, carbonate and vein-quartz cobbles imbedded in sand-rich matrix. **d** Fosfate-bearing carbonates of the Salitre Fomation, showing columnar stromatolites

The youngest Precambrian unit exposed in northern Espinhaço range, the approximately 900 m-thick Boqueirão Formation, is made up predominantly of graphitic and hematitic pelites (Fig. 6.8b), with subordinate sandstones, conglomerates and breccias (Danderfer Filho 2000; Guimarães et al. 2012). The depositional environment suggested by these authors consists of alluvial fans in transition to a shallow marine platform.

According to Danderfer Filho (2000) the lower and middle portions of the Santo Onofre Group (sensu Guimarães et al. 2012) record the development of a transtensional graben, partially superimposed on the preexistent Espinhaço rift sag basin.

6.2.2.2 São Francisco Supergroup

The São Francisco Supergroup, comprising the Una Group subdivided into the Bebedouro and Salitre formations (Fig. 6.4), occurs in the interior of large synformal structures exposed in the Chapada Diamantina (Fig. 6.3).

Bebedouro Formation

The Bebedouro Formation, with thickness varying between 1 and 200 m, is separated from the older units by a regional unconformity (Pedreira et al. 1975; Misi 1979; Montes et al. 1985; Pedreira 1997) and from the younger Salitre Carbonates by an erosional surface (Misi 1979).

Five distinct lithofacies associations have been discriminated in the formation by Guimarães (1996): (1) massive and stratified diamictites; (2) laminated pelites contain dropstones of variable sizes; (3) interbedded pelites, arkoses, lithic sandstones and diamictites; (4) arcoses showing large-scale cross beds; (5) a discontinuous bed of calcarenites and dolarenites that occurs on top of the sequence representing a cap dolomite. The diamictites contain pebbles, cobles and boulders of granitic rocks, schists, quartz, quartzites, chert, carbonates and dolerite imbedded in sand-rich matrix (Fig. 6.8c). According to Guimarães (1996), the sediments of the Bebedouro Formation were deposited by underwater debris flows, turbidity and wind currents in a glaciated marine platform.

Attempts to date the Bebedouro glacial event have been made by different authors using various methods such as Rb–Sr, K–Ar and Sr–Sr (Brito-Neves et al. 1980; Macedo and Bonhome 1981, 1984). The resulting ages fall in the interval between 1000 and 900 Ma.

Salitre Formation

The Salitre Formation (Pedreira et al. 1975) comprises a 500 m-thick sequence of carbonates and pelitic rocks. The main lithotypes of the formation are calcarenites, calcisiltites, calcilutites, pelites, and dolarenites. Columnar stromatolites (Fig. 6.8d) and subordinate conglomerates and breccias occur in different portions of the formation (Inda and Barbosa 1978; Misi 1993; Dominguez 1996; Guimarães 1996), which corresponds to an overall shallowing upward sequence, accumulated in a tide and wave influenced carbonate ramp (Leão and Dominguez 1992; Leão et al. 1992).

The Salitre Carbonates were dated by the Rb/Sr method and the results indicate deposition between 750 and 850 Ma (Macedo and Bonhome 1984; Toulkeridis et al. 1999). Trindade et al. (2004) obtained a Pb–Pb isochron age of 514 ± 33 Ma on stromatolitic limestones of the basal portion of the formation. As emphasized by these authors, this age reflects, however, an intensive fluid migration across the Salitre Formation in association with the Brasiliano deformation event. Studies of the $^{87}Sr/^{86}Sr$ isotopic system performed by Misi and Veizer (1998) led these authors to estimate a deposition age of 700 Ma for the formation. The $\delta^{13}C$ values determined by the same authors along an entire section of the formation varies between 0 and +9.4 per mil. These values differ significantly from values obtained from Cryogenian post-glacial carbonates worldwide and from the dolomites occurring on top of the glaciogenic Bebedouro Diamictites (Guimarães et al. 2012).

6.3 Structure

The regional structure of the aulacogen consists of a large, NNW-plunging basement-cored uplift, known as the Paramirim block, connected on both sides to fold and thrust systems that verge in opposite directions (Fig. 6.3).

Structural investigations performed in the basin fill units indicate that the tectonic framework of the aulacogen comprises, besides other structures, two main families of fabric elements (Danderfer Filho 1990, 2000; Cruz and Alkmim 2006; Cruz et al. 2015): (1) extensional structures related to rift nucleation and reactivation; (2) contractional structures associated to the inversion of the basin.

The first family of structures is preserved in a relatively small area of the northern Espinhaço range located near the town of Macaúbas (Fig. 6.3). Normal to normal-sinistral faults that trend NNW and dip preferentially ENE, bound major hemi-grabens filled by the Espinhaço and Santo Onofre units. Towards north and south, these structures become progressively inverted, acting mainly as oblique-slip or strike-slip shear zones (Danderfer Filho 2000; Cruz and Alkmim 2006; Figs. 6.3 and 6.9). The majority of these extensional structures formed during the Statherian rifting events documented in the aulacogen. During the subsequent rifting episodes, the preexistent fabric elements underwent reactivation and apparently only a relatively small number of new structures developed.

The second set of structures comprises NNW-trending, thick and thin-skinned thrusts, reverse and transpressional shear zones associated with folds of variable scale. These fabric elements dominate the structural picture of the aulacogen and result from an overall WNW-ENE oriented contractional stress field. Besides these structures, the southern and northern ends of the aulacogen were also affected by the orogenic front that propagated from the marginal Araçuaí, Rio Preto and Riacho do Pontal belts towards the craton interior (Fig. 6.3). The interaction of the orogenic fronts with the intracratonic rift structures give rise to locally complex interference patterns (Danderfer Filho 1990, 2000; Cruz and Alkmim 2006).

The portion of the aulacogen that involves the largest population of deformational structures and corresponds to both its least and most intensively inverted sector defines a 70–150 km-wide and 500 km-long NNW-trending zone known as the Paramirim corridor (Alkmim et al. 1993). Cored by the equally named basement block, this double-verging deformation zone comprises three distinct segments. Its northern and southern ends (Fig. 6.3) correspond to relatively complex zones, where the intracratonic NNW-trending fabric elements interact with the structures of the Rio Preto, Riacho do Pontal and Araçuaí belts that fringe the craton to the north and southeast, respectively. These segments are currently viewed as part of the marginal belts (Uhlein et al. 2012; see Caxito et al. and Alkmim et al. this volume). The central portion of the corridor preserves rift-related structures and is also affected by structures associated to the main inversion event (Figs. 6.3 and 6.9).

The Paramirim corridor is coupled to an embryonic ENE-verging fold-thrust belt that involves the basement and sedimentary units along western Chapada Diamantina. The development of this system occurs in two distinct phases of deformation. The first phase is characterized by a large number of thin-skinned and in general layer-confined structures, which include detachments coupled with ENE-verging imbricate fans, duplexes, and fold trains (Fig. 6.10a). The elements of the second phase are basement-involved, reverse-dextral to reverse shear zones, associated with large scale folds (Fig. 6.10b). The strain

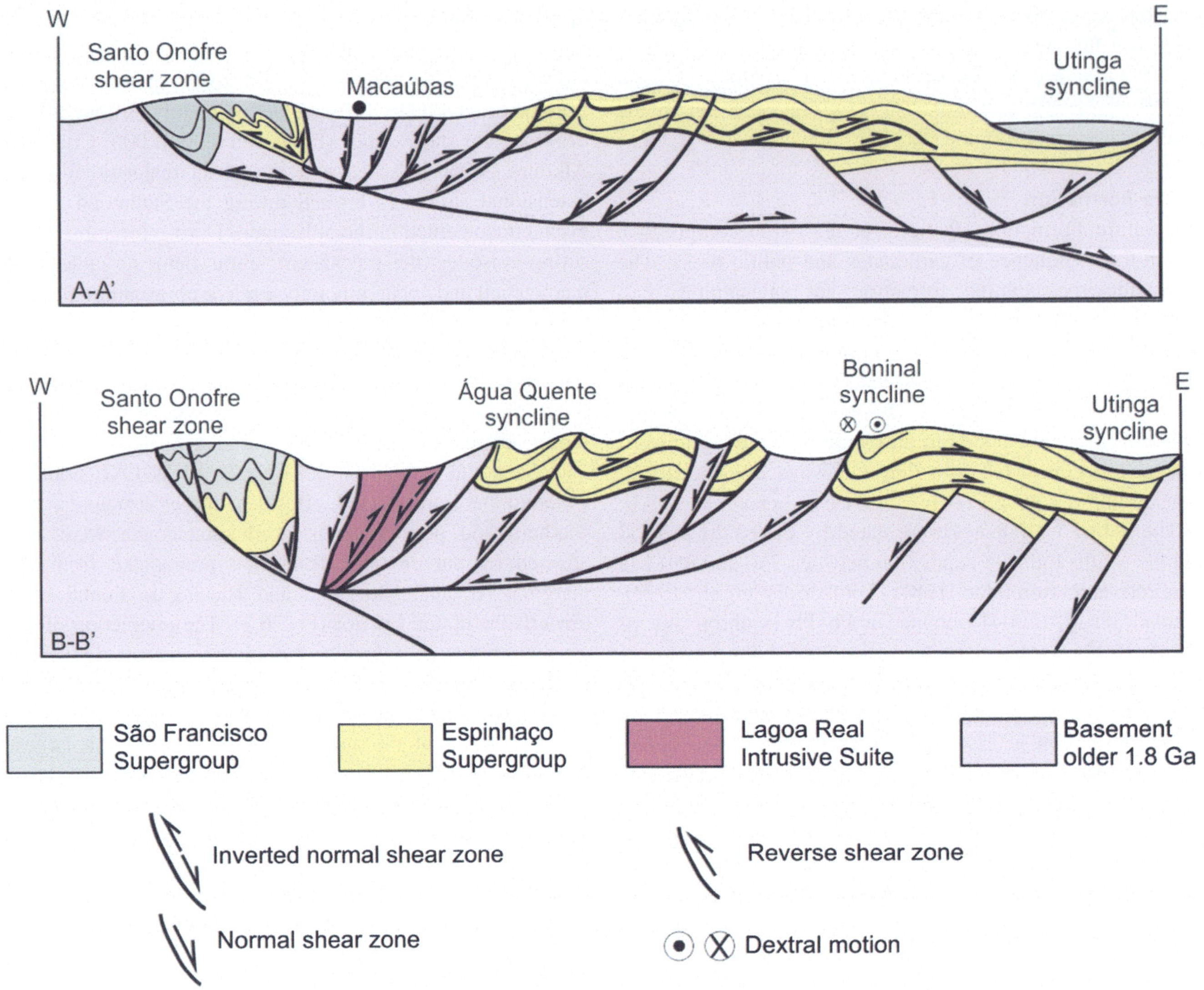

Fig. 6.9 Cross-sections of the central (AA′) and southern portion (BB′) of the Paramirim aulacogen. Their locations are indicated in the map of Fig. 6.3. See text for explanation

Fig. 6.10 Bedding confined deformation structures that characterize the first contractional deformation phase in the southwestern Chapada Dimantina. **a** A ENE-verging intra-stratal duplex structure, in the core of a F_2 anticline. **b** Sigmoildal S_1 foliation, developed around a quartzite pod, indicative of ENE-directed tectonic motion

related to both phases reaches its maximum along the southwestern border of the Chapada Diamantina (Figs. 6.3 and 6.9), where rocks of the Espinhaço Supergroup exhibit a penetrative WSW-dipping cleavage (Danderfer Filho 1990; Cruz and Alkmim 2006; Cruz et al. 2007a, b).

The structures of western half of the Paramirim corridor form a system of NNW-trending shear zones along the northern Espinhaço range (Schobbenhaus 1972; Fernandes et al. 1982; Loureiro et al. 2009; Danderfer Filho 2000). In the central segment of the corridor, this system consists essentially of ENE-dipping normal faults. Experiencing progressive inversion towards south, they are converted into WSW-directed, reverse to reverse-dextral shear zones close to the craton boundary. Likewise, these normal faults are reactivated with reverse dextral motions along northern segment of the corridor (Figs. 6.3 and 6.9; Alkmim et al. 1993). The Espinhaço and São Francisco units that occupy the hanging wall of these shear zones underwent increasing amount of shortening towards north and south, being intensively folded and brought up on top of the basement or younger units (Fig. 6.3).

The age of the aulacogen inversion, a matter intensively debated in the literature, is constrained by the age of the youngest unit involved in process, the Neoproterozoic Salitre Formation, as well as geochronological determinations on mineral phases derived from syn-kinematic metamorphic reactions. Távora et al. (1967) and Guimarães et al. (2008), using the K–Ar and Ar–Ar methods in muscovite extracted from deformed rhyolites of the Novo Horizonte Formation, obtained cooling ages in the interval of 580 and 404 Ma and suggested that their deformation occurred during the Brasiliano event. As previously mentioned, the geochronological determinations by Franz et al. (2014) and Trindade et al. (2004) on rocks of the Espinhaço and São Francisco supergroups also point toward a Neoproterozoic age for the inversion of the aulacogen.

6.4 Tectonic Evolution

Various models have been postulated for the tectonic evolution of the Paramirim Aulacogen (e.g., Jardim de Sá 1981; Costa and Inda 1982; Pedreira 1988; Danderfer Filho 1990, 2000; Dominguez 1993, 1996; Schobbenhaus 1996; Barbosa and Dominguez 1996; Brito-Neves et al. 1996; Cruz and Alkmim 2006; Loureiro et al. 2009; Guimarães et al. 2012). These models agree at least on one aspect of the geologic history of the aulacogen, namely, its origin as an intracontinental rift initiated in the Statherian period. The nature, age, regional correlations and tectonic significance of the aulacogen fill units represent issues on which the models diverge, resulting in different views of the basin evolution in the course of the Proterozoic (see Guimarães et al. 2012). Many of these differences arose as a consequence of lack of data, especially age constraints on the units involved. However, as shown in the previous sections, a considerable number of sedimentological and geochronological studies have been carried out on the basin fill units in recent years (e.g. Loureiro et al. 2009; Pedreira and De Waele 2008; Danderfer Filho et al. 2009; Alkmim and Martins-Neto 2012; Santos et al. 2013; Franz et al. 2014; Guadagnin et al. 2015; Guadagnin and Chemale Jr. 2015), so that a new view of the basin evolution emerges from the new data. According the geochronological framework presently available in the literature, the Paramirim aulacogen hosts at least eight unconformity bounded sequences, which record distinct subsidence pulses that took place during the 1.3 Ga long period that extends from the Statherian up to the and Ediacaran period (Fig. 6.4).

The Espinhaço Supergroup encompasses the Statherian I and II (STI, STII), Calymmian (C), Calymmian-Stenian (CS), and Upper Espinhaço (UE) sequences (Fig. 6.4).

The Statherian I sequence, including the Algodão and Serra da Gameleira formations, records the onset of the Espinhaço sedimentation at ca. 1775 Ga (Danderfer Filho et al. 2015). This first subsidence pulse was followed by renewed extensional tectonism at around 1740 Ma, which led to the accumulation of the Staherian II sequence, represented by the Oliveira dos Brejinhos and Rio dos Remédios groups, as well as alkaline plutons of the Lagoa Real Suite (Schobbenhaus 1996; Danderfer Filho and Dardenne 2002; Guimarães et al. 2012; Loureiro et al. 2009; Cruz and Alkmim 2006; Alkmim and Martins-Neto 2012; Guadagnin and Chemale Jr. 2015; Figs. 6.4 and 6.11). The Statherian sequences also occur in the orogenic belts along the craton margins. Examples are the basal units of Espinhaço Supergroup in its type locality, the southern Espinhaço range in the Araçuaí belt, and the Araí Group in the northern Brasilia belt comprises (Brito-Neves et al. 1996; Alkmim et al. and Valeriano et al. this book).

The Calymmian sequence encompasses the São Marcos and Paraguaçu groups, deposited between 1575 and 1420 Ma, the ages of the basal Bomba Rhyolites and tuffs of the overlying Tombador Formation, respectively (Danderfer Filho et al. 2009; Guadagnin et al. 2015; Guadagnin and Chemale Jr. 2015; Figs. 6.4 and 6.11). The Calymmian-Stenian sequence, whose deposition initiated around 1420 Ma, is represented by the Chapada Diamantina Group.

The Espinhaço basin experienced a considerable expansion during the Calymmian and subsequent periods, starting with the deposition of the Paraguaçu Group and reaching its climax during the eastward advance of the Tombador Formation, at around 1420 Ma. The facies framework and geometry of both the Calymmian and Calymmian- Stenian sequences led the authors working in the region to portray their accumulation during major sagging events (Loureiro

Fig. 6.11 Panoramic view of the hinge zone of the regional NS-trending Pai Ignácio anticline, eastern Chapada Dimantina. The cliffs expose the Calymmian Tombador Sandstone

et al. 2009; Guimarães et al. 2012). However, the acid magmatism recorded by the basal rhyolites (1582 ± 8 and 1569 ± 14 Ma, Danderfer Filho et al. 2009) and mafic intrusions [dated at 1496 ± 3.2 Ma and 1492 ± 16 Ma by Guimarães et al. (2008) and Loureiro et al. (2009)] associated with these sequences suggests that they might have been preceded by a rifting phase not preserved in the aulacogen rock record. The presence of 1580–1500 Ma detrital zircons in the Calymmian-Stenian sequence (Chemale Jr. et al. 2012; Guadagnin and Chemale Jr. 2015) suggests considerable reworking of the preexistent Calymmian sequence.

The Morro do Chapéu Formation (Fig. 6.4) is here tentatively ascribed to the uppermost portion of Espinhaço Supergroup due to the lack of geochronological data and lithological affinities with the underlying units. This unit was deposited in large stuarine channels excavated into the older units.

The São Francisco Supergroup comprises three unconformity bounded sequences, represented by the Santo Onofre Group (LM), as well as the Bebedouro (MM) and Salitre Formations (B) (Fig. 6.4), whose ages are not well constrained. These sequences are traditionally viewed as correlatives of the Macaúbas and Bambuí groups, which occur in the Araçuaí belt and São Francisco basin (see Reis et al. and Alkmim et al. this book).

The Santo Onofre Group seems to record a renewed rifting event (Danderfer Filho 2000) that took place in the Tonian Period (Fig. 6.11), probably coeval to the pre-glacial, rift-related deposition of the lower Macaúbas Group, exposed in the Araçuaí belt along the eastern margin of the craton (Kuchenbecker et al. 2015; Alkmim et al. this book).

The glaciogenic Bebedouro Formation correlates with the middle and most characteristic portion of the Macaúbas Group in the Araçuaí belt, which witnesses the development of a glacially influenced rift basin (Pedrosa-Soares et al. 2011; Uhlein et al. 2007; Kuchenbecker et al. 2015). Other potential correlatives are the Jequitaí and Poço Verde formations, exposed in the intracratonic São Francisco basin and Brasilia belt along western margin of the craton (Reis et al. this book). Attempts to date the glacial deposits and establish its connection with one of the major ice ages of the Cryogenian and Edicaran periods documented worldwide were so far unsuccessful. Two basic interpretations are present in the literature. According to Pedrosa-Soares et al. (2011) and Kuchenbecker et al. (2015), the glacial deposits of the middle Macaúbas Group and correlative units are probably Cryogenian, for its minimal age is constrained by

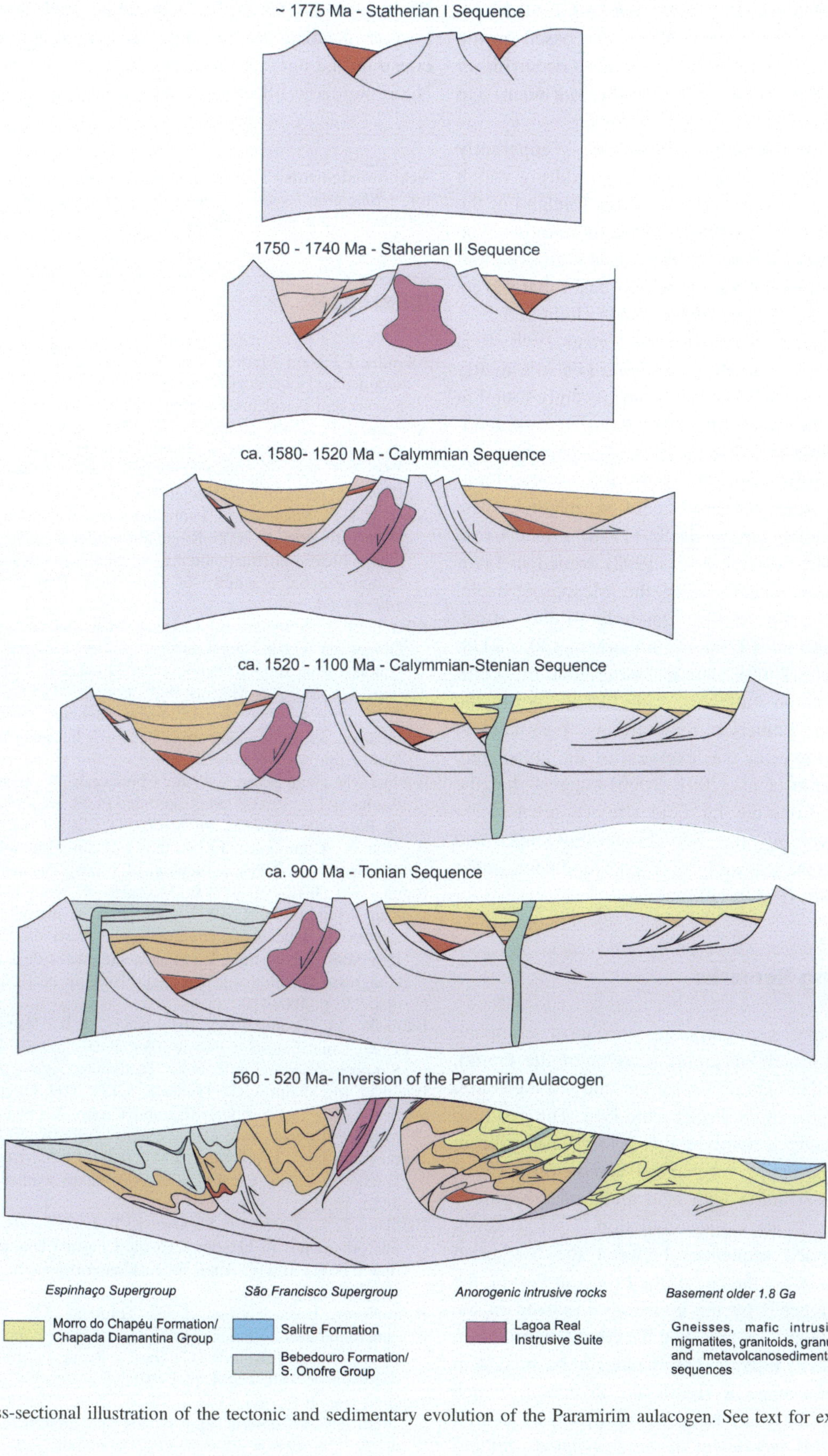

Fig. 6.12 Cross-sectional illustration of the tectonic and sedimentary evolution of the Paramirim aulacogen. See text for explanation

cap carbonates by Babinski et al. (2007) at 740 ± 40 Ma in the São Francisco basin. Caxito et al. (2012), based on the isotopic signature of associated cap carbonates occurring in the Araçuaí belt, argue in favor of a 630 Ma (Marinoan) age for these deposits.

The deposition of the Salitre Formation is apparently associated with a major marine transgression, during which the cratonic area was been converted into the foreland of the various orogenic fronts propagating across its margins. The main correlative unit of the Salitre Formation is the Bambuí Group, the type unit of the São Francisco basin that covers much of the NS-trending lobe of the craton (Inda and Barbosa 1978; Dominguez 1993; Misi and Veizer 1998; Reis et al. this book). The age of Bambuí Group is also a matter of debate in the literature. Ediacaran fossils recently found in Bambuí pelites by Warren et al. (2014) constrain its deposition age at ca. 550 Ma.

As previously mentioned, the inversion of the Paramirim aulacogen, was the subject of a major debate among various authors. Jardim de Sá (1978, 1981), Costa and Inda (1982) and Cordani et al. (1992) argued in favor of a major inversion event during the Mesoproterozoic. Danderfer Filho (1990, 2000), Danderfer Filho et al. (1993), Cruz and Alkmim (2006), Guimarães et al. (2008) and Loureiro et al. (2009), demonstrating that the compressional deformation affected (Fig. 6.11) both the Espinhaço and São Francisco Supergroups, postulate an Edicaran-Cambrian age for the inversion of the aulacogen. Furthermore, Alkmim et al. (2001, 2006) suggest that the inversion resulted from the far field stresses released by the collisions involving the São Francisco-Congo and other plates during the assembly of West Gondwana in the Late Neoproterozoic (Fig. 6.12).

6.5 Concluding Remarks

The Paramirim aulacogen, exposing the most complete section of the deformed Precambrian cover of the craton, constitutes an excellent natural lab for the study of the craton evolution during most of the Proterozoic Eon. The geologic history of the aulacogen involves first a long lasting basin stage, in which alternating mechanical and flexural subsidence pulses in association with acid and basic magmatic events led to the accumulation of seven unconformity-bounded sequences between the Statherian and Edicaran periods. By the end of the Edicaran Period, the orogenic fronts triggered by the collisions involved in the assembly of West Gondwana reached the craton margins and propagated towards its interior. The Paramirim basin system experienced in this stage a significant inversion, which basically involves the fault bounded uplift of its central basement high and development of the associated northern Espinhaço and Chapada Diamantina fold-thrusts systems. Further detailed studies are however required for better constrain the ages of the deposition and deformation of the Neoproterozoic fill units of the aulacogen.

Acknowledgments The authors acknowledge the National Council for Scientific and Technical Development-CNPq grants #308045/2013-0 and 307590/2009-7.

References

Alkmim, F.F. and Martins-Neto, M.A. 2012. Proterozoic first-order sedimentary sequences of the São Francisco craton, eastern Brazil. Marine and Petroleum Geology, 33, 27–139.

Alkmim, F.F., Brito Neves, B.B., Alves, J.A.C. 1993. Arcabouço tectônico do Cráton do São Francisco: uma revisão. In: Dominguez, J.M. and Misi, A. (eds) O Cráton do São Francisco.Sociedade Brasileira de Geologia, Núcleo BA/SE, Salvador, p. 45–62.

Alkmim, F.F., Marshak, S., Pedrosa-Soares, A.C., Peres, G.G., Cruz, S., Whittington, A. 2006. Kinematic evolution of the Araçuaí-West Congo orogen in Brazil and Africa: Nutcracker tectonics during the Neoproterozoic assembly of Gondwana. Precambrian Research, 149, 43–64.

Alkmim, F.F., Marshak, S., Fonseca, M.A. 2001. Assembling West Gondwana in the Neoproterozoic: Clues from the São Francisco craton region, Brazil. Geology, 29, 319–322.

Arcanjo, J.B., Marques-Martins, A.A., Loureiro, H.S.C., Varela, P.H.L. 2005. Projeto Vale do Paramirim, Bahia: geologia e recursos minerais. Salvador, Companhia Baiana de Pesquisa Mineral-CBPM. Série Arquivos Abertos, 22, 82p.

Babinski, M., Van Schmus, W.R., Chemale Jr., F., Brito Neves, B.B., Rocha, A.J.D. 1993. Idade isocrônica Pb/Pb em rochas carbonáticas da Formação Caboclo em Morro do Chapéu. In: Pedreira, A.J., Misi, A., Dominguez, J.M.L. (eds.), II Simpósio sobreo Craton do São Francisco. Brazilian Geological Society, Salvador, p. 160–163.

Babinski, M., Brito-Neves, B.B., Machado, N., Noce, C.M., Ulhein, A., Van Schumus, W.R. 1994. Problemas na metodologia U/Pb em zircões de vulcânicas continentais: o caso do Grupo Rio dos Remédios, Supergrupo Espinhaço, no estado da Bahia. XXXVIII Congresso Brasileiro de Geologia, Boletim de Resumos Expandidos, v.2, p. 409–410.

Babinski, M., Pedreira, A., Brito-Neves, B.B., Van-Schmus, W.R. 1999. Contribuição à geocronologia da Chapada Diamantina. 7º Simpósio Nacional de Estudos Tectônicos, Anais, p. 118–121.

Babinski, M., Vieira, L.C., Trindade, R.I.F. 2007. Direct dating of the Sete Lagoas cap carbonate (Bambuí Group, Brazil) and implications for the Neoproterozoic glacial events. Terra Nova, 19, 401–406.

Barbosa, J.S.F and Dominguez, J.M.L (eds). 1996. Mapa Geológico do Estado da Bahia. Escala: 1.000.000. Texto explicativo, Salvador, 382p.

Battilani, G.A., Gomes, N.S., Guerra, W.J. 2007. The occurrence of microdiamonds in Mesoproterozoic Chapada Diamantina intrusive rocks, Bahia, Brazil. Anais da Academia Brasileira de Ciências, 79 (2), 321–332.

Brito-Neves, B.B., Cordani, U.G., Torquato, J.R. 1980. Evolução geocronológica do Precambriano no estado da Bahia. In: Inda, H.A. D. and Duarte, F.B. (eds.) Geologia e Recursos Minerais do Estado da Bahia, 3, SME-COM, p. 1–101.

Brito-Neves, B.B., Sá, J.M., Nilson, A.A., Botelho, N.F. 1996. A tafrogênese estateriana nos blocos paleoproterozóicos da América do Sul e processos subseqüentes. Geonomos, 3, 1–21.

Caxito, F.A., Halverson, G.P., Uhlein, A., Stevensson, R., Dias, T.G., Uhlein, G.J. 2012. Marinoan glaciation in eastern Central Brazil. Precambrian Research, 203, 38–58.

Chemale Jr. F., Dussin I.A., Alkmim F.F., Martins M.S., Queiroga G., Santos M.N. 2012. Unravelling a Proterozoic basin history through detrital zircon geochronology: The case of the Espinhaço Supergroup, Minas Gerais, Brazil. Gondwana Research, 22, 200–206.

Cordani, U.G., Iyer, S.S., Taylor, P.N., Kawashita, K., Sato, K., Mcreath, I. 1992. Pb-Pb, Rb-Sr, and K-Ar sistematic of the Lagoa Real uranium province (south-central Bahia, Brazil) and the Espinhaço Cycle (ca. 1.5–1.0 Ga). Journal of South American Earth Sciences, 1, 33-46.

Costa L.A.M. and Inda H.A.V. 1982. O Aulacógeno do Espinhaço. Ciâncias da Terra 2, 13–18.

Cruz, S.C.P., Alkmim, F.F. 2006. The Tectonic interaction between the Paramirim Aulacogen and the Araçuaí Belt, São Francisco craton region, Eastern Brazil. Anais da Academia Brasileira de Ciências, 1, 151–173.

Cruz, S.C.P., Dias, V.M., Alkmim, F.F. 2007a. A história de inversão do aulacógeno do Paramirim contada pela sinclinal de Ituaçu, extremo sul da Chapada Diamantina (BA). Revista Brasileira de Geociências., 37(4, suplemento), 92–110.

Cruz, S.C.P., Dias, V.M., Alkmim, F.F. 2007b. A interação tectônica embasamento/cobertura em Aulacógenos invertidos: um exemplo da Chapada Diamantina Ocidental. Revista Brasileira de Geociências, 37(4, suplemento), 111–127.

Cruz, S. C. P., Alkmim, F. F., Leite, C. M. M., Evangelista, H. J. Cunha, J. C., Matos, E. C., Noce, C. M., Marinho, M. M., 2007c. Geologia e arcabouço estrutural do Complexo Lagoa Real, Vale do Paramirim, Centro-Oeste da Bahia. Revista Brasileira de Geociências, 37(4, suplemento), 28–146.

Cruz, S. C. P., Alkmim, F., F., Barbosa, J. S. F., Dussin, I., Corrêa-Gomes, L. C. 2015. Tectonic inversion of compressional structures in the Southern portion of the Paramirim Corridor, Bahia, Brazil. Brazilian Journal of Geology, 4: 541–567.

Danderfer Filho, A. 1990. Análise estrutural descritiva e cinemática do Supergrupo Espinhaço na região da Chapada Diamantina (BA). Unpl. Master Thesis, Universidade Federal de Ouro Preto, Ouro Preto, 169p.

Danderfer Filho, A., DeWaele, B., Pedreira, A., Nalini, H.A. 2009. New geochronological constraints on the geological evolution of Espinhaço basin within the São Francisco Craton-Brazil. Precambrian Research, 170, 116–128.

Danderfer Filho, A., Lagoeiro, L.E., Alkmim, F.F. 1993. O sistema de dobramentos e empurrões da Chapada Diamantina (BA): registro da inversão do Aulacógeno do Espinhaço no decorrer do Evento Brasiliano. In: SBG, Simpósio sobre o Cráton do São Francisco, 2, Anais, p. 197–199.

Danderfer Filho, A. 2000. Geologia sedimentar e evolução tectônica do Espinhaço Setentrional, estado da Bahia. Unpl. PhD Thesis, Universidade Federal de Brasília, Brasilia, 220p.

Danderfer Filho, A. and Dardenne, M.A. 2002. Tectonoestratigrafia da Bacia Espinhaço na porção centro-norte do Cráton do São Francisco: registro de uma evolução poliistórica descontínua. Revista Brasileira de Geociências, 4, 449–460.

Danderfer Filho, A., Lana, C., Nalini-Júnior, H.A., Costa, A.F.O. 2015. Constraints on the Statherian evolution of the intraplate rifting in a Paleo-Mesoproterozoic paleocontinent: New stratigraphic and geochronology record from the eastern São Francisco craton. Gondwana Research, 28, 668–688.

Dominguez, J.M.L. (1993). As coberturas do Cráton do São Francisco: Uma abordagem do ponto de vista da análise de bacias. In: Dominguez, J.M.L. and Barbosa, J.S.F. (eds) O Cráton do São Francisco, SGM, pp. 137–155.

Dominguez, J.M.L. 1996. As Coberturas Plataformais do Proterozóico Médio e Superior. In: Barbosa J.S.F.and Dominguez, J.M.L. (eds). Mapa Geológico do Estado da Bahia, Texto Explicativo, pp. 109–112.

Fernandes, P.E.C.A., Montes, M.L., Braz, E.R.C., Silva, L.L., Oliveira, F.L.L., Ghignone, J.I., Siga Jr., O., Castro, H.E.F. 1982. Geologia. In: BRASIL. Ministério das Minas e Energia. Secretaria Geral. Projeto RADAMBRASIL. Folha SD.23 Brasília: geologia, geomorfologia, pedologia, vegetação e uso potencial da terra. Rio de Janeiro, p. 25–204.

Fleischer, R. 1971. Observações geológicas sobre a dumortierita da Serra das Veredas – Bahia. Mineração e Metalurgia, 54, 21–24.

Franz, G.,Morteani, G., Gerdes, A., Rhede, D. 2014. Ages of protolith and Neoproterozoic metamorphism of Al-P-bearing quartzites of the Veredas Formation (Northern Espinhaço, Brazil): LA-ICP-MS age determinations on relict and recrystallized zircon and geodynamic consequences. Precambrian Research, 250, 6–26.

Guadagnin F., Chemale Jr. F., Magalhães A.J., Santana A., Dussin I., Takehara, L. 2015 Age constraints on crystal-tuff from the Espinhaço Supergoup – Insight into the Paleoproterozoic to Mesoproterzoic intracratonic basin cycles of the São Francisco craton. Gondwana Research, 27, 363–376.

Guadagnin, F. and Chemale Jr. F. 2015. Detrital zircon record of the Paleoproterozoic to Mesoproterozoic cratonic basins in the São Francisco Craton. Journal of South American Earth Sciences, 60, 104–116.

Guimarães, J.T. 1996. A Formação Bebedouro no Estado da Bahia: Faciologia, Estratigrafia e Ambiente de Sedimentação. Unpl. Master Thesis, Instituto de Geociências, Universidade Federal da Bahia, Salvador, 155p.

Guimarães J.T.; Santos R.A.S.; Melo R.C. (Orgs.). 2008. Geologia da Chapada Diamantina Ocidental. In: Guimarães J.T., Santos R.A., Melo R.C. Projeto Ibitiara – Rio de Contas. Companhia Baiana de Pesquisa Mineral-CBPM, Salvador, Arquivos Abertos 31, 64 p.

Guimarães, J.T., Alkmim, F.F., Cruz, S.C.P. 2012. Supergrupos Espinhaço e São Francisco. In: Barbosa, J.S.F., Mascarenhas, J.F. M., Corrêa-Gomes, L.C., Domingues, J.M.L. (eds.). Geologia da Bahia. Pesquisa e Atualização de Dados. Salvador, Companhia Baiana de Pesquisa Mineral-CBPM, v. 2, p. 33–86.

Inda, H.A.V., Barbosa, J.F., 1978. Texto Explicativo para o Mapa Geológico do Estado da Bahia, SME/COM. 137 p.

Jardim de Sá, E.F. 1978. Tectônica de placas vertical em ambientes intracratônicos: considerações a partir do Proterozóico médio do Cráton do São Francisco. In: 30th Congresso Brasileiro de Geologia, Anais, p. 291.

Jardim de Sá, E.F. 1981. A Chapada Diamantina e a faixa Santo Onofre: Um exemplo de tectônica intraplaca no Proterozóico Médio do Cráton São Francisco. In: Inda, H.A.A.V., Marinho, M.M., Duarte, F.B. (eds) Geologia e Recursos Minerais do Estado da Bahia, Textos Básicos, 4, SME/COM, pp. 111–120.

Jordt-Evangelista, H. and Danderfer-Filho, A. 2012. Quartzito azul com dumortierita e fosfatos de alumínio do Espinhaço setentrional, Bahia: mineralogia e petrogênese. Revista Brasileira de Geociências, 42, 363–372.

Kuchenbecker, M., Pedrosa-Soares, A.C., Babinski, M., Fanning, M. 2015. Detrital zircon age patterns and provenance assessment for pre-glacial to post-glacial successions of the Neoproterozoic Macaúbas Group, Araçuaí orogen, Brazil. Precambrian Research 266, 12–26.

Leão, Z.M.A.N., Dominguez, J.M.L., 1992. Plataformas carbonáticas precambrianas: o exemplo da Formação Salitre, Proterozóico Superior, Estado da Bahia, 37th Congresso Brasileiro de Geologia, Anais, p. 45–452.

Leão, Z.M.A.N., Dominguez, J.M.L., Camargo, S.L. 1992. Sedimentação carbonática marinha rasa no Pré-Cambriano: sobre a validade de aplicação de modelos de fácies desenvolvidos para o Fanerozóico. Simpósio de Geologia de Minas Gerais, 4, Anais, p. 103–104.

Lobato, L.M., Pimentel, M.M., Cruz, S.C.P., Machado, N., Noce, C.M., Alkmim, F.F. 2015. Geochronology of the Lagoa Real uranium district, Brazil: Implications for the age of the uranium mineralization. Journal South American Earth Sciences, 58, 129–140.

Loureiro H.S.C., Bahiense I.C., Neves J.P., Guimarães J.T., Teixeira L. R., Santos R.A., Melo R.C. (Org.). 2009. Geologia e recursos minerais da parte norte do corredor de deformação do Paramirim: (Projeto Barra - Oliveira dos Brejinhos). Companhia Baiana de Pesquisa Mineral-CBPM, Salvador, Série Arquivos Abertos 33, 113 p.

Macedo, M.H., Bonhome, M.G. 1984. Contribuição a cronoestratigrafia das Formações Caboclo, Bebedouro e Salitre na Chapada Diamantina (BA) pelos métodos RB-Sr e K-Ar. Revista Brasileira de Geociências, 3, 153–163.

Macedo, M.H.F., Bonhome, M.G. 1981. Datação Rb-Sr e K-Ar das formações Bebedouro e Caboclo na Chapada Diamantina. 1º Simpósio sobre o Cráton do São Francisco, Anais, p. 98–99.

Magalhães, A.J.C., Scherer, C.M.S., Raja Gabaglia, G.P., Bállico, M. B., Catuneanu, O. 2014. Unincised fluvial and tide-dominated estuarine systems from the Meso-proterozoic Lower Tombador Formation, Chapada Diamantina basin, Brazil. Journal South American Earth Science, 56, 68–90.

Magalhães, A.J.C., Raja Gabaglia, G.P., Scherer, C.M.S., Bállico, M. B., Guadagnin, F., Bento Freire, E., Silva Born, L.R., Catuneanu, O. 2016. Sequence hierarchy ina Mesoproterozoic interior sag basin: from basin fill to reservoir scale, the Tombador Formation, Chapada Diamantina Basin, Brazil. Basin Research, 26, 393–432

Maruèjol, P., Cuney, M., Fuzikawa, K., Maria Netto, A., Poty, B. 1987. The Lagoa Real Subalkaline Granitic Complex (South Bahia, Brazil): A Source for Uraniun Mineralizations Associated With Na-Ca Metassomatism. Revista Brasileira de Geociências, 4: 578–594.

McReath, I., Jardim de Sá, E.F., Fryer, B.J. 1981. As vulcânicas ácidas da região da Bacia do rio Paramirim-Bahia. In: Inda, H.A.V. Marinho, M.M., Duarte, F.P. (eds) Geologia e Recursos Minerais do Estado da Bahia, Textos Básicos, 4, SME/COM, pp. 120–130.

Misi, A. 1979. O Grupo Bambuí no Estado da Bahia. In: Inda, H. A. V. (eds). Geologia e Recursos Minerais do Estado da Bahia. Textos Básicos. Salvador, SME, v.1, p. 119–154.

Misi, A. 1993. A sedimentação carbonática do Proterozóico superior no Cráton do São Francisco: Evolução diagenética e estratigráfica-isotópica. In: SBG, Simpósio sobre o Cráton do São Francisco, 2, Anais, p. 192–194.

Misi, A., Veizer, J., 1998. Neoproterozoic carbonate sequences of the Una Group, Irec^e Basin Brazil: chemostratigraphy, age and correlations. Precambrian Res. 89, 87–100.

Montes, A.D.L., Gravenor, C. P., Montes, M. L. 1985. Glacial sedimentation in the Late Precambrian Bebedouro Formation, Bahia, Brazil. Sedimentary Geology, 44, 349–358.

Pedreira, A.J.C.L., Arcanjo, J. B., Pedrosa, C.J., Oliveira, J.E., Silva, B. C.E. 1975. Projeto Bahia - Geologia da Chapada Diamantina, Relatório Final. DNPM/CPRM, 225 p.

Pedreira, A.J.C.L. 1988. Seqüências deposicionais no Pré-Cambriano: exemplo da Chapada Diamantina Oriental, Bahia. In: SBG, Congresso Brasileiro de Geologia, 33, Anais, p. 648–659.

Pedreira, A.J.C.L. 1997. O limite Meso-Neoproterozóico na região central da Bahia. In: SBG/NMG, Simpósio de Geologia de Minas Gerais, 9, Anais, p. 1–2.

Pedreira, A.J. and De Waele, B. 2008. Contemporaneous evolution of the Palaeoproterozoic-Mesoproterozoic sedimentary basins of the São Francisco-Congo Craton. In: Pankhurst, R.J. Trouw, R.A.J., Brito Neves, B.B., De Wit, M.J (Ed), West Gondwana: Pre-Cenozoic Correlations Across the South Atlantic Region. Geol. Soc. Spec. Publ., 294, 33–48.

Pedrosa-Soares, A.C.; Noce, C.M.; Wiedemann, C.M., Pinto, C. P. 2001. The Araçuaí-West-Congo Orogen in Brazil: An overview of a confined orogen formed during Gondwanaland assembly. Precambrian Research, 110, 307–323.

Pedrosa-Soares, A.C., Babinski, M., Noce, C.M., Martins, M., Queiroga, G., Vilela, F. 2011. The Neoproterozoic Macaúbas Group (Araçuaí orogen, SE Brazil) with emphasis on the diamictite formations. In: Arnaud, E., Halverson, G.P., Shields-Zhou, G. (Org.). The Geological Record of Neoproterozoic Glaciations. Memoir of the Geological Society of London 36, 523–534.

Pedrosa-Soares, A. C. and Alkmim, F. F. 2011. How many rifting events preceded the development of the Araçuaí-West Congo orogen? Geonomos, 12, 244–251.

Pimentel, M.M., Machado, N., Lobato, L.M. 1994. Geocronologia U/Pb de rochas graníticas e gnáissicas da região de Lagoa Real, Bahia, e implicações para a idade da mineralização de urânio. In: SBG, Congresso Brasileiro de Geologia, 38, Boletin de Resumos Expandidos, p. 389–390.

Santos, M.N., Chemale Jr., F., Dussin, I.A., Martins, M., Assis, T.A.R., Jelinek, A.R., Guadagnin, F., Armstrong, R. 2013. Sedimentological and paleoenvironmental constraints of the Statherian and Stenian Espinhaço rift system, Brazil. Sedimentary Geology, 290, 47–59.

Schobbenhaus C. 1972. Relatório geral sobre a geologia da região setentrional da Serra do Espinhaço - Bahia Central. Recife, SUDENE/DRN/DG, 91 p. (Série Geologia Regional 19).

Schobbenhaus C., Kaul, P.T. 1971. Contribuição à Estratigrafia da Chapada Diamantina Bahia-Central. Mineração e Metalurgia, 53 (315): 116–120.

Schobbenhaus, C. 1996. As tafrogêneses superpostas Espinhaço e Santo Onofre, estado da Bahia: Revisão e novas propostas. Revista Brasileira de Geociências, 4: 265–276.

Schobbenhaus, C., Hoppe, A., Baumann, A., Lork, A. 1994. Idade U/Pb do vulcanismo Rio dos Remédios, Chapada Diamantina, Bahia. In: SBG, Congresso Brasileiro de Geologia, 38, Anais, 2, p. 397–399.

Silveira, E.M., Söderlund, U., Oliveira, E.P., Ernst, R.E., Menezes Leal, A.B. 2013. First precise U-Pb baddeleyite ages of 1500 Ma mafic dykes from the São Francisco Craton, Brazil, and tectonic implications. Lithos 174: 144 –156.

Távora, F.J., Cordani, U.G., Kawashita, K. 1967. Determinações de idades potássio-argônio em rochas da região central da Bahia. In: SBG, Congresso Brasileiro de Geologia, 21, Anais, p. 234–244.

Teixeira, L.R. 2000. Projeto Vale do Paramirim. Relatório Temático de Litogeoquímica. Convênio CPRM/CBPM, 35p.

Teixeira, L. 2005. Projeto Barra-Oliveira dos Brejinhos. Relatório Temático de Litogeoquímica. Convênio CPRM/CBPM, 29p.

Toulkeridis T., Babinski M., Buchwaldt R., Brito Neves B.B., Todt W., Santos R. 1999. Are varangian or sturtian the glacial deposits on the São Francisco cráton? Evidence from determination of sedimentary rocks and minerals of the Neoproterozoic Una Group. In: South American Symposium on Isotope Geology, 2, Cordoba, Anais, p. 453–456.

Trindade, R.I. F., D´Agrella-Filho, M.S., Babinski, M., Font, E., Brito Neves, B.B. 2004. Paleomagnetism and geochronology of the Bebebedouro Cap Carbonate: Evidence for continental-scale Cambrian remagnetization in the São Francisco craton, Brazil. Precambrian Research, 128, 83–103.

Uhlein, A., Caxito, F.A.; Egidio, M., Barbosa, J.S.F. 2012. Faixa de Dobramentos Rio Preto e Riacho do Pontal. In: Johildo Salomão Figueirêdo Barbosa, Juracy de Freitas Mascarenhas, Luiz Cesar Correa-Gomes, José Maria Landim Dominguez, Jaiilma Santos de Souza. (eds.) Geologia da Bahia. Pesquisa e Atualização. 1ed. Salvador, Bahia: CBPM-Companhia Baiana de Pesquisa Mineral, v. 2, p. 87–130.

Uhlein, A., Trompette, R., Egydio-Silva, M. and Vauchez, A. 2007. A glaciação Sturtiana (~750 Ma), a estrutura do rifte Macaúbas--Santo Onofre e a estratigrafia do Grupo Macaúbas, Faixa Araçuaí. Geonomos, 15, 45–60.

Warren LV, Quaglio F, Riccomini C, Simões MG, Poiré DG, Strikis NM, Aneli LE, Strikis, P.C. 2014. The puzzle assembled: Ediacaran guide fossil Cloudina reveals an old proto-Gondwana seaway. Geology, 5, 391–394.

The São Francisco Basin

7

Humberto L.S. Reis, Fernando F. Alkmim, Renato C.S. Fonseca, Thiago C. Nascimento, João F. Suss, and Lúcio D. Prevatti

7.1 Introduction

As a consequence of their longevity and strength to survive recycling events (O'Reilly et al. 2001; King 2005; Neill et al. 2008), cratons stand out as complex tectonic domains that have witnessed the evolutionary history of the continental lithosphere. Thanks to that, intracratonic basins, as poly-historic depocenters, generally host important records of the trajectory of the differentiated cratonic lithosphere through geological time (e.g.: Korsch and Lindsay 1989; Kaminski and Jaupart 2000; Spalletti and Limarino 2006; Kadima et al. 2011).

Focus of various studies since the nineteenth century (e.g., Derby 1879; Rimann 1917; Costa and Branco 1961; Braun 1968; Pflug and Renger 1973; Schöll 1973, Dardenne 1978, 1981; Menezes-Filho et al. 1977; Alkmim et al. 1993; Alkmim and Martins-Neto 2001, 2012; Martins-Neto 2009; Martins-Neto et al. 2001; Pimentel et al. 2011), the São Francisco basin covers a substantial portion of the São Francisco craton (Fig. 7.1) and exhibits typical attributes of intracratonic basins. Its sedimentary succession encompasses multiple and superimposed basin-fill cycles younger than 1.8 Ga., which reflect tectonic and climatic events—some of global significance—that have affected the São Francisco-Congo lithosphere after the Paleoproterozic Era (e.g., Campos and Dardenne 1997a, b; Martins-Neto 2009; Alkmim and Martins-Neto 2001; Alkmim et al. 2011; Babinski et al. 2012; Caxito et al. 2012).

Many aspects of the São Francisco basin geology are still poorly understood. The lack of good geochronological constraints hampers regional stratigraphic correlations and detailed palinspatic and paleoenvironmental reconstructions of the fill units. Moreover, the paucity of available sub-surface data has limited the studies to a few seismic lines, drill cores, and relatively scarce exposures of good quality (e.g., Coelho et al. 2008; Zalán and Romeiro-Silva 2007; Hercos 2008; Martins-Neto 2009; Alkmim and Martins-Neto 2012; Reis 2011). Nevertheless, the recent outset of hydrocarbon exploration programs and the resumption of regional mapping programs, aerogeophysical surveys, geochronological and geochemical studies are changing the scenario of São Francisco basin knowledge (e.g., Santos et al. 2000; Babinski et al. 2007; Vieira et al. 2007; Rodrigues et al. 2010, 2012; Kuchenbecker 2011; Pedrosa-Soares and Alkmim 2011; Pimentel et al. 2011; Alvarenga 2012, 2014; Lopes 2012; Reis et al. 2012, 2013).

In this chapter, we present a synthesis on the stratigraphic and tectonic framework of the São Francisco basin. Our work integrates numerous available studies performed in the region during the last decades and includes new data recently acquired during the renewal of the hydrocarbon exploration programs in the basin. Special emphasis is given on the correlation between the basin fill units and the main tectonic and climatic events affecting the São Francisco craton region during Precambrian and Phanerozoic times (e.g., Brito-Neves et al. 1999; Pedrosa-Soares et al. 2001; Alkmim and Martins-Neto 2001; Valeriano et al. 2004a, b). Our basic

H.L.S. Reis, J.F. Suss, R.C.S. Fonseca, T.C. Nascimento and L.D. Prevatti are Formerly Petra Energia S.A, Alameda Oscar Niemeyer, 119- Vila da Serra, 34.000-000 Nova Lima, MG, Brazil

H.L.S. Reis (✉)
Laboratório de Estudos Tectônicos (LESTE)/Centro de Geociências, Instituto de Ciência e Tecnologia, Universidade Federal dos Vales do Jequitinhonha e Mucuri, Campus Juscelino Kubitscheck, Diamantina, MG 39.100-000, Brazil
e-mail: humbertosiqueira@gmail.com; humberto.reis@ict.ufvjm.edu.br

F.F. Alkmim
Departamento de Geologia, Escola de Minas, Universidade Federal de Ouro Preto, Morro Do Cruzeiro, Ouro Preto, MG 35.400-000, Brazil

R.C.S. Fonseca · J.F. Suss
Interpetro Consultoria em Geologia Ltda., Rua Passa Tempo, 171-401, Carmo, Belo Horizonte, MG 30310-760, Brazil

T.C. Nascimento
IHS Energy, Rua Acre, Centro, 15, 7th floor, Rio de Janeiro, RJ 20.081-000, Brazil

L.D. Prevatti
Shell Brasil Petróleo Ltda., Av. das Americas, 4200, Bl 5 and 6, Barra da Tijuca, Rio de Janeiro, RJ 22.640-102, Brazil

M. Heilbron et al. (eds.), *São Francisco Craton, Eastern Brazil*, Regional Geology Reviews, DOI 10.1007/978-3-319-01715-0_7

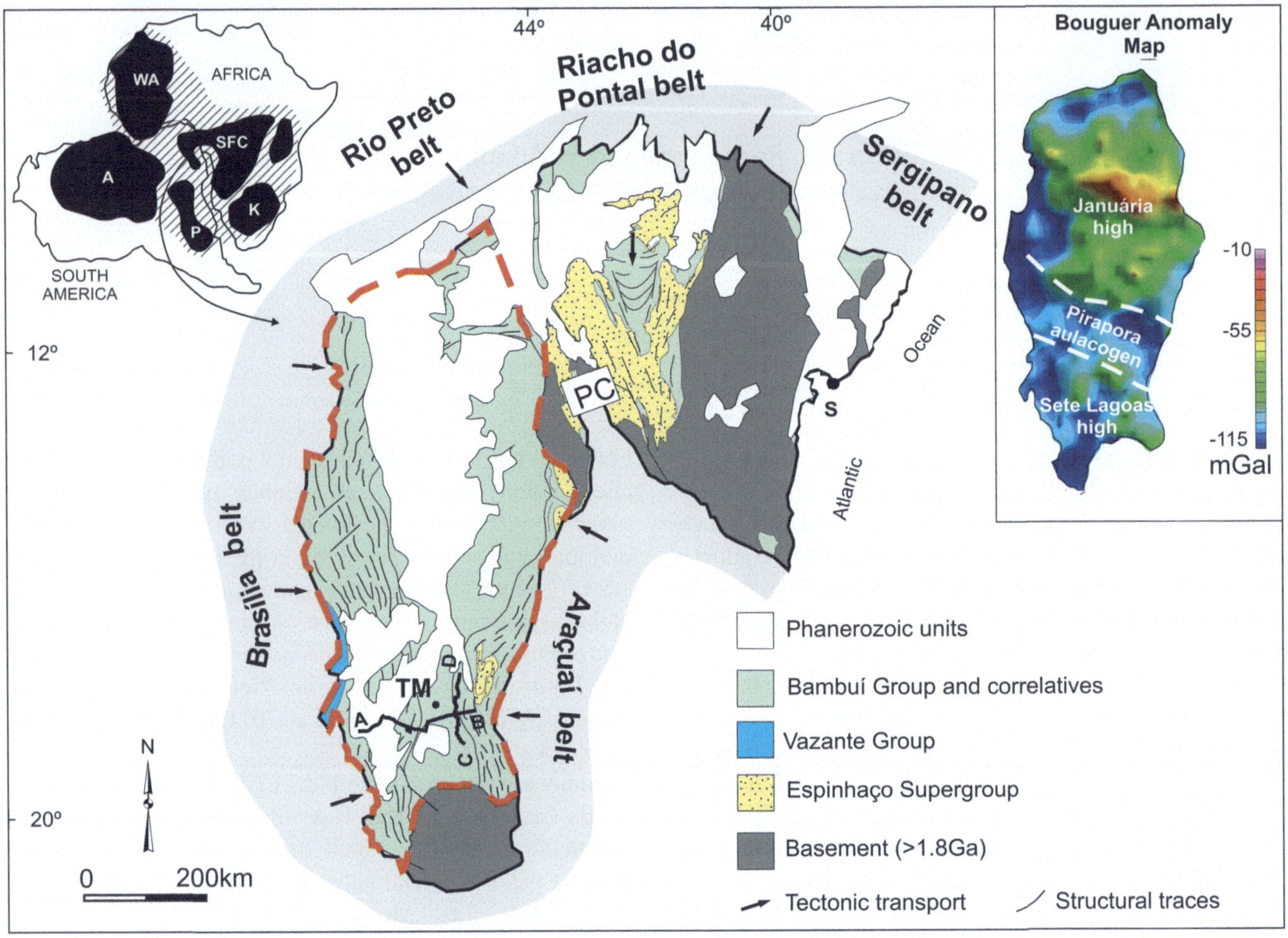

Fig. 7.1 Simplified geologic map of the São Francisco craton and the São Francisco basin (*red dashed line*) (Based on Alkmim and Martins-Neto 2001). The Bouguer anomaly map on the right shows the main basement structures of the basin (Reis 2011). The *black lines* on the geologic map indicate the location of seismic cross-sections shown on Fig. 7.4. *PC* Paramirim Corridor. West Gondwana cratons: *A* Amazonian, *P* Paranapanema, *WA* West Africa, *SFC* São Francisco-Congo, *K* Kalahari. Cities: *TM* Três Marias, *S* Salvador. Reproduced with minor modifications from Reis and Alkmim (2015), with permission from Elsevier

conclusion is that the history of São Francisco craton can be tracked not only by basement structures and surrounding orogenic belts, but also by its intracratonic cover assemblages. Additionally, the São Francisco basin rock record offers a new perspective to better understand the evolution of the adjacent orogenic belts, where metamorphism and intense deformation often overprint primary features.

7.2 Definition

The São Francisco basin occupies most of the NS-trending portion of the craton (Fig. 7.1), extending over an area of ca. 350.000 km^2 in the southeastern Brazilian highlands (Alkmim and Martins-Neto 2001). The eastern, western and northern limits of the basin, coinciding with the cratonic boundaries (see Cordani et al., this book), are given by structures of the Ediacaran Araçuaí, Brasília and Rio Preto marginal belts, respectively. The southern limit is erosional. To the northeast, the basin is bounded by the Paramirim Corridor, an Ediacaran intracratonic deformation zone that affects the neighboring Paramirim aulacogen (Cruz and Alkmim 2006 and this book) (Fig. 7.1).

According to recent drilled wells, the geothermal gradients measured in the basin area are on average 17 °C/km. As demonstrated by Rocha et al. (2011) and Assumpção et al. (this book), the basin is locally underlain by a thick cratonic lithosphere (ca. 200 km), thus corresponding to a typical cold intracratonic basin. Gravimetric inversion based on satellite surveys indicates a crust-mantle boundary at 40–45 km depths, with the deepest portions located in the basin center and adjacent to the Brasília belt, on the west (Oliveira 2009; Oliveira et al. 2012). This distribution agrees with the regional Bouguer anomalies observed in the basin area (Fig. 7.1) and coincides with the major tectonic features described further in this chapter.

Other definitions, and consequently, different delimitations for the São Francisco basin can be also found in the Brazilian geological literature (see discussion in Martins-Neto and Pinto 2001; Alkmim and Martins-Neto 2001). For instance, Domingues (1993) and Iyer et al. (1995) portrayed the São Francisco basin as the deposition site of the Neoproterozoic São Francisco Supergroup, an area that also encompasses the craton margins. The basin delimitation adopted here is quite simple and found broad acceptance in the current Brazilian petroleum industry and geological literature.

7.3 Stratigraphy

Based on the concept postulated by Catuneanu et al. (2005, 2011, 2012), at least three main Precambrian 1st-order sequences can be recognized in the São Francisco basin: (i) the Mesoproterozoic to Early Neoproterozoic Paranoá-Upper Espinhaço sequence; (ii) the Neoproterozoic Macaúbas sequence; and (iii) the Ediacaran Bambuí sequence. These successions overlie the Archean/Paleoproterozoic basement and an apparently older (unknown) Precambrian succession not exposed in the basin area and recognized on seismic sections only. The Proterozoic 1st-order sequences are locally deformed and uncomfortably overlain by discontinuous Phanerozoic strata, which comprise the late Paleozoic Santa Fé Group, as well as the Cretaceous Areado, Mata da Corda and Urucuia groups (Campos and Dardenne 1997a, b; Sgarbi et al. 2001; Fragoso 2011; Sgarbi 2011a, b). The São Francisco basin sedimentary record tracks major plate reorganizations and global events that affected the São Francisco craton and its margins in the time between the end of the Paleoproterozoic and the Upper Cretaceous (Figs. 7.2 and 7.3).

The correspondence between the stratigraphic subdivision adopted in this work and the lithostratigraphic units traditionally mapped in the basin over the last years is shown on Fig. 7.3. This figure also indicates the correlatives of these units in the orogenic belts that fringe the São Francisco craton.

7.3.1 Basement

The basement of the São Francisco basin comprises all units older than 1.8 Ga that forms a coherent NS-trending block in the western portion of the craton (Teixeira et al., this book). These units include Archean TTG-complexes (migmatites and gneisses), granitoid plutons, and greenstone belt successions, as well as Paleoproterozoic igneous and metasedimentary rocks (Dorr 1969; Alkmim and Marshak 1989; Heineck et al. 2003; Baltazar and Zucchetti 2007; Noce et al. 2007; Pinho 2008; Lana et al. 2013; Romano et al. 2013). Exposures of these assemblages occur close to the southern boundary of the basin and in relatively small statigraphic windows in the Sete Lagoas and Januaria highs, which represent major basement structures of the southern and northern portions of the basin, respectively (Fig. 7.1).

7.3.2 Paranoá-Upper Espinhaço 1st-Order Sequence

The Mesoproterozoic to Early Neoproterozoic Paranoá-Upper Espinhaço sequence represents a rift-sag basin-fill succession, which is laterally associated with deposits of a rift-passive margin basin, developed along the western boundary of the São Francisco plate. The sequence consists of a siliciclastic-dominated package that grades upward into marine to transitional pelites, sandstones and carbonates. Exposed in relatively small areas of the basin, it includes the sediments of the Paranoá Group, on the west, and the middle to upper Espinhaço Supergroup strata, on the east. Detected in various seismic sections, these units thicken westwards and within the NW–SE trending Pirapora aulacogen, a prominent structure that cuts across the basement in the central portion of the basin. In the interior of the aulacogen, the Paranoá-Upper Espinhaço sequence exhibits thickness of few kilometers and a typical steer-head geometry, overlying an apparently older and unknown Proterozoic (?) succession (Figs. 7.2 and 7.4). This succession is, however, not exposed in the basin area and more studies are needed to better constrain its age and tectono-stratigraphic significance.

The Paranoá-Upper Espinhaço sequence can be subdivided into two 2nd-order sequences: a lower rift and an upper sag/passive margin (Fig. 7.4). The rift 2nd-order sequence mainly occurs in the interior of the Pirapora aulacogen and along the western sector of the basin, where its thickness is controlled by sets of large-scale normal faults.

Few available well data from the southeastern portion of the basin reveals that the basal rift 2nd-order sequence contains poorly-sorted fluvial to deltaic sandstones, arkoses and conglomerates, mainly arranged as metric to decametric fining-upward cycles. These sediments are intruded by mafic sills and dykes and commonly contain syn-tectonic growth strata.

The rift-related units are apparently overlain by a thick package of aeolian sandstones (the Galho do Miguel Formation of the Espinhaço Supergoup), whose accumulation marks the onset of a transitional to thermally driven subsidence stage in the basin (Dupont 1995; Martins-Neto et al. 2001; Lopes 2012). This subsidence phase culminated with a marine incursion, which is recorded by the upper sag/passive margin 2nd-order sequence. This sequence can be subdivided into three 3rd-order transgressive-regressive sequences along the eastern half of the basin. These lower-rank sequences correspond to the siliciclastic and carbonatic

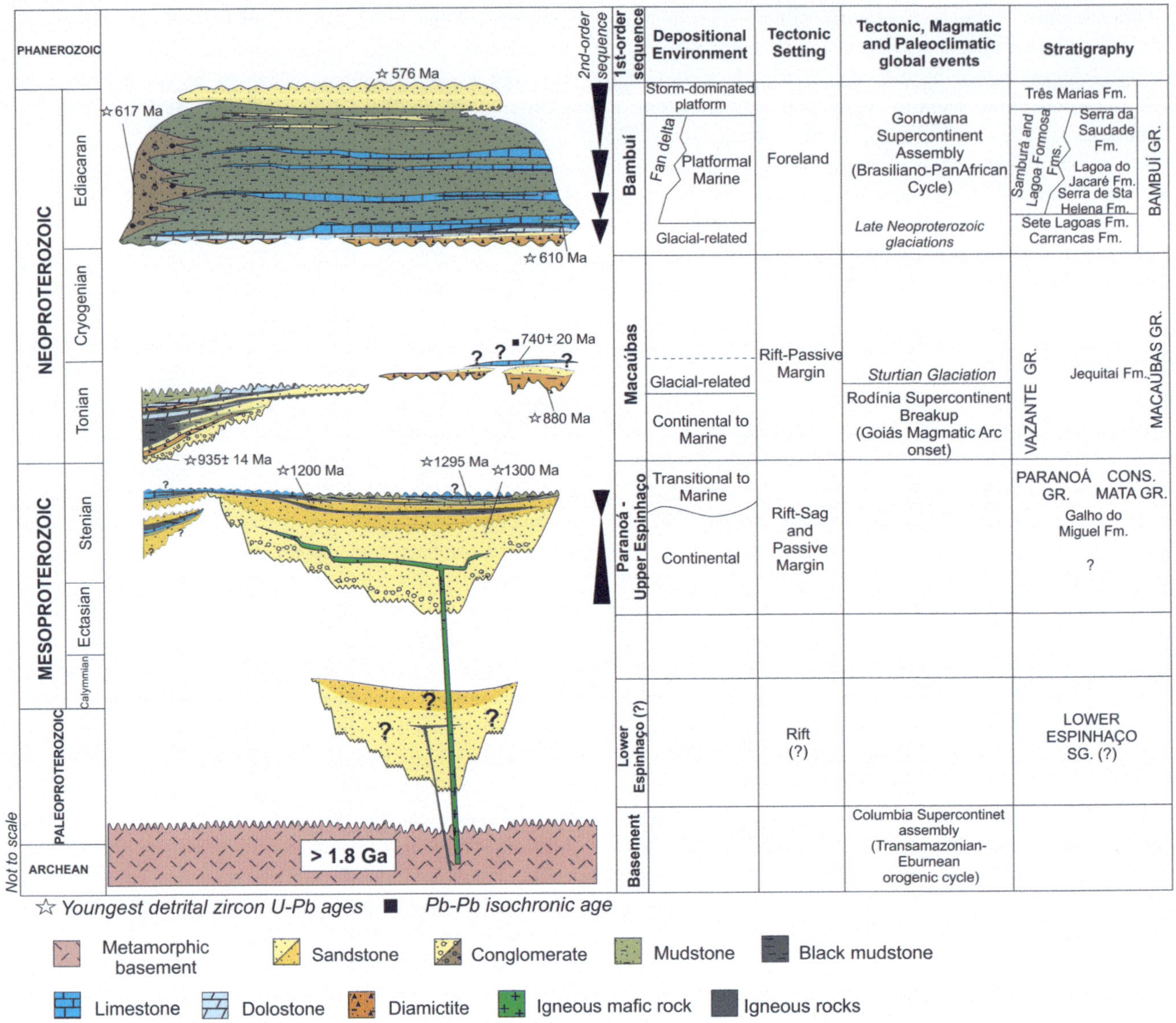

Fig. 7.2 Stratigraphic chart for the Precambrian section of the São Francisco basin. See text for explanations and references. Reproduced with minor modifications from Reis and Alkmim (2015), with permission from Elsevier

deposits of the Conselheiro Mata Group of the upper Espinhaço Supergroup (Dupont 1995; Lopes 2012) (Fig. 7.5).

In the western sector of the basin, the passive margin 2nd-order sequence also encompasses multiple 3rd-order sequences (Martins-Neto 2009). Mainly exposed along narrow deformed areas (e.g.: Serra de São Domingos area), these successions mostly consist of tidal- to storm-dominated marine and transitional siliciclastics, locally associated to platformal and stromatolite-bearing carbonates (Martins-Neto 2009; Feitosa 2012; Alvarenga et al. 2012) (Fig. 7.6).

The correlatives of the Paranoá-Upper Espinhaço sequence in the Brasília metamorphic belt are the marine packages of the Paranoá and Canastra groups, which consist of passive margin fine to coarse-grained siliciclastics, organic matter-rich shales, and carbonates with cyanobacteria mats and columnar stromatolites (Dardenne 1978, 1981, 2000; Martins-Neto 2009; Campos et al. 2013). Provenance studies carried out on these rocks indicated a sedimentary supply mainly derived from the cratonic area, with important Archean and Paleoproterozoic sources (Valeriano et al. 2004a, b; Pimentel et al. 2001; Rodrigues et al. 2012). U-Pb ages determinations on detrital zircons

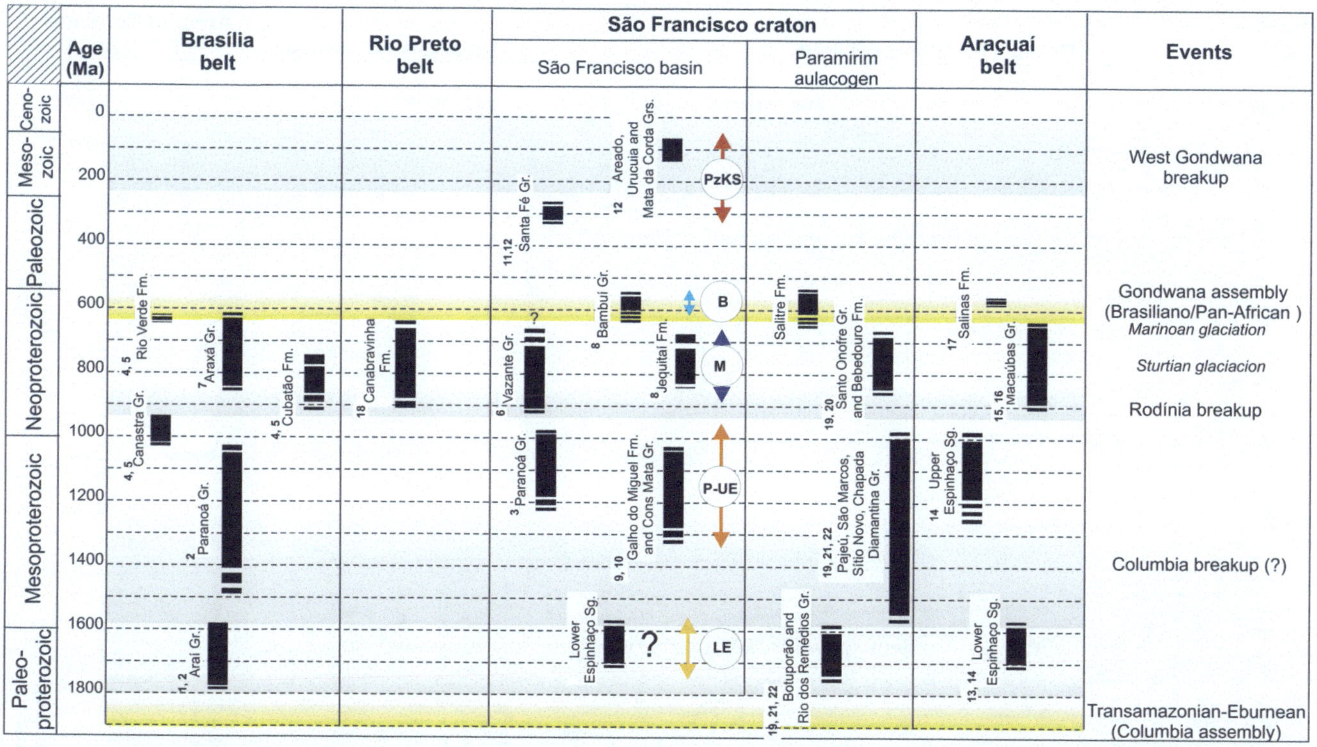

Fig. 7.3 Chronoestratigraphic correlation chart for the São Francisco basin fill successions and units exposed in the Paramirim aulacogen (northern São Francisco craton) and Neoproterozoic marginal belts. References: (1) Pimentel et al. (1999), (2) Matteini et al. (2012), (3) Alvarenga (2012), (4) Rodrigues et al. (2010), (5) Dias et al. (2011), (6) Rodrigues et al. (2012), (7) Valeriano et al. (2004a, b), (8) Pimentel et al. (2011), (9) Lopes (2012), (10) Reis et al. (no prelo), (11) Campos and Dardenne (1997a, b), (12) Sgarbi et al. (2001), (13) Machado et al. (1989), (14) Chemale et al. (2012), (15) Babinski et al. (2012), (16) Pedrosa-Soares and Alkmim (2011), (17) Pedrosa-Soares et al. (2007), (18) Caxito et al. (2014), (19) Danderfer et al. (2009), (20) Schobbenhaus (1996), (21) Pedreira and De Waele (2008), (22) Danderfer et al. . First-order sequences: *LE* Lower Espinhaço, *P-UE* Paranoá-Upper Espinhaço, *M* Macaúbas, *B* Bambuí, *PzKS* Paleozoic and Cretaceous successions

grains and diagenetic xenotime extracted from the same units yielded maximum depositional ages between 1.5 and 1.0 Ga. for the Paranoá and Canastra groups (Valeriano et al. 2004a, b; Matteini et al. 2012; Rodrigues et al. 2010; Pimentel et al. 2011; Rodrigues et al. 2012).

In the marginal Araçuaí belt, U-Pb age determinations on detrital zircons extracted from the Paranoá-Upper Espinhaço sequence correlatives (i.e.: middle to upper Espinhaço Supergroup) have indicated a maximum depositional age of ca. 1.2 Ga (Chemale et al. 2012). In the same region, their minimum age is estimated in ca. 900 Ma, as indicated by the direct dating of mafic intrusions that cut the whole succession (Machado et al. 1989).

The available U-Pb ages on detrital zircons extracted from Paranoá-Upper Espinhaço sequence exposures in the São Francisco basin (younger than ca. 1,3 Ga; Lopes 2012; Alvarenga 2012; Kuchenbecker et al. 2014; Reis et al. 2014) indicates a time span between ca 1.3 and 0.9 Ga for its deposition (Figs. 7.2 and 7.3).

7.3.3 Macaúbas 1st-Order Sequence

The Neoproterozoic Macaúbas sequence encompasses the Jequitaí Formation (Dardenne 1978) as well as siliciclastic units identified in seismic sections and drill cores along the central-western São Francisco basin (near the town of João Pinheiro). Towards the west, the sequence apparently grades into the Neoproterozoic successions of the Vazante Group. On seismic sections, this sequence thickens east and westwards. Its basal portion is often marked by the presence of normal faults, commonly associated with syn-tectonic growth strata (Fig. 7.7).

The up to 300 m-thick glaciogenic strata of the Jequitaí Formation (Karfunkel and Hoppe 1988; Uhlein et al. 2004; Hercos 2008), exposed along the rims of regional-scale anticlines in the eastern portion of basin, comprise diamictites, sandstones, and subordinated pelites, which correspond to ice-proximal till, alluvial-fan, lacustrine and proglacial fluvial (outwash plain) deposits (Martins-Neto and Hercos 2002).

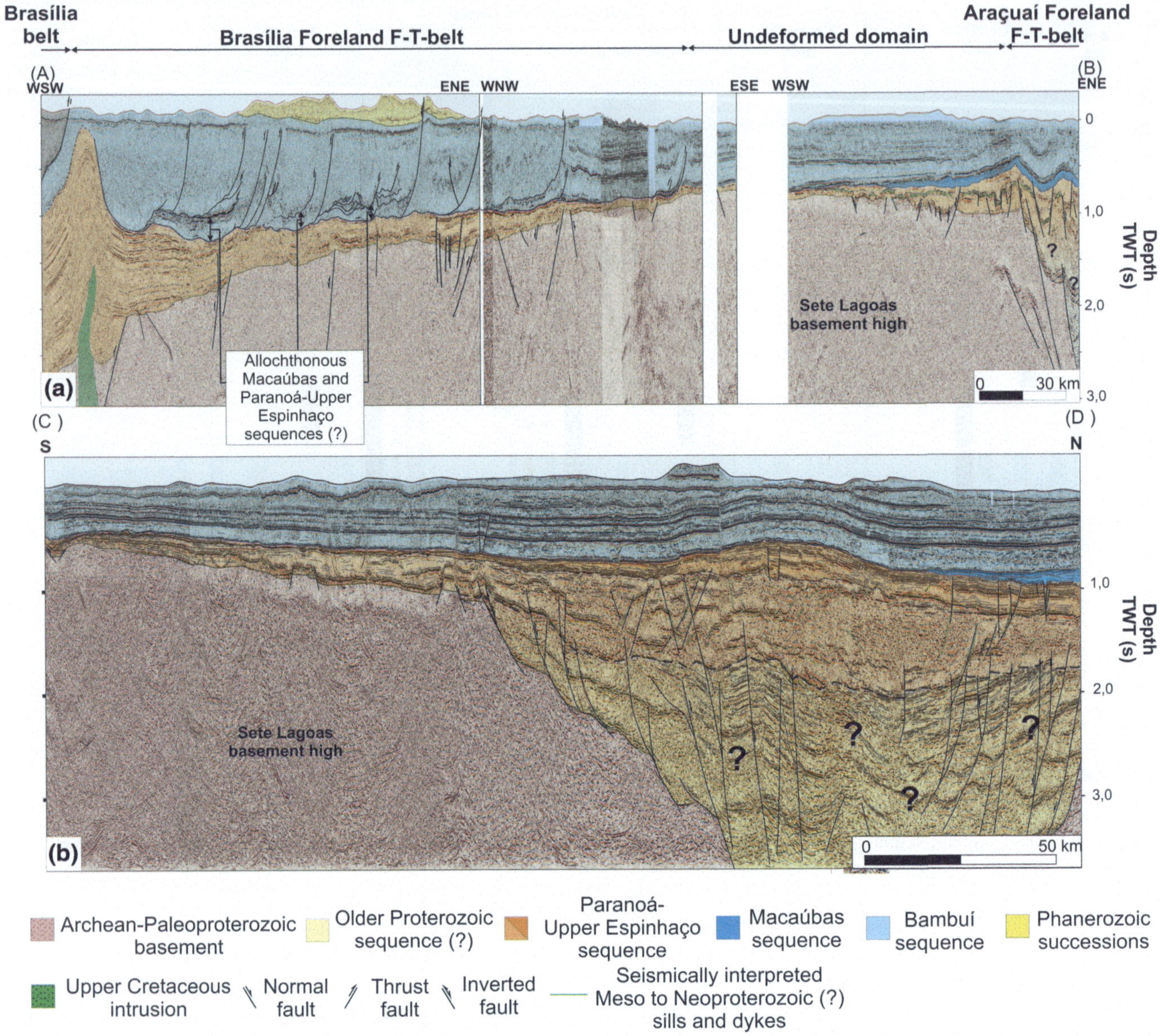

Fig. 7.4 Crooked composite seismic sections across the southern São Francisco basin showing: **a** the distribution of the main stratigraphic units of the São Francisco basin, and the expression of the Neoproterozoic Brasília and Araçuaí foreland fold-thrust belts (modified from Reis et al. 2012 and reproduced with minor modifications from Reis and Alkmim (2015), with permission from Elsevier); and **b** the Pirapora aulacogen, partially inverted in the influence area of the Araçuaí foreland f-t-belt. The rift and sag 2nd-order sequences of the Paranoá-Upper Espinhaço sequence are represented in *white* and *dark orange*, respectively. On both illustrations, the depth is shown in two-way travel time (TWT). The location of the seismic sections is indicated on Fig. 7.1

The Jequitaí Formation is traditionally interpreted as the intracratonic condensed section of rift/passive margin successions accumulated along the margins of the São Francisco plate (Uhlein et al. 2004), probably during a global-scale glacial event in the Cryogenian Period (Babinski et al. 2012).

The Macaúbas 1st-order sequence comprises an up to 800 m-thick package of transitional to shallow marine deposits in the western portion of the basin. In drill cores, it consists of radioactive shales that grade upward into deltaic well sorted, medium to fine-grained sandstones, commonly associated with thin beds of delta plain heterolithics, sandstones and conglomerates. Periodic exposures are suggested by a few beds of oxidized paleosoil-like mudstones. In its uppermost portion, the whole succession is bounded by detachment faults of the Brasília foreland fold-and-thrust belt (Fig. 7.7). These deposits apparently grade laterally into the Vazante Group, which is exposed close to the western boundary of the basin and comprises a ca. 5000 m-thick succession dominated by marine carbonates and fine-grained siliciclastics, locally interbedded with conglomerates,

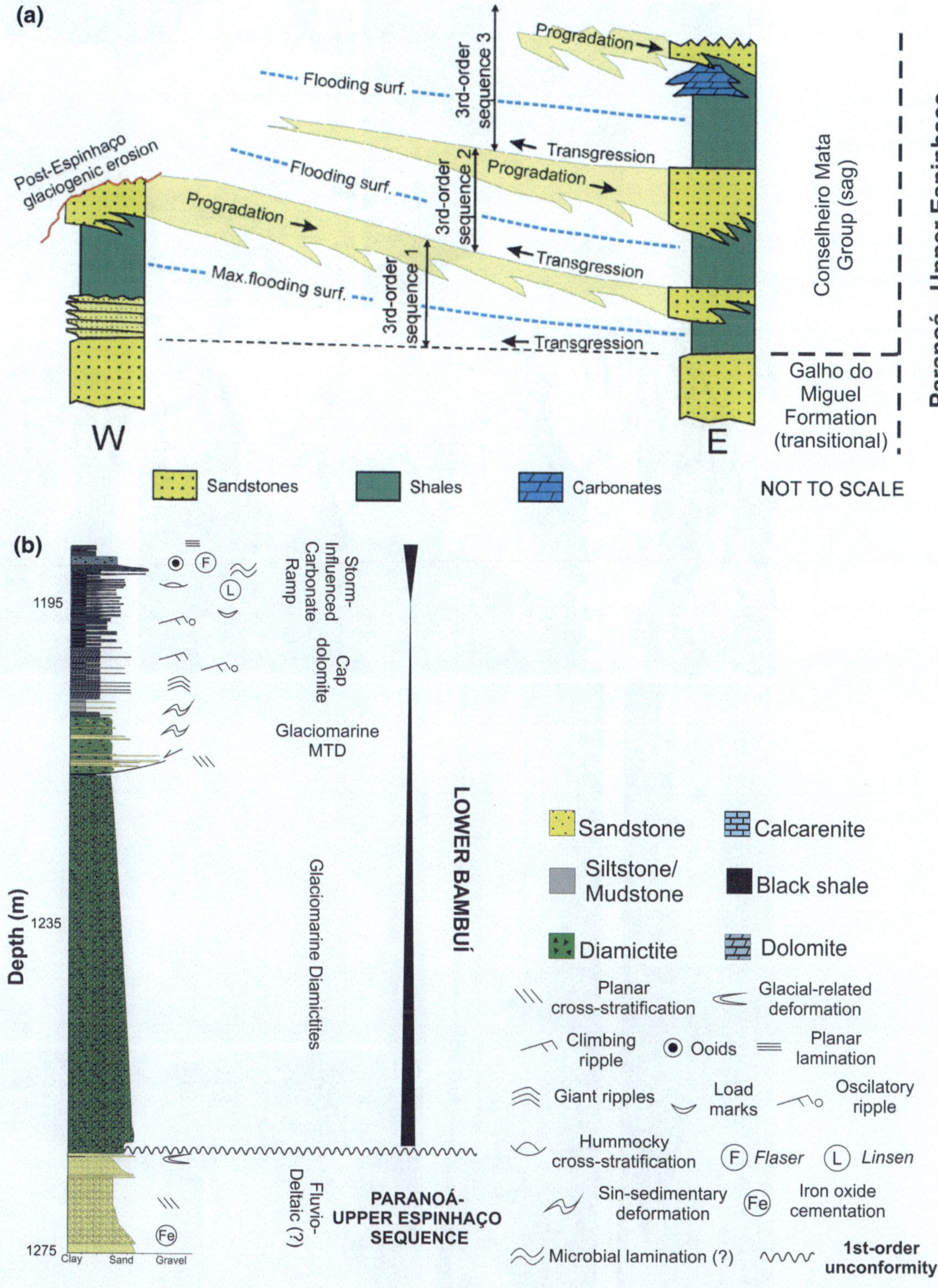

Fig. 7.5 Stratigraphic successions of the upper Paranoá-Upper Espinhaço 1st-order sequence and the lower Bambuí 1st-order sequence. **a** Schematic correlation section showing the 3rd-order transgressive-regressive sequences of the Conselheiro Mata Group (Paranoá-Upper Espinhaço sag 2nd-order sequence), which overlay the aeolian deposits of the Galho do Miguel Formation at the eastern São Francisco basin. Based on Dupont (1995) and modified from Martins-Neto (2009). **b** Stratigraphic column (based on drill cores) representing the lower portion of the Bambuí sequence at Januária basement high area. In this portion of the basin, the Ediacaran sequence overlies the Paranoá-Upper Espinhaço sequence and contains a thick package of glacial-related coarse- to fine-grained siliciclastics, which grade upwards into the carbonate ramp successions of Sete Lagoas Formation

Fig. 7.6 Rocks of the Paranoá-Upper Espinhaço sequence, exposed at the northwestern São Francisco basin: **a** Storm-dominated platformal silliciclastics, and **b** tidal flat sandstones with herring-bone cross-stratification. Lithotypes of the Vazante Group, western São Francisco basin: **c** Columnar stromatolites (Conophyton), and **d** *black* shale in drill core sample. **e** Drill core sample of the Jequitaí Formation diamictites from the eastern São Francisco basin. Rocks of the lower and upper Bambuí sequence: **f** Columnar stromatolites, and **g** sandstone showing hummocky cross-stratification

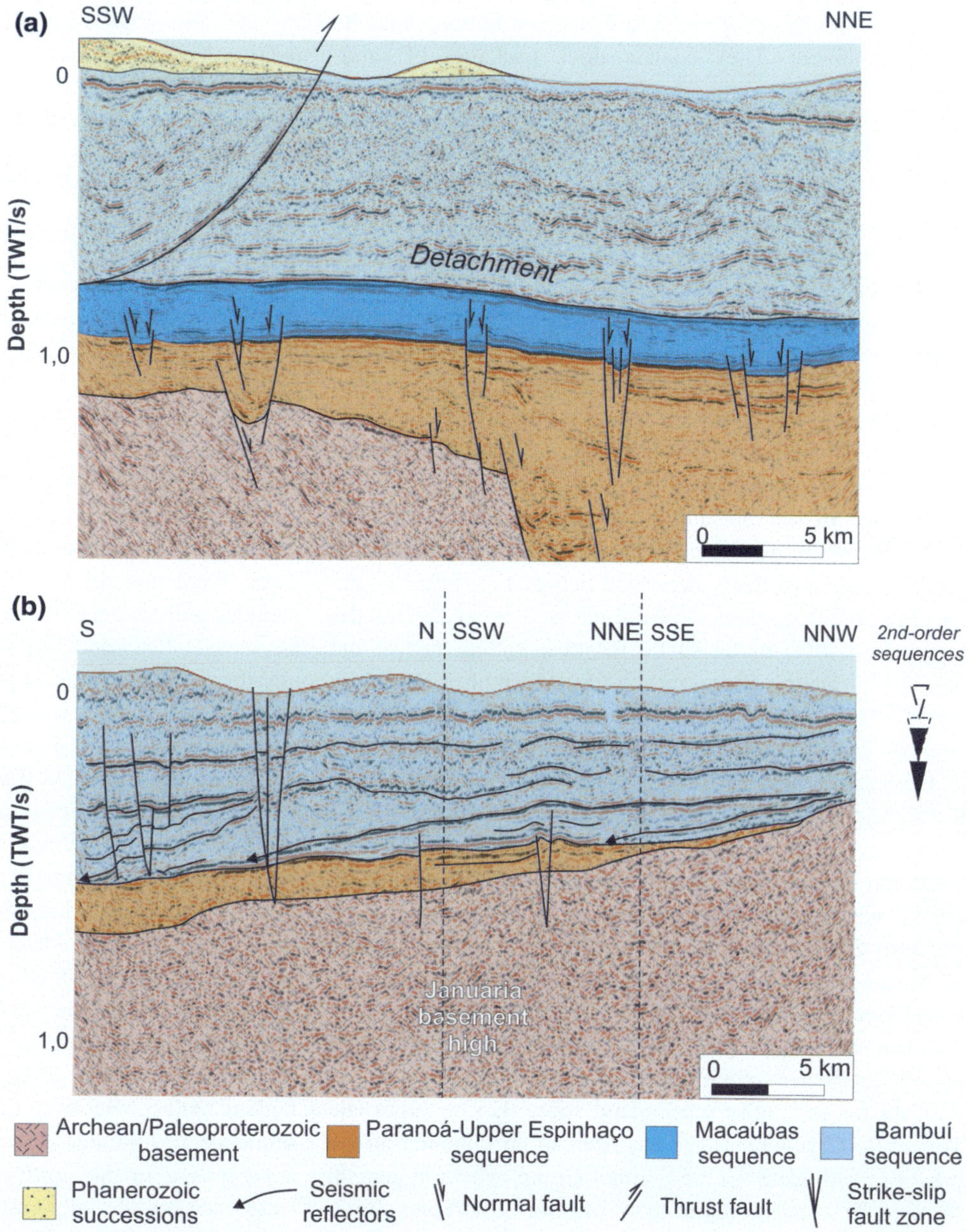

Fig. 7.7 a Single crooked seismic section of the southwestern São Francisco basin illustrating a segment of the Brasilia foreland fold-thrust belt. The contact of the Paranoá-Upper Espinhaço and Macaúbas sequences is affected by normal growth faults. The Macaúbas sequence is bounded on the top by a detachment surface, on which the Ediacaran Bambuí sequence is transported toward east. b Composite crooked seismic section of the northern portion of the basin, showing wedges of the lower Bambuí sequence prograding outward of the Januária basement high. On both illustrations the depth is shown in two-way travel time (TWT). For location of the sections see Fig. 7.1

sandstones, diamictite-bearing and phosphorite deposits (Dardenne 2000; Azmy et al. 2008; Martins-Neto 2009) (Figs. 7.1, 7.2 and 7.6). This unit is known for hosting important Pb-Zn deposits (Dardenne 2000), as well as organic-rich black shales with up to ca. 15 % of total organic content (Martins-Neto 2009).

The Vazante Group is interpreted as a passive margin basin-fill succession (Dardenne 2000), fed mainly by Archean and Paleoproterozoic cratonic sources (Pimentel et al. 2011; Rodrigues et al. 2012). Its maximum age is constrained by the youngest U-Pb detrital zircon ages of 930 Ma found in the basal Rocinha Formation (Rodrigues 2008; Rodrigues et al. 2012). This passive margin system seems to have been later converted into an active margin setting (Pimentel et al. 2011).

In the Araçuaí belt, which bounds the basin to the east, the Macaúbas 1st-order sequence is represented by the glacial to post-glacial metasedimentary successions of the Macaúbas Group (Karfunkel and Hoppe 1988; Uhlein et al. 1999; Martins-Neto et al. 2001; Martins-Neto and Hercos 2002; Martins-Neto 2009; Pedrosa-Soares et al. 2007; 2011a, b, Babinski et al. 2012). As indicated by recent provenance studies based on detrital zircon age determinations (Kuchenbecker et al. 2014; Alkmim et al., this book), these successions record a Cryogenian-Ediacaran rift-passive margin basin developed between the São Francisco peninsula (precursor of the São Francisco craton) and the Congo continent. They contain thick packages of sandstones, pelites, glaciomarine diamictites, carbonates, basic volcanics, banded iron formations and ophiolite remnants. Closure of this confined basin during the Brasiliano/PanAfricano event led to the development of Araçuaí-West Congo orogen around 570 Ma (Pedrosa-Soares et al. 2001; Alkmim et al. 2006).

The Macaúbas sequence is correlative of the rift-related successions of the Santo Onofre Group in the western Paramirim aulacogen (Schobbenhaus 1996; Danderfer et al. 2009). The diamictite-bearing Canabravinha Formation exposed in the Rio Preto belt along the northern margin of the craton is another correlative of the Macáubas sequence (Schobbenhaus 1996; Caxito et al. 2014).

The age of the Macaúbas 1st-order sequence is still a matter of debate. According to the data so far obtained in the basin, it was most likely deposited between ca. 880 Ma and 740 Ma. Its maximum depositional age is given by the youngest detrital zircon so far found in diamictites of the Jequitaí Formation (Rodrigues 2008; Pimentel et al. 2011), which is in agreement with the youngest U-Pb zircon ages obtained from siliciclastic rocks of the lower Vazante Group in the western São Francisco basin (Pimentel et al. 2011; Rodrigues et al. 2012). The minimum age of 740 Ma, on the other hand, is constrained by a Pb–Pb isochron obtained on basal cap carbonates of the Bambuí Group (Babinski et al. 2007), which unconformably cover the basement on the southern portion of basin. Unsolved issues regarding the age of Macaúbas sequence are: (1) current interpretations on correlative successions of the Macaúbas Group exposed in the Araçuaí metamorphic belt (Kuchenbecker et al. 2015) have suggested maximum depositional ages younger than 730 Ma for the entire glaciogenic strata; (2) no other late Tonian (to Cryogenian) age has been so far obtained on the basal cap carbonates of the Bambuí Group, which in general seem to be much younger than 740 Ma (Rodrigues 2008; Kuchenbecker 2011; Caxito et al. 2012); (3) at least part of the diamictite-bearing successions of the São Francisco basin commonly ascribed to the Jequitaí Formation could actually represent a late Cryogenian to Ediacaran glaciation instead of a Sturtian event, as suggested by regional stratigraphic and chemostratigraphic data (Caxito et al. 2012), as well as seismic and drill core information.

7.3.4 Bambuí 1st-Order Sequence

Covering the largest area of the São Francisco basin, the Bambuí 1st-order sequence (i.e., the Bambuí Group) comprises a package of carbonates, pelites, sandstones and subordinated conglomerates. The sequence is interpreted as the record of an Ediacaran foreland basin stage experienced by the São Francisco plate in response to lithospheric overburden caused by the uplift of the Brasília and Araçuaí belts (Barbosa et al. 1970; Chang et al. 1988; Alkmim and Martins-Neto 2001; Alkmim et al. 2011). On seismic sections, the Bambuí sequence displays a typical wedge-shaped geometry, whose thickness varies from a few hundred meters, on the east, to ca. 3000 m in the presumed foredeep area, on the west (Fig. 7.4). Remarkably, the seismic sections also reveal a slight thickening of the lower Bambuí sequence over preexisting large-scale structures (e.g.: Pirapora aulacogen), a fact that suggests reactivation of older fabric elements induced by the orogenic overburden along the craton margins.

The Bambuí 1st-order sequence encompasses four shallowing upward 2nd-order sequences, which are continuous over large areas of the basin and comprise the Paraopeba Sub-Group and the Três Marias Formation (Costa and Branco 1961; Braun 1968; Dardenne 1978, 1981). The Bambuí sequence also includes a distinct basal unit, the Carrancas Conglomerate and associated siliciclastics, which consist of a thin package of fluvial fine- to coarse-grained siliciclastics. These sediments have been interpreted as a lowstand system tract succession, restricted to some channels carved in the basement in southeastern portion of the basin (Vieira et al. 2007).

The basal 2nd-order sequence encompasses a succession of glaciogenic diamictites and lodgment tillites covered by post-glacial cap carbonates and a thick succession of shallow marine and stromatolite-bearing limestones and dolostones (Figs. 7.2, 7.5 and 7.6). These rocks correspond to the glacial-influenced units of the Carrancas Formation (Romano 2007; Kuchenbecker et al. 2013) and the overlying platformal carbonates of Sete Lagoas Formation (Nobre-Lopes 1995, 2002; Vieira et al. 2007; Iglesias and Uhlein 2009). Seismic and well data indicate that the sequence also includes diamictite-bearing successions which are exposed in the northwestern portion of the basin and have been traditionally correlated to the Jequitaí Formation, i.e., Macaúbas 1st-order sequence (e.g.: Dardenne 1978, 1981; Lima 2011).

In drill cores from the northern portion of the basin, the basal 2nd-order sequence includes diamictites and gravity flow coarse- to fine-grained siliciclastics, overlain by a ca. 100 m-thick succession of carbonate ramp deposits with subordinate interbedded organic-rich shales (Fig. 7.5). The overall δC^{13} and δO^{18} isotopic content and sedimentary facies described for the Sete Lagoas carbonates indicate typical post-glacial signatures (e.g.: Santos et al. 2000; Kuchenbecker 2011; Caxito et al. 2012; Alvarenga et al. 2014), similar to those described in Neoproterozoic glacial successions worldwide (Hoffman and Schrag 2002).

The remaining three shallowing upward 2nd-order sequences overlay the glacial-related deposits and encompass the upper Sete Lagoas, the Serra de Santa Helena,

Lagoa do Jacaré, Serra da Saudade and Três Marias formations (Costa and Branco 1961; Dardenne 1978, 1981). In the eastern and central sectors of the basin, the second and third lower-rank sequences consist of 100 ms-thick cycles with pelite-dominated packages at the base that grade upwards into shallow platformal carbonates. The uppermost 2nd-order sequence, on the other hand, consists mostly of marine fine-grained siliciclastic deposits overlain by the storm-influenced and sand-dominated Três Marias Formation (Chiavegatto 1992) (Fig. 7.6).

The 2nd-order sequences grade laterally into the fan-deltaic and submarine fan siliciclastics of the Samburá and Lagoa Formosa formations in the western half of the basin (Castro and Dardenne 2000; Dardenne et al. 2003; Baptista 2004; Fragoso et al. 2011; Uhlein et al. 2011). In this portion of the basin, the Bambuí sequence also contains deep marine glauconite- and phosphorite-bearing sediments of the Serra da Saudade Formation (Lima et al. 2007). The facies changes in the Bambuí sequence reflect the transition from shallower forebulge to foredeep deposits, accumulated, respectively on the central-eastern and western sectors of the basin (Alkmim and Martins-Neto 2001; Alkmim et al. 2011; Martins-Neto 2009).

Provenance studies and facies distribution point toward two main source areas for the Bambuí sequence: (i) Neoproterozoic orogenic sources associated with the Brasília belt on the west, and (ii) Archean and Proterozoic sources, represented by the craton basement and older basin-fill units. The orogenic source is indicated by the spatial distribution and provenance of the Samburá and Lagoa Formosa rudites, as well as consistent eastward paleocurrent indicators measured in the uppermost Três Marias sandstones (Chiavegatto 1992; Castro and Dardenne 2000; Uhlein 2014). The expressive occurrence of shallow carbonate facies in the central and eastern portions of the basin (Nobre-Lopes 1995, 2002; Vieira et al. 2007; Iglesias and Uhlein 2009; Costa 2011a, b), associated with progradational seismic patterns outward from the large Januária and Sete Lagoas basement highs (Fig. 7.7) suggest that the craton basement has also contributed with sediments during the deposition of the Bambuí sequence. An additional and late sourcing associated with the Araçuaí belt is represented by thin beds of coarse-grained siliciclastics exposed in a few areas along the eastern São Francisco basin (Chavegatto et al. 1997; Kuchenbecker et al. 2014).

The detrital zircon U-Pb age spectrum obtained from Bambuí Group rocks also point toward bimodal (cratonic and orogenic) sources, constraining, in addition, a maximum deposition age of ca. 610 Ma for most of the sequence (Rodrigues 2008; Lima 2011; Pimentel et al. 2011; Reis et al. 2012). This age has been, however, matter of debate. The recent discover of *Cloudina sp.* fossil remnants (Warren et al. 2014) and late Ediacaran zircon grains in the basal sediments of the Sete Lagoas Formation (Paula-Santos et al. 2015; Pimentel et al. 2012; Kuchenbecker et al. 2014) have indicated ages younger than ca. 550 Ma for almost the entire Bambuí sequence and thus a mismatch between the Late Ediacaran basin-cycle and the southern Brasília belt evolution, whose main collisional stage was dated at ca. 630 Ma (Valeriano et al. 2004a, b; Pimentel et al. 2011). Since the whole Bambuí sequence is involved in the external portion of the southern Brasília belt in the western São Francisco basin, it seems that the orogenic stage in the belt has lasted longer than previously thought. Nevertheless, further studies are required to better evaluate the real significance of the mentioned ages and their tectono-stratigraphic implications.

The Bambuí sequence has correlatives in other portions of the craton and marginal orogenic belts. The Salitre Formation, exposed in the Paramirim aulacogen, consists of shallow marine carbonates and subordinated fine-grained siliciclastics and phosphate-bearing successions (Souza et al. 1993; Misi and Veizer 1998; Misi 2001). In the marginal Brasília and Araçuaí belts, the chronological equivalents of the Bambui sequence are the Ediacaran syn-orogenic assemblages of the Rio Verde (with the underlying Cubatão diamictites?) and Salinas formations, respectively (Martins-Neto et al. 2001; Lima et al. 2002; Pedrosa-Soares et al. 2007; Santos et al. 2009; Rodrigues et al. 2010; Dias et al. 2011). The regional distribution of the units suggests, at least partially, the coexistence of the foreland intracratonic system and marginal orogenic basins during the Ediacaran-Cambrian boundary.

7.3.5 Late Paleozoic Santa Fé Group

The Santa Fé Group is composed of an up to 180 m-thick package of diamictites, dropstone-bearing shales, sandstones and subordinated varvites and tillites of glaciolacustrine, glaciofluvial and periglacial origin (Campos and Dardenne 1997a, b). Occurrences of the Santa Fé sedimentary rocks are scarce and limited to the central and northwestern portions of the basin, where they fill large valleys carved on upper Bambuí strata and are often associated with striated pavements. The Santa Fé deposits record the Permo-Carboniferous glaciation that has affected the West Gondwana during the Late Paleozoic (Sgarbi et al. 2001; Limarino et al. 2014; Torsvik and Cocks 2013; Vesely and Assine 2006).

7.3.6 The Cretaceous Areado, Mata da Corda and Urucuia Groups

The youngest units of the São Francisco basin are the Cretaceous Areado, Mata da Corda and Urucuia groups, which occur in a series of isolated mesas and large plateaus in its southwestern and northern portions.

The Lower Cretaceous Areado Group is composed of an up to ca. 300 m-thick sand-dominated succession, whose main exposures are located at the southwestern sector of the basin (e.g., Campos and Dardenne 1997a, b; Sgarbi et al. 2001; Kattah 1991; Fragoso 2011; Pedrosa-Soares and Alkmim 2011). The group is subdivided into three formations. The basal Abaeté Formation consists of alluvial medium- to coarse-grained sedimentary successions, locally interbedded with sandstones and mudstone layers. These deposits are overlain by the discontinuous lacustrine strata of the Quiricó Formation, which encompasses distal fine-grained siliciclastics associated with turbidite deposits and thinner black shale layers (Kattah 1991; Sgarbi et al. 2001; Fragoso 2011). Conchostracan remnants, plant fragments and fresh water fossils are commonly described in this succession (Sgarbi 2000; Sgarbi et al. 2001). The upper Três Barras Formation corresponds to fluvial and fluvio-deltaic sandstones, as well as widespread and thicker aeolian strata. Restricted thin layers of silexite with bathyal/abyssal radiolarian fauna strongly contrasts with the overall continental record of the Areado Group (e.g.: Kattah 1991; Sgarbi 2000; Sgarbi et al. 2001; Arai 2009; Fragoso 2011). These continental/marine Early Cretaceous fossil-bearing layers represent a challenge for understanding the paleogeography of the basin during the Early Cretaceous.

The sedimentary facies of the Areado Group indicate accumulation under semi-arid and arid conditions (Sgarbi et al. 2001), partially coeval to sauropoda and theropoda dinosaur faunae dispersion (Domingues 2009; Zaher et al. 2011). Considering their depositional systems, geographic distribution and common relationship with extensional structures along the southwestern São Francisco basin, Fragoso (2011) grouped the early Cretaceous sediments into two unconformity-bounded sequences. The lower sequence contains almost the entire Areado Group sediments and corresponds to a rift-related succession, which shows the record of initial, climax and subsequent post-rift stages. This succession, filling the Abaeté graben (Alkmim and Martins-Neto 2001; Sgarbi et al. 2001), is overlain by aeolian sediments, which could be associated either with a late flexural post-rift stage or another, unrelated Upper Cretaceous subsidence phase (Fig. 7.8).

The Upper Cretaceous Mata da Corda Group is exposed in the southwestern São Francisco basin and consists of a succession of alkaline volcanic/sub-volcanic, volcaniclastic and epiclastic rocks (Campos and Dardenne 1997a, b; Sgarbi et al. 2001; Sgarbi 2011a, b). It is related to alkaline-carbonatitic-phosphoritic igneous intrusions and a system of NW-trending dykes that cut extensive areas along southeastern and central Brazil (Borges and Drews 2001; Silva 2006; Grasso 2010). This magmatism is generally associated with a late uplift event of the Cretaceous Paranaíba arch, which separates the Paraná and São Francisco basins (Hasui and Haraliy 1991; Sgarbi et al. 2001; Sgarbi 2011a, b).

The Cretaceous Urucuia Group covers a large area along the central and northern portions of the basin (Campos and Dardenne 1997a, b; Sgarbi et al. 2001 and references therein). The group consists of an up to 360 m-thick succession of aeolian sandstones, locally associated with fluvial fine- to coarse-grained sediments. These deposits extend further to the north of São Francisco basin, where they overlie the Lower Cretaceous units of the Parnaíba basin (Campos and Dardenne 1997a, b; Sgarbi et al. 2001). Commonly viewed as a correlative unit of the Upper Cretaceous Mata da Corda Group, the age and stratigraphic correlations of this group are still unclear (Sgarbi et al. 2001).

The Areado, Mata da Corda and Urucuia groups are interpreted as continental interior manifestations of the Cretaceous rifting event that led to the development of the South Atlantic (e.g., Hasui and Haralyi 1991; Sgarbi et al. 2001; Mohriak and Leroy 2012).

7.4 Tectonic Framework

Three families of large-scale fabric elements occur in the São Francisco basin (Alkmim and Martins-Neto 2001; Reis and Alkmim 2015): (i) Proterozoic rift structures; (ii) Neoproterozoic foreland fold-thrust belts; and (iii) Cretaceous rift structures. The Precambrian structures comprise the main structural grain of the basin (Fig. 7.9) and generally show evidence of multiple reactivations, including partial inversion. Their expressions vary along the basin and a large number of fabric elements described below are only detected in seismic sections.

7.4.1 Proterozoic Rift Structures

The largest Proterozoic rift structure preserved in the basin is the NW-trending Pirapora aulacogen. Partially inverted, this graben is defined by a system of NW-striking normal faults and conjugate NE-oriented structures. It hosts the Paranoá-Upper Espinhaço sequence, which covers an apparently older and unknown succession confined to its central portion (Fig. 7.4). The aulacogen separates two large basement highs located in the southern and northern sectors of the basin, the Sete Lagoas and Januária highs, respectively. Exposed in a few windows along the basin, these basement highs stand out on geophysical maps and seismic sections (Figs. 7.1 and 7.4) and seem to have acted as positive structures during the whole Proterozoic evolution of the basin. During the Neoproterozoic, the onset of the Macaúbas basin-cycle was responsible for the nucleation of a new generation of rift-related structures (Fig. 7.7). In the eastern portion of the basin, seismic sections suggest that these structures formed through the extensional reactivation of pre-existing Pirapora aulacogen faults. The overall

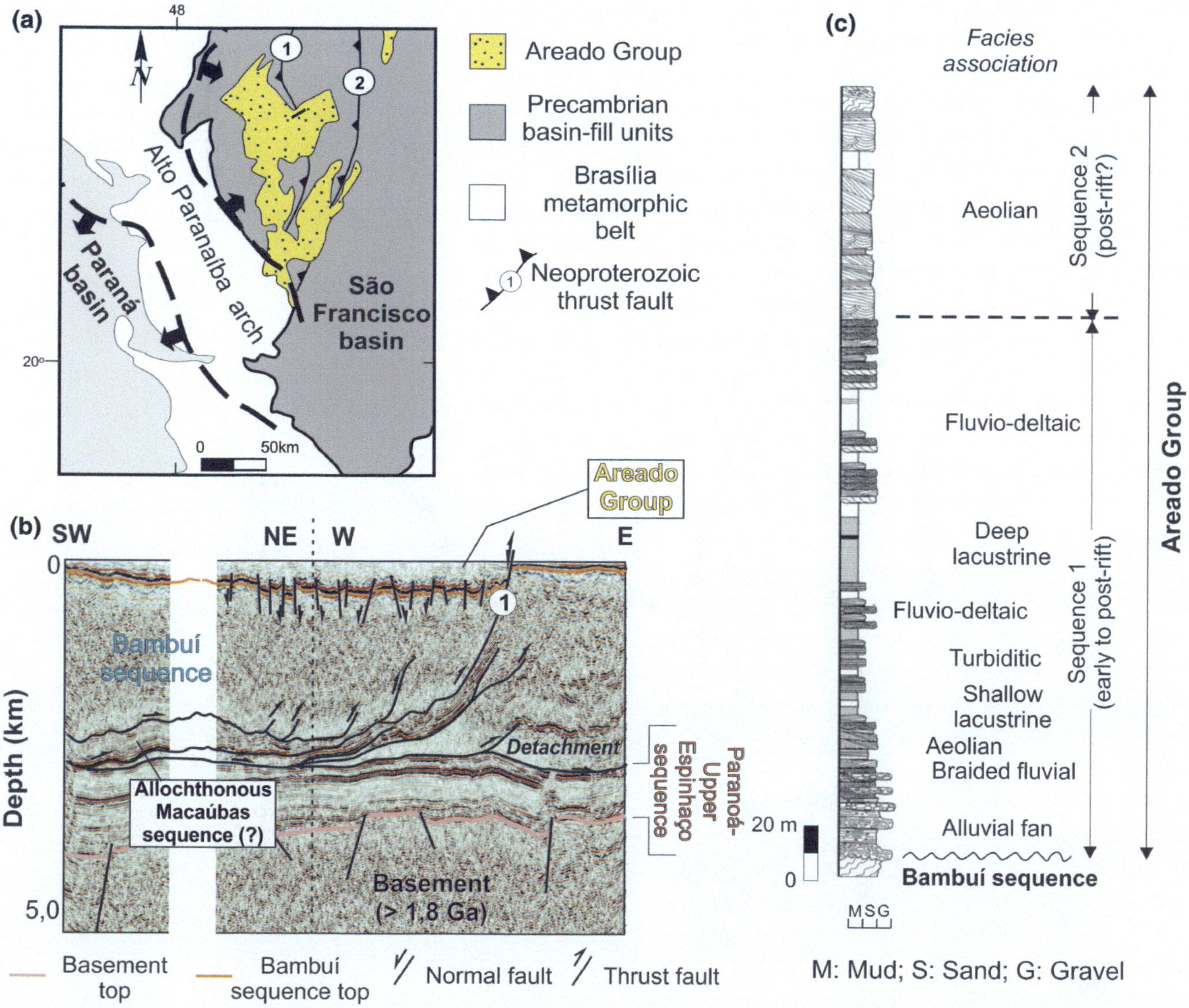

Fig. 7.8 The Lower Cretaceous Areado Group in the southwestern São Francisco basin. **a** Simplified geological map showing the distribution of the Areado Group and its relationship with the main Precambrian and Phanerozoic tectonic elements. **b** Interpreted seismic section showing the Areado Group filling a half-graben in the Presidente Olegário area. Note that the graben has formed through the extensional reactivation of a Neoproterozoic thrust fault (1: João Pinheiro Fault; 2: São Domingos Fault). **c** Stratigraphic column and depostional environments of the Areado Group, near the town of Presidente Olegário, southwestern São Francisco basin (Fragoso 2011)

configuration of the Pirapora aulacogen indicates that its nucleation dates back to the early evolutionary stages of the São Francisco basin (Late Paleoproterozoic?).

7.4.2 Foreland Fold-Thrust Belts

The Precambrian fill units of the São Francisco basin were caught by the Brasiliano orogenic fronts, which propagated from the Brasília, Rio Preto and Araçuaí belts towards the craton interior during the Ediacaran Period (Alkmim et al. 1996; Brito-Neves et al. 1999; Brito-Neves 2004; Valeriano et al. 2004a, b; Alkmim et al. 2006; Pedrosa-Soares et al. 2001, 2007; Caxito 2010). Two roughly NS-trending foreland fold-thrust belts of opposite vergences developed along the borders of the basin (Alkmim et al. 1996; Alkmim 2004) (Fig. 7.9), the Brasília on the west, and the Araçuaí on the east. Separated by a central undeformed sector, these foreland f-t-belts show significant differences in tectonic style, as shown on Table 7.1.

7.4.2.1 The Brasília Foreland Fold-Thrust Belt

The Brasília foreland fold-thrust belt is thin-skinned, fold-dominated (Figs. 7.10 and 7.11), and coupled to detachments located near the base of the Bambuí sequence (Coelho 2007; Zalán and Romeiro-Silva 2007; Reis 2011;

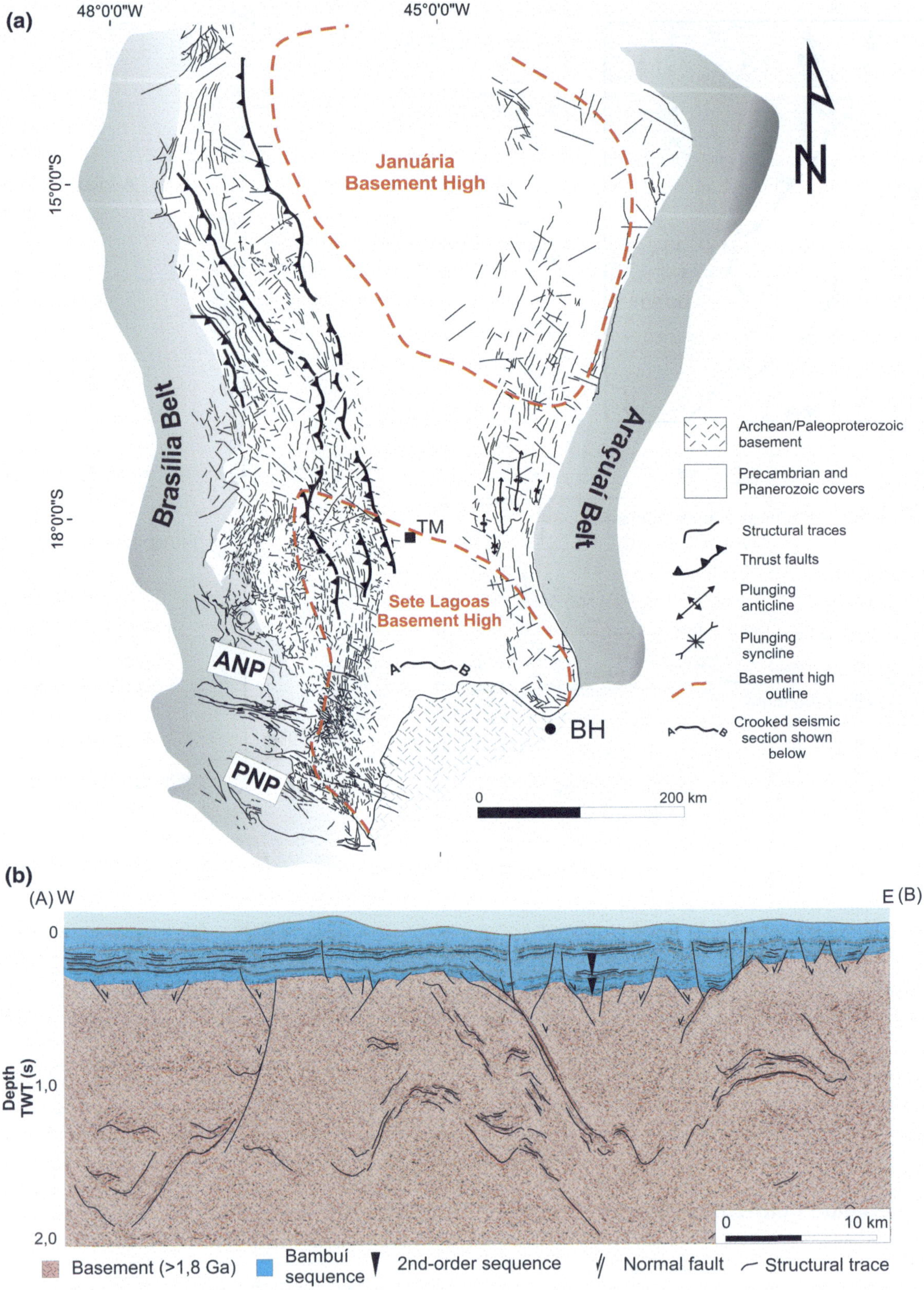

Fig. 7.9 **a** Simplified tectonic map of the southern São Francisco basin highlighting the Neoproterozoic Brasília and Araçuaí foreland fold-and-thrust belts, respectively on the west and on the east. The map also shows the relationship between these belts and the main basement structures. **b** Crooked seismic section showing the extensional reactivation of Archean compressional structures during the deposition of the lower Bambuí sequence on the Sete Lagoas basement high. Depth is shown in two-way travel time (TWT)

Table 7.1 Main attributes of the Brasília and Araçuaí foreland fold-thrust belts

	Brasília Foreland f-t-belt	Araçuaí Foreland f-t-belt
Main Structures	Folds and major thrusts associated with subordinate duplexes coupled to a regional detachment zone. Late stage NW-trending left-lateral and NE-trending right-lateral strike-slip faults affect the previous structures respectively in the southern and northern sectors of the belt	Thin-skinned, W-verging thrusts and folds associated with duplexes and imbricate fans. At its central sector, large-scale double plunging folds developed as cover structures of inverted preexisting (buried) rift structures. NE and NW-trending conjugate fracture sets are widespread throughout the whole belt
Polarity	Marked by a few west-dipping thrusts and decreasing strain intensity towards east	Given by the systematic west vergence of the fabric elements and progressive decrease in the strain intensity in the same direction
Architecture	Wedge-shaped and fold-dominated thin-skinned belt, coupled to detachments located near the base of the Bambuí sequence	Segmented in two thin-skinned domains, separated by a central thick-skinned sector
Metamorphism	Absent (sharp contact with the Brasília metamorphic belt on the west)	Gradual increase towards east, reaching the greenschist facies conditions close to the eastern boundary of the basin
Age[a]	ca. 630–540 Ma (?)	ca. 580–530 Ma

Based on Schöll (1973), Bonhomme (1976), Magalhães (1988), Chang et al. (1988), Muzzi Magalhães (1989), Oliveira (1989), Alkmim et al. (1993), D'Arrigo (1995), Fonseca et al. 1995, Souza-Filho (1995), Alkmim et al. (1996), Costa-Neto (2006), Coelho (2007), Hercos (2008), Reis (2011) and Reis et al. (2011, 2012)

[a]Based on the approximated age of the main collisional events recorded in the Brasília (Pimentel et al. 2004; Valeriano et al. 2004a, b) and Araçuaí (Pedrosa-Soares et al. 2001, 2007) belts, as well as the younger detrital zircon and fossil remnants found within the Bambuí sequence (Warren et al. 2014; Paula-Santos et al. 2015)

Reis et al. 2012; Reis and Alkmim 2015). Corresponding to the external domains of the Brasília and Rio Preto marginal belts, it exhibits recesses and salients that culminate in central portion of the basin (Reis 2011; Reis et al. 2011) (Fig. 7.9). In seismic sections, the belt displays a typical wedge shape, involving strata of the upper Paranoá-Upper Espinhaço and Macaúbas (?) sequences in the western proximal domain. The detachment zone becomes progressively shallower towards the east, where a few thrusts separate large sectors affected by chevron folding of decreasing intensity (Fig. 7.4). Remarkably, neither a penetrative cleavage nor metamorphic features have been observed along its extension (Alkmim et al. 1996; Alkmim and Martins-Neto 2001).

The evolution of the foreland belt was partially accompanied by the extensional reactivation of pre-existing structures (e.g.: Pirapora aulacogen) and nucleation of a system of NE-trending grabens in the southern part of the basin (Fig. 7.9). These grabens formed through the reactivation of Archean contractional structures and host complete sections of the lower Bambuí sequence. They developed in the outer arc of the Sete Lagoas basement high, which seem to have acted as the forebulge during the imposition of the orogenic loads along the western margin of the basin.

The Brasília foreland fold-thrust belt can be subdivided into three segments, southern, central and northern, which exhibit remarkable differences in tectonic style. The southernmost segment mostly involves rocks of the Bambuí sequence and contains structures of two distinct generations. First-stage chevron folds with NNE-trending hinges are locally bounded by west-dipping thrusts in the interior of duplexes and imbricate fans. These structures are cut by a system of late stage sinistral strike-slip faults, which affect the basement and are responsible by an overall counterclockwise rotation of preexistent fabric elements along narrow NW- trending corridors. The southern foreland belt segment is bounded to the west by the metamorphic nappes of the Brasília belt (Figs. 7.9 and 7.10).

The late stage strike-slip faults are practically absent in the central segment of the foreland belt, where upright chevron folds of the first phase frankly predominate (Fig. 7.11). In this segment, the belt exhibits its maximum cratonward advancing, culminating as the Três Marias salient, a 130 km long and 80 km wide antitaxial curve (Reis 2011; Reis and Alkmim 2015; Reis et al. 2011) (Figs. 7.9 and 7.10). The salient corresponds to a basin-controlled curve, whose apex roughly coincides with the sector of the

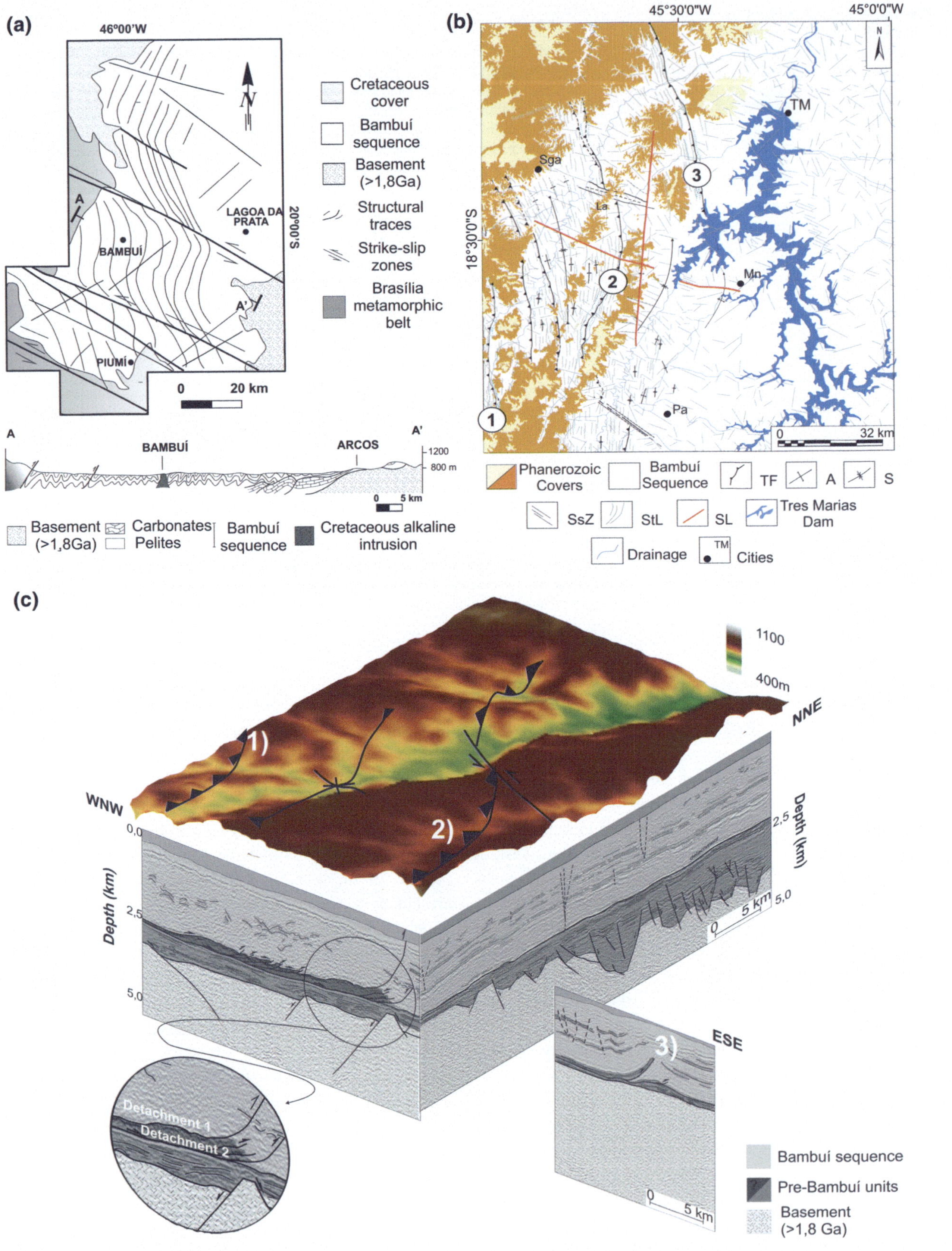

Fig. 7.10 **a** Schematic structural map of the southern Brasília foreland fold-thrust belt (Modified from Magalhães 1989). **b** Simplified structural map of the Três Marias Salient in the culmination of the Brasília fold-thrust belt, central-western São Francisco basin (Adapted from Reis 2011). *TF* thrust faults (1—João Pinheiro. 2—Borrachudo. 3—São Domingos). *A* anticline, *S* syncline, *SsZ* strike-slip zone, *StL* structural trend-lines, *SL* location of the seismic lines shown on (**c**). Cities: *TM* Três Marias, *Pa* Paineiras, *Mn* Morada Nova de Minas. **c** Block-diagram of the Três Marias salient, made through a combination of seismic sections and a digital elevation model (Reproduced with minor modifications from Reis and Alkmim 2015, with permission from Elsevier)

Fig. 7.11 Structures of the Brasília foreland f-t-belt: **a** West-dipping thrust, juxtaposing pre-Bambuí siliciclastics with the Ediacaran Bambuí sequence in the northwestern São Francisco basin; **b** *Box* and chevron folds affecting sandstones and mudstones of the Vazante Group in the northwestern São Francisco basin; **c** *Upright* chevron fold in glauconite-bearing pelites of Bambuí sequence near the town of Cedro do Abaeté (western São Francisco basin). Structures of the Araçuaí foreland f-t-belt (eastern São Francisco basin): **d** East-dipping cleavage in Bambuí sequence carbonates; **e** Large-scale quartz-veins cutting Macaúbas sequence siliciclastics at Araçuaí foreland f-t-belt culmination zone, central São Francisco basin

basin that contains the thickest pre-orogenic Bambuí strata (Reis and Alkmim 2015).

The main structures of the northern Brasília foreland belt are NNW-trending folds and subordinated thrusts, locally overprinted by sets of NE-oriented strike-slip faults (Fonseca et al. 1995; Alkmim et al. 1996; Alkmim and Martins-Neto 2001). The strike-slip system affects the Archean/Paleoproterozoic basement and continues westwards into the northern Brasília marginal belt. Structural analyses carried out by Araújo Filho (2000) in this region indicate that the northern segment of the Brasília marginal belt is significantly younger than its southern counterpart.

In the northernmost São Francisco basin, the Brasília foreland fold-thrust belt curves towards northeast and merges with the external domain of the of Rio Preto belt, a cratonward verging system of folds and thrusts, which roots in right-lateral strike-slip system (Egydio-Silva et al. 1989; Caxito et al. this book).

7.4.2.2 The Araçuaí Foreland Fold-Thrust Belt

The Araçuaí foreland fold-thrust belt extends along the eastern edge of the São Francisco basin and represents the external portion of the equally named marginal belt that bounds the craton to the east (Fig. 7.9). Differently from the Brasilia foreland belt, folds of the Araçuaí foreland belt are asymmetric, show a clear W-directed vergence and are associated with a regionally penetrative axial plane cleavage (Fig. 7.11). Furthermore, close to the eastern boundary of the basin, the rocks involved in the belt have experienced metamorphism under lower greenschist facies conditions (Schöll 1973; Magalhães 1988; Oliveira 1989; Souza-Filho 1995). Conjugate NE- and NW-trending fractures are observed along its whole extension (e.g.: Oliveira 1989; Alkmim and Martins-Neto 2001).

The southern segment of the Araçuaí foreland belt is thin-skinned and linked to a regional detachment developed along the basal contact of the Bambuí Group with older units (Magalhães 1988; D'Arrigo 1995). The main structures of the segment are NNW-trending duplexes and W-verging imbricate fans. These structures are locally affected by late stage co-axial folds, which seem to have been generated during the late reactivation of basement-involved faults along the eastern craton boundary (D'Arrigo 1995).

The central segment of the belt is associated with the partial inversion of the Pirapora aulacogen structures (Figs. 7.4 and 7.9). The most striking surface expression of the compressional reactivation of buried extensional structures is a set of regional and double-plunging drape folds (Fig. 7.9). The interaction between the foreland belt and the preexistent NW-oriented graben produced a salient, whose development was probably influenced by along-strike thickness variations of the pre-orogenic strata (Marshak and Wilkerson 1992; Souza-Filho 1995). Some oroclinal bending was likely active during the generation of the southern salient limb, as indicated by fabric rotation along WNW-trending left-lateral strike-slip faults, which coincide with the Pirapora aulacogen boundary (Souza-Filho 1995). The widespread occurrence of large quartz and calcite veins in the culmination zone of the salient denotes an expressive fluid migration episode, probably related to the reactivation of older basement-involved structures (Fig. 7.11). Deep-seated faults become less significant in the northern segment of the belt and thin-skinned structures seem to predominate over the entire sector (e.g. Oliveira 1989).

7.4.3 Cretaceous Rift Structures

Although less expressive, Cretaceous rift structures are well exposed in the southwestern São Francisco basin. These structures control the deposition of the Lower Cretaceous Areado Group, as well as the emplacement and extrusion of the Upper Cretaceous Mata da Corda igneous rocks.

In the southwestern sector of the basin, NNW-trending normal faults, formed through the extensional reactivation of the Brasilia foreland fold-thrust belt structures, control the thickness and sedimentary dispersal of the Lower Cretaceous Areado Group (Sawasato 1995; Fragoso 2011; Reis 2011) in the interior of a major structure, the Abaeté graben (Fig. 7.8). These faults are genetically related to an assembly of NW-striking tensile fractures, NS-oriented normal faults and shear fracture sets, which were likely formed under extensional stress field with sub-vertical σ_1 and horizontal NE-trending σ_3 (Sawasato 1995; Reis 2011).

Upper Cretaceous intrusions, as well as a swarm of NW-trending dykes associated with the Mata da Corda Group often deform the Areado Group sediments and produce a variety of folds and faults (Sawasato 1995; Borges and Drews 2001). A younger family of structures consists of normal-sinistral NE-striking faults, which affect the Cretaceous strata and apparently control the morphology of the southwestern sector of the basin (Alkmim and Martins-Neto 2001). The precise age of these structures remains unknown.

The most prominent structure of the southwestern edge of the São Francisco basin corresponds to a ca. 350 km long and ca. 80 km wide arch, the Alto Paranaíba arch that separates the Paraná and São Francisco sedimentary and hydrographic basins (Hasui and Haralyi 1991; Alkmim and Martins-Neto 2001) (Fig. 7.8). The main uplift stage of this structure seems to have occurred in the Cretaceous period, simultaneously to the deposition and emplacement of the Areado and Mata da Corda successions (Hasui and Haralyi 1991; Sgarbi et al. 2001; Hackspacher et al. 2007).

7.5 Evolutionary Synthesis

Stabilized after the 2.1–2.0 Ga Transamazonian-Eburnean event, the São Francisco craton and its African counterpart were probably incorporated in the Columbia supercontinent (see D´Agrella et al. this book, Roger and Santosh 2002; Zhao et al. 2004; Meert 2012; Nance et al. 2014). Involved in the subsequent supercontinent cycles—except for Rodinia -, the craton underwent multiple tectonic events and experienced climatic changes of global significance (Brito-Neves et al. 1999; Martins-Neto et al. 2001; Alkmim and Martins-Neto 2012). Due to its long-term intracratonic residence, the São Francisco basin records a complex evolutionary history briefly discussed in the next sections.

The first tectonic stage so far identified in the São Francisco basin corresponds to the opening of the NW-trending Pirapora aulacogen (Fig. 7.4). Cutting across the central portion of the basin, this structure hosts the Paranoá-Upper Espinhaço sequence, which unconformably overlies an apparently older unknown succession. This succession is a potential correlative of the lower Espinhaço Supergroup exposed in the Araçuaí belt and Paramirim aulacogen. The basal Espinhaço Supergroup consists of 1775 and 1740 Ma continental sediments and volcanic rocks associated with anorogenic plutons (e.g.: Machado et al. 1989; Dussin and Dussin 1995; Schobbenhaus 1996; Knauer 2007; Chemale et al. 2012; Danderfer Filho et al. 2015). Stratigraphic and tectonic studies have demonstrated that the Lower Espinhaço accumulated in a system of Statherian rifts, developed in a crustal segment presently represented by the São Francisco-Congo craton and its margins (Dussin and Dussin 1995; Pedreira and De Waele 2008; Danderfer et al. 2009; Chemale et al. 2012; Danderfer Filho et al. 2015; Cruz and Alkmim, this book (Fig. 7.12). In the northern Brasília belt, the Paleoproterozoic metassedimentary successions and associated rocks of the Araí Group (e.g., Dardenne 2000; Matteini et al. 2012) are also potential correlatives of these unknown strata. Nevertheless, additional analyses are needed to better evaluate the age and tectono-stratigraphic significance of this succession.

The manifestation of the second tectonic event documented in the basin corresponds to the reactivation of the Pirapora aulacogen coupled with the onset of the ca. 1.3–0.9 Ga Paranoá-Upper Espinhaço basin-cycle. Although no correlative magmatic ages have been so far reported in the São Francisco basin, Mesoproterozoic igneous zircons (age peaks around 1.5 and 1.4 Ga) are widespread in its entire sedimentary record (e.g.: Pimentel et al. 2011; Matteini et al. 2012; Rodrigues et al. 2012). Documented worldwide, the Mesoproterozoic anorogenic magmatism associated with multiple rifting episodes might represent a manifestation of the Columbia supercontinent breakup, which started around 1.6 Ga and ended between 1.3 and 1.2 Ga (Zhao et al. 2004). The continents generated by the dispersal of Columbia reassembled to form the Rodinia supercontinent at very end of the Mesoproterozoic (Brito-Neves et al. 1999; Hoffman 1991; Nance et al. 2014). Most of the Rodinia reconstructions (e.g.: Kroener and Cordani 2003; Pisarevsky et al. 2003; Tohver et al. 2006; Nance et al. 2014) shows the São Francisco-Congo in the periphery of the supercontinent or as an isolated landmass. In fact, no clear evidence for collisional tectonism affecting the São Francisco craton in time between 1.2 and 1.0 (Grenvillian event) has been yet found.

Sometime during the Neoproterozoic Era, the São Francisco-Congo continent experienced a renewed rifting phase, which is recorded by the accumulation of the Macaúbas 1st-order sequence. The broad tectonic scenario and the exact chronology of this and the subsequent tectonic events are still poorly understood. In the São Francisco basin, the Neoproterozoic Macaúbas basin-cycle is represented by the sedimentary successions of the Jequitaí Formation, Vazante Group (?) and correlatives (Figs. 7.2 and 7.3).

Two main diachronic events that affected the São Francisco-Congo plate during the Neoproterozoic are chronologically associated with the Macaúbas basin-cycle. The development of a convergent margin between 900 and 600 Ma on the western portion (present-day coordinates) of the plate culminated with the development of an extensive arc-related basin partially floored by the oceanic remnants of the Araxá Group of Brasilia belt (Pimentel et al. 2004, 2011; Valeriano et al. 2004a, b; Rodrigues et al. 2012). The São Francisco continental margin received the sediments of the Vazante Group and its landward sand-dominated correlatives (Figs. 7.2 and 7.12).

Along the eastern margin of the craton, the Macaúbas sequence records a rifting event followed by the development of a passive margin system that defined a large gulf separating a continent, represented by the Congo craton, from a peninsula, represented by the São Francisco craton and its margins (Fig. 7.12). This basin-cycle apparently took place after the Tonian extensional episode that culminated with the deposition of the lower Macaúbas Group (exposed in the Araçuaí metamorphic belt) and is characterized by an assembly of Cryogenian to Ediacaran anorogenic intrusions, bimodal volcanics, ophiolite slices and rift-passive margin sedimentary successions (Pedrosa-Soares et al. 2001, 2007; Queiroga et al. 2007; Silva et al. 2008; Kuchenbecker et al. 2015; Alkmim et al. this book).

The Macaúbas basin-cycle, recorded in the eastern São Francisco basin by the Jequitaí formation, developed in the course of a glacial event, often ascribed to the Sturtian ice age (e.g.: Babinski et al. 2007, 2012). As previously mentioned, the precise age of the glacial units remains unclear (e.g.: Schobbenhaus 1996; Figueiredo 2008; Caxito et al. 2014; Babinski et al. 2012; Pedrosa-Soares and Alkmim 2011).

The convergence of São Francisco-Congo and other plates, including Rio de la Plata, Paranapanema, Amazonia, West Africa, and Kalahari, led to the assembly of West

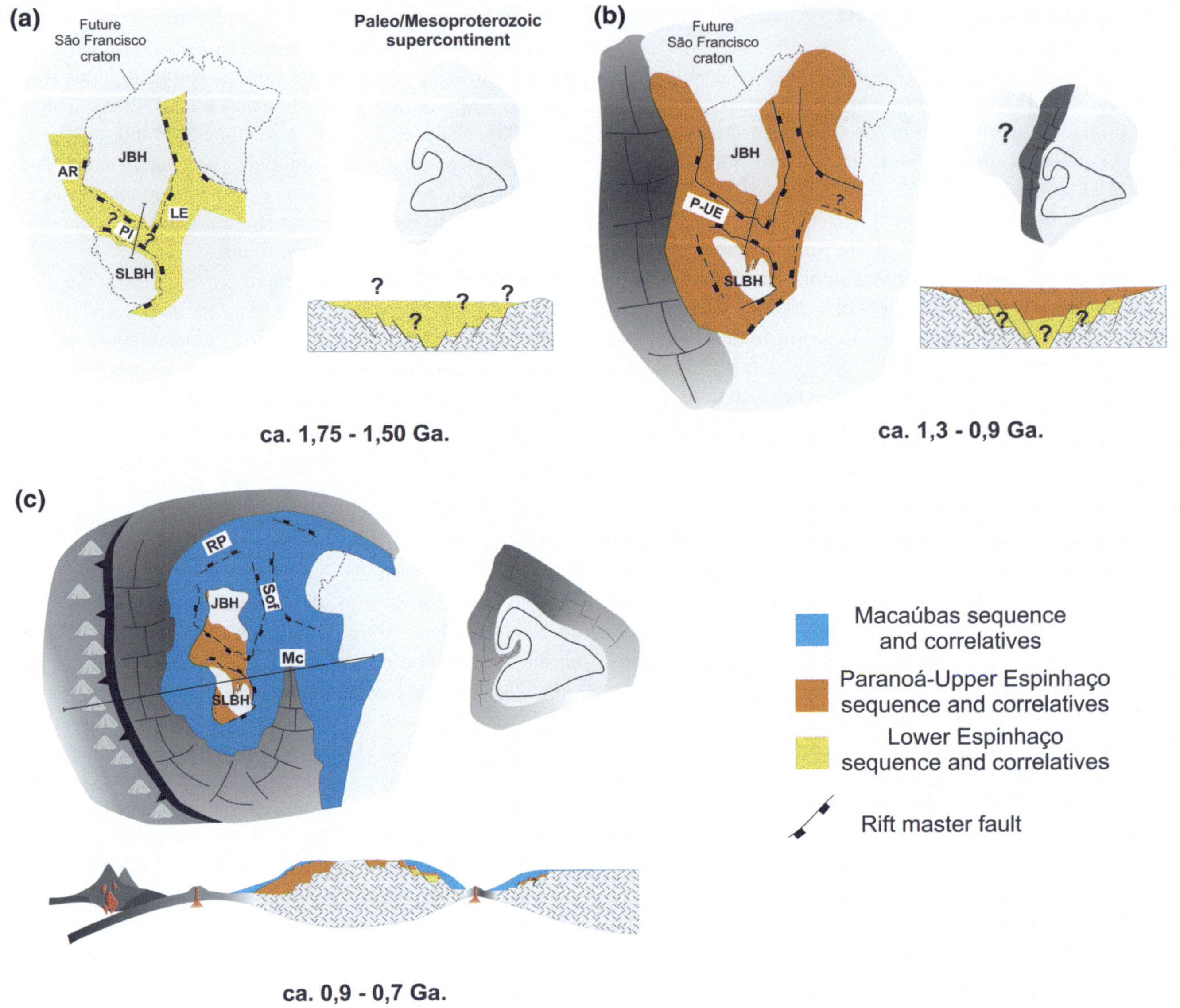

Fig. 7.12 Cartoons illustrating the evolutionary history of the São Francisco craton between ca. 1.8 and 0.7 Ga. Paleoproterozoic rifts: *AR* Araí, *Pi* Pirapora (?), *LE* Lower Espinhaço. Mesoproteroic rift-passive margin system: *P-UE* Paranoá-Upper Espinhaço. Neoproterozoic rifts: *Mc* Macaúbas, *Sof* Santo Onofre, *RP* Rio Preto. Basement highs: *SLBH* Sete Lagoas basement high, *JBH* Januária basement high. (See text for explanation)

Gondwana by the end of the Neoproterozoic and beginning of the Cambrian. The margins of the São Francisco-Congo were then converted into the Brasiliano orogenic belts (Almeida 1977). The craton interior subsided in response to the orogenic loads imposed along its margins, receiving sediments shed from the newly uplifted areas around. The sediments deposited in the craton interior were later locally caught by the Neoproterozoic/early Paleozoic orogenic fronts (Fig. 7.13).

The oldest Neoproterozoic orogenic event documented in the margins of São Francisco craton, representing the Paranapanema-São Francico/Congo collision, was responsible for the uplift of the southern Brasília belt around 630 Ma (Alkmim and Martins-Neto 2001; Pimentel et al. 2004; 2011; Valeriano et al. 2004a, b). After this event, a major transgression affected the São Francisco peninsula that started to behave as a downwarp basin. The mixed carbonate-siliciclastic sediments of the Ediacaran Bambuí 1st-order sequence started to fill the foreland basin-system (e.g., Chang et al. 1988; Alkmim and Martins-Neto 2001, 2012; Martins-Neto 2009; Pimentel et al. 2011). The glacial deposits of the basal Bambuí sequence might represent the record of one of the ice ages that covered low latitudes worldwide in the late Cryogenian/early Ediacaran (Hoffman and Schrag 2002) or even younger climatic events.

Closure of the gulf that separated the São Francisco mega-peninsula from the Congo continent culminated with the development of the Araçuaí-West Congo orogen between

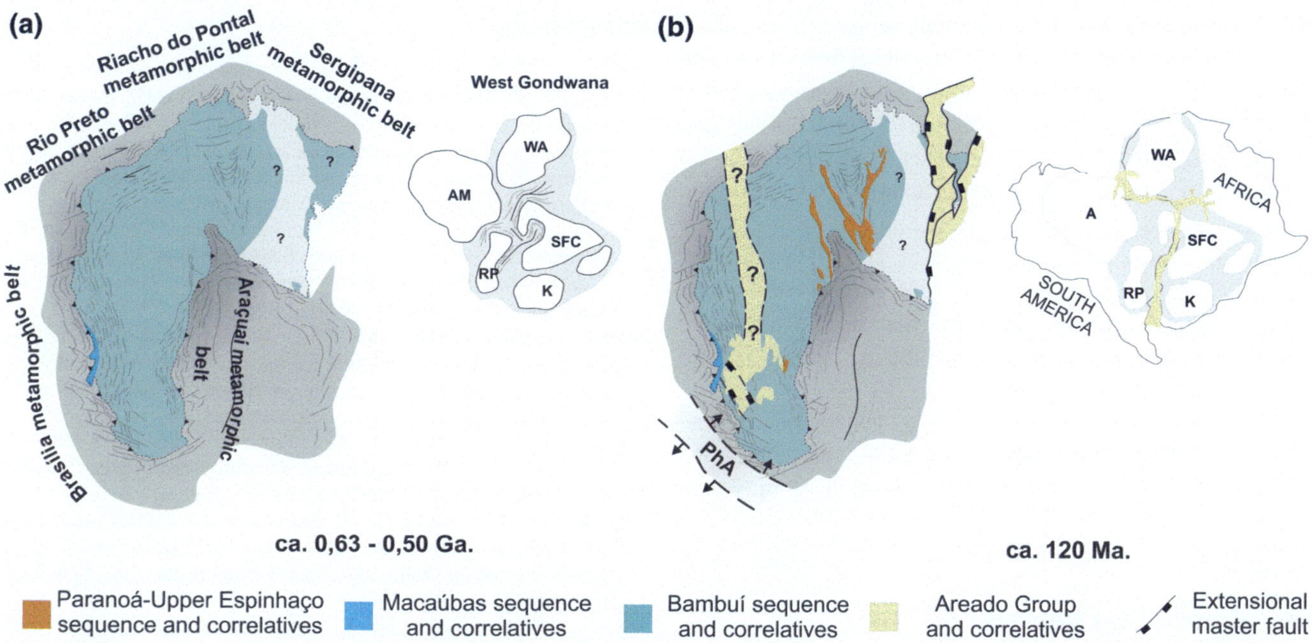

Fig. 7.13 Cartoons illustrating the tectonic evolution of the São Francisco craton between ca. 630 and 120 Ma (See text for explanation)

ca. 580 and 540 Ma. (Pedrosa-Soares et al. 2001, 2007; Alkmim et al. 2006). The Araçuaí orogenic front propagated towards the São Francisco basin, thereby affecting the Bambuí strata and causing partial inversion of the Pirapora aulacogen within the foreland foreland fold-thrust belt (Figs. 7.4, 7.9 and Table 7.1). Once formed, the Araçuaí belt has also fed the restricted continental deposits of the uppermost Bambuí, exposed in few areas along the eastern São Francisco basin (Kuchenbecker et al. 2014).

In the final stages of West Gondwana assembly, the collision between Amazonia and São Francisco triggered the development of the northern Brasilia belt (Tohver et al. 2006; Araujo-Filho 2000), and a second contractional deformation phase along the preexistent southern Brasilia belt. As a consequence of that, the package of Bambuí sediments deposited on the craton interior became involved in the deformation, giving rise to the northern Brasília foreland fold-thrust belt (Figs. 7.4, 7.9 and 7.10). Simultaneously, an extensive right-lateral strike-slip system, which includes the Rio Preto belt and its external foreland portions, formed along the northern margin of the craton (Caxito et al. 2014).

From its residence in West Gondwana and Pangea, the São Francisco basin preserves only the Santa Fé glaciogenic sediments. These sediments accumulated around the Carboniferous-Permian boundary, when the supercontinent wandered along polar latitudes (Campos and Dardenne 1997a, b; Sgarbi et al. 2001; Limarino et al. 2014).

The Gondwana breakup in the early Cretaceous resulted in the opening of the South Atlantic and separation of the São Francisco and Congo cratons (Porada 1989; Pedrosa-Soares et al. 2001). The main segments of this large rift system evolved into the present-day passive and transform margins of the South America and Africa (e.g.: Zalán 2004; Mohriak and Leroy 2012). Other branches of the system, such as the Abaeté graben in the São Francisco basin, failed and became important depocenters in interior of the new continent (Figs. 7.8 and 7.13).

The opening of the Abaeté graben took place through the extensional reactivation of the underlying Precambrian structures (Sawasato 1995; Reis 2011) (Fig. 7.8). The graben received the sediments of the early Cretaceous Areado Group under arid to semi-arid climatic conditions (Sgarbi 2000; Sgarbi et al. 2001; Fragoso 2011). The radiolarian fauna found in silexites interbedded with the Areado continental deposits led Kattah (1991) and Arai (2009) to postulate that a marine incursion coming from the north reached the Abaeté graben.

Intraplate stress rearrangements in the course of the South Atlantic event caused the uplift of the Alto Paranaíba arch that affected the southwestern portion of the São Francisco basin (Alkmim and Martins-Neto 2001; Alkmim 2004) (Figs. 7.8 and 7.13). Despite of the paucity of precise paleotectonic/paleothtermal reconstructions, the arch uplift most likely started prior to the Cretaceous, and was successively amplified afterwards (Hasui and Haralyi 1991; Sgarbi et al. 2001; Hackspacher et al. 2007; Sgarbi 2011a, b). This scenario indicates that the Alto Paranaíba arch uplift is, at least partially, coeval to the generation of the Abaeté graben.

In the Upper Cretaceous, the Alto Paranaíba arch experienced a new reactivation. This uplift phase is associated with the extensive intraplate magmatic episode that culminated with the intrusion of NW-trending dykes, alkaline plutons, and the extrusion of kamafugitic lavas and pyroclastics rocks of the Mata da Corda Group (Borges and Drews 2001; Silva 2006; Hackspacher et al. 2007; Grasso 2010; Sgarbi 2011a, b). For some authors, this magmatic episode occurred in response to the action of a mantle plume beneath the craton border (Sgarbi et al. 2001; Hackspacher et al. 2007; Franco-Magalhães 2009). Regardless of the lack of precise age constraints, the sedimentation of the Urucuia Group might have occurred, at least in part, during the Alto Paranaíba arch uplift.

The mechanism involved in the generation of the prominent NE-striking oblique-slip faults that affect the Cretaceous units in the southwestern portion of the basin remains unclear.

7.6 Concluding Remarks

Covering a large portion of the craton, the São Francisco basin is underlain by a thick and cold lithosphere and corresponds to a typical intracratonic basin. Its multiple and superimposed basin-cycles (younger than 1.8 Ga) record the main tectonic and climatic events, which affected the São Francisco/Congo plate in the time between the Late Paleoproterozoic and the Upper Cretaceous. The following tectonic events have been so far recognized in the basin: (i) multiple Paleoproterozoic (?) to Early Neoproterozoic rifting and magmatic events; (ii) the Late Neoproteroic to Early Paleozoic Brasiliano-PanAfrican collisions (West Gondwana assembly) and (iii) the Lower Cretaceous South Atlantic rifting. These events and their geological record in most cases are also described in other parts of the São Francisco craton and along its marginal belts.

The Neoproterozoic basin-forming events took place contemporaneously to glacial ages of global significance. Similar successions found worldwide have been interpreted as the record of the Neoproterozoic Snowball Earth glaciations (Hoffman and Schrag 2002). More studies are required to better constrain the age and distribution of the glacial strata that occur within the São Francisco craton and along its marginal belts. The few Permo-Carboniferous glaciogenic deposits of the Santa Fé Group mark the passage of the Gondwana through polar latitudes in the Late Paleozoic.

Acknowledgments This work was partially developed during hydrocarbon exploration programs conducted by Petra Energia S.A., which has provided technical and financial support and allowed the use of seismic and well data. F.F. Alkmim received support from CNPq, grant # 308045/2013-0. U.G. Cordani, M. Kuchenbecker, D.G.C Fragoso, and J.C.D. Sanglard are thanked for the helpful discussions and reviews, which greatly improved the original manuscript.

References

Alkmim FF (2004) O que faz de um craton um cráton? O Cráton do São Francisco e as revelações almeidianas ao delimitá-lo. In: Mantesso-Neto V, Bartorelli A, Carneiro CDR, Brito-Neves BB de (orgs) Geologia do Continente Sul Americano: Evolução da obra de Fernando Flávio Marques de Almeida. Beca, São Paulo.

Alkmim FF, Marshak S (1989) Transamazonian Orogeny in the Southern São Francisco Craton Region, Minas Gerais, Brazil: evidence for Paleoproterozoic collision and collapse in the Quadrilátero Ferrífero. Precambrian Resarch 90: 29–58.

Alkmim FF, Martins-Neto M (2001) Bacia intracratonica do São Francisco: Arcabouço estrutural e cenários evolutivos. In: Pinto CP, Martins-Neto M A (eds) Bacia do São Francisco: Geologia e Recursos Naturais. SBG/MG, Belo Horizonte.

Alkmim FF, Martins-Neto M (2012) Proterozoic first-order sedimentary sequences of the São Francisco craton, eastern Brazil. Marine and Petroleum Geology 33(1): 127–139.

Alkmim FF, Brito-Neves BB, Castro-Alves JA (1993) Arcabouço tectônico do Cráton do São Francisco: Uma revisão. In: Dominguez JML, Misi A (eds) O Cráton do São Francisco. SBG/BA-SE, Salvador.

Alkmim FF, Chemale F, Endo I (1996) A deformação das coberturas proterozoicas do Cráton do São Francisco e o seu significado tectônico. Revista Escola de Minas 49(1): 22–38.

Alkmim FF, Marshak S, Pedrosa-Soares AC, Peres GG, Cruz SCP, Whittington A (2006) Kinematic evolution of the Araçuaí-West Congo orogeny in Brazil and Africa: Nutcracker tectonics during the Neoproterozoic assembly of Gondwana. Precambrian Res 149: 43–64.

Alkmim FF, Pedrosa-Soares AC, Cruz SCP, Silva CMT (2011) Map-view curves of the Brasiliano/PanAfrican Araçuaí and West Congolian belts: Products of craton-orogen interactions during the assembly of West Gondwana. In: Schmitt RS, Trouw R, Carvalho IS, Collins A (eds) Gondwana 14, Abstracts. Universidade Federal do Rio de Janeiro (UFRJ), Rio de Janeiro.

Almeida FFM (1977) O Cráton do São Francisco. Revista Brasileira de Geociências 7(4): 349–364.

Alvarenga CJS (2012) Estratigrafia da Bacia do São Francisco: Uma discussão sobre novos dados. In: Proceedings of the 46th Congresso Brasileiro de Geologia, Sociedade Brasileira de Geologia (SBG), Santos.

Alvarenga CJS, Dardenne MA, Vieira LC, Martinho CT, Guimarães EM, Santos RV, Santana RO (2012) Estratigrafia da borda ocidental da Bacia do São Francisco. Boletim de Geociências da Petrobras 20 (1/2): 145–164.

Alvarenga CJS, Santos RV, Vieira LC, Lima BAF, Mancini LH (2014) Meso-Neoproterozoic isotope stratigraphy on carbonate platforms in the Brasília Belt of Brazil. Precambrian Research 251: 164–180.

Azmy K, Kendall B, Creaser RA, Heaman L, Oliveira TFO (2008) Global correlation of the Vazante Group, São Francisco Basin, Brazil: Re-Os and U-Pb radiometric age constraints. Precambrian Res 164: 160–172.

Arai M (2009) Paleogeografia do Atlântico Sul no Aptiano: um novo modelo a partir de dados micropaleontológicos recentes. Boletim de Geociências da Petrobras 17(2): 331–351.

Araújo Filho JO (2000) The Pirineus Syntaxis: An example of the intersection of two Brasiliano Fold-Thrust Belts in central Brazil and its implications for the tectonic evolution of Western Gondwana. Revista Brasileira de Geociências 30(1): 144–148.

Babinski M, Vieira LC, Trindade RIF (2007) Direct dating of the Sete Lagoas cap carbonate (Bambuí Group, Brazil) and implications for the Neoproterozoic glacial events. Terra Nova 19: 401–406.

Babinski M, Pedrosa-Soares AC, Trindade RIF, Martins M, Noce CM, Liu D (2012) Neoproterozoic glacial deposits from the Araçuaí orogen, Brazil: Age, provenance and correlations with the São Francisco craton and West Congo belt. Gondwana Res 21: 451–465.

Baltazar OF, Zucchetti M (2007) Lithofacies associations and structural evolution of the Archean Rio das Velhas greenstone belt, Quadrilátero Ferrífero, Brazil: A review of the setting of gold deposits. Ore Geology Reviews 32: 471–499.

Baptista MC (2004) Estratigrafia e evolução geológica da região de Lagoa Formosa (MG). Dissertation, Universidade Federal de Minas Gerais (UFMG).

Barbosa O, Braun OPG, Dyer RC, Cunha CABR (1970) Geologia do Triângulo Mineiro, Boletim 136. DNPM/DFPM, Rio de Janeiro.

Bonhomme MG (1976) Mineralogie des fractions fines et datation rubidium-strontium das le Group Bambuí, Brésil. Revista Brasileira de Geociências 6(4): 39–43.

Borges AJ, Drews MGP (2001) Características magnetométricas da Bacia do São Francisco em Minas Gerais. In: Pinto C P, Martins-Neto MA (eds) Bacia do São Francisco: Geologia e Recursos Naturais. SBG/MG, Belo Horizonte.

Braun OPG (1968) Contribuição à estratigrafia do Grupo Bambuí. In: Proceedings of the 12th Congresso Brasileiro de Geologia, Sociedade Brasileira de Geologia (SBG), Belo Horizonte.

Brito-Neves BB de (2004) A história dos continentes: trajetórias e tramas tectônicas. In: Mantesso-Neto V, Bartorelli A, Carneiro CDR, Brito-Neves BB de (orgs) Geologia do Continente Sul Americano: Evolução da obra de Fernando Flávio Marques de Almeida. Beca, São Paulo.

Brito-Neves BB de, Campos-Neto M da C, Fuck RA (1999) From Rodinia to Western Gondwana: Na approach to the Brasiliano-Pan African Cycle and orogenic collage. Episodes 22(3): 155–166.

Campos JEG, Dardenne MA (1997a) Estratigrafia e sedimentação da Bacia Sanfranciscana: Uma revisão. Revista Brasileira de Geociências 27(3): 269–282.

Campos JEG, Dardenne MA (1997b) Origem e evolução tectônica da Bacia Sanfranciscana. Revista Brasileira de Geociências 27(3): 283–294.

Campos JEG, Dardenne MA, Freitas-Silva FH, Martins-Ferreira MAC (2013) Geologia do Grupo Paranoá na porção externa da Faixa Brasília. Braz J Geol 43(3): 461–476.

Castro PTA, Dardenne MA (2000) The sedimentology, stratigraphy and tectonic context of the São Francisco Supergroup at the southern boundary of the São Francisco craton, Brazil. Revista Brasileira de Geociências 30: 345–437.

Catuneanu O, Martins-Neto MA, Eriksson PG (2005) Precambrian sequence stratigraphy. Sedimentary Geology 176: 67–95.

Catuneanu O, Galloway W E, Kendall CGStC, Miall AD, Posamentier HW, Strasser A, Tucker ME (2011) Sequence stratigraphy: Metodology and Nomenclature. Newsletter on Stratigraphy (Special Issue) 44(3): 173–245.

Catuneanu O, Martins-Neto MA, Eriksson PG (2012) Sequence stratigraphic framework and application to the Precambrian. Marine and Petroleum Geology 33: 26–33.

Caxito F de A (2010) Evolução tectonica da Faixa Rio Preto, noroeste da Bahia/sul do Piauí. Dissertation, Universidade Federal Minas Gerais (UFMG).

Caxito F de A, Uhlein A, Sanglard JCD, Gonçalves-Dias T, Mendes M de CO (2012) Depositional systems and stratigraphic review proposal of the Rio Preto Fold Belt, northwestern Bahia/Southern Piauí. Revista Brasileira de Geociências 42(3): 523–538.

Caxito F de A, Dantas EL, Stevenson R, Uhlein A (2014) Detrital zircon (U-Pb) and Sm-Nd isotope studies of the provenance and tectonic setting of basins related to collisional orogens: The case of the Rio Preto fold belt on the northwest São Francisco Craton margin, NE Brazil. Gondwana Research 26: 741–754.

Chang HK, Miranda FP, Magalhães L, Alkmim FF (1988) Considerações sobre a evolução tectônica da bacia do São Francisco. In: Proceedings of the 35th Congresso Brasileiro de Geologia, Sociedade Brasileira de Geologia (SBG), Belém, vol 5, pp 2076–2090.

Chiavegatto JRS (1992) Análise estratigráfica das sequencias tempestíticas da Formação Três Marias (Proterozóico Superior), na porção meridional da Bacia do São Francisco. Dissertation, Universidade Federal de Ouro Preto (UFOP).

Chavegatto JRS, Gomes NS, Dardenne MA (1997) Conglomerados oligomíticos da Formação Três Marias na Serra do Gurutuba, norte de Minas Gerais. Boletim SBG/Núcleo MG 14:83–84.

Chemale Jr F, Dussin IA, Alkmim FF, Martins MS, Queiroga G, Armstrong R, Santos, MN (2012) Unravelling a Proterozoic basin history through detrital zircon geochronology: The case of the Espinhaço Supergroup, Minas Gerais, Brazil. Gondwana Res 22: 200–206.

Coelho JCC (2007) Estilos estruturais e evolução tectônica da borda oeste da Bacia do São Francisco, com base na integração de dados de superfície, sub-superfície, litogeoquímica e isótopos. Dissertation, Universidade Federal de Ouro Preto (UFOP).

Coelho JCC, Martins-Neto MA, Marinho MS (2008) Estilos estruturais e evolução tectônica da porção mineira da bacia proterozoica do São Francisco. Revista Brasileira de Geociências 38(2-suplemento): 149–165.

Costa DA (2011) Controle lito-estrutural e estratigráfico na Hidrogeoquímica e nas concentrações de fluoreto no sistema aquífero carstico-fissural do Grupo Bambuí, norte de Minas Gerais. Dissertation, Universidade Federal de Minas Gerais (UFMG).

Costa MT da, Branco JJR (1961) Introdução. *In*: Branco JJJR (ed) Roteiro para excursão Belo Horizonte-Brasília. SBG/Congresso Brasileiro de Geologia 15: 1–119.

Costa-Neto SF da (2006) Ritmito superior do Grupo Paranoá e fim da deposição de margem passiva. Dissertation, Universidade de Brasília (UnB).

Cruz SCP, Alkmim FF (2006) The tectonic interaction between the Paramirim Aulacogen and the Araçuaí belt, São Francisco craton region, Eastern Brazil. Anais da Academia Brasileira de Ciências 78 (1): 151–173.

D'Arrigo HBP (1995) O descolamento basal do Grupo Bambuí e o Alto de Sete Lagoas. Dissertation, Universidade Federal de Ouro Preto (UFOP).

Danderfer A, De Waele B, Pedreira AJ, Nalini HA (2009) New geochronological constraints on the geological evolution of Espinhaço basin within the São Francisco Craton – Brazil. Precambrian Res 170: 116–128.

Danderfer Filho, A., Lana C.C., Nalini Junior, H.A., Costa, A.F.O. (2015). Constraints on the Statherian evolution of the intraplate rifting in a Paleo-Mesoproterozoic paleocontinent: New stratigraphic and geochronology record from the eastern São Francisco craton. Precambrian Research, 28, 668–688.

Dardenne MA (1978) Síntese sobre a estratigrafia do Grupo Bambuí no Brasil Central. In: Proceedings of the 30th Congresso Brasileiro de Geologia, Sociedade Brasileira de Geologia (SBG), Recife, vol 2, pp 507–610.

Dardenne MA (1981) Os grupos Paranoá e Bambuí na Faixa Dobrada Brasília. In: Proceedings of the 1st Simpósio sobre o Cráton do São Francisco, Sociedade Brasileira de Geologia (SBG) - Núcleo BA, Salvador, pp 140–157.

Dardenne MA (2000) The Brasília fold belt. In: Cordani UG, Milani EJ, Thomaz Filho A, Campos DA (eds) Tectonic Evolution of South America. 31st International Geological Congress, Rio de Janeiro.

Dardenne MA, Pimentel MM, Alvarenga CJS (2003) Provenance of conglomerates of the Bambuí, Jequitaí, Vazante and Ibiá groups: Implications for the evolution of The Brasília Belt. In: Boletim de

Resumos of the 9th Simpósio Nacional de Estudos Tectônicos, pp 47–49.

Derby OA (1879) Contribuição para o estudo da geologia do Valle do Rio S. Francisco. Archivos do Museu Nacional, vol 4, Rio de Janeiro.

Dias PHA, Noce CM, Pedrosa-Soares AC, Seer HJ, Dussin IA, Valeriano C de M, Kuchenbecker M (2011) O Grupo Ibiá (Faixa Brasília Meridional): evidências isotópicas Sm-Nd e U-Pb de bacia colisional do tipo *flysch*. Geonomos 19(2): 90–99.

Domingues JML (1993) As coberturas do Cráton do São Francisco: uma abordagem do ponto de vista da análise de bacias. In: Dominguez JML, Misi A (eds) O Cráton do São Francisco. Sociedade Brasileira de Geologia-Núcleo BA-SE/SGM, Salvador.

Domingues RAP (2009) Paleogeografia do Alto de Paracatu: O registro geológico dos *bone-beds* de dinossauros da Bacia Sanfranciscana. Dissertation, Universidade de São Paulo (USP).

Dorr II JVN (1969) Physiographic, stratigraphic and structural development of the Quadrilátero Ferrífero, Minas Gerais, Brazil. US Geological Surrvey Professional Paper 641(A): 1–110.

Dupont H (1995) O Grupo Conselheiro Mata no seu quadro paleogeográfico e estratigráfico. Boletim da Sociedade Brasileira de Geologia, Núcleo Minas Gerais 13: 9–10.

Dussin IA, Dussin TM (1995) Supergrupo Espinhaço: Modelo de Evolução geodinâmica. Geonomos 3(1):19–26.

Egydio-Silva M, Karmann I, Trompette RR (1989) Litoestratigrafia do Supergrupo Espinhaço e Grupo Bambuí no noroeste do estado da Bahia. Revista Brasileira de Geociências 19(2): 101–112.

Feitosa IL (2012) Litofácies, ambientes deposicionais e ciclicidade do Grupo Paranoá: Exemplo da região da Serra de São Domingos, nos municípios de Buritis e Formoso, MG. Dissertation, Universidade de Brasília (UnB).

Figueiredo FT (2008) Fácies sedimentares e proveniência da Formação Bebedouro, Neoproterozoico (BA). Dissertation, Universidade de São Paulo (USP).

Fonseca MA, Dardenne MA, Uhlein A (1995) Faixa Brasília setor setentrional: estilos estruturais e arcabouço tectônico. Revista Brasileira de Geociências 25(4): 267–278.

Fragoso DGC (2011) Geologia da região de Presidente Olegário e evolução tectono-sedimentar do Grupo Areado, Eocretácio da Bacia Sanfranciscana, Minas Gerais. Dissertation, Universidade Federal de Minas Gerais (UFMG).

Fragoso DGC, Uhlein A, Sanglard JCD, Suckau GL, Guerzoni HTG, Faria PH (2011). Geologia dos grupos Bambuí, Areado e Mata da Corda na Folha Presidente Olegário (1:100.000), MG: Registro deposicional do neoproterozoico ao neocretácio da Bacia do São Francisco. Geonomos 19(1): 28–38.

Franco-Magalhães AOB (2009) Exumação tectônica e evolução associada do relevo no Arco de Ponta Grossa, Sul-Sudeste do Brasil. PhD thesis, Universidade Estadual Paulista (Unesp).

Grasso CB (2010) Petrologia do Complexo Alcalino-Carbonatítico de Serra Negra, MG. Dissertation, Universidade de Brasília (UnB).

Hackspacher PC, Godoy DF de, Ribeiro LFB, Hadler-Neto JC, Franco AOB (2007) Modelagem térmica e geomorfologia da borda sul do Cráton do São Francisco: termocronologia por traços de fissão em apatita. Revista Brasileira de Geociências 37(4): 78–86.

Hasui Y, Haralyi NLE (1991) Aspectos lito-estruturais e geofísicos do soerguimento do Alto Paranaíba. Geociências 10: 57–77.

Heineck CA, Leite CAS, Silva MA da, Vieira VS (2003) Mapa Geológico de Minas Gerais, Escala 1:1.000.000. CPRM-Serviço Geológico do Brasil, Belo Horizonte.

Hercos CM (2008) Arcabouço estrutural da Bacia do São Francisco nos arredores da Serra da Água Fria e da Onça, porção centro-norte do estado de Minas Gerais. Dissertation, Universidade Federal de Ouro Preto (UFOP).

Hoffman PF (1991) Did the breakout of Laurentia turn Gondwanaland inside out? Science 252: 1409–1412.

Hoffman PF, Schrag DP (2002) The snowball Earth hypotesis: testing the limits of global change. Terra Nova 14: 129–155.

Iglesias M, Uhlein A (2009) Estratigrafia do Grupo Bambuí e coberturas fanerozoicas no vale do rio São Francisco, norte de Minas Gerais. Revista Brasileira de Geociências 39(2): 256–266.

Iyer SS, Babinski M, Krouse HR, Chemale Jr F (1995) Highly ^{13}C enriched carbonate and organic matter in the Neoproterozoic sediments of the Bambuí Group, Brazil. Precambrian Research 73: 271–282.

Kadima E, Delvaux D, Sebagenzi SN, Tack L, Kabeya SM (2011) Structure and geological history of the Congo Basin: an integrated interpretation of gravity, magnetic and reflection seismic data. Basin Research 23: 499–527.

Kaminski E, Jaupart C (2000) Lithosphere structure beneath the Phanerozoic intracratonic basins of North America. Earth and Planetary Science Letters 178: 139–149.

Karfunkel J, Hoppe A (1988) Late Proterozoic Glaciation in Central-Eastern Brazil: synthesis and model. Paleogeogr., Paleoclim., Paleoecol. 65: 1–21.

Kattah SS (1991) Análise faciológica e estratigráfica do Jurássico Sup./ Cretáceo Inf. na porção meridional da Bacia do São Francisco, oeste do estado de Minas Gerais. Dissertation, Universidade Federal de Ouro Preto (UFOP).

King SD (2005) Archean cratons and mantle dynamics. Earth and Planetary Science Letters 234: 1–14.

Knauer LG, Costa, RD, Freimann M, Ferreira MP (2011) Folha Morada Nova de Minas, SE.23-Y-B-VI, escala 1:100.000: nota explicativa. In: Pedrosa-Soares AC, Noce CM, Voll E, Kuchenbecker M, Reis HLS, Fragoso DGC (eds) Projeto Alto Paranaíba. CODEMIG-UFMG, Belo Horizonte.

Korsch RJ, Lindsay JF (1989) Relationships between deformation and basin evolution in the intracratonic Amadeus Basin, central Australia. Tectonophysics 158: 5–22.

Kroener, A, and Cordani, U. G. (2003). African, southern Indian and South American cratons were not part of the Rodinia supercontinent: evidence from field relationships and geochronology. Tectonophysics, 375, 325–352.

Kuchenbecker M (2011) Quimioestratigrafia e proveniência sedimentar da porção basal do Grupo Bambuí em Arcos (MG). Dissertation, Universidade Federal de Minas Gerais (UFMG).

Kuchenbecker M, Babinski M, Pedrosa-Soares AC, Costa RD da, Lopes-Silva L, Pimenta F (2013) Proveniência e análise sedimentar da porção basal do Grupo Bambuí em Arcos (MG). Geologia USP 13(4): 49–61.

Kuchenbecker M, Pedrosa-Soares AC, Babinski M, Fanning M (2015) Detrital zircon age patterns and provenance assessment for pre-glacial to post-glacial successions of the Neoproterozoic Macaúbas Group, Araçuaí orogen, Brazil. Precambrian Research 266: 12–26.

Kuchenbecker M, Reis HLS, Silva, LC da, Costa RD da, Fragoso DGC, Knauer LG, Pedrosa-Soares AC, Dussin I (2014) Age constraints for deposition and sedimentary provenance of the Espinhaço Supergroup and Bambuí Group in eastern São Francisco craton. Geonomos (in the press).

Lana C, Alkmim FF, Armstrong R, Scholz R, Romano R (2013) The ancestry and magmatic evolution of Archean TTG rocks of the Quadrilátero Ferrífero province, southeast Brazil. Precambrian Research 231: 157–173.

Lima OMB (2011) Estratigrafia isotópica e evolução sedimentar do Grupo Bambuí na borda ocidental do Cráton do São Francisco: implicação tectônica e paleo-ambiental. PhD thesis, Universidade de Brasília (UnB).

Lima ONB, Uhlein A, Britto W de (2007) Estratigrafia do Grupo Bambuí na Serra da Saudade e geologia do depósito fosfático de Cedro do Abaeté, Minas Gerais. Revista Brasileira de Geociências 37(4-suplemento): 204–215.

Lima SA de A, Martins-Neto MA, Pedrosa-Soares AC, Cordani UG, Nutman A (2002) A Formação Salinas na área-tipo, NE de Minas Gerais: Uma proposta de revisão da estratigrafia da Faixa Araçuaí com base em evidências sedimentares, metamórficas e idades U-Pb SHRIMP. Revista Brasileira de Geociências 32(4): 491–500.

Limarino CO, Césari SN, Spalletti LA, Taboada AC, Isbell JL, Geuna S, Gulbranson EL (2014) A paleoclimatic review of Southern South America during the late Paleozoic: A record from icehouse to extreme greenhouse conditions. Gondwana Research 25 (4): 1396–1421.

Lopes TC (2012) O Supergrupo Espinhaço na Serra do Cabral, Minas Gerais: contribuição ao estudo de proveniência sedimentar. Dissertation, Universidade Federal de Minas Gerais (UFMG).

Machado N, Schrank A, Abreu FR, Knauer LG, Abreu PAA (1989) Resultados preliminares da geocronologia U-Pb na Serra do Espinhaço meridional. In: Proceedings of the 5th Simpósio de Geologia de Minas Gerais, Sociedade Brasileira de Geologia-Núcleo MG, Belo Horizonte, 10: 171–174.

Magalhães L (1988) Análise estrutural qualitativa dos sedimentos do Grupo Bambuí, região sudeste da Bacia do São Francisco (Faixa Sete Lagoas-Serra do Cipó). Dissertation, Universidade Federal de Ouro Preto.

Marshak S, Wilkerson MS (1992) Effect of overburden thickness on thrust belt geometry and development. Tectonics 11(3): 560–566.

Martins-Neto MA (2009) Sequence stratigraphic framework of Proterozoic successions in eastern Brazil. Marine and Petroleum Geology 26: 163–176.

Martins-Neto MA, Hercos CM (2002) Sedimentation and tectonic setting of Early Neoproterozoic glacial deposits in southern Brazil. In: Altermann W, Corcoran PL (eds) Precambrian Sedimentary Environments: A Modern Approach to Ancient Depositional Systems, 33. Special Publication of the International Association of Sedimentologists.

Martins-Neto M A, Pinto CP (2001) Bacia do São Francisco: Definição e base de dados. In: Pinto CP, Martins-Neto MA (eds) Bacia do São Francisco: Geologia e Recursos Naturais. Sociedade Brasileira de Geologia-Núcleo MG, Belo Horizonte.

Martins-Neto MA, Pedrosa-Soares AC, Lima SAA (2001) Tectono-sedimentary evolution of sedimentar basins from late Paleoproterozoic to late Neoproterozoic in the São Francisco craton and Araçuaí fold belt, eastern Brazil. Sedimentary Geology 141–142: 343–370.

Matteini M, Dantas EL, Pimentel MM, Alvarenga CJS, Dardenne MA (2012) U-Pb and Hf isotope study on detrital zircons from Paranoá Group, Brasília Belt, Brazil: constraints on depositional age at Mesoproterozoic-Neoproterozoic transition and tectono-magmatic events in the São Francisco craton. Precambrian Research 206/207: 168–181.

Meert JG (2012) What's the name? The Columbia (Paleopangea/Nuna) supercontinent. Gondwana Research 21: 987–993.

Menezes-Filho MR, Mattos GMM, Ferrari PG (1977) Projeto Três Marias: Relatório final, vol 6. DNPM/CPRM-Serviço Geológico do Brasil, Brasília.

Misi A (2001) Estratigrafia isotópica das sequencias do Supergrupo São Francisco, coberturas neoproterozoicas do Cráton do São Francisco: idade e correlações. In: Pinto CP, Martins-Neto MA (eds) Bacia do São Francisco: Geologia e Recursos Naturais. Sociedade Brasileira de Geologia-Núcleo MG, Belo Horizonte.

Misi A, Veizer J (1998) Neoproterozoic carbonate sequences of the Una Group, Irecê Basin: chemostratigraphy, age and correlations. Precambrian Research 89: 87–100.

Mohriak WU, Leroy S (2012) Architecture of rifted continental margins and break-up evolution: insights from the South Atlantic, North Atlantic and Red Sea-Gulf of Aden conjugate margins. In: Mohriak WU, Danforth A, Post PJ, Brown DE, Tari GC, Nemcok M, Sinha ST (eds) Conjugate Divergent Margins. Geological Society of London, Special Publication 369. doi:10.1144/SP369.17.

Muzzi Magalhães P (1989) Análise estrutural qualitativa das rochas do Grupo Bambuí, na porção sudoeste da Bacia do São Francisco. Dissertation, Universidade Federal de Ouro Preto (UFOP).

Nance RD, Murphy JB, Santosh M (2014) The supercontinent cycle: A retrospective essay. Gondwana Research 25(1): 4–29.

Neill CJ, Lenardic A, Griffin WL, O'Reilly SY (2008) Dynamics of cratons in an evolving mantle. Lithos 102: 12–24.

Nobre-Lopes J (1995) Faciologia e gênese dos carbonatos do Grupo Bambuí na região de Arcos, estado de Minas Gerais. Dissertation, Universidade de São Paulo (USP).

Nobre-Lopes J (2002) Diagenesis of the dolomites hosting Zn/Ag mineral deposits in the Bambuí Group at Januária region-MG. PhD thesis, Universidade Estadual de Campinas (Unicamp).

Noce CM, Pedrosa-Soare AC, Silva LC da, Alkmim FF (2007) O embasamento arqueano e paleoproterozoico do Orógeno Araçuaí. Geonomos 15(1): 17–23.

Oliveira JRP (1989) Comportamento estrutural dos grupos Macaúbas e Bambuí na porção centro-norte de Minas Gerais. Dissertation, Universidade Federal de Ouro Preto (UFOP).

Oliveira LGS (2009) A Missão Grace e a estrutura da listosfera na regiaõ do Cráton do São Francisco. PhD thesis, Universidade Federal de Ouro Preto (UFOP).

Oliveira NV de, Endo I, Shukowsky W (2012) Interpretation of the geomagnetic anomalies in the São Francisco craton region based on CHAMP Mission data. Brazilian Journal of Geophysics 30(1): 93–101.

O'Reilly SY, Griffin WL, Djomani YHP, Morgan P (2001) Are lithospheres forever? Tracking changes in subcontinental lithospheric mantle through time. GSA Today 11(4): 4–10.

Paula-Santos GM, Babinski M, Kuchenbecker M, Caetano-Filho S, Trindade RI, Pedrosa-Soares AC (2015). New evidence of na Ediacaran age for the Bambuí Group in Southern São Francisco craton (eastern Brazil) from zircon U-Pb data and isotope chemostratigraphy. Gondwana Research 28(2): 702–720.

Pedreira AJ, de Waele B (2008) Contemporaneous evolution of the Paleoproterozoic-Mesoproterozoic sedimentary basins of the São Francisco-Congo Craton. In: Pankhurst RJR, Trow AJ, Brito-Neves BB, de Wit MJ (eds) West Gondwana: Pre-Cenozoic Correlations Across the South Atlantic Region. Geological Society of London Special Publication 294:33–48.

Pedrosa-Soares AC, Alkmim FF (2011) How many rifting events preceded the development of the Araçuaí-West Congo orogeny? Geonomos 19(2): 244–251.

Pedrosa-Soares AC, Noce CM, Wiedemann CM, Pinto CP (2001) The Araçuaí-West-Congo Orogen in Brazil: na overview of a confined orogen formed during Gondwanaland assembly. Precambrian Res. 110: 307–323.

Pedrosa-Soares AC, Noce CM, Alkmim FF, Silva LC da, Babinski M, Cordani UG, Castañeda C (2007) Orógeno Araçuaí: síntese do conhecimento 30 anos após Almeida 1977. Geonomos 15(1): 1–16.

Pedrosa-Soares AC, Babinski M, Noce CM, Martins M, Queiroga G, Vilela F (2011a). The Neoproterozoic Macaúbas Group (Araçuaí orogen, SE Brazil) with emphasis on the diamictite formations. In: Arnaud E, Halverson GP, Shields-Zhou G (orgs) The Geological Record of Neoproterozoic Glaciations. Memoir of the Geological Society of London 36: 523–534.

Pedrosa-Soares AC, Noce CM, Voll E, Kuchenbecker M, Reis HLS, Fragoso DGC (2011b) Projeto Alto Paranaíba. CODEMIG-UFMG, Belo Horizonte.

Pflug R, Renger F (1973) Estratigrafia e evolução geológica da margem sudeste do Cráton Sanfranciscano. *In*: Proceedings of the 27th Congresso Brasileiro de Geologia, Sociedade Brasileira de Geologia, Aracaju, vol 1(2), pp 5–19.

Pimentel MM, Fuck RA, Botelho NF (1999) Granites and the geodynamic history of the Neoproterozoic Brasília belt, Central Brazil: a review. Lithos 46: 463–483.

Pimentel MM, Dardenne MA, Fuck RA, Viana MG, Junges SL, Fischel DP, Seer HJ, Dantas EL (2001) Nd isotopes and the provenance of detrital sediments of the Neoproterozoic Brasília Belt, central Brazil. South American Earth Sciences 14: 571–585.

Pimentel MM, Jost H, Fuck RA (2004) O embasamento da Faixa Brasília e o Arco Magmático de Goiás. In: Mantesso-Neto V, Bartorelli A, Carneiro CDR, Brito-Neves BB de (orgs) Geologia do Continente Sul Americano: Evolução da obra de Fernando Flávio Marques de Almeida. Beca, São Paulo, pp 356–368.

Pimentel MM, Rodrigues JB, DellaGiustina MES, Junges S, Matteini M, Armstrong R (2011) The tectonic evolution of the Neoproterozoic Brasília Belt, central Brazil, based on SHRIMP and LA-ICPMS U-Pb sedimentary provenance data: a review. Journal of South American Earth Sciences 31: 345–357.

Pimentel M, Della Giustina MES, Rodrigues JB, Junges SL (2012) Idades dos grupos Araxá e Bambuí: Implicações para a evolução da Faixa Brasília. In: Proceedings of the 46th Congresso Brasileiro de Geologia, Sociedade Brasileira de Geologia (SBG), Santos.

Pinho JMM (2008) Mapa Geológico da Folha Belo Horizonte, Escala 1:100.000. CPRM-Serviço Geológico do Brasil, Belo Horizonte.

Pisarevsky, S.A., Wingate, M.T.D., Powell, C.M., Johnson, S., Evans, D.A.D., 2003.Models for Rodinia assembly and fragmentation. In: Yoshida, M., Windley, B.F., Dagupta, S. (Eds.), Proterozoic East Gondwana: Supercontinent Assembly and Breackup. Geological Society, London, Special Publication, 206, p. 35–55.

Porada H (1989) Pan-African rifting and orogenesis in southern to equatorial Africa and Eastern Brazil. Precambrian Research 44: 103–136.

Queiroga GN, Pedrosa-Soares AC, Noce CM, Alkmim FF, Pimentel MM, Dantas E, Martins M, Castañeda C, Suita MT de F, Prichard H (2007) Age of the Ribeirão da Folha Ophiolite, Araçuaí Orogen: The U-Pb zircon (LA-ICPMS) dating of a plagiogranite. Geonomos 15(1): 61–65.

Reis HLS (2011) Estratigrafia e tectônica da Bacia do São Francisco na zona de emanações de gás natural do baixo Rio Indaiá (MG). Dissertation, Universidade Federal de Ouro Preto (UFOP).

Reis HLS, Alkmim FF (2015) Anatomy of a basin-controlled foreland fold-thrust belt curve: The Três Marias salient, São Francisco basin, Brazil. Marine and Petroleum Geology 66: 711–731.

Reis HLS, Alkmim FF, Pedrosa-Soares AC, Barbosa MSC (2011) The Três Marias Salient of the Neoproterozoic Brasília Foreland Fold-Thrust Belt, São Francisco Basin, Brazil. In: Schmitt RS, Trouw R, Carvalho IS, Collins A (eds) Gondwana 14, Abstracts. Universidade Federal do Rio de Janeiro, Rio de Janeiro.

Reis HLS, Alkmim FF, Silva LC (2012) O Cinturão neoproterozoico de antepaís da Faixa Brasília, Bacia do São Francisco (Brasil) – Características e principais traços tectônicos. In: Proceedings of the 46th Congresso Brasileiro de Geologia, Sociedade Brasileira de Geologia (SBG), Santos.

Reis HLS, Costa RD da, Prezotti FPS, Tedeschi M, Fonseca HAM da, Kuchenbecker M (2014) Geologia e recursos minerais da Folha Andrequicé (SE.23-Z-A-I). CPRM-Serviço Geológico do Brasil, Belo Horizonte.

Reis HLS, Fonseca RCS, Alkmim FF, Nascimento TC, Suss J (2013) A Bacia do São Francisco (MG): registro de uma longa história de ativações e reativações em domínio cratônico. In: Proceedings of the 17th Simpósio de Geologia de Minas Gerais and 13th Simpósio de Geologia do Sudeste, Sociedade Brasileira de Geologia (SBG) – Núcleo MG, Juiz de Fora.

Rimann F (1917) A kimberlita no Brasil. In: Proceeding of the Escola de Minas-Universidade Federal de Ouro Preto, Ouro Preto, vol 15, pp 27–32.

Rocha MP, Schimmel M, Assumpção M (2011) Upper-mantle seismic structure beneath SE and Central Brazil from P- and S-wave regional travel time tomography. Geophysical Journal International 184: 268–286.

Rodrigues JB (2008) Proveniência dos sedimentos dos grupos Canastra, Ibia, Vazante e Bambui. Um estudo de zircões detriticos e idades modelo Sm-Nd. PhD thesis, Universidade de Brasília (UnB).

Rodrigues JB, Pimentel MM, Dardenne MA, Armstrong RA (2010) Age, provenance and tectonic setting of the Canastra and Ibiá Groups (Brasília Belt, Brazil): Implications for the age of a Neoproterozoic glacial event in central Brazil. Journal of South American Earth Sciences 29: 512–521.

Rodrigues JB, Pimentel MM, Buhn B, Matteini M, Dardenne MA, Alvarenga CJS, Armstrong RA (2012) Provenance of the Vazante Group: New U-Pb, Sm-Nd, Lu-Hf isotopic data and implications for the tectonic evolution of the Neoproterozoic Brasília Belt. Gondwana Research 21: 439–450.

Roger JJW, Santosh M (2002) Configuration of Columbia, a Mesoproterozoic Supercontinent. Gondwana Research 5(1): 5–22.

Romano AW (2007) Geologia da Folha Pará de Minas (SE.23-Z-C-IV), escala 1:100.000: nota explicativa. UFMG/CPRM-Serviço Geológico do Brasil, Belo Horizonte.

Romano R, Lana C, Alkmim FF, Stevens G, Armstrong R (2013) Stabilization of the southern portion of the São Francisco craton, SE Brazil, through a long-lived period of potassic magmatism. Precambrian Research 224: 143–159.

Santos RV, Alvarenga CJS, Dardenne MA, Sial AN, Ferreira VP (2000) Carbon and oxygen isotope profiles across Meso-Neoproterozoic limestones from central Brazil: Bambuí and Paranoá groups. Precambrian Research 104: 107–122.

Santos RF dos, Alkmim FF, Pedrosa-Soares AC (2009) A Formação Salinas, Orógeno Araçuaí (MG): história deformacional e significado tectônico. Revista Brasileira de Geociências 39(1): 81–100.

Sawasato EY (1995) Estruturação da porção meridional da Bacia Alto-SanFranciscana, Cretáceo do Oeste de Minas Gerais. Dissertation, Universidade Federal de Ouro Preto (UFOP).

Schobbenhaus C (1996) As tafrogêneses superpostas Espinhaço e Santo Onofre, estado da Bahia: revisão e novas propostas. Revista Brasileira de Geociências 26(4): 265–276.

Schöll WU (1973) Sedimentologie und Geochemie der Bambuí Gruppe am SE-Rand des São Francisco Beckens, Minas Gerais, Brasilien. PhD thesis, Universität Heidelberg.

Sgarbi GNC (2000) The Cretaceous Sanfranciscan basin, eastern plateau of Brazil. Revista Brasileira de Geociências 30(3): 450–452.

Sgarbi GNC (2011a) Sedimentação do Cretáceo Inferior na Bacia Sanfranciscana: O Grupo Areado. In: Pedrosa-Soares AC, Noce CM, Voll E, Kuchenbecker M, Reis HLS, Fragoso, DGC (eds) Projeto Alto Paranaíba. CODEMIG-UFMG, Belo Horizonte.

Sgarbi GNC, Sgarbi PB de A, Campos JEG, Dardenne MA, Penha UC (2001) Bacia Sanfranciscana: O registro Fanerozóico da Bacia do São Francisco. In: Pinto CP, Martins-Neto MA (eds) Bacia do São Francisco: Geologia e Recursos Naturais. Sociedade Brasileira de Geologia (SBG) – Núcleo MG, Belo Horizonte.

Sgarbi PBA (2011b) Magmatismo do Cretáceo na Região Sudoeste da Bacia Sanfranciscana: O Grupo Mata da Corda.. In: Pedrosa-Soares AC, Noce CM, Voll E, Kuchenbecker M, Reis HLS, Fragoso, DGC (eds) Projeto Alto Paranaíba. CODEMIG-UFMG, Belo Horizonte.

Silva LG da (2006) Metodologia geofísica para discriminação de corpos intrusivos na Província Alcalina do Alto Paranaíba – MG. Dissertation, Universidade de Brasília (UnB).

Silva LC, Pedrosa-Soares AC, Teixeira LR, Armstrong R (2008) Tonian rift-related, A-type continental plutonism in the Araçuaí Orogen, eastern Brazil: New evidence for the breakup stage of the São Fransico-Congo Paleocontinent. Gondwana Research 13(4): 527–537.

Souza SL, Brito PCR, Silva RWS (1993) Estratigrafia, Sedimentologia e Recursos Minerais da Formação Salitre na Bacia de Irecê, Bahia (Série Arquivos Abertos 2). Companhia Baiana de Pesquisa Mineral (CBPM), Salvador.

Souza-Filho RG (1995) O Arcabouço Estrutural da porção externa da Faixa Aracuaí na Serra do Cabral (MG) e o contraste de estilos deformacionais entre os supergrupos Espinhaço e São Francisco. Dissertation, Universidade Federal de Ouro Preto (UFOP).

Spalletti LA, Limarino CO (2006) Comparison between intracratonic and active margin basins: Examples from the Late Paleozoic of western Gondwana. Journal of South American Earth Sciences 22 (3–4): 131–254.

Tohver E, D'Agrella-Filho MS, Trindade RIF (2006) Paleomagnetic record of Africa and South America for the 1200–500 Ma interval, and evaluation of Rodinia and Gondwana assemblies. Precambrian Research 147: 193–222.

Torsvik TH, Cocks LRM (2013) Gondwana from top to base in space and time. Gondwana Research 24: 999–1030.

Uhlein GJ (2014) Proveniência sedimentar e estratigrafia das formações Carrancas e Lagoa Formosa e a evolução do Grupo Bambuí (635-570 Ma) em Minas Gerais. Dissertation, Universidade Federal de Minas Gerais (UFMG).

Uhlein A, Trompette RR, Alvarenga CJS (1999) Neoproterozoic glacial and gravitational sedimentation on a continental rifted margin: The Jequitaí-Macaúbas sequence (Minas Gerais, Brazil). Journal of South American Earth Sciences 12: 435–451.

Uhlein A, Alvarenga CJS, Trompette R, Dupont HSJB, Egydio-Silva M, Cukrov N, Lima ONB (2004) Glaciação neoproterozóica sobre o Cráton do São Francisco e faixas dobradas adjacentes. In: Mantesso-Neto V, Bartorelli A, Carneiro CDR, Brito-Neves BB de (orgs) Geologia do Continente Sul Americano: Evolução da obra de Fernando Flávio Marques de Almeida. Beca, São Paulo.

Uhlein A, Baptista MC, Seer HJ, Caxito FA, Uhlein GJ, Dardenne MA (2011) A Formação Lagoa Formosa, Grupo Bambuí (MG): Sistema deposicional de leque submarino em bacia de antepais. Geonomos 19(2): 163–172.

Valeriano C de M, Dardenne MA, Fonseca MA, Simões LSA, Seer HJ (2004a) A evolução tectônica da Faixa Brasília. In: Mantesso-Neto V, Bartorelli A, Carneiro CDR, Brito-Neves BB de (orgs) Geologia do Continente Sul Americano: Evolução da obra de Fernando Flávio Marques de Almeida. Beca, São Paulo.

Valeriano C de M, Machado N, Simonetti A, Valadares CS, Seer HJ, Simões LSA (2004b) U-Pb geochronology of the southern Brasília belt (SE-Brazil): sedimentary provenance, Neoproterozoic orogeny and assembly of West Gondwana. Precambrian Research 130: 27–55.

Vesely FF, Assine ML (2006) Deglaciation sequences in the Permo-Carboniferous Itararé Group, Paraná Basin, southern Brazil. Journal of South American Earth Sciences 22: 156–168.

Vieira LC, Almeida RP, Trindade RIF, Nogueira ACR, Janikian L (2007) A Formação Sete Lagoas em sua área tipo: Fácies estratigrafia e sistemas deposicionais. Revista Brasileira de Geociências 37(4-suplemento): 1–14.

Warren LV, Quaglio F, Riccomini C, Simões MG, Poiré DG, Strikis NM, Aneli LE, Strikis PC (2014) The puzzle assembled: Ediacaran guide fossil Cloudina reveals an old proto-Gondwana seaway. Geology 42(5): 391–394.

Zaher H, Pol D, Carvalho AB, Nascimento PM, Riccomini C, Larson P, Juarez-Valieri R, Pires-Domingues R, Silva Jr. NJ de, Campos D de A (2011) A complete skull of an Early Cretaceous Sauropod and the evolution of advanced titanosaurians. PLoS One 6(2): e16663. doi:10.1371/jornal.pone.0016663.

Zalán PV (2004) A evolução fanerozoica das bacias sedimentares brasileiras. In: Mantesso-Neto V, Bartorelli A, Carneiro CDR, Brito-Neves BB de (orgs) Geologia do Continente Sul Americano: Evolução da obra de Fernando Flávio Marques de Almeida. Beca, São Paulo.

Zalán PV, Romeiro-Silva PC (2007) Bacia do São Francisco. Boletim de Geociências Petrobrás 15(2): 561–571.

Zhao G, Sun M, Wilde AS, Li S (2004) A Paleo-Mesoproterozoic supercontinent: assembly, growth and breakup. Earth-Science Reviews 67: 91–123.

Mafic Dykes: Petrogenesis and Tectonic Inferences

8

Vicente A.V. Girardi, Wilson Teixeira, Maurízio Mazzucchelli, Elson Paiva de Oliveira, and Paulo César Corrêa da Costa

Abstract

The main mafic dyke swarms of the São Francisco craton range in age from Late Archean to the Early Neoproterozoic, and record extensional and transpression tectonic events. The Uauá and Lavras regions include norite (2.73 and 2.66 Ga) and tholeiite swarms (2.62 and 1.97 Ga) respectively. Other swarms are made up of tholeiites: Paraopeba (2.10 Ga), Pará de Minas (1.71 Ga), Curaçá and Chapada Diamantina (1.5 Ga), Diamantina (0.93 Ga) and Salvador-Olivença (0.92 Ga). Geochemical and Sr–Nd isotopic data indicate variable scales of mantle heterogeneity, caused by recurrent metasomatic effects since Archean times. The main processes are attributed to the action of slab fluids from recycling of oceanic crust and overlying sediments on the sub-lithospheric mantle in continental settings, and by upwelling of OIB-like material. These processes are consistent with the protracted geologic history of the country rocks, given by episodic juvenile accretion and successive continental amalgamations from Archean to ca. 2.0 Ga. The adopted mixing geochemical model permits to evaluate the variation range of N-MORB melts, OIB and slab fluids and melts, which are considered to represent the main components of the parental mantle composition of the studied dykes.

Keywords

Mafic dykes • São Francisco craton • Mantle • Geochemistry • Sr–Nd isotopes

V.A.V. Girardi (✉) · W. Teixeira
Instituto de Geociências, Universidade de São Paulo, Rua do Lago 562, Cidade Universitária, São Paulo, 05508-080, Brazil
e-mail: girardi@usp.br

M. Mazzucchelli
Dipartimento di Scienze Chimiche e Geologiche, Università degli Studi di Modena e Reggio Emilia, Via Campi 103, 41125 Modena, Italy

E.P. de Oliveira
Departamento de Geologia e Recursos Naturais, Universidade de Campinas, Campinas, 13083-970, Brazil

P.C.C. da Costa
Instituto de Ciências Exatas e da Terra, Universidade Federal de Mato Grosso, Av. Fernando Costa s/n, Cuiabá, 78060-900, Brazil

M. Heilbron et al. (eds.), *São Francisco Craton, Eastern Brazil*, Regional Geology Reviews, DOI 10.1007/978-3-319-01715-0_8

8.1 Introduction

The investigation of Precambrian mafic dyke swarms in cratonic areas provides valuable information on the timing of the lithosphere breakup through geologic time. As ties for the geodynamic evolution, they also offer important clues to understand the formation of supercontinents and large igneous provinces (LIPs) (e.g., Bleeker and Ernst 2006; Bryan and Ernst 2008; Ernst et al. 2013a; Nance et al. 2014). If the age and geochemistry of dyke swarms are well constrained, their petrogenetic aspects can be inferred. For instance, the following effects on magma sources can evaluated: (i) influence of subducted material on the mantle composition; (ii) degree of mantle melting; (iii) contribution of deep mantle fluids, either caused by the rise of hot plumes, or due to local mantle thermal anomalies; (iv) degree of crustal contamination during the ascent of the mafic melts.

Archean and Proterozoic dyke swarms are conspicuous within the northern and southern portions of the São Francisco craton (SFC) crosscutting the basement complexes and the so-called Espinhaço rift system (e.g. Chaves 2011; Silveira et al. 2013; Oliveira et al. 2013; Menezes Leal et al. 2012; Girardi et al. 2013). During the last decades these swarms have been investigated by means of geochemistry, petrology and geochronology. In the northern part of the craton, they crop out along the Atlantic coast, between the cities of Salvador and Olivença, as well as in northeastern portion of Bahia State (Fig. 8.1). Dykes of the so-called Diamantina swarm occur in the Araçuaí marginal belt, intruding the Espinhaço Supergroup and the basement outside of the craton (Mazzucchelli et al. 2000). The similar geochronological, geochemical and isotopic features of these and the Salvador-Olivença swarm (see next sections) provide additional support for its inclusion into the group of mafic dykes from the southern SFC.

This review deals with the geochemical-isotopic signatures of the main mafic dyke swarms of the SFC, according to their distinct temporal and spatial distribution. The integrated information is used to address petrogenetic processes and infer the composition of their mantle sources. The regional correlation of dyke swarms are also drawn based on age-equivalent swarms, whereas the tectonic inferences for each particular swarm are used for discussing paleogeographic implications.

8.1.1 Geological Setting

The SFC has been considered to be an extension of the West Congo craton of Africa in the pre-Atlantic configuration (Fig. 8.1). At Neoproterozoic times this continental landmass acted as a tectonically stable cratonic remnant for the orogenic belts of West Gondwana. The basement rocks of the SFC, which have undergone medium- to high grade metamorphism and polycyclic evolution, are attributed to distinct Archean and Paleoproterozoic sialic provinces. From the geodynamic perspective these provinces were formed by episodic juvenile accretion processes of different blocks with concomitant shortening and collision due to a long-lived Paleoproterozoic accretionary regime.

Subsequent tectonic stability of the Archean/ Paleoproterozoic provinces at ca. 2.0–1.9 Ga formed the so-called São Francisco paleocontinent, which was linked to its African counterpart, the Congo craton (e.g., Teixeira et al. 2000; Barbosa and Sabaté 2004; Alkmim and Noce 2006; Noce et al. 2007; Oliveira et al. 2010; Ávila et al. 2010). The Espinhaço rift system (Espinhaço Supergroup) developed on the São Francisco paleocontinent between ca. 1.78 and 1.20 Ga. Extensive portions of the Espinhaço Supergroup were overprinted by Neoproterozoic deformation and metamorphism due to the marginal collisional dynamics, for instance in the case of the Paramirim aulacogen (Bahia state) and region around Diamantina (Minas Gerais state) (e.g., Alkmim 2004; Alkmim and Martins-Neto 2012; Chemale et al. 2012; Guadagnin et al. 2015). These overprints are due to the tectonic activities of the Araçuaí marginal belt. Neoproterozoic and Phanerozoic covers are also present (Fig. 8.1).

This work aims to characterize the local geologic framework of the most representative mafic dyke swarms and sills of the SFC, given by the isotopic and geochemical information reassessed here. We evaluate the reliable data for each of them according to the methods applied and materials dated, in order to support the petrogenetic-tectonic interpretation. From this evaluation eight dyke swarms aging from ca. 2.7 to 0.9 Ga can be distinguished within the SFC and its margins. These swarms are: (1) Uauá; (2) Curaçá; (3) Chapada Diamantina (Paramirim aulacogen); (4) Salvador-Olivença; (5) Lavras; (6) Paraopeba; (7) Pará de Minas; (8) Diamantina (southern Espinhaço system).

8.1.2 Mafic Dykes of the Northern São Francisco Craton

These dykes are mainly emplaced into medium- to high grade gneiss terrains, Archean and Paleoproterozoic in age (e.g., Uauá, Curaçá, Salvador-Olivença); but the northern branch of the Chapada Diamantina swarm also crosscuts the Espinhaço sedimentary rocks (Menezes Leal et al. 2012 and references therein).

8.1.2.1 Uauá Dyke Swarms

Two mafic dyke swarms crosscut the so-called Uauá Block, which consists mostly of Mesoarchean juvenile gneisses with NW-trending bands intruded by layered anorthosite,

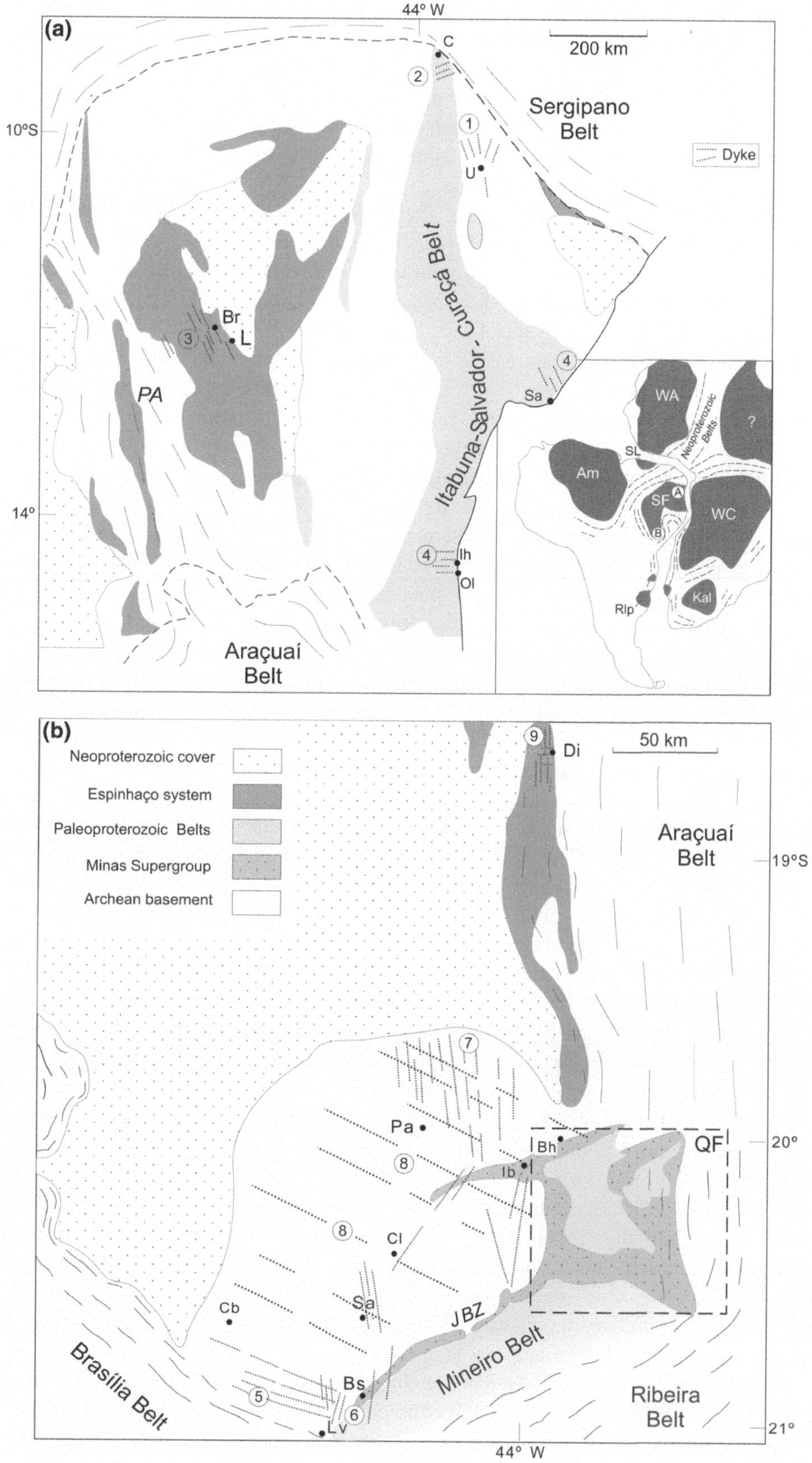

Fig. 8.1 Distribution of Precambrian dyke swarms in the northern (1a) and southern (1b) portions of the SFC, bounded by Neoproterozoic belts. The Paramirim Aulacogen (PA) and Jeceaba-Bom Sucesso shear zone (JBZ) are also shown. Keys: Sa (Salvador), U (Uauá), C (Curaçá), L (Lençóis), Br (Brotas de Macaúbas), Ih (Ilhéus), Ol (Olivença), Di (Diamantina), Pa (Pará de Minas), Bh (Belo Horizonte), Ib (Ibirité), Cl (Cláudio), Sa (Santo Antônio do Amparo), Cb (Campo Belo), Lv (Lavras) Bs (Bom Sucesso). Dyke swarms: 1 (Uauá), 2 (Curaçá), 3 (Chapada Diamantina), 4 (Salvador-Olivença), 5 (Lavras; Archean), 6 (Lavras; Paleoproterozoic), 7 (Paraopeba), 8 (Pará de Minas), 9 (Diamantina). See text and Table 8.1 for details. *Inset*: tectonic framework of Gondwana, highlighted by Neoproterozoic belts and cratons: SF (São Francisco), WC (West Congo), Kal (Kalahari), Rlp (Rio de la Plata), Am (Amazonian), WA (West Africa)

peridotite and diorite complexes, and tonalite-granodiorite bodies (Oliveira et al. 2013). Most of these rocks were metamorphosed under granulite facies conditions and later retrogressed to amphibolite grade. Paixão and Oliveira (1998) obtained a 3161 ± 65 Ma whole-rock Pb–Pb isochron for the Lagoa da Vaca layered anorthosite intrusive complex and zircon Pb-evaporation age of 3072 ± 20 Ma for host rocks (orthogranulites). On the other hand, Cordani et al. (1999) presented zircon U–Pb SHRIMP ages between 3120 and 3130 Ma for the so-called Capim tonalite. Several other Archean felsic igneous bodies occur in the Uauá Block, some of which, i.e. thc Uauá quarry enderbitic granulite and a gneissic granodiorite to the southeast of Uauá, yielded U–Pb ages of 2933 ± 3 Ma and 2991 ± 22 Ma (Oliveira et al. 2002). K-Ar mineral ages in Uauá-type gneisses yielded ages between 2.2 and 1.8 Ga, and mark the time of regional cooling of this Block (Bastos Leal et al. 1994). These data reveal the polycyclic history of the Archean basement in the Paleoproterozoic orogen.

The older dykes of the Uauá swarm trend mainly NW–SE and N–S. They are composed of unmetamorphosed or weakly metamorphosed norites and metamorphosed tholeiites (amphibolites). The norite dykes are composed mainly of plagioclase and orthopyroxene, sometimes also olivine, brown amphibole, biotite-phlogopite, pyrrhotite and ilmenite. The norites are classified into two groups: one group is represented by dykes that are no more than 5 m wide. They are fine-grained and show quench textures, with olivine, pyroxene, and radiate plagioclase needles. The other group is composed by medium to coarse-grained thicker dykes, which are up to 100 m in width, and very often shows well preserved poikilitic texture with centimeter-size plagioclase oikocrysts with enclosed bronzite chadacrysts. Ilmenite aggregates occur in some samples of this group (Oliveira et al. 2013).

In contrast, the younger dyke swarm trends NE–SW and is made up of unmetamorphosed tholeiites (diabases) and metabasic dykes (Bastos Leal et al. 1994; Bellieni et al. 1995; Menezes Leal et al. 1995). In general, the dykes have fine to coarse grain size and their thicknesses vary from few centimeters to several dozen of meters (Oliveira 2011; Oliveira et al. 2013). These tholeiite dykes are the most abundant among all the Uauá dykes. Their margins are occasionally sheared, but in the centre the primary mineralogy (calcic pyroxene, plagioclase and magnetite) and intergranular texture are all preserved. The pyroxene is often replaced by green to brown amphibole at the rim.

Some samples from the NE–SW trending unmetamorphosed dykes have been analyzed by the Rb/Sr method (internal mineral isochrones), and yielded apparent ages between 2384 ± 114 Ma and 1983 ± 31 Ma (Bastos Leal. 1992, Bastos Leal et al. 1994). Recently, Oliveira et al. (2013) dated one norite dyke from the NW–SE and N–S trending swarm at 2726.2 ± 3.2 Ma, and a tholeiite dyke from the younger NE–SW swarm at 2623.8 ± 7.0 Ma, using U-Pb on baddeleyite (ID-TIMS) and on zircon (SIMS) techniques respectively. These ages, from a tectonic perspective, are interpreted, as a large scale fracturing episode, coeval with the ultimate stages of the Neoarchean evolution.

8.1.2.2 Curaçá Dyke Swarm

The Curaçá dyke swarm (Fig. 8.1a) intrudes Archean to Paleoproterozoic rocks of the Caraíba metamorphic complex (2580–2027 Ma; Oliveira et al. 2010), and it is overlain in places by Neoproterozoic rocks of the Canudos Group (Bastos Leal et al. 1995; Oliveira and Tarney 1995). The dykes trend preferentially NE–SW. Their widths vary from a few centimeters to several tens of meters, and are undeformed. They are classified as dolerites (Silveira et al. 2013), composed of plagioclase (An_{64} to An_{45}), and augite ($En_{36}Wo_{44}Fs_{20}$ to $En_{28}Wo_{42}Fs_{30}$) with interstitial quartz, K-feldspar, and sometimes green-brown amphibole, biotite, and the accessories: ilmenite, magnetite, apatite, pyrite and pyrrhotite (Oliveira and Tarney 1995). The dykes vary in grain size from aphanitic to coarse-grained. Textures are subophitic to ophitic, with >2 mm plagioclase laths in the dyke center but <1 mm at the margin. Calcic pyroxene phenocrysts are often observed in the inner part of individual dykes. Vesicular and flow structures are locally observed, particularly at dyke margins, where vesicles (<2 mm) are bordered by tangential plagioclase laths and elongate augite. The vesicles are filled with chlorite, carbonate and sericite. These minerals, along with epidote and uralite, fill post-emplacement fractures and locally replace groundmass plagioclase and pyroxene. Xenoliths of the host rocks may be also observed.

Previous Rb/Sr mineral isochrones for two Curaçá dykes yielded apparent ages of 704 ± 56 and 650 ± 94 Ma that were tentatively interpreted as coeval with the onset of the Neoproterozoic Sergipano marginal belt (Bastos Leal 1992). However, a recently published U-Pb (ID-TIMS) baddeleyite age for one of the Curaçá dykes yielded 1506.7 ± 6.9 Ma which is considered the probable emplacement age of the Curuçá dyke swarm (Silveira et al. 2013). This indicates that the Rb–Sr system was isotopically reset.

8.1.2.3 Chapada Diamantina Swarm (Paramirim Aulacogen)

The Chapada Diamantina dyke swarm comprises mafic dykes and sills (Fig. 8.1a). They are intrusive into the sedimentary rocks of both the Paraguaçu and Chapada Diamantina groups of the Espinhaço Supergroup (e.g., Pedreira and De Waele 2008), and also into the Archean rocks of the so-called Gavião Block (Menezes Leal et al. 2010). The dykes are neither deformed, nor metamorphosed, and trend preferentially in the NNW–SSE direction. Their widths vary from less than a meter to tens of meters. They are fine to medium grained, aphyric to phyric diabases. Detailed petrographic description in Silveira et al. 2013 (and references therein) indicates textures varying from intergranular to ophitic, subophitic, and porphyritic. The mineralogy is composed mainly of augite and plagioclase (andesine to labradorite), with titanite, ilmenite, pyrrhotite, zircon, apatite, and magnetite as accessories. Plagioclase and augite are the common phenocryst phases. Rarely, plagioclase is partially replaced by epidote and calcite, and pyroxene by uralitic amphibole.

The Lagoa de Dentro gabbro sill (at least 30 m-thick) crosscuts the Mangabeira Formation (Paraguaçu Group) in the Brotas de Macaúbas region of Chapada Diamantina. U–Pb zircon dating of this gabbro yielded an upper intercept age of 1514 ± 22 Ma in the Concordia diagram (Babinski et al. 1999). Guimarães et al. (2005) obtained a more precise U–Pb zircon age of 1496 ± 3.2 Ma for another mafic dyke that intrudes the Mangabeira Formation in the Lagoa do Dionísio area. More recently, Silveira et al. (2013) obtained another precise U-Pb (ID-TIMS) age of 1501 ± 9.1 Ma on baddeleyites from a mafic dyke intrusive into quartzites of the Mangabeira Formation near Brotas de Macaúbas. This high-quality data is taken as the emplacement age of the Chapada Diamantina dykes, which are probably coeval to the Curaçá swarm (Silveira et al. 2013), although occurring 400 km apart and displaying perpendicular trends. In addition, the so-called muscovite–martite (iron-oxide) NNW–SSE trending dykes that cut the sandstones of the lowermost unit of the Chapada Diamantina Group (Tombador Formation) could also be coeval, as they yielded Ar–Ar ages of 1512 ± 6 and 1514 ± 5 Ma (Battilani et al. 2005).

8.1.2.4 Salvador-Olivença Swarm

The Salvador and Ilhéus-Olivença dykes are amongst the best studied dyke swarms in the São Francisco craton (Fig. 8.1a). The dykes are intrusive into Archean to Paleoproterozoic country rocks. Since both groups have approximately the same age, according to U-Pb datings (see below), and are considered branches of a single swarm, they will be hereafter called the Salvador-Olivença mafic swarm. Their thicknesses vary from a few centimeters up to 50 m. In the north (Salvador region), the dykes have N–S and E–W orientation, whereas in the south (Ilhéus and Olivença region) they are mostly NE–SW oriented (Bellieni et al. 1998). They are composed of unmetamorphosed and undeformed diabases and amphibolites. This study refers to the diabases, which are mainly composed of zoned plagioclase (An_{74-43}), augite (Wo_{44-29}), pigeonite (Wo_{3-15}), rare orthopyroxene (Wo_4), Ti-magnetite and ilmenite (Bellieni et al. 1998). Olivine can occur in some samples. Amphibole and biotite may replace pyroxene borders. Coarse-grained dykes may contain quartz-feldspar intergrowths. Slightly to moderate porphyritic textures prevail over holocrystalline textures found in the central part of the dykes thicker than at least 10 m (Bellieni et al. 1991).

In the Salvador branch, U-Pb baddeleyite dating indicate that the Ondina dyke emplaced at 921.5 ± 4.3 Ma and the Meridian dyke at 924.2 ± 3.8 Ma (Evans et al. 2010). At Ilhéus, the age of a third dyke, from the Olivença branch is 926 ± 5 Ma. (Oliveira et al. 2012). These accurate ages for distinct dykes in coastal Bahia support the opinion that previous published Ar/Ar analyses on those dykes, with significantly older plateau ages at 1080 and 1020–1000 Ma (Renne et al. 1990), probably reflect the presence of excess argon (Evans et al. 2010).

8.1.3 Mafic Dykes of the Southern São Francisco Craton

Mafic dyke swarms are very common in the southern part of the SFC, especially in areas located to the northwest and southwest of the city of Belo Horizonte and in the vicinity of Lavras, where they intrude the Archean granitoid basement partly remobilized in Paleoproterozoic times (Fig. 8.1b). Similar to the dykes in the northern portion of the craton, they yielded ages from the Neoarchean (Lavras norites) to the Tonian (Diamantina swarm).

Different names have been proposed for groups of dykes, according to their orientation, emplacement mechanisms and radiometric ages. For instance, Chaves (2001), Chaves and Correa Neves (2005), and Chaves (2011) distinguished four Proterozoic swarms within the southern cratonic portion, highlighted by field relationships, aeromagnetic data, petrography, geochemistry, and geochronology. In addition other dyke swarms have also been described on the basis of geologic mapping carried out over particular areas (e.g., Campo Belo, Santo Antônio do Amparo, Lavras; (Fig. 8.1b; Pinese 1997; Carneiro et al.1998; Teixeira et al. 1998; Oliveira 2004; Corrêa da Costa et al. 2006). Overall, these successive mafic activities demonstrate the important role of episodic crustal extension and rifting episodes affecting the SFC, particularly during the Proterozoic Eon.

8.1.3.1 Lavras Dyke Swarms

The Lavras dyke swarms occur in the SW cratonic edge. The oldest swarm is intrusive into Meso- to Neoarchean (3.20–2.90; 2.70–2.65 Ga) medium- to high-grade facies rocks of the cratonic basement (Teixeira et al. 1998), including granite-greenstone terranes, which have been traditionally attributed to the Neoarchean Rio das Velhas orogeny (e.g., Alkmim and Noce 2006). They also transect mafic-ultramafic layered metamorphic complexes (2.75–2.71 Ga) that locally occur in the region (Carneiro et al. 2004; Goulart et al. 2013).

Two mafic dyke swarms occur in the vicinity of Lavras-Perdões-Santo Antônio do Amparo, distinguished by their compositions and ages. These dykes show diverse orientations (N20 W, N50 W, N–S, N20E, N60E, and E–W), always in contrast with the structural orientation of the Archean country rocks, which ranges between N60 and 70 W, overprinted in places by an E–W-trending foliation (Pinese et al. 1995).

The first group, comprising about 90 % of the dykes, is mainly made up of norites and subordinate gabbronorites (Fernandes 2001). It has a wide distribution and the individual dykes are up to 30 km long and up to 100 m wide. Collectively they show a consistent WWN-ESE direction (Corrêa da Costa et al. 2006). They are made up of plagioclase, ortho and clinopyroxenes, minor quartz-feldspar intergrowths, opaque minerals, amphibole, apatite, and biotite. Cumulate, subophitic and intergranular textures characterize these rocks. According to Pinese (1997), the norite dykes emplaced at 2658 ± 44 Ma, as suggested by a Sm/Nd whole rock isochron using several nearby samples. As such, these dykes have a broad age match with the emplacement of the mafic ultramafic complexes of Ribeirão das Motas and Carmópolis de Minas (2.75–2.71 Ga), which permits to hypothesize that they may have been part of the same magmatic system originated shortly after the Neoarchean stabilization of the Rio das Velhas orogen (Goulart et al. 2013).

The second group of dykes comprises coeval amphibolites, metabasites and diabases, and exhibit NNW–SSE, N–S and NE–SW trends. The diabase dykes show sub-ophitic to intergranular textures and are composed of andesine-labradorite, augite, rare pigeonite, and minor hornblende, magnetite-ilmenite, quartz, apatite and epidote. Hornblende, frequently replacing pyroxene, and plagioclase constitute the main paragenesis of the metadiabases. Textures are blasto-subophitic and blasto-intergranular. Nematoblastic and granoblastic amphibolites are made up of amphibole and plagioclase, and small amounts of quartz, titanite, opaque minerals and apatite (Pinese 1997). These dykes show generally mean widths from 2 to 10 m, and have sharp intrusive contacts with both the Archean basement and Early Paleoproterozoic metasedimentary rocks that crop out along the NE–SW Bom Sucesso-Jeceaba lineament (Fig. 8.1b). These metasediments have been ascribed to the Minas Supergroup, interpreted to be a Neoarchean/Early Paleoproterozoic foreland sequence, outboard of which a Paleoproterozoic (2.45–2.00 Ga) orogenic belt was developed (Alkmim and Noce 2006; Teixeira et al. 2015).

The age of this second dyke swarm is attributed to the Paleoproterozoic due to an Ar–Ar amphibole plateau age for one metabasite with 1970 ± 7 Ma, which indicates a minimum age for the time of emplacement (Pinese 1997). Some of the metabasites and amphibolites around Lavras-Bom Sucesso-Oliveira show a foliation parallel to the sharp contact with the Archean host rocks. This suggests that the emplacement of these dykes took place during a late transpressive episode in association with regional shear zones. Particularly, a metagabbro at São Tiago (not shown in Fig. 8.1) to the northeast of Lavras yields a K–Ar amphibole age of 2049 ± 61 (Teixeira 1985), which is roughly comparable with the Ar–Ar plateau age of the Lavras metabasite indicated above.

8.1.3.2 Paraopeba Swarm

The Paraopeba swarm crops out to the northwest of Belo Horizonte, along the Paraopeba river valley (Fig. 8.1b), and crosscuts the 2.80–2.65 Ga gneissic and granitoid rocks of the Belo Horizonte Complex (e.g., Lana et al. 2013; Romano et al. 2013 and references therein), which was formed during the Neoarchean Rio das Velhas orogeny (see above).

The Paraopeba dykes are made up of scattered, NNW–SSE trending linear bodies (rare NE–SW ones), which vary from hundreds of meters to 5 km in length and from 2 to 10 m in width. They are composed of diabases, metadiabases and amphibolites. Many of them show strongly sheared margins with amphibolite facies mineralogy, and less deformed, massive metaigneous cores, with blastoporphyritic and granoblastic. The dykes are composed of plagioclase (andesine-labradorite), augite, hornblende, occasional garnet corona between plagioclase and pyroxene, and minor ilmenite, biotite, chlorite, epidote, carbonate, zircon, and quartz (Chaves 2011). The field observations suggest an emplacement mechanism associated with transpression, along regional NW–SE strike-slip shear zones that developed under medium- to high amphibolite facies conditions over the area. This tectonic setting is similar to that envisaged for the younger Lavras dykes (Pinese 1997).

According to Chaves (2001) one amphibolite dyke of the Paraopeba swarm yielded a Rb–Sr mineral isochron of 2189 ± 45 Ma, interpreted to be the time of emplacement. Teixeira et al. (1998) reported comparable ages for amphibolite dykes, metabasites and diabases from the same swarm with scattered distribution to the northwest of Belo Horizonte, as supported by field evidence and petrographic features. The amphibolite and metaigneous dykes yield K–Ar ages of 2116 ± 46, 2054 ± 37, 2049 ± 61, 2005 ± 26 and 1995 ± 31 Ma on amphibole concentrates. Taking into account these data, the Paraopeba swarm probably intruded the Archean continental crust at around 2100 Ma. Additional U-Pb work is needed for better constrain the emplacement age because both the K–Ar and Rb–Sr techniques in basic rocks are usually prone to isotopic disturbance.

Another set of dykes of amphibolite, gabbro and gabbronorite with parallel to subparallel (NNW) orientation to the Paraobeba swarm, crops out between the towns of Campo Belo and Santo Antonio do Amparo. It is similarly emplaced under transpressive conditions into the Campo

Belo metamorphic complex (Corrêa da Costa et al. 2006). These dykes were not dated, and they could be associated either with the Paraopeba dykes, or the younger Lavras dykes located to the south. Moreover, we cannot rule out the hypothesis of successive fissural igneous episodes between ca. 2.10 and 1.97 Ga, given that tectonic reactivations has affected the entire crystalline crust of the southern portion of the SFC at the time of the Paleoproterozoic orogeny (Teixeira et al. 2015).

8.1.3.3 Pará de Minas Swarm

This is the most prominent dyke swarm in the area, crosscutting the Archean Belo Horizonte metamorphic complex. The swarm is made up of NW-trending diabase dykes, which are ca. 50 m wide and may reach up to several hundred kilometers in length, as suggested by aeromagnetic data (Chaves 2001 and references therein).

These dykes consist of fine- to coarse grained diabase to gabbro of tholeiitic affinity. Most of the dykes are massive, but occasional porphyritic dykes also occur. They are mainly composed of labradorite-andesine and augite, with minor amounts of ilmenite, interstitial K-feldspar, quartz, apatite, and zircon or baddeleyite. In some outcrops the dykes are crosscut by narrow fractures at the margins of which fine-grained chlorite, quartz, clay minerals and sericite may crystallize as alteration products. Sometimes augite is partially replaced by tremolite-actinolite. The predominant texture is sub-ophitic, but some dykes are porphyritic, characterized by plagioclase phenochysts parallel to their walls.

One of these dykes located to the north of the town of Pará de Minas yielded a Rb–Sr mineral age of 1740 ± 54 Ma (Chaves 2001), whereas the N–S trending Ibirité gabbro to the east (Quadrilátero Ferrífero area) gives a U-Pb baddeleyite upper intercept age of 1714 ± 5 Ma (Silva et al. 1995; Alkmim and Noce 2006). A well preserved NW–SE trending amphibolite dyke, occurring near Carmópolis de Minas has yielded a Ar–Ar plateau age of 1752 ± 15 Ma (Oliveira 2004). Another NE–SW trending amphibolite dyke exposed to the east of this town, from the so-named Alto Candeias mafic suite, transects the Neoarchean Bonfim metamorphic complex. This dyke yielded a K–Ar amphibole apparent age of 1707 ± 64 Ma (Carneiro et al. 1998). The U-Pb age of 1.71 Ga can be thus considered as the representative of this swarm, which matches with the onset of the Espinhaço rift system (Guadagnin et al. 2015 and references therein).

8.1.3.4 Diamantina Swarm

The Diamantina dyke swarm occurs mainly to the south of the Diamantina town and exhibits E–W, N–S and NE–SW trends. These dykes crosscut Archean basement rocks, as well as metasedimentary rocks of the Espinhaço Supergroup involved in the Aruaçaí belt. The dykes display variable deformation and metamorphism due to the Neoproterozoic overprint related to the Araçuaí marginal belt (Mazzucchelli et al. 2000). Sills and plugs are also associated with this fissural igneous event (e.g. Silva et al. 1995).

Metabasite and amphibolite, the main lithologic types of this swarm, are commonly ten meters wide. Metabasite, the dominant lithotype, shows plagioclase and pyroxene relics with subophitic texture in a metamorphic assemblage of chlorite, actinolite, epidote, sodic plagioclase, and quartz. The amphibolite dykes are foliated and contain chlorite, plagioclase, epidote, quartz and opaque minerals (Mazzucchelli et al. 2000).

Machado et al. (1989) reported a U–Pb baddeleyite-zircon age of 906 ± 2 Ma for one porphyritic metadiabase sill near the village of Pedro Lessa. More recently, a gabbro dyke from the same region yielded a U–Pb zircon age of 933 ± 20 Ma, which is considered as the representative emplacement age for the swarm (Girardi et al. 2013 and references therein). Therefore, within error, this rock has a U–Pb age match with the Salvador-Olivença swarm (see previous section).

A K–Ar hornblende age of 1006 ± 43 Ma was reported by Carneiro et al. (1998) for a metabasic dyke from the so-called Conceição do Itaguá suite, which crops out adjacent to the Quadrilátero Ferrífero, crosscutting the Minas Supergroup. This local suite is similarly composed of metadiabases that have NNW–SSE trend (subordinately NNE-SSW) with preserved igneous texture. Secondary overprint characterized by uralitization and saussuritization is also present. To the west of this area two gabbroic dykes located between Carmo da Mata and Claudio gave Ar–Ar plateau amphibole (alteration from the pyroxene) ages of 990 ± 30 and 952 ± 7 Ma (Oliveira 2004). These gabbros are here tentatively ascribed to a large N60-70 E trending dyke that has been previously attributed to the so-called Januária swarm (Chaves 2001).

The main geologic data of the mafic dyke swarms of the SFC are summarized on Table 8.1.

8.1.4 Geochemistry

In this section, the mafic dyke swarms of the São Francisco craton are described in two groups. The Archean and Early Paleoproterozoic Uauá, Lavras and Paraopeba dykes belong to the older group, whereas the younger group includes the Late Paleoproterozoic to Neoproterozoic Pará de Minas, Curaçá, Chapada Diamantina, Salvador-Olivença, and Diamantina dykes (Table 8.1).

8.1.4.1 Chemical Classification

The Late Paleoproterozoic to Neoproterozoic mafic dykes are mostly basalts in the total alkalis-silica (TAS) diagram.

Table 8.1 Geologic-tectonic characteristics for mafic dyke swarms of the SFC, in space and time

Swarm	Rock	Material & method	Best/mean ages (Ga)	Dyke's trends	Tectonics; (location)
Uauá	Norite#, Diabase* Amphibolite	U/Pb bd#., zr.*	2.73#; 2.62*	NW; NE*	Late to Post orogenic; (NSFC)
Lavras (og)	Norite	Sm/Nd min. isochron	2.66	NW	Failed arm rift (SSFC)
Paraopebas	Amphibolite, Metadiabase, Diabase	Rb/Sr min isochron;	2.10	NNW, NS	Transpression/shearing (SSFC)
Lavras (yg)	Amphibolite, Diabase, Metabasite	Ar/Ar amph. (plateau)	1.97	NE (NS)	Transpression/ shearing (SSFC)
Pará de Minas	Diabase	U/Pb bd.	1.71	NW	Bimodal, Intraplate (SSFC).
Curaçá	Dolerite	U/Pb bd.	1.50	NE	Intraplate (NSFC)
Chapada Diamantina-Paramirim	Dolerite	U/Pb bd.	1.50	NW	Intraplate (NSFC)
Diamantina	Metabasite, Amphibolite	U/Pb bd.	0.93	EW (NW, NS, NE)	Intraplate (SSFC)
Salvador-Olivença	Diabase	U/Pb bd.	0.92	N-NW, EW	Intraplate (NSFC)

Dashed lines distinguish coeval dykes (probably swarm branches). See text for details. Keys: og = oldest generation; yg (youngest generation); bd (baddeleyite); zr. (zircon); min. (mineral); amph. (amphibole). Dyke locations: NSFC (Northern São Francisco Craton); SSFC (Southern São Francisco Craton). The right column shows the subordinate dyke's orientations in parenthesis

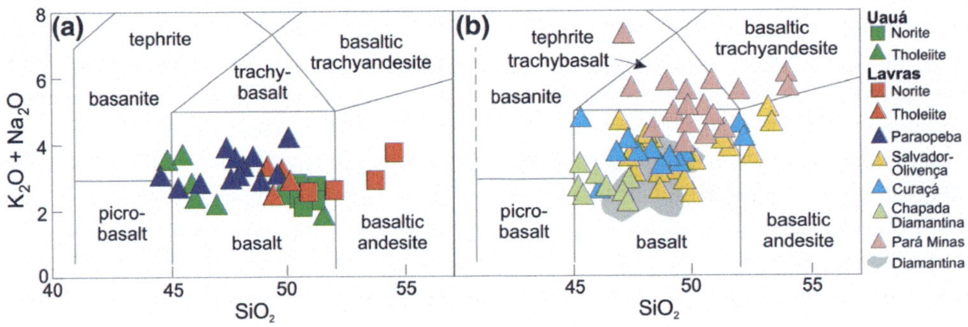

Fig. 8.2 Major element classification of dyke swarms after LeMaitre et al. (1989). Data from Oliveira et al. (2013) (Uauá swarm); Pinese (1997) (Lavras swarm); Chaves (2011) (Paraopeba swarm); Silveira et al. (2013) (Curaçá and Chapada Diamantina swarms); Mazzucchelli et al. (2000) (Diamantina swarm); Bellieni et al. (1998) (Salvador-Olivença swarm)

Samples from the Pará de Minas dykes have the most variable composition and plot in the basalt, trachybasalt and basaltic trachyandesite fields. Some Curaçá and Salvador-Oliveira dykes plot also in the basaltic andesite field (Fig. 8.2). On the least mobile trace elements Nb/Y versus $Zr/TiO_2*0.0001$ diagram (Fig. 8.3) most dykes are classified as subalkaline basalts, but several samples from the Diamantina and Curaçá, as well as two of them from Uauá, and one from Salvador-Olivença plot in the alkaline basalt field.

8.1.4.2 Major and Trace Elements

Variation major and trace elements diagrams are currently used on the study of the origin and evolution of igneous rocks, to elucidate petrogenetic processes such as melting, fractionation, crystallization and assimilation. Besides fractional crystallization and open-system processes, such as assimilation and mixing, different amounts of partial melting may obscure the mantle source characteristics inferred from the major element chemical composition of basaltic melts. Incompatible trace element diagrams and ratios in basic melts are instead currently used to identify source characteristics, because they coincide with those in their source in case of: a) highly incompatible elements whose bulk composition coefficients are similar (i.e., element partition coefficient $D \sim 0$); b) their partition coefficient is much smaller than the melt fraction ($D \ll F$). In both cases the ratios of element pairs do not change during fractional crystallization and change very little during batch melting. This seems to be a reasonable assumption for most elements and element ratios used in the following discussion. Moreover it is important to bear in mind that, in addition to incompatible element ratios whose bulk partition coefficients are similar (e.g., Ba/Nb, La/Nb), other element ratios, whose partition coefficients are not very close, such as La/Yb, Ti/Zr, Zr/Y and Zr/Nb can also be useful as petrogenetic indicators.

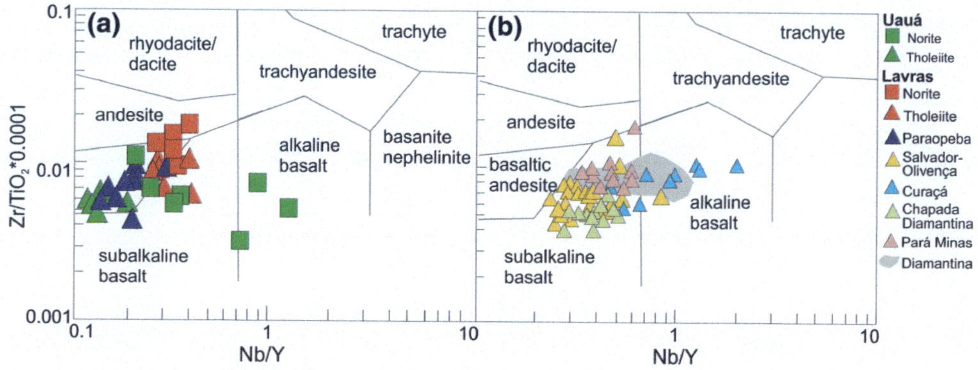

Fig. 8.3 Classification of dyke swarms after Winchester and Floyd (1976). Data from Oliveira et al. (2013) (Uauá swarm); Pinese (1997) (Lavras swarm); Chaves (2011) (Paraopeba swarm); Silveira et al. (2013) (Curaçá and Chapada Diamantina swarms); Mazzucchelli et al. (2000) (Diamantina swarm); Bellieni et al. (1998) (Salvador-Olivença swarm)

The variation ranges of the 107 representative chemical analyses used in this study were compiled from the literature and are shown in Table 8.2. They are considered to represent melts of mafic intrusions. Results corresponding to cumulates and crustal contaminated rocks were rejected. Particularly, chemical and Sr–Nd isotopic analyses from the Pará de Minas swarm will not be discussed because the available data indicate significant crustal contamination (Chaves 2001). In Table 8.2 the trace element list was compiled aiming to select the incompatible elements useful for one of the main purposes of this study, which is to compare the geochemical characteristics of the swarms to get inferences on their mantle sources. All selected samples are igneous, except those from the Paraopeba (Chaves 2011) and Diamantina (Mazzucchelli et al. 2000) swarms, which are represented by their metamorphic derivatives, metadiabases, metabasites and amphibolites. In these cases MPR (molecular proportion ratio) diagrams were used to assess metamorphic mobility. The applied method did not show evidence of mobility, except for K and Rb, which were not considered for petrogenetic interpretations, in these swarms. To discuss the geochemical features of the swarms, only two general terms will be employed: tholeites for basic tholeiitic suites including diabases, gabbros and their metamorphic derivatives and norites.

The Uauá norites and tholeiites show significant geochemical differences in evolution degree and trends. The mg# values vary from 0.63 to 0.68, and from 0.50 to 0.62, in norites and tholeiites, respectively. Norites have higher values of MgO, Na_2O FeO_t, TiO_2, P_2O_5 and lower contents of CaO and Al_2O_3 (Fig. 8.4; Table 8.2). These rocks are also more enriched in most incompatibles, with the exception of Yb and Lu (Table 8.2). In both rocks Ba displays positive spikes, while the Nb-Ta and Th anomalies are negative (Fig. 8.5). Tholeiites exhibit positive Rb anomaly, REE flat pattern, whereas norites display steeper patterns characterized by Sr, Zr, LREE and MREE enrichment, and Yb and Lu depletion. In consequence the swarms have distinct incompatible element ratios (Table 8.3), which, except for similar Ti/Zr and La/Nb, show significant differences of all values, mainly in Ba/La, Zr/Y, Ce/Y, Ti/Y and Ba/Nb. La/Nb and Ba/Nb are higher in tholeiites.

Lavras norites have higher mg# (0.50–0.72, Table 8.2), SiO_2, Rb, Ba and K values and lower contents of TiO_2, FeO_t, MnO, Na_2O, Nd, Sm, Zr, Eu, Y, Yb and Lu in comparison with tholeiites (mg# = 0.42–0.51, Table 8.2). The evolution of both suites is also distinct. In norites with mg# decreasing, Al_2O_3 and CaO increase (Fig. 8.4), and FeOt decreases (not shown), whereas in the tholeiitic suite with mg# decreasing, FeOt increases (not shown), and CaO and Al_2O_3 decrease (Fig. 8.4). Increasing of Al_2O_3 in norites reflects probably the importance of olivine and orthopyroxene fractionation, whereas decrease of Al_2O_3 and CaO in tholeiites is due to plagioclase and clinopyroxene fractionation (Pinese 1997). Spidergrams of both suites reveal several important geochemical differences (Fig. 8.5). Norites display steeper patterns, negative Nb anomaly, positive Rb, K and Zr spikes, and larger MREE and HREE depletion; whereas in the tholeiitic suite LILE elements (Ba, K, Sr) and Eu display negative anomalies, whereas Th, Zr and Y show positive spikes, and Nb is enriched with respect to LREE (La and Ce). In consequence Rb/Sr, Ce/Y, Ba/La, La/Yb, La/Nb and Ba/Nb ratios are higher in norites than in tholeiites (Table 8.3).

Variation diagrams of the Paraopeba tholeiites show that with mg# increasing CaO, Al_2O_3 (Fig. 8.4) and Ni (not shown) increase, whereas FeO_t (not shown) TiO_2 and incompatible elements, Ba, Nb, La (Fig. 8.4), Ce, Nd, Sr, Sm, Eu, Gd, Dy, Y and Yb (not shown) decrease (Chaves 2011). These trends are compatible with the gabbro fractionation. The spidergram is characterized by high contents of LILE elements, REE patterns enriched in LREE and strong negative anomalies of Nb and Sr (Fig. 8.5). Trace element ratios are mainly distinguished by high Ba/Nb and La/Nb values (Table 8.3).

Table 8.2 Variation range of chemical analyses of selected samples from mafic dykes

Swarm	Uauá	UaUá	Lavras	Lavras	Paraopeba	Curuçá	Chapada Diamantina	Diamantina	Salvador-Olivença
Rock Type	Norite	Tholeiite	Norite	Tholeiite	Tholeite	Tholeiite	Tholeiite	Tholeiite	Tholeiite
No. samples	Min–Max (6)	Min–Max (7)	Min–Max (9)	Min–Max (10)	Min–Max (10)	Min–Max (7)	Min–Max (11)	Min–Max (15)	Min–Max (32)
SiO_2 (%)	42.54–46.08	50.09–51.28	50.05–55.49	49–51.24	49.97–50.24	46.3–50.4	45.47–47.56	47.07–49.97	47.15–53.33
TiO_2	1.94–3.26	0.8–1.02	0.59–0.87	1.08–2.28	1.16–2.54	2.3–2.67	0.93–1.5	0.91–2.42	1.22–3.59
Al_2O_3	5.5–9.62	13.28–15.6	12.62–16.66	13.01–16.12	10.94–14.18	14.4–15.6	14.83–18.82	12.05–15.84	14.43–17.07
FeOT	14.25–17.9	10.44–13.36	9.39–11.01	11.70–14.80	13.5–17.62	12.4–14.9	9.72–16.93	12,23–14,62	10.61–17.01
MnO	0.2–0.27	0.18–0.214	0.14–0.18	0.21–0.26	0.21–0.25	0.19–0.24	0.141–0.173	0.16–0.20	0.14–0.23
MgO	14.36–16.02	6.43–8.3	4.83–13.58	4.97–6.76	3.52–6.24	4.9–7.1	7.23–10.65	6.03–12.48	3.79–12.22
CaO	7.48–9.37	10.1–11.57	8.26–10.17	8.32–11.17	8.24–10.79	8.1–10.1	8.68–12.04	9.24–11.33	8.24–11.50
Na_2O	1.81–2.96	1.6–2.54	1.81–2.62	2.66–3.65	1.98–3.20	1.8–3.6	1.72–2.43	1.59–3.04	1.95–4.85
K_2O	0.35–0.76	0.12–0.39	0.54–1.32	0.25–0.55	0.37–0.73	0.58–1.33	0.28–0.93	0.23–0.84	0.10–1.67
P_2O_5	0.14–0.25	0.08–0.12	0.07–0.16	0.18–0.52	0.09–0.48	0.33–0.64	1,72–2.43	0.10–0.54	0.08–0.41
LOI	0.16–1.43	0–1.15	0.71–2.2	0.82–2.06	0.15–1.62			1.17–2.69	0.82–3.0
mg#	0.63–0.68	0.50–0.62	0.5–0.72	0.41–0.52	0.42–0.49	0.43–0.49	0.57–0.69	0.44–0.66	0.35–0.69
Rb (ppm)	9.5–18.5	4.2–31.2	14–46	7.00–14.0	12–35	9.2–10.33	4.06–17.87	3.0–19.0	2.0–33.0
Sr	224.1–380.5	107.0–134.7	96–234	127–110	92–130	274.3–620.2	209.4–310.9	276–444	193–442
Y	16.6–22.2	17.1–21.5	16–24	30.0–66.0	43–50	35.5–40.3	10.9–21.8	12.0–32.0	19.0–72.0
Zr	105.7–205.5	45.8–65.6	70–135	82.0–236	65–197	149.5–283.7	41.1–90.8	85.0–230	66.0–416
Nb	5.2–19.85	2.10–3.24	5–10	8.00–16.0	4.3–11.8	20.19–72.16	4.88–8.69	11.0–22.0	5.0–35.0
Ba	140.5–711.6	56.5–768.8	152–305	51.0–106	96–215	385.1–577.3	109.5–298.7	92.0–403	155–708
La	10.75–23.63	3.05–5.14	6.7–16.3	7.00–20.0	9.79–18.4	22.28–50.09	4.62–10.33	11.0–22.0	7.0–45.0
Ce	27.43–45.24	7.90–12.06	15–33	16.0–46.0	17.9–42.8	44.32–105.9	10.21–17.53	22.5–51.5	115.6–97.0
Pr	3.96–6.91	1.23–1.69				5.88–12.84	1.39–3	3.10–7.10	
Nd	18.74–31.49	6.49–8.20	4–16	12.0–30.0	11.1–26.9	25.76–52.61	6.43–13.71	13.0–32.0	11.4–37.4
Sm	4.63–6.11	2.0–2.47	1.7–3.37	2.29–5.55	3.38–6.48	6.18–9.81	1.61–3.18	2.80–7.10	3.37–9.40
Eu	1.52–2.33	0.78–0.90	0.45–0.96	0.73–1.42	1.15–2.15	2.07–2.7	0.83–1.33	1.00–2.60	1.20–2.80
Gd	4.29–5.96	2.47–3.28			4.87–7.89	6.49–9.1	1.85–3.44	2.60–7.80	3.98–9.07
Tb	0.54–1.01	0.48–0.60	<0.5–0.6	0.77–1.29		1.16–1.41	0.33–0.66	0.40–1.10	0.50–1.20
Dy	3.77–4.55	3.14–4.08			6.03–7.82	6.82–7.82	2.1–4.06	2.30–6.20	3.10–8.51
Ho	0.71–0.84	0.69–0.88				1.37–1.59	0.44–0.85	0.40–1.30	
Er	1.81–2.41	2.0–2.56				3.84–4.35	1.28–2.48	1.30–3.50	2.51–3.87
Tm	0.24–0.30	0.29–0.37				0.52–0.59	0.18–0.38	0.10–0.50	

(continued)

Table 8.2 (continued)

Swarm	Uauá	UaUá	Lavras	Lavras	Paraopeba	Curuçá	Chapada Diamantina	Diamantina	Salvador-Olivença
Yb	1.46–1.79	1.9–2.4	1.25–1.88	2.07–4.48	2.99–4.29	3.46–3.84	1.22–2.37	1.00–3.00	1.44–4.10
Lu	0.20–0.23	0.27–0.36	0.19–0.27	0.31–0.67		0.48–0.56	0.19–0.35	0.20–0.50	0.18–0.67
Hf	3.6–5.73	1.37–1.86	1.2–2.5	1.6–4.3	10.2–14.7	4.04–6.86	1.22–2.36		
Ta	0.35–1.14	0.17–0.25				1.16–5.85	0.34–0.59		
Pb	3.11–3.65	0.47–5.44				5.32–10.88	1.31–10.41	4.0–16.0	
Th	1.41–2.50	0.30–0.74	1.4–3.5	0.8.0–3.30		2.25–5.01	0.44–0.98		
U	0.35–0.65	0.07–0.18				0.48–1.64	0.1–0.26		
Cr	810–1576	155–388	200–1974	102–766		16–132	15–103	31–707	18–899
Ni	562–854	64–431	113–453	49–215	44–98	41–89	106–235	59–167	20–322

Uauá (Oliveira et al. 2013), Lavras (Pinese 1997), Paraopeba (Chaves 2011), Curaçá and Chapada Diamantina (Silveira et al. 2013), Diamantina (Mazzucchelli et al. 2000), Salvador–Olivença (Bellieni et al. 1998)
mg# = Mg/(Mg +Fe_2) for Fe_2O_3/FeO = 0.15

The Mesoproterozoic swarms of Curaçá and Chapada Diamantina (Silveira et al. 2013) also display some similarities regarding variation diagrams. CaO and Al_2O_3, increase, with mg# increasing, whereas the incompatible elements Ba, La and Zr decrease (Fig. 8.6). However, their chemical compositions show significant differences. Curaçá tholeiites have lower mg# (0.43–0.49), CaO and Al_2O_3 (Table 8.2; Fig. 8.6) values and higher contents of Na_2O, FeO_t, MnO, K_2O and all incompatible elements (Table 8.2; Figs. 8.5 and 8.6) when compared with those from Chapada Diamantina (mg# 0.57–0.59). Their respective incompatible element patterns and ratios are very distinct (Fig. 8.5; Table 8.3). Positive anomalies of Ba, Nb, Ta, negative of K and Sr, characterizes the Curaçá swarm, whereas Chapada Diamantina dykes display pronounced positive Ba, K, Sr and Eu anomalies, modest ones of La and Ce, and Nb and Ta negative peaks. Consequently several incompatible ratios are very different: Zr/Y, Ce/Y, La/Yb of Curaçá dykes are higher and Ba/Nb, La/Nb, Ti/Zr, Ba/La, Zr/Nb lower when compared with the Chapada Diamantina swarm (Table 8.3).

The Neoproterozoic swarms of Diamantina and Salvador-Olivença display similarities in terms of evolution, incompatible element patterns and ratios (Girardi et al. 2013). The Salvador-Olivença swarm was divided in two groups, LTi ($TiO_2 < 2$ wt%) and HTi ($TiO_2 > 3$ wt%) (Bellieni et al. 1998). The HTi group is more enriched in incompatible elements. However major and trace elements variation trends of both groups and their respective spidergrams are similar. For these reasons both groups will be represented as a single swarm. Diamantina and Salvador-Olivença variation diagrams show that with mg# increasing CaO and Al_2O_3 increase, whereas FeO_t (not shown), TiO_2 and incompatible elements decrease (Fig. 8.6), which indicates the gabbro fractionation. In terms of incompatible element they are similar as shown by several ratios (La/Nb, Zr/Nb, Ce/Y, Zr/Y) and respective patterns (Fig. 8.5; Table 8.3). According to Mazzucchelli et al. (2000) the variation trends of the Diamantina swarm suggest that the most primitive samples were dominated by fractional crystallization of olivine in the mg# range 0.65–0.58, followed or accompanied by plagioclase and clinopyroxene in more advanced crystallization stages. In the other samples fractional crystallization is dominated by plagioclase and clinopyroxene in all the mg# range.

8.1.4.3 Sr–Nd Isotopes

Radiogenic isotopes of some elements, as widely known, provided the main basis for its study and to characterize the so-called mantle reservoirs. Zindler and Hart (1986) used the compositional isotopic variation of Sm, Nd and Pb to propose five end-members compositions: DMM (depleted mantle), EM1 (enriched mantle type 1), EMII (enriched mantle type 2), HIMU and PREMA. Later on several other mantle reservoirs were proposed (e.g., FOZO, PHEM, BSE) taking into account the compositional range of these elements as well as the influence of other isotopes of U, Hf and He (Hanan and Graham 1996; Farley et al. 1992; Hofmann 2005). Most authors have considered the mixing of the end-members proposed by Zindler and Hart (1986), except for PREMA, which is in disuse, to characterize the isotopic mantle composition. DMM is defined as the most depleted component and the main component of the MORBs, especially of the N-MORB, which was intended to define "normal" MORBs, but is usually referred to "depleted" MORB (Hofmann 2005). EMI and EMII are distinguished by their $^{87}Sr/^{86}Sr$ and $^{143}Nd/^{144}Nd$ contents, and are considered by the many authors as originated, respectively, from subducted pelagic and terrigenous continental material (e.g. Weaver 1991;

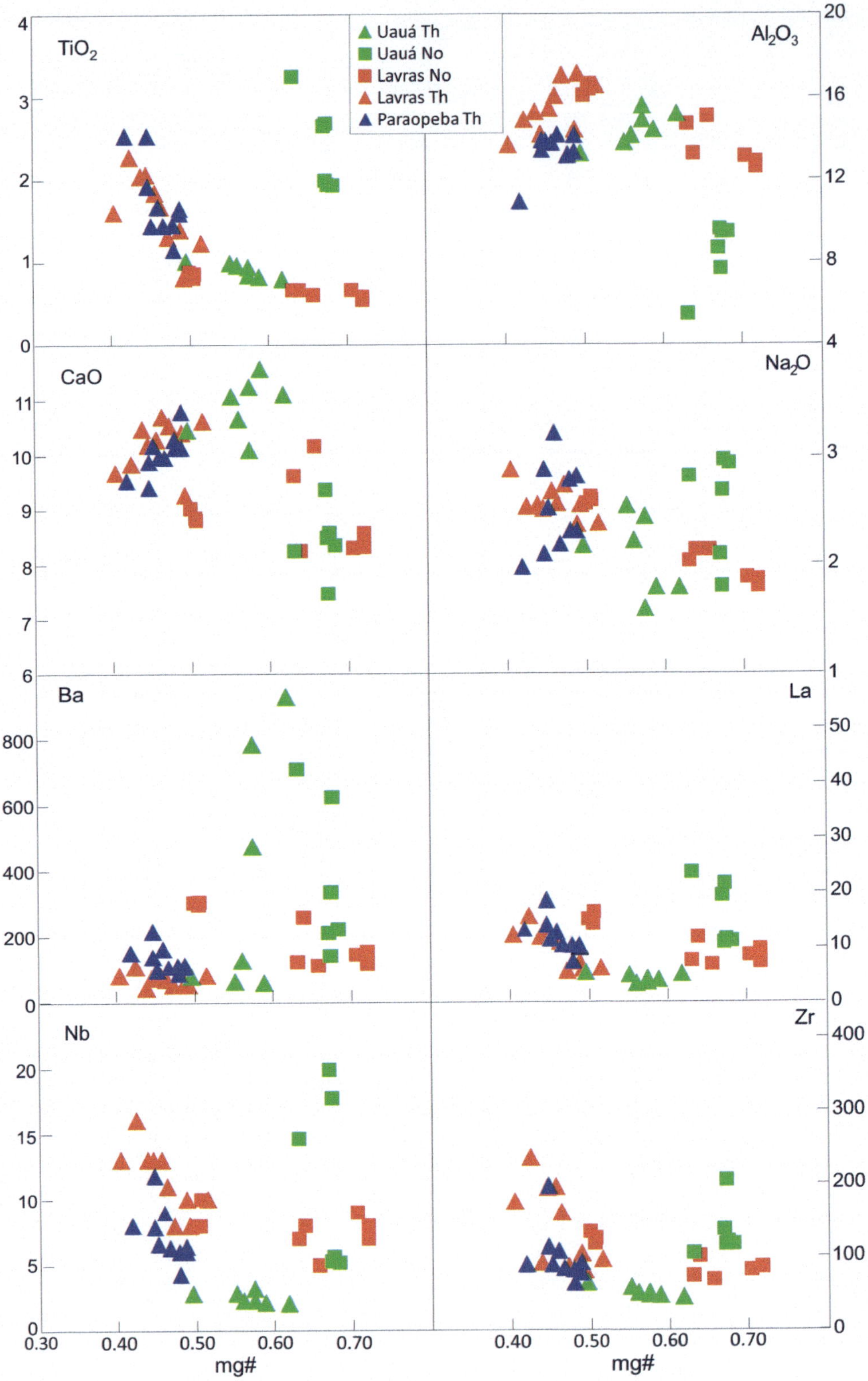

Fig. 8.4 Selected major and trace elements *vs* mg# values for Archean and Paleoproterozoic dyke swarms

Woodhead and Devey 1993; Hauri et al. 1996; Hofmann 1997, 2005; Blichert-Toft et al. 1999; Eisele et al. 2002). On the other hand, Plank and Langmuir (1998) attributed the EMs compositional variation to the nature of the lithological constituents (clastic, biogenic, hydrothermal), whereas the EMI origin was related to recycling of subcontinental metasomatized lithosphere from craton margins (Tatsumoto and Nakamura 1991; Rollinson 1993). The origin of HIMU is

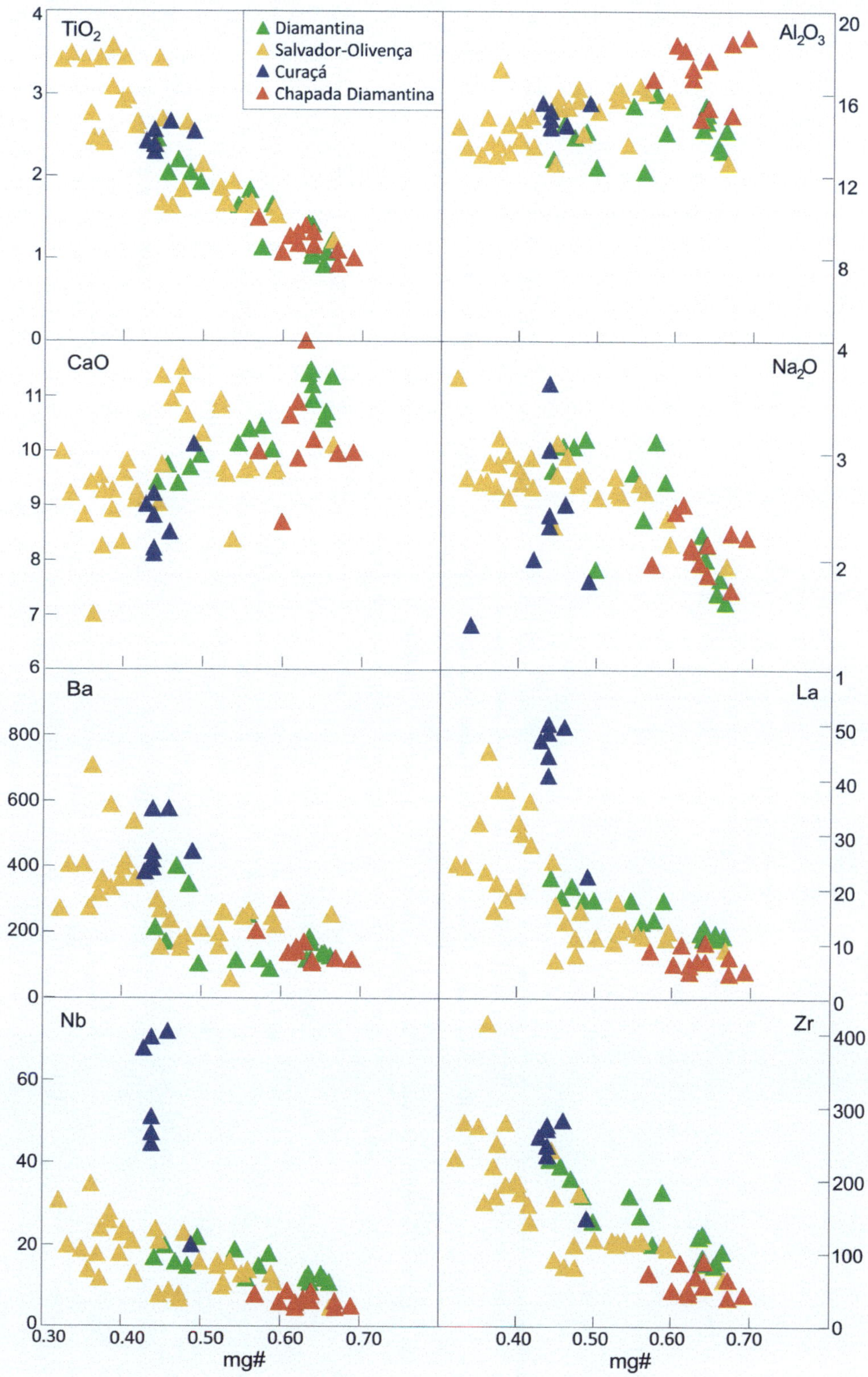

Fig. 8.5 Primitive mantle normalized diagrams of incompatible elements from dyke swarms

generally attributed to the role of subducted oceanic crust, whose main characteristics are the high $^{206}Pb/^{204}Pb$ and $^{208}Pb/^{204}Pb$ ratios.

Table 8.4 shows the results of 53 Rb–Sr and Sm–Nd isotopic analyses of the SFC dyke swarms, which were plotted on Figs. 8.7 and 8.8. The absence of Pb isotopic data in the pertinent literature hampers to evaluate the role of the HIMU component. Therefore the isotopic variation of the swarms will be considered, in this section, in terms of EMI, EMII and DMM. The $\varepsilon_{Sr(T)}$ and the $\varepsilon_{Nd(T)}$ values were

Table 8.3 Average of incompatible element ratios of selected samples from mafic dykes compared with N-MORB, E-MORB, OIB (Sun and McDonough 1989), average values of SW Amazonian mafic dykes and related intrusions (Girardi et al. 2013), and Gough, Tristan da Cunha and Samoa islands (Willbold and Stracke 2006)

	Uauá	UaUá	Lavras	Lavras	Paraopeba	Curuçá	Chapada Diamantina	Diamantina	Salvador–Olivença	N–MORB	OIB	SW Amazonian Dykes	St. Helena	Tristan da Cunha	Samoa
	Norite	Tholeiite	Norite	Tholeiite	Tholeiite	Tholeiite	Tholeiite	Tholeiite	Tholeiite			Tholeiite			
Rb/Sr	0.05	0.08	0.18	0.08	0.19	0.04	0.04	0.02	0.05	0.01	0.05	0.05	0.03	0.05	0.08
Zr/Y	7.23	2.86	4.60	3.18	2.80	6.74	3.89	7.14	6.31	2.64	9.66	3.09	8.74	10.24	8.45
Ce/Y	1.91	0.52	1.05	0.70	0.74	2.49	1.00	1.60	1.14	0.26	2.75	0.63	3.03	4.61	2.82
Ba/Nb	32.9	140	25.8	6.31	17.0	8.68	23.6	12.0	20.6	2.70	7.29	42.7	4.86	9.34	7.73
La/Yb	10.5	1.95	7.10	3.04	3.33	11.8	4.16	7.70	5.29	0.82	17.1	2.70	18.5	23.7	21.0
Ti/Zr	108	104	43.7	69.2	100	60.0	116	64	85.4	103	61.0	105	78.6	70.2	85.9
La/Nb	1.43	1.61	1.38	0.87	1.64	0.81	1.06	1.00	1.05	1.07	0.77	2.39	0.64	0.83	0.78
Ti/Y	780	297	201	220	281	405	452	456	539	271	593	312	687	719	726
Ba/La	23.0	87.0	18.7	7.2	10.4	10.7	22.2	11.7	19.8	2.5	9.5	17.9	7.6	11.3	9.9
Zr/Nb	11.8	20.7	12.4	12.6	14.2	4.6	9.4	9.8	9.8	31.8	5.8	31.5	3.9	4.0	4.7

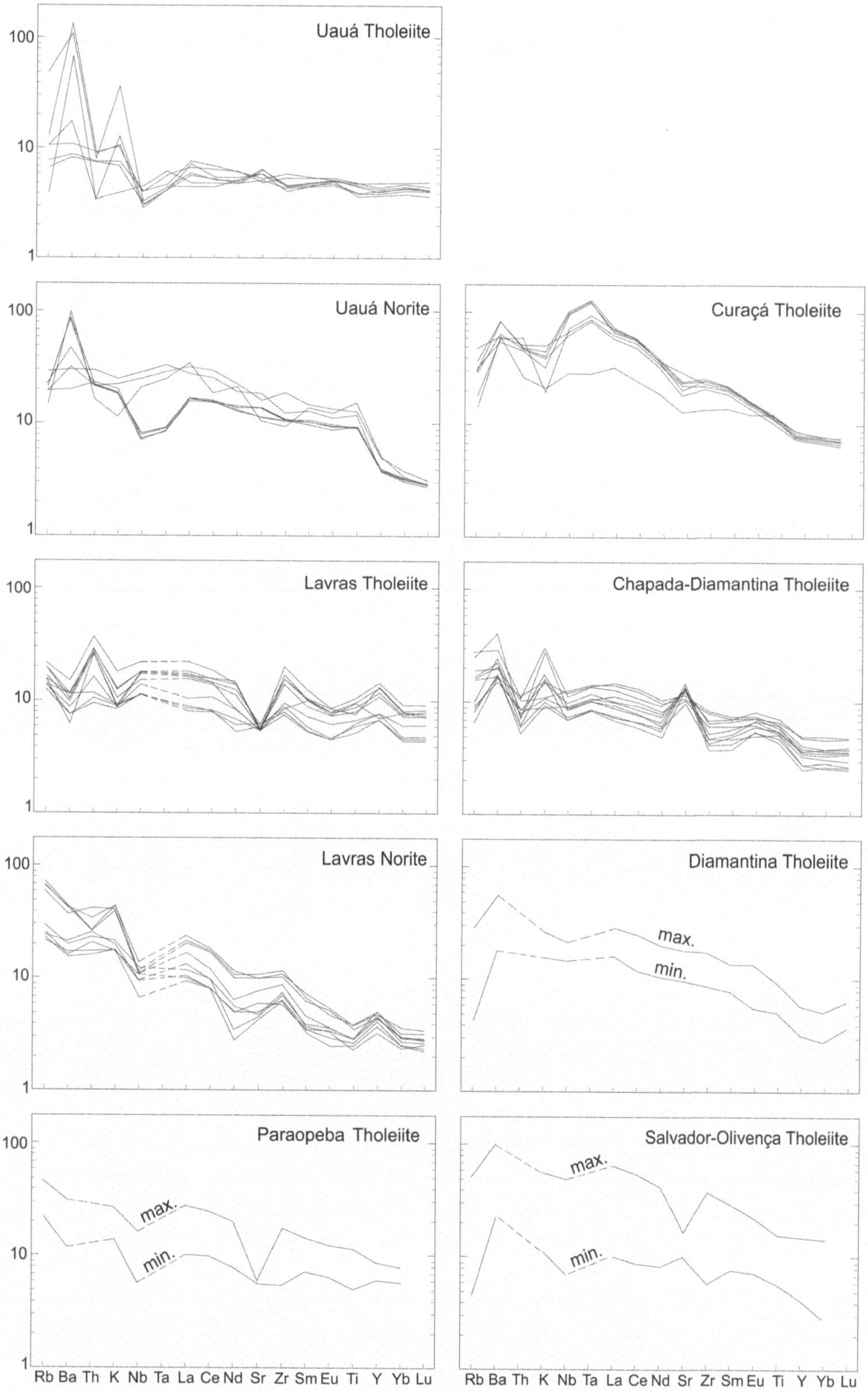

Fig. 8.6 Selected major and trace elements *vs* mg# values for Mesoproterozoic and Neoproterozoic dyke swarms

calculated assuming ages of 2.73 Ga (norites), 2.62 Ga (tholeiites) and 2.66 Ga (norites), 1.97 Ga (tholeiites) for the Uauá and Lavras swarms respectively. The samples from Curaçá and Chapada Diamantina were plotted considering an age of 1.5 Ga, whereas those from Diamantina and Salvador-Olivença were calculated assuming ages of 0.93 and 0.92 respectively.

The Uauá norites reveal small variation of $\varepsilon_{Nd(T)}$ positive values (+2.4 to +3.7) and a larger $\varepsilon_{Sr(T)}$ variation (−0.6 to +18.5). The tholeiites, except for one sample, which is very

Table 8.4 Sr–Nd isotopic values of dyke swarms

Swarm and Age	Sample	$^{87}Rb/^{86}Sr$	$^{87}Sr/^{86}Sr$	$^{147}Sm/^{144}Nd$	$^{143}Nd/^{144}Nd$	$\varepsilon^{T}(Sr)$	$\varepsilon^{T}(Nd)$
Uauá Norite 2.76 Ga	CL11–D3	0.1409	0.707030 (20)	0.13390	0.511620(14)	2.7	2.4
	EO-119.1	0.1473	0.707050(20)	0.15048	0.511957(14)	−0.6	2.9
	MP-07	0.0935	0.705474(20)	0.13561	0.511727(14)	7.2	3.7
	MP-34	0.1635	0.709025(20)	0.13274	0.512413(13)	18.5	2.9
	EO-47.2	0.1435	0.706985(20)	0.15046	0.511970(16)	0.6	3.2
	UAEO-126.1	0.1344	0.707440(20)	0.15070	0.511951(8)	12.2	2.7
Uauá Tholeiite 2.62 Ga	EO-148.3	0.1164	0.705890(20)	0.18268	0.512466(14)	1.0	1.4
	EO-45.1	0.1755	0.707360(20)	0.16201	0.512221(14)	−10.0	3.6
	MP-02	0.1707	0.705748(20)	0.19707	0.512766(14)	−30.4	2.4
	UA96–3.2	0.0969	0.704645(20)	0.19367	0.512684(13)	−6.2	1.2
	LR-34F	0.1020	0.704240(20)	0.19174	0.512668(19)	−14.7	2.2
	LR-38F	0.0898	0.704740(20)	0.18680	0.512581(13)	−1.0	2.2
Lavras Norite 2.66 Ga	54	0.4542	0.720730(90)	0.1479	0.511731(23)	26.6	−1.6
	55	0.3212	0.714790(100)	0.1539	0.511789(38)	15.2	−2.5
	64	0.3962	0.719170(70)	0.1349	0.511527(23)	36.5	−1.1
	25	0.6534	0.726800(50)	0.1410	0.511586(44)	4.1	−2.1
	28	0.3228	0.713880(90)	0.1501	0.511778(25)	1.4	−1.5
	112	0.4628	0.717920(80)	0.1326	0.511793(33)	−17.9	4.8
	70	0.4656	0.717940(70)	0.1330	0.511842(34)	−19.1	5.6
Lavras Tholeiite 1.97 Ga	60	0.1736	0.707930(100)	0.1851	0.512444(53)	11.7	−1.2
	45	0.1794	0.707360(80)	0.1833	0.512523(33)	0.9	1.1
	19	0.1608	0.706930(80)	0.1761	0.512311(28)	2.6	−1.2
	27	0.2827	0.710980(90)	0.1749	0.512287(36)	1.1	−4.8
	34	0.2005	0.707340(80)	0.1797	0.512235(99)	−7.6	−3.6
Curuçá Tholeiite 1.5 Ga	DB-1	0.0971	0.710819(56)	0.1451	0.511745(8)	85.2	−7.5
	DB-6	0.1375	0.710789(51)	0.1126	0.511746(8)	72.4	−1.3
	DB-10A	0.1209	0.710372(59)	0.1104	0.511741(5)	100.0	−1.0
	DB-8A	0.1429	0.710585 (53)	0.1137	0.511681(6)	67.8	−2.8
	DB-9A	0.0772	0.710159(46)	0.1075	0.511613(6)	81.9	−2.9
Chapada Diamantina Tholeiite 1.5 Ga	CHD-05	0.1116	0.706959(58)	0.1581	0.512237(6)	25.8	−0.4
	CHD-07	0.0566	0.705997(43)	0.1520	0.512215(15)	28.9	0.3
	CHD-103	0.1164	0.707488(49)	0.1497	0.512339(5)	31.9	3.2
	CHD-110	0.0512	0.706265 (33)	0.1442	0.512239(9)	34.4	5.2
Diamantina Tholeiite 0.93 Ga	DI-11	0.0233	0.70657(12)	0.1539	0.512337(17)	40.4	−0.8
	DI-30	0.0164	0.70708(25)	0.1412	0.512314(20)	49.0	0.3
	DI-59	0.0400	0.70720(12)	0.1485	0.512237(11)	46.2	−2.1
	DI-48	0.1623	0.70803(8)	0.1341	0.512281(7)	34.9	0.5
	DI-50	0.1182	0.70805(17)	0.1499	0.512254(10)	43.5	−1.9
	DI-35	0.0101	0.70680(9)	0.1217	0.512163(12)	46.2	−0.4

(continued)

Table 8.4 (continued)

Swarm and Age	Sample	$^{87}Rb/^{86}Sr$	$^{87}Sr/^{86}Sr$	$^{147}Sm/^{144}Nd$	$^{143}Nd/^{144}Nd$	$\varepsilon^T(Sr)$	$\varepsilon^T(Nd)$
Salvador–Olivença Tholeiite 0.92 Ga	6099	0.1550	0.70608(2)	0.1577	0.51234(4)	8.7	−1.2
	6091	0.0463	0.70443(2)	0.1784	0.51246(2)	5.6	−1.3
	6383	0.0837	0.70549(4)	0.1781	0.51247(3)	13.7	−1.1
	6067	0.0629	0.70401(2)	0.1740	0.51262(3)	−3.5	2.3
	6142	0.0989	0.70386(2)	0.1485	0.51247(9)	−12.3	2.4
	6380	0.0833	0.70585(3)	0.1642	0.51217(3)	18.1	−5.3
	6131	0.0699	0.70445(2)	0.1659	0.51262(3)	1.5	3.3
	6391	0.1650	0.70770(3)	0.1814	0.51234(2)	29.8	−4.0
	6346	0.0842	0.70527(4)	0.1766	0.51250(3)	10.5	−3.4
	6063	0.1348	0.70662(2)	0.1499	0.512010(2)	20.2	−6.7
	6078	0.2390	0.70809(3)	0.1887	0.51234(9)	21.5	−3.1
	6125	0.0391	0.70384(2)	0.1887	0.51234(4)	−1.4	−3.1
	6059	0.0676	0.70630(2)	0.1803	0.51213(3)	28.2	−8.0
	6115	0.1421	0.70335(3)	0.1522	0.51243(8)	0.7	0.8

Sources of Uauá, Lavras, Diamantina and Salvador-Olivença as in Table 8.2. Samples from Curaçá and Chapada Diamantina (this work)

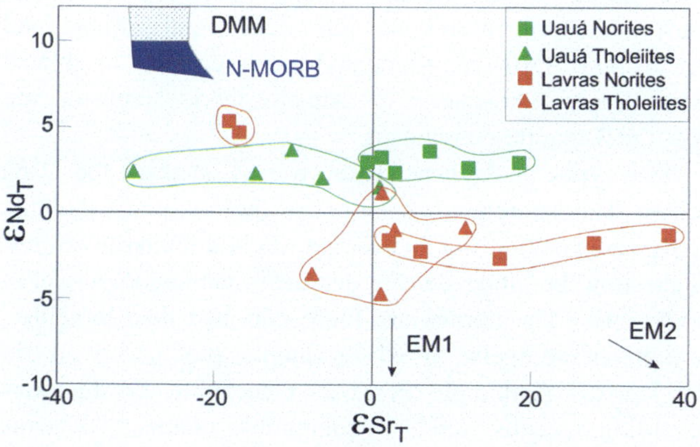

Fig. 8.7 $\varepsilon_{Sr(T)}$ *vs* $\varepsilon_{Nd(T)}$ compositional fields of Archean and Paleoproterozoic dyke swarms. DMM, MORB, EMI and EMII after Rollinson (1993)

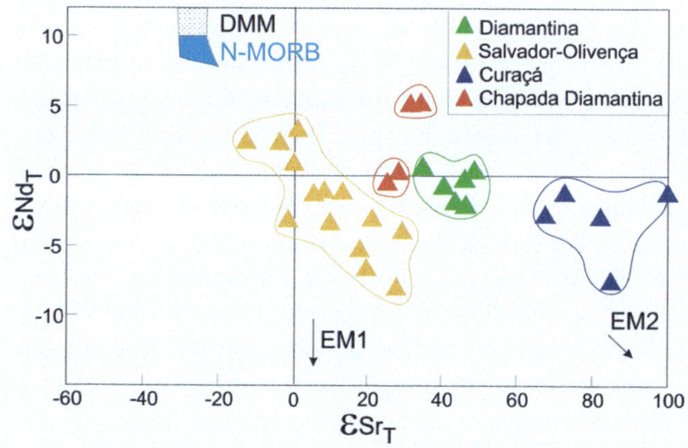

Fig. 8.8 $\varepsilon_{Sr(T)}$ *vs* $\varepsilon_{Nd(T)}$ compositional fields of Mesoproterozoic and Neoproterozoic dyke swarms. DMM, MORB, EMI and EMII as Fig. 8.7

close to the "Bulk Earth", plot in the depleted quadrangle ($\varepsilon_{Sr(T)} = +1.0$ to -30.4; $\varepsilon_{Nd(T)} = +1.2$ to $+3.6$) (Table 8.4 and 8.7). The isotopic ranges show relatively small differences in $\varepsilon_{Nd(T)}$ values, in average, slightly higher in norites, but a large difference in the $\varepsilon_{Sr(T)}$ ranges.

The norites from Lavras show two distinct isotopic fields. The smaller group of samples ($\varepsilon_{Sr(T)} = -17.9$ to -19.1; $\varepsilon_{Nd(T)} = +4.8$ to $+5.8$) lies in the depleted quadrangle, which indicates a stronger influence of the DMM component, whereas the larger group ($\varepsilon_{Sr(T)} = +1.4$ to $+36.54$; $\varepsilon_{Nd(T)} = -1.1$ to -2.5) plots in the enriched quadrangle, suggesting mixing between the depleted mantle and an EM end-member. Samples from the Lavras tholeiite compositional field ($\varepsilon_{Sr(T)} = -7.6$ to $+11.7$; $\varepsilon_{Nd(T)} = -4.8$ to $+1.1$; Table 8.4; Fig. 8.7) display less isotopic variation, and their average plot close to the "Bulk Earth".

The Curaçá tholeiite field ($\varepsilon_{Sr(T)} = +67.8$ to $+100.0$; $\varepsilon_{Nd(T)} = -1{,}0$ to -7.5) lie in the enriched quadrangle (Table 8.4; Fig. 8.8), whereas samples from Chapada Diamantina apparently constitute two compositional fields ($\varepsilon_{Sr(T)} = +31.9$ to $+34.4$; $\varepsilon_{Nd(T)} = +3.2$ to $+5.2$) and ($\varepsilon_{Sr(T)} = +25.8$ to $+28.9$; $\varepsilon_{Nd(T)} = -0.4$ to $+0.3$).

The Diamantina samples exhibit high $\varepsilon_{Sr(T)}$ values and small $\varepsilon_{Nd(T)}$ variation ($\varepsilon_{Sr(T)} = +34.9$ to $+49.0$; $\varepsilon_{Nd(T)} = -2.1$ to $+0.5$). The Salvador-Olivença dyke swarm has a large compositional field ($\varepsilon_{Sr(T)} = +28.9$ to -12.3; $\varepsilon_{Nd(T)} = -8.0$ to $+3.3$) (Table 8.4; Fig. 8.8), which extends from the enriched to the depleted quadrangle.

In sum Sr–Nd isotopic data of the studied swarms evidence that the parental mantles of all swarms are heterogeneous in variable scale. Their integrated isotopic and geochemical studies are a fundamental tool to understand the related petrogenetic processes.

8.1.5 Petrogenesis

Mantle heterogeneity is attributed to the effects caused by several geological and geochemical processes, as subduction, lithospheric recycling and action of deep lithospheric or asthenospheric fluids. In this respect, subduction of oceanic crust, overlaid by variable amounts of sediments, beneath the sub-lithospheric mantle in island arcs and intra-continental settings, plays an important role to explain mantle enrichment, recurrent through geological time. In this scenario, fluids generated from the slab induce metasomatism in the overlying mantle. The observed LILE and LREE enrichment can be attributed to overlying sediments (e.g., Weaver 1991), whereas retention by rutile in the slab eclogites may be attributed the observed negative Nb-Ta anomalies (e.g., Cordery et al. 1997; Takahashi et al. 1998; Ayers 1998; Rivalenti et al.1998; Leitch and Davis 2001). Small amounts of rutile (~ 2 %) would be enough to prevent Nb and Ta enrichment in the mantle wedge (Brenan et al. 1994), and this would result in high LILE and LREE contents and low Nb values, as shown by high Ba/Nb, La/Nb, Ba/La and Zr/Nb ratios in the studied tholeiitic dykes of the SW Amazonian craton (Girardi et al. 2013). In that case contribution of the slab fluids in the mantle source was estimated to reach 30 %, which reflects the importance of the sedimentary contribution on the mantle source. In this respect the high Ba contents and ratios are significant, probably due to the presence of pelagic sediments (Weaver 1991). Table 8.3 compares incompatible element ratios of the studied dykes of SFC with average ratios of swarms from the SW Amazonian craton (Girardi et al. 2013), assuming N-MORB and OIB values from Sun and McDonough (1989). It also shows average ratios from Santa Helena, Samoa and Tristan da Cunha oceanic islands.

The OIB source is debatable. According to Willbold and Stracke (2006) recycling of oceanic crust, containing variable proportion of upper and lower continental crust material, is responsible for isotopic and trace element variations, and permit to distinguish OIB HIMU from OIB EMI and EMII types. For instance, the origin of the Hawaiian lavas, made up of tholeiitic basalts and picrites, is attributed to a heterogeneous mantle plume which resulted from mixing of recycled oceanic crust containing HIMU and EMI components and the peridotite matrix from the lower mantle (Ren et al. 2006, 2009). However, Niu and O'Hara (2004) and Niu et al. (2012), stated that recycled oceanic crust is too depleted to give rise to OIB melts, although not ruling out the contribution of continental sediments as an enriched OIB component. They distinguished OIB and MORB sources, and advocated that MORB sources are more depleted, both isotopically and in terms of trace element abundance, with respect to OIB. Finally, these Authors attributed the OIB enrichment to mantle metasomatism in the deeper portions of the oceanic lithosphere. The OIBs patterns have, in general, common characteristics, as enrichment in the more incompatible elements, especially Nb and Ta, MREE and mainly HREE depletion. These characteristics evidence a remarkable contrast of OIBs patterns and ratios in relation to the average values of arc and intraplate continental tholeiites, as shown by the mafic swarms from SW Amazonian craton. Lower Ba/Nb, La/Nb, Zr/Nb and Ba/La, and higher La/Yb, Ce/Y, Ti/Y and Zr/Y distinguish both sources and characterize OIBs ratios (Table 8.3).

The origin of the parental source of continental mafic dyke swarms is complex, because it may involve mixing in several proportions of the so-called "mantle components", and, in consequence can display geochemical characteristics of basalts from different environments. This is the case of the dykes of the São Francisco craton. In order to consider the petrogenesis of these dykes we will take into account the influence of the slab fluids and melts resulted from oceanic crust subduction in arcs and intra-continental settings, and the action of OIB-type fluids and melts, possibly originated from deeper lithospheric or asthenospheric levels on the N-MORB mantle component.

Two types of Archean dyke swarms occur in the Uauá region. The isotopic $\varepsilon_{Nd(T)}$ values of the norite and tholeiite groups are positive and, in average, slightly higher in norites, suggesting the influence of a depleted mantle component in both suites. The norites are more enriched than tholeiites, in terms of radiogenic Sr, whose samples plot, except one, in the depleted quadrangle (Fig. 8.7). Geochemical distinctions regarding igneous trends, incompatible elements contents and ratios highlight the differences between the two suites. LILE/HSFE and LREE/HSFE ratios are high in both lithologies, but higher in tholeiites, suggesting a more significant influence of fluids from subducted ocean crust in their source, whereas high LREE/HREE (Ce/Y and La/Yb) ratios and HSFE/HREE (Zr/Y and Ti/Y) ratios, similar to those from OIB islands (Table 8.3), suggest that uprising of OIB-type fluids could have an important role on the norite parental mantle. This scenario is consistent with different lithospheric thicknesses and distinct parental mantles. According to Oliveira et al. (2013) the origin of tholeiite melts is ascribed to shallower source with respect to the deeper garnet-bearing mantle, which originated the norite swarm. Moreover the norite melts derived from lower melting degree of an enriched refractory parental mantle, whereas the tholeiites resulted from higher melting degrees of a more depleted source.

The available data of both swarms permit an additional speculation. If different enrichment degrees at different depths are admitted for the origin of the respective mantle sources of both swarms, and taking into account the different geochemical and isotopic features, and crystallization ages, it is reasonable to suppose that the metasomatic effects occurred at different times during the Archean.

The Archean norites and Paleoproterozoic tholeiites dyke swarms of the Lavras region have isotopic and geochemical differences. The isotopic plots of the norite swarm indicate a heterogeneous mantle composed by two sources, one close to the N-MORB component, and another, in which the depleted component was enriched mainly by a radiogenic Sr-bearing contaminant (Fig. 8.7). The average of isotopic Sr–Nd values of the Lavras tholeiites indicates composition close to the Bulk Earth. The distinction between the two types of dykes is also shown by geochemical data. Variation diagrams distinguish melts evolution and fractionation processes (Fig. 8.4). Incompatible elements diagrams and ratios are very different (Fig. 8.5; Table 8.3). Consequently La/Nb, Ba/Nb, Ce/Y and La/Yb ratios are much higher in norites. Low Ba/Nb and La/Nb (<1) ratios do not favor slab fluids metasomatism in the tholeiite parental source, whereas this process is suggested in the norite source by these higher element ratios. In short, geochemical and isotopic data indicate distinct mantle sources of Archean norites and Paleoproterozoic tholeiites.

Variation diagrams indicate that the Paleoproterozoic Paraopeba tholeiites evolved by gabbro fractionation, characterized by removal mainly of plagioclase and clinopyroxene. Trace elements patterns and ratios reveal modest fractionation (La/Yb = 3.3), as well as strong positive anomalies of Ba, K and La in relation to Nb, giving rise to high Ba/Nb and La/Nb ratios (Table 8.3). These geochemical features are compatible with mantle enrichment by subduction of oceanic crust and overlying sediments in arc and intraplate continental settings. The absence of isotopic data hampers a more detailed discussion about this swarm.

The coeval Mesoproterozoic Chapada Diamantina and Curaçá swarms (1.5 Ga) have distinct geochemical and isotopic characteristics. The Curaçá melts are more evolved and more enriched in all incompatible elements (Table 8.2). Lower LREE/HSFE and LILE/HSFE ratios, due to the Nb and Ta contents with respect to Ba, K, La and Ce (Fig. 8.5; Table 8.3), of the Curaçá dykes, favors the larger influence of OIB-type fluids on the genesis of the parental source of the these dykes in comparison with those of the Chapada Diamantina swarm. The isotopic variation fields of the Curaçá and Chapada Diamantina dykes (Fig. 8.8), are in accordance with the geochemical differences of both swarms.

The Neoproterozoic Diamantina and Salvador-Olivença swarms can be considered coeval, taking into account the respective age errors. Incompatible element patterns are similar (Fig. 8.5), as well as several element ratios (Ce/Y, Zr/Y, Ti/Zr, La/Nb and Zr/Nb) (Table 8.3). The similarity of their heterogeneous mantle sources was supported by the application of the mixing geochemical model, whose compositions vary from an almost depleted N-MORB to an N-MORB component enriched by significant amounts of OIB- like and slab derived fluids (Girardi et al. 2013).

8.1.6 Mantle Source Inferences

In this section we use the mixing model reported in Girardi et al. (2013) in order to give to the readers a flavor of the variability of the mantle sources involved in the petrogenesis of the various dyke swarms, as well as to verify if estimated values of the source components are consistent with the geochemical and isotopic data of the studied mafic rocks.

The model considers parent melts derived from mantle reservoirs: N-MORB, OIB (Sun and McDonough 1989) and a slab component representative of fluids released from the subducted oceanic mafic crust and overlying sediments. The slab component is that reported by Rivalenti et al. (2007, Table 8.3), which is representative of the fluid released from the Nazca plate (Kilian and Behrmann 2003; Klein and Karsten 1995) and was estimated from the fluid/eclogite partition coefficients from Kessel et al. (2005) and batch melting equation. The use of the Nazca plate fluids, whose the main components are represented by the Chile Ridge basalts and by the LEG 141 sediments, as one of the model end-members, is supported by the respective similarity of its basalt and sediment compositions to those of the N-MORB, and GLOSS, which represents the average of the global subducting sediments (Plank and Langmuir 1998).

The model concerns ratios of highly incompatible elements, which, as previously mentioned, in closed-system process remain broadly constant during fractional crystallization if variable, but not large melting degrees of the mantle source occur. In order to minimize eventual effects of fractionation, we selected, for the various mafic intrusions, representative samples of the most primitive melts. The selected element ratios for each dyke swarm are: Rb/Sr, Zr/Y, Ce/Y, Ba/Nb, La/Yb, Ti/Zr, La/Nb, Ti/Y, Ba/La and Zr/Nb (Table 8.5).

Table 8.5 shows the estimated end-members for the following dyke swarms: Uauá tholeiites and norites, Lavras tholeiites and norites, Paraopeba, Curaçá, Chapada Diamantina, Diamantina and Salvador-Olivença tholeiites.

As far as the Uauá dykes are concerned, the element ratios are consistent with a mixing model between the following end-members: 100 % N-MORB to 40 % N-MORB + 10 % OIB + 50 % slab fluids for the tholeiites and 92 % N-MORB + 8 % OIB and 40 % N-MORB + 50 % OIB + 10 % slab fluids for the norites. The Ti/Y and Ti/Zr ratios for the Uauá norites are slightly higher than the maximum ratios of the model, maybe because of some ilmenite cumulus effect.

Table 8.5 Incompatible element ratios of the mafic dykes from the São Francisco Craton and minimum and maximum values of the mixing models. N-MORB, OIB as in Table 8.3 and slab fluid composition from Rivalenti et al. (2007)

	N-MORB	OIB	Slab Fluids	Rb/Sr	Zr/Y	Ce/Y	Ba/Nb	La/Yb	Ti/Zr	La/Nb	Ti/Y	Ba/La	Zr/Nb
	Uauá tholeiite			0.05	2.91	0.52	31.1	1.88	102	1.63	297	19.1	21.9
Min	100	0	0	0.01	2.64	0.27	2.7	0.82	75	1.07	271	2.5	9.5
Max	40	10	50	0.06	5.07	4.16	75.6	20.56	103	3.78	378	20.0	31.8
	Uauá norite			0.05	7.99	2.09	18.9	10.12	96	1.30	768	14.5	12.1
Min	92	8	0	0.02	3.22	0.47	5.6	1.77	66	0.88	298	6.4	8.2
Max	40	50	10	0.06	8.35	2.62	20.9	13.96	92	1.22	549	17.1	15.1
	Lavras tholeiite			0.07	2.90	0.48	6.7	2.60	81	0.68	236	10.0	10.3
Min	97	3	0	0.02	2.88	0.41	5.6	1.45	76	0.76	272	6.6	7.6
Max	70	30	0	0.04	4.13	0.95	7.0	3.99	95	0.85	314	9.2	15.6
	Lavras norite			0.18	4.60	1.05	25.8	7.10	44	1.38	201	18.7	12.4
Min	95	5	0	0.02	3.01	0.40	5.1	1.41	63	0.92	288	5.6	8.6
Max	40	10	50	0.18	5.92	5.09	72.2	24.69	96	3.50	373	20.6	18.3
	Paraopeba tholeiite			0.19	2.80	0.74	17.0	3.33	100	1.64	281	10.4	14.2
Min	99	1	0	0.01	2.72	0.29	3.5	0.94	68	1.02	275	3.4	8.9
Max	50	30	20	0.09	6.82	2.63	29.8	13.09	101	1.62	466	18.4	27.3
	Curuçá tholeiite			0.04	6.74	2.49	8.7	11.76	60	0.81	405	10.7	4.6
Min	70	30	0	0.04	4.80	1.03	6.8	4.62	62	0.80	370	8.5	4.6
Max	10	90	0	0.04	9.81	4.16	11.6	24.22	77	0.89	609	13.0	8.5
	Chapada Diamantina tholeiite			0.04	4.92	1.27	15.8	5.45	90	1.16	442	13.6	10.2
Min	90	10	0	0.02	3.37	0.52	5.9	2.01	70	0.86	305	6.8	9.0
Max	50	30	20	0.06	6.69	2.48	28.7	12.40	90	1.59	467	18.1	13.7
	Diamantina tholeiite			0.02	7.08	1.64	10.1	8.10	65	0.97	460	10.4	8.8
Min	92	8	0	0.02	3.22	0.47	5.6	1.77	62	0.88	298	6.4	6.7
Max	30	40	30	0.16	7.54	3.68	25.6	19.68	92	1.54	470	16.6	15.1
	Salvador–Olivença tholeiite			0.03	3.55	0.91	17.2	4.38	92	0.92	328	18.7	8.8
Min	92	8	0	0.02	3.22	0.47	5.6	1.77	74	0.88	298	6.4	9.0
Max	62	18	20	0.16	4.80	2.03	30.0	9.35	92	1.76	355	17.0	15.1

The best end-members of the mixing model for the element ratios of the Lavras mafic swarms are the following: 97 % N-MORB + 3 % OIB and 70 % N-MORB + 30 % OIB for the tholeiites and 95 % N-MORB + 5 % OIB and 40 % N-MORB + 10 % OIB + 50 % slab fluids for the norites. Similarly to the Uauá norites, the Ti/Y and Ti/Zr ratios of the Lavras norites are slightly higher than the maximum ratios of the model, due probably also to cumulus of ilmenite.

The incompatible element ratios of the Paraopeba dykes indicate a quite heterogeneous source, starting from a "low" end-member composition of 99 % N-MORB + 1 % OIB to a "high" component consisting of 50 % N-MORB + 30 % OIB + 20 % slab fluids. In this case, in order to reconcile the high La/Nb ratio of this swarm, the selected OIB component in the source was of Gough-type (Sun and McDonough 1989).

The mixing model of the Curaçá tholeiites ranges from a "low" source of 70 % N-MORB + 30 % OIB to an enriched source, represented by 10 % of N-MORB and 90 % of OIB. In this case, in order to reconcile the Ba/Nb and La/Nb ratios of this swarm, the selected OIB component in the source is of Tristan da Cunha-type (Sun and McDonough 1989).

Regarding the Chapada Diamantina dyke swarm the best end-members of the mixing model are the following: 90 % N-MORB + 10 % OIB and 50 % N-MORB + 30 % OIB + 20 % slab fluids. In this case, in order to reconcile the high Ti/Y ratio of this swarm, the selected OIB component in the source is of Gough-type (Sun and McDonough 1989).

The mixing model for the Diamantina and Salvador-Olivença dyke swarms requires the same "low" end-member (92 % N-MORB + 8 % OIB). The "high" component for the Diamantina swarm is 30 % N-MORB + 40 % OIB + 30 % slab fluids; and for the Salvador-Olivença dykes it is 62 % N-MORB + 18 % OIB + 20 % slab fluids.

The results obtained by the application of mixing model are coherent with the inferences from geochemical and isotopic data (see petrogenesis section), and estimate the proportional contribution of the various mantle components of the SFC swarms.

8.1.7 Regional Correlations and Tectonic Inferences

This section presents a summary of the main tectonic events linked with the emplacement of mafic dykes of the SFC. In addition, the age matches with mafic magmatism on the African counterpart are presented, using the barcode concept (e.g., Bleeker and Ernst 2006; Ernst et al. 2013b) to address some paleogeographic inferences.

The Archean mafic dykes of Uauá (ca. 2.70, 2.62 Ga) and Lavras (older generation; ca. 2.66 Ga) are the oldest swarms, and crop out in the northern and southern portion of the craton, respectively. Their emplacement can be spatially and temporally related with ductile/brittle extensional tectonics linked with the final orogenic phases of the Neoarchean evolution. From a tectonic perspective, these fissural activities are roughly contemporary or slightly younger than the greenstone belt sequences such as the 2.74 Ga Umburanas of the Gavião Block in Bahia (Bastos Leal et al. 2003) and the 2.79–2.75 Ga Rio das Velhas, in Minas Gerais (Noce et al. 2007), therefore suggesting a genetic link.

The Paleoproterozoic Paraopeba (2.10 Ga) and Lavras (1.97 Ga) swarms at the southern cratonic portion emplaced under alternate episodes of transtension and transpression across the Archean crust. Such geologic conditions are tectonically linked to a recognized thrust-fold belt which formed as deformation and metamorphism propagated outwards the Paleoproterozoic Minas accretionary orogeny at the time (Teixeira et al. 2015 and references therein).

In contrast, the 1.70 Ga Pará de Minas dyke swarm is the earliest anorogenic magmatism that followed the cratonization of the Minas orogeny. This interpretation is consistent with comparable ages (ca. 1.7 Ga) for both basal felsic volcanics of the Espinhaço Supergroup and A-type granitic plutonism such as the Borrachudos Granite. These magmatic activities have been collectively associated with an intraplate event accompanying the opening of the Espinhaço rift within the Paleoproterozoic continental crust (e.g., Babinski et al. 1994; Dussin and Dussin 1995).

From a geodynamic point of view, the Mesoproterozoic Curaçá and Chapada Diamantina dyke swarms (1.50 Ga) are representative of an intra-plate extensional episode over old and thick cratonized continental crust, accompanying basin processes associated with the polycyclic history of the Espinhaço basin. These swarms are contemporary with the Humpala sill (1502 ± 5 Ma), which is part of a wider sill province in SW Angola portion of the Congo craton, thus suggesting a much larger scale for such a phenomenon (Ernst et al. 2013b). According to these authors, given the age match (within the errors), all these 1.50 Ga mafic units represent the initial magmatic barcode as parts of a LIP (large igneous province) system. This gives support to the idea of a coherent São Francisco-Congo landmass from at least the late Mesoproterozoic (Silveira et al. 2013). These Authors suggested that Curaçá and Chapada Diamantina parental mantles could have had contributions from a plume. The high content of the OIB component and the absence of subducted slab fluids, as indicated by the geochemical model (Table 8.5), favor the suggested influence of a plume on the Curaçá parental source. However, if the plume model is considered, the significant difference between the Chapada Diamantina and Curaçá parental mantles needs the intervention of a geochemically and isotopically heterogeneous plume, and/or by the interaction of the plume with the lithospheric mantle previously contaminated, similarly to the model proposed for the origin of the Hawaiian lavas (Ren et al. 2006, 2009). Alternatively, the geochemical and isotopic differences between the Chapada Diamantina and Curaçá sources could be explained by different geodynamic scenarios, as suggested by the adopted geochemical model. In this case, Curaçá parental mantle underwent only the action of uprising of OIB-like fluids, probably due to a local thermal anomaly. In contrast, the source of Chapada Diamantina dykes shows the probable influence of fluids and melts from an ancient oceanic subducted crust and also the action of OIB-like fluids (Table 8.5).

The Salvador-Olivença and Diamantina dyke swarms of Tonian age are the youngest Precambrian intraplate mafic magmatism so far recognized in the São Francisco craton. They are well-preserved markers of a prominent anorogenic magmatic episode (ca 0.93–0.91 Ga) akin to the break-up of a coherent lithosphere – the São Francisco- Congo landmass (e.g., Corrêa-Gomes and Oliveira 2000). For instance, these dykes are coeval with Gangila basalts in the Congo craton (Tack et al. 2001). Moreover, a newly identified NNW–NNE trending gabbro-norite dyke swarm in southeastern Angola yields a U/Pb TIMS baddeleyite age of 1.10 Ga (Ernst et al. 2013b). According to these authors, this event has also a precise match with the Umkondo Large Igneous Province (LIP) of the Kalahari craton and mafic intraplate magmatism on other blocks such as the Bundelkhand craton (India). This suggests a long-lived extensional regime over the lithosphere, worldwide in scale. Finally, if all this landmasses shared roughly contemporary magmatic events between 1.10 and 0.91 Ga the magmatic barcode record suggests that former nearest neighbors to the proto-São Francisco-Congo craton at Late Mesoproterozoic may be envisaged.

8.1.8 Final Remarks

The main dyke swarms of the SFC craton show age span from 2.7 to 0.9 Ga. The great majority of analyzed samples plot in the basalt field in the chemical classification diagrams. As consequence of significant differences of geochemical and isotopic features Uauá and Lavras swarms were distinguished into two groups: norite and tholeiite. The latter includes diabases, gabbros and metamorphic derivatives. Geochemical and isotopic data point out that metasomatic effects from subducted oceanic crusts and OIB-like fluids and melts are present in almost all parental mantles of the swarms. The exceptions are the Lavras and Curaçá tholeiites, which do not show effects of subducting oceanic slabs. These data are consistent with the complexity of processes related to the origin of the parental mantles. The adopted geochemical model indicated the variable contribution of the metasomatic agents on the N-MORB component of the SFC dykes mantle sources.

Available paleomagnetic data from granulites of the Jequié complex in the northern São Francisco craton retrogressed at 2.0 Ga ago suggest that crustal shortening processes between Archean blocks that included slab subduction operated at that time, which is consistent with Minas accretionary orogeny (Chap. 5, this book). Since then the SFC landmass was not subjected to collisional episodes until the Neoproterozoic, and therefore mantle enrichment from slab fluids in Meso and Neoproterozoic dyke swarms should originate from ancient subduction systems.

The slab fluid component is part of the Archean Uauá norites and tholeiites, and Lavras norite parental sources (Table 8.5) in the SFC, similarly to the inferred for dykes of Crixás-Goiás (2.49 Ga) block, in Central Brazil (Girardi et al. 2013). This scenario is consistent with recycling of oceanic lithosphere since the Late Archean, as advocated by several Authors (e.g., Condie 2005). The parental mantle compositions of Uauá and Lavras norites and tholeiites, which are geographically close to each other in the respective regions, indicate small scale mantle heterogeneity.

From a tectonic perspective, the 2.73 Ga Uauá norites, the nearby 2.62 tholeiites, and the 2.66 Ga Lavras norites, mark late to post-tectonic extensional episodes (failed rifts?) at regional scale related to the Neoarchean assembly of the sialic core of the proto-São Francisco craton. As widely recognized, rifting of continental areas may cause thinning of the lithospheric mantle and consequent uprising of deep OIB-type fluids and melts, probably from the asthenosphere, favoring metasomatic processes in several different levels of the peridotitic mantle. This model is consistent with local thermal anomalies (e.g., Comin Chiaramonti et al. 2013), and can be considered as a possibility to explain the small scale compositional variation of the parental mantle of the Archean dykes.

In contrast, the tholeiitic dyke swarms of Paraopeba (2.10 Ga) and Lavras (1.97 Ga) were emplaced under a transtension-transpression regime as response from contraction tectonics across the Archean landmass, coeval with the latest stages of the Minas accretionary orogeny. According to Teixeira et al. (2015) the tectonic stability of this orogeny formed the São Francisco paleocontinent at ca. 2.0 Ga.

The striking Pará de Minas (1.71 Ga) swarm can be interpreted as the first regional episodic attempt of rupture of the thick, stabilized Paleoproterozoic lithosphere in association with the opening of the Espinhaço rift.

The emplacement of the Curaçá and Chapada Diamantina dykes took place at 1.5 Ga ago linked with a particular extension episode that is also recorded in the Espinhaço rift related basin and in the magmatism of the African counterpart. Although coeval, the parental mantle sources of these swarms exhibit different geochemical and isotopic characteristics (see previous section).

Finally, the Diamantina (0.93 Ga) and Salvador-Olivença (0.92 Ga) swarms are considered coeval, taking into account the U-Pb age errors. The similarity of their mantle source compositions suggests a common origin related to a pervasive intraplate episode. Regional correlations with mafic magmatic events onto the Congo craton support that such anorogenic activities were widespread at that time, and are probably related to large rift-related basins. Extensional tectonic was recurrent through time, starting at 1.0–0.9 Ga in different cratonic domains of the South America continent (Cordani et al. 2010).

Acknowledgments The authors acknowledge the financial support by Brazilian agencies FAPESP (Fundação de Pesquisa do Estado de São Paulo) and CNPq (Conselho Nacional de Desenvolvimento Científico e Tecnológico).

References

Abreu, P.A.A., 1995. O Supergrupo Espinhaço da Serra do Espinhaço Meridional (Minas Gerais): o rifte, a bacia e o orógeno. Geonomos, 3, 1–18.

Alkmim, F.F., 2004. O que faz de um cráton um cráton? O Cráton do São Francisco e as revelações almeidianas ao delimitá-lo. In: Mantesso-Neto, V., Bartorelli, A., Dal Ré Carneiro, C., Brito-Neves, B.B. (org.) Geologia do Continente Sul-Americano. Evolução da obra de Fernando Fávio Marques de Almeida. Becca, São Paulo, pp. 17–35.

Alkmim, F.F., Noce, C.M., 2006. Outline of the geology of Quadrilátero Ferrífero. In: Alkmim, F.F., Noce, C.M. (eds.). The Paleoproterozoic Record of São Francisco Craton. IGCP Field Workshop, Bahia and Minas Gerais. Field Guide and Abstracts, pp. 37–73.

Alkmim, F.F., Martins-Neto, M.A., 2012. Proterozoic first-order sedimentary sequences of the São Francisco Craton, eastern Brazil. Marine and Petroleum Geology 33, 127–139.

Ávila, C.A., Teixeira, W., Cordani, U.G., Moura, C.A.V., Pereira, R. M., 2010. Rhyacian (2.23–2.20 Ga) juvenile accretion in the southern São Francisco craton, Brazil: geochemical and isotopic evidence from the Serrinha magmatic suite, Mineiro belt. Journal of South American Earth Science, 29, 464-482.

Ayers, J., 1998. Trace element modeling of aqueous fluid-peridotite interaction in the mantle wedge of subduction zones: Contributions to Mineralogy and Petrology, 132, 390–404.

Babinski, M., Brito-Neves, B.B., Machado, N., Noce, C.M., Uhlein, A., Van Schmus, W.R., 1994. Problema na metodologia U/Pb em zircões de vulcânicas continentais: o caso do Grupo Rio dos Remédios, Supergrupo Espinhaço, no Estado da Bahia. Boletim de Resumos Expandidos, SBG, Congresso Brasileiro de Geologia, 2, pp. 409–410.

Babinski, M., Pedreira, A.J., Brito Neves, B.B., Van Schmus, W.R., 1999. Contribuição à Geocronologia da Chapada Diamantina. VII Simpósio Nacional de Estudos Tectônicos, Lençóis, pp. 118–120.

Barbosa, J.S.F., Sabaté, P., 2004. Archean and Paleoproterozoic crust of the São Francisco Craton, Bahia, Brazil: geodynamic features. Precambrian Research, 133, pp. 1–27.

Bastos Leal, L.R., 1992. Geocronologia Rb–Sr e K–Ar, evolução isotópica e implicações tectônicas dos enxames de diques máficos de Uauá e Vale do Rio Curaçá, Bahia. IGc-USP, Dissertação de Mestrado, 118 p.

Bastos Leal, L.R., Teixeira, W., Menezes Leal, A.B., Girardi, V.A.V., Piccirillo, E.M., 1994. Geocronologia Rb/Sr e K/Ar do enxame de diques máficos de Uauá, Bahia (Brasil). Geochimica Brasiliensis, 8 (1), 99–114.

Bastos Leal, L.R., Teixeira, W., Bellieni, G., Petrini, R., Piccirilo, E.M., 1995. Geocronologia e geoquímica isotópica de Sr e Nd nos diques máficos do Curaçá Craton do São Francisco (Brasil): registro de um evento distensivo Neoproterozóico associado à evolução da Faixa Colisional Sergipana. Geochimica Brasiliensis, 9, 141–159.

Bastos Leal, L.R., Cunha, J.C., Cordani, U.G., Teixeira, W., Nutman, A.P., Menezes Leal, A.B., Macambira, M.J.B., 2003. SHRIMP U–Pb, 207Pb/206Pb zircon dating, and Nd isotopic signature of the Umburanas greenstone belt, northern São Francisco craton, Brazil, Journal of South American Earth Sciences, 15(7), 775–785.

Battilani, G.B., Vasconcellos, P.M., Gomes, N.S., Guerra, W.J., 2005. Geochronological data of dykes and sills intruding Proterozoic sequences of the Tombador Formation, Bahia, Brazil. III Simpósio sobre o Cráton do São Francisco, Salvador, pp. 139–142.

Bellieni, G., Petrini, R., Piccirillo, E.M., Cavazzini, G., Civetta, L., Ccmin-Chiaramonti P., Melfi, A.J., Bertolo, S. and De Min, A., 1991. Proterozoic mafic dyke swarms of the São Francisco Craton (SE-Bahia State, Brazil): petrology and Sr–Nd isotopes. European Journal of Mineralogy, 3, 429–449.

Bellieni, G., Piccirillo, E.M., Petrini, R., Girardi, V.A.V., Menezes Leal, A.B., Teixeira, W., Bastos Leal, L.R., De Min, A., Comin Chiaramonti, P., Tanner de Oliveira, M.A.F., 1995. Petrological and Sr–Nd evidence bearing on Early Proterozoic magmatic events of the subcontinental mantle: São Francisco craton (Uaua, NE-Brazil). Contribution to Mineralogy and Petrology, 122, 252–261.

Bellieni, G., Petrini, R., Piccirillo, E.M., Brito, C.M., Figueiredo, A.M. G., Marques, L.S., De Min, A., Melfi, A.J., 1998. Petrogenesis and tectonic significance of the Late Proterozoic unmetamorphosed mafic dyke swarms from the Salvador area (NE Brazil). Neues Jahrbuch für Mineralogie, Abhandlungen, 173, 327–350.

Bleeker, W., Ernst, R., 2006. Short-lived mantle generated magmatic events and their dyke swarms: The key unlocking Earth´s paleogeographic pole. In: Hanski, S., Mertanen, T., Rämö, T., Vuollo, J. (eds.). Dyke swarms – tie markers of crustal evolution, pp. 3–26.

Blichert-Toft, J., Frey, F.A., Albarède, F., 1999. Hf isotope evidence for pelagic sediments in the source of Hawaiian basalts. Science, 285, 879–882.

Brenan, J.M., Shaw, H.F., Phinney, D.L., Ryerson, F.J., 1994. Rutile-aqueous partitioning of Nb, Ta, Hf, Zr, U and Th: implications for high field strength element depletions in island-arc basalts. Earth and Planetary Science Letters, 128, 327–339.

Bryan, S.E., Ernst, R.E., 2008. Revised definition of large igneous provinces (LIPs). Earth Science Reviews, 86, 175–202.

Carneiro, M.A., Carvalho Jr. I. M., Teixeira, W., 1998. Petrologia, geoquímica e geocronologia dos diques máficos do Complexo Metamórfico Bonfim Setentrional (Quadrilátero Ferrífero) e suas implicações na evolução crustal do Craton do São Francisco Meridional. Revista Brasileira de Geociências, 29 (1), 29–44.

Carneiro M.A., Teixeira W., Carvalho Jr. I. M., Pimentel M. M., Oliveira A. H. 2004. Comportamento dos Sistemas Sm-Nd e Rb–Sr da Seqüência Acamadada Máfico-Ultramáfica Ribeirão dos Motas (Arqueano), Cráton São Francisco Meridional: Evidências de Enriquecimento Mantélico e Fracionamento Isotópico. Revista do Instituto de Geociências, Geol. USP Série Científica, São Paulo, 4 (2), 13–26.

Chaves, A.O., 2001, Enxames de diques máficos do setor sul do Craton do São Francisco (MG, Brasil). Tese de Doutorado, Instituto de Geociências USP, São Paulo, 154 pp.

Chaves, A.O., 2011. O enxame de diques de anfibolito do Cráton São Francisco meridional. Revista Brasileira de Geociências, 41(3), 509–524.

Chaves, A.O., Correa Neves, J.M., 2005, Radiometric ages, aeromagnetic expression, and general geology of mafic dykes from southeastern Brazil and implications. Journal of South American Earth Sciences, 19, pp. 387–397.

Chemale Jr., F., Dussin, I.A., Alkmim, F.F., Martins, M.S., Queiroga, G., Armstrong, R., Santos, M.N., 2012. Unravelling a Proterozoic basin history through detrital zircon geochronology: the case of the Espinhaço Supergroup, Minas Gerais, Brazil. Gondwana Research, 22, 200–206.

Comin-Chiaramonti P., De Min A., Cundari, A., Girardi V.A.V., Ernesto, M., Gomes, C. B., Riccomini, C., 2013. Magmatism in Assuncion- Sapucai-Villarica Graben (Eastern Paraguay) revised: petrological, geophysical, geochemical and geodynamical inferences. Juornal of Geological Research, article ID 590835 22 pp http://dx.doi.org/10.1155/2013/590835.

Condie, K.C., 2005. High field strength elements in Archean basalts: a window evolving sources of mantle plumes? Lithos, 79, 491–504.

Cordani, U.G., Sato, K., Nutman, A., 1999. Single zircon SHRIMP determination from Archean tonalitic rocks near Uauá, Bahia, Brazil. Proceedings II South American Symposium on Isotope Geology, Cordoba, 27–30

Cordani, U.G., Fraga L.M., Reis, N., Tassinari, C.C.G., Brito Neves, B., 2010. On the origin and tectonic significance of the intra-plate eventsof Grenvillian-type age in South America: A discussion. Journal of South American Earth Sciences, 29, 143–159.

Cordery, M.C., Davies, J.F., Campbell, I.H., 1997. Genesis of flood basalts from eclogite-bearing mantle plumes. Journal of Geophysical Research, 102, 20179–20198.

Corrêa da Costa, P.C., Carneiro, M.A., Teixeira, W., Girardi, V.A.V. Nalini Júnior, H.A., Oliveira, A.H., Fernandes, R.A. 2006. Estudo Geoquímico e Petrológico dos Diques Máficos da Região de Candeias-Campo Belo-Santo Antônio do Amparo (MG), Porção Meridional do Craton São Francisco. Revista Brasileira de Geociências USP Série Científica, 5(2), 65–84.

Corrêa-Gomes, L.C., Oliveira, E.C. 2000. Radiating 1.0 Ga mafic dyke swarms of Eastern Brazil and Western Africa: evidence of post-assembly extension in the Rodinia Supercontinent? Gondwana Research 3, 325–322.

Dussin, I.A., Dussin, T.M., 1995. Supergrupo Espinhaço: Modelo de Evolução Geodinâmica. Geonomos, 1, 19–26.

Eisele, J., Sharma, M., Galer, S. J. G., Blichert-toft, J., Devey, C. J., Hofmann, A. W. 2002. The role of sediment recycling in EM-1 inferred from OS, PB, Hf, Nd, Sr isotope and trace element systematics of the Pitcarn hotspot. Earth Planetary Science Letters 196, 197–212.

Ernst, R.E., Bleeker, W., Söderlund, U., Kerr, A.C., 2013a. Large Igneous Provinces and supercontinents: Toward completing the plate tectonic revolution. Lithos, 174, 1–14.

Ernst, R. E., Pereira, E., Hamilton, M.A., Pisarevsky, S.A., Rodrigues, J., Tassinari, C.C.G., Teixeira, W., Van-Dunem, V., 2013b. Mesoproterozoic intraplate magmatic barcode record of the Angola portion of the Congo craton: newly dated magmatic events at 1500 and 1110 Ma and implications for Nuna (Columbia) supercontinent reconstructions. Precambrian Research, 230, 103–118.

Evans, D.A.D., Heaman, L.M., Trindade, R.I.F., D'Agrella-Filho, M. S., Smirnov, A.V., Catelani, E.L., 2010. Precise U-Pb Baddeleyite Ages from Neoproterozoic Mafic Dykes in Bahia, Brazil, and their Paleomagnetic/Paleogeographic Implications, AGU Brazil Abstract, GP31E-07 American Geophysical Union, Joint Assembly, Meeting of the Americas, Iguassu Falls, August 2010, CD-ROM.

Farley, K. A., Natland, J. H., Craig, H., 1992. Binary mixed of enriched and undergassed (and primitive?) mantle components (He, Sr, Nd, Pb) in Samoan lavas. Earth and Planetary. Science Letters, 111, 183–199.

Fernandes, R. A., 2001. Etapas de formação de Crosta Continental (do Mesoarqueano ao Mesoproterozóico) no Craton São Francisco Meridional. Dissertação de Mestrado, Departamento de Geologia, Universidade Federal de Ouro Preto, Ouro Preto.

Girardi, V.A.V., Teixeira, W., Mazzucchelli, M., Corrêa da Costa, P.C. 2013. Sr–Nd constraints and trace-elements geochemistry of selected Paleo and Mesoproterozoic mafic dykes and related intrusions from the South American Platform: insights into their mantle sources and geodynamic implication. Journal of South American Earth Sciences, 415–82.

Goulart, L., E., A., Carneiro, M. A., Endo, I., Suita, M. T. F., 2013. Novas evidências de crescimento crustal neoarqueano no Cráton São Francisco Meridional: a Suíte Acamadada Carmópolis de Minas, Minas Gerais, Brasil. Brazilian Journal of Geology, 43(3), 445–459.

Guadagnin F., Chemale Jr. F., Magalhães, A.J.C., Santana, A., Dussin, I., Takehara, L. (2015). Age constraints on crystal-tuff from the Espinhaço Supergroup - Insight 2 into the Paleoproterozoic to Mesoproterozoic intracratonic basin cycles 3 of the Congo – São Francisco Craton. Gondwana Research 27: 363–376.

Guimarães, J.T., Teixeira, L.R., Silva, M.G., Martins, A.A.M., Filho, E. L.A., Loureiro, H.S.C., Arcanjo, J.B., Dalton de Souza, J., Neves, J. P., Mascarenhas, J.F., Melo, R.C., Bento, R.V., 2005. Datações U–Pb em rochas magmáticas intrusivas no Complexo Paramirim e no Rifte Espinhaço: uma contribuição ao estudo da evolução geocronológica da Chapada Diamantina. III Simpósio sobre o Cráton do São Francisco, Salvador, pp. 159–161.

Hanan, B., Graham, D. 1996. Lead and helium isotope evidence from oceanic basalts for a common deep source of mantle plumes. Science, 272, 991–995.

Hauri, F.H., Shimizu, N., Dieu, J.J., Hart, S.R., 1996. Evidences of hotspot-related carbonatite metasomatism in the oceanic upper mantle. Nature, 365, 221–227.

Hofmann, A.W., 1997. Mantle geochemistry: the message from oceanic volcanism. Nature, 356, 210–229.

Hofmann, A.W., 2005. Sampling mantle heterogeneity through oceanic basalts: isotopes and trace elements. In: Carlson, R.W. (Ed.) Mantle and Core. Treatise of Geochemistry, 2, 61–102.

Kessel, R., Schmidt, M.W., Ulmer, P., Pettke, T., 2005. Trace element signature of subduction zone fluids, melts and supercritical liquids at 120-180 km depth. Nature 437, 724–727.

Kilian, R., Behrmann, J.H., 2003. Geochemical constraints on the sources of Southern Chile trench sediments and their recycling in arc magmas of the Southern Andes. Journal of Geological Society of London 160, 57–70.

Klein, E.M., Karsten, J.L., 1995. Ocean-ridge basalts with convergent margin geochemical affinities from the Chile ridge. Nature, 374, 52–57.

Lana C., Alkmim F.F., Armstrong R., Scholz, R., Romano, R., Nalini, H.A. 2013. The ancestry and magmatic evolution of Archaean TTG rocks of the Quadrilátero Ferrífero province, southeast Brazil. Precambrian Research, 230, 1–30.

Leitch, A.M., Davis, G.F., 2001. Mantle plumes and flood basalts: enhanced melting from plume ascent and an eclogite component: Journal of Geophysical Research, 106, 2047–2059.

LeMaitre, R.W., Bateman, P., Dudek, A., Keller, J., Lameyre, J., LeBas, M.J., Sabine, P.A., Schmid, R., Sorensen, H., Streckeisen, A., Wooley, A.R., Zanettin, B., 1989. A classification of igneous rocks and glossary of terms. Blackwell, Oxford.

Machado, N., Schrank, A., Abreu, F.R., Knauer, L.G., Almeida-Abreu, P.A., 1989. Resultados preliminares da geocronologia U-Pb na Serra do Espinhaço Meridional - Boletim do Núcleo Minas Gerais. Sociedade Brasileira de Geologia, 10, pp. 171–174.

Mazzucchelli, M., Rivalenti, G., Menezes Leal, A.B., Girardi, V.A.V., Brito Neves, B.B., Teixeira, W., 2000. Petrology of metabasic dykes in the Diamantina region, Minas Gerais, Brazil. Periodico di Mineralalogia, 70, 231–254.

Menezes Leal, A. B., Girardi, V.A.V., Bellieni, G., Bastos Leal, L.R., Teixeira, W., Piccirillo, E.M. 1995. Contribuição ao estudo de diques máficos de Uauá, Bahia, Brasil. Geochimica Brasiliensis, 9 (1), 61–90.

Menezes Leal, A.B., Brito, D.C., Girardi, V.A.V., Corrêa-Gomes, C., Cruz, S.C., Bastos Leal, L.R., 2010. Petrology and geochemistry of the tholeiitic mafic dykes from the Chapada Diamantina, Northeastern São Francisco Craton, Brazil. 6th International Dyke Conference, Varanasi, India, p. 79.

Menezes Leal, A. B, Barbosa, J. S. F., Corrêa-Gomes, L. C, 2012. Corpos Máficos-Ultramáficos. In: Barbosa, J. S. F., Marscarenhas, J. F., Corrêia-Gomes, L. C. Dominguez, J. M. L., Souza, J. S. Geologia da Bahia: Pesquisa e atualização. Salvador-Bahia: CBPM, VII, pp. 443–483.

Nance, R.D., Murphy, J.B., Santosh, M., 2014. The supercontinent cycle: A retrospective essay, Gondwana Research, 25(1), 4–29.

Niu, Y., and O'Hara, M.J., 2004. Mantle plumes are not ancient oceanic crust. In Oceanic Hotspots, chapter 7, Ed. R. Hékinian and P. Stoffers, Springer- Verlag, New York, 239–252.

Niu, Y., Wilson, M., Humphreys, O'Hara, M.J., 2012. A trace element perspective on the source of ocean island basalts (OIB) and fate of subducted ocean crust (SOC) and mantle lithosphere (SML). Episodes, 35(2), 311–327.

Noce, C.M., Pedrosa-Soares, A.C., Silva, L.C., Armstrong, R., Piuzana, D., 2007. Evolution of polycyclic basement complexes in the Araçuaí orogen, based on U– Pb SHRIMP data: Implication of Brazil-Africa links in Paleoproterozoic time. Precambrian Research 159, 60–78.

Oliveira, A.H., 2004. Evolução Tectônica de um fragmento do Cráton São Francisco Meridional com base em aspectos estruturais, geoquímicos (rocha total) e geocronológicos (Rb–Sr, Sm-Nd, Ar-Ar, U-Pb). Tese de Doutorado, Escola de Minas, Universidade Federal de Ouro Preto. 134 p.

Oliveira, E.P., 2011. The Late Archaean Uauá Mafic Dyke Swarm, São Francisco Craton, Brazil, and Implications for Palaeoproteozoic extrusion tectonics and Orogen Reconstruction. In: Srivastava, R.K. (ed.). Dyke Swarms: Keys for Geodynamic Interpretation. Springer-Verlag, Berlin-Heidelberg, pp. 19–31.

Oliveira, E.P., Tarney, J., 1995. Petrogenesis of the Late Proterozoic Curaçá mafic dyke swarm, Brazil: asthenospheric magmatism

associated with continental collision. Mineralogy and Petrology 53, 27–48.
Oliveira, E.P., Mello, E.F., McNaughton, N., 2002. Reconnaissance U–Pb geochronology of Precambrian quartzites from the Caldeirão belt and their basement, NE São Francisco Craton, Bahia, Brazil: implications for the early evolution of the Paleoproterozoic Itabuna-Salvador-Curaçá orogen. Journal of South American Earth Sciences, 15, 349–362.
Oliveira, E.P., McNaughton, N.J., Armstrong, R., 2010. Mesoarchaean to Palaeoproterozoic growth of the northern segment of the Itabuna Salvador Curaçá orogen, São Francisco craton, Brazil. Geological Society, London, Special Publications, 338, 263–286.
Oliveira E.P., Silveira E., Söderlund U., Ernst R., Evans D.A.V., 2012. New U-Pb zircon/baddeleyite ages on Archaean to Neoproterozoic LIPS (mafic dykes) of the São Francisco Craton, Brazil, and their potential use for palaeocontinent reconstruction. In: Mertanen S., Pesonen L.J., Sangchan P. (eds) Supercontinent Symposium 2012. University of Helsinki, Finland. Programme and Abstracts, pp. 94–95.
Oliveira, E.P., Silveira, E.M., Söderlund U., Ernst R.E., 2013. U–Pb ages and geochemistry of mafic dyke swarms from the Uauá Block, São Francisco Craton, Brazil: LIPs remnants relevant for Late Archaean break-up of a supercraton. Lithos 174, 308–322.
Paixão, M.A., Oliveira, E.P., 1998. The Lagoa da Vaca Complex: an Archean layered anorthosite body on the western edge of the Uauá Block, Bahia, Brazil. Revista Brasileira de Geociências 28, 201–208.
Pedreira, A.J., and De Waele, B., 2008. Contemporaneous evolution of the Palaeoproterozoic-Mesoproterozoic sedimentary basins of the São Francisco-Congo Craton. In: Pankhurs, R.J., Trouw, R.A.J., Brito Neves, B.B. & De Wit, M.J. (eds) West Gondwana: PreCenozoic Correlations Across the South Atlantic Region. Geological Society, London, Special Publications, 294, 33–48.
Pinese, J.P.P., 1997. Geoquímica, geologia isotópica e aspectos petrológicos dos diques máficos Pré-Cambrianos da região de Lavras (MG), porção sul do Cráton São Francisco. Tese de Doutorado, Instituto de Geociências, Universidade de São Paulo, São Paulo.
Pinese, J.P.P., Teixeira,W., Piccirillo, E.M., Quéménéur, J.J.G., Bellieni, G., 1995. The Precambrian Lavras mafic dykes, southern São Francisco Craton, Brazil: preliminary geochemical and geochronological results. In: Baer, G., Heimann, A. (eds.). Physics and Chemistry of Dykes, 205–218.
Plank, T., Langmuir C.H., 1998. The chemical composition of subducting sediment and its consequences for the crust and mantle. Chemical Geology, 145, 325–394.
Ren, Z-Y., Shibata, T., Yoshikava, M., Johnson, K. T. M., Takahashi, E. 2006. Isotope composition of the submarine Hana ridge lavas, Haleakala volcano, Hawaii. Journal of Petrology, 47, 255–275.
Ren, Z-Y., Hanyu, T., Miazaki, T., Chan, Q., Kawabata, H., Takahashi, T., Hirahara, Y., Nichols, A. R. L., Tatsumi, T. 2009. Geochemical differences of the Hawaiian shield lavas: implications for melting process in the heterogeneous Hawaiian plume. Journal of Petrology, 50(8), 1556–1573.
Renne, P. R., Tullis C. Onstott. T.C., D'Agrella-Filho. M.S., Pacca, I. G., Teixeira,W., 1990. 40Ar/39Ar dating of 1.0–1.1 Ga magnetizations from the Francisco and Kalahari cratons: tectonic implications for Pan-African and Brasiliano mobile belt. Earth and Planetary Science Letters, 101, 349-366.
Rivalenti, G., Mazzucchelli, M., Girardi, V.A.V., Cavazzini, G., Finatti, C., Barbieri, M.A., Teixeira, W., 1998. Petrogenessis of the Paleoproterozoic basalt-andesit-rhyolite dyke association in the Carajás region - Amazonian Craton. Lithos, 43, 235–265.
Rivalenti, G., Mazzucchelli, M., Zanetti, A., Vannucci, R., Bollinger, C., Hémond, C., Bertotto, G.W., 2007. Xenoliths from Cerro de los Chenques (Patagonia): an example of slab-related metasomatism in the backarc lithospheric mantle. Lithos, 99, 45–67.
Rollinson, H., 1993. Using geochemical data: evaluation, presentation, interpretation. Pearson Education Limited, Longman Group, UK.
Romano, R., Lana, C., Alkmim, F.F., Stevens, G., Armstrong, R., 2013. Stabilization of the Southern São Francisco Craton, SE Brazil, through a long-lived and episodic period of potassic magmatism. Precambrian Research: 224, 1–20.
Silva, A.M., Chemale Jr., F., Kuyjimian, R. Heaman, L. 1995. Mafic dyke swams of the Quadrilátero Ferrífero and Espinhaço range, MG, Brazil. Revista Brasileira de Geociências, 25, 124–137.
Silveira, E.M., Söderlund, U., Oliveira, E.P., Ernst, R.E., Menezes Leal, A.B., 2013. First precise U–Pb baddeleyite ages of 1500 Ma mafic dykes from the São Francisco Craton, Brazil, and tectonic implications, Lithos, 174, 144–156.
Sun, S.S., McDonough, W.F., 1989. Chemical and isotopic systematics of oceanic basalts: implications for mantle composition and process. In: Saunders, A.D. and Norry, M.J. (Eds.), Magmatism in the Ocean Basins. Geological Society Special Publication, 42, 313–345.
Tack, L., Wingate, M.T.D., Liégeois, J.P., Fernandez-Alonso, M., Deblond, A., 2001. Early Neoproterozoic magmatism (1000–910 Ma) of the Zadinian and Mayumbian Groups (Bas-Congo): onset of Rodinia rifting at the western edge of the Congo craton. Precambrian Research, 110, 277–306.
Takahashi, E., Nakajima, K., Wright, T.L., 1998. Origin of the Columbia River basalts: melting model of a heterogeneous plume head. Earth and Planetary Science Letters, 162, 63–80.
Tatsumoto, M., Nakamura, Y., 1991. Dupal anomaly in the sea of Japan, Pb, Nd and Sr isotopic variation at the Eurasian continental margin. Geochimica et Cosmochimica, Acta, 55, 3697–3708.
Teixeira, W., 1985. A Evolução geotectônica da porção meridional do Craton do São Francisco, com base em interpretações geocronológicas. Tese de Doutorado. Instituto de Geociências, USP, 207 pp.
Teixeira, W., Cordani, U.G., Nutman, A.P., Sato, K. 1998. Polyphase Archean evolution in the Campo Belo metamorphic complex, Southern São Francisco Craton, Brazil: SHRIMP U-Pb zircon evidence. Journal of South American Earth Sciences, 11 (3), 279–289.
Teixeira, W., Sabatè, P., Barbosa, J., Noce, C.M., Carneiro, M.A., 2000. Archean and Paleoproterozoic evolution of the São Francisco Craton, Brazil. In: Cordani, U.G., Milani, E.J., Thomaz Filho, A., Campos, D.A. (eds.). Tectonic Evolution of South America. 31st International Geological Congress, 101–137.
Teixeira, W., Ávila, C.A., Dussin, I.A., Corrêa Neto, A.V., Bongiolo, E.M., Santos, J.O.S., Barbosa, N. 2015. Zircon U-Pb-Hf, Nd-Sr constraints and geochemistry of the Resende Costa Orthogneiss and coeval rocks: new clues for a juvenile accretion episode (2.36–2.33 Ga) in the Mineiro belt and its role to the long-lived Minas accretionary orogeny. Precambrian Research: 256: 148–169.
Weaver, B.L., 1991. The origin of ocean island basalt end-member compositions: trace element and isotopic constraints. Earth and Planetary Science Letters, 104, 381–397.
Willbold, M., Stracke, A., 2006. Trace element composition of mantle end-members: Implications for recycling of oceanic and upper and lower continental crust. Geochemical Geophysics Geosystems, 7 (4) doi:10.1029/2005GC001005.
Winchester, J.A., Floyd, P.A., 1976. Geochemical magma type discrimination: application to altered and metamorphosed basic igneous rocks. Earth and Planetary Science Letters 28, 459–469.
Woodhead, J.D., Devey, C.W., 1993. Geochemistry of the Pitcairn seamount I: Source character and temporal trends. Earth Planetary and Science Letters, 116, 81–99.
Zindler, A., Hart, S., 1986. Chemical geodynamics. Annual Review Earth and Planetary Science Letters, 14, 493–571.

The Recôncavo-Tucano-Jatobá Rift and Associated Atlantic Continental Margin Basins

9

Andres Gordon, Nivaldo Destro, and Monica Heilbron

Abstract

During the opening of the South Atlantic, in the Lower Cretaceous, the Recôncavo-Tucano-Jatobá aborted rift, together with the Camamu-Almada and Jacuípe continental margin basins, developed on the São Francisco cratonic cold lithosphere. This tectonic event finally dismembered the São Francisco-Congo cratonic bridge. The rifting tectonics reactivated Precambrian fabrics of the São Francisco Craton (the Paleoproterozoic Salvador-Itabuna-Curaçá belt), as well as the Neoproterozoic Araçuaí and Sergipano belts. The pre-existing basement structures governed the nucleation of the new rift zones with the generation of intra-basin transfer zones, master listric fault systems, horsts, grabens/hemi-grabens. The listric (crustal scale) master fault systems were able to accommodate a large crustal extension decoupled from the lithospheric thinning, thus preventing magmatic activity in the Recôncavo, Camamu and Almada basins. During the rift stage large wedges of coarse siliciclastic sediments (fluvial and lacustrine) accumulated along syn-sedimentary master faults systems. The syn-rift lacustrine sediments are the most prolific source rocks and responsible for most of the oil produced in the Brazilian passive margin basins. The post-rift evaporite deposits were recorded in both Camamu and Almada basins but are absent in Jacuípe basin. The drift stage sediments display a similar evolution in Jacuípe, Camamu and Almada basins, initiating the cycle with platform carbonates, from Albian to Turonian, evolving later into open marine deposits.

Keywords

West Gondwana breakup • Brazilian continental margin basins • Recôncavo-Tucano-Jatobá interior rift • Jacuípe-Camamu-Almada basins • Cretaceous

9.1 Introduction

The cratonic bridge linking the São Francisco and Congo cratons was only dismembered during the Cretaceous tectonic activity that resulted in the break-up of West Gondwana and generation of the Atlantic Ocean. During the opening of the South Atlantic in the Lower Cretaceous, the Recôncavo-Tucano-Jatobá aborted rift, together with the Camamu-Almada and Jacuípe continental margin basins, developed on the northern São Francisco cratonic lithosphere (Fig. 9.1). Because of that, these basins differ in many aspects from the other eastern Brazilian margin basins. The study of these basins provides an opportunity to understand the behavior of the cratonic lithosphere under extensional tectonism and the role of the São Francisco craton structural inheritance on the architecture of the newly developed interior rift and continental margin basins.

A. Gordon (✉) · M. Heilbron
Tektos Research Group, Faculdade de Geologia, Universidade do Estado Rio de Janeiro (UERJ), Rua São Francisco Xavier, 524, Bloco a-4020, Rio de Janeiro, RJ 20550-900, Brazil
e-mail: acgordon@ymail.com

N. Destro
Petrobras Exploration and Production, Avenida República do Chile, 330, Centro, Rio de Janeiro, RJ 20031-170, Brazil

M. Heilbron et al. (eds.), *São Francisco Craton, Eastern Brazil*, Regional Geology Reviews, DOI 10.1007/978-3-319-01715-0_9

In this section, after summarizing the stratigraphy and structural framework of the Recôncavo-Tucano-Jatobá, Camamu, Almada and Jacuípe basins, we discuss their evolutionary history, emphasizing the cratonic nature of the lithosphere, on which they nucleated in the Lower Cretaceous.

9.2 The Recôncavo-Tucano-Jatobá Rift

The Recôncavo-Tucano-Jatobá rift (RTJ) is a Late Jurassic to Early Cretaceous system of continental half-grabens connected to the eastern Brazilian margin basins (Fig. 9.2). Cutting across the boundary between two major provinces of

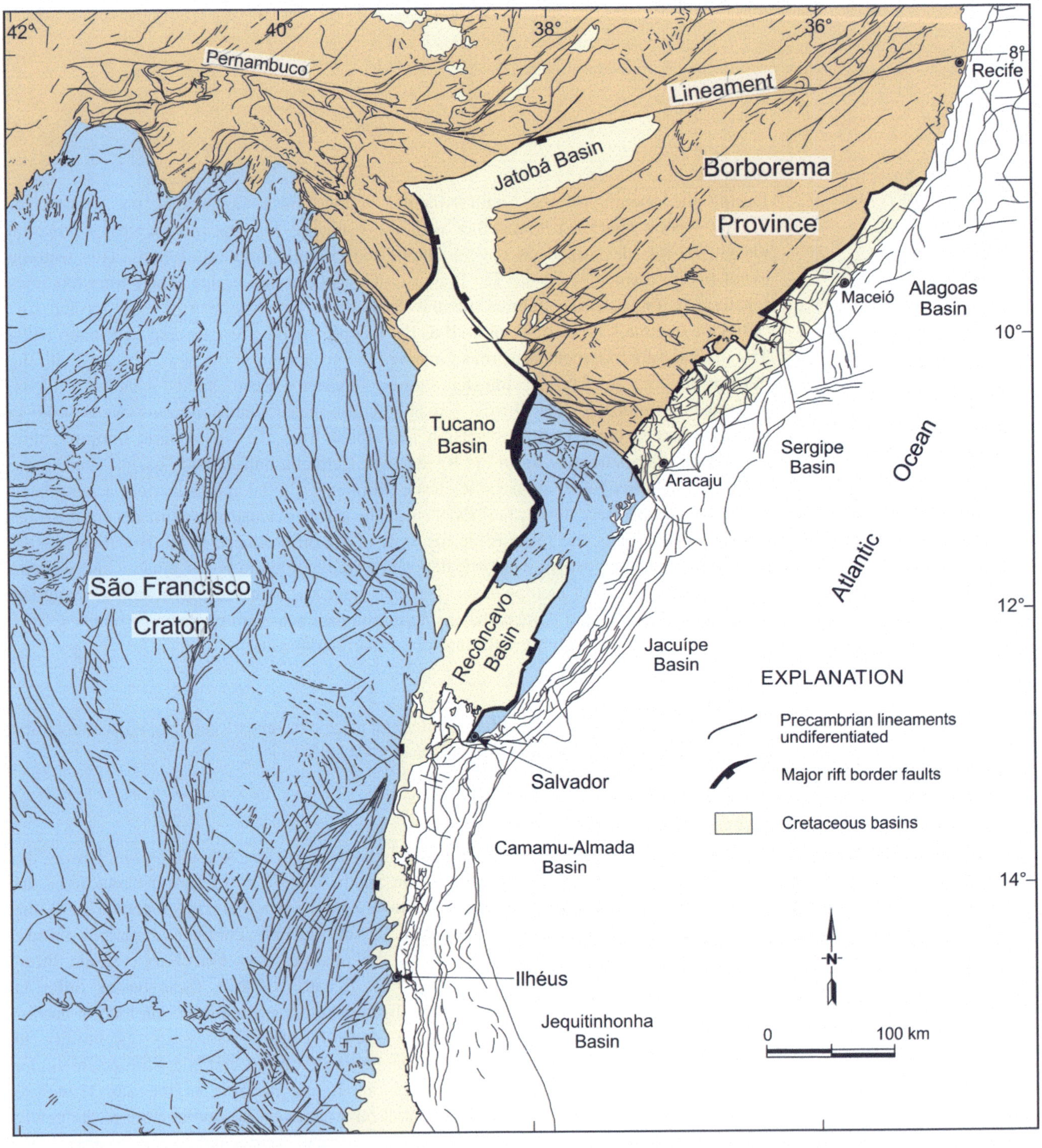

Fig. 9.1 Cretaceous basins developed within the São Francisco craton (adapted from Destro 2002)

the Brazilian shield, the São Francisco craton (SFC) and the Borborema province (Almeida et al. 1981), the RTJ basin evolved as a failed rift arm during the opening of the South Atlantic (e.g. Szatmari et al. 1985, 1987; Milani et al. 1988; Szatmari and Milani 1999) (Fig. 9.1).

The northern SFC crust consists of roughly NS-trending Archean and Paleoproterozoic rock assemblages (see Teixeira et al. and Barbosa and Barbosa, this book) not affected by the Neoproterozoic Brasiliano/Pan-African tectonism. The Borborema province, on the other hand, is locally represented by the NW-oriented Sergipano fold-thrust belt, which roots in a major strike-slip system generated in the course of the Neoproterozoic Brasiliano/Pan-African event.

9.2.1 Stratigraphy

The RTJ basin is filled with strata deposited during the pre-, syn-, and post-rift phases, whose average thickness along the main depocenters exceeds 10 km (Fig. 9.3). The Upper

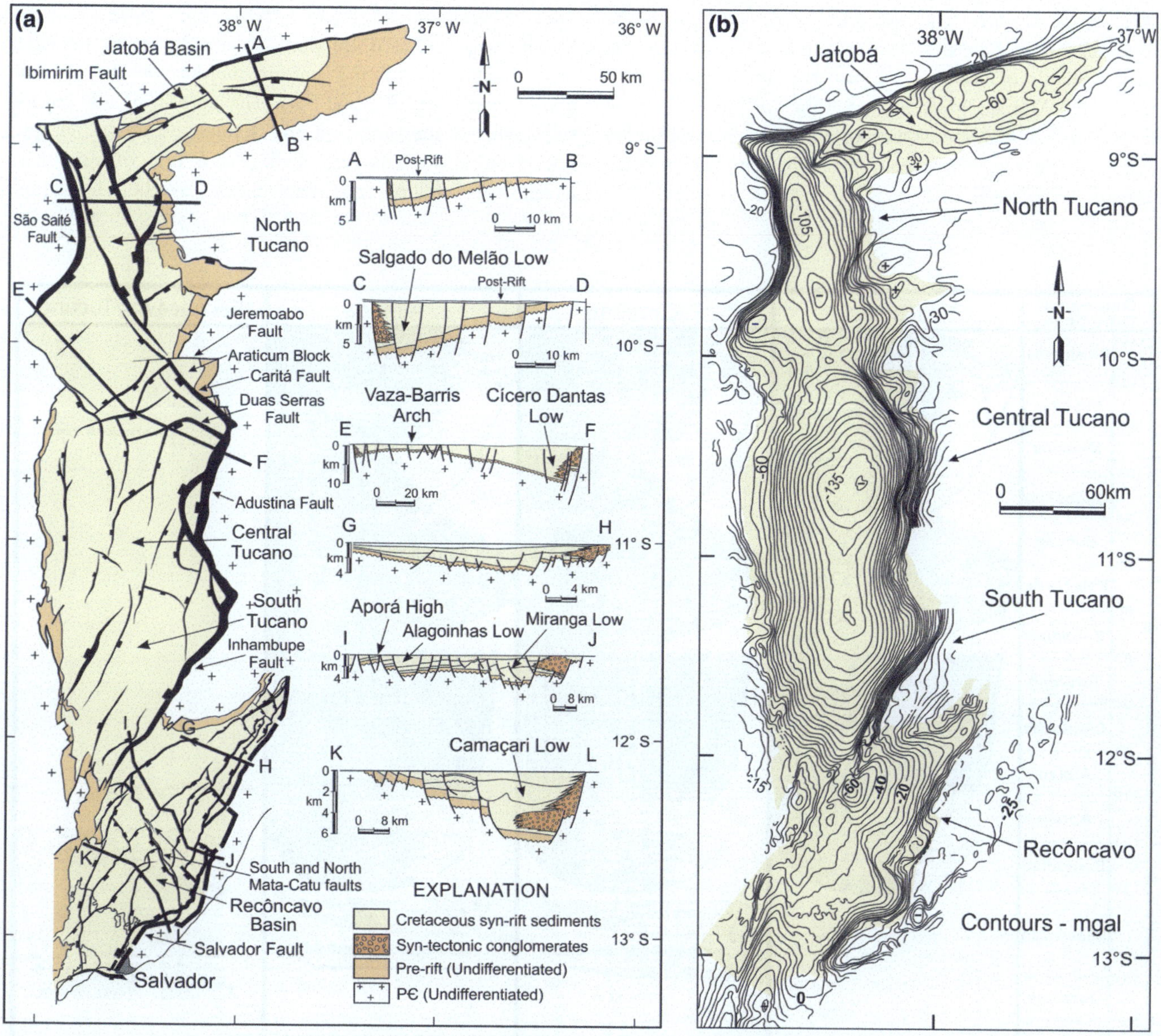

Fig. 9.2 **a** Simplified tectonic map of the Recôncavo-Tucano-Jatobá rift (Destro 2002; adapted from Aragão and Peraro 1994). Cross-sections adapted from Magnavita 1992). Major release faults are the South and North Mata-Catu faults in the center of the Recôncavo basin and the Duas Serras fault in Central Tucano. Major transfer faults are the Caritá and Jeremoabo faults in the transition between the Central and North Tucano rift. **b** Map of the residual gravimetric anomalies of the Recôncavo-Tucano-Jatobá rift. The alternating asymmetry of the half-grabens depicts from the contour lines near the major border faults. Contours are in mgal (Adapted from Magnavita 1992)

Paleozoic to Jurassic pre-rift deposits consists mainly of red beds (Aliança Formation) and coarse-grained fluvial deposits (Sergi Formation). The syn-rift strata accumulated during the Neocomian, when active faulting along the borders of half-grabens led to the deposition of fanglomerates that exceeds 4 km in thickness (Salvador Formation). Syn-rift shales of the Candeias Formation (representing the main hydrocarbon source rocks of the basin) deposited in the deep lakes along with occasional turbidity influxes and sandstone fan incursions (Netto and Oliveira 1985). By the end of the Neocomian, in the Hauterivian time, the subsidence rate declined and a prograding system of delta fans filled the lakes (Ilhas Group). The rift phase terminated with the Barremian São Sebastião fluvial sediments, which are unconformably overlain by the post-rift Aptian conglomerates of the Marizal formation. A thin (100 m) veneer of alluvial and fluvial sandstones of the Barreiras formation represents the Late Cretaceous through Cenozoic post-rift deposits in the basin.

9.2.2 Structure

The most remarkable feature of the RTJ rift is the reversal in the asymmetry of the half-grabens along strike (Fig. 9.2). The border fault is located on the east in the southern portion of the system. In contrast, the northern segment of the basin is bounded by a master fault located on the west. A broad feature named Vaza-Barris arch separates these two half-grabens (Fig. 9.2, section E–F). The Vaza-Barris arch interacted in a complex way with the sinistral transpressional Jeremoabo transfer fault (Magnavita 1992; Destro 2002; Destro et al. 2003b) and the dextral-normal Caritá transfer fault (Magnavita 1992 and Vasconcellos 2003). The ca. 20 km-long and EW-trending Jeremoabo fault connects offset segments of the eastern border of the Tucano half-graben (Figs. 9.2 and 9.4). The portion of the Tucano sub-basin to the south of the Jeremoabo fault, an area referred to as the Araticum block (Fig. 9.2a), corresponds to a relatively shallow basement block, overlain by Upper

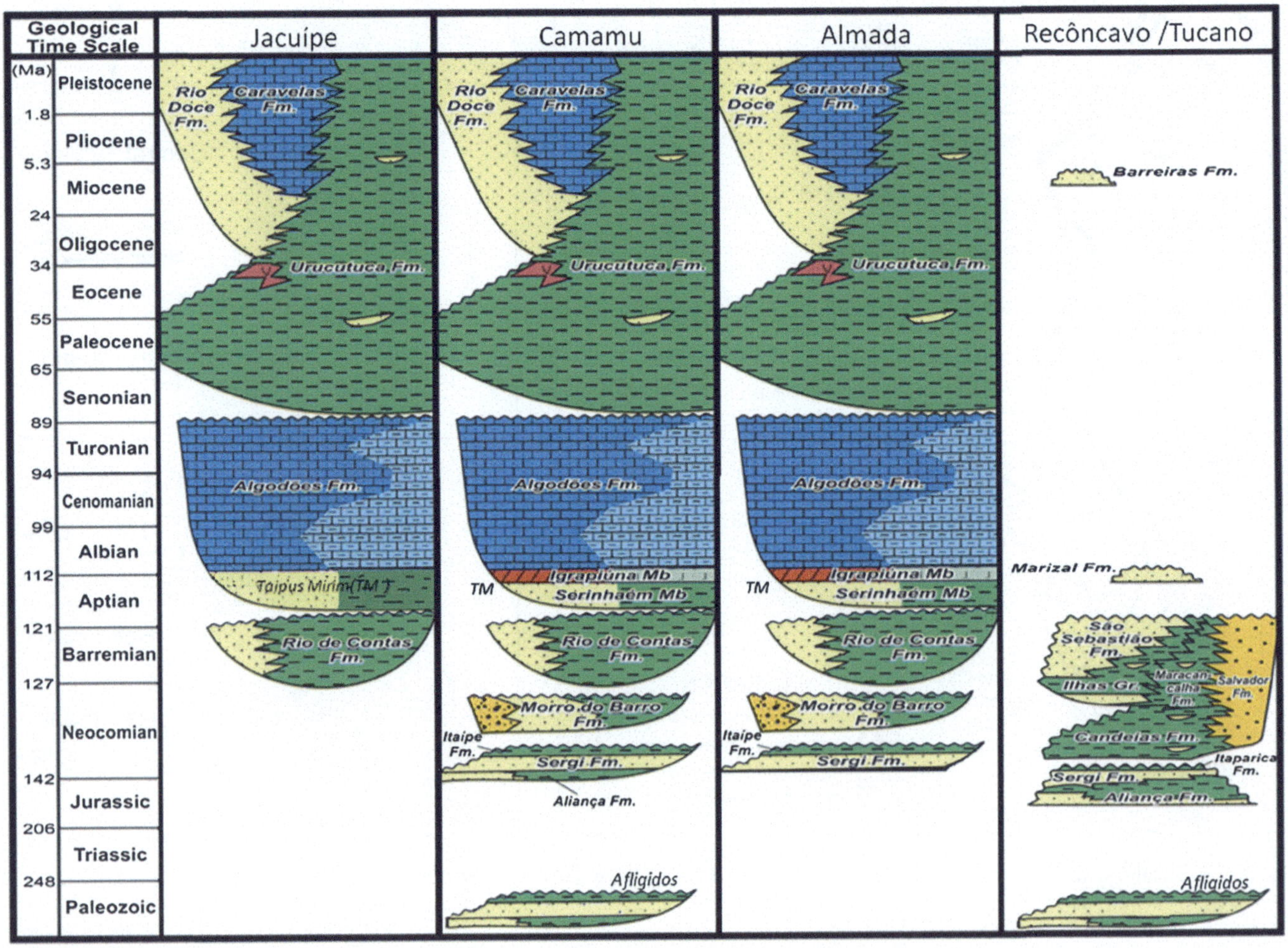

Fig. 9.3 Stratigraphic chart of the offshore Jacuípe, Camamu and Almada basins, and Recôncavo-Tucano-Jatobá rift (after Gontijo et al. 2007; Caixeta et al. 2007; Graddi et al. 2007)

Jurassic pre-rift sediments and Lower Cretaceous rift sequence.

An overall NW–SE extension direction has been determined for the RTJ rift (e.g. Szatmari et al. 1985, 1987; Milani and Davison 1988; Milani et al. 1988; Szatmari and Milani 1999; Destro 2002). From a paleostress investigation conducted in the system, Magnavita (1992) concluded that an EW-oriented extension direction acted during the Early Neocomian, whereas a NW–SE-oriented extension predominated during the Late Barremian/Early Aptian time. All previously mentioned authors agree on an overall extensional setting for generation of the RTJ rift. The NW–SE extension direction has also been inferred in other Cretaceous rift basins of northeastern Brazil, such as the Sergipe–Alagoas (Destro 1995) and Camamu basins (Mércio 1996). Since the RTJ basin exhibits an overall NS orientation, it can be viewed as an oblique rift system.

The major structural features of the RTJ system are (Figs. 9.2a and 9.4):

- the Camaçari and Alagoinhas lows representing the main depocenters in the Recôncavo sub-basin, with average thicknesses up to 6 and 4 km, respectively;
- the Cícero Dantas and Salagado do Melão lows, in the Tucano sub-basin, with average thickness up to 15 and 6 km, respectively;
- The Salvador main border fault in the Recôncavo half-graben;
- The Inhambupe-Adustina normal fault system, in the south and central Tucano sub-basin, and the São Saité normal fault in the northern Tucano sub-basin;
- Transfer and accommodation zones, exemplified by the Aporá high and Vaza-Barris arch;
- Cross-faults including transfer and release faults.

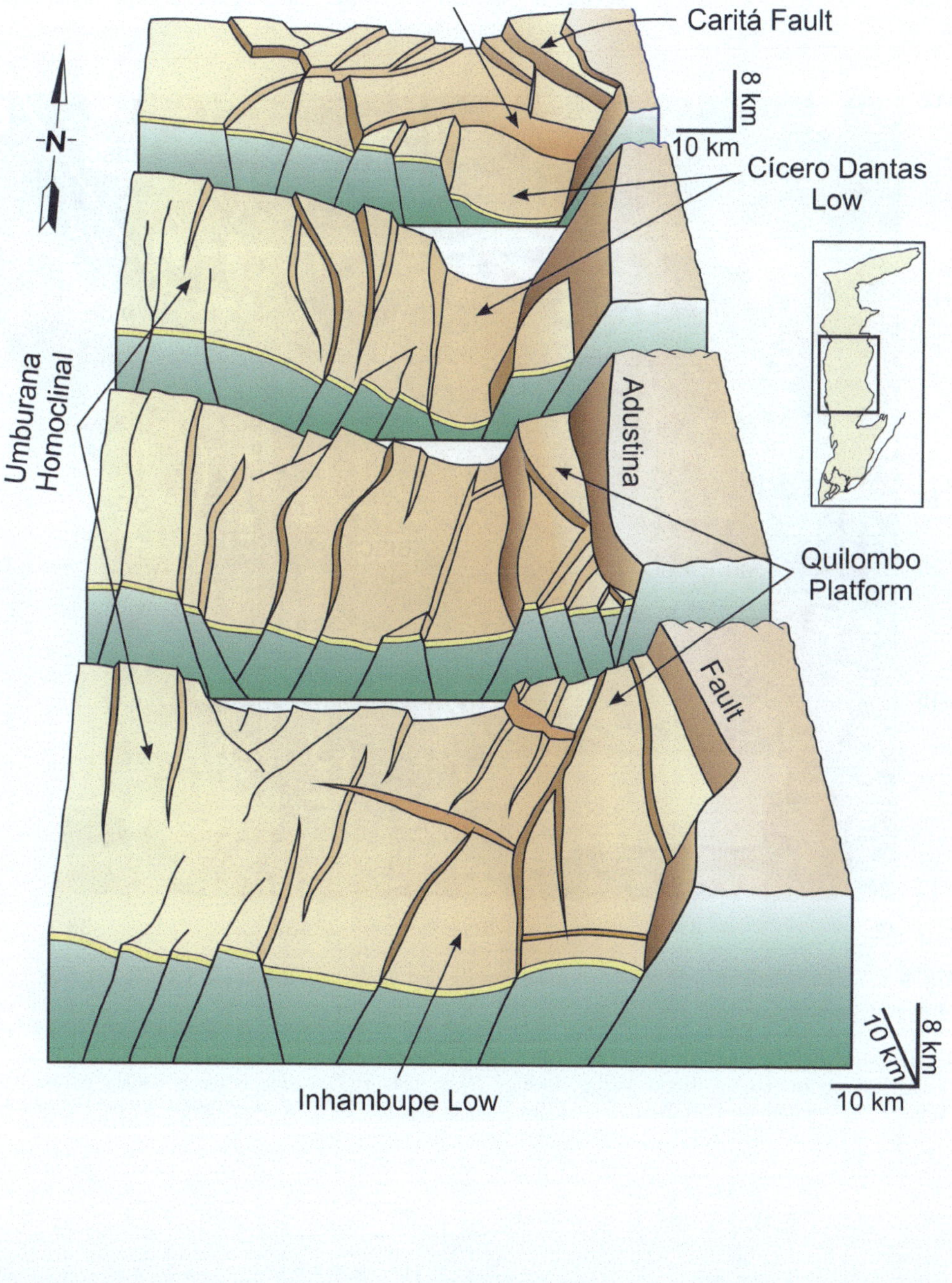

Fig. 9.4 Block-diagram of the South and Central Tucano sub-basins (Destro 2002). Displacement increases from south to north along the Adustina fault, and from west to east along the Duas Serras release fault. These faults join at the Cícero Dantas low, where the *top* of the pre-rift sequence reaches depths of ca. 15 km

Transfer faults are cross-faults that connect at least two distinct normal faults and show predominance of strike-slip displacement. Release faults, on the other hand, are associated to a single normal fault and form to accommodate the variation of vertical displacement along the strike of the parent normal fault (Destro 1995; Destro et al. 2003a). Differently from transfer faults, release faults show predominance of normal component of displacement. Figure 9.4 shows the example of Duas Serras release fault associated to the Adustina parent normal fault. At their contact, they show throws respectively of 6 and 15 km (see Fig. 9.2a for the total length of these faults in map view).

9.3 The Camamu-Almada and Jacuípe Basins

The Jacuípe, Camamu and Almada (CAJ) are narrow basins located along the eastern boundary of the São Francisco craton on the Brazilian northeast Atlantic coast, approximately between 11°30′ to 14° 37′S Lat. and 37°30′ to 39°W Long (Figs. 9.1 and 9.5). The northernmost Jacuípe basin is bounded to the north by the Vaza-Barris fault, and separated from the Camamu basin on the south by the Salvador transfer zone. The Taipus basement high separates the Camamu basin from the southernmost Almada basin, which is turn bounded to the south by the Olivença basement high (Fig. 9.5). A regional NE-trending normal fault system inherited from the Precambrian São Francisco cratonic fabric bounds the Atlantic margin basins on the west (Fig. 9.5).

These coastal basins are located mainly offshore, although Camamu and Almada basins exhibit restricted onshore outcrops of Upper Paleozoic, Jurassic and Cretaceous sediments. Onshore portions of the basins occur in the Camamu embayment, in the Maraú Peninsula, and Almada Canyon (Bruhn and Moraes 1989; Fig. 9.5).

Petroleum exploration is more extensive in the Camamu and Almada basins than in the Jacuípe basin. The Camamu and Almada large database includes more than 87 boreholes

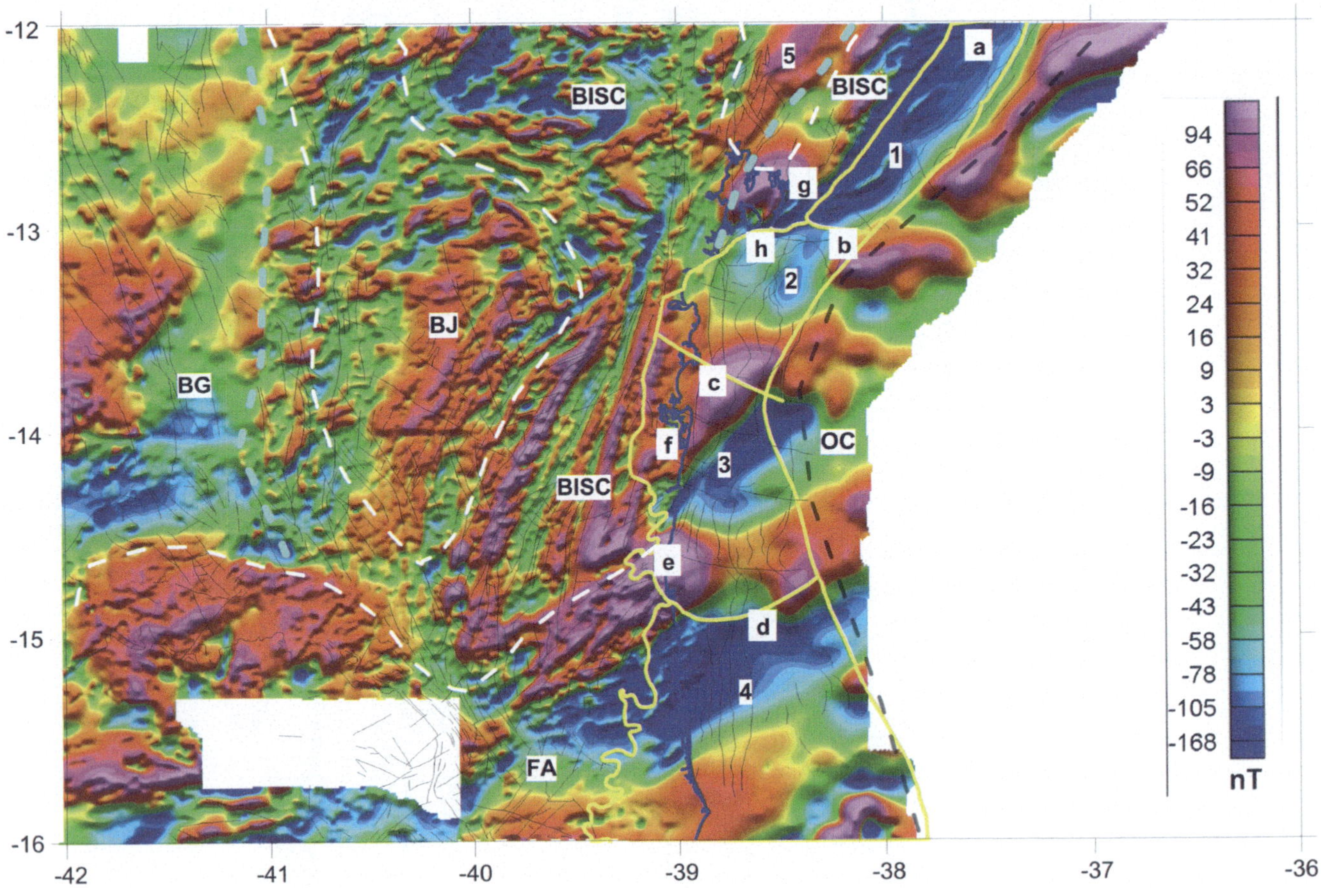

Fig. 9.5 Aeromagnetic map of the continental margin and adjacent onshore basement exposures (total magnetic field reduced to IGRF) (After Arcanjo 1997). São Francisco craton units: Gavião block (*BG*), Jequié block (*BJ*), Itabuna-Salvador-Curaçá Belt (*BISC*). Araçuaí Belt (*FA*). Ocean crust (*OC*). Offshore basins: Jacuípe (*1*), Camamu (2), Almada (*3*), Jequitinhonha (*4*). Basin boundaries, tectonic features and localities: Vaza-Barris fault (*a*), Salvador Transfer zone (*b*), Taipus basement high (*c*), Olivença basement high (*d*), Almada Canyon (*e*), Maraú Peninsula (*f*), Salvador city (*g*), Itaparica Island (*h*). *White dashed lines* are the São Francisco unit boundaries. *Light-blue dashed line* is the boundary of the Itabunas-Salvador-Curaçá Orogen. *Black dashed line* is the ocean/crust boundary. *Blue line* is the coastline

drilled to date, and a broad 2D and 3D seismic coverage. On the contrary, the Jacuípe Basin is less known and data include only one well, drilled in shallow water, and a limited number of 2D seismic lines. The discovery of the large Manati gas field has proven the hydrocarbon potential of the Camamu basin. This gas field is currently in production and encompasses a series of the smaller oil and gas accumulations. Several exploratory wells have demonstrated the hydrocarbon potential of Almada basin but the discoveries are not yet economically viable.

9.3.1 Stratigraphy

Regionally, the stratigraphic record of these basins is divided into five units: basement, pre-rift, syn-rift, post-rift, and drift supersequences (Chang et al. 1992). However, when examined in detail, the fill successions of each individual basin exhibit remarkable differences (Fig. 9.3).

9.3.1.1 Basement

The basement of the Jacuípe, Camamu and Almada basins is made up of granulite grade gneisses and granitoids of the Paleoproterozoic Eastern Bahia orogenic domain that forms the substratum of the northern São Francisco craton (see Barbosa and Barbosa and Teixeira et al. this book; Almeida 1977; Delgado et al. 2003; Barbosa et al. 2003; Barbosa and Sabate 2003). The Bouguer anomaly map of Fig. 9.6 (Sandwell and Smith 1997) shows the Archean blocks, the Itabuna-Salvador-Curaçá belt, and the Neoproterozoic Araçuaí belt, which constitute the substratum of the focused basins.

The great lithological heterogeneity observed in the outcrops of the SFC extends into the basement of the Jacuípe, Camamu and Almada basins. The structural domains of the SFC and the Araçuaí belt correlate well with the structural domains of the Jacuípe, Camamu and Almada basement. This relationship is evident in the magnetic, composite aero-gamma-spectrometric, and seismic maps shown on Figs. 9.5 and 9.7, respectively (see also Table 9.1). Several

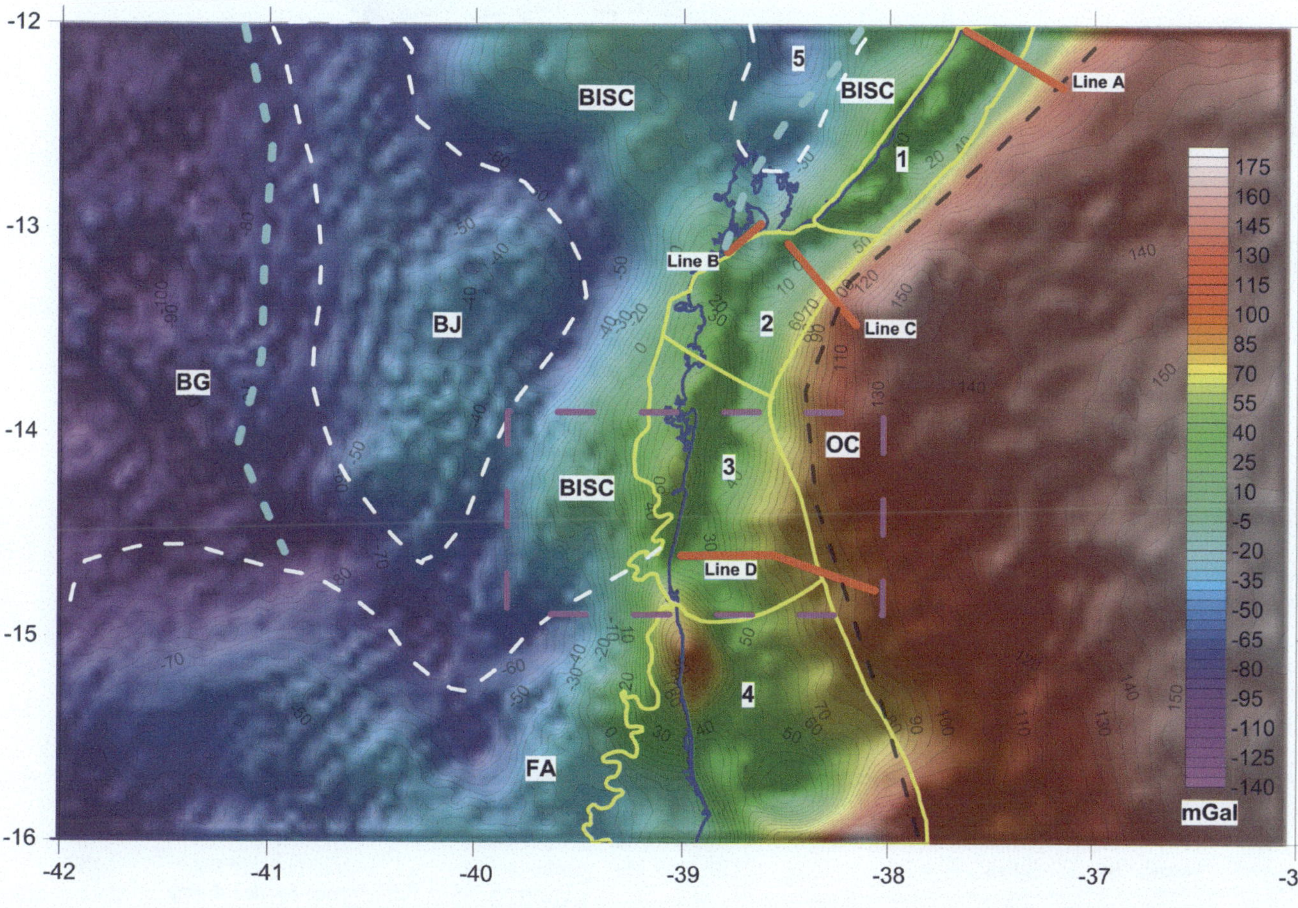

Fig. 9.6 Satellite Bouguer anomaly map (after Sandwell and Smith 1997) of the western part of the São Francisco craton and adjoining Atlantic margin basins. Offshore basins: Jacuípe (*1*), Camamu (*2*), Almada (*3*), Jequitinhonha (*4*). *White dashed lines* are the São Francisco unit boundaries and *light-blue dashed line* is the boundary of the Itabunas-Salvador-Curaçá Orogen. Craton blocks: Gavião block (*BG*), Jequié block (*BJ*), Itabuna-Salvador-Curaçá Belt (*BISC*). Araçuaí Belt (*FA*). *Black dashed line* is the ocean/crust boundary, and ocean crust (OC). *Blue line* is the coastline. *Red thick lines* are the locations of the seismic examples A to D, and *magenta dashed box* is the location of the map of Fig. 9.7

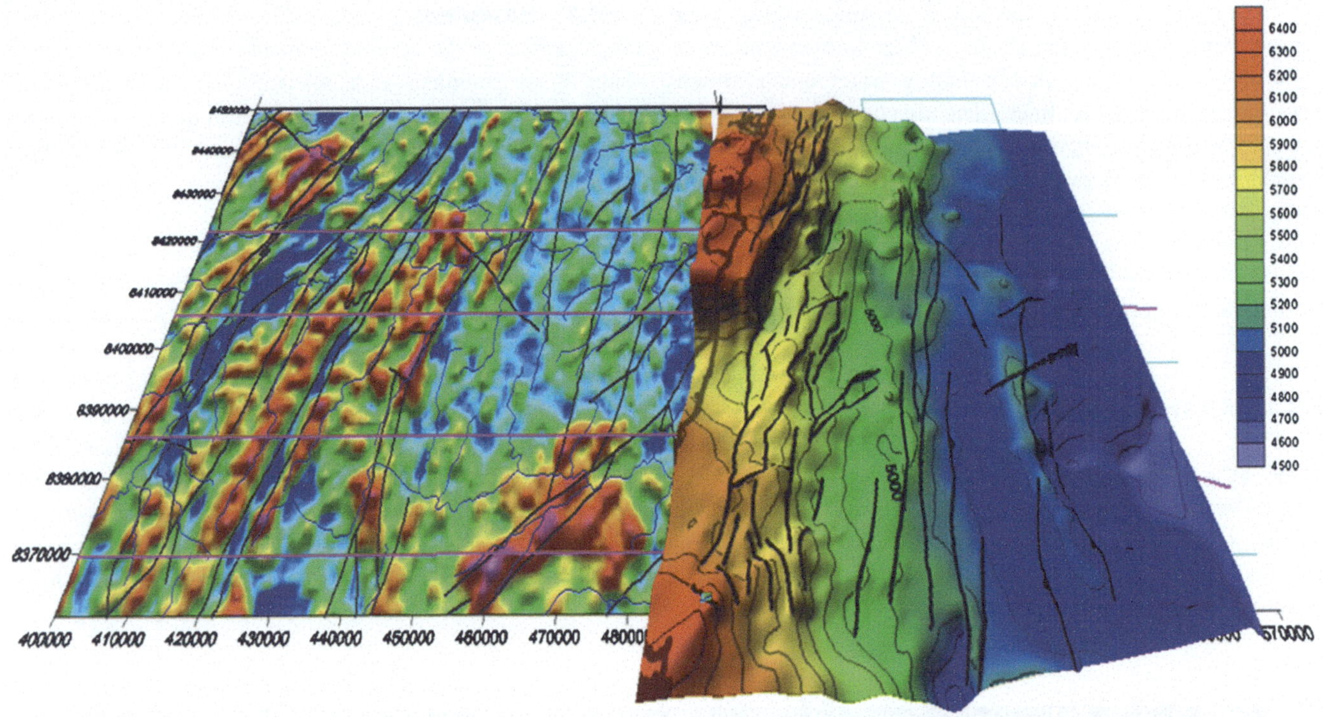

Fig. 9.7 São Francisco craton-Almada basin composite map. Composite aero-gamma spectrometric map (*left*) and seismic map (*right*), of the western part of the São Francisco craton and the offshore Almada basin. Note the continuity of the NE- and WNW-trending domains (Itajú do Colonia shear zone) from the SFC into the Almada basin

Table 9.1 Correlation between the basement of the Jacuípe, Camamu and Almada basins with the SFC and Araçuaí belt units (see Chaps. 3, 4 and 14 of this book)

Basin/Sub Basin	Basement unit	Craton/orogenic belts	Age	Composition	Structure	References
Southern Almada	Itajú do Colônia belt	Araçuaí belt	Neoproterozoic (696–676 Ma)	Alkaline basic to acid rocks	NE-trending shear zone and anarogenic alkaline stocks	Corrêa Gomes et al. (1998)
Northern Almada and southern Camamu	São José Complex	Itabuna-Salvador-Curaçá belt	Paleoproterozoic (~2151 Ma)	Granulite facies tholeiitic and calc-alkaline basic rocks Pre-to syn-collisional plutons	NNE-(primary) and NE-trending (secondary) transpressive shear belts	Corrêa Gomes et al. (1998)
Northern Camamu	Salvador shear zone	Itabuna-Salvador-Curaçá belt	Paleoproterozoic (~2151 Ma)		NW-striking shear zone	Corrêa Gomes et al. (2005)
Southern Jacuipe	Salvador-Ilheus block	Itabuna-Salvador-Curaçá block	Archean (2719–2516 M.a)	Granulite facies orthogneisses		Silva et al. (2007)
Northern Jacuipe	Eastern Bahia coastal belt	Itabuna-Salvador-Curaçá belt	Paleoproterozoic (2230–2169 Ma)	Arc-related orthogneisses reworked during the Brasiliano orogeny (585–500 Ma)	NE-trending transpressive shear zone formed during the Paleoproterozoic collision	Silva et al. (2007)

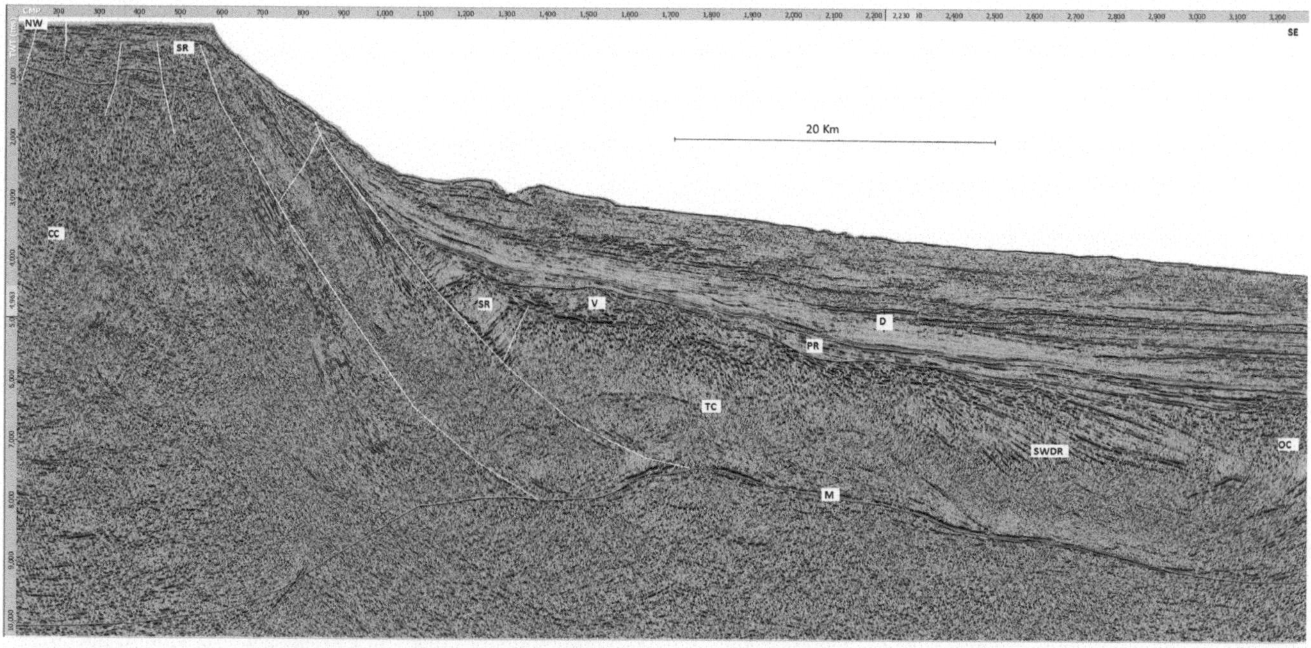

Fig. 9.8 Seismic line A of the Jacuípe basin: Mohorovicic discontinuity (*M*), Continental crust (*CC*), Transitional crust (*TC*), Oceanic crust (*OC*), Seaward dipping reflectors (*SWDR*), Volcanic (*V*), Syn-rift deposits (*SR*), Post-rift deposits (*PR*), Drift deposits (*D*). (After Mohriak et al. 1998)

of the wells drilled in these basins sampled gneisses and granulites characteristic of the Eastern Bahia orogenic domain (ANP–Agência Nacional de Petróleo well files).

9.3.1.2 Pre-rift Supersequence

The pre-rift supersequence, comprising the oldest fill units, shows important changes among the Jacuípe, Camamu and Almada basins (Fig. 9.3). The pre-rift units are absent in the Jacuípe basin, where the well BAS-53 recorded Barremian to Aptian rift deposits overlaying the Precambrian basement, a fact also observed in the deep seismic crustal profiles. The seismic record of the Jacuípe basin shows rift sediment wedges deposited on top of the continental and transitional crust without the pre-rift sequences (Mohriak et al. 1998; Fig. 9.8).

The Camamu basin exhibits the most complete record of the pre-rift units, which consist of Permian and Neo-Jurassic sediments. The Permian sequence is composed of continental and marine deposits known as the Afligidos Formation (Aguiar and Matos 1990; Silva et al. 2012). The Afligidos Formation accumulated in a large continental sag and correlates with similar deposits of the Brazilian intracratonic Paraná, Parnaiba and Recôncavo basins, as well as the Sergipe-Alagoas passive margin basin (Caixeta et al. 2007). The Upper Jurassic/Lower Cretaceous sequence is composed of fluvial, aeolian and lacustrine sediments with an extensive regional distribution. This sequence is known as the Brotas Group, which includes the Aliança, Sergi and Itaipe formations (Viana et al. 1971; Netto et al. 1994). Recent geochronological studies carried out on samples of the Aliança and Itaparica formations in the Recôncavo basin yielded Neotriassic ages for these units (Silva et al. 2012), indicating that the duration of the pre-rift phase probably extends from the Neotriassic to the Lower Cretaceous. The Sergi Formation is composed of fluvial and aeolian sandstones with local intercalations of shales and siltstones containing non-marine ostracodes (Viana et al. 1971). The Sergi Formation is the main reservoir in the gas fields of the Camamu basin (Fig. 9.3). Equivalent deposits are also found in the Recôncavo, Tucano, Jatobá, Sergipe and Alagoas basins (Viana et al. 1971; Netto et al. 1994; Gontijo et al. 2007).

The pre-rift unit in the Almada basin is incomplete. Neither the Paleozoic sequence nor the Aliança Formation occur and the Sergi Sandstones lie directly on the basement (Netto et al. 1994; Gontijo et al. 2007) (Fig. 9.3).

9.3.1.3 Syn-rift Supersequence

The syn-rift sequences of the Jacuípe, Camamu and Almada basins are composed of siliciclastic sediments deposited in fluvio-deltaic and lacustrine environments during Berriasian and Aptian times (Fig. 9.3). These units appear in the lithostratigraphic chart as Morro de Barro, Rio de Contas, and lower Taipus Mirim (Itararé Member) formations (Netto et al. 1994; Gontijo et al. 2007). Major listric fault systems, horsts and grabens nucleated during this phase, controlling the deposition of a thick package of organic matter-rich shales in deep anoxic lakes. The Morro de Barro Formation is the main source rock in the rift sequence (Mello et al. 1995; Mohriak 2003). This unit locally shows intercalations of siliciclastic rocks with reservoir capacity in fan delta, debris flows and lake turbiditic deposits. The Morro de Barro Formation was not sampled in a well drilled in the Jacuípe basin (Graddi et al. 2007). Regionally, the Morro de Barro Formation correlates with the Maracangalha and the Candeias (Gomo Member) formations in the Recôncavo and Tucano basins, and with the Feliz Deserto Formation in the Sergipe-Alagoas basin (Gontijo et al. 2007).

9.3.1.4 Post-rift Supersequence

This unit, also referred to as the Taipus Mirim Formation (Fig. 9.3), consists of siliciclastic sediments (Serinhaém Member) and evaporites (Igrapiúna Member) deposited in a sag basin during the Upper Aptian (Netto et al. 1994; Gontijo et al. 2007). During the post-rift phase, important salt structures, such as domes and pillows developed mainly in the Almada basin and, to a lesser degree, in the Camamu basin (Gordon et al. 2013). Salt deposits are absent in the Jacuípe basin (Graddi et al. 2007).

9.3.1.5 The Drift Supersequence

Marine sediments deposited from the Albian to Recent form the drift supersequence (Fig. 9.3). During the drift phase, the Jacuípe, Camamu and Almada basins evolved from the final thermal subsidence to the marine passive divergent margin stage. The drift supersequence comprises three distinct sequences (Chang et al. 1992; Netto et al. 1994; Gontijo et al. 2007):

- a carbonate platform (Algodões Formation) developed in a restricted, shallow and narrow sea from the Albian to Cenomanian;
- a transgressive marine system (Urucutuca Formation), characterized by siliciclastic and marls deposited from the Turonian to the Eocene;
- a regressive marine system made up of coarse-grained sandstones, platform carbonates and distal mudstones (Caravelas, Rio Doce and Urucutuca formations), accumulated from the Oligocene to the present.

Salt and shale tectonics greatly affected the sediments of the drift supersequence (Gontijo et al. 2007).

9.3.2 Structure

Deep seismic profiles reveal the crustal structure and the rift-related magmatism of the Jacuípe, Camamu and Almada basins. The study of this sections show in the first place that different tectono-magmatic settings are recorded in the depocenters located to the north and south of the so-called Salvador transfer zone (STZ) (Figs. 9.8, 9.9, 9.10 and 9.11). The Salvador transfer zone marks the boundary between the Jacauípe, Camamu, and Recôncavo basins. It is in continuity with the NW-trending fabric elements of the Itabuna-Salvador-Curaça belt of the SFC basement (Fig. 9.5) and the N´Komi fracture zone in the conjugated West African margin (Gordon et al. 2013). Corrêa Gomes et al. (2005) describe a wrench tectonic reactivation in both onshore and offshore segments of this transfer zone. Seismic profiles and wells drilled north and south of the STZ in the Itaparica Island indicate great differences in the sedimentary thickness and the existence of positive flower structures along its trace (Fig. 9.9). In the West African counterpart, the

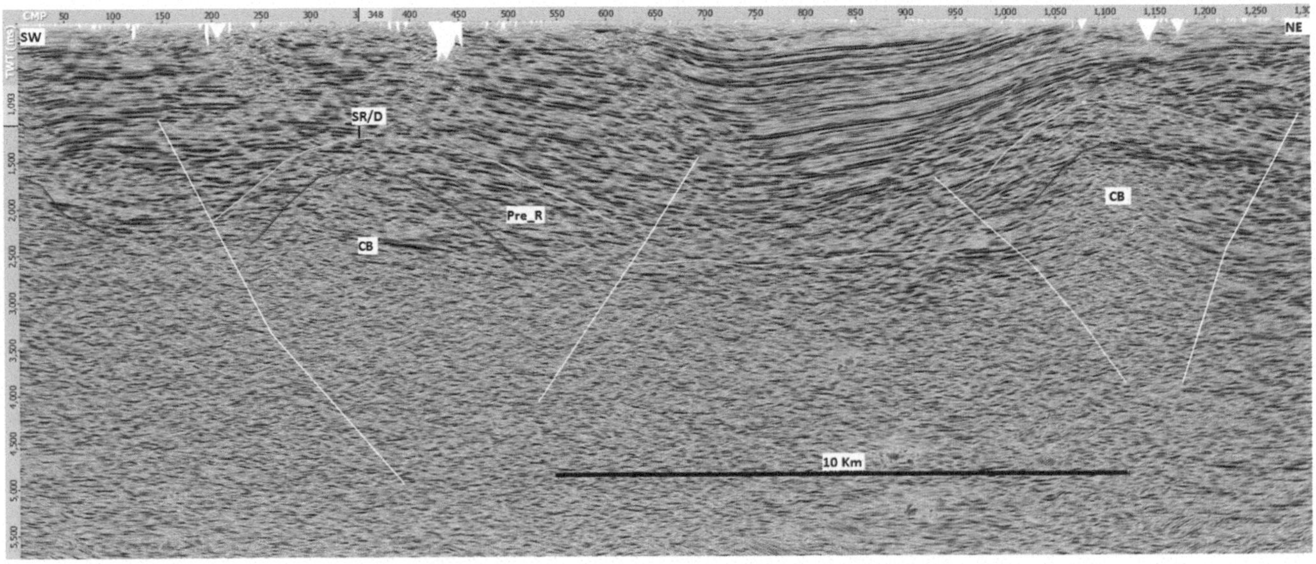

Fig. 9.9 Seismic line B, Itaparica Island: *CB* crystalline basement, *Pre-R* pre-rift deposits, and *SR/D* syn-rift-drift deposits

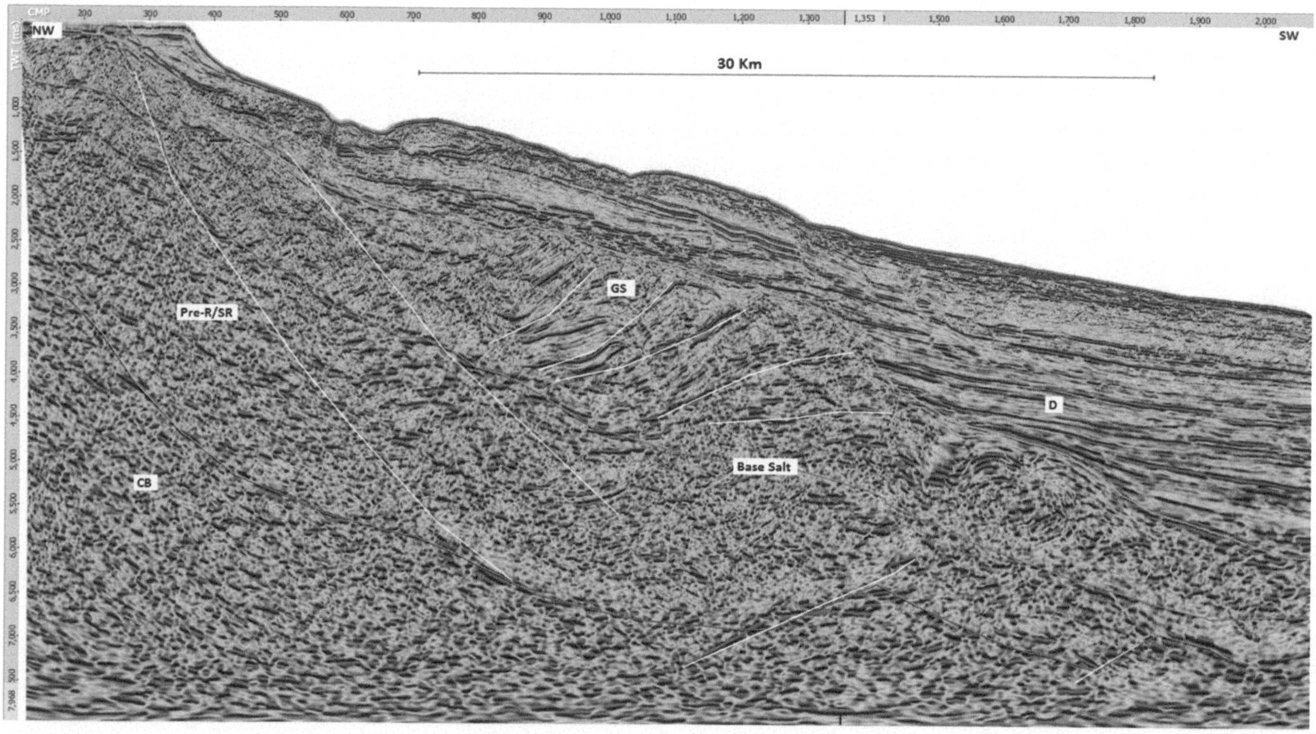

Fig. 9.10 Seismic Line C, Camamu basin: *CB* crystalline basement, *Pre-R/SR* Pre-rift/Syn-rift deposits, Gravitational slide (Albian to Eocene), *D* Eocene to recent marine section. (After Cobbold et al. 2010)

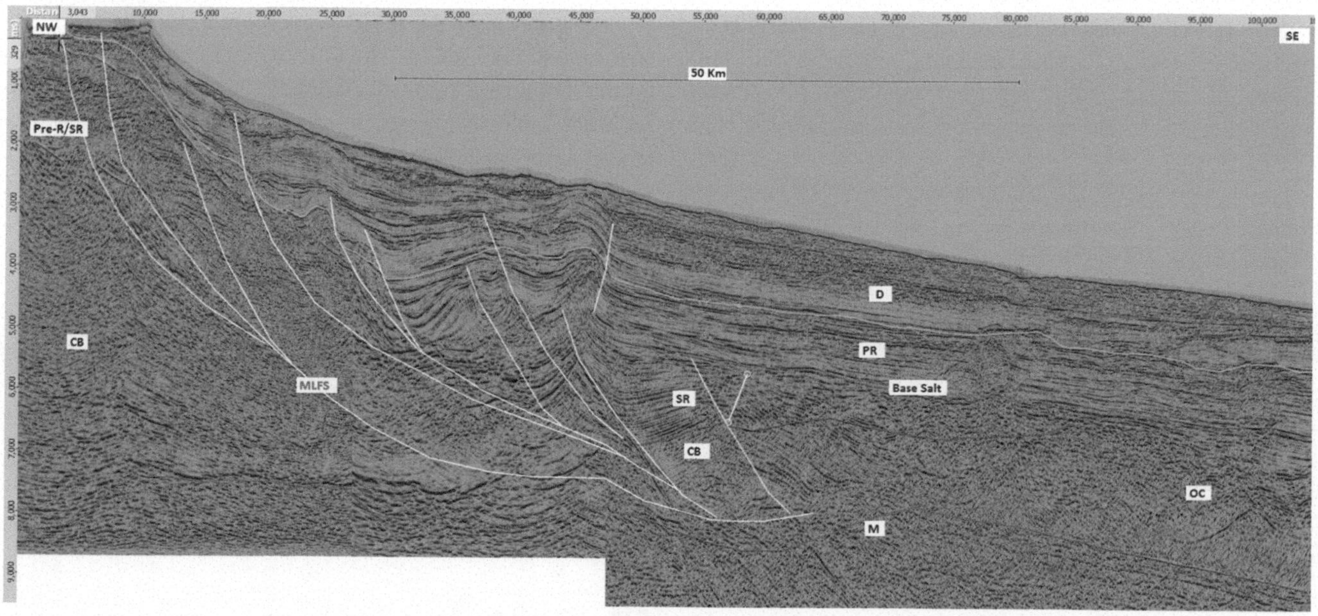

Fig. 9.11 Seismic line D, Almada basin (after Gordon et al. 2013). *M* Mohorovicic discontinuity, *CB* Continental crust, *OC* Oceanic crust, *Pre-R/SR* Pre-rift/syn-rift deposits, *PR* post-rift deposits, *D* drift deposits, *MLFS* master listric fault system

dextral N′Komi wrench fault separates the North and South Gabon basins, which correspond to domains of different geological evolution. According to refraction seismic profiles, the depths of basement in areas to the north and south of the N′Komi wrench fault are respectively 12 and 9 km (Teisserenc and Villemin 1989).

From the Paleocene to Recent, the STZ acted as weak zone, controlling the development of the Salvador canyon, which was the main feeder system for the deep-water sand deposits in the Camamu basin (Ferreira et al. 2009).

9.3.2.1 The Tectono-Magmatic Setting North of the Salvador Transfer Zone

Deep seismic profiles indicate the presence of an important mantle rise in the Jacuípe basin located to the north of the STZ, as attested by the geometry of the MOHO on the section of Fig. 9.8 and by the Bouguer anomaly map of Fig. 9.6. The Jacuípe basin is narrow (less than 100 km-wide), exhibits a crustal β factor of approximately 3, and records a syn-rift magmatic event in form of seaward dipping reflectors in the transitional crust (Mohriak 2003). The magmatic activity during the rift/post-rift stage contrasts with the non-volcanic character of Camamu and Almada basins located to the south of the STZ (Gordon et al. 2013).

9.3.2.2 Tectono-Magmatic Setting South of the Salvador Transfer Zone

No evidence for magmatic activity during the rift stage was detected in seismic or borehole data available for the Camamu and Almada basins (Gordon et al. 2013). Their magma-poor nature is in remarkable contrast with the adjacent Eastern Brazilian margin basins. Deep seismic profiles of these basins show an important crustal thinning, which records crustal β factors ranging from 5 to 8 before reaching the true oceanic crust (Figs. 9.10 and 9.11; Gordon et al. 2013). 2D gravity forward modeling confirms the high amount of crustal stretching. No evidence for mantle exhumation was found in the seismic sections (Gordon et al. 2013).

The basement of the southern Camamu and northern Almada basins corresponds to the Paleoproterozoic Itabuna-Salvador-Curaçá belt (Figs. 9.5, 9.6; Table 9.1), which locally involves a series NNE-oriented high-angle oblique-slip faults (Barbosa et al. 2003). This Precambrian cratonic fabric continues further into the offshore Camamu and Almada basins (Fig. 9.6).

The most remarkable tectonic features observed in gravity and magnetic anomaly maps, onshore outcrops, and seismic sections of the Camamu and Almada basins are:

- The presence of a NNE-trending master listric fault system that cuts the upper crust, links a series of half-graben sub-basins and detaches into the lower crust (Gordon et al. 2013; Figs. 9.10 and 9.11).
- Three systems of NW-, NNE-, and WNW-trending faults (Destro et al. 1994; Mercio and Alkmin 1995; Fig. 9.5). The NW-oriented system consists of normal faults, probably more recent in age (Mércio and Alkmim 1995). The dominant NNE system comprises normal faults that are parallel to the basin boundary and to the main foliation in the basement. Destro et al. (1994) interpreted the WNW-trending system as a transfer faults with a strike-slip displacement probably of the same nature of the Salvador shear zone (Fig. 9.5). Reactivation of this system affected the basement and the syn-rift section onshore, as well as the Lower Cretaceous section offshore. Motion along this system caused a displacement in the Camamu basin border fault (Destro et al. 1994).
- Basement highs such as the Taipus high (Fig. 9.5).
- A giant thin-skinned gravitational sliding in the northern Camamu basin (Cobbold et al. 2010; Fig. 9.10). Cobbold et al. (2010) interpret this structure as detaching into the Aptian evaporites and Neocomian shales, affecting the Campanian to Eocene sedimentary section and creating folds and thrusts. Slope instability after reactivation and steepening of the margin is a possible causative mechanism for the sliding (Cobbold et al. 2010).

9.3.2.3 The Itajú do Colônia Shear Zone

This NE-trending mega shear zone of the Neoproterozoic Araçuaí Orogen (Fig. 9.5) exhibits dextral strike-slip displacement and several reactivations during the Precambrian and Mesozoic times (Corrêa Gomes et al. 1988). The Itajú do Colônia shear zone is intruded by Paleoproterozoic amphibolite dykes, Mesoproterozoic tholeiitic dykes, and Neoproterozoic alkaline dykes and stocks. Its expression in the offshore Almada basin corresponds to various horst structures (Perez 2004). The Olivença basement high, which is the boundary between Almada and the southernmost Jequitinhonha basin (Figs. 9.5 and 9.6) is likely to be related to this shear zone (Perez 2004).

A reactivation of the northeast segment of the Itajú do Colônia shear zone during the Lower Cretaceous created the onshore Almada Canyon (Ferreira et al. 2009; Fig. 9.5). The Almada Canyon is well exposed and filled by the turbiditic deposits of the Urucutuca Formation (Bruhn and Moraes 1989). The canyon has been delivering sands into the Southern Almada basin since the Cenomanian (Arcanjo 1997). Recent drilling in the Almada basin proved the existence of numerous confined turbiditic systems with good reservoir quality, mainly in deep water locations in front of the Almada Canyon.

9.4 Evolutionary Synthesis

After the Brasiliano/PanAfrican orogeny (Neoproterozoic to Early Paleozoic), a series of intracontinental and active margin basins developed in and around West Gondwana (Limarino and Spalletti 2006; Silva et al. 2012). Shallow marine to continental Paleozoic sediments found in the Reconcavo-Tucano and Camamu basins represent a subsidence pulse in the large sag basin known as the "Afro-Brazilian depression" (Cesero and Ponte 1997; Silva et al. 2007), which was connected with the Paraná, Parnaíba and Sergipe-Alagoas interior basins (Caixeta et al. 2007).

The geologic evolution continues with extensive continental deposits of the pre-rift supersequence (Aliança, Sergi, Itaípe and Itaparica formations) in the Recôncavo-Tucano, Sergipe-Alagoas, Camamu, and Almada basin areas in Brazil, and in the conjugated West African basins of Congo and Gabon. The pre-rift sedimentation in the Afro-Brazilian sag basin records slow subsidence rates (Silva et al. 2012).

The onset of rifting tectonics in the Lower Cretaceous reactivated the Paleoproterozoic fabrics of the São Francisco craton (Itabuna-Salvador-Curaçá belt) and the Neoproterozoic Araçuaí belt. The pre-existing basement structures governed the nucleation of master listric fault systems, horsts, grabens, hemi-grabens, and transfer zones. The listric (crustal scale) master fault systems were able to accommodate a large crustal extension during the rift development decoupled from the lithospheric thinning, thus preventing magmatic activity in the Recôncavo, Camamu and Almada basins (Gordon et al. 2013). The cratonic nature of the basement resulted in the relatively small width of these basins and consequent location of the continental/ocean boundary closer to the coastline in comparison to the other eastern Brazilian margin basins.

The rifting process affected the western São Francisco Craton area in two different tectonic domains: (1) the eastern rift branch, represented by the Jacuípe, Camamu and Almada basins; and (2) the aborted western rift branch (aulacogen) of the Recôncavo-Tucano-Jatobá basins (Silva et al. 2007). The abortion of the western rift branch ended the stratigraphic evolution of the Recôncavo-Tucano-Jatobá basins (Fig. 9.3), while the eastern rift branch continued its evolution into a passive margin since the Aptian with post-rift and drift deposits.

Syn-rift wedges of coarse siliciclastic sediments (fluvial and lacustrine) were deposited along the syn-sedimentary master faults systems. The syn-rift lacustrine sediments are the most prolific source rocks and are responsible for most of the oil produced in the Brazilian passive margin basins.

Post-rift evaporite deposits occur in both the Camamu and Almada basins and are absent in Jacuípe Basin (Fig. 9.3). The drift stage sediments display a similar evolution in Jacuípe, Camamu and Almada basins, initiating the cycle with platform carbonates, from Albian to Turonian, evolving later into open marine deposits.

9.5 Final Remarks

The Camamu, Almada and Jacuípe (CAJ) basins differs significantly from other Eastern Brazilian passive margin basins (Espírito Santo, Campos, Santos and Pelotas, the ESCSP basins) in tectonic and stratigraphic aspects such as basin width, crustal thinning, structural style, pre-rift sedimentation, and syn-rift magmatism. These differences, which basically reflect the cratonic nature of the lithosphere of the CAJ basins, are:

- CAJ basins are narrower (less than 100 km in width), display an important seismic crustal thinning ($ß \geq 3$), and show an irregular distribution of their main structures. Deformation is accommodated mainly along large crustal listric fault systems, which characterize several half grabens sub-basins. In contrast, the ESCSP basins are larger (more than 250 km in width), show smaller crustal thinning in seismic ($ß < 3$), and a symmetric distribution of the deformation. Deformation is accommodated along horst and grabens associated with planar faults systems that affect mainly the upper crust. Thermo-mechanical numerical models of lithosphere extension with a strong crust/mantle coupling and frictional-plastic strain softening (Huismans and Beaumont 2005) are able to explain the tectonic style observed in the CAJ basins. Rifting of old, strong and cold cratonic lithosphere, such as the São Francisco craton, is likely to provide the required strong crust/mantle coupling.
- The Camamu and Almada basins exhibit an important pre-rift sedimentation that extends northwards into the Sergipe and Alagoas basins. The ESCSP basins do not show pre-rift sedimentation, with the exception of the Paleozoic pre-rift sediments of the Pelotas Basin. The presence/absence of pre-rift sedimentation is controlled by the occurrence of a Paleozoic/Jurassic sag basin.
- The syn-rift magmatism, in form of intra-basinal volcanism and seaward dipping reflectors, seems to be absent in Camamu and Almada basins. Syn-rift lavas, dykes and sills have been documented in Espirito Santo, Campos and Santos basins. Seaward dipping reflectors are observed in the seismic record of Pelotas, Santos, Campos, Jacuípe and Sergipe-Alagoas basins.

References

Aguiar G.A. and Matos L.F. (1990). Definição e relações estratigráficas da Formação Afligidos nas Bacias do Recôncavo, Tucano Sul e Camamu, Bahia, Brasil. In: Congresso Brasileiro de Geologia, 36, 1990, Natal. Anais. Natal: SBG, 1990. V. 1, p. 157–170.

Almeida F.F.M. (1977). O Cráton do São Francisco. Revista Brasileira de Geociências, São Paulo, v. 7, n. 4, p. 349–364.

Almeida F.F.M., Hasui Y., Brito Neves B.B., Fuck R.A. (1981). Brazilian Structural Provinces. Earth Sci. Rev., Special Issue 17:1–29.

Aragão M.A.N.F. (1994). Arquitetura, estilos tectônicos e evolução da Bacia do Recôncavo, Brasil. In: Simp. Cret. Brasil, 3, Rio Claro, Boletim, 165–167.

Aragão M.A.N.F. and Peraro A.A. (1994). Elementos estruturais do rifte Tucano/Jatobá. In: Simp. Cret. Brasil, 3, Rio Claro, Boletim, 161–164.

Arcanjo J.B.A. (1997). Programa de Levantamentos Geológicos Básicos do Brasil: Itabuna Folha SD 24-Y-B-Vl. Brasília: CPRM, 276 p., escala 1:100.000.

Barbosa J.S.F. and Sabate P. (2003). Colagem paleoproterozoica de placas arqueanas do Cráton de São Francisco na Bahia. Revista Brasileira de Geociências, São Paulo, v. 33, n. 1, p. 7–14, mar. Suplemento.

Barbosa J.S.F., Sabaté P., Marinho M.M. (2003). O Cráton do São Francisco na Bahia: uma síntese. Revista Brasileira de Geociências, São Paulo, v.33, n. 1, p. 3–6.

Bruhn C.H.L. and Moraes M.A.S. (1989). Turbiditos da Formação Urucutuca na Bacia de Almada, Bahia: um laboratório de campo para estudos de reservatórios canalizados. Separata de: Boletim de Geociências da Petrobras, Rio de Janeiro, v. 3, p. 235–267.

Cobbold P.R., Gilchrist I.S., Chiossi D.N., Fonseca F., Gomes De Souza F., Lilletveit R. (2010). Large submarine slides on a steep continental margin (Camamu Basin, NE Brazil). Jorunal of the Geological Society of London, V 167, P. 1–9.

Caixeta J.M., Milhomem P.S., Witzke R.E., Dupuy I.S.S., Gontijo G.A. (2007). Bacia de Camamu. In: Boletim de Geociências da Petrobras, Rio de Janeiro, v. 15, n. 2, p. 455–461.

Chang H.K., Kowsmann R.O., Figueiredo, A.M.F., Bender A.A. (1992). Tectonic and stratigraphy of the east Brazil rift system: an overview. Tectonophysics, Amsterdam, v. 213: 97–138.

Cesero P. D and Ponte F.C. (1997). Análise comparativa da paleogeologia dos litorais atlânticos brasileiros e africano. Boletim de Geociências da Petrobras, V. 11 (1), P. 1–18.

Corrêa Gomes C.G.L., Oliveira P.E., Barbosa, J.F.S., Silva, F.P.C. (1988). Tectônica associada à colocação de diques alcalinos félsicos a máficos Neoproterozóicos da Zona de cisalhamento de Itabuna-Itajú do Colônia, Bahia, Brasil. Revista Brasileira de Geociências, Rio de Janeiro, v. 28, p. 449–458, 1988.

Corrêa Gomes L.C., Dominguez, J.M.L., Barbosa, J.S.F., Silva, I.C. (2005). Padrões de orientação dos campos de tensão, estruturas, herança do embasamento e evolução tectônica das bacias de Camamu e porção sul do recôncavo, costa do dendê, Bahia. Revista Brasileira de Geociências, São Paulo, v. 35, n. 4, p. 117–128, dez. Suplemento.

Delgado, I.M., Souza, J.D., Silva L.C., Silveira Filho N.C., Santos R. A., Pedreira A.J., Guimarães J.T., Angelim L.A.A., Vasconcelos A. M., Gomes I.P., Lacerda Filho J.V., Valente, C.R., Perrota M.M., Heineck C.A. (2003). Geotectônica do Escudo Atlântico. In: Bizzi L.A., Schobbenhaus, C., Vidotti R.M., Gonçalves J.H. (eds.). Geologia, tectônica e recursos minerais do Brasil: texto, mapa & SIG. Brasília: Companhia de Pesquisa de Recursos Minerais, p. 227–334.

Destro N., Amorim J.L., Witzke R.E. (1994). Identificação de falhas de transferência na Bacia de Camamu. In: Congresso Brasileiro de Geologia 38, Camboriú, SBG. V2, p. 275–277.

Destro N. (1995). Release fault: a variety of cross fault in linked extensional fault systems in the Sergipe-Alagoas Basin, NE Brazil. J. Struct. Geol., 17(5): 615–629.

Destro N. (2002). Falhas de alívio e de transferência: O significado tectônico e econômico no Rifte do Recôncavo-Tucano-Jatobá, NE Brasil. Federal University of Ouro Preto, Brazil, PhD, 173p.

Destro N., Szatmari P., Alkmim F.F., Magnavita L.P. (2003a). Release faults, associated structures, and their control on petroleum trends in the Recôncavo Rift, northeast Brazil. Bull. Am. Ass. Petrol. Geol., 87(7):1123–1144.

Destro N., Alkmim F.F., Magnavita L.P., Szatmari P. (2003b). The Jeremoabo transpressional transfer fault, Recôncavo-Tucano Rift, NE Brazil. J. Struct. Geol., 25 (8):1263–1279.

Ferreira T.S., Caixeta J.M., Lima F.D. (2009). Controle do embasamento no rifteamento das bacias de Camamu e Almada. In: Boletim de Geociências da Petrobras, Rio de Janeiro, v. 17, n. 1, p. 69–88.

Graddi J.C.S.V., Campos Neto O.P.A., Caixeta J.M. (2007). Bacia de Jacuípe. In: Boletim de Geociências da Petrobras, Rio de Janeiro, v. 15, n. 2, p. 417–421.

Gontijo G.A., Milhomem O.S., Caixeta J.M., Dupuy I.S.S., Menezes P. E.L. (2007). Bacia de Almada. Boletim de Geociências Petrobras, Rio de Janeiro, v. 15, n. 2, p. 463–473.

Gordon A.C., Mohriak W.U., Barbosa V.C.F. (2013). Crustal Architecture of the Almada Basin, Ne Brazil: an example of a non-volcanic rift segment of the South Atlantic passive margin. Geological Society, London, Special Publications, v. 369, p. 215–234.

Huismans, R.S., Beaumont, C. (2005). Effect of litospheric stratificationon extensional styles and rift basin geometry. In: Post, P. (ed). Petroleum Systems of Divergent Continental Margin Basins, 25th Annual GCSSEPM Foundation Bob F. Perkins Research Conference proceedings.

Limarino C.O. and Spalletti L. (2006). Paleogeography of the upper Paleozoic basins of southern South America: An overview. Journal of South American Earth Sciences, V. 22, P. 134–155.

Mello M.R., Gonçalves F.T.T., Netto A.S.T., Amorim J.L., Witzke R.E. (1995). Application of the Petroleum System concept in the assessment of exploration risk: the Camamu Basin example, offshore Brazil. In: Congresso Internacional da Sociedade Brasileira de Geofísica, 40, 1995, Rio de Janeiro, Resumos…Rio de Janeiro: SBGF.

Mercio S.R. and Alkmin F.F. (1995). Duas Famílias de Falhas na Bacia de Camamu (BA) e seu Significado Tectônico. In: 5° Simpósio Nacional de Estudos Tectônicos, Gramado, Boletim de Resumos, v. 5, p. 290–292.

Magnavita L.P. (1992). Geometry and kinematics of the Recôncavo--Tucano-Jatobá Rift, NE Brazil. Oxford University, PhD Thesis, 493p.

Milani E.J. and Davison I. (1988). Basement control, and transfer tectonics in the Recôncavo–Tucano–Jatobá rift, Northeast Brazil. Tectonophysics, 18:41–70.

Milani E.J., Lana MC, Szatmari P. (1988). Mesozoic rift basins around the northeast Brazilian microplate (Recôncavo-Tucano-Jatobá, Sergipe-Alagoas). In: W. Manzpeizer (ed.) Triassic-Jurassic rifting: Continental breakup and the origin of the Atlantic Ocean and passive margins. Amsterdam, Elsevier, 833–858.

Mohriak W.U., Basseto M., Vieira I.S. (1998). Crustal architecture and tectonic evolution of the Sergipe-Alagoas and Jacuipe basins, offshore northeastern Brazil. Tectonophysics, Amsterdam, v. 288, p. 199–220.

Mohriak, W.U. (2003). Bacias sedimentares da margem continental Brasileira. In: Bizzi L.A, Schobbenhaus C., Vidotti, R.M., Gonçalves J.H. (Ed.). Geologia, Tectônica e Recursos Minerais do Brasil. Brasília, DF: CPRM, p. 87–94.

Netto A.S.T. and Oliveira J.J. (1985). O preenchimento do rift valley na Bacia do Recôncavo: Revista Brasileira de Geociências, v. 15, p. 97–102.

Netto A.S., Wanderley Filho J.R.E., Feijo F.J. (1994). Bacias de Jacuípe, Camamu e Almada. Boletim de Geociências da Petrobras, Rio de Janeiro, v. 8, n. 1, p. 173–175.

Perez C.F.E. (2004). Emprego de Técnicas de Sensoriamento Remoto e Métodos Potenciais na Caracterização Estrutural do Embasamento da Bacia de Camamu-Almada. PhD Thesis, COOPE, Federal University of Rio de Janeiro, Brazil, 93p.

Sandwell D.T. and Smith W.H.F. (1997). Marine gravity anomaly from Geosat and ERS 1 satellite altimetry. Journal of Geophysical Research, Washington, v. 102, n. B, 10054 p.

Silva O.B.D., Caixeta J.M., Milhomem P.D.S. (2007). Bacia de Recôncavo. Boletim Geociências da Petrobras, Rio de Janeiro, v. 15, n. 2, p. 423–432.

Silva D.R., Mizusaki, A.M.P., Milani E.J., Pimentel, M. (2012). Depositional ages of Paleozoic and Mesozoic pre-rift supersequences of the Recôncavo Basin in northeastern Brazil: A Rb-Sr radiometric study of sedimentary rocks. Journal of South American Earth Sciences, V. 37, P. 13–24.

Szatmari P. and Milani E.J. (1999). Microplate rotation in northeast Brazil during South Atlantic rifting: Analogies with the Sinai Microplate. Geology, 27(12):1115–1118

Szatmari P., Milani E. J., Lana M. C., Conceição J. C., Lobo A. (1985). How South Atlantic rifting affects Brazilian oil reserves distribution. Oil and Gas Journal, 14: 107–113.

Szatmari P., Françolin J. B. L., Zanotto O., Wolff S. (1987). Evolução tectônica da margem equatorial brasileira. Revista Brasileira de Geociências, 17(2): 180–188.

Teisserenc P. and Villemin J. (1989). Sedimentary basins of Gabon-geology and oil systems. In: Edwards D. and Santogrossi A. (eds). Divergent/Passive Margin Basins. AAPG, Memoir, 48, 117–199.

Vasconcellos D.V.F. (2003). Falha de transferência de Caritá: O significado tectônico no Rifte do Recôncavo-Tucano-Jatobá, NE Brasil. MsC Thesis, Federal University of Ouro Preto, Brazil, 66p.

Viana C.F, Junio G.R., Simões I., Moura J.A., Fonseca J.R., Alves R. J. (1971). Revisão estratigráfica da Bacia do Recôncavo/Tucano. Boletim Técnico da Petrobras, Rio de Janeiro, v. 14, n. 3–4, p. 157-192

Part IV
Marginal Belts

The Southern Brasília Belt

10

Claudio de Morisson Valeriano

Abstract

This chapter presents a review of the main tectonic features of the Southern Brasilia belt with emphasis on its relationships with the southwestern margin of the São Francisco craton in the context of the Neoproterozoic Brasiliano orogenic collage of West Gondwana. The belt is characterized by a stack of W-vergent nappes of passive margin units bounded to the west by the Archean-Paleoproterozoic Goiás microcontinent and the Neoproterozoic Mara Rosa, Goiás and Arenápolis magmatic arcs. The protracted history of the passive margin development initiated with Mesoproterozoic rifting events that led to a final Tonian continental break up and development of a wide continental margin basin. Minimum sedimentation ages of the sedimentary units involved are given by subduction-related regional metamorphism between 650 and 610 Ma. An older sedimentary phase of the passive margin is represented by the Paranoá Group, the Jequitaí Formation (Bambuí Group), the Vazante and Canastra groups, the Cubatão Formation (Ibiá Group) and part of the Araxá Group, whose youngest detrital zircons were dated between 1.0 and 0.9 Ga. The orogenic phase of the southern Brasilia belt started with progressive subduction of distal to proximal passive margin units at 650–630 Ma and development of recumbent folds associated with a regionally penetrative foliation and medium to high-pressure metamorphic parageneses. Post-metamorphic peak nappe exhumation took place around 610–605 Ma (monazite U-Pb) and final cooling at 600–580 Ma, as indicated by K-Ar ages on white mica and biotite. The syn-orogenic Neoproterozoic sedimentary units show detrital zircons as young as 600 Ma, with mixed cratonic/arc-related sedimentary provenance, represented by the Bambuí Group, Rio Verde Formation (Ibiá Group) and upper thrust sheets of the Araxá Group.

Keywords

Brasiliano orogeny • Continental margin • Ocean continent transition • Tocantins orogenic system • Southern Brasília belt

C. de Morisson Valeriano (✉)
TEKTOS Research Group, Faculdade de Geologia, Universidade do Estado Rio de Janeiro (UERJ), Rua São Francisco Xavier, 524, Bloco A-4026, Rio de Janeiro, RJ 20.550-900, Brazil
e-mail: valeriano.claudio@gmail.com

M. Heilbron et al. (eds.), *São Francisco Craton, Eastern Brazil*,
Regional Geology Reviews, DOI 10.1007/978-3-319-01715-0_10

10.1 Introduction

The Southern Brasilia belt (SBB) is a wide orogenic belt essentially composed of metasedimentary rocks that stretches for ca. 800 km along the southwestern margin of the São Francisco craton. Although telescoped by the 650–610 Ma Brasiliano orogeny, representative sections of the stratigraphy of one of the former Neoproterozoic passive margins developed around the São Francisco paleocontinent are exposed across the SBB .

The Brasilia belt was first outlined in the classic article by Almeida (1967). Afterwards, successive geodynamic models have evolved following the transition from geosynclinal to plate tectonics concepts (e.g., Fuck et al. 1993; Dardenne 2000; Pimentel et al. 2004; Valeriano et al. 2004a, b; Valeriano et al. 2008). Current geodynamic models for the belt embrace a complete Wilson Cycle, starting with the development of a passive margin and opening Goianides ocean in theTonian Period.

Subduction of distal units of the passive margin, accretionary tectonics and associated nappe exhumation took place relatively early (~650–630 Ma) in the SSB when compared to other orogenic belts that surround the São Francisco craton. The SSB provides thus the oldest records of the tectonic processes involving the continental growth around the São Francisco craton (Valeriano et al. 2008).

In view of the significant advances regarding SBB geology in of the last decade, this chapter presents a descriptive synthesis on the lithostratigraphy, sedimentary provenance, radiometric age constraints, and structural style of the rock assemblages that characterize the different tectonic domains of southwestern margin of the São Francisco craton (SFC). The last section of the chapter presents a brief evolutionary summary of tectonic processes from the Tonian extensional tectonics and passive margin development, until the Cryogenian accretionary and collisional processes that resulted in the present exposure of one of the most complete Precambrian orogenic edifices of South America.

10.2 Geodynamic Context of the Brasilia Belt

The Brasilia belt as a whole (Fig. 10.1) extends for more than 1200 km along the western margin of the SFC. It resulted from subduction of São Francisco paleoplate oceanic lithosphere towards west (current coordinates), followed by the accretion of several exotic terranes onto the Neoproterozoic passive margin of the former São Francisco paleocontinent. Started in the Tonian, this processes involved the Goiás Jost et al. (2005) and Paranapanema micro-continents (Mantovani et al. 2005), as well as magmatic arcs, such as the Goiás, Mara Rosa, and Guaxupé arcs (Pimentel and Fuck 1992; Pimentel et al. 2000; Laux et al. (2005). The collision of the São Francisco and the Amazon paleoplates took place much later, at around 540 Ma, leading to the development of the Araguaia and Paraguay belts of the western Tocantins orogenic system (Valeriano et al. 2008).

The protuberant shape of the western São Francisco paleocontinental margin, as inferred from gravimetric data (Lesquer et al. 1981; Ussami 1993), acted as an indentor during the orogenic process, dividing the belt into a NE-trending segment, the Northern Brasilia belt (Fonseca et al. 1995), and the SE-trending Southern Brasilia belt. The northern and southern segments of the Brasilia belt merge along a highly deformed E-W trending zone, currently referred to as the Pirineus syntaxis (Araujo Filho 2000) (Fig. 10.1).

Large portions of the Neoproterozoic passive margin of the southwestern São Francisco paleocontinent are now exposed along the fold-thrust systems of the external and part of the internal Southern Brasilia belt.

10.3 Tectonic Framework of the Southern Brasilia Belt

The overall structure of the SSB consists of a fold-thrust belt involving Neoproterozoic metasedimentary rocks (locally containing detached basement slivers), covered by a system of sub-horizontal spoon-shaped metamorphic nappes (upper greenschist to granulite facies) (Fig. 10.2). According to structural style, metamorphic grade, and lithostratigraphic content, the tectonic zoning of the Southern Brasilia belt and adjoining São Francisco craton can be defined as follows, from the base to the top (or from east to west): (i) the cratonic zone; (ii) the external metamorphic fold-thrust belt; (iii) the upper nappe complex (Fig. 10.3).

In the next sections, the lithostratigraphy, age constraints, sedimentary provenance patterns (Table 10.1) and structure of the Neoproterozoic units involved in the SBB and adjacent SFC are briefly described from cratonic area to the upper and westernmost nappe units.

10.4 The Cratonic Zone

The cratonic zone is characterized by basement assemblages older than 1.8 Ga unconformably overlain by the Neoproterozoic strata of the Vazante and Bambuí groups as pointed

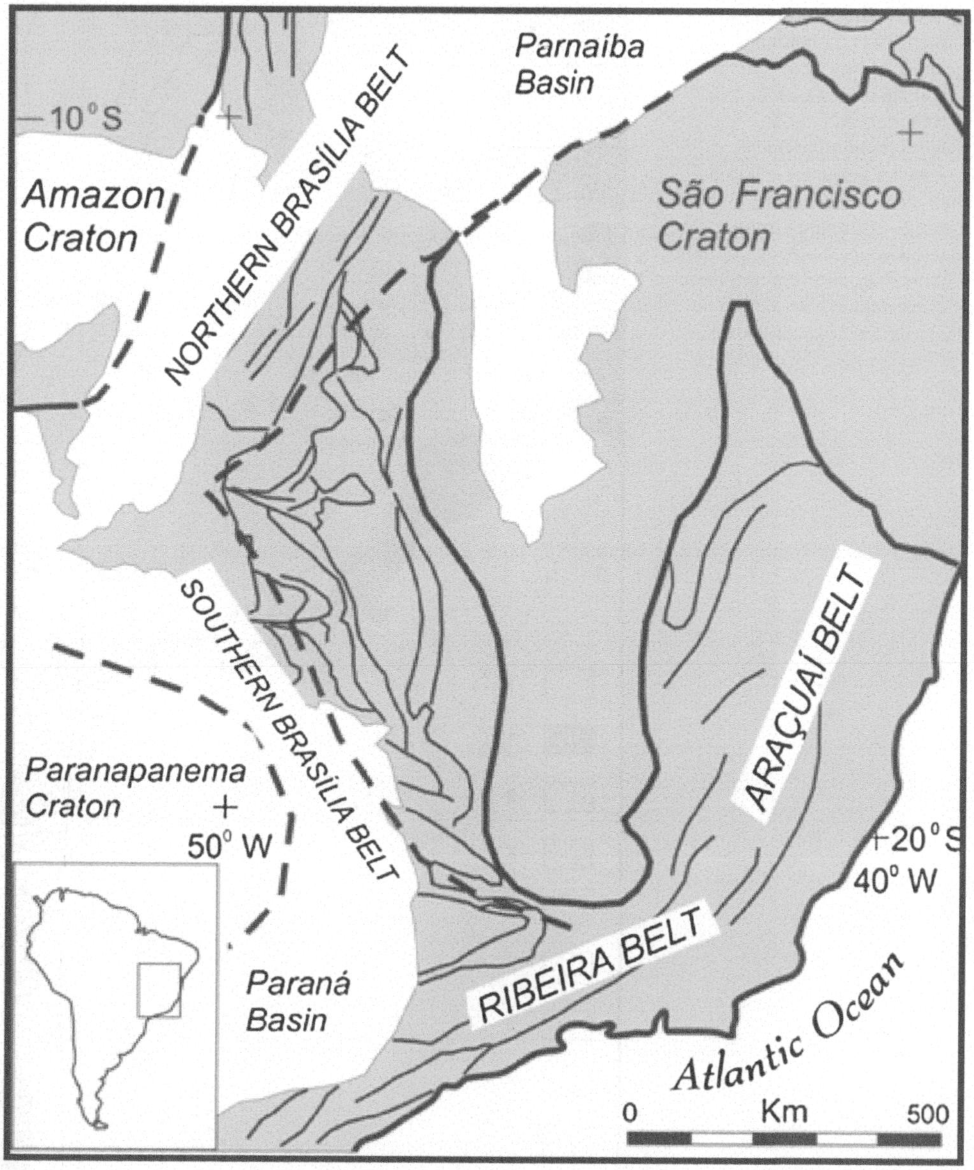

Fig. 10.1 The Northern and Southern Brasilia belts as orogenic domains developed by the indentation of the western São Fancisco paleocontinental margin against accretionary terranes during the Brasiliano orogeny (modified from Valeriano et al. 2008). The *box* indicates the area of the map shown on Fig. 10.2

out by Alkmim and Martins-Neto (2012). The truly cratonic zone, where the Neoproterozoic cover is virtually undeformed and unmetamorphosed is restricted to the central portion of SFC, located in the eastern extremity of the sections shown on Fig. 10.2. Almeida (1981) modified his original delimitation of the SFC (presented in Almeida 1967) to include the ca. 100-wide thin-skinned foreland fold-thrust belt of the SBB in the western cratonic zone (see Cordani et al., this book). This concept is also adopted in this chapter. Thus, the craton boundary, in map view, is here placed along the thrust fault that brings the older Meso- to Neoproterozoic units of the Paranoá, Canastra, and Vazante groups (see next sections) on top of the parautochthonous Bambuí strata (Figs. 10.2, 10.3 and 10.5a).

10.4.1 Stratigraphy

10.4.1.1 Basement

The basement in the cratonic zone reached tectonic stability at around 1.8 Ga, following the Rhyacian-Orosirian amalgamation of Archean blocks and juvenile Paleoproterozoic magmatic arcs documented in Eastern Bahia orogenic domain (Barbosa and Barbosa, this book) and Mineiro belt (Alkmim and Marshak 2001; Alkmim and Teixeira, this book), located in the northern and southern SFC, respectively. The craton basement, exposed around the town of Campo Belo (Fig. 10.2), consists of Archean granite-greenstone complexes (e.g., the Campo Belo Complex, Teixeira et al., this book) bounded to the east and southeast by the Quadrilátero

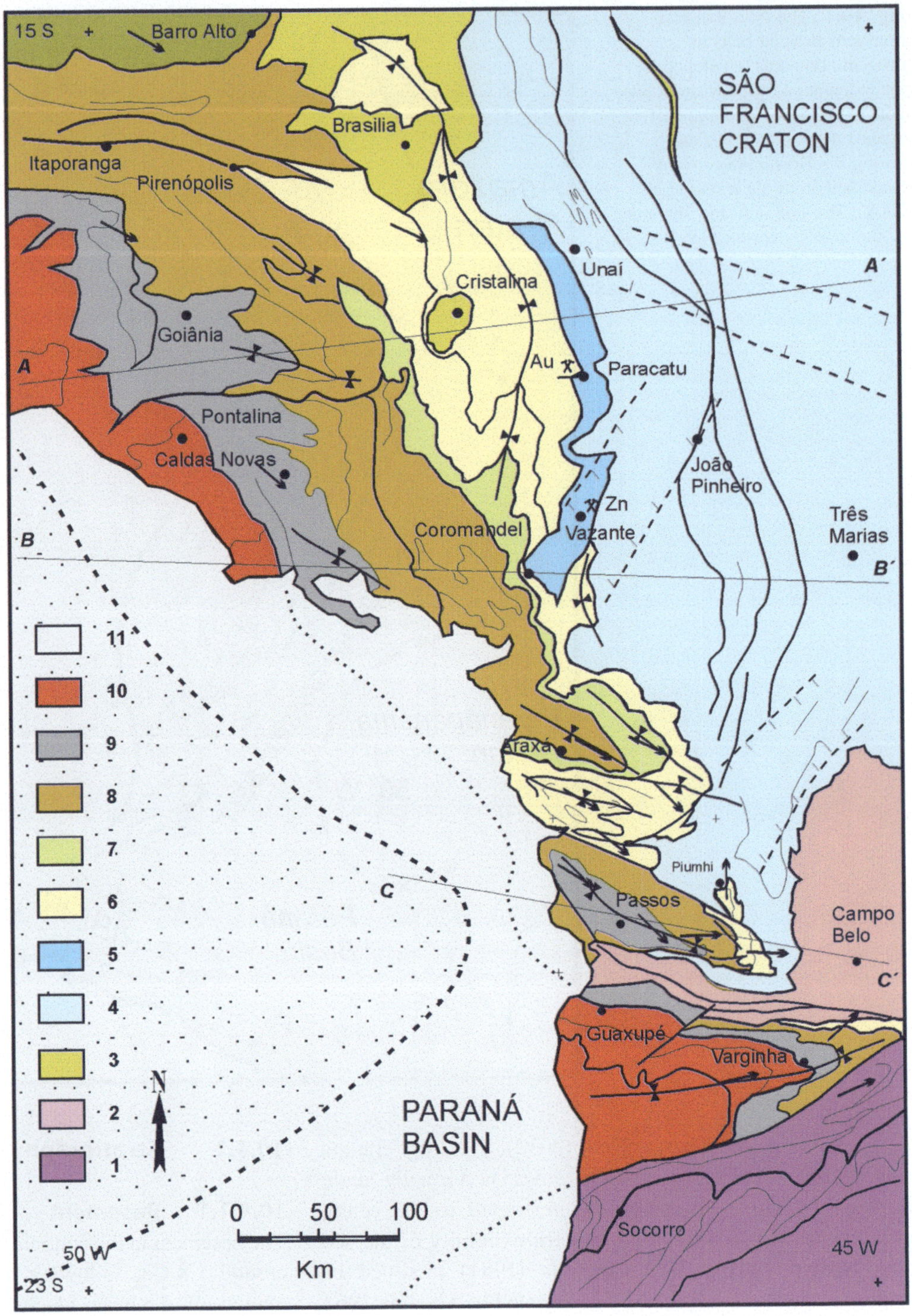

Fig. 10.2 Tectonic map of Southern Brasilia belt and adjoining southwestern margin of the São Francisco craton. Key: *1*—Ribeira belt; *2*—São Francisco craton basement; *3*—Paranoá Group; *4*—Bambuí group cratonic cover; *5*—Vazante Group; *6*—Canastra Group; *7*—Ibiá Group; *8*—Araxá Group; *9*—Araxá Group metamorphosed under granulite facies conditions; *10*—Magmatic arc complexes; *11*—Phanerozoic cover (Paraná Basin)

Ferrífero province and the Mineiro belt, respectively (Ávila et al. 2010; Teixeira et al. 2015). Besides Archean TTG-complexes and a greenstone belt sequence, the Quadrilátero Ferrífero province also exposes Paleoproterozoic rift, passive margin and syn-orogenic sedimentary successions. The NE-trending Mineiro belt is made up by mainly of accreted juvenile 2.42–2.10 Ga granitoids of (Ávila et al. 2010; Teixeira et al. 2015; Alkmim and Teixeira, this book). Ages of regional metamorphism in the Mineiro belt fall between ca. 2.25 and 2.05 Ga, followed by regional cooling and uplift between ca. 1.9 and 1.8 Ga. Quartzites and associated siliciclastic successions, well exposed in the Mineiro belt along the southern craton boundary, are close correlatives to the 1.78–1.00 Ga Espinhaço Supergroup that crops out in

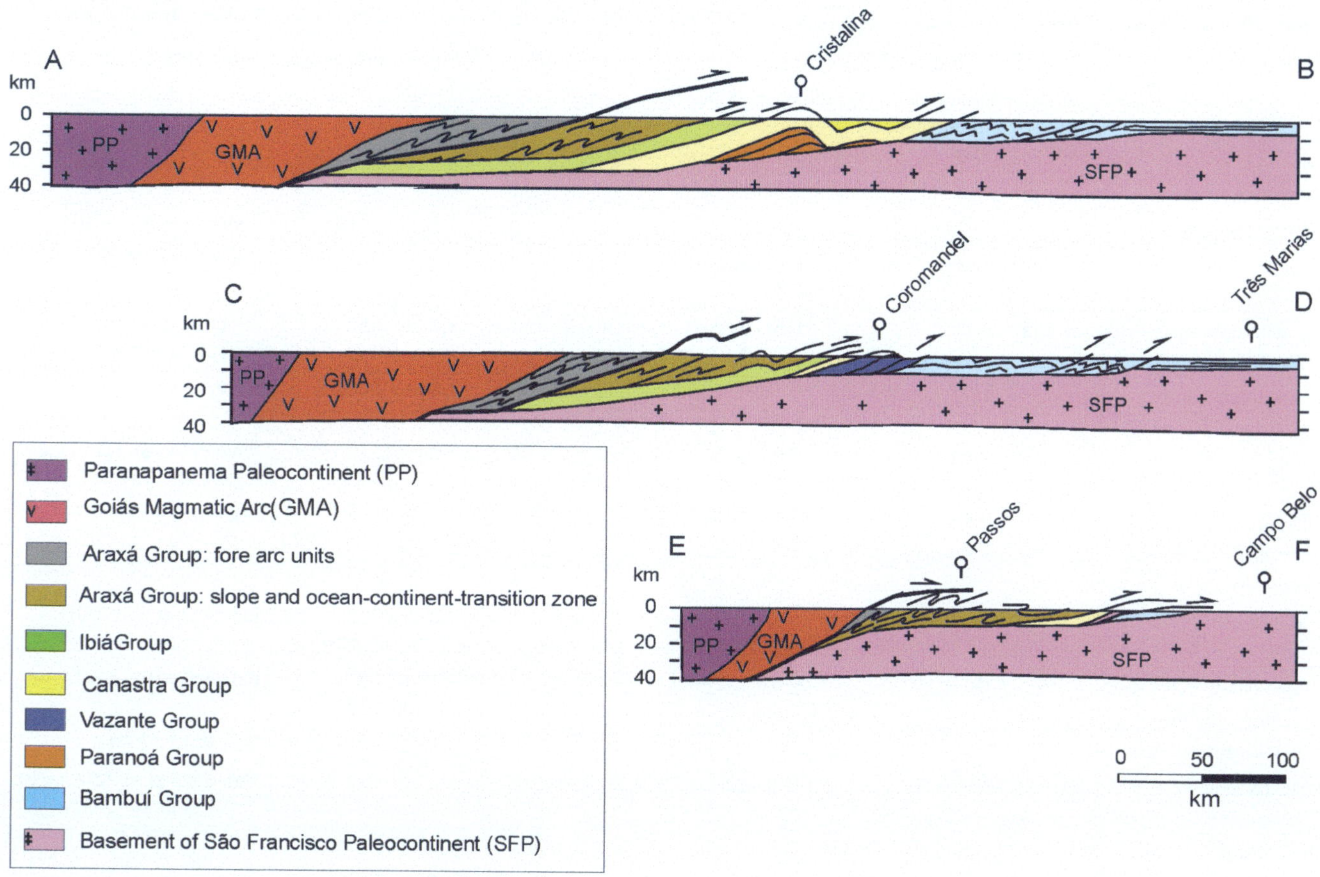

Fig. 10.3 Schematic sections (locations indicated on Fig. 10.2) showing main tectonic elements of the Southern Brasilia belt

the craton interior and Neoproterozoic marginal belts (see Heilbron et al., Alkmim et al. and Cruz et al., this book).

10.4.1.2 Vazante Group

The Vazante Group occurs as an individual thrust sheet sandwiched between the deformed Bambuí cratonic cover on the east and the Canastra-Ibiá thrust sheet on the west. The 250 km-long and 25 km-wide outcrop belt of the Vazante group is mostly composed of metapelites (slates) and meta-dolomites, which host the widely known mineral deposits of Vazante (Zn), Paracatu-Morro Agudo (Zn-Pb) and Lagamar (phosphate) (Dardenne et al. 1997).

The Vazante Group was originally regarded as belonging to the Bambuí Group (Cloud and Dardenne 1973). Dardenne (1978) first called attention to the fact that the Vazante Group contains well preserved Conophyton columnar stromatolites (Fig. 10.5c), which would be possibly Mesoproterozoic and thus older than the typical Neoproterozoic structures of the same nature recognized in the Bambuí Group (Dardenne 2005.

The best described section of the Vazante Group (Fig. 10.4) is exposed along the Lagamar-Vazante road (Dardenne 2000), allowing the definition by Dardenne (2000) of seven formations. The basal Santo Antonio do Bonito and Rocinha formations comprise metapelites and metassiltites with conglomeratic intercalations. The latter formation contains important phosphorite beds (Dardenne et al. 1997). The overlying Lagamar stromatolitic limestones are covered by the Serra do Garrote carbonaceous metapelites (Fig. 10.5f) with psammitic intercalations. The Serra do Poço Verde and Morro do Calcário formations are made up dominantly of dolomites, and host most of the Zinc deposits of the Vazante Group. These dolomitic domains are typically low and flat areas with common karstic features such as caves and sinkholes (Fig. 10.5d).

The uppermost Lapa formation is composed of marly metapelites with few quartzitic intercalations.

Based on stromatolite associations found in the carbonatic rocks of the Vazante Group (Moeri 1972), Dardenne et al. (1976, 2005a, b) suggested an age of sedimentation between 1350 and 950 Ma. Re–Os ages presented by Azmy et al. (2008) point to an age in the 1100–1000 Ma interval. However, detrital zircons dated by the U-Pb method (Rodrigues 2008; Rodrigues et al. 2012) point to a maximum depositional age of 935 ± 14 Ma, placing the sedimentation of the Vazante group most likely in the same range of other

Table 10.1 U-Pb ages of detrital zircons and whole-rock Sm–Nd ages for the main metasedimentary units of the Southern Brasilia belt

Unit	U-Pb detrital zircon ages		Sm–Nd model age T_{dm} (Ga) (n)[a]	References
	Youngest zircon/xenotime (Ma)	Main modes (Ga)		
Bambuí Group				
Três Marias Fm			1.53 to 1.81 (5)	Pimentel et al. (2001)
	616	0.56–0.63–0.68–0.78–1.15–1.55–1.70–2.60		Rodrigues (2008)
S. da Saudade, L. do Jacaré and Serra Sta. Helena Fms	612	0.61–0.66–0.79–1.06–1.30–1.56–1.78–2.17–2.58–2.70	1.42 to 1.87 (17)	Pimentel et al. (2001) and Rodrigues (2008)
Sete Lagoas Fm	537 ± 4, 505 ± 7	0.54–0.56–0.58–0.63–0.66–0.87–0.93–0.97–1.07–1.27–1.94	1.80 to 2.00 (3)	Santos (2012)
	610	0.65–0.79	1.60 to 2.19 (11)	Pimentel et al. (2001) and Rodrigues (2008)
Jequitaí Fm	880	1.23–1.54–1.72–2.12–2.27–2.75–2.86–3.01		Rodrigues (2008)
Araxá group				
Metapelites			1.0 to 1.3, 1.8 to 2.1 (26)	Pimentel et al. (2001)
Anápolis-Itauçu nappe	643	0.67–1.05–1.5–2.0	1.19 to 1.37 1.76 to 2.18 (8)	Piuzana et al. (2003a, b)
Araxá nappe			1.9 (1)	Seer et al. (2001)
Passos nappe	607 ± 1	0.95–1.25–1.75–2.15–2.40–2.95–3.0		Valeriano et al. (2004a, b)
Ibiá group				
Rio Verde Fm			1.16 to 1.33 1.93 to 2.01 (7)	Pimentel et al. (2001)
			1.16 to 1.33 (3)	Seer et al. (2001)
	636 ± 21	0.64–0.67–0.72–0.76–0.83–0.88–0.96–1.07	1.46 (1)	Rodrigues (2008)
	639 ± 15	0.69–0.88–1.57–1.72–1.95–2.53	1.20 to 1.24 (3)	Dias et al. (2011)
			1.58 to 2.69 (2)	Klein (2008)
Cubatão Fm (pebbles)			2.20 to 2.47 (3)	Pimentel et al. (2001)
	935 ± 11	0.94–0.99–1.19–1.37–1.60–1.84–1.99–2.14–2.54	1.77 to 1.89 (2)	Rodrigues (2008)
Cubatão Fm (matrix)	922 ± 16	1.20–1.54–1.87–2.10–2.50–2.72		Dias et al. (2011)
Canastra group				
Metapelites			1.90 to 2.34 (6)	Pimentel et al. (2001)
Passos nappe	1011	1.3		Valeriano et al. (2004a, b)
	1030		1.47 to 1.87 (4)	Rodrigues (2008)
Vazante group				
			1.70 to 2.10 (13)	Pimentel et al. (2001)
	935 ± 14	0.94–1.16–1.21–1.53 –1.81–2.08–2.15	1.67 to 2.76 1.90 to 2.08 (17)	Rodrigues (2008) and Rodrigues et al. (2012)
			1.41 to 2.40 (9)	Santana (2011)
Paranoá group				
			2.03 to 2.27 (6)	Pimentel et al. (2001)
Northern Brasilia belt	1042 (diagenetic xenotime)	1,54–1,78–2.15–2,70–2,85–3,10–3,43		Matteini et al. (2012)

[a](n) = number of Sm–Nd samples

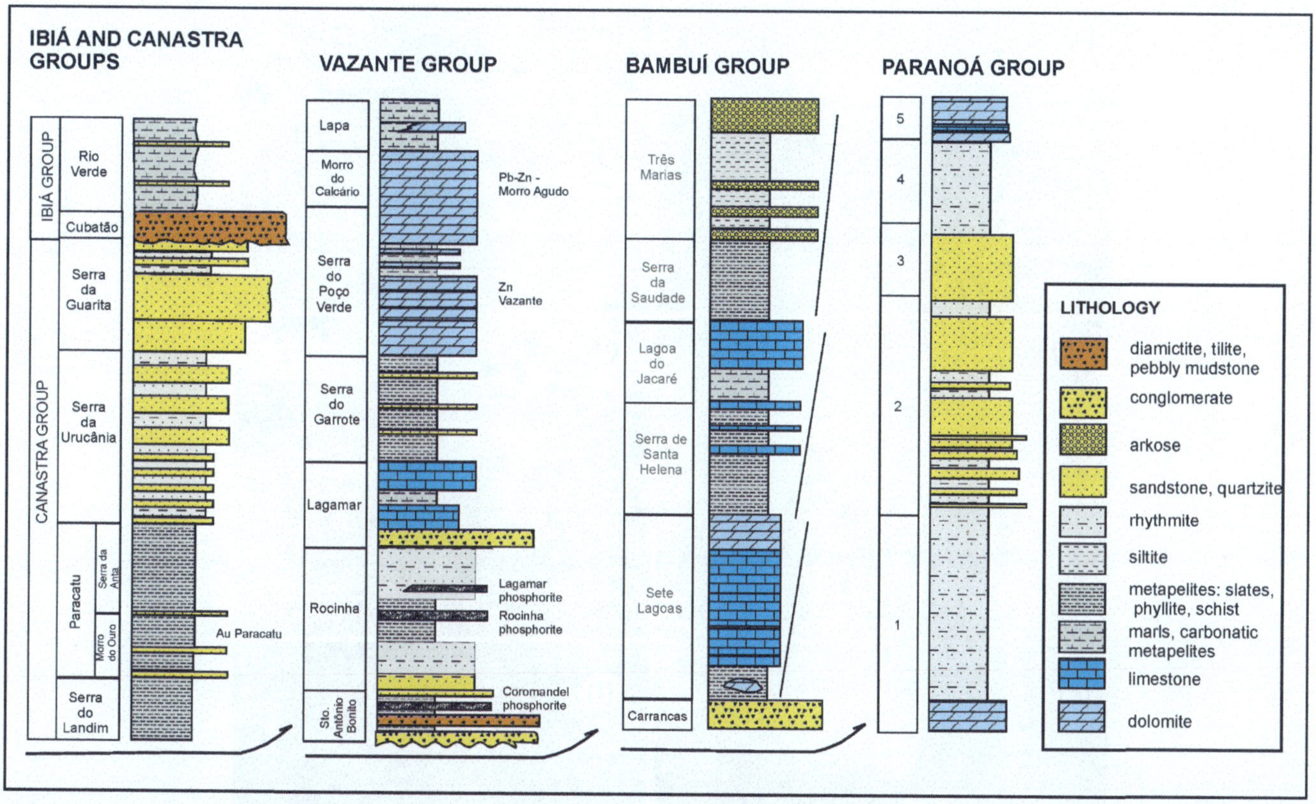

Fig. 10.4 Representative lithostratigraphic columns for the main passive margin metasedimentary units of the Southern Brasilia: Ibiá and Canastra groups (Dias et al. 2011); Vazante, Bambuí (Dardenne 2000) and Paranoá (Alvarenga et al. 2012) groups

passive margin units, such as the Canastra and Paranoá groups (Table 10.1). Main modes of U-Pb ages of detrital zircons extracted from the group are 0.94, 1.16, 1.21, 1.53, 1.81, 2.08 and 2.15 Ga.

10.4.1.3 Bambuí Group

The stratigraphic succession of the Bambuí Group (Fig. 10.4) is fairly continuous along the cratonic zone. The most referenced section of the group crops out along the BR-040 highway, which connects the Belo Horizonte city on the south to Brasilia on the northwest, via the towns of João Pinheiro and Paracatu (Fig. 10.2).

According to Dardenne (1978), the Bambuí Group encompasses three shallowing-upward cycles. In the first cyle, the basal calcilutites of the Sete Lagoas Formation are covered by stromatolitic dolomites and by calcarenites. The second cycle starts with the metapelitic Serra de Santa Helena Formation, which displays intercalations of calcarenites and oolithic or stromatolitic limestones of the Lagoa do Jacaré Formation towards the top (Zalán and Romeiro-Silva 2007). The third and uppermost cycle starts with the metapelites of Serra da Saudade Formation and ends with the rhythmites and arkoses of the Três Marias Formation (Chiavegatto 1992). The feldspathic composition of the Três Marias psammites has been interpreted as indicative of syn-orogenic sedimentation, in a foredeep originated by east-vergent thrust stacking (Chiavegatto and Dardenne 1997).

Rudaceous intercalations of the Samburá and Lagoa Formosa formations have led to interpretations pointing to the syn-compressional (foreland) basin nature of all (Chang et al. 1988) or at least of the upper part of the Bambuí Group (Castro and Dardenne 2000; Valeriano et al. 2000; Rodrigues 2008). The Samburá (Fig. 10.5b) and Lagoa Formosa rudites consist of polymict conglomerates containing pebbles of the basement, older metasedimentares units involved in the SSB, abundant limestone, sandstone and mudstone clasts, which may have sourced by the upper passive margin units. Castro and Dardenne (2000) interpreted the decimetric to metric fining upward cycles (coarse conglomerates at the base, mudstone at the top) as fan-delta deposits.

The interpretation that most of the section of the Bambuí Group was deposited in a foreland basin was recently corroborated by the expressive occurrence of young ($\sim$620 Ma) detrital zircons from the uppermost portion of the Sete Lagoas Formation up to the Três Marias Formation (Rodrigues 2008; Pimentel et al. 2011; Reis et al., this book).

The age of the Bambuí Group has since long been regarded as Neoproterozoic, although the suggestion of a Cambrian age has also appeared in very early studies

Fig. 10.5 **a** View to NE of the Pimenta range, east of the town of Piumhi, where the allochthonous Canastra Group quartzites are thrust onto the foreland Bambui sequence; **b** Fan delta Samburá polymict conglomerates overlying slates and carbonatic rocks of the foreland Bambuí basin, MG-50 highway, east of Piumhi; **c** Conophyton stromatolites of the Lagamar Formation, Vazante Group, Neoproterozoic proximal passive margin; **d** Sinkhole developed within dolomitic domains of the Vazante Group; **e** Typical waterfall developed on fractured Araxá group sub-horizontal laminated quartzites of the Furnas formation in the Passos Nappe, turbidite slope deposits; **f** carbonaceous metapelites of the Garrote formation, Vazante Group, with graded bedding in drill core of Votorantim Metais

focusing this unit. The first direct Rb-Sr ages determinations performed by Bonhomme (1976) and Thomaz Filho and Bonhomme (1979) in pelitic rocks of the group indicated sedimentation (or metamorphism) around 600 Ma.

An isotopic study of the carbonatic rocks of the group (Babinski et al. 1999) indicate that Pb isotope system was disturbed in the 690–500 Ma interval, which broadly overlaps the orogenic evolution of the Brasiliano belts surrounding the southern SFC. A Pb–Pb isochronic age of 740 ± 22 Ma was presented by Babinski et al. (2007) as the sedimentation age of the basal Sete Lagoas Formation.

U-Pb ages of detrital zircons from the upper Sete Lagoas Formation presented by Rodrigues (2008) and Pimentel et al. (2011) yield an expressive mode at ~640 Ma, placing the sedimentation age of the Bambuí Group as coeval to the orogenesis in the Brasilia belt, corroborating thus the foreland basin hypothesis. More recently, detrital zircons from the Sete Lagoas Formation in the Lagoa Santa region (southern cratonic area, Santos 2012), have yielded puzzling young ages (537 ± 4 and 505 ± 7 Ma), with an expressive age mode at ca. 557 Ma, placing the sedimentation of the Bambuí Group at a much younger interval than the K-Ar cooling ages (600–580 Ma, Valeriano et al. 2000) of the external thrust sheets of the of SSB in the reagion of Piumhi town (Fig. 10.2).

10.4.2 Structure

As previously mentioned, the cratonic zone encompasses a thin-skinned foreland fold-thrust belt and a central domain, where the Neoproterozoic strata are undeformed. Regional seismic sections crossing these domains (Coelho et al. 2008; Reis 2011) reveal that the foreland fold-thrust belt involving the Bambuí Group rocks is detached over of a sub-horizontal surface. Beneath this E-verging thin-skinned fold-thrust zone, conspicuous reflectors in the autochthonous substratum have been interpreted as unconformities separating the Paleo- to Mesoproterozoic sag megasequences from a succession filling a series of grabens. In surface exposures, the slates and carbonatic rocks of the Bambuí Group in general display upright chevron and box folds, which are progressively replaced by isolated kink bands towards the central SFC (Reis and Alkmim 2015, Reis et al. this book).

10.5 The External Metamorphic Fold-Thrust Belt

An important thrust front separates the cratonic zone from the external fold-thrust belt (ETFB), in which allochtonous metasedimentary associations display greenschist facies metamorphic parageneses, locally reaching the garnet zone.

The predominant slates and phyllites of this domain present typical hilly landscapes with quartzite cuestas covered by savanna-like pastures (Fig. 10.6g).

Involvement of the basement in the E-verging deformation structures is characteristically absent in areas north of Piumhi and abundant south of this town, as demonstrated by the increasing proportion of basement thrust slices (Figs. 10.2 and 10.3).

The external fold-thrust belt is composed from base to top of the Paranoá, Canastra and Ibiá groups. With the exception of the Ibiá-Canastra contact, which is regarded as an erosional unconformity (Pereira et al. 1994; Dias et al. 2011), all units are bounded by important thrust faults.

10.5.1 Stratigraphy

10.5.1.1 Paranoá Group

The stratigraphic succession of the Paranoá Group (Campos et al. 2013) is well exposed east of the city of Brasília, along the Serra de São Domingos thrust-sheet (Alvarenga and Dardenne 1978; Alvarenga et al. 2012), where it is overlain by a 1500 m-thick succession of the Bambuí Group. Most of the Paranoá Group rocks crop out from the vicinity of Brasilia towards north. In the SSB, exposures of this unit are restricted to the core of the Cristalina dome, a prominent structure portion belt focused here (Fig. 10.1).

According to Alvarenga et al. (2012), the Paranoá Group underlies the Bambuí Group in the cratonic domain, showing a sag-like structure characterized by horizontal internal reflectors in seismic sections.

The Paranoá Group was divided by Alvarenga et al. (2012) into five units or sequences (Fig. 10.4), which consist of a basal conglomerate, followed by metasiltites, metarhythmites and quartzites, with subordinated stromatolitic dolomites and limestones. Sedimentation environments are interpreted as varying from tidal, with sporadic sub-aereal exposures, to intermediate platform environments, with frequent storm-related facies.

The only detrital zircon U-Pb ages available for the Paranoá Group were obtained from outcrops located in the Northern Brasilia belt (Matteini et al. 2012), where the group uncomformably overlies the Paleoproterozoic Araí Group and is thrust over the Canastra-Ibiá thrust sheet. The minimum and maximum ages of sedimentation (Table 10.1) are ~1.04 and 1.54 Ga, as indicated respectively by U-Pb ages of diagenetic xenotime overgrowth on detrital zircons and the youngest detrital zircon population (Matteini et al. 2012). Main modes of detrital zircon ages are: 1.54, 1.78, 2.15, 2.70, 2.85, 3.10 and 3.43, indicating a provenance from the São Francisco craton or basement of the Northern Brasilia belt.

10.5.1.2 Canastra and Ibiá Groups

The Canastra Group rocks make up most of the outcrop area of the Canastra-Ibiá thrust sheet, reaching thicknesses of ca. 2000 m, with distinct quartzite-phyllite (and minor carbonatic) lithology, leading most authors (e.g. Campos Neto 1984a, b; Pereira et al. 1994) to envisage a shelf environment for this unit.

The Paracatu section, well exposed along the BR-040 highway (Fig. 10.2), contains basal carbonaceous phyllites and sericite phyllites containing quartzite intercalations that grade up into thick quarztite successions with conglomeratic lenses. This section, starting with distal marine and ending with proximal platform and shoreface deposits, suggests a regressive evolution for the Canastra Group (Pereira et al. 1994). Discontinuous carbonatic intercalations are common in the basal zones of the Canastra-Ibiá thrust-sheet.

The contact between the Canastra and Ibiá groups is consensually interpreted as an unconformity (Barbosa et al. 1970; Pereira et al. 1994; Seer et al. 2000; Dias et al. 2011). Over this surface, the basal metadiamictites of the Cubatão Formation are capped by the homogeneous, feldspar-rich sericite phyllites of the Rio Verde Formation.

Detrital zircon U-Pb age distribution of the Canastra Group indicates a cratonic provenance, with main modes of Paleoproterozoic to Archean ages, but also containing an important contribution of 1.2–1.5 Ga Mesoproterozoic sources. The youngest zircons, dated at ca. 1.0 Ga (Valeriano et al. 2004a, b; Rodrigues 2008), suggest a Neoproterozoic sedimentation age for this group. This is corroborated by Nd whole-rock T_{dm} model ages of 1.47–2.34 Ga, provided by Pimentel et al. (2001) and Rodrigues (2008).

Fig. 10.6 Common deformational features found in the Southern Brasilia belt. **a** recumbent D2 fold and **b** stretching lineation paralel to fold hinge, in Neoproterozoic basal quartzites of the Araxá group in the northern portion of the Passos Nappe; **c** Isoclinal D2 folds in quartzitic layers of the Araxá Group, Passos Nappe near Capitólio; **d** East-verging sin-D2 thrust-related sigmoidal quartz veins in muscovite schist at Paredão waterfalls near Guapé town; **e** Slight inclined D3 fold affecting limestones of the Lagamar Formation (Vazante Group; **f** two directions of crenulation on weathered slate of the Vazante group, typical overprinting of D3 and D4 weak foreland zone deformation; **g** Canastra Group quartzite beds with anticlinal structure thrust to the east on top of Vazante Group dolomites and slates

The Ibiá Group seems to be composed of two contrasting units, in terms distribution of detrital zircon ages. The basal Cubatão Formation diamictites show similar pattern to the Canastra Group, with youngest dated zircons at 935 ± 11 (Rodrigues 2008; Rodrigues et al. (2010) and 922 ± 16 Ma (Dias et al. 2011). The upper Rio Verde Formation, on the other hand, yields a much younger detrital zircons age spectrum (Table 10.1), with the youngest grains dated at 636 ± 21 Ma (Rodrigues 2008) and at 639 ± 15 (Dias et al. 2011). The expressive detrital zircon age mode of ca.

639 Ma is interpreted as derived from the magmatic arc terranes that were approaching from the west (Dias et al. 2011), rendering the upper Ibiá Group as syn-orogenic.

10.5.2 Structure and Metamorphism

The Canastra-Ibiá thrust sheet advanced to the east, covering large areas underlain by the Paranoá, Vazante and Bambuí groups. The sole thrust of this allochthon very often truncates the structures of the underlying units, as exemplified by a region near the town of Coromandel (Fig. 10.2), where Vazante Group rocks disappear beneath an anticline of the overriding Canastra thrust sheets.

The deformation style of the Canastra and Vazante thrust sheets is typically polyphasic. Although the number and hierarchy of deformational phases vary in the descriptions of different authors (Campos Neto 1984a, b; Pinho and Dardenne 1989; Rostirolla et al. 2002), four deformational phases can be easily characterized.

The first two (D1 and D2) are syn-metamorphic and resulted in pervasive low-angle shear deformation with the development of penetrative (S1 and S2) subhorizontal foliations associated to tight to isoclinal recumbent folds. Due to the intense transposition and rotation of D1 structures by the subsequent D2 phase, it is rare to observe and unambiguously identify these early (D1) structures.

The structures imprinted by D2, the main deformation phase, developed pervasive crenulation cleavage and axial plane schistosity associated with recumbent folds (Fig. 10.6a). Fold axes are mostly sub-parallel to a conspicuous NW-plunging L2 stretching lineation (Fig. 10.6b).

The two younger deformation phases are post-metamorphic and generated faults and typically upright folds. Although a wide dispersion of orientation is observed, D3 folds axes and crenulation lineations plunge preferentially NW (Fig. 10.6f, g). The axes and crenulation lineations associated with the D4 folds plunge gently NW and NE. D3 and D4 folds in general do not show axial plane foliations, except for local spaced cleavage in thick quartzite beds. Small-scale folding is frequently chevron style, in association with kink bands (Fig. 10.6e).

10.6 The Nappe Complex

The upper nappes, much more pervasively deformed than the underlying units and exhibiting parageneses of the upper greenschist, amphibolite and (high-P) granulite facies, are mainly composed of rocks of the Araxá Group. On the west, these nappes are in tectonic contact with the Goiás magmatic arc, locally exposed in the Pontalina region and Guaxupé nappe (Fig. 10.2).

10.6.1 Stratigraphy

10.6.1.1 Araxá Group

The dominant lithotypes of the Araxá Group are in marked contrast with the previously described units. The group contains larger proportions of metapelitic schists and paragneisses, and less quartzites.

Rock associations of oceanic affinity are always present, with graphitic schists, calc-silicate rocks and minor iron and manganese formations, associated to abundant metabasic schists, amphibolites and ultramafic lenses (Strieder and Nilson 1992; Seer and Dardenne 2000; Seer et al. 2001; Valeriano et al. 2004a, b).

The Araxá Group schists and paragneisses frequently contain lenses and pods of metabasic rocks (chlorite schists and amphibolites) studied by Valeriano and Simões (1996) in the Passos Nappe. The lower strata of the nappe contain frequent high-TiO_2 rocks and the upper levels contain mostly low-TiO_2 metabasic rocks. The distribution of these two geochemical types were interpreted as indicative of progressive lithospheric thinning that took place in the distal shelf and slope environments of the São Francisco paleocontinental margin.

Lacking on unequivocal syn-sedimentary volcanic intercalations, the Araxá Group depositional ages are loosely bracketed by the youngest detrital zircons and ages of the oldest metamorphism. The maximum depositional age of the group is given by the U-Pb age of 907 $\pm$ 1 Ma of the youngest concordant detrital zircon from the Furnas Quartzite (Fig. 10.5e) in the Passos nappe (Valeriano et al. 2004a, b). The minimum age is given by the oldest metamorphism affecting the unit, dated at ca. 630 Ma (Seer et al. 2005).

Younger detrital zircons of 600 Ma have been found in some units attributed to the Araxá group, but these will be discussed below, in the context of the syn-collisional units of the SSB.

10.6.1.2 Fore-Arc and Syn-collisional Sedimentary Units

Some of the uppermost slices of the nappe stack contain abundant mica-schists, paragneisses, and even granulites, traditionally ascribed to the Araxá Group. However, the presence of detrital zircon populations of younger ages in these rocks are interpreted as evidence for a provenance from the Goiás magmatic arc (Piuzana et al. 2003a, b; Valeriano et al. 2004a, b; Rodrigues 2008; Pimentel et al. 2011), which accreted to the São Francisco continental margin between 650 and 600 Ma. The top sections of the upper metamorphic nappes ("Araxá Group") and the Rio Verde Formation of the upper Ibiá Group seem to represent syn-orogenic sedimentation in a fore arc tectonic (accretionary complex) setting. Therefore, the name Araxá Group for these successions should be discontinued.

10.6.2 Structure and Metamorphism

Contrasting with the imbricated fan geometry of the underlying tectonic domains, the nappes are associated with recumbent isoclinal folds and a regionally pervasive foliation.

The most conspicuous small-scale structure of the nappes of the internal zone of the SSB is the subhorizontal S2 foliation, associated to the L2 stretching (or mineral) lineation. In thin sections, the S2 foliation is often represented by a crenulation cleavage, generated by transposition of the S1 foliation. In some outcrops, the superimposition of D2 over D1 isoclinal folds is observed. As described in the external fold-thrust belt, the L2 stretching lineation is sub-parallel to D2 fold hinges and interpreted as indicative of tectonic transport towards east and southeast.

The upper nappe complexes truncate all underlying structures, often with a "metamorphic jump" along the sole thrust, where the phyllites of the Ibiá and Canastra groups give way to schists and paragneisses of the Araxá Group, well within the biotite zone of upper greenschist facies.

The Araxá Group display higher metamorphic grades towards the top of the rock pile. In the area around Goiânia (Fig. 10.2), Piuzana et al. (2003b) describes granulite facies schists, paragneisses and calc-silicate rocks (the Anápolis-Itauçu complex), overlying amphibolites facies rocks of the same unit. Simões (1995) described an inverted metamorphic gradient with rocks of the upper greenschist facies (biotite zone) in the lower levels of the Passos nappe, evolving upwards to garnet, staurolite and kyanite zones, reaching the transition to high-pressure granulites at the top. Further south, the Andrelândia Group in the Três Pontas nappe, broadly equivalent to the Araxá Group, is also made up of high-pressure granulitic rocks (Campos Neto and Caby 1999). Contrasting with several other Brasiliano orogenic belts (e.g., the Ribeira and Araçuaí belts), sillimanite has never been described in rocks of the Araxá Group, an indication of the high pressure of the metamorphic regime affecting this unit.

10.7 Evolutionary History

The protracted history of the passive margin represented by the Southern Brasilia belt initiates in the Tonian, with the deposition of the Vazante, Canastra and Araxá passive margin units, evolving up to the Cryogenian with the onset of the orogeny.

Based on the available U-Pb ages of detrital zircons, two sets of metasedimentary units can be discriminated in the SBB. The older successions, containing as the youngest zircons grains dated at 1.0–0.9 Ga, seem to constitute the southwestern passive margin of the São Francisco paleocontinent. The Paranoá, Vazante and Canastra groups, the Cubatão Formation (Ibiá Group) and part of the Araxá Group derive from São Francisco cratonic sources and show a minimum sedimentation age of 650–610 Ma, given by regional metamorphism associated to west-directed subduction. Younger units, such as the Bambuí Group, the Rio Verde Formation (Ibiá group) and part of the Araxá Group, with detrital zircons as young as 0.6 Ga, seem to have a mixed sedimentary provenance, both from cratonic sources and presumably from magmatic arc terranes that were colliding or at least approaching from the west, as subduction of the oceanic lithosphere of the São Francisco paleoplate progressed. Except for the Bambuí Group, these units apparently were deposited in tectonic settings such as fore-arc basins or accretionary prisms. The Bambuí Group was deposited on the lower plate in a foreland basin, as evidenced by its young detrital zircon content and increasing depth of the basement towards west.

The subduction-related main metamorphism is dated between 650 and 630 Ma, with the generation of S1 penetrative foliation and development of garnet staurolite, kyanite mica schists, as well as high-P facies in the Araxá Group rocks of the Passos Nappe, and in the Andrelândia Group rocks just below the Guaxupé nappe in the south, where kyanite/K-feldspar parageneses occur in garnet and biotite schists and paragneisses.

Late D2 exhumation of metamorphic nappes is indicated by a second generation of U-Pb ages of ca. 605 Ma on monazite (Valeriano et al. 2004a, b), which is considered coeval with retrogressive metamorphism associated to the development of S2 foliation (Simões 1995).

Post-metamorphic peak final thrusting caused telescoping of the various sectors of the original passive margin with the emplacement of distal over medium units, all of them underlying by proximal deposits. Telescoping also affected the metamorphic zones, leading to frequent metamorphic jumps along an overall inverted metamorphic gradient. Cooling and closure of biotite and white mica K-Ar systems by 600–580 Ma (Hasui and Almeida 1970; Valeriano et al. 2000) could be coeval with the exhumation of oldest foreland basin sedimentary units, such as the Samburá Rudites of the Bambuí Group.

Acknowledgments Helpful comments by Fernando Alkmim and by Monica Heilbron brought substantial improvement to an earlier version of the manuscript.

References

Alkmim F.F. &, Martins-Neto M.A. 2012. Proterozoic first-order sedimentary sequences of the São Francisco craton, eastern Brazil. Marine and Petroleum Geology 33: 12–139.

Alkmim F.F.& Marshak S. 2001. The Transamazonian orogeny in the Quadrilátero Ferrífero, Minas Gerais, Brazil: Paleoproterozoic collision and collapse in the southern São Francisco Craton region. Precambrian Research 90: 29–58.

Alkmim F.F., Fonseca M.A., Marshak S. 2001. Assembling West Gondwana during the Neoproterozoic. Clues from the São Francisco Craton region. Geology 29: 319–322.

Almeida F.F.M. 1967. Origem e evolução da Plataforma Sul-Americana. Boletim N.° 241, Departamento Nacional de Produção Mineral-Divisão de Geologia e Mineralogia, Rio de Janeiro, 36 p.

Almeida F.F.M. 1981. O Cráton do Paramirim e suas relações com o do São Francisco. In: SBG, Simpósio sobre o Cráton do São Francisco e suas faixas marginais. Salvador, 1981. Anais..., pp. 1-10.

Alvarenga C.J.S. & Dardenne M.A. 1978. Geologia dos grupos Bambuí e Paranoá na Serra de São Domingos, MG. In: Anais do 30° Congresso Brasileiro de Geologia, Recife, Sociedade Brasileira de Geologia, 2: 546–556.

Alvarenga C.J.S., Dardenne M.A., Vieira L.C., Martinho C.T., Guimarães E.M., Santos R.V., Santana R.O. 2012. Estratigrafia da borda ocidental da Bacia do São Francisco. Boletim de Geociências da Petrobras 20: 145–164.

Araujo Filho J.O. 2000. The Pirineus Syntaxis: an example of the intersection of two Brasiliano fold thrust belts in Central Brazil and its implications for the tectonic evolution of Western Gondwana. Revista Brasileira de Geociências 30: 144–148.

Ávila C.T., Teixeira W., Cordani U.G., Moura C.A.V., Pereira R.M. 2010. Rhyacian (2.23 2.20 Ga) juvenile accretion in the southern São Francisco craton, Brazil: Geochemical and isotopic evidence from the Serrinha magmatic suite, Mineiro belt. Journal of South American Earth Sciences 29: 464–482.

Azmy K., Kendall B., Creaser R.A., Heaman L., Oliveira T.F. 2008. Global correlation of the Vazante Group, São Francisco Basin, Brazil: Re-Os and U-Pb radiometric age constraints. Precambrian Research 164: 160–172.

Babinski M.,Vieira L.C., Trindade, R.I.F. 2007. Direct dating of the Sete Lagoas cap carbonate (Bambuí Group, Brazil) and implications for the Neoproteozoic glacial events. Terra Nova 19: 401–406.

Babinski, M., Van Schmus W.R., Chemale Jr. F. 1999. Pb-Pb dating and Pb isotope geochemistry of Neoproterozoic carbonate rocks from the São Francisco basin, Brazil: Implications for the mobility of Pb isotopes during tectonism and metamorphism. Chemical Geology 160: 175–199.

Barbosa O., Braun, O.P.C., Dyer R.C., Cunha C.A.B.R. 1970. Geologia da região do Triangulo Mineiro. Boletim N.° 136, Departamento Nacional de Produção Mineral-Divisão de Fomento à Produção Mineral, Rio de Janeiro, 140 p.

Bonhomme M. 1976. Mineralogie des fractions fines et datations Rubidium-Strontium dans le groupe Bambuí, MG, Brésil. Revista Brasileira de Geociências 6: 211–222.

Campos J.E.G, Dardenne M.A., Freitas-Silva F.H., Martins Ferreira M. A.C. 2013. Geologia do Grupo Paranoá na porção externa da Faixa Brasília. Revista Brasileira de Geociências 43: 461–476.

Campos Neto M.C. 1984a. Geometria e fases de dobramentos brasilianos superpostos no oeste de Minas Gerais. Revista Brasileira de Geociências 14: 60–68.

Campos Neto M.C. 1984b. Litoestratigrafia, relações estratigráficas e evolução paleogeográfica dos Grupos Canastra e Paranoá (região de Vazante-Lagamar, MG). Revista Brasileira de Geociências 14: 81–91.

Campos Neto M.C. & Caby R. 1999. Tectonic constraint on Neoproterozoic high-pressure metamorphism and nappe system south of the São Francisco Craton, southeast Brazil. Precambrian Research 97: 3–26.

Castro P.T.A. & Dardenne M.A. 2000. The sedimentology stratigraphy and tectonic context of the São Francisco Supergroup at the southwestern domain of the São Francisco Craton, Brazil. Revista Brasileira de Geociências 30: 439–441.

Chang H.K., Miranda F.P., Magalhães E., Alkmim F.F. 1988. Considerações sobre a evolução tectônica da bacia do São Francisco. In: SBG, Congresso Brasileiro de Geologia, 35, Anais, 5: 2076–2090.

Chiavegatto J.R.S. 1992. Análise estratigráfica das sequencias tempestíticas da Formação Três Marias (Proterozóico Superior), na porção meridional da Bacia do São Francisco. PhD Thesis, Universidade Federal de Ouro Preto, 196p.

Chiavegatto J.R.S., Dardenne M.A. 1997. Contribuição à sedimentologia e estratigrafia do Grupo Bambuí no norte de Minas Gerais. In: Simpósio de Geologia de Minas Gerais, 9, Anais do..., Belo Horizonte, pp. 81–82.

Cloud P. and Dardenne M.A. 1973. Proterozoic age of the Bambui Group in Brazil. Geological Society of America Bulletin 84: 1673–1676.

Coelho J.C.C., Martins-Neto M.A., Marinho M.S. 2008. Estilos estruturais e evolução tectônica da porção mineira da bacia proterozoica do São Francisco. Revista Brasileira de Geociências 38: 149–165.

Dardenne M.A. 1978. Síntese sobre a estratigrafia do Grupo Bambuí no Brasil Central. Anais do XXX Congresso Brasileiro de Geologia. Recife, Sociedade Brasileira de Geologia, 2: 597–602.

Dardenne M.A. 2000. The Brasília Fold Belt. In: Cordani U.G., Milani E.J., Thomaz Filho A., Campos D.A. (Eds.), Proceedings of the 31st International Geological Congress on the Tectonic Evolution of South America. Rio de Janeiro, pp. 231–263.

Dardenne M.A. 2005. Conophytons de Cabeludo, Grupo Vazante (MG) - Construções dolomíticas por ciano-bactérias no Proterozóico. In: Winge, M., Schobbenhaus, C., Berbert-Born M., Queiroz E.T., Campos D.A., Souza, C.R.G. (Eds.) Sítios Geológicos e Paleontológicos do Brasil. Published August 3 2005 at http://www.unb.br/ig/sigep/sitio073/sitio073.pdf.

Dardenne M.A., Faria A., Andrade G.F. 1976. Ocurrence de stromatolithes columnaires dans le Groupe Bambuí (Goiás, Brésil). Anais da Academia Brasileira de Ciências 48: 555–566.

Dardenne M.A., Freitas-Silva F.H., Nogueira G.M.S., Souza J.F.C. 1997. Depósitos de fosfato de Rocinha e Lagamar, Minas Gerais. In: Schobbenhaus C., Queiroz E.T., Coelho, C.E.S. (Eds.) Principais depósitos minerais do Brasil, DNPM/CPRM, v. IV C, pp. 113–122.

Dardenne M.A., Campos J.E.G., Campos Neto M.C. 2009. Estromatólitos colunares no Sumidouro do Córrego Carrapato, Lagamar (MG) - Registro de construções dolomíticas cilindro-cônicas por ciano-bactérias no Proterozóico do Brasil. In: Winge,M.; Schobbenhaus,C.; Berbert-Born,M.; Queiroz,E.T.; Campos,D.A.; Souza,C.R. G. ; Fernandes,A.C.S. (Eds.) Sítios Geológicos e Paleontológicos do Brasil. 1st Ed. Rio de Janeiro, CPRM- CEDOC, 2: 311–320.

Dias P.H.A., Noce C.M., Pedrosa-Soares A.C., Seer H.J., Dussin I.A., Valeriano C. M., Kuchenbecker M. 2011. O Grupo Ibiá (Faixa Brasília Meridional): evidências isotópicas Sm-Nd e U-Pb de bacia colisional tipo flysch. Geonomos 19: 90–99.

Fonseca M.A., Dardenne M.A., Uhlein A. 1995. Faixa Brasília Setor Setentrional: estilos estruturais e arcabouço tectônico. Revista Brasileira de Geociências 25: 267–278.

Fuck R.A., Jardim de Sa E.F., Pimentel M.M., Dardenne M.A., Pedrosa Soares A.C. 1993. As faixas de dobramentos marginais do Craton do São Francisco. In: J.M.L. Dominguez & A. Misi (eds.) O Craton

do São Francisco. Salvador, Sociedade Brasileira de Geologia, pp. 161–185.
Hasui Y. & Almeida F.F.M. 1970. Geocronologia do centro-oeste brasileiro. Boletim da Sociedade Brasileira de Geologia 19: 5-26.
Jost H., Fuck R.A., Dantas E.L., Rancan C.C., Rezende D.B., Santos E., Portela J.F., Mattos L., Chiarini M.F.N., Oliveira R.C., Silva S.E. 2005. Geologia e geocronologia do Complexo Uvá, Bloco Arqueano de Goiás. Revista Brasileira de Geociências 35: 559–572.
Klein P.B.W. 2008. Geoquímica de Rocha Total, Geocronologia de U-Pb e Geologia Isotópica de Sm-Nd das Rochas Ortognáissicas e Unidades Litológicas Associadas da Região Ipameri Catalão (Goiás). PhD Thesis, Universidade de Brasília, 154 p.
Laux J.H., Pimentel M.M., Dantas E.L., Armstrong R.A., Junges S.L. 2005. Two Neoproterozoic crustal accretion events in the Brasília belt, central Brazil. Journal of South American Earth Sciences 18: 183–198.
Lesquer A., Almeida F.F.M., Davino A., Lachaud J.C., Mailard P. 1981. Signification structurale des anomalies gravimétriques de la partie sud du craton de São Francisco (Brésil). Tectonophysics, 76: 273–293.
Mantovani M.S.M., Quintas M.C.L., Shukowsky W., Brito Neves B.B. 2005. Delimitation of the Paranapanema Proterozoic block: a geophysical contribution. Episodes 28: 18–22.
Matteini M., Dantas E.L., Pimentel M.M., Alvarenga C.J.S., Dardenne M.A. 2012. U–Pb and Hf isotope study on detrital zircons from the Paranoá Group, Brasília Belt Brazil: Constraints on depositional age at Mesoproterozoic – Neoproterozoic transition and tectono-magmatic events in the São Francisco craton. Precambrian Research 206: 168–181.
Moeri E. 1972. On a columnar stromatolite in the Precambrian Bambui Group of Central Brazil, Eclog Geol Helv, 65: 185–195.
Pereira L.F., Dardenne, M.A., Rosiére, C.A., Pedrosa-Soares A.C. 1994. Evolução geológica dos Grupos Canastra e Ibiá na região entre Coromandel e Guarda-Mor. Geonomos 2: 22–32.
Pimentel M.M., Fuck R.A., 1992. Neoproterozoic crustal accretion in central Brazil. Geology 20: 375–379.
Pimentel, M.M., Dardenne, M.A., Fuck, R.A., Viana M.G., Junges S.L., Fischel D.P., Seer H.J., Dantas E.L. 2001. Nd isotopes and the provenance of detrital sediments of the Neoproterozoic Brasília Belt, central Brazil. Journal of South American Earth Sciences 14: 571–585.
Pimentel M.M., Fuck R.A., Jost H., Ferreira Filho C.F., Araujo S.M. 2000. The basement of the Brasília Fold Belt and the Goiás Magmatic Arc. In: U.G. Cordani, E.J. Milani, A. T. Filho, D.A. Campos (eds): Tectonic evolution of South America, 31st International Geological Congress, Rio de Janeiro, pp. 195–230.
Pimentel M.M., Jost H., Fuck R.A. 2004, O Embasamento da Faixa Brasília e o Arco Magmático de Goiás. In: V. Mantesso Neto, A. Bartorelli, C.D.R. Carneiro, B.B. Brito Neves (Organizadores), Geologia do Continente Sul-Americano – Evolução da obra de Fernando Flávio Marques de Almeida. Beca Ed, pp 355–368.
Pimentel M.M., Rodrigues J.B., Della Giustina M.E.S., Junges S.; Matteini M., Armstrong R. 2011. The tectonic evolution of the Neoproterozoic Brasília Belt, central Brazil, based on SHRIMP and LA-ICPMS U Pb sedimentary provenance data: A review. Journal of South American Earth Sciences 31: 345–357.
Pinho J.M.M. & Dardenne M.A. 1989. Deformação no Grupo Canastra na Região de Coromandel, MG. In: SBG, Simposio Nacional de Estudos Tectônicos, 4, Anais, p. 12.
Piuzana D., Pimentel M.M., Fuck R.A., Armstrong R.A. 2003a. SHRIMP U Pb and Sm Nd data for the Araxá Group and associated magmatic rocks: constraints for the age of sedimentation and geodynamic context of the southern Brasília Belt, central Brazil. Precambrian Research 125: 139–160.
Piuzana D., Pimentel M.M., Fuck R.A., Armstrong R.A. 2003b. Neoproterozoic magmatism and high-grade metamorphism in the Brasília Belt, central Brazil: regional implications of SHRIMP U-Pb and Sm-Nd geochronological studies. Precambrian Research 125: 245–273.
Reis H.L.S. 2011. Estratigrafia e tectônica da Bacia do São Francisco na zona de emanações de gás natural do Baixo Indaiá (MG). Ms Thesis Universidade Federal de Ouro Preto, 127 p.
Reis H.L.S. & Alkmim F.F. 2015. Anatomy of a basin-controlled foreland fold-thrust belt curve: the Três Marias salient, São Francisco basin, Brazil. Marine and Petroleum Geology 66: 711–731.
Rodrigues J. B. 2008. Proveniência de sedimentos dos grupos Canastra, Ibiá, Vazante e Bambuí – Um estudo de zircões detríticos e Idades Modelo Sm-Nd. PhD Thesis, Universidade de Brasília, 128p.
Rodrigues J.B., Pimentel M.M, Dardenne M.A, Armstrong R.A. 2010. Age, provenance and tectonic setting of the Canastra and Ibiá Groups (Brasília Belt, Brazil): Implications for the age of a Neoproterozoic glacial event in central Brazil. Journal of South American Earth Sciences 29: 512–521.
Rodrigues J.B., Pimentel M.M., Buhn B., Matteiini M., Dardenne M. A., Alvarenga C.J.S., Armstrong R.A. 2012. Provenance of the Vazante Group: New U-Pb, Sm-Nd, Lu-Hf isotopic data and implications for the tectonic evolution of the Neoproterozoic Brasília Belt. Gondwana Research 21: 439–450.
Rostirolla S.P., Mancini F., Reis Neto J.M., Figueira E.G., Araújo E.C. 2002. Análise estrutural da mina de Vazante e adjacências: Geometria, cinemática e implicações para a hidrogeologia. Revista Brasileira de Geociências 32: 59–68.
Santana A.V.A. 2011. Estratigrafia, sedimentologia e proveniência das unidades superiores do Grupo Vazante na região da fazenda Fagundes, Paracatu, MG. Ms. Dissertation, Instituto de Geociências, Universidade de Brasília, Brasília, 160 p.
Santos G.M.P. 2012. Quimioestratigrafia isotópica (C,O,Sr) e geocronologia das rochas da Formação Sete Lagoas, Grupo Bambuí. PhD Thesis, Universidade de São Paulo. 132 p.
Seer H.J. & Dardenne M.A. 2000. Tectonostratigraphic terrane analysis on neoproterozoic times: the case study of Araxa Synform, Minas Gerais State, Brazil; implications to the final collage of the Gondwanaland. Revista Brasileira de Geociências 30: 78–81.
Seer H.J., Dardenne M.A., Pimentel M.M., Fonseca M.A., Moraes L.C. 2000. O Grupo Ibiá na sinforma de Araxá, um terreno tectonoestratigráfico ligado à evolução de arcos magmáticos. Revista Brasileira de Geociências 30:734–744.
Seer H.J., Brod J.A., Valeriano C.M., Fuck R.A. 2005. Leucogranitos intrusivos no Grupo Araxá: registro de um evento magmático durante colisão Neoproterozóica na porção meridional da Faixa Brasília. Revista Brasileira de Geociências 35: 33–42.
Seer H.J., Brod J.A., Fuck R.A., Pimentel M.M., Boaventura G.R.; Dardenne M.A. 2001. Grupo Araxá em sua área tipo: um fragmento de crosta oceânica Neoproterozóica na Faixa de Dobramentos Brasília. Revista Brasileira de Geociências 31:389–400.
Simões L.S.A. 1995. Evolução tectonometamórfica da nappe de Passos, sudoeste de Minas Gerais. Unpubl. Doct. Thesis, Universidade de São Paulo, 149 p.
Strieder A.J. & Nilson A.A. 1992. Melange ofiolítica nos metassedimentos Araxá de Abadiania (GO) e implicações tectônicas regionais. Revista Brasileira de Geociências 22: 204–215.
Teixeira, W. Ávila C.A., Dussin I.A., Correa Neto, A.V., Bongiolo E. M. Santos J.O.S., Barbosa N. 2015. A juvenile accretion episode (2.35–2.32 Ga) in the Mineiro belt and its role to the Minas accretionary orogeny: zircon U-Pb-Hf and geochemical evidences. Precambrian Research 256: 148-169.
Thomaz Filho A. & Bonhomme M. 1979. Datations isotopiques Rb/Sr et K/Ar dans le Groupe Bambuí, a São Francisco (M.G.), au Brésil. C. R. Acad. Sc. Paris, t. 289, Série D, pp. 1221–1224.
Ussami N. 1993. Estudos Geofísicos no Cráton do São Francisco: Estágio atual e Perspectivas. In: Dominguez, J.M.L. & Misi, A, O

Cráton do São Francisco, Sociedade Brasileira de Geologia, Salvador, Bahia, pp. 35–43.

Valeriano C.M. & Simões L.S.A. 1996. Geochemistry of Proterozoic mafic rocks from the Passos Nappes (Minas.Gerais, Brazil): tectonic implications to the evolution of the Southern Brasilia belt. Revista Brasileira de Geociências 27: 99–110.

Valeriano C.M., Simões L.S.A., Teixeira W., Heilbron M. 2000. Southern Brasilia belt (SE Brazil): tectonic discontinuities, K-Ar data and evolution during the Neoproterozoic Brasiliano orogeny. Revista Brasileira de Geociências 30:195–199.

Valeriano C.M., Machado N., Simonetti A., Valladares C.S., Seer, H.J., Simões L.S.A. 2004a. U-Pb geochronology of the southern Brasilia belt(SE-Brazil): sedimentary provenance, Neoproterozoic orogeny and assembly of West-Gondwana. Precambrian Research 130: 27–55.

Valeriano C.M., Dardenne M. A., Fonseca M.A., Simões L.S.A., Seer H.J. 2004b. A Evolução Tectônica da Faixa Brasília. In: V. Mantesso Neto, A. Bartorelli, C.D.R. Carneiro, B.B. Brito Neves (Organizadores), Geologia do Continente Sul-Americano – Evolução da obra de Fernando Flávio Marques de Almeida. Beca Ed, pp 575–592.

Valeriano, C.M., Pimentel M.M., Heilbron M., Almeida J.C.H., Trouw R.A.J. 2008. Tectonic evolution of the Brasília Belt, Central Brazil, and early assembly of Gondwana. In: Pankhurst, R.J., Trouw, R.A.J., Brito Neves, B.B. & de Wit, M.J. (eds) West Gondwana: Pre-Cenozoic Correlations Across the South Atlantic Region. Geological Society, London, Special Publications 294: 197–210.

Zalán P.V., Romeiro-Silva P.C. 2007. Bacia do São Francisco. Boletim de Geociências da Petrobras 15: 561–571.

11 The Northern Brasília Belt

Reinhardt A. Fuck, Márcio M. Pimentel, Carlos J.S. Alvarenga, and Elton L. Dantas

Abstract

The Brasília belt is a well-preserved Neoproterozoic orogenic belt within the Tocantins Province, central Brazil. Its northern segment strikes N–S, and verges eastwards to the São Francisco craton. The northern Brasília belt external zone is a fold-thrust belt of low-grade passive margin metasedimentary rocks as well as syn-orogenic sedimentary sequences related to a magmatic arc. The internal zone includes deep sea sediments associated to an ophiolite mélange, arc-type calc-alkaline volcanics and intrusives, and S-type collisional granites. Metamorphic grade increases westwards from non-metamorphic and low-grade rocks, in the east, to high-temperature amphibolite facies rocks, culminating in ultrahigh-T granulites in the metamorphic core. The belt is the result of convergence of the Amazonian, São Francisco and Paranapanema paleocontinents and involved the subduction of a wide oceanic lithosphere and development of primitive island arc systems. Convergence also entrapped the Goiás massif, an exotic continental fragment, comprising Archean granite-greenstone terrains, Paleoproterozoic granite-gneiss, Neoproterozoic layered complexes, and Proterozoic cover rocks, exposed between the magmatic arc and the metamorphic core in the central part of the belt, and the external zone farther north.

Keywords

Northen Brasília belt • Tocantins Province • Neoproterozoic • Magmatic arc

11.1 Introduction

The Tocantins Province (Almeida et al. 1981), in central Brazil, borders the western margin of the São Francisco craton. The province is a large Neoproterozoic orogen that resulted from convergence and collision of the Amazonian and São Francisco paleocontinents. Final convergence and collision took place during the Brasiliano orogeny in the late Neoproterozoic, as part of West Gondwana amalgamation. Phanerozoic sedimentary rocks of the Parnaíba and Paraná basins cover the province on the north and south, respectively (Fig. 11.1). These basins also cover older stable continental blocks, such as the so-called Parnaíba and Paranapanema blocks, both of which also involved in the Brasiliano orogeny.

The province comprises three fold belts: Brasília, Araguaia and Paraguay. The Araguaia and Paraguay belts formed along the eastern and southeastern margins of the Amazonian craton, respectively, whereas the Brasília belt (Figs. 11.1 and 11.2), developed along the western and southern margins of the São Francisco craton.

The Brasília belt extends for more than 1100 km roughly in the N–S direction, and may be structurally subdivided into three segments. The northern segment of the belt, focus of the present chapter, strikes NE and verges eastwards to the São Francisco craton. In the central part of the belt, south of the Pireneus syntaxis (Ferreira Filho 2000), the structural grain turns to NW–SE, and the main vergence of the nappe stack is to the ESE. Further to the south, the belt wraps around the southern tip of the São Francisco craton, and verges mostly eastwards (Fig. 11.2).

R.A. Fuck (✉) · M.M. Pimentel · C.J.S. Alvarenga · E.L. Dantas
Instituto de Geociências, Universidade de Brasília, Brasília, DF, Brazil
e-mail: reinhardt@unb.br

M. Heilbron et al. (eds.), *São Francisco Craton, Eastern Brazil*, Regional Geology Reviews, DOI 10.1007/978-3-319-01715-0_11

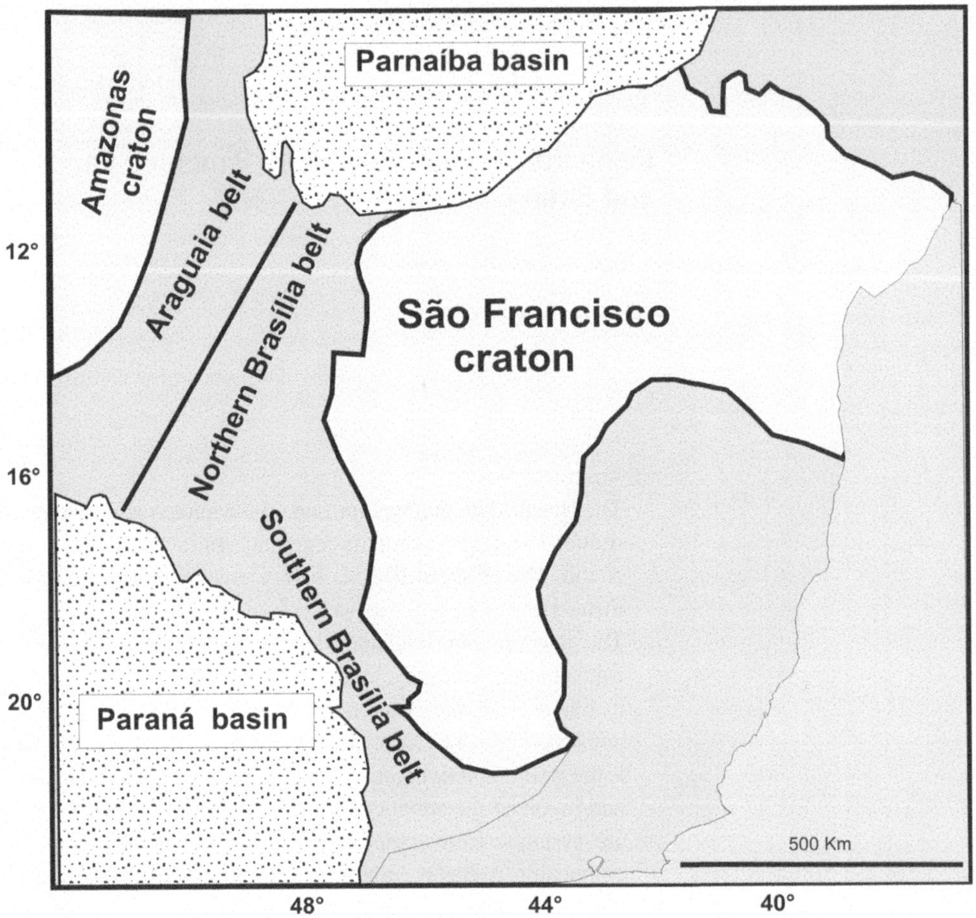

Fig. 11.1 The northern segment of the Brasília belt in context of the Tocantins Province

The Brasília belt is a well preserved orogenic belt, comprising a thick package of passive margin sedimentary deposits, younger deep sea sediments associated to volcanic arc rocks as well as an ophiolite mélange, arc-type calc-alkaline volcanics and intrusives, and S-type collisional granites. Metamorphic grade increases westwards from non-metamorphic and low-grade rocks, at the cratonic border in the east, to high-temperature amphibolite facies rocks, culminating in ultrahigh-T granulites in the metamorphic core, and decreases again westwards to amphibolite and greenschist facies within the Goiás Arc. Structurally, the Brasília belt is a fold-thrust belt that verges towards the São Francisco craton.

This chapter aims to provide the reader with an overview of tectonic zoning and stratigraphic framework of the northern Brasilia belt. After presenting the various tectonic compartments of belt, we discuss in more detail their stratigraphic assemblages and paleogeographic significance. We conclude with an evolutionary synthesis.

11.2 Tectono-Stratigraphic Zoning

Distinct petrotectonic associations are recognized the northern Brasília belt and adjoining craton margin, allowing their subdivision into three main compartments, which from east to west are: the cratonic, external and internal zones (Pimentel et al. 1999, 2000, 2004; Dardenne 2000; Valeriano et al. 2004, 2008) (Fig. 11.2).

The cratonic zone encompasses the portion of the craton covered by Neoproterozoic strata and currently referred to as the São Francisco basin. In areas adjacent to the craton boundary, the Neoproterozoic fill units of the basin were captured by the Neoproterozoic deformation front. The external zone is a thick-skinned fold-thrust belt, which besides the basement also involves low-grade Paleo- to Neoproterozoic metasedimentary rocks. These units are normally in tectonic contact with the Neoproterozoic cratonic cover, mainly through reverse faults. The internal zone is made up of three distinct components: the metamorphic core, the Goiás massif and the Goiás magmatic arc.

11.3 The Cratonic Zone

The cratonic zone encompasses two tectonic domains, currently viewed as sectors of the São Francisco basin (see Reis et al. this book): an eastern domain, in which the Neoproterozoic units are undeformed, and a thin-skinned foreland fold-thrust belt developed along the craton boundary.

Fig. 11.2 Geological map of the Brasília belt

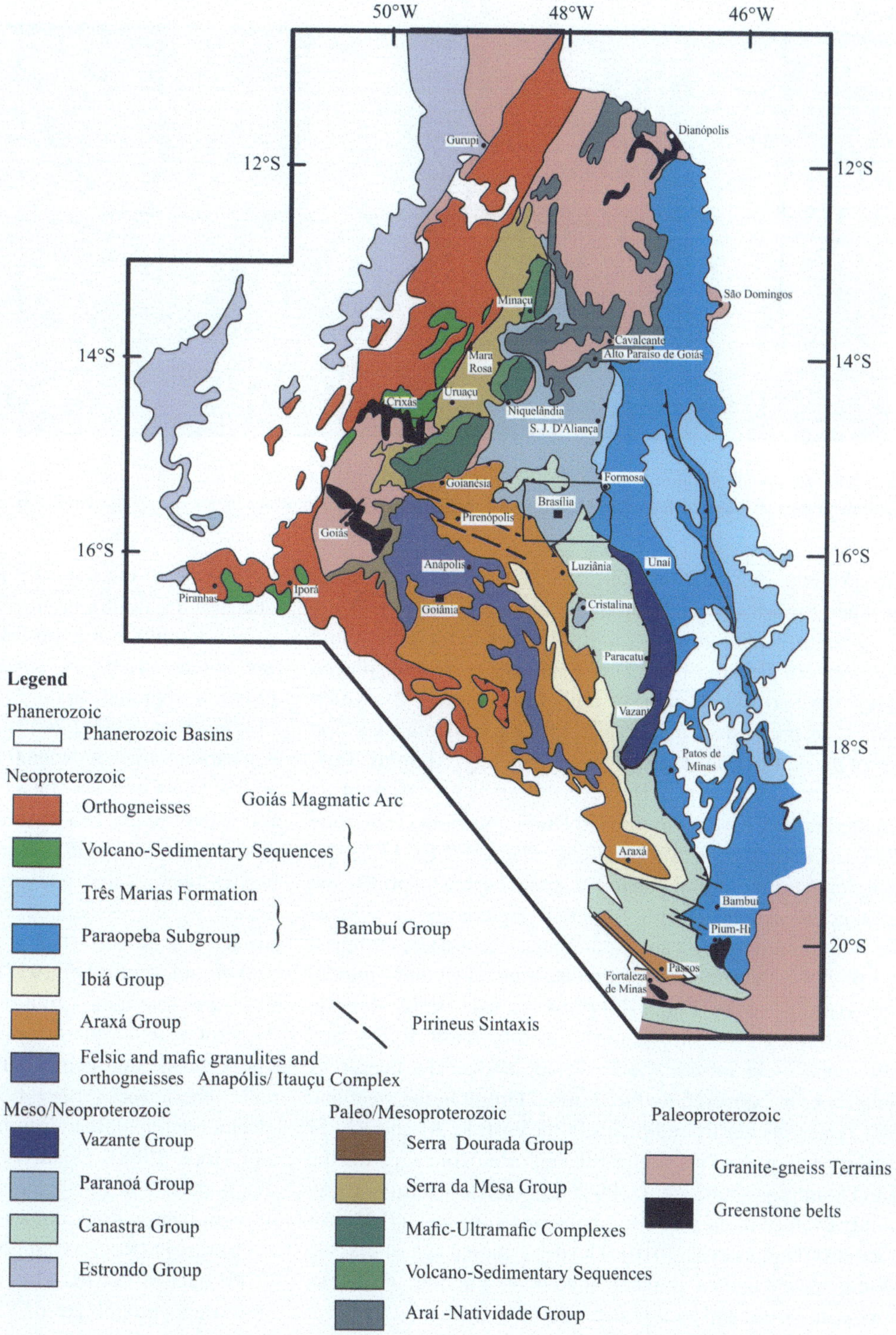

The undeformed domain corresponds to the central portion of the São Francisco basin, bounded on the west by the São Domingos fault. Local deformation is recognized only close to the fault.

The thin-skinned foreland fold-thrust belt occupies the area located between the São Domingos fault (Dardenne 2000) and the major thrust that brings the Mesoproterozoic Paranoá Group on top of the Neoproterozoic strata of the São Francisco basin (Fig. 11.2). The São Domingos fault is a 1-km wide reverse fault zone, marked by a whole series of subsidiary faults, which ramp up and exposes different stratigraphic units at the surface. A seismic section of the southern portion of the belt shows that this fault zone corresponds to frontal ramp of the basal detachment of the

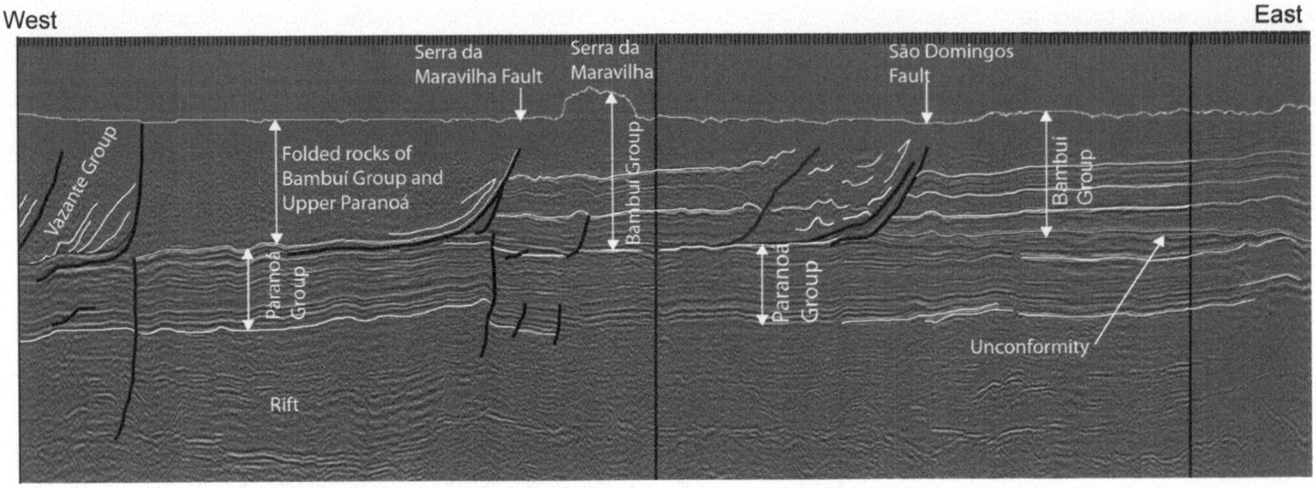

Fig. 11.3 Regional seismic profile across the external Brasília belt and adjacent cratonic zone showing deformed layer of Upper Paranoá and Bambuí groups dethatched from the Middle and Lower Paranoá Group strata (from Alvarenga et al. 2012)

thin-skinned deformation zone, located along the contact between the Meso and Neoproterozoic units (Fig. 11.3).

A controversial point is the limit between the craton and the northern Brasília belt. This boundary is usually placed along the Serra Geral reverse fault zone (Fig. 11.2), which places, as previously mentioned, Mesoproterozoic rocks on top of the deformed Neoproterozoic strata (e.g. Alkmim 2004, and references therein). However, Brasiliano deformation reached farther east, up to the São Domingos fault (Dardenne 2000), apparently not involving the basement (Fig. 11.3). An NS-elongated structural window exposes basement ortho- and paragneiss associated with low-grade metamorphic rocks of the São Domingos volcanic-sedimentary sequence, comprising phyllites, biotite schist, carbonaceous schist and bimodal metavolcanic rocks (Fig. 11.2) (Teixeira et al. 1982; Rendon Dávila 2002). Tonalite, granite, gabbro, mafic dykes and ultramafic rocks intrude the low-grade rocks. The metamorphic rocks are covered by the Bambuí Group in the west and by Cretaceous clastic sediments in the east. K–Ar and Rb–Sr isotope determinations suggest Paleoproterozoic ages for the São Domingos sequence. U–Pb zircon data of intrusive rocks yielded values of ca. 590 Ma for a granite and ca. 540 Ma for a gabbro (Dantas et al. 2010). The depositional age of the São Domingos sequence remains unknown, but the intrusions appear to be related to a transtensional–extensional event at the end of the Neoproterozoic and Early Cambrian.

11.3.1 Stratigraphic Assemblages

The exposures in the cratonic zone are frankly dominated by the Neoproterozoic São Francisco megasequence, including glaciogenic deposits of the Jequitaí Formation and pelite-carbonate rocks of the Bambuí Group, which correspond to the main fill units of the São Francisco basin (see Reis et al. this book) (Fig. 11.4).

The Jequitaí Formation is a succession of diamictites with rare intercalations of sandstone and siltstone, deposited above a regional unconformity (Fig. 11.4, Walde 1976; Uhlein et al. 1999, 2011). The thickness of the Jequitaí Formation varies between 0 and 20 m in the northern portion of the São Francisco basin (Alvarenga et al. 2007, 2014), and it may reach 180 m in the southern portion of the basin (Cukrov et al. 2005). The glacial origin of the Jequitaí Formation was first recognized ca. 100 years ago and confirmed by the associated occurrence of striated clasts and striated pavements (Isotta et al. 1969; Walde 1976; Karfunkel and Karfunkel 1977). The diamictite is mostly massive and contains granite, gneiss, quartzite, limestone, dolostone and quartz clasts in a matrix of silt and fine-grained sand cemented by diagenetic carbonate. Originally, these rocks have been interpreted as deposits of a glacial continental environment (Isotta et al. 1969; Karfunkel and Hope 1988). However the scarcity of clasts, stratified diamictites, and fine-grained intercalations and the absence of significantly thick out-wash facies indicate a glaciomarine environment (Rocha Campos et al. 1996; Uhlein et al. 1999). U–Pb data of detrital zircon in diamictites suggest a maximum depositional age of 850 Ma (Pimentel et al. 2011).

The Bambuí Group comprises five formations (Fig. 11.4): Sete Lagoas, Serra de Santa Helena, Lagoa do Jacaré, Serra da Saudade and Três Marias. Three transgressive/regressive cycles were described along the stratigraphic succession of the Bambuí Group (Dardenne et al. 1978 and references therein). The lower Sete Lagoas Formation consists of dolostone, limestone and claystone and includes boundary sequences in its upper portion, which mark an unconformity defined by dolostones with a

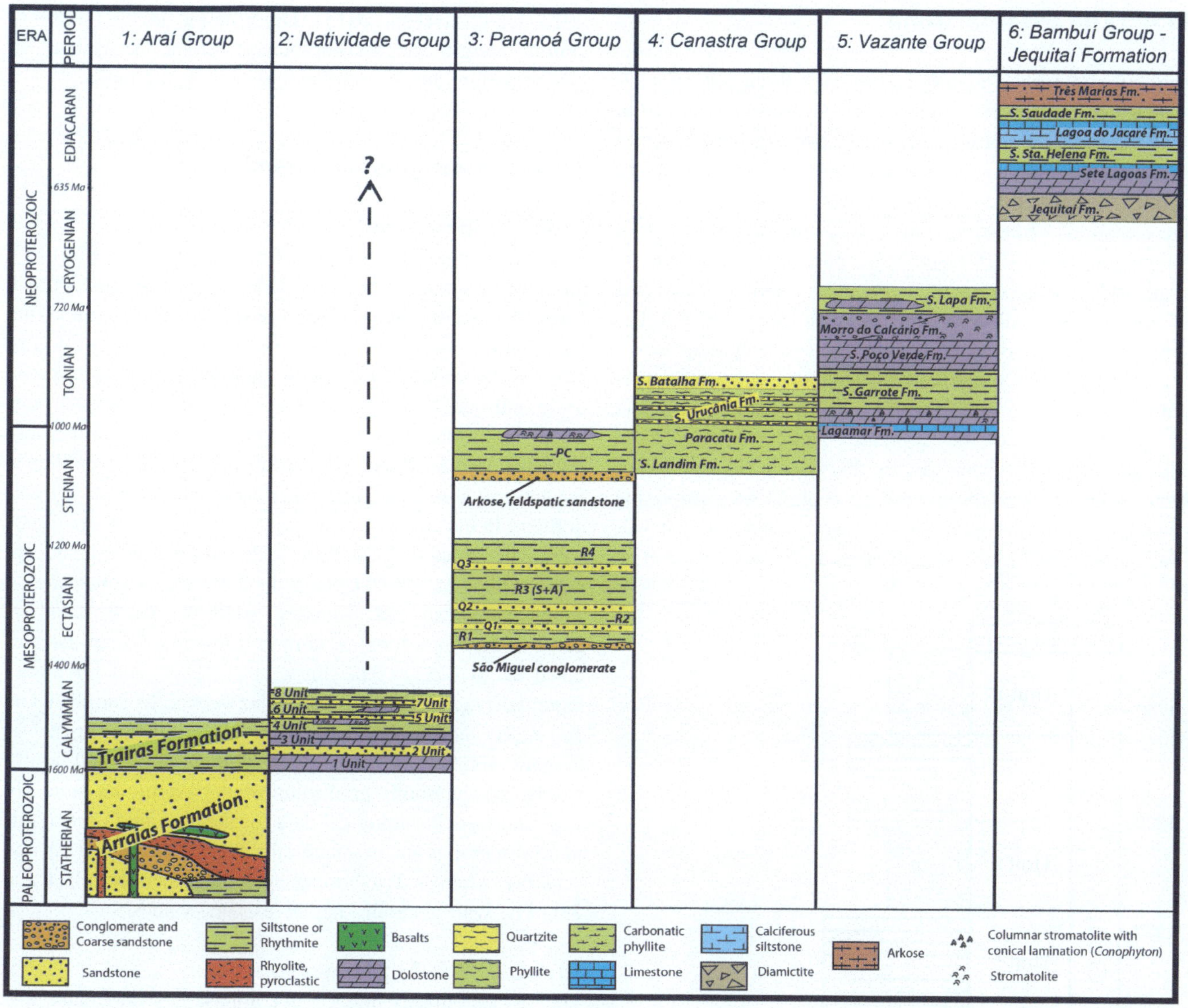

Fig. 11.4 Comparative stratigraphic columns of the different supracrustal units of the external Brasília belt

paleokarst horizon exposed at the border of the basin (Martins and Lemos 2007; Vieira et al. 2007). This unconformity has not been identified in the inner parts of the basin. The Serra de Santa Helena Formation consists of mudstone with thin layers of fine-grained sandstone beds and lenses. The upper carbonate succession of the Bambuí Group is composed of claystone, marl, siltstone, and lenses and layers of dark gray limestone of the Lagoa do Jacaré Formation. The assemblage of the three lower formations of the Bambuí Group has been referred to as the Paraopeba Subgroup. The uppermost succession of the Bambuí Group is composed of siliciclastic mudstones of the Serra da Saudade Formation, followed by mudstone intercalated with layers of arkosic sandstone of the Três Marias Formation, deposited on a shallow platform dominated by storm currents with locally tidal facies. The Bambuí Group is interpreted as foreland-type basin deposits, accumulated at the front of the Brasília mountain chain.

The lower portion of the Sete Lagoas Formation has an estimated age of 740 ± 22 Ma based on Pb–Pb isochron data in dolostone (Babinski et al. 2007; Vieira et al. 2007). In contrast, the upper portion of this formation has a maximum age of ca. 620 Ma, or perhaps even younger than that, based on detrital zircon U–Pb data (Pimentel et al. 2011). These two ages indicate an interval larger than 100 Ma, in which a positive excursion of $\delta^{13}C$ values from carbonates is observed. This excursion, marked by values greater than +7.5 ‰, is located in the upper portion of the Sete Lagoas Formation, near the previously mentioned sequence boundary (Santos et al. 2004; Martins and Lemos 2007; Alvarenga et al. 2007, 2014).

11.4 The External Zone

The external zone is mainly formed of a thick pile of sedimentary sequences deformed against the western margin of the São Francisco craton, as well as rather large exposures of their sialic basement. The external zone corresponds thus to a thick-skinned fold-thrust belt, which includes a thick succession of sedimentary and metasedimentary rocks of Proterozoic age (1750–600 Ma), partially equivalent to stratigraphic units occurring in the interior and other margins of the craton (Alkmim and Martins Neto 2012). The supracrustal succession is divided into two megasequences: (1) Paleo-Mesoproterozoic rift-sag megasequence represented by the Araí and Natividade groups (Figs. 11.5 and 11.6), correlatives of the Espinhaço Supergroup (see Cruz and Alkmim, and Alkmim et al., this book); (2) late Mesoproterozoic Paranoá megasequence (Fig. 11.4), comprising siliciclastic deposits with dolomite and limestone intervals (Fuck et al. 1988; Martins and Lemos 2007; Alvarenga et al. 2012; Campos et al. 2013). Deformation in this sector of craton margin is accommodated along a system of NNE-trending and E-verging folds and thrust faults.

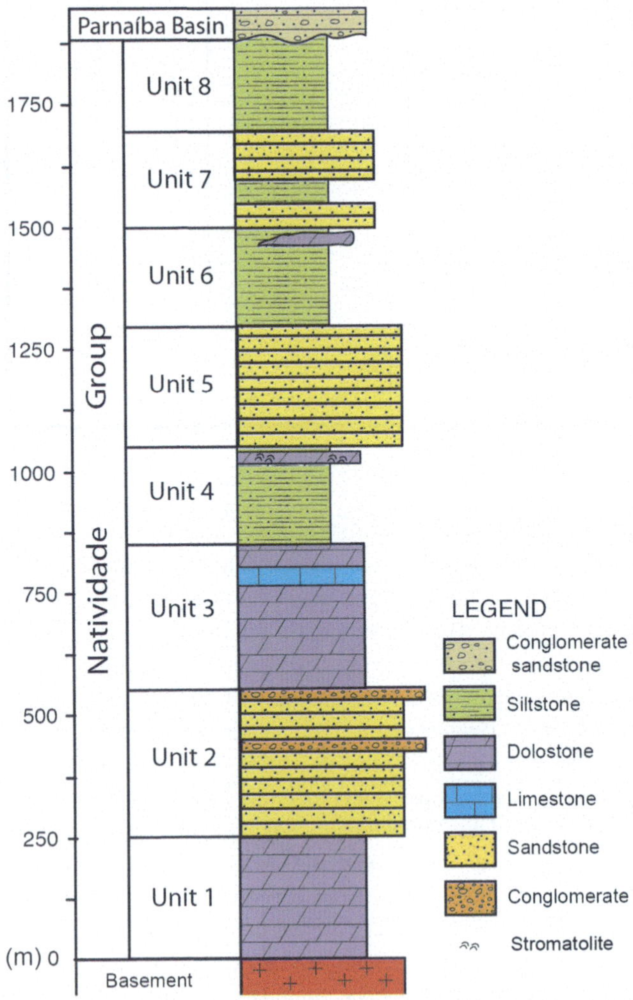

Fig. 11.5 Stratigraphic column of the Natividade Group (modified from Dardenne and Saboia 2007)

11.4.1 Stratigraphic Assemblages

11.4.1.1 Basement

The basement of the northern Brasília belt is exposed in a large area known as the Natividade-Cavalcante crustal block (Fig. 11.6). This is a Paleoproterozoic terrain comprising two main domains, the limits of which are not well constrained due to the lack of geological maps at an appropriate scale and geochronological data. Narrow supracrustal belts wrapping around gneiss domes characterize the northeastern domain. At least three rock-forming events are recorded in the calc-alkaline, metaluminous basement gneisses of this domain. In the Almas-Dianópolis area (Fig. 11.6), granitoids were dated at ca. 2.2 Ga. Early Paleoproterozoic magmatism was dated at ca. 2.3–2.4 Ga in the Natividade-Conceição do Tocantins area, and younger additions were dated at 2144 ± 21 Ma; a tonalite sample from close to Arraias was dated at 2042 ± 12 Ma (Fuck et al. 2014, and references therein). T_{DM} model ages range between 2.24 and 3.11 Ga, and $\varepsilon_{Nd}(t)$ values range from +2 to close to zero, suggesting juvenile sources. Bordering this domain west- and southwards, peraluminous granitoids of the Aurumina suite are exposed. Geochronological data for the Aurumina granitoids of this sector point to magmatic events between 2.13 and 2.18 Ga, whereas T_{DM} model ages range between 2.21 and 2.92 Ga, with generally negative $\varepsilon_{Nd}(t)$ values.

11.4.1.2 Rift-Sag Megasequence

The rift-sag megasequence encompasses sedimentary and volcanic rocks of the Araí and Natividade groups (Figs. 11.4 and 11.5).

The Araí Group deposits may be subdivided into four stages: pre-rift, sin-rift, transitional and flexural stages (Alvarenga et al. 2007; Marques 2010; Tanizaki et al. 2015). The pre-rift stage is composed of polymictic conglomerate, aeolian quartzite and oligomictic conglomerate. Pyroclastic rocks of rhyolite composition and basalt flows represent the syn-rift stage. This continental sedimentary sequence and associated volcanic rocks correspond to the Arraias Formation. The transitional stage, representing an association of continental and marine rocks, comprises quartzite, oligomictic conglomerate and locally graywacke, followed by a succession of phyllite, quartzite and some oligomictic conglomerate included in the Traíras Formation. The flexural stage, corresponding to the upper Traíras Formation, consists of carbonate and siliciclastic sediments formed in a marine shelf environment, represented by two fining-upward depositional sequences. Rocks of the flexural stage may be

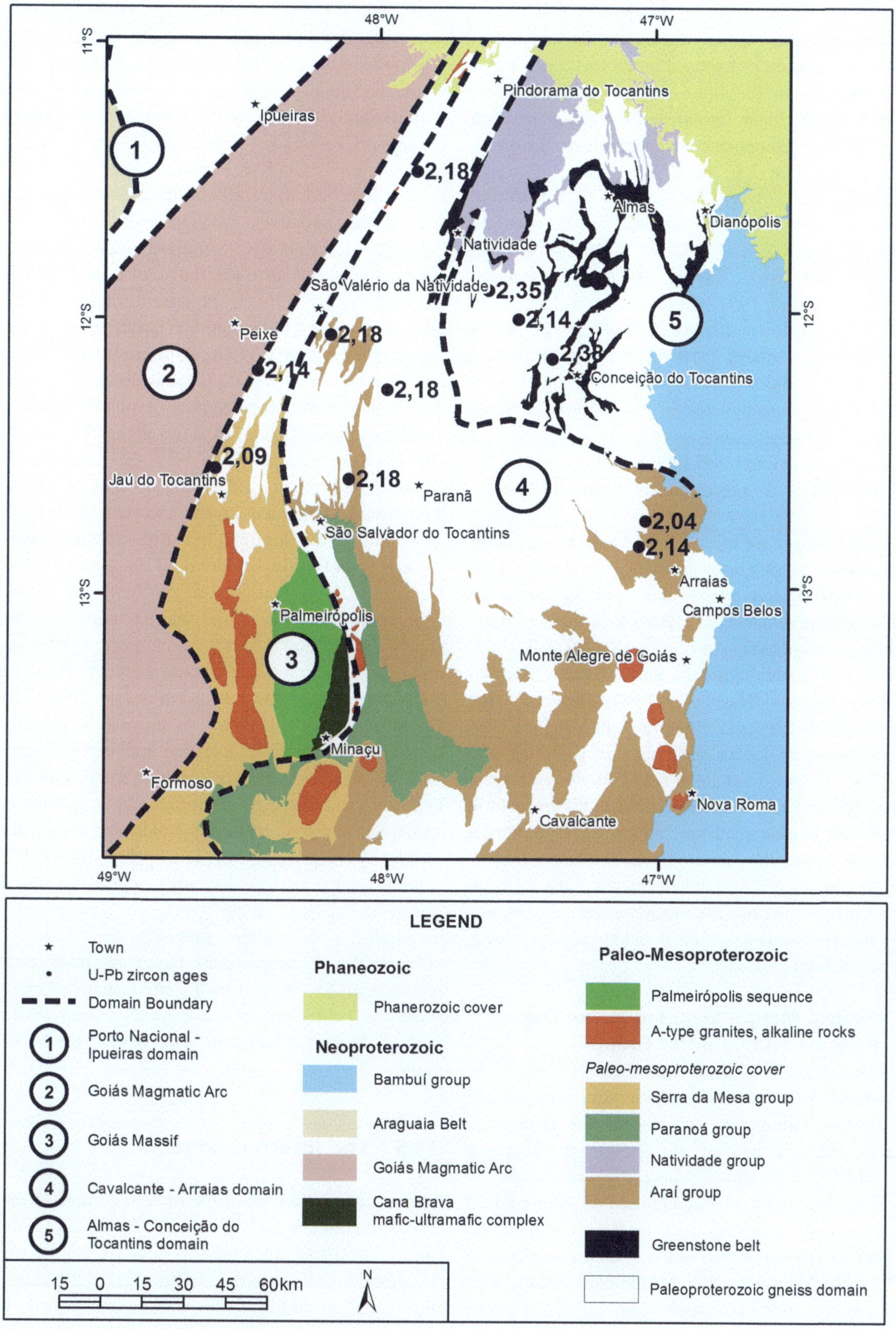

Fig. 11.6 Geological map of the Almas-Dianópolis terrain with available U–Pb data (from Fuck et al. 2014)

correlated with the Serra da Mesa Group, which is also a marine sequence formed mainly of amphibolite facies siliciclastic and carbonate rocks exposed in the internal zone of the belt. In addition to the lithostratigraphic similarities, rocks of the Serra da Mesa Group and Traíras Formation have similar isotopic compositions: carbonate rocks show $\delta^{13}C_{PDB}$ values between 2.6 and 2.93 ‰. Sm–Nd model ages in both units are between 1.85 and 2.4 Ga, indicating Paleoproterozoic sources and detrital zircon U–Pb ages are between 2.4 and 1.55 Ga. The 1.77 Ga U–Pb age for volcanic and pyroclastic rocks intercalated in the basal rift sediments (Pimentel et al. 1991a) indicates equivalence with rift deposits from the lower Espinhaço Supergroup, in the eastern margin and northern parts of the São Francisco craton (see Cruz and Alkmim and Alkmim et al., this book). The Araí Group is not recognized in the southern Brasília belt.

Open folds mark deformation of the Araí Group rocks and metamorphism took place under greenschist facies conditions.

The Natividade Group is exposed in the northeastern corner of the external zone, resting unconformably on Paleoproterozoic basement gneiss and relicts of volcanic-sedimentary sequences of the Natividade-Cavalcante block. The group is thought to be equivalent to the Araí Group (Marini et al. 1984; Gorayeb et al. 1988), although correlation with the Bambuí Group was also suggested (Hasui et al. 1990). The base of the succession comprises low-grade dolomite, quartzite and conglomerate, followed by layers and lenses of dolomite and limestone, siltstone and quartzite (Fig. 11.5). According to Dardenne and Saboia (2007), the deposition of this group took place in a mixed marine platform, which could represent extension of the sedimentary environment characterizing the Araí Group flexural stage. All units of the Natividade Group underwent strong deformation under grenschist facies conditions, along the Transbrasiliano lineament strike-slip system (Dardenne and Saboia 2007).

11.4.1.3 Passive Margin Megasequence: The Paranoá and Canastra Groups

The Paranoá Group (Fig. 11.4), exposed between 13° and 17°S, has been suggested to be a stratigraphic equivalent of the Conselheiro Mata Group, the upper unit the Espinhaço Supergroup in the Araçuaí belt and Paramirim aulacogen (Dardenne 2000). The Paranoá Group comprises a sequence of shallow-water marine deposits composed of mature siliciclastic craton-sourced sediments made up mainly of quartzite and rhythmite. The rhythmic succession of quartzite, siltstone and slate also includes occasional carbonate lenses with stromatolites (Faria 1995; Campos et al. 2013). Unconformably overlying the Araí Group, the ca. 1000 m-thick Paranoá succession has been subdivided into 12 lithofacies (Faria 1995): São Miguel conglomerate (SM), rhythmite (R1), quartzite (Q1), rhythmite (R2), quartzite (Q2), siltstone (S), slate (A), rhythmite (R3), quartzite (Q3), rhythmite (R4), feldspathic quartzite (QF), and pelite-carbonate (PC). Dolostone, limestone, and stromatolitic dolostone lenses are found within some of the siltstone successions. The upper portion the Group (PC) includes stromatolitic dolostones, in which Conophyton metula Kirichenko has been described, a species that is believed to represent the 0.9–1.2 Ga age interval (Cloud and Dardenne 1973; Dardenne et al. 1976). Stromatolite and microfossils (Stratifera undata) of the Paranoá Group suggest a narrower range of ages of 1170–950 Ma (Fairchild et al. 1996). U–Pb data from detrital zircon along the typical section of the Paranoá Group indicates a maximum depositional age of 1.54 Ga. A minimum depositional U–Pb age of approximately 1.04 Ga was obtained from diagenetic xenotime overgrowth on detrital zircon (Matteini et al. 2012).

The Canastra Group, exposed along the central-southern area of the Brasília belt (Fig. 11.2, see Valeriano, this book), is composed mainly of psammite and pelite strata with subordinate carbonate layers and lenses. Due to intense deformation, the stratigraphic organization of the Canastra Group is not fully understood. In previous studies, the Canastra Group was correlated with the Araxá Group, exposed to the western and internal portion of the belt. In recent studies, however, they have been mapped as separate units despite similar lithological contents (Dardenne 2000; Valeriano et al. 2004). In northwestern Minas Gerais, the Canastra Group represents a regressive megacycle (Dardenne 2000). The lower part, comprising sedimentary rocks rich in organic matter and pyrite, represents deep water deposits, overlain by turbidite, with gravitational flow structures, covered in turn by platform sediments, characterized by storm structures. The uppermost part is formed of shallow platform deposits, with tidal structures indicating sediment transport to the west. The Canastra Group was metamorphosed under greenschist facies conditions, although locally amphibolite facies assemblages have been reported. In the present study, the Canastra Group is interpreted as equivalent to the Paranoá Group and, therefore, part of the passive margin sequence of the São Francisco craton.

11.5 The Internal Zone

The metamorphic core of the orogen comprises the Neoproterzoic metasedimentary rocks of the Araxá Group and the high-grade rocks of Anápolis–Itauçu Complex (Fig. 11.2). Metamorphism increases from greenschist facies on the east to ultrahigh-T granulitic facies rocks on the west. The Goiás massif includes Archean granite-greenstone terrains and Paleoproterozoic orthogneiss, covered by the Paleoproterozoic Serra da Mesa Group. Eastwards these terrains are

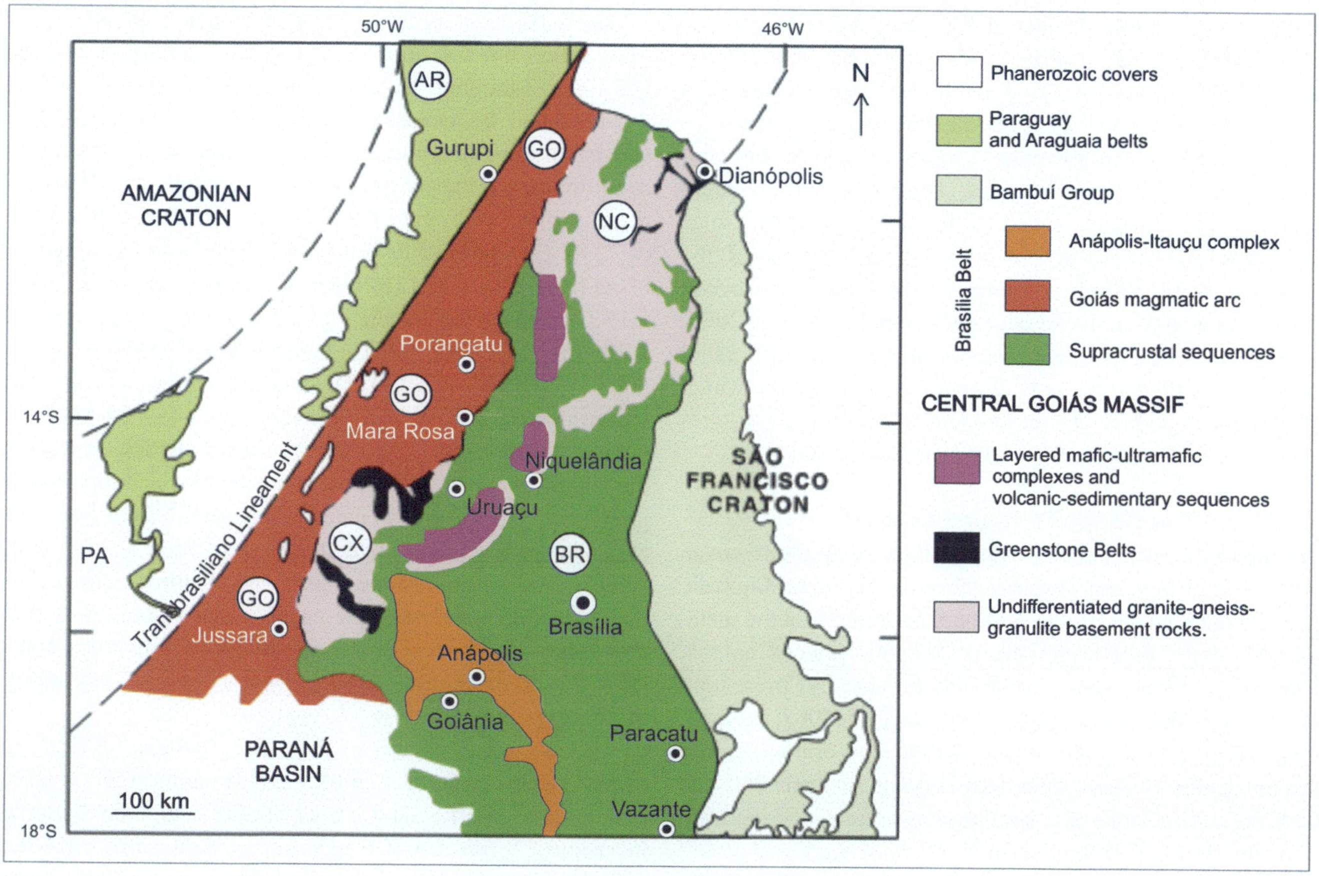

Fig. 11.7 Map of the central part of the Tocantins Province. *AR* Araguaia Belt, *PA* Paraguay Belt, *GO* Goiás Magmatic Arc, *NC* Natividade-Cavalcante Block, *BR* Brasília Belt, *CX* Crixás granite-greenstone terrain. From Cordani and Teixeira (2007)

juxtaposed to Mesoproterozoic volcanic-sedimentary sequences and associated gabbro-anorthosite complexes, as well as the Neoproterozoic layered Barro Alto, Niquelândia and Cana Brava mafic-ultramafic complexes, which make up the eastern border of the massif (Fig. 11.2). The Goiás Massif appears to be an exotic continental fragment taken up between the Goiás Magmatic Arc and the external zone during the Brasiliano orogeny (Fig. 11.7). The Goiás Magmatic Arc comprises Neoproterozoic orthogneiss terrains and volcanic-sedimentary sequences.

11.5.1 Stratigraphic Assemblages

11.5.1.1 The Araxá Group

The Araxá Group comprises mainly mica-schist and subordinate quartzite layers, mostly metamorphosed under amphibolite facies conditions. The structural pattern is represented by flat-lying foliation associated with thrust sheets verging east-southeast towards the São Francisco craton, and limited by lateral ramps due to differential movement of the thrust fronts. A distinct feature of the Araxá Group is the presence of a number of small (a few hundred meters to several km long) serpentinite bodies, many of which containing podiform chromite. Os isotopic composition of chromite suggests crystallization at ca. 0.8 Ga (Walker et al. 2002). The serpentinite bodies may be isolated or accompanied by amphibolite either derived from ocean-floor basalts or from mafic plutonic rocks, associated with meta-chert, manganese and iron formation, calc-silicate rocks, and carbon-rich schist (Brod et al. 1992). The association is interpreted as an ophiolite mélange, which extends for about 200 km along strike (Strieder and Nilson 1992). In the Araxá synform, southern part of the belt, where the unit was defined originally, the Araxá Group is dominated by amphibolite, mafic schist and interbedded chemical deposits, covered by pelitic schist (Seer and Dardenne 2000; Seer et al. 2001). Metamorphism is of amphibolite facies, which affected also c. 630 Ma syn-collision S-type granites intruded into the supracrustal rocks. Geochemical signature of the mafic rocks is similar to that of E-MORB tholeiite, and the metapelite is typical of deep-sea deposits. One of the mafic slivers within micaschist attributed to the Araxá Group has a SHRIMP U–Pb age of ca. 838 Ma (Piuzana et al. 2003).

The Araxá metasedimentary rocks present a bimodal pattern of Sm–Nd model ages, suggesting independent sources, one of which related to the Neoproterozoic Goiás Magmatic Arc. In its type-area (southern segment of the belt), the Araxá Group is interpreted as remnant of ocean crust from a back arc basin setting (Seer and Dardenne 2000; Seer et al. 2001). The interpretation has been corroborated by recent U–Pb age patterns of detrital zircon grains, which display an abundant Neoproterozoic population, with ages between ca. 900 and 630 Ma (Pimentel et al. 2011).

11.5.1.2 The Anápolis–Itauçu Complex

The high-grade rocks of the Anápolis–Itauçu Complex form an NW–SE elongated zone, exposed between the Goiás Magmatic Arc and occurrences of Araxá Group. Interpreted as the metamorphic core of the Brasília belt, the Anápolis–Itauçu Complex consists of ortho- and paragranulites, locally presenting ultrahigh-T mineral assemblages (Moraes et al. 2007; Baldwin et al. 2005; Baldwin and Brown 2008). The orthogranulites display U–Pb crystallization ages varying between 760 and 630 Ma and metamorphic overgrowths around 650–640 Ma (Piuzana et al. 2003; Baldwin et al. 2005; Moraes et al. 2007; Baldwin and Brown 2008). Most of the orthogranulites in the Anápolis–Itauçu complex are represented by granite intrusions and layered mafic–ultramafic complexes. The granite intrusions took place between ca. 630 and 650 Ma (Piuzana et al. 2003). U–Pb (LA-MC-ICPMS, SHRIMP and ID-TIMS) zircon analyses of mafic-ultramafic complexes reveal a spread of concordant ages spanning an interval of ~80 Ma with an upper intercept age of ~670 Ma (Giustina et al. 2011a, b).

11.5.1.3 The Goiás Massif

The Goiás Archean Block

This is a small (approximately 18,000 km^2) oval-shaped Archean terrain located between the supracrustal units of the Brasília belt, to the east, and the Goiás Magmatic Arc, to the north, west and south (Figs. 11.2 and 11.7). It comprises typical Archean TTG-greenstone belt terrains, with komatiites exhibiting well-developed spinifex textures. Paleoproterozoic supracrustal rock units are also recognized, especially in the upper parts of the greenstone belts (Jost et al. 2010). The Goiás block, formerly labeled Goiás massif, has been considered to be an allochthonous terrane accreted to the western margin of the Brasília Belt during the latest stages of evolution of the Brasiliano orogeny (Pimentel et al. 2000; Jost et al. 2013).

The TTG terrains form different blocks, separated from each other by the supracrustal belts, among others the Hidrolina, Moquém, Caiamar and Anta blocks, studied in greater detail by Queiroz et al. (2008). In summary, they consist of tonalite, granodiorite and granite gneisses and subordinate porphyritic tonalite intrusions. U–Pb SHRIMP ages for these rocks vary from ca. 2.84 to 2.71 Ga. Recently, ca. 2.82 Ga charnockite intrusions have also been recognized (Beghelli Jr. 2012). Inherited zircon grains dated at ca. 3.11 Ga suggest the presence of older continental crust in the northern part of the Goiás Archean block (Queiroz et al. 2008). Beghelli Jr. (2012) confirmed this assumption, reporting the LA-ICPMS U–Pb age of 3.144 ± 35 Ma for a gneiss sample from the Caiçara Complex, which represents the oldest rock of the Goiás Archean block. Amphibolite grade metamorphism was dated using titanite from gneiss of the Caiamar block and yielded the age of ca. 2.71 Ga. Also, some rocks in this terrane show calc-alkaline signature as well as positive and negative $\varepsilon_{Nd}(T)$ values, suggesting that its evolution was marked by polydiapiric intrusions, representing both generation of new juvenile crust and reworking of older continental crust.

The southernmost part of the Goiás Archean block (Uvá Complex, Jost et al. 2005, 2013) comprises mainly tonalite-granodiorite, and minor diorite and monzogranite intrusions. U–Pb LA-ICP-MS zircon data show that the complex formed during two magmatic stages. The older stage includes tonalite and granodiorite gneisses as well as a diorite stock with crystallization ages between ca. 3.04 and 2.93 Ga, whereas rocks of the younger event have crystallization ages between ca. 2.88 and 2.85 Ga.

In general, the greenstone belt sequences comprise a lower metavolcanic sequence dominated by komatiite, with minor BIF and carbonaceous schist, an intermediate unit formed mainly of metabasalt, komatiitic metabasalt and chemical sediments, and an upper clastic sequence with carbonaceous phyllites and marbles (Jost and Oliveira 1991, and references therein). These rocks were metamorphosed under conditions of upper greenschist to amphibolite facies. In the Crixás greenstone belt, the upper metasedimentary sequence is not Archean in age, as indicated by the U–Pb zircon provenance work of Jost et al. (2010), which identified an important zircon population showing ages of ca. 2.22 Ga.

The only geochronology data for the metavolcanic sections of the greenstone belts is a poorly defined Sm–Nd whole-rock isochron including samples of komatiite and komatiitic basalt of the Crixás greenstone belt. It indicated the age of ca. 3.0 Ga, interpreted as the crystallization of the original magmas (Fortes et al. 2003). Therefore, available geochronological data indicate a large age gap between the lower metavolcanic units and the upper metasedimentary package.

Barro Alto, Niquelândia and Canabrava Complexes and Associated Volcanic-Sedimentary Sequences

Three composite Neoproterozoic layered complexes define the eastern boundary of the Goiás massif (Fig. 11.7). Due to their geological and geochronological similarities, they have been interpreted as representative of an originally single regional-scale structure, which was disrupted during the Brasiliano event, forming three individual bodies (Wernick and Almeida 1979; Danni et al. 1982; Ferreira Filho et al. 1998). The three complexes consist of two magmatic systems, distinguished by petrological characteristics (Danni et al. 1982; Suita 1996; Ferreira Filho et al. 1994; Ferreira Filho and Pimentel 2000). The composite intrusions are known as the Barro Alto, Niquelândia and Canabrava complexes, from south to north, although their nomenclature was recently reviewed, suggesting that these terms should be assigned only to the granulite-facies units exposed in the eastern parts of the complexes (Ferreira Filho et al. 2010). Accordingly, the western amphibolite facies layered units, exposed in the southern and central intrusions were re-named as the Serra da Malacacheta and Serra dos Borges complexes, respectively (Ferreira Filho et al. 2010). In previous studies the latter were considered to be Mesoproterozoic intrusions (ca. 1.25 Ga, U–Pb SHRIMP zircon ages, Pimentel et al. 2004a, 2006), however recent work by Correia et al. (2012) suggests that both magmatic systems are Neoproterozoic. The western parts of the layered complexes are composed of interlayered leucotroctolite, leucogabbro, anorthosite and occasional pyroxenite, whereas the easternmost parts ($\sim$0.79 Ga, U–Pb SHRIMP zircon ages, Pimentel et al. 2004a, 2006) consist of several cyclic units of dunite, pyroxenite and gabbronorite, within which slices of supracrustal rocks and granite intrusions metamorphosed under high grade conditions are common. Mineral chemistry, whole-rock trace element compositions and Sm–Nd isotopic data reveal contrasting geochemical signature for both units; in the western units, analyses attest to a depleted-mantle source for the original magmas (Ferreira Filho and Pimentel 2000; Moraes et al. 2003), whereas the eastern association suggests large degrees of crustal contamination with older sialic crust (Suita 1996; Pimentel et al. 2004a, 2006). To the west, the three composite bodies are in contact with Mesoproterozoic bi-modal volcanic-sedimentary sequences, namely the Juscelândia, Indaianópolis and Palmeirópolis sequences, from south to north (Fig. 11.7). The three sequences display similar stratigraphy and consist of metapelite, calc-silicate rocks and metachert interbedded with metavolcanic rocks, represented by fine-to-medium grained amphibolite and felsic gneiss (Brod and Jost 1991; Araujo 1996; Moraes and Fuck 1994, 1999; Ferreira Filho et al. 1998; Moraes et al. 2003). Amphibolites of the Juscelândia and Palmeirópolis sequences show positive ε_{Nd} values and trace element signatures typical of MORB magmas, suggesting that the volcanic-sedimentary sequences represent a continental rift that evolved towards an oceanic basin (Araujo 1996; Moraes et al. 2003, 2006).

The composite mafic-ultramafic complexes, together with the associated volcanic-sedimentary sequences, underwent amphibolite to granulite facies metamorphism, with P and T conditions increasing progressively from west to east, in contrast with the generally westwards increase in metamorphic grade of the Brasília belt. Locally, ultrahigh temperature mineral assemblages are recorded (Moraes and Fuck 1994, 2000; Ferreira Filho et al. 1998). SHRIMP U–Pb zircon analyses performed on metamorphic overgrowths constrain the age of the high-grade metamorphism in the composite layered intrusions at approximately 760–750 Ma (Pimentel et al. 2004a, 2006; Moraes et al. 2006), contrasting with the ca. 630 Ma peak metamorphism generally recorded in the Brasília belt. Additionally, rutile U–Pb ages reveal a younger metamorphic event around 610 Ma, which is interpreted as related to the uplift of the complexes during the Brasiliano orogeny (Ferreira Filho et al. 1994). The current structural configuration of the three composite layered complexes is ascribed to this final tectono-metamorphic episode.

11.5.1.4 The Goiás Magmatic Arc

The Goiás Magmatic arc is one of the most important tectonic features of the Brasília belt (Fig. 11.2). It comprises two sectors, separated by the Archean terranes of the Goiás massif. The southern sector is known as the Arenópolis Arc, and the northern sector, as the Mara Rosa arc (Fig. 11.7, Pimentel and Fuck 1992; Pimentel et al. 2000b; Junges et al. 2002; Laux et al. 2005). In both sectors, early evolution of the arc took place in intraoceanic island arcs with the crystallization of very primitive tholeiitic to calc-alkaline volcanics and associated tonalites/granodiorites dated at ca. 890–860 Ma. The rocks display positive $\varepsilon_{Nd}(t)$ values (up to approximately +5.0) and Nd model ages mostly between 0.8 and 1.1 Ga (Pimentel et al. 1991, 1997, 2000a; Pimentel and Fuck 1992). Geochemical and isotopic data suggest that some of the original tonalitic/andesitic magmas were similar to modern adakites (Pimentel et al. 1991a, 1997). A second cycle of volcanic and plutonic activity took place between ca. 670 and 600 Ma in both sectors (Dantas et al. 2001; Laux et al. 2004, 2005), followed by bi-modal intrusions of gabbro, diorite, tonalite to granite, and accompanied by metamorphism and deformation (Laux et al. 2005). $\varepsilon_{Nd}(T)$ values are more varied and T_{DM} model ages vary between 1.0 and 2.0 Ga, suggesting that this second phase of magmatic arc development took place in the presence of older continental crust, most likely in a continental arc setting.

Fig. 11.8 Tonalitic orthogneisses of the Goiás Magmatic Arc

The Arenópolis Arc

The Arenópolis segment of the Goiás magmatic arc comprises N–S trending volcano-sedimentary sequences (from west to east, the Bom Jardim de Goiás, Arenópolis, Iporá, Jaupaci and Anicuns-Itaberaí sequences), separated from each other by calcic- to calc-alkaline orthogneiss ranging in composition from tonalite to granite. The supracrustal and orthogneiss units are juxtaposed along important NNE–NNW strike-slip faults, some of which are part of the transcontinental Transbrasiliano strike-slip fault system.

The volcanic sequences comprise mainly low-K tholeiitic metabasalts, calc-alkaline metabasalts, metandesites, metadacites, and metarhyolites in some of these sequences (Bom Jardim de Goiás and Arenópolis). Other sequences comprise bimodal associations with metabasalts and rhyolites (e.g. Iporá and Jaupaci). The metavolcanic rocks typically present very primitive geochemical and isotopic characteristics, with low initial $^{87}Sr/^{86}Sr$ ratios (0.7032–0.7050) and positive $\varepsilon_{Nd}(t)$ values. Felsic metavolcanic rocks have U–Pb zircon ages between ca. 0.9 and 0.64 Ga (Pimentel et al. 1991b; Rodrigues et al. 1999). Fine-grained amphibolites of the Arenópolis volcano-sedimentary sequence are probably the most primitive of the metavolcanic rocks, representing low-K tholeiites and calc-alkaline metabasalts with very primitive isotopic compositions (initial $^{87}Sr/^{86}Sr$ of 0.7026, $\varepsilon_{Nd}(t)$ of +6.9; Pimentel 1991), which represent the early stages of development of an intraoceanic island arc system. Small metamorphosed gabbro-diorite intrusions are also recognized in these sequences; one of these, dated at 890 ± 9 Ma (SHRIMP U–Pb zircon age; Pimentel et al. 2003), corresponds to a plutonic/subvolcanic equivalent of the Arenópolis volcanic sequence. The Anicuns-Itaberaí sequence, exposed along the contact between the eastern part of the Goiás magmatic arc and the Anápolis–Itauçu high-grade terrain, is represented dominantly by amphibolites (metavolcanic and metaplutonic) and metapelitic rocks, with subordinate iron formations, cherts, marbles, and ultramafic rocks of uncertain age. Mafic rocks in this sequence crystallized in two main periods of time: 0.89–0.82 and 0.63–0.60 Ga (Laux et al. 2005 and references therein).

The orthogneisses exposed between the volcanic-sedimentary belts are dominantly hornblende- and biotite-bearing metadiorites, metatonalites and metagranodiorites (Fig. 11.8). They show mineral assemblages indicative of metamorphism under epidote amphibolite facies conditions, locally display relict igneous textures and structures, and can also show features of strong deformation and metamorphism such as mylonitic textures and migmatitic structures. Major and trace element data suggest that the igneous protoliths were metaluminous, calcic- to calc-alkaline, with typically low K contents (Pimentel and Fuck 1986). The Arenópolis, Sanclerlândia and Firminópolis gneisses are typical examples of these rocks and have been compared with primitive M-type granitoids of intraoceanic island arcs, whereas the Matrinxã granodiorite presents characteristics of both I- and M-type rocks (Pimentel and Fuck 1986, 1987, 1992; Rodrigues et al. 1999).

U–Pb, Sm–Nd and Rb–Sr isotopic determinations indicate ages between ca. 890 and 630 Ma. Initial $^{87}Sr/^{86}Sr$ ratios are low (<0.705) and $\varepsilon_{Nd}(T)$ values are positive, indicating the juvenile character of the original magmas (Pimentel and Fuck 1986, 1992; Rodrigues et al. 1999). Similarly to the mafic rocks in the volcanic-sedimentary rocks, these intrusions crystallized in two main periods: 890–750 and 660–600 Ma (Laux et al. 2005).

Hornblende-bearing gneisses in the easternmost part of the Arenópolis arc have T_{DM} model ages between ca. 0.94 and 1.13 Ga. To the southeast, however, orthogneisses display Paleoproterozoic model ages (1.89–2.27 Ga), indicating the presence of older sialic crust between the Neoproterozoic juvenile rocks. Examples of involvement or presence of small blocks of older rocks in between the juvenile rock units have been described in previous studies in some areas of the Arenópolis Arc (Pimentel and Fuck 1986; Rodrigues et al. 1999).

The Mara Rosa Arc

This segment of the arc also comprises narrow strips of metavolcanic rocks (mostly low-K metatholeiite and calc-alkaline metabasalt). Two supracrustal sequences are recognized: (1) the Mara Rosa Sequence is mainly made of amphibolites and schists with isotopic characteristics suggesting affinities with island arc rocks (Pimentel et al. 1997; Junges et al. 2002); (2) the Santa Terezinha de Goiás Sequence, in the southernmost area of the arc terrain, is a volcanic-sedimentary sequence including schist and quartzite intercalated with metavolcanic rocks, with U–Pb zircon age of 670 Ma (Dantas et al. 2001; Fuck et al. 2007).

The limits of the Mara Rosa Arc are mainly tectonic. To the south it is marked by the low-angle Mandinópolis shear zone where the arc rocks thrust over the Archean TTG terrains and greenstone belts (Jost et al. 2001). To the east, the Mara Rosa rocks were thrusted over the Serra da Mesa Group and its basement as well as the Paleoproterozoic Campinorte Complex (Pimentel et al. 1997; Giustina et al. 2009), along the Rio dos Bois fault system (Kuyumjian et al. 2004). To the west, the Transbrasiliano Lineament, a remarkable regional structure, joins the arc rocks and the Archean Serra Azul complex with the 570–530 Ma felsic ortogneisses and mafic intrusions of the Porto Nacional-Porangatu terrain (Dantas et al. 2006). To the north, between Porto Nacional and Natividade, Tocantins state, two samples of tonalite gneiss and leucogranite have similar Nd model ages (ca. 1.3 Ga) and zircon U–Pb SHRIMP ages of ca. 860 Ma, extending the Mara Rosa Arc far to the north (Fuck et al. 2014) where it is covered by the Paleozoic sedimentary rocks of the Parnaíba Basin.

Similarly to the Arenópolis Arc, the available geochronological data for the Mara Rosa Arc indicate that the arc terrain comprises two magmatic phases: (1) the older one, between ca. 915 and 800 Ma, including metavolcanic rocks and tonalite orthogneisses with very primitive geochemical and isotopic signature; (2) the younger, represented by the Santa Terezinha de Goiás sequence and granitoid rocks, crystallized between ca. 0.67 and 0.63 Ga.

11.6 Summary of the Tectonic Evolution of the Northern Brasília Belt

The Brasília belt show many of the tectonic features of modern mountain belts, such as an ophiolite association, a large juvenile magmatic arc, sedimentary sequences which have been interpreted as passive margin sequences, arc-related sedimentary piles and a foreland basin. The convergence of the Amazonian, São Francisco and Paranapanema paleocontinents involved the subduction of a wide ocean lithosphere and development of primitive island arc systems within the so-called Goiás-Pharusian Ocean. During the late Neoproterozoic, oceanic lithosphere was consumed along possibly west-dipping subduction zones, and finally the island arc systems were accreted to the western margin of the São Francisco paleocontinent. This was followed by the development of a continental arc between ca. 670 and 630 Ma. During this process, the development of the ophiolite mélange and deposition of syn-orogenic sedimentary sequences (Araxá and Ibiá groups) took place. Final collision most likely took place at ca. 630 Ma and was marked by the emplacement of S-type granites, especially into the Araxá Group metasedimentary rocks. Thickening of the lithosphere was followed by orogen collapse and the emplacement of large volumes of mafic magmas and roughly coeval high grade-metamorphism recorded in the Anápolis–Itauçu Complex. The final stages of evolution of the Brasília belt were marked by uplift, erosion and the emplacement of late- to post-tectonic granite intrusions associated with small gabbro-diorite bodies, especially within the Goiás Magmatic Arc.

Acknowledgments Authors are thankful to CNPq for continuous support through INCT-ET and INCT-PETROTEC, and authors' research fellowships.

References

Almeida, F.F.M., Hasui, Y., Brito Neves, B.B., Fuck, R.A. Brazilian structural provinces: An introduction. Earth-Science Reviews 17(1), 1–29, 1981.

Alkmim, F.F. O que faz de um cráton um cráton? O Cráton do São Francisco e as revelações almeidianas ao delimitá-lo. In: V. Mantesso Neto, A. Bartorelli, C.D.R., Carneiro, B.B. Brito Neves (Org.). Geologia do Continente Sul Americano. Evolução da obra de Fernando flavio Marques de Almeida. São Paulo, Beca, 17–35, 2004.

Alkmim, F.F., Martins Neto, M.A. Proterozoic first-order sedimentary sequences of the São Francisco Craton, eastern Brazil. Marine and Petroleum Geology 33, 127–139, 2012.

Alvarenga, C.J.S., Dardenne, M.A., Vieira, L.C., Martinho, C.T., Guimarães, E.M., Santos, R.V., Santana, R.O. Estratigrafia da Borda Ocidental da Bacia do São Francisco. Boletim de Geociências da Petrobrás 20, 145–164, 2012.

Alvarenga, C.J.S., Giustina, M.E.S.D., Silva, N.G.C., Santos, R.V., Gioia, S.M.C.L., Guimarães, E.M., Dardenne, M.A., Sial, A.N., Ferreira, V.P. Variações dos isótopos de C e Sr em carbonatos pré- e pós-glaciação Jequitaí (Estuartiano) na região de Bezerra-Formosa, Goiás. Revista Brasileira de Geociências 37(supl), 147–155, 2007.

Alvarenga, C.J.S., Santos, R.V., Vieira, L.C., Lima, B.A.F., Mancini, L.H. Meso-Neoproterozoic isotope stratigraphy on carbonates platforms in the Brasília Belt of Brazil. Precambrian Research 251, 164–180, 2014.

Araujo, S.M. Geochemical and isotopic characteristics of alteration zones in highly metamorphosed volcanogenic massive sulfide deposits and their potential application to mineral exploration. PhD Thesis. University of Toronto, 1996.

Babinski, M., Vieira, L.C., Trindade, R.I.F. Direct dating of the Sete Lagoas cap carbonate (Bambuí Group, Brazil) and implications for the Neoproterozoic glacial events. Terra Nova 19, 401–406, 2007.

Baldwin, J.A., Brown, M. Age and duration of ultrahigh-temperature metamorphism in the Anápolis-Itauçu complex, southern Brasília Belt, central Brazil – constraints from U-Pb geochronology, mineral rare earth element chemistry and trace-element thermometry. Journal of Metamorphic Geology 26, 213–233, 2008.

Baldwin, J.A., Powell, R., Brown, M., Moraes, R., Fuck, R.A. Modelling of mineral equilibria in ultrahigh-temperature metamorphic rocks from the Anapolis-Itaucu Complex, central Brazil. Journal of Metamorphic Geology, v. 23, p. 511–531, 2005.

Beghelli Jr., L.P. Charnockitos e ortognaisses da porção centro-oeste do bloco arqueano de Goiás: dados geoquímicos e isotópicos. MSc Thesis, Universidade de Brasilia, 77p., 2012.

Brod, J.A., Leonardos, O.H., Meneses, P.R., Albuquerque, M.A.C., Almeida, R., Blanco, S.B., Cardoso, F.B.F., Romão, P.A., Tallarico, F. H.B., Thomsen, F.P.R. Geoquímica da Sequência Vulcano-Sedimentar de Abadia dos Dourados, Grupo Araxá, Triângulo Mineiro, MG. Revista da Escola de Mina 1–2, 164–166, 1992.

Brod, J.A. Jost, H. Características estruturais, litológicas e magmáticas da zona de cisalhamento dúctil do Rio Traíras, bloco do Complexo de Niquelândia, Goiás. Revista Brasileira de Geociências, BRASIL, v. 21, n. 3, p. 205–217, 1991.

Campos, J.E.G, Dardenne, M.A., Freitas-Silva, F.H., Martins Ferreira, M.A.C. Geologia do Grupo Paranoá na porção externa da Faixa Brasília. Revista Brasileira de Geociências 43, 461–476, 2013.

Cloud, P., Dardenne, M.A. Proterozoic age of the Bambuí Group in Brazil. Geological Society of America Bulletin 84, 673–1676, 1973.

Cordani, U.G. and Teixeira, W. Proterozoic Accretionary belts in the Amazonian Craton. Geological Society of America, v. 200, p. 297–320, 2007.

Correia, C.T., Sinigoi, S., Girardi, V.A.V., Mazzucchelli, M., TassinariI, C.C.G., Giovanardi, T. The growth of large mafic intrusions: Comparing Niquelândia and Ivrea igneous Complexes. Lithos (Oslo. Print), v. 155, p. 167–182, 2012.

Cukrov, N., Alvarenga, C.J.S., Uhlein, A., 2005. Litofácies da glaciação neoproterozóica nas porções sul do Cráton do São Francisco: exemplos de Jequitaí, MG e Cristalina, GO. Revista Brasileira de Geociências 35, 69–76.

Danni, J.C.M., Fuck, R.A.,Leonardos, O.H. Archean and lower Proterozoic units in central Brazil. Geologische Rundschau 71(1), 291–317, 1982.

Dantas, E.L., Jost H., Fuck, R.A., Brod, J.A., Pimentel, M.M., Meneses, P.R. Proveniência e idade deposicional de seqüências metavulcano-sedimentares da região de Santa Terezinha de Goiás, baseada em dados isotópicos Sm-Nd e U-Pb em monocristal de zircão. Revista Brasileira de Geociências 31, 329–334, 2001.

Dantas, E.L., Oliveira, C.G., Fuck, R.A. Nota explicativa e Mapa Geológico da Folha Porangatu (Folha SD.22-X-D-IV, 1:100.00). Brasília: CPRM, 2006. 100 p.

Dantas, E.L., Santos, R.V., Alvarenga, C.J.S., Pimentel, M.M., Armstrong, R., Fuck, R.A., Laux, J.H. What is the São Domingos Sequence? A Meso-Neoproterozoic rift margin or an intra- cratonic Paleoproterozoic basin? New constraints on the western boundary of the São Francisco Craton. In: VII SSAGI - South American Symposium on Isotope Geology, 2010, Brasília. Abstracts, 2010.

Dardenne, M.A. Síntese sobre a estratigrafia do Grupo Bambuí no Brasil Central. In: 30º Congresso Brasileiro de Geologia, 1978, Recife. Anais, v. 2. p. 597–610, 1978.

Dardenne, M.A. The Brasília Fold Belt. In U.G. Cordani, E.J. Milani, A. Thomaz Filho, D.A. Campos (eds.). Tectonic Evolution of South America. Rio de Janeiro, 31st International Geological Congress, p. 231–263, 2000.

Dardenne, M.A., Faria, A., Andrade, G.F. Occurrence de stromatolites columnaires dans le Group Bambuí, Goiás, Brésil. Anais da Academia Brasileira de Ciências 48, 555–566, 1976.

Dardenne, M.A., Saboia, A.M. Litoestratigrafia do Grupo Natividade na região de Natividade-Pindorama, Tocantins. In: Horbe, A.M.C., Souza, V.S. (Org.). Contribuições à Geologia da Amazônia. Belém: Sociedade Brasileira de Geologia,, v. 5, 29–38, 2007.

Fairchild, T.R., Schopf, J.W., Shen-Miller, J., Guimarães, E.M., Edwards, M.D., Lagstein, A., Li, X., Pabst, M., Melo-Filho, L.S., Recent discoveries of Proterozoic microfossils in south-central Brazil. Precambrian Research 80, 125–152, 1996.

Faria, A., 1995. Estratigrafia e sistemas deposicionais do Grupo Paranoá nas áreas de Cristalina, Distrito Federal e São João da Aliança-Alto Paraíso de Goiás. Thesis, Universidade de Brasília, Brazil.

Ferreira Filho, C.F., Kamo, S., Fuck, R.A., Krogh, T., Naldrett, A. J. Zircon and rutile U-Pb geochronology of the Niquelândia layered mafic-ultramafic complex, Brazil. Precambrian Research 68, 241–255, 1994.

Ferreira Filho, C.F., Naldrett, A.J., Gorton, M.P. REE and pyroxene compositional variation across the Niquelândia mafic and ultramafic layered intrusion, Brazil: petrological and metallogenetic implications. Transaction of the Institution of Mining and Metallurgy 107, 1–22, 1998.

Ferreira Filho, C.F., Pimentel, M.M. Sm-Nd isotope systematics and REE data of troctolites and their metamorphic equivalents of the Niquelândia Complex, central Brazil: further constraints for the timing of magmatism and metamorphism. Journal of South American Earth Sciences 13(7), 647–659, 2000.

Ferreira Filho, C.F., Pimentel, M.M., Araujo, S.M., Laux, J.H. Layered intrusions and volcanic sequences in central Brazil: Geological and geochronological constraints for Mesoproterozoic (1.25 Ga) and Neoproterozoic (0.79 Ga) igneous associations. Precambrian Research 183, 617–634, 2010.

Fortes, P.T.F.O., Pimentel, M.M., Santos, R.V., Junges, S.L., 2003. Sm-Nd study at the Mina III gold deposit, Crixás greenstone belt, central Brazil: implications for the depositional age of the upper metasedimentary rocks and associated Au mineralization. Jour. South Am. Earth Sci, 16:503–512

Fuck, R.A., Dantas, E.L, Pimentel, M.M., Botelho, N.F., Armstrong, R.A., Laux, J.H., Junges, S.L., Soares, J.E.P., Praxedes, I.F. Paleoproterozoic crust-formation and reworking events in the Tocantins Province, central Brazil: A contribution for Atlantica supercontinent reconstruction. Precambrian Research 244, 53–74, 2014.

Fuck, R.A., Marini, O.J., Dardenne, M.A., Figueiredo, A.N. Coberturas metassedimentares do Proterozóico Médio: Os grupos Araí e Paranoá na região de Niquelândia-Colinas, Goiás. Revista Brasileira de Geociências 18(1), 54–62, 1988.

Giustina, M.E.S.D., Oliveira, C.G., Pimentel, M.M., Melo, L.V., Fuck, R.A., Dantas, E.L., Bühn, B.M. U-Pb and Sm-Nd constraints on the nature of the Campinorte sequence and related Palaeoproterozoic juvenile orthogneisses, Tocantins Province, central Brazil. Geological Society Special Publication 323, 255–269, 2009.

Giustina, M.E.S.D., Pimentel, M.M., Ferreira Filho, C.F., Fuck, R.A., Andrade, S. U-Pb-Hf-trace element systematics and geochronology of zircon from a granulite-facies metamorphosed mafic-ultramafic layered complex in central Brazil. Precambrian Research 189, 176–192, 2011a.

Giustina, M.E.S.D., Pimentel, M.M., Ferreira Filho, C.F., Hollanda, M. H.B.M. Dating coeval mafic magmatism and ultrahigh temperature metamorphism in the Anápolis Itauçu Complex, Central Brazil. Lithos 124, 82–102, 2011b.

Gorayeb, P.S.S., Costa, J.B.S., Lemos, R.L., Gama Jr., T., Bemerguy, R.L., Hasui, Y. O Pré-Cambriano da Região de Natividade-GO. Revista Brasileira de Geociências, 18(4), 391–397, 1988.

Hasui, Y., Costa, J.B.S., Saad, A.R., Campanha, V.A. O Grupo Natividade e sua correlação com o Grupo Bambuí. Geociências (UNESP), v. Esp., 299–316, 1990.

Isotta, C.A.L., Rocha Campos, A.C., Yoshida, R. Striated pavement of the Upper Precambrian glaciations in Brazil. Nature 222, 466–468, 1969.

Jost, H., Chemale Jr., F., Dussin, I.A., Martins, R. A U-Pb zircon Paleoproterozoic age for the metasedimentary host rocks and gold mineralization of the Crixás greenstone belt, Goiás, Central Brazil. Ore Geology Reviews 37, 127–139, 2010.

Jost, H., Chemale Jr., F., Fuck, R.A., Dussin, I.A. Uvá complex, the oldest orthogneisses of the Archean-Paleoproterozoic terrane of central Brazil. Journal of South American Earth Sciences 47, 201–212, 2013.

Jost, H., Fuck, R.A., Brod, J.A., Dantas, E.L. Meneses, P.R., Assad, M.L.L. Pimentel, M.M., Silva, A.M. Geologia dos terrenos arqueanos e proterozóicos da região de Crixás-Cedrolina, Goiás. Revista Brasileira de Geociências, São Paulo, v. 31, n. 3, p. 315–328, 2001.

Jost, H., Fuck, R.A., Dantas, E.L., Rancan, C.C., Rezende, D.B., Santos, E., Portela, J.F., Mattos, L., Chiarini, M.F.N., Oliveira, R.C. Geologia e geocronologia do Complexo Uvá, bloco arqueano de Goiás. Revista Brasileira de Geociências 35(4), 563–576, 2005.

Jost. H., Oliveira, A.M. Stratigraphy of the greenstone belts, Crixás region, Goiás, Central Brazil. Journal of South American Earth Sciences 4, 201–214,1991.

Junges, S.L., Pimentel, M.M., Moraes, R. Nd Isotopic study of the Neoproterozoic Mara Rosa Arc, central Brazil: implications for the evolution of the Brasilia Belt. Precambrian Research 117, 101–118, 2002.

Karfunkel, J., Hope, A. Late Proterozoic glaciation in central eastern Brazil: synthesis and model. Palaeogeography, Palaeoclimatology, Palaeoecology 65, 1–21, 1988.

Karfunkel, J., Karfunkel, B. Fazielle Entwicklung der mittleren Espinhaço-Zone mit besonderer Beruecksichtigung des Tillit-Problems, Minas Gerais, Brasilian. Geologisches Jahrbuch 24, 3–91, 1977.

Kuyumjian, R. M., Oliveira, C.G., Campos, J.E.G., Queiroz, C.L. Geologia do limite entre os terrenos arqueanos e o Arco Magmático de Goiás na região de Chapada-Campinorte, Goiás. Revista Brasileira de Geociências 34(3), 329–334, 2004.

Laux, J.H., Pimentel, M.M., Dantas, E.L., Armstrong, R., Armele, A. Mafic magmatism associated with the Goiás Magmatic Arc in the Anicuns-Itaberaí region, Goiás, Brazil: Sm-Nd isotopes and new ID-TIMS and SHRIMP U-Pb data. Journal of South American Earth Sciences 16(7), 599–614, 2004.

Laux, J.H., Pimentel, M.M., Dantas, E.L., Armstrong, R.A., Junges, S.L. Two Neoproterozoic crustal accretion events in the Brasília belt, central Brazil. Journal of South American Earth Sciences 18, 183–198, 2005.

Marini, O.J., Fuck, R.A., Dardenne, M.A., Danni, J.C.M. Província Tocantins, setores Central e Sudeste. In: F.F.M. Almeida, Y. Hasui (Org.). O Pré-Cambriano do Brasil. São Paulo, E. Blücher, 205–264, 1984.

Marques, G.C. Geologia dos grupos Araí e Serra da Mesa e seu embasamento no sul do Tocantins. MSc thesis, Universidade de Brasília, 199p, 2010.

Martins, M., Lemos, V.B. Análise estratigráfica das seqüências neoproterozóicas da Bacia do São Francisco. Revista Brasileira de Geociências 37 (suplemento), 156–167, 2007.

Matteini, M., Dantas, E.L., Pimentel, M.M., Alvarenga, C.J.S., Dardenne, M.A. U-Pb and Hf isotope study on detrital zircons from the Paranoá Group, Brasília Belt, Brazil: Constraints on depositional age at Mesoproterozoic-Neoproterozoic transition and tectono-magmatic events in the São Francisco Craton. Precambrian Research 206–207, 168–181, 2012.

Moraes, R., Fuck, R.A. Trajetória horária para o metamorfismo da seqüência Juscelândia, Goiás: condições do metamorfismo e implicações tectônicas. Revista Brasileira de Geociências, São Paulo, v. 29, n. 4, 1999.

Moraes, R., Fuck, R. A. Deformação e metamorfismo das sequências Juscelandia e Serra da Malacacheta, Complexo Barro Alto, Goiás. Revista Brasileira de Geociências 24(3), 189–197, 1994.

Moraes, R., Fuck, R.A., Brown, M., Piccoli, P.M., Baldwin, J., Dantas, E.L., Laux, J.H., Junges, S.L. Wollastonite-scapolite-clinopyroxene marble of the Anápolis-Itauçu Complex, Goiás: more evidence of ultrahigh-temperature metamorphism. Revista Brasileira de Geociências 37, 877–883, 2007.

Moraes, R., Fuck, R.A., Pimentel, M.M., Gioia, S.M.C.L., Figueiredo, A.M.G. Geochemistry of bimodal rocks of Juscelândia, Goiás, Brazil: Mesoproterozoic transition from continental rift to oceanic ridge. Precambrian Research 125, 317–336, 2006.

Pimentel, M.M., Ferreira Filho, C.F., Armele, A. Neoproterozoic age of the Niquelândia Complex, central Brazil: Further ID-TIMS U-Pb and Sm-Nd isotopic evidence. Journal of South American Earth Sciences 21(3), 228–238, 2006.

Pimentel, M.M., Ferreira Filho, C.F., Armstrong, R.A. SHRIMP U-Pb and Sm-Nd ages of the Niquelândia layered complex: Meso- (1.25 Ga) and Neoproterozoic (0.79 Ga) extensional events in central Brazil. Precambrian Research 132(1), 133–153, 2004.

Pimentel, M.M., Fuck, R.A. Geologia da seqüência vulcano-sedimentar de Arenópolis (GO). Revista Brasileira de Geociências 16, 217–223, 1986.

Pimentel, M.M., Fuck, R.A. Origem e evolução das rochas metavulcânicas e metaplutônicas da região de Arenópolis (GO). Revista Brasileira de Geociências 17, 2–14, 1987.

Pimentel, M.M., Fuck, R.A. Neoproterozoic crustal accretion in central Brazil. Geology 20(4), 375–379, 1992.

Pimentel, M.M., Fuck, R.A., Botelho, N.F. Granites and the geodynamic history of the Neoproterozoic Brasília Belt, Central Brazil: a review. Lithos 46(3), 463–483, 1999.

Pimentel, M.M., Fuck, R.A., Gioia, S.M. The Neoproterozoic Goiás Magmatic Arc: a review and new Sm-Nd isotopic data. Revista Brasileira de Geociências, São Paulo, v. 30, n. 1, p. 35–39, 2000a.

Pimentel, M.M., Fuck, R.A, Jost, H., Ferreira Filho, C.F., Araujo, S.A. The basement of the Brasília Fold Belt and the Goiás Magmatic Arc. In: Cordani, U.G., Milani, E.J., Thomaz Filho, A., Campos, D. A. (Org.). The Tectonic Evolution of South America. Rio de Janeiro, 31st International Geological Congress, 195–229, 2000b.

Pimentel, M.M., Heaman, L., Fuck, R.A., Marini, O.J. U-Pb zircon geochronology of Precambrian tin-bearing continental-type acid magmatism in central Brazil. Precambrian Research 52, 321–335, 1991a.

Pimentel, M.M., Heaman, L., Fuck, R.A. U-Pb zircon and sphene geochronology of late Proterozoic volcanic arc rock units from southwestern Goiás, Central Brazil. Journal of South American Earth Sciences 4, 329–339, 1991b.

Pimentel, M.M., Hollanda, M.H.B.M., Armstrong, R. U-Pb SHRIMP age and Sm-Nd isotopic characteristics of the Morro do Baú gabbro: implications for the evolution of the Arenópolis volcano-sedimentary sequence, Goiás Magmatic Arc. Anais da Academia Brasileira de Ciências 75(3), 331–339, 2003.

Pimentel, M.M, Jost, H., Fuck, R.A. O Embasamento da Faixa Brasília e o Arco Magmático de Goiás. In: V. Mantesso Neto, A. Bartorelli, C.D.R. Carneiro, B.B. Brito Neves (Org.). Geologia do Continente Sul-Americano: Evolução da Obra de Fernando Flávio Marques de Almeida. São Paulo Beca, 355–368, 2004.

Pimentel, M.M., Rodrigues, J.B., Giustina, M.E.S.D., Junges, S.L., Matteini, M. The Tectonic evolution of the Neoproterozoic Brasília Belt, central Brazil, based on SHRIMP and LA-ICPMS U-Pb sedimentary provenance data: A review. Journal of South American Earth Sciences 31, 345–357, 2011.

Pimentel, M.M., Whitehouse, M.J., Viana, M.G., Fuck, R.A, Machado, N. The Mara Rosa arc in the Tocantins Province: Further evidence for Neoproterozoic crustal accretion in central Brazil. Precambrian Research 81, 299–310, 1997.

Piuzana, D., Pimentel, M.M., Fuck, R.A., Armastrong, R.A. SHRIMP U-Pb and Sm-Nd data for the Araxá Group and associated magmatic rocks: constraints for the age of sedimentation and tectonic evolution of the southern Brasília Belt, central Brazil. Precambrian Research 125, 139–160, 2003.

Queiroz, C.L., Jost, H., Silva, L.C., McNaughton, N.J., 2008. U-Pb SHRIMP and Sm-Nd geochronology of granite-gneiss complexes and implications for the evolution of the central Brazil Archean terrain. Journal of South American Earth Sciences, 26: 100–124.

Rendon Dávila, C.A. Ambiente geotectônico, geocronologia e mineralizações de ouro nas janelas erosivas de São Domingos e Correntina, MSc Thesis, Universidade de Brasília, 80 p., 2002.

Rocha Campos, A.C., Young, G.M., Santos, P.R., 1996. Re-examination of a striated pavement near Jequitaí, MG: implications for proterozoic stratigraphy and glacial geology. Anais da Academia Brasileira de Ciências 68, 593

Rodrigues, J. B., Gioia, S.M.L.C., Pimentel, M.M. Geocronologia e geoquímica de ortognaisses da região de Iporá e Firminópolis: Implicações para evolução do Arco Magmático de Goiás. Revista Brasileira de Geociências 29(2), 207–216, 1999.

Santos, R.V., Dardenne, M.A., Sial, A.N., Ramos, M.L.S., Fonseca, M. A. The Mesoproterozoic Neoproterozoic Transition in the Southeast portion of the São Francisco Craton, Brazil. Journal of South American Earth Sciences 18(1), 27–39, 2004.

Seer, H.J., Brod, J.A., Fuck, R.A., Pimentel, M.M., Boaventura, G.R., Dardenne, M. A. Grupo Araxá em sua área tipo: um fragmento de crosta oceânica neoproterozóica na Faixa de Dobramentos Brasília. Revista Brasileira de Geociências, Brasil 31(3), 385–396, 2001.

Seer, H.J., Dardenne, M.A. Tectonostratigraphic terrane analysis on Neoproterozoic times: the case study of the Araxá Synform, Minas Gerais, Brazil - Implications to the final collage of the Gondwanaland. Revista Brasileira de Geociências 30(1), 78–81, 2000.

Strieder, A.J., Nilson, A.A. Estudo petrológico de alguns fragmentos da mélange ofiolítica em Abadiânia (GO): I - O protolito dos corpos de serpentinito. Revista Brasileira de Geociências 22(3), 338–362, 1992.

Suita, M.T.F. Geoquímica e metalogenia de platinoides em complexos máfico-ultramáficos do Brasil: Critérios e guias com ênfase no complex acamadado de alto grau Barro Alto (CBA, Goiás). UFGRS, Doctorate Thesis, 1996.

Tanizaki, M.L.N., Campos, J.E.G., Dardenne, M.A. Estratigrafia do Grupo Araí: registro de rifteamento paleoproterozoico no Brasil Central. Brazilian Journal of Geology 45(1), 95–108, 2015.

Teixeira, A.N., Poli, C.J. Barbosa, F.M., 1982, Contribuição à geologia e petrologia da região de São Domingos. Ver. Brasileira de Geociências, 12(4): 562–571.

Uhlein, A., Alvarenga, C.J.S., Dardenne, M., Trompette, R.R. The glaciogenic Jequitaí Formation, southeastern Brazil. Geological Society, London, Memoirs 36, 541–546, 2011.

Uhlein, A., Trompette, R., Alvarenga, C.J.S. Neoproterozoic glacial and gravitational sedimentation on a continental rifted margin: The Jequitaí-Macaúbas sequence (Minas Gerais, Brazil). Journal of South American Earth Sciences 12, 435–451, 1999.

Valeriano, C.M., Machado, N., Simonetti, A., Valladares, C.S., Seer, H. J., Simões, L.A. U-PB geochronology of the southern Brasilia belt (SE-Brazil): sedimentary provenance, Neoproterozoic orogeny and assembly of West-Gondwana. Precambrian Research 130, 27–55, 2004.

Valeriano, C.M., Pimentel, M.M., Heilbron, M., Almeida, J.C.H., Trouw, R. Tectonic evolution of the Brasília Belt, Central Brazil, and early assembly of Gondwana. Geological Society of London Special Publication 294, p. 197–210, 2008.

Vieira, L.C., Almeida, R.P., Trindade, R.I.F., Nogueira, A.C.R., Janikian, L. A, 2007. Formação Sete Lagoas em sua área-tipo: fácies, estratigrafia e sistemas deposicionais. Revista Brasileira de Geociências 37 (suplemento), 168–181.

Walde, D.H.G. Neue Hinweise für eine Proterozoische Vereisung in Ostbrasilien. Münstersche Forschungen zur Geologie und Paläontologie 38/39, 17–59, 1976.

Walker, R.J., Prichard, H., Ishiwatari, A., Pimentel, M.M. The osmium isotopic composition of the convecting upper mantle deduced from ophiolite chromites. Geochimica et Cosmochimica Acta 66(2), 329–345, 2002.

Wernick, E., Almeida, F.F.M. The geotectonic environments of Early Precambrian granulites in Brazil. Precambrian Research 8(1), 1–17, 1979.

The Rio Preto and Riacho do Pontal Belts

12

Fabrício A. Caxito, Alexandre Uhlein, Elton Dantas, Ross Stevenson, Marcos Egydio-Silva, and Silas S. Salgado

Abstract

Together, the Rio Preto and Riacho do Pontal belts form a 600 km-long orogenic system developed along the northwestern and northern margins of the São Francisco craton during the Neoproterozoic Brasiliano orogeny. Involving the Paleoproterozoic ($\sim$1.9 Ga) Formosa Formation (schist, quartzite, greenschist and amphibolite) and the Neoproterozoic (900–600 Ma) Canabravinha Formation (metadiamictite, metawacke, metaturbidite), the Rio Preto fold belt, exposed in Bahia and Piauí states, borders the craton to the northwest. Neoproterozoic deformation between 600 and 540 Ma originated a complex, asymmetrical and double-verging thrust wedge, whose southern branch propagated for over 100 km into the craton interior in form of a thin-skinned deformation front. The Rio Preto belt probably represents an inverted Neoproterozoic hemi-graben developed along the northern margin of the craton. The Riacho do Pontal fold belt occupies the northern margin of the craton. Its external zone is made up of a south-verging thin-skinned nappe system detached along the basement-cover contact. Ages of syn- to late-collisional granitic intrusions suggest that the main deformation phase in the Riacho do Pontal belt occurred between 667 and 555 Ma. The Barra Bonita Formation (quartzite, schist and marble), a correlative of the Una Group in craton interior (Paramirim aulacogen), represents a platformal unit, deposited on the northern São Francisco passive margin. The Monte Orebe metabasalts, exposed further north in the central sector of the belt, might represent remnants of a Neoproterozoic oceanic crust.

Keywords

Rio Preto belt • Riacho do Pontal belt • Brasiliano orogeny • Neoproterozoic • Basin inversion • Passive margin basin

F.A. Caxito (✉) · A. Uhlein · S.S. Salgado
Instituto de Geociências, Departamento de Geologia, Centro de Pesquisas Manoel Teixeira da Costa Universidade Federal de Minas Gerais, Campus Pampulha. Av. Antônio Carlos 6627, Belo Horizonte, MG 31270-901, Brazil
e-mail: boni@ufmg.br

E. Dantas
Instituto de Geociências, Universidade de Brasília, Campus Universitário, Asa Norte, Brasília, DF 70910-900, Brazil

R. Stevenson
GEOTOP, Université du Québec à Montréal, P.O. Box 8888 Station Centre Ville, Montreal, QC H3C 3P8, Canada

M. Egydio-Silva
Instituto de Geociências, Universidade de São Paulo, Rua do Lago 562, São Paulo, SP 05508-080, Brazil

M. Heilbron et al. (eds.), *São Francisco Craton, Eastern Brazil*, Regional Geology Reviews, DOI 10.1007/978-3-319-01715-0_12

12.1 Introduction

The Rio Preto and Riacho do Pontal belts form together a 600 km-long orogenic system developed along the north and northwest margins of the São Francisco craton (Fig. 12.1). In the context of the fold-thrust belts that surround the craton and Brasiliano systems in general, the Rio Preto and Riacho do Pontal represent the least studied orogenic zones.

In this chapter, we present a synthesis on the state-of-the art of the geological understanding of each of these fold belts, exploring the stratigraphic and structural relations between them and the adjoining cratonic area (Table 12.1). In order to establish links to the adjacent cratonic domain, the descriptions provided here focus mainly the external zones of each belt, rather than their internal sectors.

12.2 The Rio Preto Belt

12.2.1 Stratigraphy

12.2.1.1 Basement Units

The basement (Figs. 12.2 and 12.3), exposed in the northern and more internal portions of the belt, is represented by the Cristalândia do Piauí Complex (Arcanjo and Braz Filho 1994), composed of biotite and hornblende orthogneisses with amphibolite intercalations. The regional dominant structure is a southeast-dipping gneissic foliation, which transposes older structures and parallels the axial-plane of isoclinal folds. Available geochronological data suggest either crystallization or metamorphism around 2.1 Ga (Rb-Sr whole-rock isochron). K-Ar geochronological deter-

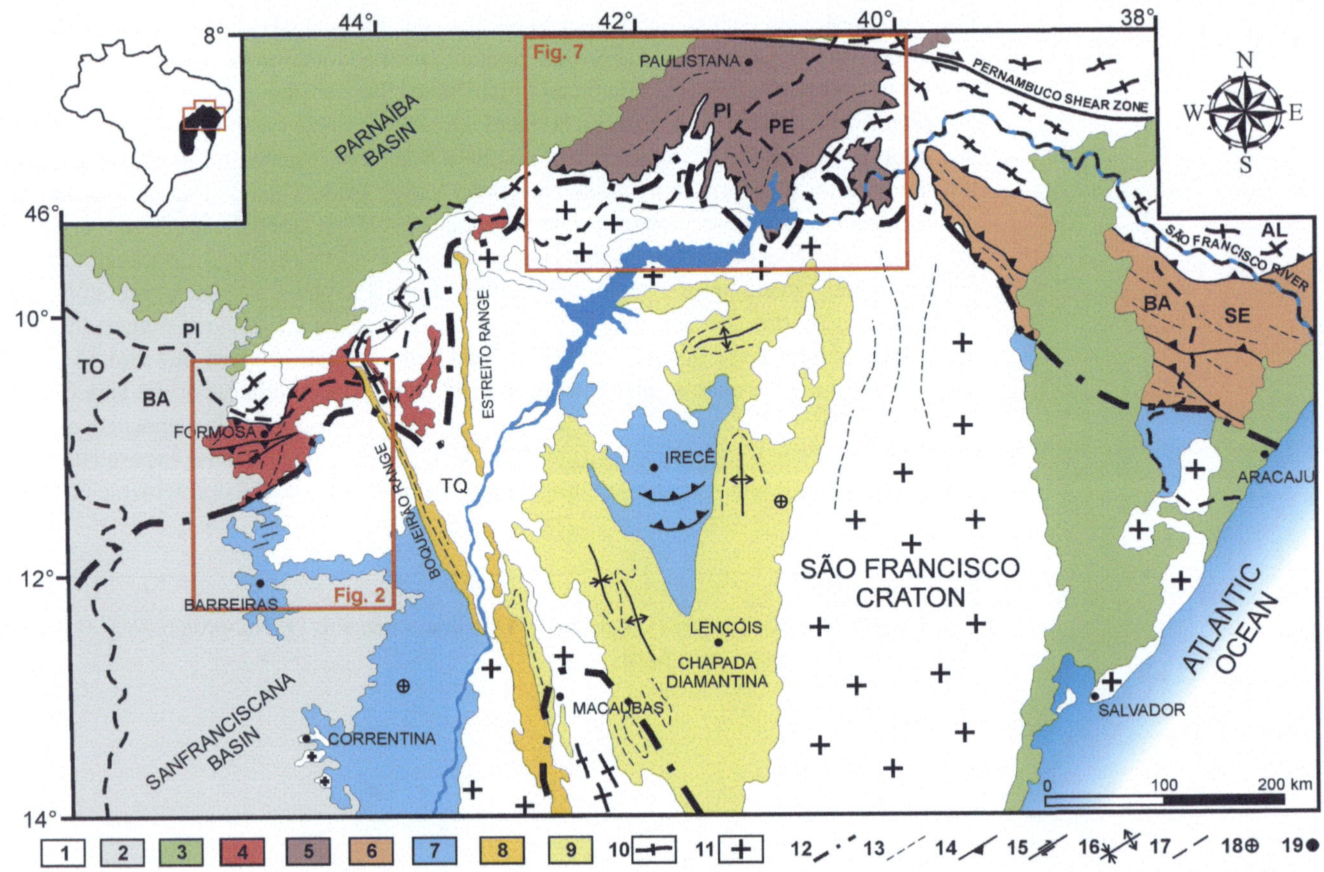

Fig. 12.1 Simplified geologic map of the northern São Francisco craton and its marginal fold belts (modified from Uhlein et al. 2011; 2012). *1*—Cenozoic covers; *2*—Cretaceous Urucuia Group; *3*—Paleo and Mesozoic basins; Northern São Francisco Craton marginal fold belts: *4*—Rio Preto; *5*—Riacho do Pontal; *6*—Sergipana; Proterozoic covers of the São Francisco Craton: *7*—Bambuí Group and correlatives; *8*—Santo Onofre Group; *9*—Espinhaço Supergroup; Archean/Paleoproterozoic basement: *10*—Reworked basement within the orogenic domain; *11*—São Francisco Craton basement; *12*—Craton/fold belt boundary; *13*—Structural lineaments; *14*—Thrust; *15*—Strike-slip shear zone; *16*—Synclinal/anticlinal axis; *17*—Interstate border; *18*—Horizontal bedding; *19*—City

Table 12.1 Comparative stratigraphy of the Rio Preto and Riacho do Pontal fold belts and adjoining cratonic areas

Age	Rio Preto belt	NW São Francisco Craton	Riacho do Pontal Belt		Northern São Francisco Craton
			Volcanosedimentary sequences	*Plutonic suites*	
Ediacaran/Cambrian (ca. 630–530 Ma)		Bambuí Group	Mandacaru Formation	Serra da Aldeia/Caboclo Suite	Una Group
				Serra da Esperança Suite	
				Rajada Suite	
Cryogenian/Early Ediacaran (ca. 820–630 Ma)	Canabravinha Formation		Barra Bonita Formation		
			Monte Orebe Complex		
Late Tonian/Early Cryogenian (ca. 900–820 Ma)			Paulistana Complex	Brejo Seco Complex	
				São Francisco de Assis Complex	
Early Tonian (ca. 1000–960 Ma)			Morro Branco Complex Santa Filomena Complex (?)	Afeição Suite	
Archean/Paleoproterozoic	Formosa Formation	Basement in the Correntina region	Morro do Estreito Complex		São Francisco craton basement and Paleoproterozoic covers
	Cristalândia do Piauí Complex				

minations on biotite crystals yielded ages around 540 Ma (Egydio-Silva 1987). Recent published Sm–Nd model ages (T_{DM}) between ca. 2.8 and 2.6 Ga suggest that an Archean crust is the main component of this segment of the basement (Caxito 2010; Caxito et al. 2014a). Altogether, the data suggests the imprinting of a Rhyacian tectono-metamorphic event upon an Archean continental crust, followed by the resetting of the K-Ar clock during the Brasiliano orogeny.

The Formosa Formation (Figs. 12.2 and 12.4) corresponds to the Rio Preto Group of Egydio-Silva et al. (1989), as redefined by Caxito et al. (2012a). This unit crops out in the northern portion of the belt, alongside the Preto river valley, extending for ca. 20 km towards north, reaching the region of the boundary between the states of Bahia and Piauí (Figs. 12.2 and 12.3). It is composed of muscovite schists, often garnet-rich, with layers of micaceous quartzite, sandy metarhythmite, iron-manganese metachert (the guide-layer of the unit), chlorite-actinolite-epidote schist (greenschist), and, locally, ortho-amphibolite intercalations. The latter are particularly well exposed at the Angico farm, to the west of Formosa do Rio Preto town, where a 200 m-thick ortho-amphibolite layer is concordantly intercalated with the mica schists of the Formosa Formation. Caxito (2010) and Caxito et al. (2015) conducted a petrographic, geochemical, isotopic and geochronological study on these amphibolites and interpreted them as tholeiitic gabbros metamorphosed under epidote-amphibolite facies conditions (around 500 °C and 2–5 kbar). Flat chondrite-normalized rare earth elements (REE) patterns and slight enrichment of light REE compared to heavy REE (La_N/Yb_N: 1.35–2.97; Eu/Eu*: 0.94–1.14), negative Nb and positive Sr anomalies, as well as high LIL/HFS ratio suggest an hybrid of mid-ocean ridge and island arc environments, probably reflecting a back-arc setting (Caxito 2010; Caxito et al. 2015). U-Pb analyses of magmatic zircon crystals (Th/U: 0.11–1.56) yielded a precise age of 1958.3 ± 16 Ma, with $\varepsilon Nd_{(1.96)}$ ranging from slightly negative to positive values (−0.3 to +1.0), suggesting variable mixing of a depleted mantle source and older continental crust (Caxito et al. 2015).

Studies on the detrital zircon record indicates a relatively monotonous sedimentary provenance for the Formosa Formation, with bimodal distribution of detrital zircon U-Pb ages between 1.9–2.2 and 2.5–2.6 Ga. T_{DM} model ages between 1.9 and 2.6 Ga corroborate the provenance data (Caxito 2010; Caxito et al. 2014a), indicating that the

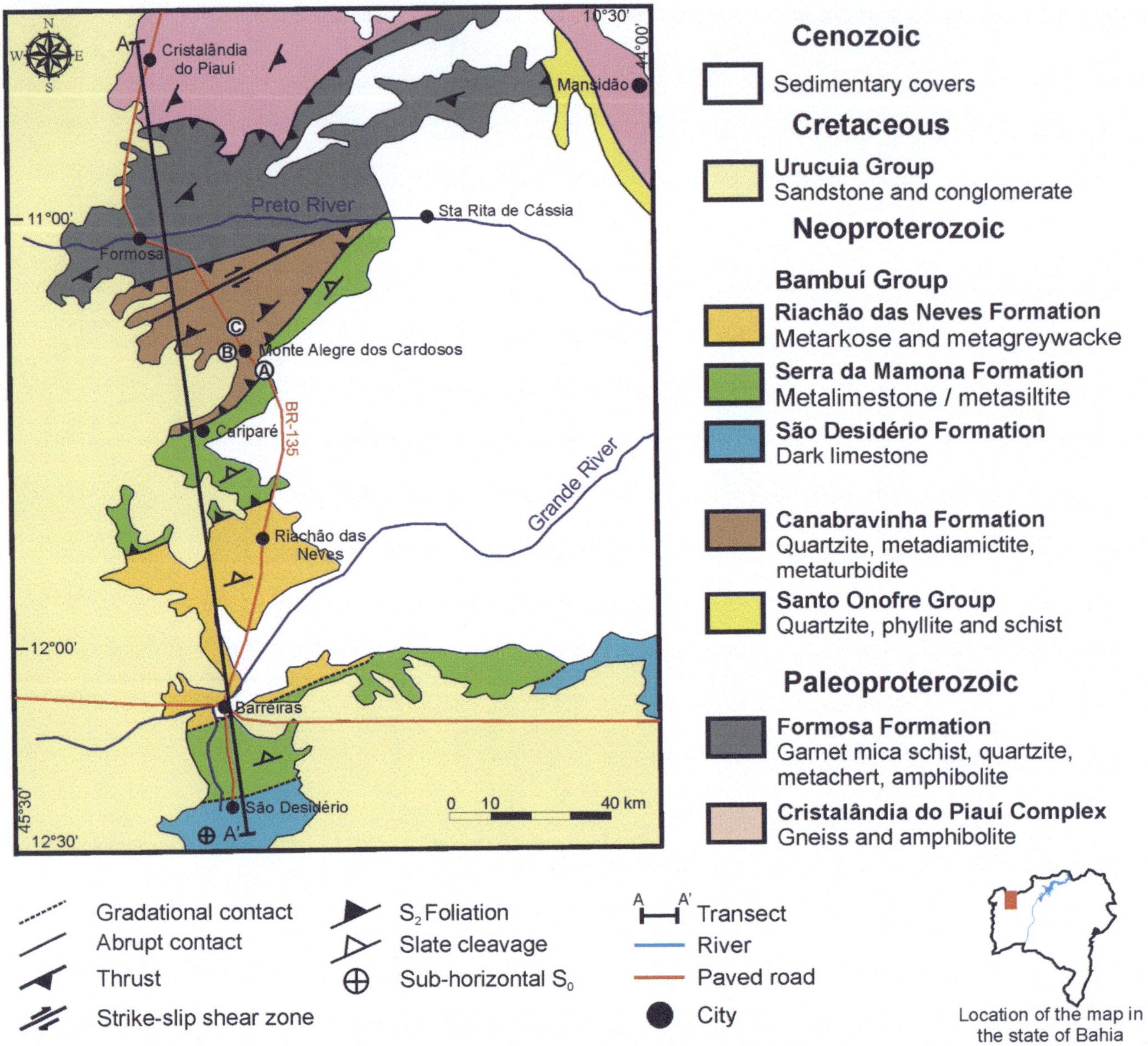

Fig. 12.2 Simplified geologic map of the Rio Preto fold belt and adjacent northwestern São Francisco craton (modified from Egydio-Silva 1987; Egydio-Silva et al. 1989; Caxito 2010)

basement assemblage of the Cristalândia do Piauí Complex was the main source for the Formosa Formation. The absence of Mesoproterozoic and Neoproterozoic zircons and the intercalation of Paleoproterozoic mafic rocks suggest that at least part of the Formosa Formation was deposited during the Paleoproterozoic ($\sim$1.96 Ga ago), probably in a basin related to a magmatic arc.

The Formosa Formation it is thrust northward over the gneisses of the Cristalândia do Piauí Complex. The south contact with the Canabravinha Formation is made by a reverse-dextral shear zone (Malhadinha-Rio Preto Shear Zone, Gonçalves Dias and Mendes 2008; Caxito et al. 2014b). The Formosa Formation is thrust on top of the quartzites of the Santo Onofre Group along the Boqueirão ridge on the east (Egydio-Silva 1987) (Figs. 12.2 and 12.3). The main Neoproterozoic regional structure affecting this unit is a Neoproterozoic crenulation cleavage (S_2). Rocks exhibiting this cleavage have yielded muscovite K-Ar ages between 600 and 540 Ma (Egydio-Silva 1987).

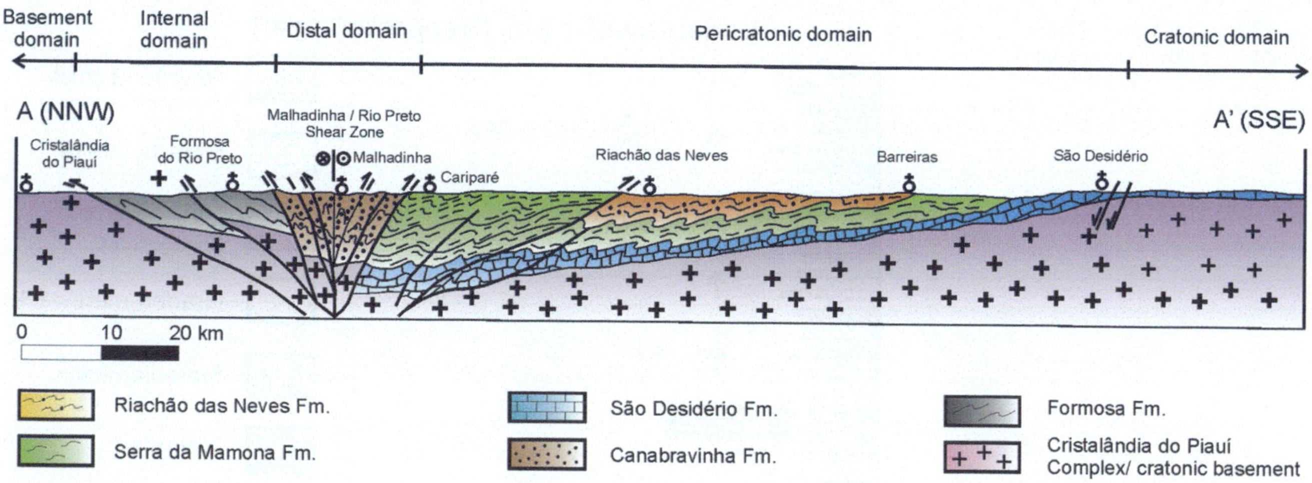

Fig. 12.3 Simplified transect of the Rio Preto fold belt and northwestern São Francisco craton. See Fig. 12.2 for location of the cross-section (modified from Egydio-Silva 1987; Egydio-Silva et al. 1989; Caxito 2010)

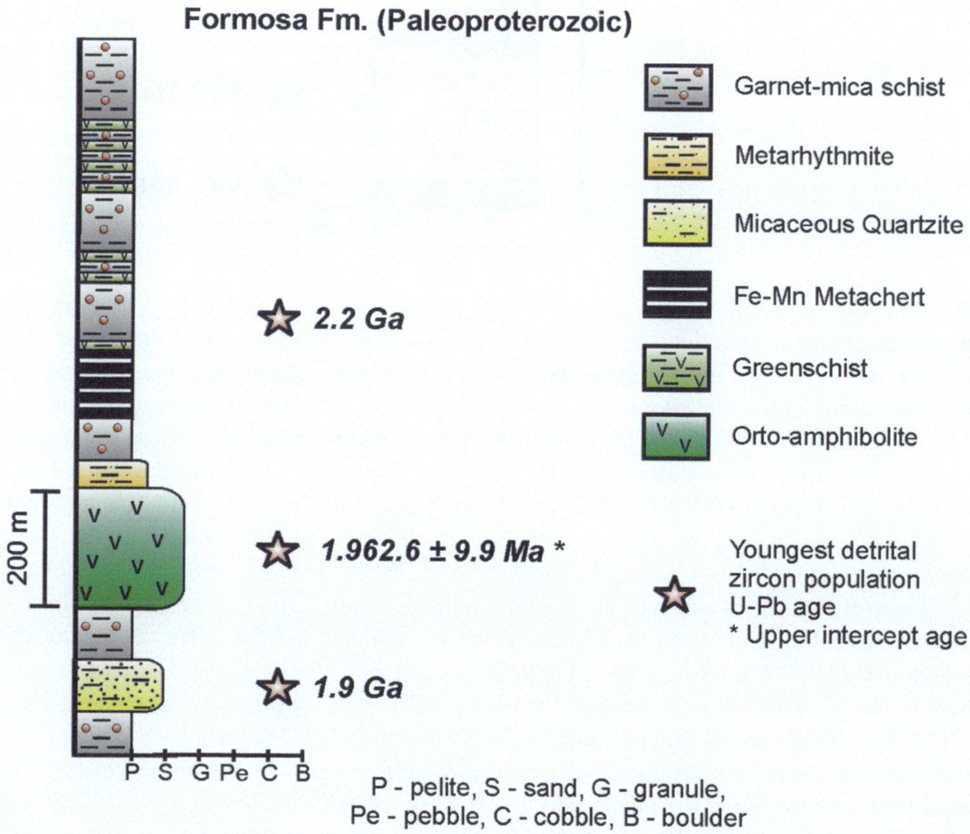

Fig. 12.4 Stratigraphic column of the Formosa Formation, Rio Preto belt (after Caxito et al. 2012a, 2014a)

12.2.1.2 Neoproterozoic Units

Canabravinha Formation

The Canabravinha Formation (Figs. 12.2 and 12.5) occurs in the southern portion of the Rio Preto belt, extending from the village of Cariparé towards north for almost 40 km. It is composed of texturally and mineralogically immature quartzites and metawackes (Fig. 12.6a), phyllite (locally carbonaceous), sand-pelite metarhythmite, metadiamictite (Fig. 12.6b), and, locally, metamarl (Fig. 12.5). The quartzites and metawackes show lithic, feldspathic, carbonate-rich and micaceous varieties. The conglomeratic varieties show graded (Fig. 12.6a), plane-parallel or cross-bedding, locally displaying climbing ripples. The vertical and lateral facies changes observed in the formation characterize a transition from coarse-grained rocks, on the south (metadiamictite,

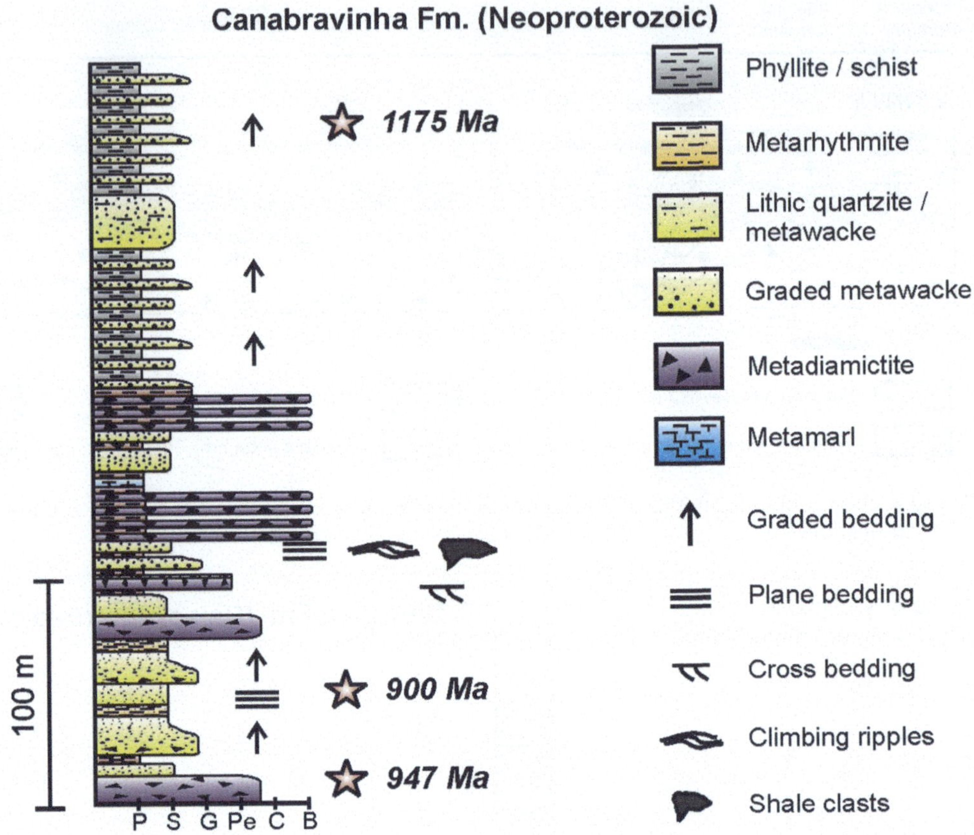

Fig. 12.5 Stratigraphic column of the Canabravinha Formation, Rio Preto belt (after Caxito et al. 2012a, 2014a)

metawackes), to medium and fine-grained facies, on the north (metarhytmite, phyllite). Caxito et al. (2012a) interpret the deposition of the Canabravinha Formation as occurring in a submarine gravel-rich slope-apron environment.

The spectrum of U-Pb ages of the detrital zircons of the Canabravinha Formation differs from the Formosa Formation, spreading from 3000 to 900 Ma. The Nd isotopes also indicate a larger variety of sources, with T_{DM} model ages between 1.5 and 2.7 Ga (Caxito 2010; Caxito et al. 2014a).

Caxito et al. (2012b) analyzed the isotopic composition of twelve carbonate clasts from the Canabravinha Formation, which yielded $\delta^{13}C$ values from −4.5 to 0 ‰, and $\delta^{18}O$ values of a more limited range, between −13.1 and −10.6 ‰. These results suggest erosion of a pre-glacial carbonate platform, for the fact that pre-glacial carbonate platforms are in general characterized by negative $\delta^{13}C$ anomalies, such as the Islay (~750 Ma) and Trezona (~635 Ma) anomalies documented worldwide (Halverson et al. 2010). No direct evidence for glacial influence on the deposition of the Canabravinha Formation such as dropstones and striated clasts were yet found. Likewise, a cap carbonate unit was never described in the Rio Preto fold belt. However, glacial periods favor the development of submarine slope-apron systems, as well as the associated lowering of the sea-level, causing exposure and erosion of the continental platform. Thus, the Canabravinha Formation could represent submarine gravitational re-sedimentation of glacially-related sediments.

The Canabravinha Formation is characterized by low to medium-grade metamorphism and complex structural evolution related to the Brasiliano orogeny, as indicated by K-Ar ages in muscovite around 590 Ma (Egydio-Silva 1987). The Canabravinha strata are thrust southward on top of the Serra da Mamona Formation of the Bambuí Group along the Cariparé shear zone and juxtaposed with the Formosa Formation by a reverse-dextral shear zone (Malhadinha-Rio Preto Shear Zone). Based on gravimetric data, Egydio-Silva (1987) estimates a thickness of almost 7500 m for the Canabravinha Formation. However, this estimation must be carefully interpreted, due to the high ductile deformation accommodated in these rocks, which are isoclinally folded in the central portion of the fold belt.

Bambuí Group

Egydio-Silva et al. (1989) subdivided the Bambuí Group in western Bahia state into three formations, including from the

Fig. 12.6 **a** Graded bedding in coarse metagreywacke of the Canabravinha Formation, near Monte Alegre dos Cardosos, Bahia; **b** Granite boulder in metadiamictite of the Canabravinha Formation, Canabravinha creek; **c** Open fold with a E–W trending axis affecting limestones of the São Desidério Formation in the cratonic domain, São Desidério town, Bahia; **d** cross-cutting relationships between north-dipping bedding (S_0) and cleavage (S_1) in metapelite of the Serra da Mamona Formation, near Barreiras, Bahia

base to the top the São Desidério, Serra da Mamona and Riachão das Neves formations. Although this author considered the previously described Canabravinha Formation as the basal unit of the Bambuí Group, Caxito et al. (2012a) suggests that it should be better treated as a separate unit.

The São Desidério Formation is composed of dark gray limestones with intercalations of marls and siltstones, with an estimated thickness of 450 m (Egydio-Silva et al. 1989). The basal contact is not observed. The São Desidério carbonates are potential correlatives of the Sete Lagoas Formation that unconformably cover the basement gneisses in the intracratonic São Francisco basin (see Reis et al. 2017), as suggested by the upward increasing of $\delta^{13}C$ values exhibited by both units (Sial et al. 2010). The upper contact of the formation is characterized by a progressive increase in the proportion of clastic material in the carbonates, which grade into the metamarls and carbonatic slates of the Serra da Mamona Formation.

The Serra da Mamona Formation is marked by the intercalation of carbonate and pelite layers, affected by incipient to low grade metamorphism. Its thickness is estimated in 3000 m (Egydio-Silva et al. 1989), a value that must be carefully considered, for the intensively deformed unit.

On top of the Serra da Mamona Formation, the lithotypes of the Riachão dos Neves Formation show a gradual increase in grain size and a great amount of feldspar and lithic fragments, characterizing meta-arkoses and metagraywackes. This formation, with an estimated thickness of 4000 m, can be correlated to the Três Marias Formation (Egydio-Silva et al. 1989), exposed in the central São Francisco basin (see Reis et al. 2017).

12.2.2 Structure

In the northwestern Bahia state, S-N transect along route BR-135 between São Desidério and Cristalândia do Piauí highlights an example of the transition between the cratonic and the orogenic belt domains (Fig. 12.3). The undeformed flat laying Bambuí Group strata in the cratonic domain become progressively affected by flexural and concentric folds associated to an axial plane disjunctive cleavage in the pericratonic domain between the towns of Barreiras and Riachão das Neves (Fig. 12.6c). In the proximities of Cariparé, the folds become progressively tighter, becoming similar and isoclinal folds. In this region, the deformation history recorded by the Neoproterozoic rocks is quite complex and took place in three distinct phases (Egydio-Silva 1987; Caxito 2010; Caxito et al. 2014b).

The overall architecture of the Rio Preto belt is characterized by a double thrust wedge, in which the southern and wider portion displays a clear vergence towards the São Francisco craton, whilst the northern and narrower branch verges north, as exemplified by the low angle thrust bringing the Formosa Formation over the Cristalândia do Piauí gneisses (Fig. 12.3) (Egydio-Silva 1987; Caxito et al. 2014b).

As demonstrated by the K-Ar age determinations carried out by Egydio-Silva (1987), the development of the Rio Preto belt probably results from a polyphase evolution related exclusively to the Brasiliano orogeny (600–540 Ma). The main deformational foliation recognized in the Rio Preto fold belt rocks (S_2) corresponds to the axial plane cleavage documented in the Bambuí Group exposed in the craton interior.

Only one deformational phase, marked by SSE-verging folds, can be recognized in the cratonic cover domain between the towns of Cariparé and São Desidério. An axial plane slate cleavage preferentially oriented at N60–70E/50°NW is associated with these folds (Fig. 12.6d), which exhibit nearly vertical short limbs and shallow plunges toward NE or SW. Reverse, south-verging faults can locally invert the stratigraphy within the cratonic cover. The folds become progressively symmetrical further south, but still showing a weak south-vergence near São Desidério (Fig. 12.6c). South of this town, the Bambui strata are undeformed and flat lying. The deformation in the cratonic domain is typically thin-skinned and associated with a regional detachment located along the basement-cover contact. Egydio-Silva (1987) estimates a shortening of about 15–20 % for this domain. The Cariparé shear zone, marking the boundary between the craton and the fold belt, is a NE-trending fault zone, along which the Canabravinha Formation are thrust towards SE on top of Serra da Mamona Formation.

Three main deformation phases can be recognized in the Rio Preto fold belt (Egydio-Silva 1987; Caxito 2010; Caxito et al. 2014b):

- Phase D_1 generated a penetrative S_1 foliation, which is in general parallel to the sedimentary bedding (S_0). The tectonic meaning of this phase is obscure due to intensive transposition in the subsequent phases.
- Phase D_2 is the main deformational phase, responsible for development of the double-verging large-scale structure of the belt. Gentle and concentric folds affecting the Bambuí Group in the external zone grade progressively into tight to isoclinal folds in the internal portions of the fold belt. This phase also generated the large ductile/brittle shear zones, among them the Cariparé and Malhadinha/Rio Preto faults (Fig. 12.3).
- Phase D_3 generated gentle to open folds associated with a south-dipping crenulation cleavage (S_3), generated in the course of a final hinterland-verging compression.

Caxito (2010) and Caxito et al. (2014c) proposed that the double thrust wedge that dominates the structural picture of the Rio Preto belt was produced by the oblique convergence between the São Francisco craton on the south and the Cristalândia do Piauí block on the north. This convergence generated frontal thrusts followed by back-thrusts. The continuation of the process led to an overall right-lateral transpressional modification of the system, thereby generating the prominent Malhadinha-Rio Preto shear zone in the central portion of the belt (Fig. 12.3).

12.3 The Transition Between the Rio Preto and Riacho do Pontal Belts

Sparse occurrences of supracrustal rocks (mainly mica schists) mark the transition zone between the Rio Preto and Riacho do Pontal belts located between the Boqueirão range and Campo Alegre de Lourdes town in Bahia state (Fig. 12.1). In this region, the supracrustal units are strongly deformed and cut by a system of curved, SSE-verging thrusts (Arcanjo and Braz Filho 1999) in a similar way as further south in the Irecê basin, where the Neoprotezoic cratonic cover was also transported southwards along a system of arcuated thrusts (Fig. 12.1) (Danderfer Filho et al. 1993; Arcanjo and Braz Filho 1999). The Boqueirão and Estreito ranges, underlain by quartzites of the Santo Onofre Group (Egydio-Silva 1987), might have acted as lateral ramps linked to the curved SSE-verging thrusts. The craton boundary exhibits a complex and irregular geometry in map-view in this region, which is not yet covered with detailed geological mapping.

12.4 The Riacho do Pontal Belt

The Riacho do Pontal fold belt (Brito Neves 1975; Caxito 2013) borders the São Francisco craton to the north in the states of Bahia, Pernambuco and Piauí (Figs. 12.1 and 12.7) and is bounded on the north by the western branch of the Pernambuco lineament, a continental-scale dextral strike-slip shear zone. On the eastern sector, the belt grades discontinuously into the Sergipano belt (Oliveira et al. 2017); on northwest sector, it is covered by the Phanerozoic strata of the Parnaíba basin (Fig. 12.1).

The original concept of the Riacho do Pontal as a Brasiliano orogenic belt was challenged in the eighties by some authors, who interpreted the deformation and metamorphism in the region as a manifestation of the Paleoproterozoic Transamazonian event (Jardim de Sá and Hackspacher 1980). The first Rb-Sr whole rock geochronological data of syn- to late-collisional plutons performed in the nineties yielded ages around 555 Ma (Jardim de Sá et al. 1992, 1996), confirming thus the original assumption by Brito-Neves (1975).

The Riacho do Pontal belt is subdivided in three sectors: the internal, central and external zones (Oliveira 1998; Caxito 2013; Caxito and Uhlein 2013; Caxito et al. 2016) (Fig. 12.8).

12.4.1 Stratigraphy

12.4.1.1 Basement

The basement of the belt in the Sobradinho dam area (Fig. 12.7) is represented by the Gavião/Sobradinho block of the São Francisco craton (Barbosa and Sabaté 2004; Dantas et al. 2010; Teixeira et al. 2017; Barbosa et al. 2017). TTG-orthogneisses with tonalitic, granodioritic and leucogranite bands predominate in the region. Supracrustal sequences, composed mainly of quartzite and calc-silicatic

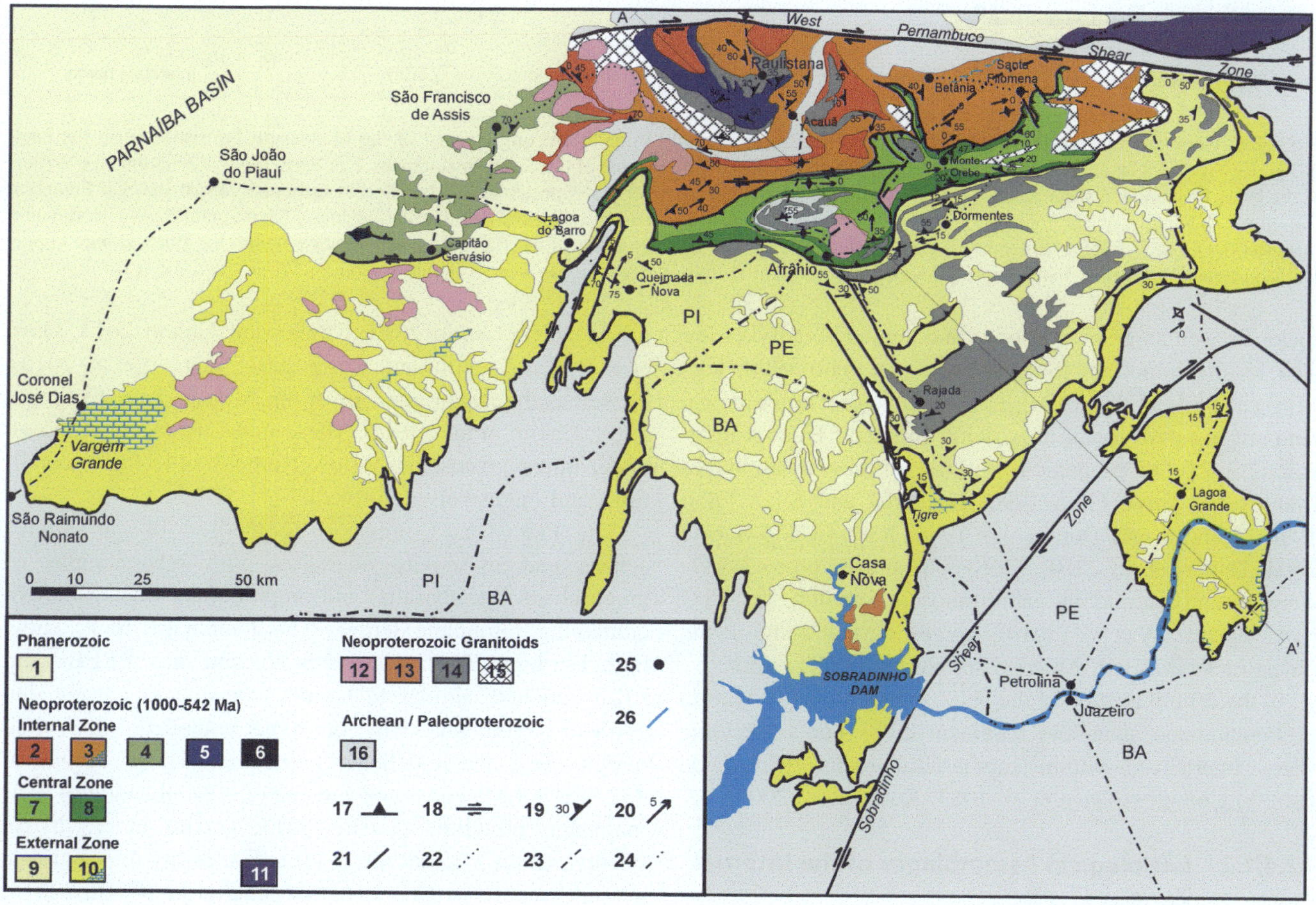

Fig. 12.7 Geologic map of the Riacho do Pontal fold belt (after Angelim and Kosin 2001 and Caxito 2013). *1*—Phanerozoic covers; *2*—Afeição Suite (1000–960 Ma); *3*—Santa Filomena Complex (brick pattern: marble); *4*—Morro Branco Complex; *5*—Paulistana Complex; *6*—Brejo Seco Complex; *7*—Monte Orebe Complex (metasedimentary rocks); *8*—Monte Orebe Complex (metavolcanic rocks); Casa Nova Group: *9*—Mandacaru Formation; *10*—Barra Bonita Formation (brick pattern: marble); *11*—Santana dos Garrotes Formation; *12*—Serra da Aldeia Suite; *13*—Serra da Esperança Suite (ca. 555 Ma); *14*—Rajada Suite (650-575 Ma); *15*—Undifferentiated granitoids; *16*—Archean/Paleoproterozoic basement; *17*—Thrust; *18*—Strike-slip shear zone; *19*—S_2 foliation; *20*—L_2 lineation; *21*—Interpreted photolineament; *22*—Land road; *23*—Paved road; *24*—Interstate border; *25*—City; *26*—São Francisco river

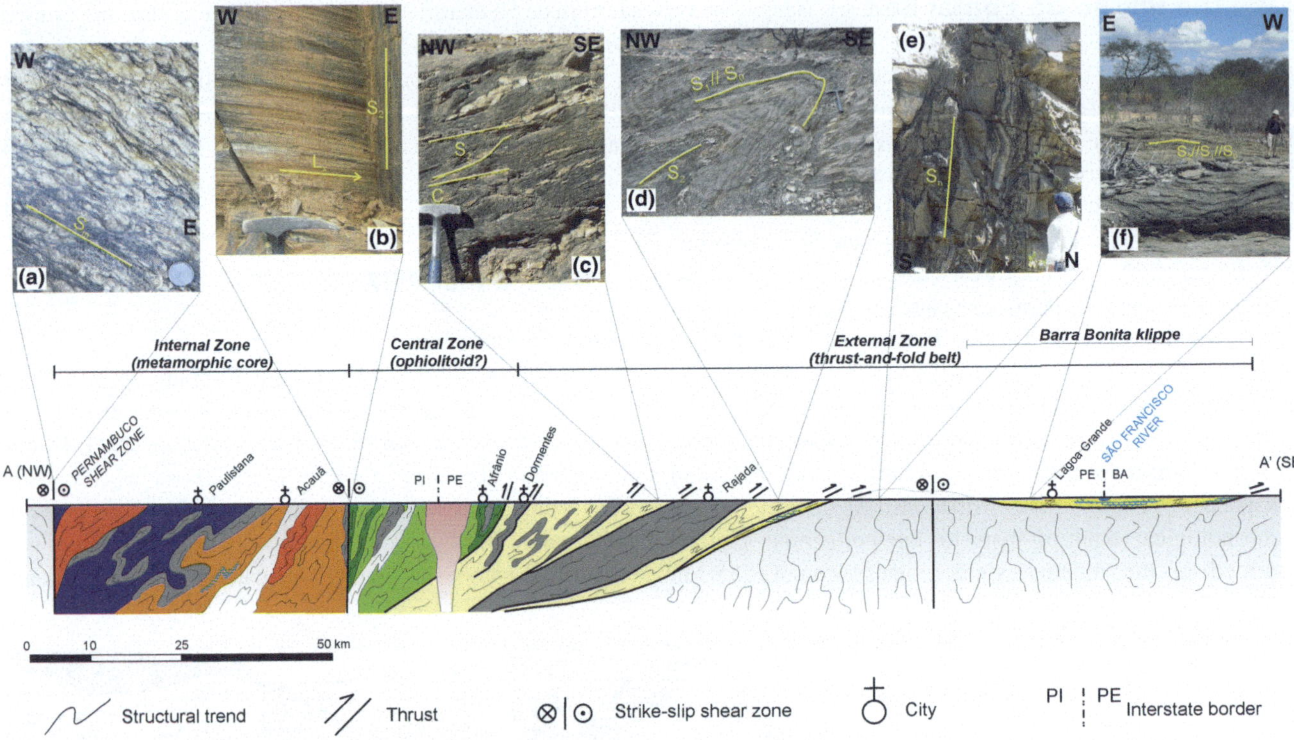

Fig. 12.8 Geologic transect of the Riacho do Pontal fold belt (Caxito 2013). Key is the same as in Fig. 7. **a** Mylonitic gneiss of the basement in the West Pernambuco shear zone; **b** greenschist of the Monte Orebe Complex, showing vertical foliation (S_2) and horizontal stretching lineation (Lx), near Afrânio, Pernambuco; **c** S-C structure in a schist of the Mandacaru Formation, indicating top-to-southeast motion; **d** tight F_2 fold affecting a schist of the Mandacaru Formation, near the basal thrust of the Casa Nova nappes, Pau Ferro town, Pernambuco; **e** vertical Sn (Paleoproterozoic?) banding in orthogneisses of the São Francisco basement, near Petrolina, Pernambuco; **f** horizontal $S_2//S_1//S_0$ foliation in schist of the Barra Bonita Formation within the Barra Bonita klippe, near Lagoa Grande, Pernambuco

rocks, as well as granitic plutons and amphibolite dykes are also locally important (Santos and Silva Filho 1990; Figueirôa and Silva Filho 1990). The available geochronological data suggests Archaean ages, with important Paleoproterozoic (2.2–2.0 Ga) orogenic reworking for these assemblages (Santos and Silva Filho 1990; Figueirôa and Silva Filho 1990; Barbosa and Dominguez 1996; Barbosa and Sabaté 2004; Dantas et al. 2010). Paleoarchean ages up to 3.5 Ga were recently found in gabbro-dioritic xenoliths in TTG rocks by Dantas et al. (2010), suggesting that some of the oldest rocks of South America are present in this region.

In the central portion of the belt, near the town of Afrânio in Pernambuco state, the basement crops out as tectonic slices interleaved within supracrustal rocks (Morro do Estreito Complex; Gava et al. 1983; Kosin et al. 2004).

12.4.1.2 Lithological Assemblages of the Internal Zone

The internal zone of the Riacho do Pontal belt (Figs. 12.7 and 12.8) involves volcanosedimentary sequences (Paulistana, Santa Filomena and Morro Branco complexes), intruded by large volume of igneous rocks, including the mafic-ultramafic complexes of Brejo Seco and São Francisco de Assis (Gomes and Vasconcelos 1991; Angelim and Kosin 2001; Caxito 2013; Caxito and Uhlein 2013; Gava et al. 1983; Angelim and Kosin 2001; Salgado et al. 2016), as well as the augen-gneisses of the Afeição Suite (Angelim 1988; Caxito et al. 2014c). These units are metamorphosed under upper greenschist to lower amphibolite facies conditions and intensively deformed.

The Paulistana Complex is composed of garnet-mica schists and muscovite quartzites with intercalations of greenschists (metabasalts) and amphibolites (metagabbros), containing ultramafic lenses. The metamafic rocks show high-Ti, high Th/Yb and Nb/Yb and are LREE- and LILE-enriched, similar to basalts extruded in present day evolved continental rifts or hyper-extended continental margins (e.g. the Red Sea rift margins). The U-Pb age of 882.8 ± 4.4 Ma constrains the age of crystallization of their magmatic protoliths (Caxito 2013; Caxito et al. 2016). Positive $\varepsilon Nd_{(882\ ma)}$ around +4.0 indicates involvement of a juvenile mantle source in their generation (Caxito 2013; Caxito et al. 2016).

The Santa Filomena Complex comprises coarse muscovite, biotite, garnet, kyanite, staurolite, cordierite and sillimanite schists, with local calcitic marble intercalations.

The Morro Branco Complex (Caxito 2013; Caxito and Uhlein 2013) consists of fine-grained quartz-mica schist and phyllite, metachert, quartzite, metabasalts, intermediate to acid metavolcanics, basic to felsic metatuffs, as well as subordinate graphite-schists. The intermediate to acid metavolcanics comprise rhyodacite, dacite, rhyolite and crystal tuffs. The metabasalts locally preserves amygdaloidal structures, suggesting eruption in a shallow, low-pressure environment. No geochronological data is available for the Morro Branco Complex.

Angelim and Kosin (2001) postulated that the above mentioned volcanosedimentary complexes are older than the intrusive 1000–960 Ma Afeição Augen Gneiss (Jardim de Sá et al. 1988, 1992; Van Schmus et al. 1995; Caxito et al. 2014c). Caxito et al. (2014c) presented a systematic petrographic, lithogeochemical, geochronological and isotopic study on the plutonic rocks of the Afeição Suite, suggesting that they represent the southwesternmost edge of the Tonian Cariris Velhos magmatic arc of the Transversal Zone of the adjacent Borborema Province (Kozuch 2003; Santos et al. 2010). U-Pb zircon ages of 1000–960 Ma, εNd(t) between −1.0 and +3.1 and T_{DM} model ages of 1.2–1.5 Ga corroborate this correlation (Van Schmus et al. 1995; Caxito et al. 2014c). However, at least the Paulistana Complex seems to be younger than the Afeição Suite, as indicated by the aforementioned U-Pb metagabbro age. The Morro Branco Complex and part of the Santa Filomena Complex might, on the other hand, be older than the Afeição Suite and probably representing volcanosedimentary sequences related to the Cariris Velhos event, a hypothesis that demands further tests.

The mafic-ultramafic Brejo Seco Complex is an approximately 10 km-long intrusion, tectonically interleaved within the Morro Branco volcanosedimentary sequence (Fig. 12.7). It is composed of a thin basal mafic unit (gabbros and troctolites), followed by variably serpentinized dunite, layered troctolite, minor olivine gabbro, layered gabbro, leucogabbro, minor anorthosite, ilmenite gabbro and ilmenite-magnetitite. The whole complex, with a maximum thickness of ca. 3 km, is tectonically inverted, with the ultramafic units (on the north) sitting on top of the mafic units (on the south). Basic dykes, represented by amygdaloidal diabase, crosscut the upper-layered gabbros. Both the northern and southern contacts are marked by EW-trending reverse shear zones. Preliminary geochemical data points to a tholeiitic geochemical affinity for the plutonic rocks of the Brejo Seco Complex, which was then suggested to be characteristic of island arc rocks (Marimon 1990). However, a systematical study on the petrogenetic and lithochemical evolution of the Brejo Seco Complex (Salgado 2014) suggests instead that it represents a classic mafic-ultramafic layered intrusion emplaced in a continental rift setting around 900 Ma ago (Salgado et al. 2016). A thick lateritic cover containing nickel deposits is developed above the ultramafic assemblage (Santos 1984). Approximately 40 km to the northeast of the Brejo Seco mafic-ultramafic exposures (Fig. 12.7), sparse outcrops of mafic and ultramafic rocks can be found. The region is highly weathered, and the rare outcrops that can be found are of coarse gabbro and serpentinite. If a correlation of the São Francisco de Assis and Brejo Seco complexes is admitted, then the zone of influence of mafic-ultramafic magmatism in the western portion of the Riacho do Pontal fold belt would extend up to 60 km in the NE–SW direction.

12.4.1.3 Lithological Assemblages of the Central Zone

The central zone of the Riacho do Pontal belt is characterized by a 100 km-long, EW-trending synformal structure, known as the Monte Orebe synform (Kreysing et al. 1973; Angelim 1988; Moraes 1992), which extends from the town of Afrânio, in the state of Pernambuco to Paulistana, in the state of Piauí (Figs. 12.7 and 12.8). Moreover, the central zone is characterized by a paired positive-negative linear Bouguer anomaly typical for suture zones (Gibb and Thomas 1976), which shows a difference of ca. 50 mgal from peak-to-peak (Oliveira 1998).

The dominant unit in the central zone, the Monte Orebe Complex, is made up mainly of basic metavolcanics (actinolite schists, amphibolites and metatuffs), interleaved within deep-sea pelagic metasedimentary rocks, mainly metachert (locally iron-rich) and garnet-mica schist, with local metagreywacke and quartz-schist. Locally, millimetric vesicular structures can be found in otherwise massive metabasalts (Fig. 12.9c). Medium to coarse grained ortho-amphibolites are also common and generally concordant to the actinolite-plagioclase greenschists.

Preliminary lithochemistry data of major and selected trace elements suggests a tholeiitic, MORB-type affiliation for the igneous protoliths of the Monte Orebe Complex (Moraes 1992). Caxito et al. (2014d) present new lithogeochemical and Sm–Nd isotope data on the metabasalts. Trace and rare earth elements data confirm a T-MORB chemistry, and Sm–Nd isotope data yields a whole-rock isochron age of 819 ± 120 Ma with an initial εNd(t) = +4.4, indicating derivation from a depleted mantle source. The Monte Orebe Complex might thus contain oceanic crust remnants, thereby marking a suture zone within the central portion of the belt (Caxito et al. 2014d, 2016).

12.4.1.4 Lithological Assemblages of the External Zone

The external zone of the Riacho do Pontal belt comprises the Casa Nova Group (Souza et al. 1979; Santos and Silva Filho 1990; Figueirôa and Silva Filho 1990; Bizzi et al. 2007) subdivided into two units: the Barra Bonita and Mandacaru formations.

The Barra Bonita Formation, composed mainly of fine-grained metapelitic rocks and muscovite quartzite (Fig. 12.9a) with marble lenses (Fig. 12.9b), is interpreted as

Fig. 12.9 **a** Muscovite quartzite of the Barra Bonita Formation, Ponta da Serra, Piauí; **b** folded marble of the Barra Bonita Formation, near Afrânio, Pernambuco; **c** amygdaloidal metabasalt of the Monte Orebe Complex, Serra do Trancelim, Dormentes—Santa Filomena road; **d** south-verging folds affecting the limestones of the Salitre Formation, Una basin, Chapada Diamantina region (photo by Daniel G.C. Fragoso, UFMG)

deposited in a shallow marine, platformal setting (Santos and Silva Filho 1990; Caxito 2013; Caxito et al. 2016). Fine to medium grained grey mica schists and phyllites predominate, with quartz, biotite, muscovite, garnet porfiroblasts, and minor detrital feldspar as the main mineral phases. The quartzites are commonly whitish and schistose, with muscovite and minor detrital feldspar. They occur mainly near the base of the Barra Bonita formation, in contact with the basement migmatites and gneisses. The marble lenses can reach thicknesses of up to 200 m, locally preserving original sedimentary structures (Fig. 12.7).

The Mandacaru Formation is composed mainly of garnet mica schists with centimetric intercalations of metagraywacke with detrital feldspar, quartz, muscovite, garnet and chlorite. Based on the composition and preliminary geochemical data, a deep-sea turbiditic, syn-orogenic sedimentation is inferred for the Mandacaru Formation (Santos and Silva Filho 1990).

Ongoing sedimentary provenance studies suggests that the Barra Bonita Formation represents the continuation of the northern São Francisco craton passive margin, with a detrital zircon age spectra of 1.6–2.1 Ga and T_{DM} values between 2.4 and 1.5 Ga (Caxito 2013; Caxito et al. 2016). These patterns are very similar to those of the Una Group of the Paramirim aulacogen (Chapada Diamantina region) (Santos et al. 2012), and might reflect sources of the São Francisco craton basement, along with the Proterozoic volcanosedimentary units of the Paramirim aulacogen. $\delta^{13}C$ data and $^{87}Sr/^{86}Sr$ around 0.7074–0.7080 for the marble intercalations within the Barra Bonita Formation (Caxito 2013; Caxito et al. 2016) are also very similar from those of the Salitre Formation of the Una Group (Misi and Veizer 1998), suggesting a broad Neoproterozoic carbonatic platform in this area.

The Mandacaru Formation, on the other hand, presents quite distinct provenance patterns, with T_{DM} model ages around 1.6–1.4 Ga (Van Schmus et al. 2011; Caxito 2013; Caxito et al. 2016). U-Pb detrital zircon data for the Mandacaru Formation and for the upper portion of the Monte Orebe Complex indicates a peak at 1.0 Ga, with the youngest detrital zircons around 640–665 Ma (Caxito 2013; Brito Neves et al. 2015; Caxito et al. 2016). These data suggests assemblages of the internal zone of the belt, including the Afeição Suite augen-gneisses and Stage I Brasiliano granites of the Borborema Province (Van Schmus et al. 2011), as source for these units. Thus, an important shift in sedimentary provenance is marked by the transition from the Barra Bonita Formation (passive margin with provenance from the craton) into the Mandacaru Formation (syn-orogenic sedimentation with provenance from the internal zone (Caxito 2013; Caxito et al. 2016).

12.4.1.5 Neoproterozoic Granitoids

The Riacho do Pontal belt contains a large number of Neoproterozoic granitoid plutons emplaced from the syn- to the post-collisional development stages (Fig. 12.7).

The syn-collisional magmatism is represented by tabular bodies of mesocratic, fine to medium-grained two-mica orthogneisses (Rajada Suite; Siqueira Filho 1967; Santos and Caldasso 1978; Gomes and Vasconcelos 1991). Based on the geochemical characteristics of the Rajada Suite syn-collisional orthogneisses, Angelim (1988) suggested that they generated by melting of the Casa Nova metagreywackes, due to the heating produced by crustal thickening during the main deformation phase recorded in the belt. Preliminary U-Pb zircon data indicates crystallization ages between 620 and 575 Ma (Caxito 2013; Brito Neves et al. 2015; Caxito et al. 2016).

The Serra da Esperança Suite represents the syn- to late-collisional magmatism, and comprises grey to pinkish syenites, quartz-syenites and associated granites, as well as pegmatite and syeno-granite dykes, which intrudes the Casa Nova metasedimentary rocks in the Sobradinho Dam area (Pla Cid et al. 2000) (Fig. 12.7). A Rb-Sr whole-rock isochron age of 555 ± 10 Ma ($Ri = 0.7068 \pm 1$) reported by Jardim de Sá et al. (1996) is the best available estimation for emplacement and deformation of the Serra da Esperança Suite.

The last expression of magmatic activity in the Riacho do Pontal belt is represented by the late- to post-collisional Serra da Aldeia/Caboclo Suite. This unit occurs as circular to oval plutons, concentrated in the northwestern part of the belt (Gava et al. 1984). It is composed of grey to pink medium to coarse-grained syenite and K-feldspar granite, locally affected by late-stage shear zones. A limited number of chemical analyses performed in this unit indicates an alkaline affinity, with peralkaline/shoshonitic/potassic terms (Gava et al. 1984; Pla Cid et al. 2000).

12.4.2 Structure

The Brasiliano deformation took place in two main phases in the Riacho do Pontal belt: the contractional D_n and the strike-slip D_{n+1} phases (Gomes 1990; Gomes and Vasconcelos 1991; Caxito 2013; Caxito and Uhlein 2013).

The D_n phase corresponds to a progressive regional deformation, assisted by magmatism and metamorphism, during which a system of nappes were transported southwards along low-angle detachment surfaces. The frontal thrust zone of this system is marked by blastomylonites exhibiting a variety of kinematic indicators (Fig. 12.8c). The D_n phase led to the nucleation of three generations of folds and associated axial plane foliations. The most prominent among them is the S_2 foliation associated with the dominant south-verging tight to isoclinal F_2 folds (Fig. 12.8d) coeval to the Casa Nova nappes. A down-dip stretching lineation, plunging between NW and NNW, is normally associated with the S_2 foliation.

The metamorphism associated D_n phase reaches the amphibolite facies conditions and generated the Rajada Suite syn-collisional intrusions through crustal melts. The metamorphic isogrades show a inverse pattern within the nappes, the upper structures showing higher grade parageneses.

The manifestations of the second D_{n+1} phase are dextral strike-slip motions accommodated along EW-trending shear zones. The West Pernambuco shear zone is the master structure related to the D_{n+1} phase (Figs. 12.1 and 12.7). This continental scale shear zone is associated with a high-grade mylonitic foliation that overprints all units of the internal zone of the belt. Away from the Pernambuco shear zone, the D_{n+1} deformation is represented by a train of open folds, whose axial plane is marked by a crenulation cleavage (Fig. 12.7).

A number of subsidiary EW-trending sub-vertical shear zones related to this phase can be observed along the road connecting the towns Afrânio and Paulistana. These structures transpose preexistent contractional fabric elements, generating a penetrative sub-vertical mylonitic foliation with an associated E–W oriented stretching lineation (Fig. 12.8b). Syn-kinematic garnet porphyrclasts attests the overall dextral sense of shear along these structures.

Vauchez and Egydio-Silva (1992) estimated the P-T conditions for the development of the West Pernambuco shear zone at around 630–700° C and 6 kbar, which are consistent with syn-kinematic partial melting of the units involved. Towards east, the West Pernambuco shear zone terminates in form of a horsetail structure, which resolves the strike-slip deformation in a wide transpressional zone (Vauchez and Egydio-Silva 1992).

12.4.2.1 Effects of the Riacho do Pontal Deformation Within the Northern São Francisco Craton

Near the Sobradinho Dam area (external zone), a large number of isolated schist outcrops characterize klippen structures (Figs. 12.7 and 12.8f), which testify the original extension of the Neoproterozoic nappes. The basement outcrops between the Casa Nova nappe front and these klippen commonly show no sign of Brasiliano deformation, preserving the original Paleoproterozoic NS-trending vertical foliation characteristic of the Eastern Bahia orogenic domain (Barbosa and Barbosa 2017) (Fig. 12.8e). In contrast to the central and internal zones, the external domain corresponds to a thin-skinned deformation front, which propagates even further south, affecting the cratonic cover within the northern Paramirim aulacogen (see Cruz and Alkmim 2017). The contraction D_n deformation evolved, thus, from a thick-skinned domain on the north, to a thin-skinned zone on south, in which tectonic displacements are in the order of 30–60 km, as estimated by Jardim de Sá et al. (1992).

Proterozoic cover units of northern Paramirim aulacogen in the Chapada Diamantina region are affected by S-verging thrusts and folds that represent a manifestation of the Brasiliano thin-skinned deformation front 250 km south of the craton boundary (Figs. 12.1 and 12.9d) (Danderfer et al. 1993; Cruz and Alkmim 2017).

12.5 A Tectonic Model for the Northern São Francisco Craton Margin Belts

During the early Neoproterozoic (~900–820 Ma), the northern sector of the São Francisco paleocontinent was the site of intense crustal stretching, which led to the development of rift to passive margin basins. The widespread occurrence of mafic dyke swarms in the craton interior (Girardi et al. 2017) marks this extensional period. The emplacement of mantle plumes beneath the future cratonic margins might have been one of the causes for crustal warping and stretching. One of these plumes was probably located in the northern São Francisco craton margin and responsible for the emplacement of mafic-ultramafic complexes such as the Brejo Seco and São Francisco de Assis involved in the Riacho do Pontal belt (Salgado et al. 2016; Caxito et al. 2016). The Paulistana Complex, extruded in a continental rift environment at around 882 Ma (Caxito et al. 2016), also marks this phase of crustal stretching. The positive $\varepsilon Nd_{(882\ ma)}$ values around +4.0 suggests the involvement of large portions of a juvenile mantle source, such as in mantle upwelling areas, in the transition of active-type continental rifts to passive margins (Caxito 2013).

The development of the Rio Preto basin is probably related to this rifting event, which might extend throughout the Santo Onofre basin of the São Francisco craton and towards the Macaúbas basin of the Araçuaí Orogen further east (Fig. 12.10; Schobbenhaus 1996; Alkmim et al. 2017). The Canabravinha Formation was then deposited in a hemi-graben basin located within the Archean/Paleoproterozoic basement of the region (Fig. 12.10).

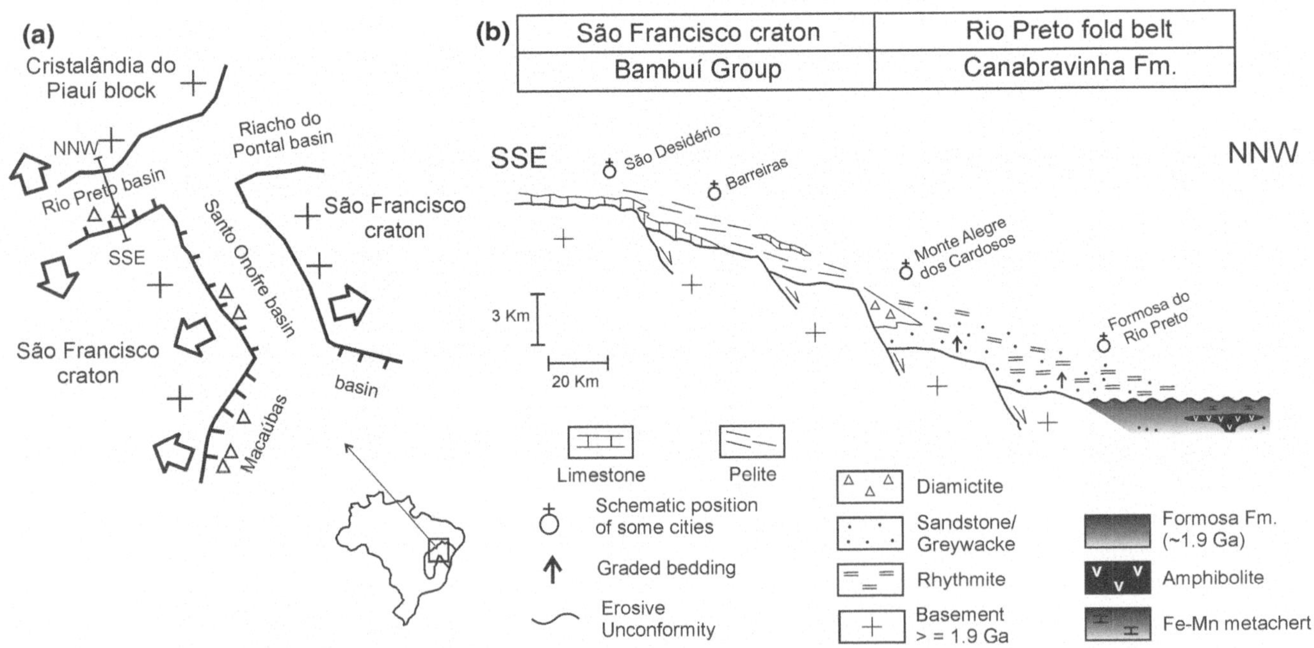

Fig. 12.10 Paleogeographic model for the Rio Preto area. See text for explanation (from Caxito et al. 2012a)

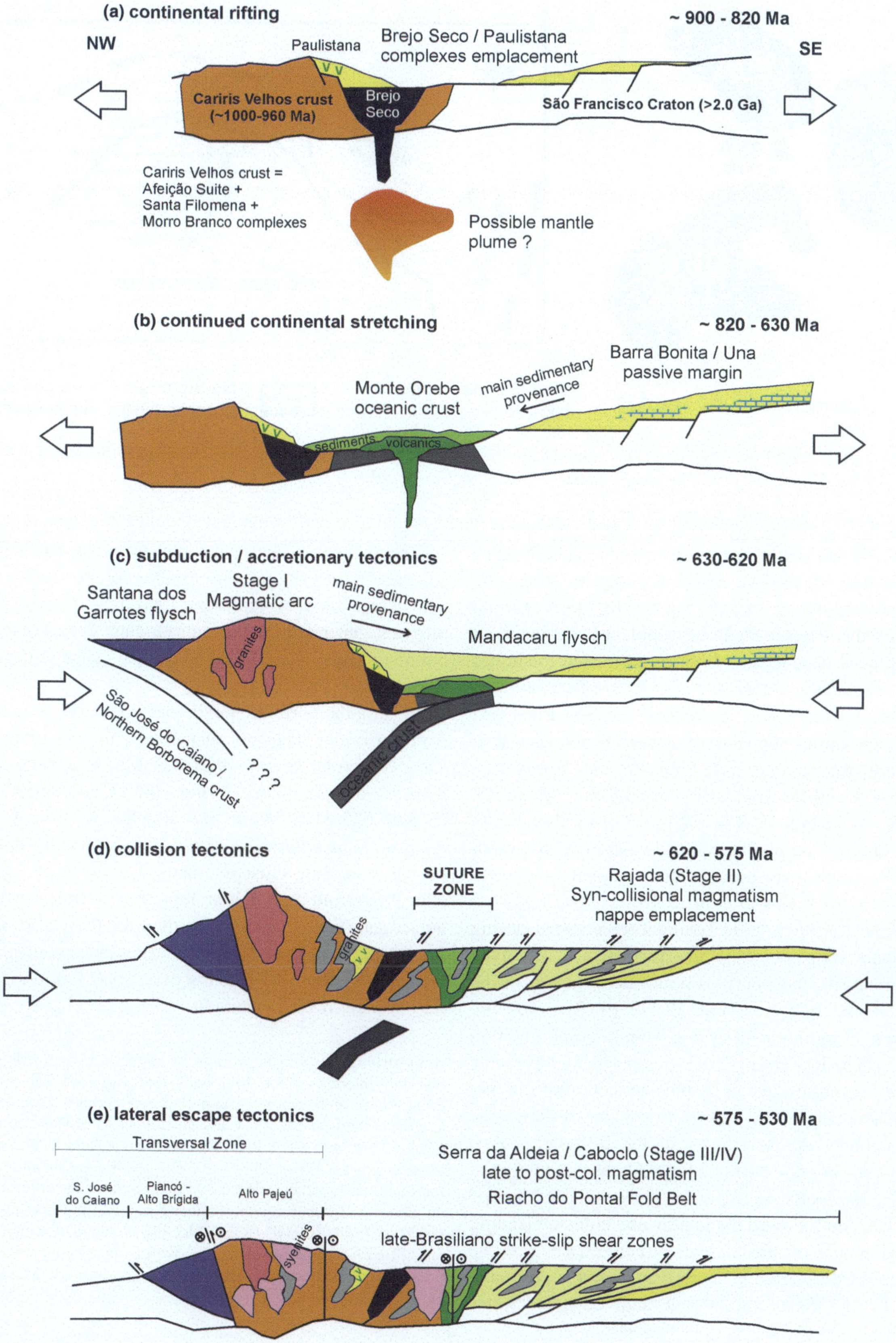

Fig. 12.11 Paleogeographic and tectonic model for the Riacho do Pontal belt. See text for discussion (from Caxito 2013; Caxito et al. 2016)

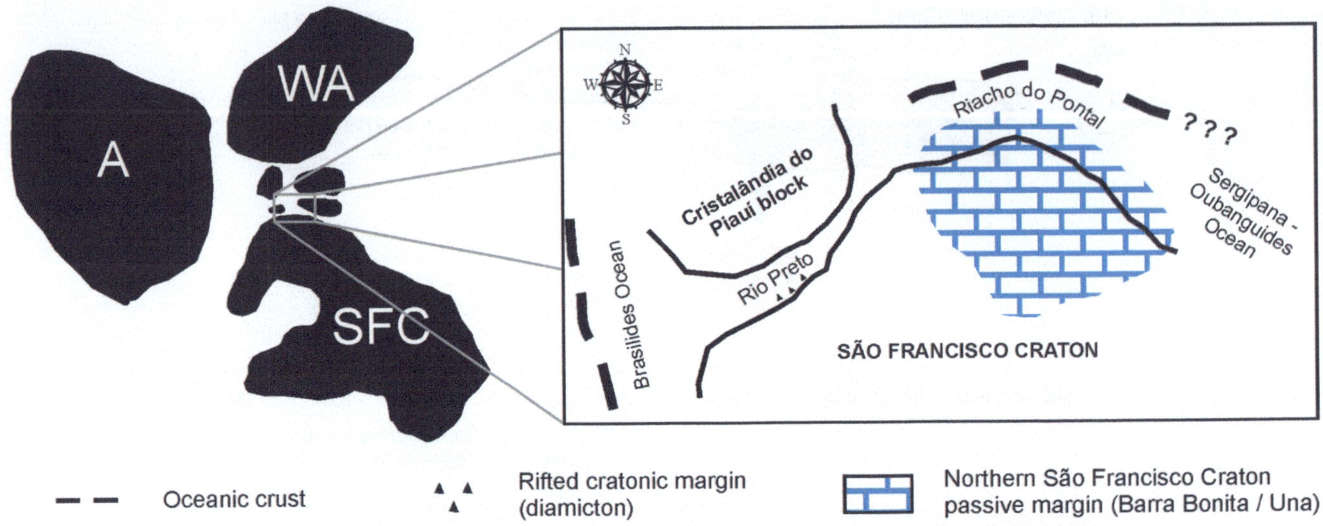

Fig. 12.12 Paleogeography of the northern São Francisco Craton margin and adjoining areas during the Late Neoproterozoic (ca. 650–630 Ma). *A* Amazon, *WA* West Africa, *SFC* São Francisco-Congo (modified from Caxito 2013)

Between 820 and 630 Ma, with progressive stretching of the continental crust, passive margins started to fringe the São Francisco continent on all sides. The Barra Bonita Formation of the Riacho do Pontal fold belt represents a typical platformal unit, which might be correlated to similar units within the cratonic area (e.g. the Salitre Formation of the Una Group). A tectonic model for the São Francisco northern margin during the Neoproterozoic would involve a broad passive margin represented by the Una Group carbonate platform and the Barra Bonita platformal quartzites with marble intercalations (Fig. 12.11a, b). Further north, the Monte Orebe Complex metabasalts seem to represent relicts of a Neoproterozoic oceanic crust coupled the passive margin (Caxito 2013; Caxito et al. 2016).

The rift and passive margin basins were inverted during the Brasiliano orogeny (630–530 Ma), which involved intense magmatism, metamorphism and deformation in the Riacho do Pontal belt area (Fig. 12.11c–e). Syn-orogenic sedimentation is represented by the Mandacaru Formation (Fig. 12.11c). The Riacho do Pontal belt probably represents a collisional orogen originated by the subduction of the São Francisco paleoplate towards north, below the western edge of the Pernambuco-Alagoas block of the Borborema Province, with consumption of the Monte Orebe oceanic crust (Fig. 12.11c, d). Finally, an important phase of lateral escape at ca. 575–530 Ma followed the continental collision leading to the development of crustal scale structures, such as the West Pernambuco shear zone, as well as granite and syenite intrusions of the Serra da Aldeia and Caboclo Suites (Fig. 12.11e).

The lack of evidence for oceanic crust and a Brasiliano subduction zone suggests an ensialic evolution for the Rio Preto belt, which seem to represent a strongly inverted rift basin. Stresses generated in the surrounding Tocantins and Borborema provinces could have been the cause of the inversion (Fig. 12.10).

In sum, the northern São Francisco craton margin evolved from an initial stage of crustal stretching and rifting in the early Neoproterozoic ($\sim$900–820 Ma) to a stage of passive margin development during the Cryogenian-Ediacaran ($\sim$820–635 Ma). While the Riacho do Pontal fold belt records the development of a full passive margin, including probable oceanic crust remnants (Caxito et al. 2014d), the Rio Preto fold belt might represent a failed rift arm, or aulacogen (Fig. 12.12). Altogether, the Rio Preto, Riacho do Pontal and Sergipano fold belts represents a continental scale orogenic system that extends over 1000 km along northern São Francisco craton margin.

Acknowledgments This work is the result of a sustained and ongoing research interest in the fold belts that surround the northern São Francisco craton margin, supported by various Brazilian funding agencies throughout the years, mainly FAPEMIG through grants CRA-505-06 and CRA-RDP-00120-10 (alongside VALE) and CNPq through grant 475510/2008-9. We also thank the Canadian Bureau for International Education and the Ministère de l'Éducation, du Loisir et du Sport du Québec for the financial support for short term research stages at the GEOTOP-UQàM-McGill, Montréal, through the ELAP and PBEEE programs. An original version of the manuscript was greatly improved after comments and suggestions by M. Heilbron.

References

Alkmim, F.F., Kuchenbecker, M., Reis, H.L.S., and Pedrosa-Soares, A. C. 2017. The Araçuaí Belt. In: U.G. Cordani et al. (eds.), São Francisco Craton, Eastern Brazil, Regional Geology Reviews, pp. xxx–xxx. doi: 10.1007/978-3-319-01715-0_14

Angelim, L.A.A. 1988. Programa Levantamentos Geológicos Básicos do Brasil-PLGB, carta geológica, carta metalogenética, Escala 1:100000 Folha SC.24-V-A-III, Santa Filomena, Estados de Pernambuco e Piauí. DNPM/CPRM, 146 p.

Angelim, L.A.A. and Kosin, M. 2001 (Org.) Folha Aracaju – NW. Nota Explicativa. CPRM – Serviço Geológico do Brasil. Programa Levantamentos Geológicos do Brasil, CD-Rom.

Arcanjo J.B.A. and Braz Filho P.A. 1994. O mapeamento geológico das folhas Curimatá/Corrente – Uma abordagem sobre os principais aspectos estruturais, estratigráficos e geomorfológicos. In: SBG, Congresso Brasileiro de Geologia, 38, Balneário Camboriú, Bol. de Res. Exp., 2:106-107.

Arcanjo J.B.A. and Braz Filho P.A. (ed.) 1999. Programa levantamentos geológicos básicos do Brasil. Folhas Curimatá (SC.23-Z-A), Corrente (SC.23-Y-B-Parcial) e Xique-Xique (SC.23-Z-B-Parcial), escala 1:250.000. CPRM, Brasília: 84 p.

Barbosa, J.S.F. and Dominguez, J.M.L. 1996. Mapa Geológico do Estado da Bahia, Texto Explicativo. SGM – UFBA, Salvador, 188 p.

Barbosa J.S. and Sabaté P. 2004.Archean and Paleoproterozoic crust of the São Francisco Craton, Bahia, Brazil: geodynamic features. Precambrian Reseach. 133:1–27.

Barbosa, J.S.F. and Barbosa, R.G. 2017. The Paleoproterozoic Eastern Bahia Orogenic Domain. In: U.G. Cordani et al. (eds.), São Francisco Craton, Eastern Brazil,Regional Geology Reviews, pp. xxx–xxx. doi: 10.1007/978-3-319-01715-0_4

Bizzi, L.A., Schobbenhaus, C., Gonçalves, J.H., Baars, F.J., Delgado, I. M., Abram, MD., Neto, R.L., Matos, G.M.M., Santos, J.O.S. 2007. Mapa Geológico do Brasil, escala 1:2 500 000. MME-CPRM.

Brito Neves, B.B. 1975. Regionalização Geotectônica do Pré-Cambriano Nordestino. Instituto de Geociências da Universidade de São Paulo, Tese de Doutoramento, São Paulo, 198 p.

Brito Neves, B.B., Van Schmus, W.R., Angelim, L.A.A., 2015. Contribuição ao conhecimento da evolução geológica do Sistema Riacho do Pontal – PE, BA, PI. Geologia USP - Série Científica, 15 (1):57-93.

Caxito, F.A., 2010. Evolução tectônica da Faixa Rio Preto, Noroeste da Bahia/ Sul do Piauí. Msc Dissertation, Universidade Federal de Minas Gerais, Belo Horizonte, 151 p.

Caxito, F.A., 2013. Geotectônica e evolução crustal das faixas Rio Preto e Riacho do Pontal, estados da Bahia, Pernambuco e Piauí. PhD Thesis, Universidade Federal de Minas Gerais, Belo Horizonte, 288 p.

Caxito, F.A., Uhlein, A., Sanglard, J.C.D., Gonçalves Dias, T., Mendes, M.C.O., 2012a. Depositional systems and stratigraphic review proposal of the Rio Preto Fold Belt, northwestern Bahia/southern Piauí. Revista Brasileira de Geociências 42(3), 523–538.

Caxito, F.A., Halverson, G.P., Uhlein, A., Stevenson, R., Gonçalves Dias, T., Uhlein, G.J. 2012b. Marinoan glaciation in east central Brazil. Precambrian Research 200, 38–58. doi:10.1016/j.precamres.2012.01.005

Caxito, F.A., and Uhlein, A., 2013. Arcabouço tectônico e estratigráfico da Faixa Riacho do Pontal, divisa Pernambuco-Piauí-Bahia. Geonomos, 21(2), 19–37. doi:10.18285/geonomos.v21i2.269

Caxito, F.A., Dantas, E.L., Stevenson, R., Uhlein, A., 2014a. Detrital zircon (U-Pb) and Sm-Nd isotope studies of the provenance and tectonic setting of basins related to collisional orogens: The case of the Rio Preto fold belt on the northwest São Francisco Craton margin, NE Brazil. Gondwana Research, 26(2):741–754, doi:10.1016/j.gr.2013.07.007

Caxito, F.A., Uhlein, A., Morales, L.F.G., Egydio-Silva, M., Sanglard, J.C.D., Gonçalves Dias, T., Mendes, M.C.O., 2014b. Structural analysis of the Rio Preto fold belt (northwestern Bahia/ southern Piauí), a doubly vergent asymmetric fan developed during the Brasiliano Orogeny. Anais da Academia Brasileira de Ciências, 86 (3):151–163. doi:10.1590/0001-3765201420130173

Caxito, F.A., Uhlein, A., Dantas, E.L., 2014c. The Afeição augen-gneiss Suite and the record of the Cariris Velhos Orogeny within the Riacho do Pontal fold belt, NE Brazil. Journal of South American Earth Sciences, 51:12–27, doi:10.1016/j.jsames.2013.12.012.

Caxito, F.A., Uhlein, A., Stevenson, R., Uhlein, G.J., 2014d. Neoproterozoic oceanic crust remnants in northeast Brazil. Geology, 42 (5):387–390. doi:10.1130/G35479.1

Caxito, F.A., Uhlein, A., Dantas, E.L., Stevenson, R., Pedrosa-Soares, A.C., 2015. Orosirian (ca. 1.96 Ga) mafic crust of the northwestern São Francisco Craton margin: Petrography, geochemistry and geochronology of amphibolites from the Rio Preto fold belt basement, NE Brazil. Journal of South American Earth Sciences, 59:95–111. doi: 10.1016/j.jsames.2015.02.003

Caxito, F.A., Uhlein, A., Dantas, E.L., Stevenson, R., Salgado, S.S., Dussin, I.A., Sial, A.N., 2016. A complete Wilson Cycle recorded within the Riacho do Pontal Orogen, NE Brazil: Implications for the Neoproterozoic evolution of the Borborema Province at the heart of West Gondwana. Precambrian Research, 282:97-120. doi: 10.1016/j.precamres.2016.07.001

Cruz, S.C.P., and Alkmim, F.F. 2017. The Paramirim Aulacogen. In: U. G. Cordani et al. (eds.), São Francisco Craton, Eastern Brazil, Regional Geology Reviews, pp. xxx–xxx. doi: 10.1007/978-3-319-01715-0_6

Danderfer Filho, A., Lagoeiro, L.E., and Alkmim, F.F. 1993. O sistema de dobramentos e empurrões da Chapada Diamantina (BA): registro da inversão do aulacógeno do Espinhaço no decorrer do evento Brasiliano. In: SBG, Núcleo Bahia-Sergipe, Simpósio sobre o Cráton do São Francisco, 2, Salvador, Anais, 197–199.

Dantas, E.L., Brito-Neves, B.B., Fuck, R.A. 2010. Looking for the oldest rocks of South America: Paleoarchean orthogneiss of the Sobradinho Block, northernmost foreland of the São Francisco Craton, Petrolina, Pernambuco, Brazil. VII SSAGI - South American Symposium on Isotope Geology, Brasília, p. 137–140.

Egydio-Silva, M., 1987. O sistema de dobramentos Rio Preto e suas relações com o Cráton São Francisco. PhD Thesis, Universidade de São Paulo, São Paulo, 95 p.

Egydio-Silva, M., Karmann, I., Trompette, R.R., 1989. Litoestratigrafia do Supergrupo Espinhaço e Grupo Bambuí no noroeste do estado da Bahia. Revista Brasileira de Geociências 19(2), 101–112.

Figueirôa, I. and Silva Filho, M.A. 1990. Programa Levantamento Geológicos Básicos do Brasil. Carta geológica, carta metalogenética, Escala 1:100 000 Folha SC.24-V-C-III, Petrolina, Estados de Pernambuco e Bahia.DNPM/CPRM, 108 p.

Gava, A., Nascimento, D.A., Vidal, J.L.B. et al., 1983. Geologia. In: BRASIL DNPM. Projeto RADAMBRASIL. Folha SC-24/25 – ARACAJU/RECIFE. Rio de Janeiro, 1983. 826p. il. p. 27 – 376.

Gava, A., Montes, A.S.L., Oliveira, E.P. 1984. Granitos alcalinos no sudeste do Piauí. Caracterização geológica, petrográfica e geoquímica. In: SBG, Congresso Brasileiro de Geologia, 33, Rio de Janeiro, anais, p. 2767–2786.

Gibb, R.A., Thomas, M.D., 1976. Gravity signature of fossil plate boundaries in the Canadian Shield. Nature 262, 199–200.

Girardi, V.A.V., Teixeira, W., Mazzucchelli, de Oliveira, M.,E.P., and da Costa. P.C.C. 2017. Mafic Dykes: Petrogenesis and Tectonic Inferences. In: U.G. Cordani et al. (eds.), São Francisco Craton,

Eastern Brazil, Regional Geology Reviews, pp. xxx–xxx. doi: 10.1007/978-3-319-01715-0_8

Gomes, F.E.M. 1990. Relações litoestratigráfico-estruturais e evolução tectônica na Faixa Riacho do Pontal – Região de Paulistana (PI). In: SBG, XXXVI Congresso Brasileiro de Geologia, Natal, vol. 6: 2843–2857.

Gomes, F.E.M. and Vasconcelos, A.M. 1991. Programa Levantamento Geológicos Básicos do Brasil. Carta geológica, carta metalogenética, Escala 1:100 000 Folha SC.24-V-A-II, Paulistana, Estados de Pernambuco e Piauí.DNPM/CPRM, 146 p.

Gonçalves Dias, T., Mendes, M.C.O., 2008. Geologia da Faixa Rio Preto entre Formosa do Rio Preto e Malhadinha, Bahia. Trabalho de Graduação, Universidade Federal de Minas Gerais, Belo Horizonte, 62 p.

Halverson, G.P., Wade, B.P., Hurtgen, M.T., Barovich, K.M., 2010. Neoproterozoic chemostratigraphy. Precambrian Research 182, 337–350.

Jardim de Sá, E.F. and Hackspacher, P.C. 1980. Reconhecimento estrutural na borda noroeste do Cráton São Francisco. In: SBG, Congresso Brasileiro de Geologia, 31, Balneário Camboriú, SC, Anais, 5:2719–2731.

Jardim de Sá, E.F., Macedo, M.H.F., Torres, H.H.F., Kawashita, K. 1988. Geochronology of metaplutonics and evolution of supracrustal belts in the Borborema Province, NE Brazil. In: Cong. Latino-Americano de Geologia, 7, Belém, Anais, p. 49–62.

Jardim de Sá, E.F., Macedo, M.H.F., Fuck, R.A., Kawashita, K. 1992 Terrenos proterozóicos na província Borborema e a margem norte do Cráton do São Francisco. Rev. Brás. Geoc., 22(4), 472–480.

Jardim de Sá, E.F., Macedo, M.H.F., Kawashita, K., Peucat, J.J., Leterrier, J., Fuck, R.A.. 1996. A suíte Serra da Esperança: intrusões alcalinas sintectônicas aos nappes brasilianos na Faixa Riacho do Pontal, NE do Brasil. In: 39 Congresso Brasileiro de Geologia, SBG, Salvador, 6: 499–501.

Kosin, M.D., Angelim, L.A.A., Souza, J.D., Guimarães, J.T., Teixeira, L.R., Martins, A.A.M., Bento, R.V., Santos, R.A., Vasconcelos, A. M., Neves, J.P., Wanderley, A.A., Carvalho, L.M., Pereira, L.H.M., Gomes, I.P. 2004. Folha SC.24-Aracaju. In: Schobbenhaus, C., Gonçalves, J.H., Santos, J.O.S., Abram, M.B., Leão Neto, R., Matos, G.M.M., Vidotti, R.M., Ramos, M.A.B., Jesus, J.D.A. de (eds.). Carta Geológica do Brasil ao Milionésimo, Sistema de Informações Geográficas. Programa Geologia do Brasil. CPRM, Brasília. CD-ROM.

Kozuch, M., 2003.Isotopic and trace element geochemistry of early Neoproterozoic gneissic and metavolcanic rocks in the Cariris Velhos orogen of the Borborema Province, Brazil, and their bearing on tectonic setting.PhD thesis, University of Kansas, USA.

Kreysing, K., Lenz, R., Ribeiro, G.F. 1973. Salinização das águas subterrâneas do centro do polígono das secas do nordeste brasileiro. Recife:SUDENE. 69p. il.

Marimon, M.P.C. 1990. Petrologia e litogeoquímica da seqüência plutono-vulcanosedimentar de Brejo Seco, Município de São João do Piauí. Universidade Federal da Bahia, Curso de Pós-Graduação, Dissertação de Mestrado, Salvador, 102 pg.

Misi A. & Veizer J. 1998. Neoproterozoic carbonate sequences of the Una Group, Irece Basin, Brasil: chemostratigraphy, age and correlations. Precambrian Research, 89:87–100.

Moraes, J.F.S. 1992. Petrologia das rochas máficas-ultramáficas da seqüência vulcanosedimentar de Monte Orebe, PE-PI. Universidade Federal da Bahia, Curso de Pós-Graduação, Dissertação de Mestrado, Salvador, 98 pg.

Oliveira, R.G. 1998. Arcabouço geotectônico da região da Faixa Riacho do Pontal, Nordeste do Brasil: dados aeromagnéticos e gravimétricos. Dissertação de Mestrado, IG-USP, 157 p.

Oliveira, E.P., Windley, B.F., McNaughton, N.J., Bueno, J.F., Nascimento, R.S., Carvalho, M.J., and Araújo, M.N.C. 2017. The Sergipano Belt. In: U.G. Cordani et al. (eds.), São Francisco Craton, Eastern Brazil, Regional Geology Reviews, pp. xxx–xxx. doi: 10.1007/978-3-319-01715-0_13

Pla Cid, J., Nardi, L.V.S., Conceição, H., Bonin, B., Jardim de Sá, E.F. 2000. The alkaline sílica-satured ultrapotassic magmatism of the Riacho do Pontal fold belt, NE Brazil. Journal of South American Earth Science, 13 (7):661–683.

Reis, H.L.S., Alkmim, F.F., Renato, Fonseca, C.S., Nascimento, T.C., Suss, J.F., and Prevatti, L.D. 2017. The São Francisco Basin. In: U. G. Cordani et al. (eds.), São Francisco Craton, Eastern Brazil, Regional Geology Reviews, pp. xxx–xxx. doi: 10.1007/978-3-319-01715-0_7

Salgado, S.S., 2014. Contexto Geológico, Petrologia, Geoquímica e Potencial Metalogenético do Complexo Máfico-Ultramáfico de Brejo Seco, Sudeste do Piauí. Unpublished Masters Dissertation, Universidade Federal de Minas Gerais, Belo Horizonte, Brazil.

Salgado, S.S., Ferreira Filho, C.F., Caxito, F.A., Uhlein, A., Dantas, E. L., Stevenson, R., 2016. The Ni-Cu-PGE mineralized Brejo Seco mafic-ultramafic layered intrusion, Riacho do Pontal Orogen: Onset of Tonian (ca. 900 Ma) continental rifting in Northeast Brazil. Journal of South American Earth Sciences, 70:324–339. doi: 10.1016/j.jsames.2016.06.001

Santos, J.F., 1984. Depósito de Níquel de São João do Piauí, Piauí. In: Schobbenhaus, C. and Coelho, C.E.S. (eds), Principais Depósitos Minerais do Brasil: Brasília-DF, DNPM/CVRD, v. 2, p. 341–345.

Santos, E.J., Caldasso, A.L.S., 1978. Síntese dos conhecimentos e ensaio interpretativo da área do Riacho do Pontal, Nordeste do Brasil. In: Simpósio sobre o Cráton do São Francisco e suas faixas marginais, Salvador, Anais, p. 399–426.

Santos, C.A and Silva Filho, M.A. 1990. Programa Levantamentos Geológicos Básicos do Brasil. Riacho do Caboclo. Folha SC.24-V-A-VI, Estados de Pernambuco e Bahia. Secretaria Nacional de Minas e Metalurgia, Texto e Mapas, 113 p.

Santos, E.J., Van Schmus, W.R., Kozuch, M., Brito Neves, B.B. 2010. The Cariris Velhos tectonic event in northeast Brazil. Journal of South American Earth Sciences 29, 61–76.

Santos, T.C., Pimentel, M.M., Brito Neves, B.B., Gruber, L., Rodrigues, J.B., 2012. Proveniência do Grupo Bambuí/Una na região da Chapada Diamantina e comparação com unidades cronocorrelatas na porção oeste do Cráton do São Francisco. In: SBG, Congresso Brasileiro de Geologia, 46, Santos, CD-ROM.

Sial, A.N., Misi, A., Pedreira, A.J., Gaucher, C., Ferreira, V.P., Cezario, W.S., 2010. Carbon-isotope stratigraphy of the Neoproterozoic Serra do Paraíso (RioPardo Basin) and São Desiderio (Rio Preto Belt) formations, Bahia, Brazil. In: VII South American Symposium on Isotope Geology (SSAGI), Brasília, CD-ROM.

Schobbenhaus C. 1996. As tafrogêneses superpostas Espinhaço e Santo Onofre, estado da Bahia: Revisão e novas propostas. Revista Brasileira de Geociências, 26(4):265–276.

Souza J.D., Fernandes Filho J.; Guimarães J.T., Lopes J.N. 1979. Projeto Colomi. Relatório Final, Geologia da Região do Médio São Francisco, texto e mapas. Escala 1:250 000. Salvador, DNPM-CPRM, 389 p.

Siqueira Filho, J., 1967. Geologia da folha de Jutaí - Pernambuco. Recife: SUDENE, 52p. (Brasil SUDENE. Série Geologia Regional, 7).

Teixeira, W., Oliveira, E.P., and Marques L.S. 2017. Nature and Evolution of the Archean Crust of the São Francisco Craton. In: U. Cordani et al. (eds.), São Francisco Craton, Eastern Brazil, Regional Geology Reviews, pp. xxx–xxx. doi: 10.1007/978-3-319-01715-0_3

Uhlein, A., Caxito, F.A., Sanglard, J.C.D., Uhlein, G. J., Suckau, G.L. 2011. Estratigrafia e Tectônica das Faixas Neoproterozóicas da Porção Norte do Craton do São Francisco. Geonomos 19 (2): 8–31.

Uhlein, A., Caxito, F.A., Egydio-Silva, M., Barbosa, J.F.S. 2012. Faixas de Dobramentos Rio Preto e Riacho do Pontal. Capítulo IX. In: Geologia da Bahia. Editor Geral: Barbosa, J.F.S. Série Publicações Especiais, 13, CBPM, 2 vol.: 87 – 130.

Van Schmus, W.R., Brito Neves, B.B., Hackspacher, P., Babinski, M. 1995. U/Pb and Sm/Nd geochronolgic studies of eastern Borborema Province, northeastern Brazil: initial conclusions. Journal of South American Earth Sciences 8, 267–288.

Van Schmus, W.R., Kozuch, M., Brito Neves, B.B. 2011. Precambrian history of the Zona Transversal of the Borborema Province, NE Brazil: Insights from Sm-Nd and U-Pb geochronology. Journal of South American Earth Sciences 31: 227–252

Vauchez, A. and Egydio-Silva, M. 1992. Termination of a continental-scale strike-slip fault in partially melted crust: the West Pernambuco shear zone, northeast Brazil. Geology 20, 1007–1010.

The Sergipano Belt

13

Elson P. Oliveira, Brian F. Windley, Neal J. McNaughton, Juliana F. Bueno, Rosemery S. Nascimento, Marcelo J. Carvalho, and Mario N.C. Araújo

Abstract

The Neoproterozoic Sergipano is the main orogenic belt between the Borborema province in the north and the São Francisco craton in the south. The belt is divisible from north to south into the Canindé, Poço Redondo-Marancó, Macururé, Vaza Barris, and Estância lithostratigraphic domains; the first three are composed of plutonic, volcanic and sedimentary rocks, and the last three of sedimentary rocks. Field relationships combined with zircon geochronology of igneous and sedimentary rocks and whole-rock elemental and isotopic data allow us to propose the following evolution for the Sergipano belt. An early Neoproterozoic ($\sim$980–960 Ma) continental arc (Poço Redondo tonalitic gneisses) developed on the margin of the Paleoproterozoic-Neoproterozoic Pernambuco-Alagoas block in the north. Extension of this continental block gave rise to: (i) A-type crustal melt granites and associated sedimentary rocks on the stretched, rifted margin of the Poço Redondo-Marancó domain; (ii) the Canindé rift sequence between the Pernambuco-Alagoas block and the Poço Redondo/Marancó domain; (iii) a passive margin on the southern boundary of the Pernambuco-Alagoas block, upon which sediments were deposited soon after 900 Ma; and (iv) a second sedimentary shelf on the passive margin of the São Francisco craton. In the Canindé domain, rifting continued until ca. 640 Ma and led to emplacement of a bimodal association of A-type granites (715 Ma) and continental mafic volcanic rocks, a continental-type layered gabbroic complex (ca. 700 Ma), magma-mingled gabbro/quartz–monzodiorite (688 Ma), and rapakivi granites (684 and 641 Ma). Deformed pillow basalts and interleaved marble lenses are likely ocean floor relicts in the Canindé domain. Closure of the Canindé oceanic basin began at ca. 630 Ma with the intrusion of arc-type granitic plutons in the Macururé (628–625 Ma), Canindé (ca. 621 Ma) and Poço Redondo-Marancó (ca. 625 Ma) domains. Convergence of the Pernambuco-Alagoas block and the São Francisco craton led to deformation on the passive margins and granite emplacement (590–570 Ma) mainly in the Macururé domain. Exhumation of the Pernambuco-Alagoas block and the Canindé, Poço Redondo-Marancó, and Macururé domains in the north led to deposition of uppermost clastic sediments in the Estância and Vaza Barris domains in the south, interpreted as a foreland basin, and to final thrusting of the continental margin sedimentary rocks onto the São Francisco craton.

E.P. Oliveira (✉) · J.F. Bueno · M.J. Carvalho
Department of Geology and Natural Resources, Institute of Gesciences, University of Campinas, P.O. Box 6152 Campinas, 13.083-970, Brazil
e-mail: elson@ige.unicamp.br

B.F. Windley
Department of Geology, University of Leicester, Leicester, LE1 7RH, UK

N.J. McNaughton
John de Laeter Centre of Mass Spectrometry, School of Applied Physics, Curtin University of Technology, Perth, WA 6845, Australia

R.S. Nascimento
Faculty of Geology, Institute of Geosciences, Federal University of Pará, Belém, 66.075-110, Brazil

M.N.C. Araújo
CENPES, Petrobras, Rio de Janeiro, 21.941-598, Brazil

M. Heilbron et al. (eds.), *São Francisco Craton, Eastern Brazil*, Regional Geology Reviews, DOI 10.1007/978-3-319-01715-0_13

Keywords
Sergipano belt • Brasiliano event • Neoproterozoic • Fold-thrust belt • Sediment provenance

13.1 Introduction

The Neoproterozoic Sergipano belt, along with the Riacho do Pontal and Rio Preto belts defines the N-NE limit of the São Francisco craton (SFC) (Fig. 13.1). The Sergipano Belt represents the western segment of the major Oubanguide orogen that extends into NW Africa (Trompette 2000). It consists of a triangular-shaped, E-SE- to W-NW-trending belt formed by the collision of the Pernambuco-Alagoas block in the north with the SFC in the south. The Sergipano belt was first interpreted as a geosyncline (Humprey and Allard 1969; Santos and Silva Filho 1975), later as a collage of tectono-stratigraphic terranes or microplates (Davison and Santos 1989), or a fold-and-thrust belt (D'el-Rey Silva 1999), and more recently as representative of a complete plate tectonic cycle in West Gondwana (Oliveira et al. 2010a).

Here we present a synthesis of current knowledge of the Sergipano Belt with data on its structure, geochronology, and whole-rock and isotopic geochemistry.

13.2 Internal and External Tectonic Domains

From north to south the Sergipano belt is organized into five lithotectonic domains, namely: Canindé, Poço Redondo-Marancó, Macururé, Vaza Barris and Estância (Fig. 13.1). The shear zones of Macururé, Belo Monte-Jeremoabo, São Miguel do Aleixo, and Itaporanga form the boundaries between the various domains (Fig. 13.2) (Davison and Santos 1989; Silva Filho 1998; Oliveira et al. 2010a). Geophysical data obtained by Oliveira (2008) and Oliveira and Medeiros (2012) show that the Belo Monte-Jeremoabo shear zone (BMJSZ) marks the northern limit of the SFC crust. On this basis we suggest here that this shear zone separates the internal from the external tectonic domains in the Sergipano belt.

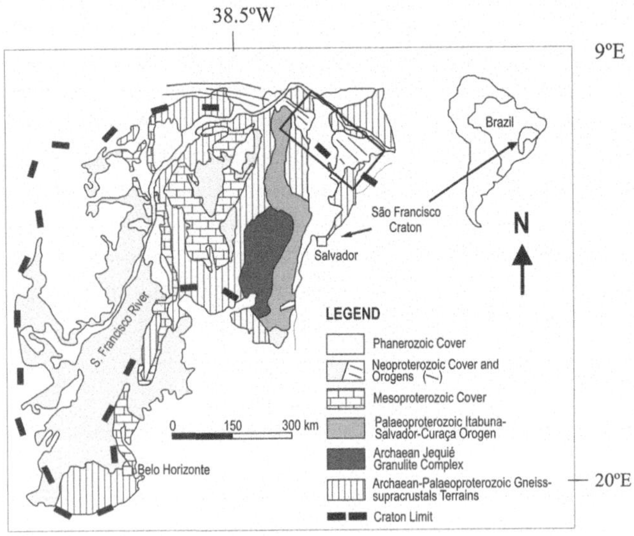

Fig. 13.1 The São Francisco craton and its margins (adapted from Oliveira and Tarney 1995). Figure 13.2 shows location of the Sergipano belt

Located to the north of the BMJSZ are the allochthonous domains/terranes (Marancó-Poço Redondo and Canindé) accreted during the Neoproterozoic. The three domains to the south of the BMJSZ (Estância, Vaza Barris and Macururé) largely consist of metasedimentary rocks (Oliveira et al. 2010a).

13.3 Lithotypes and Age of Igneous and Metamorphic Rocks

13.3.1 External Tectonic Domains

13.3.1.1 Estância Domain

This domain is located in the southern portion of the belt (Fig. 13.2) and composed of sub-horizontal, undeformed to weakly deformed platform-type sedimentary rocks of the Estância Group (Davison and Santos 1989). The base of the group is represented by autochthonous unit of conglomerates, argillites, sandstones and diamictites of the Juetê Formation, succeeded upwards by limestones and dolomites, sometimes with stromatolites (Acauã Formation), overlain by feldspathic sandstones, siltstones, and argillites that contain well-preserved ripple marks, mudcracks and hummocky structures (Lagarto Formation; Saes and Vilas Boas 1989; Basilici et al. 2012). These platform-type sediments are overlain by molasse-type, clastic conglomerates, sandstones, siltstones and argillites of the Palmares Formation (Saes and Vilas Boas 1989). Units of this domain rest unconformably upon Archean to Paleoproterozoic rocks of the SFC (Oliveira et al. 2010b; Oliveira 2012).

13.3.1.2 Vaza Barris Domain

The Vaza Barris domain is more deformed than the Estância domain and for this reason its stratigraphy is more contentious. The domain contains several formations that are grouped into the Miaba and Vaza Barris Groups (Humprey and Allard 1969; Silva Filho et al. 1978; D'el-Rey Silva and McClay 1995) with the modifications suggested by Oliveira et al. (2010a) (Fig. 13.3).

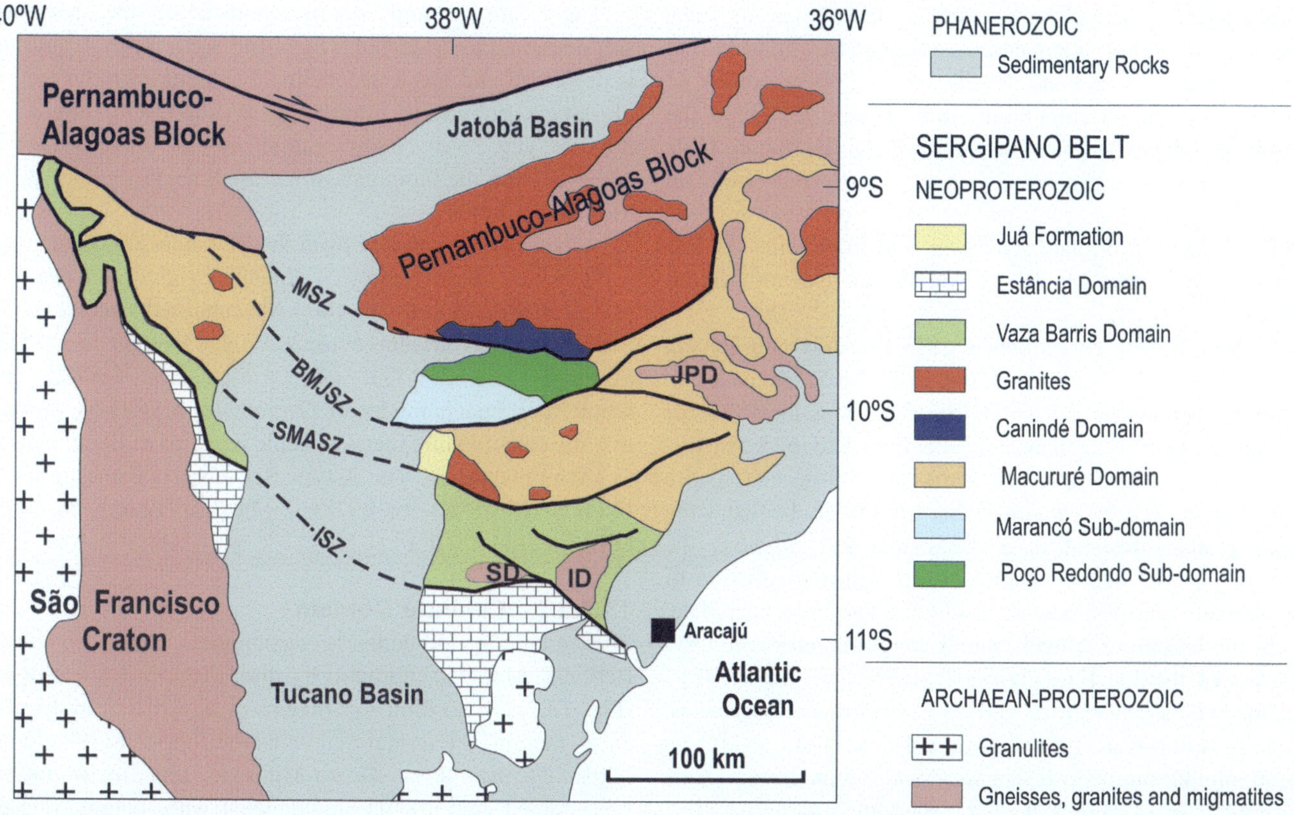

Fig. 13.2 The Sergipano belt and its division into domains (modified after Oliveira et al. 2010a). The Poço Redondo-Marancó domain is divided into two sub-domains. MSZ, BMJSZ, SMASZ and ISZ stand, respectively, for Macururé, Belo Monte-Jeremoabo, São Miguel do Aleixo and Itaporanga shear zones. *ID* Itabaiana dome, *SD* Simão Dias dome, *JPD* Jirau do Ponciano dome

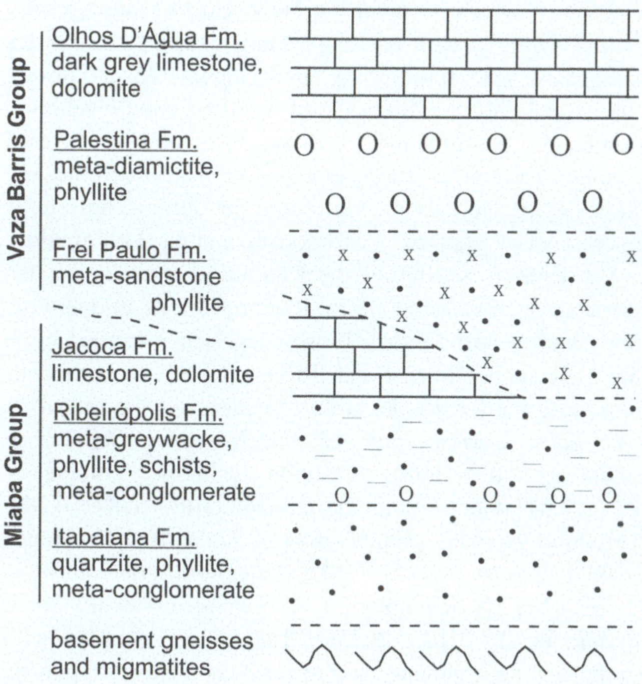

Fig. 13.3 Stratigraphy of the Vaza Barris domain after Oliveira et al. (2010a). Contacts between formations are tectonic or unconformities. Thicknesses are not to scale

At the type locality (Capitão Farm), the lower Miaba Group comprises a basal unit of quartzite (Itabaiana Fm), overlain by poorly sorted meta-conglomerate with pebbles of gneiss and quartzite (Ribeirópolis Fm) and then by intercalated stromatolite-bearing dolomite and limestone (Jacoca Fm). Basal quartzites of the Miaba Group rest unconformably on basement gneisses and migmatites of the Itabaiana and Simão Dias domes (Fig. 13.2); in places the unconformity has been transposed into a thrust (Oliveira et al. 2010). The migmatite paleosome of the Simão Dias dome yields a SHRIMP U-Pb zircon age of 2868 ± 25 Ma (Oliveira 2012).

The overlying Vaza Barris Group was thrust southwards over rocks of the Miaba Group. According to Oliveira et al. (2010a), it starts with siltstones, phyllites, and meta-sandstones (Frei Paulo Fm) deposited over the Ribeirópolis Fm or over the Jacoca Fm, succeeded upwards by phyllites and diamictites of the Palestina Fm, and then by grey to black limestones and dolomites, occasionally with phyllite intercalations (Olhos D'Água Fm).

13.3.1.3 Macururé Domain

According to Oliveira et al. (2010a, and references therein) the Macururé domain (Fig. 13.2) contains amphibolite facies garnet-bearing meta-turbidites, feldspathic-aluminous mica schists with minor intercalations of quartzite, marble and meta-volcanic rocks, and lenses up to 200 m across of amphibolite, garnet-amphibolite and chlorite schist. In the northeast of this domain, a quartzite of the Santa Cruz Formation rests unconformably on the basement (Jirau do Ponciano dome, Fig. 13.2) and is considered to be the base of the Macururé domain. Gneisses and migmatites of this dome have a 2000 Ma Rb-Sr whole-rock isochron (Brito Neves et al. 1978).

Many granites have intruded the Macururé domain. Recent U-Pb age of these granites (Long et al. 2005; Bueno et al. 2009; Oliveira et al. 2010a, 2015a) constrains their time of emplacement in two main age intervals: 628–625 Ma and 590–570 Ma. Bueno (2008), Bueno et al. (2009), and Oliveira et al. (2010a, 2015a) suggested that from their chemistry the older granites formed in a continental arc, whereas the younger, crustal-derived granites formed during collision of the Pernambuco-Alagoas block with the São Francisco plate.

In the Macururé domain, minor conglomerates and greywackes of the Juá Fm (Fig. 13.2) underwent minor deformation and metamorphism, and were not intruded by granites. Menezes Filho et al. (1988) interpreted these clastic sediments as an alluvial fan deposit that is possibly correlative in time with the Palmares Fm of the Estância domain farther south.

13.3.2 Internal Domains

13.3.2.1 Poço Redondo-Marancó Domain

This domain is separated from the Macururé domain by the Belo Monte Jeremoabo shear zone (Fig. 13.2). It is divisible into two major sub-domains, namely Marancó and Poço Redondo.

The Poço Redondo sub-domain is a migmatitic gneiss complex dominated by granodioritic-tonalitic rocks that represent the basement of the Marancó sub-domain. Migmatitic gneisses from this sub-domain have zircon U-Pb SHRIMP ages of 980 ± 4 Ma and 961 ± 38 Ma (Carvalho et al. 2005; Oliveira et al. 2010a). Several granitoids intrude the sub-domain, namely the Sitios Novos, Poço Redondo and Queimada Grande; their zircon U-Pb SHRIMP ages are: 631 ± 4 Ma (Sitios Novos), 623 ± 7 Ma (Poço Redondo), and 618 ± 4 Ma (Queimada Grande) (Oliveira et al. 2015a).

The Marancó sub-domain comprises greenschist- to amphibolite-facies, pelitic to psammitic metasedimentary rocks, rhythmites interleaved with calc-alkaline andesite to dacite beds, and intercalations of basalt, andesite, gabbro and serpentinites. Peridotites and gabbros with variable degrees of serpentinization mainly occur as lenses in metasedimentary rocks or as intrusions in the south of the sub-domain; they may be slices of lithospheric mantle from beneath the orogen, or ophiolite fragments (Silva Filho 2006).

There are several granite bodies in the Marancó sub-domain, the largest of which (the Serra Negra batholith) is deformed and has a geochemical signature similar to that of A-type granites (Carvalho et al. 2005; Oliveira et al. 2010a). The Serra Negra granite and the Poço Redondo gneisses form the basement of metasedimentary rocks of the Marancó sub-domain.

Geochronological data from the Marancó sub-domain are as follows. The Marancó volcanic rocks (dacite) are 602 ± 4 Ma-old with ca. 1000 Ma inherited zircons (U-Pb SHRIMP data, Carvalho et al. 2005). Nd isotope data (TDM ages of 1.12–1.74 Ga; $\varepsilon_{Nd(t)} = -1.1$ and -8.62) and whole-rock geochemistry of Carvalho et al. (2005) indicate that the calc-alkaline meta-volcanic rocks most likely belong to a continental arc. The A-type Serra Negra batholith has a U-Pb SHRIMP age of 952 ± 2 Ma (Carvalho et al. 2005; Oliveira et al. 2010a).

13.3.2.2 Canindé Domain

The Canindé domain is separated from the Poço Redondo-Marancó domain by the Macururé shear zone (Fig. 13.2). According to Oliveira et al. (2010a, and references therein) the domain contains the following lithodemic units: (i) the Novo Gosto-Mulungu unit made up of fine-grained amphibolites intercalated with metamorphosed pelites, siltstones, cherts, graphite schists, calc-silicate rocks and marbles, crosscut by mafic and felsic dykes, granites and Fe–Ti-rich gabbros; (ii) the Garrote unit, which is a continuous, up to 2 km-wide, strongly deformed, granite sheet the protolith of which intruded the Novo Gosto-Mulungu unit; (iii) the Gentileza unit, made up of amphibolites and diorites intercalated with porphyritic quartz-monzonite and minor dolerite and gabbroic bodies; and (iv) the Canindé gabbroic complex comprising massive and layered olivine-gabbronorite, leucogabbro, anorthosite, troctolite, and minor pegmatitic gabbro, norite and peridotite. These units are crosscut by granites, granodiorites, and rapakivi granites.

The tectonic setting of the Canindé domain is controversial since Silva Filho (1976) interpreted it as an ophiolite. Later, Jardim de Sá et al. (1986) suggested it formed in an island arc, and Oliveira and Tarney (1990) by intracontinental magmatism. Rocks of this domain have the following U-Pb ages: Garrote unit—ca. 715 Ma (Van Schmus in Santos et al. 1998); Canindé gabbroic complex—701 ± 8 Ma, Gentileza quartz-monzodiorite—688 ± 6 Ma; Curralinho rapakivi granite—684 ± 7 Ma, Boa Esperança rapakivi granite—641 ± 5 Ma; Lajedinho granodiorite—621 ± 9 Ma (Nascimento 2005; Nascimento et al. 2005; Oliveira et al. 2010a, 2015a). These ages, coupled with whole-rock geochemistry and mineral chemistry, supporting a continental signature of the rocks (Oliveira and Tarney 1990), led Oliveira et al. (2010a) to interpret the Canindé domain as a rift sequence that was later deformed and accreted to the Poço Redondo-Marancó domain. The rift is likely to have evolved into an ocean basin owing to fragments of amphibolite with interleaved marble lenses and relics of deformed pillow basalts in the Novo Gosto-Mulungu unit.

13.4 Structural Relationships and Timing of Deformation Phases

Oliveira et al. (2010a) presented a thorough description of field relationships, deformation phases, and ages of deformation in the Sergipano belt. Accordingly, four deformation events (D1–D4) affected the supracrustal rocks in the Macururé and Vaza Barris domains, probably in the time period of 628–570 Ma (Ar–Ar, U-Pb, and Sm–Nd ages). Deformation also reworked earlier structures in the basement domes and/or inliers throughout the belt. A summary of Oliveira's et al. (2010a) structural description is presented in the following sections and a representative cross-section across the metasedimentary domains is shown in Fig. 13.4.

13.4.1 Deformation of Metasedimentary Rocks—Macururé, Vaza Barris and Estância Domains

Deformation that affected the metasedimentary rocks of the Sergipano belt developed during a progressive sequence of four events, namely D1, D2, D3, and D4.

13.4.1.1 D1 Event

D1 event is hardly observed in the belt owing to the strong effects of the D2 event. D1 is recognized only as cleavage in phyllites and meta-greywackes of the Vaza Barris domain and as schistosity in garnet-mica schists of the Macururé domain. F1 folds may be inclined, horizontal, tight or asymmetric with their axial plane foliations parallel or oblique to bedding.

In the Macururé domain, Bueno et al. (2009) described a post-D1, pre-D2 tonalite (Camará Tonalite) with xenoliths of garnet mica schist; the tonalite yielded a U-Pb SHRIMP zircon age of 628 ± 12 Ma. This data and that of the Coronel João Sá granodiorite (625 ± 2 Ma, zircon U-Pb age

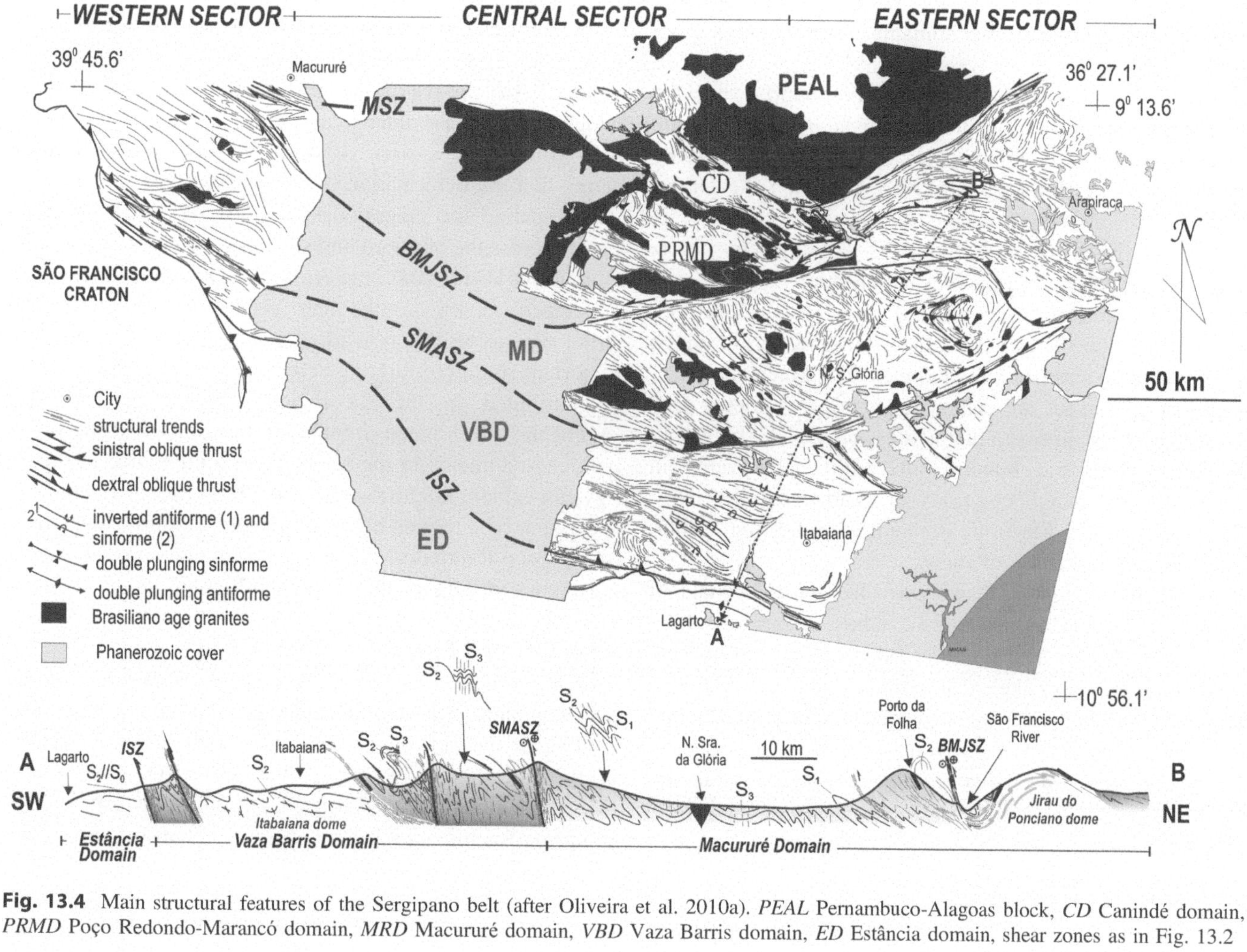

Fig. 13.4 Main structural features of the Sergipano belt (after Oliveira et al. 2010a). *PEAL* Pernambuco-Alagoas block, *CD* Canindé domain, *PRMD* Poço Redondo-Marancó domain, *MRD* Macururé domain, *VBD* Vaza Barris domain, *ED* Estância domain, shear zones as in Fig. 13.2

after Long et al. 2005), which is intrusive into the Macururé mica schists and affected by D2, set a minimum age limit for the D1 event.

13.4.1.2 D2 Event

The very important D2 event is the most penetrative in the Sergipano belt and was associated with the main phase of collision between the SFC and the Pernambuco-Alagoas block. It gave rise to pervasive south-verging thrusts and nappes throughout most of the metasedimentary domains in the belt.

Structures associated with this event formed under different strain magnitudes from the margins to the centre of the belt. Anchizonal sandstones of the Estância domain (Lagarto Formation) contain well-preserved sedimentary structures weakly affected by open to gentle, asymmetric, F2 folds. South-verging fault-bends and thrusts (Fig. 13.5a) confirm the south-directed shear sense of D2. In the western part of the belt, meta-carbonates of the Vaza Barris domain were thrust-stacked onto meta-siltstones of the Estância domain, and low-T ultramylonites were preferentially nucleated in the hanging wall of south-verging imbricate fans coeval with a D2 compressive shear zone that marks the limit between the Vaza Barris and Estância domains (the Itaporanga shear zone—ISZ, Fig. 13.2).

The D2 deformation is most pronounced in the high-grade central portion of the orogen in the Macururé domain. It may be comparable to the foreland-verging imbricate fan of Huiqi et al. (1990), in which the largest displacements are accommodated.

The L2x lineation coeval with D2 is defined by quartz rods and stretched feldspars. This lineation is very pronounced in quartzites intercalated with metasedimentary rocks of the Vaza Barris and Macururé domains, in which L2x evolved to form L > S tectonites. The kinematics related to the D2 event are demonstrated by asymmetric tails around pebbles in meta-diamictites of the Vaza Barris domain, asymmetric boudins affecting quartz veins in quartzites, and recumbent folds in quartzose mica schist (Fig. 13.5b) of the Macururé domain, all indicating a dominant top-to-south sense of shear.

Age constraints for D2 are provided by U-Pb data of granites emplaced along the S2 fabric of mica schists (584 ± 10 Ma and 571 ± 9 Ma; Bueno et al. 2009); muscovite oriented along the D2 foliation in quartzite and mica schist (591 ± 4 Ma; ^{40}Ar–^{39}Ar age; Oliveira et al. 2010a), and a Sm–Nd isochron of garnet mica schist in the Macururé domain (573 ± 1 Ma; Oliveira et al. 2010a). In summary, D2 lasted at least 20 million years, from approximately 590–570 Ma.

13.4.1.3 D3 Event

Oliveira et al. (2010a) suggested that the D3 event occurred in kinematic continuity with the D2 thrust tectonics by activation of lateral displacement along frontal thrust ramps. Structures formed prior to D3 were strongly transposed and refolded coevally with the development of steeply dipping strike-slip shear zones.

The nucleation of D3 shear zones occurred at the expense of the early-formed structures. This is demonstrated at outcrop scale by the development of localized shearing on the attenuated limbs of F3 folds (Fig. 13.5c, d) that affected originally flat-lying S2 foliation that carries a down-dip L2x lineation at a high angle to L3x.

Folds formed during D3 range from open to tight and verge to the SE, S and SW, depending on their location in the northwestern, central and northeastern sectors of the belt respectively. Asymmetric, rotated boudins, rotated garnet porphyroblasts, S-C and C′ structures, shear bands, and syn-mylonitic folds (Fig. 13.5e, f) indicate dextral and sinistral movements, respectively on the northwestern and northeastern boundaries of all the shear zones that bound the Estância, Vaza Barris, Macururé, Marancó-Poço Redondo and Canindé domains.

In the western extension of the Macururé shear zone, muscovite flakes oriented along the S3 mylonitic foliation yield a ^{40}Ar/^{39}Ar age of 581 ± 2 Ma (Oliveira et al. 2010a), which is somewhat older than the Sm–Nd age of the metamorphism in the eastern sector of the belt, probably indicating that the deformation events varied across it.

Comparing the magnitudes of strain of the D2 and D3 events across the northern and southern segments of the belt, Oliveira et al. (2010a) suggested that D2 fabrics are better developed in the south than in the north, where D3 strike-slip tectonics prevail. Such a contrast illustrates important strain partitioning across the belt that can, in part, be attributed to the rheological contrast between the sedimentary-dominated domains in the south and the more magmatic domains in the north. Such a difference in rheology can explain the formation of the Belo Monte-Jeremoabo shear zone as a terrane boundary developed during accretion of the Poço Redondo-Marancó and Canindé domains to the northernmost edge of the Sergipano belt.

13.4.1.4 D4 Event

This event marks the end of Neoproterozoic deformation in the Sergipano belt. Continued shearing during uplift and cooling of the belt developed, either new ductile-brittle to brittle structures, or reactivated the D2 and D3 ductile fabrics. The main D4 structures are kink folds, shear and extension fractures, faults and en-echelon tension gashes, all

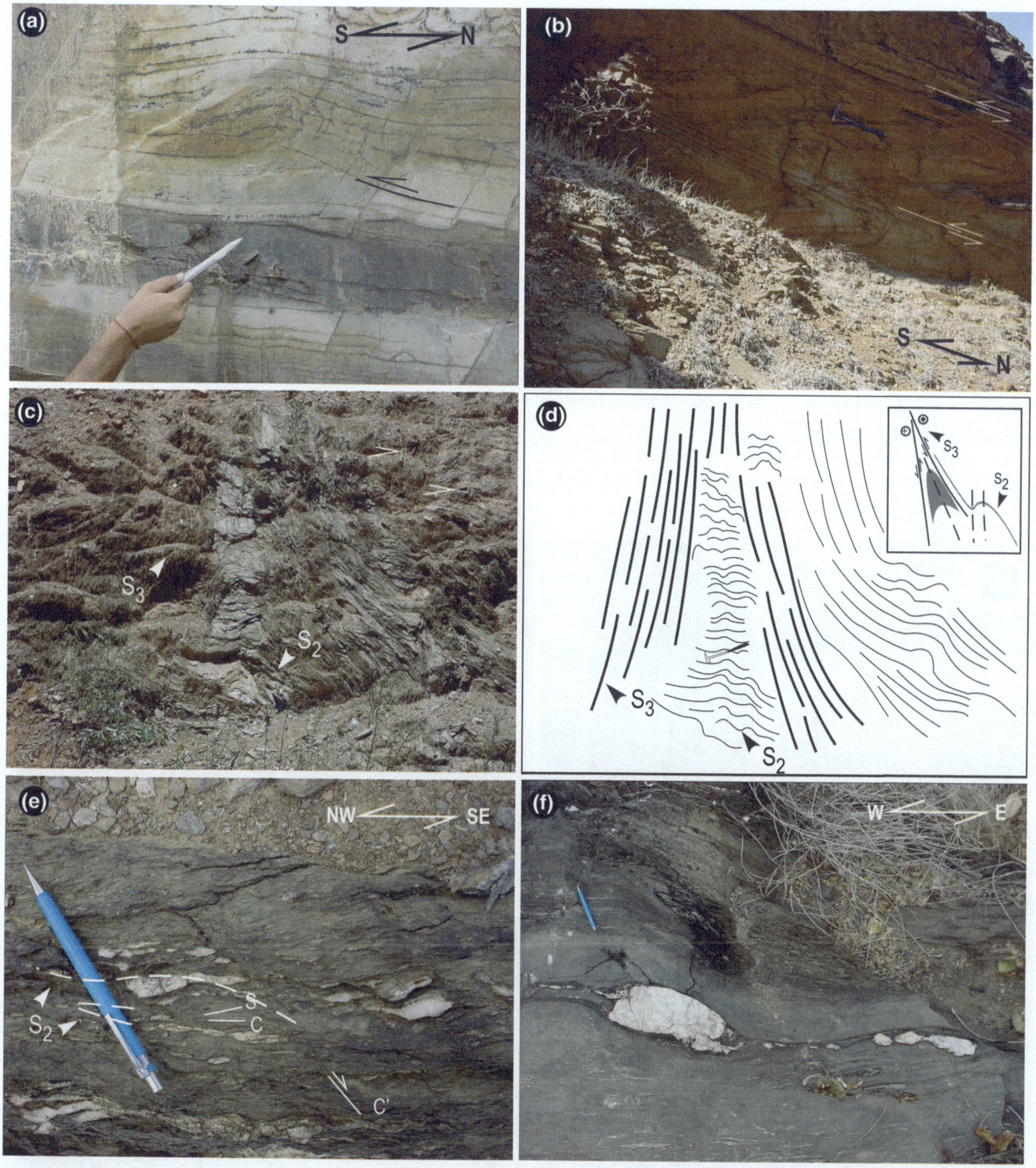

Fig. 13.5 Field relationships in sedimentary domains of the Sergipano belt. **a** Fault-bends and thrust faults affecting the bedding of low-grade sandstones of the Lagarto Formation, Estância domain. **b** F2 recumbent folds affecting the S0 bedding of meta-pelites from the western segment of the Macururé domain. Note the nucleation of D2 shear zones in the attenuated folds limbs. **c**, **d** Nucleation of the S3 foliation along attenuated limbs of F3 folds that have deformed the S2 foliation of mica schists in the western Macururé domain. **e** asymmetric boudins, S-C and C′ fabrics defining dextral shear movement in the western segment of the Macururé shear zone, and **f** asymmetric trails of quartz boudins showing sinistral shear sense in the eastern segment of the Macururé shear zone

showing a shear sense towards the south. In the Estância domain several strike-slip faults indicate southward displacement of the whole sedimentary pile.

13.4.2 Deformation in the Poço Redondo/Marancó Domain

This domain comprises a late Mesoproterozoic/early Neoproterozoic basement (migmatites and gneisses), a supracrustal cover, and igneous bodies. Due to the fact that timing of deformation and migmatization in the basement is not yet well constrained, the sequence of structural events is numbered with the youngest recorded in the basement rocks, labeled as D_n. Therefore, all structures developed prior to this event are referred to in alphanumeric indexes, such as D_n, D_{n-1}, etc. (Oliveira et al. 2010a).

This domain preserves relicts of a complex deformation history (D_{n-1}) that was coeval with a period of anatexis that generated the migmatitic gneisses of the Poço Redondo region. F_n isoclinal folds subsequently affected the migmatitic banding S_{n-1} (Fig. 13.6a). The axial planar S_n foliation mostly dips steeply and trends E–W.

The timing of the D_n event can be constrained by the syntectonic emplacement of the Queimada Grande Granite, which shows flow structures (oriented mafic enclaves), pre-terminal crystallization features (ductile deformed feldspar, but weakly deformed or undeformed quartz) and granodiorite apophyses parallel to the migmatite banding D_n (Fig. 13.6b). This granite has a U-Pb age of ca. 624 Ma (Brito et al. 2006), a SHRIMP age of 618 ± 4 Ma (Oliveira et al. 2015a), and a continental arc geochemical signature (Bueno 2008; Oliveira et al. 2015a).

Structures older than D_{n-1} occur within amphibolite xenoliths in migmatites of the Poço Redondo/Marancó domain. These are small, gentle to open, asymmetric F_{n-2} folds in a differentiated banding S_{n-3} (Fig. 13.6c).

Fig. 13.6 Main deformation features in the Marancó/Poço Redondo and Canindé domains (after Oliveira et al. 2010a). **a** Isoclinal Fn folds with steep Sn axial planes and vertical axes, Poço Redondo migmatite. **b** Syn-Dn conformable apophysis of the Queimada Grande granite. **c** Earlier structures (F_{n-2}) preserved in mafic enclaves in migmatites. **d** Metasedimentary rocks of the Mulungú-Novo Gosto unit of the Canindé domain showing S_0 affected by D1

Investigating deformation in metasedimentary and meta-igneous rocks of the Marancó sub-domain, Carvalho (2005) documented two ductile deformation events D_{n+1} and D_{n+2} that succeeded the D_n event observed in the migmatites. D_{n+1} developed a penetrative, NW-trending axial plane foliation marked by oriented micas, garnet and sillimanite in aluminous schists and quartzites, and by amphibole in meta-basalts; this deformation partially converted the Serra Negra batholith into augen gneiss. The D_{n+1} structural event developed SSW-verging, asymmetric folds associated with shallow to steeply-dipping, as well as SW-directed oblique thrusts. In the northern part of the sub-domain the fold vergence is opposite, possibly a result of back-thrusting. D_{n+2} is a continuum of D_{n+1}. F_{n+1} folds were coaxially refolded and transposed along left-lateral, strike-slip shear zones linked to further displacements of D_{n+1} thrust sheets. Apparently, D_{n+2} did not significantly deform felsic volcanic rocks in the southernmost part of this domain, suggesting that the volcanic rocks are syn- to late D_{n+2}. A third deformation event affected this domain, though its effects were restricted to areas close to the Belo Monte-Jeremoabo shear zone that delimits the Poço Redondo-Marancó and Macururé domains to the south. Small-scale chevron folds and kink bands are the main structures associated with D_{n+3}. A fourth D_{n+4} event, correlative with D_4 in the sedimentary domains of the belt, is distinguished by a set of left-lateral NE-trending strike-slip faults.

Geochronological constraints for D_{n+1} come from ^{40}Ar–^{39}Ar dating of amphiboles (625 ± 3 Ma; Oliveira et al. 2010a) extracted from a meta-basalt close to a D_{n+1} thrust shear zone. Muscovite plates from a garnet-mica schist give a younger ^{40}Ar–^{39}Ar age of 612 ± 7 Ma, which was interpreted by Oliveira et al. (2010a) as the time of exhumation of the Poço Redondo-Marancó domain during D_{n+2}. This may have lasted until at least ca. 602 ± 4 Ma, i.e. the U-Pb SHRIMP age of dacite dated by Carvalho (2005, see also Oliveira et al. 2010a), if we take into account the fact that the felsic volcanic rocks were weakly deformed during D_{n+2}.

13.4.3 Deformation in the Canindé Domain

The Canindé domain is separated from the Poço Redondo-Marancó domain by the Macururé shear zone. According to Oliveira et al. (2010a) deformation in this domain was similar to that observed in supracrustal rocks of the Poço Redondo-Marancó domain, and well preserved in metasedimentary rocks and amphibolites of the Novo Gosto-Mulungu unit. D_1 is commonly transposed by D_2, but is recognizable in metasedimentary rocks by minor folds parallel or oblique to S_0 (Fig. 13.6d). Increase in deformation intensity produced a metamorphic banding S_1 that is usually folded by D_2 structures. D_2 is the main deformation event in the Canindé domain, and is characterized by open to tight folds with N-NW-dipping S_2 that evolved to a near-upright mylonitic foliation as the Macururé shear zone is approached. During D_{2m}-thick sheets of pink deformed leucogranite were emplaced parallel to S_2; they are syn-collision/accretion crustal melt granites.

A third D_3 event, correlative with D_4 in the sedimentary domains of the Belt, is distinguished by NE- and NW-trending conjugate shear fractures and by a set of NE-trending faults; the latter show a dominant sinistral shear sense and offset the mylonite fabric S_{2m} of the Macururé shear zone.

Oliveira et al. (2010a) suggested the following constraints for the timing of deformation in the Canindé domain. The Canindé gabbroic complex (701 ± 8 Ma) and the Gentileza porphyritic quartz-monzodiorite (688 ± 6 Ma) are not as intensively deformed as the Canindé supracrustal rocks; they behaved more competently during deformation and were affected by D_1 and D_2 close to their contact with the Mulungu-Novo Gosto unit and by the D_3-related faults—therefore D_1 and D_2 are younger than the ages referred to above. The Lajedinho granodiorite (Seixas and Moraes 2000) entrains elongate mafic enclaves that parallel S_1–S_2 of the host Gentileza amphibolite; our U-Pb SHRIMP date for this granodiorite is 621 ± 9.5 Ma (Oliveira et al. 2015a). The Curituba monzogranite (617 ± 7 Ma, zircon U-Pb—Silva Filho et al. 2005) crosscuts the Macururé shear zone and thus sets an upper age limit for D_2. No age information is available for D_3. However, given that D_3 correlates with D_{n+4} in the Poço Redondo-Marancó domain and with D_4 in the metasedimentary domains, Oliveira et al. (2010a) suggested that it might be younger than 581 ± 2 Ma, i.e. the ^{40}Ar–^{39}Ar age of muscovite plates oriented along the D_3 mylonitic foliation in the western extension of the Macururé shear zone in the Macururé domain.

13.5 Provenance of the Metasedimentary Rocks

The provenance of clastic sedimentary rocks in ancient basins is a robust geological tool for reconstructing orogenic belts; the Sergipano belt is no different. In this section, we first present previous interpretations of sediment provenance in the belt and then review our published detrital zircon ages and whole-rock Sm–Nd isotope data.

Brito Neves et al. (1977), Silva Filho et al. (1978), and Dominguez (1993) suggested that the uppermost Estância Group sediments, mainly the Lagarto and Palmares Formations were deposited in a foreland basin and formed by erosion of the Sergipano belt during the Neoproterozoic orogeny. This would imply a syn- to late-collisional sedimentary influx towards the SFC. On the other hand, D'el-Rey Silva (1999) suggested that the sediments were derived by erosion from the SFC to the south. This model implies that all sediments in the Estância domain should yield depleted-mantle Nd model ages (T_{DM}), or detrital zircon grains no younger than Paleoproterozoic, for the SFC country rocks are older than 2.0 Ga (e.g., Oliveira et al. 2010b).

Oliveira et al. (2005, 2006, 2010a, 2015b) and Oliveira (2012) resolved this controversy using depleted-mantle Sm–Nd T_{DM} model ages and detrital zircon U-Pb dates. As illustrated in Fig. 13.7, which is a plot of Sm–Nd model ages for the three sedimentary domains of the Sergipano belt and their potential sources, samples from the two uppermost formations of the Estância domain show Sm–Nd T_{DM} model ages in the time range 1.7–1.4 Ga, whereas three samples from the basal Juetê Formation indicate model ages between 2.9 and 2.1 Ga. These ages suggest that the São Francisco craton may have supplied clasts to sediments of the Juetê Formation, but that younger sources are required to explain the Sm–Nd T_{DM} model ages of the Lagarto and Palmares Formations. U-Pb SHRIMP detrital zircon ages of a sandstone from the Lagarto Formation yield clusters at 570, 634, and 958 Ma with a few Paleoproterozoic and Archean grains (Fig. 13.8), this implying deposition after 570 Ma, which is more consistent with the foreland basin model, at least for the two younger formations. On the other hand, detrital zircons from the basal Juetê Formation are all older than 2073 Ma (Oliveira 2012; Oliveira et al. 2015b), thus poorly constraining sediment deposition to any time between 2.0 and 0.6 Ga.

A detrital zircon study of the Vaza Barris units (Oliveira et al. 2006, 2010a, 2015b; Oliveira 2012) indicates the following ages for the youngest zircon populations within each clastic formation: 2000 Ma (quartzite—Itabaiana Formation), 780 Ma (metagreywacke—Ribeirópolis Formation), 657 Ma (meta-sandstone—Frei Paulo Formation), and 653 Ma (diamictite—Palestina Formation). These results support a model of sedimentary provenance from the São Francisco Craton for the basal Itabaiana Formation, which rests upon basement gneisses, but not for the other formations. It is more likely that the sedimentary provenance of the uppermost formations of this domain was largely controlled by uplift of sources farther north, probably in other domains of the Sergipano belt or in other parts of the Borborema Province.

Detrital zircon ages of quartzite and mica schist in the Macururé domain indicate dominantly Mesoproterozoic (~1000 Ma) sources in the west, and early Neoproterozoic to Paleoproterozoic (~950 and 2100 Ma) sources in the east (Oliveira 2012; Oliveira et al. 2010a, 2015b), suggesting that the Borborema Province was the main provenance of the Macururé sediments. No zircons younger than 856 Ma were found, also implying that clast contribution was from areas older than Brasiliano/Pan-African orogenic event.

Detrital zircon SHRIMP data indicate that metasedimentary rocks of the Marancó sub-domain were mainly derived from sources with ages in the interval 980–1100 Ma, and less often from Paleoproterozoic and Archean sources (Carvalho et al. 2005; Oliveira 2012; Oliveira et al. 2015b). The former age group is found in gneisses and migmatites of the Cariris Velhos orogenic belt to the north (Brito Neves et al. 1995; Santos et al. 2010) and from the Poço Redondo sub-domain.

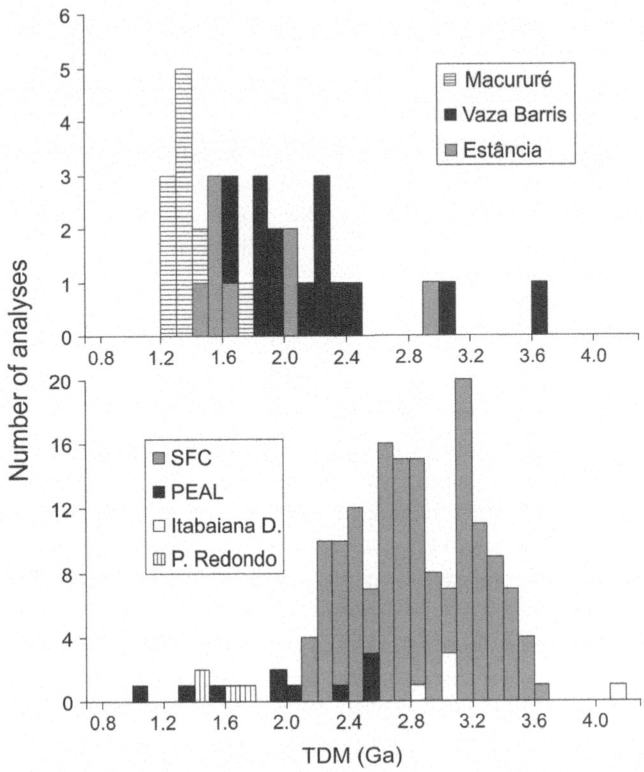

Fig. 13.7 Depleted-mantle Nd-model ages for clastic sedimentary domains of the Sergipano belt and their potential sources—*SFC* São Francisco craton, *PEAL* Pernambuco-Alagoas block, Itabaiana dome and Poço Redondo migmatites (after Oliveira et al. 2006)

13.6 Tectonic Evolution

The evolutionary history presented here corresponds to a synthesis of the model by Oliveira et al. (2010a). Accordingly, the evolution of the Sergipano belt began with the breakup of a Paleoproterozoic continent (Fig. 13.9a) followed by development of an early Neoproterozoic (~980–960 Ma) continental arc (Poço Redondo gneisses) possibly on the margin of the Paleoproterozoic Pernambuco-Alagoas block (Fig. 13.9b). Extensional tectonism acting upon this continental block (Fig. 13.9c, d) gave rise to the following geological units: (i) the Serra Negra A-type granites and associated sedimentary rocks formed on the stretched margin of the Poço Redondo-Marancó domain; (ii) the volcanic-sedimentary sequence of the Canindé domain started to form between the Pernambuco-Alagoas Block and

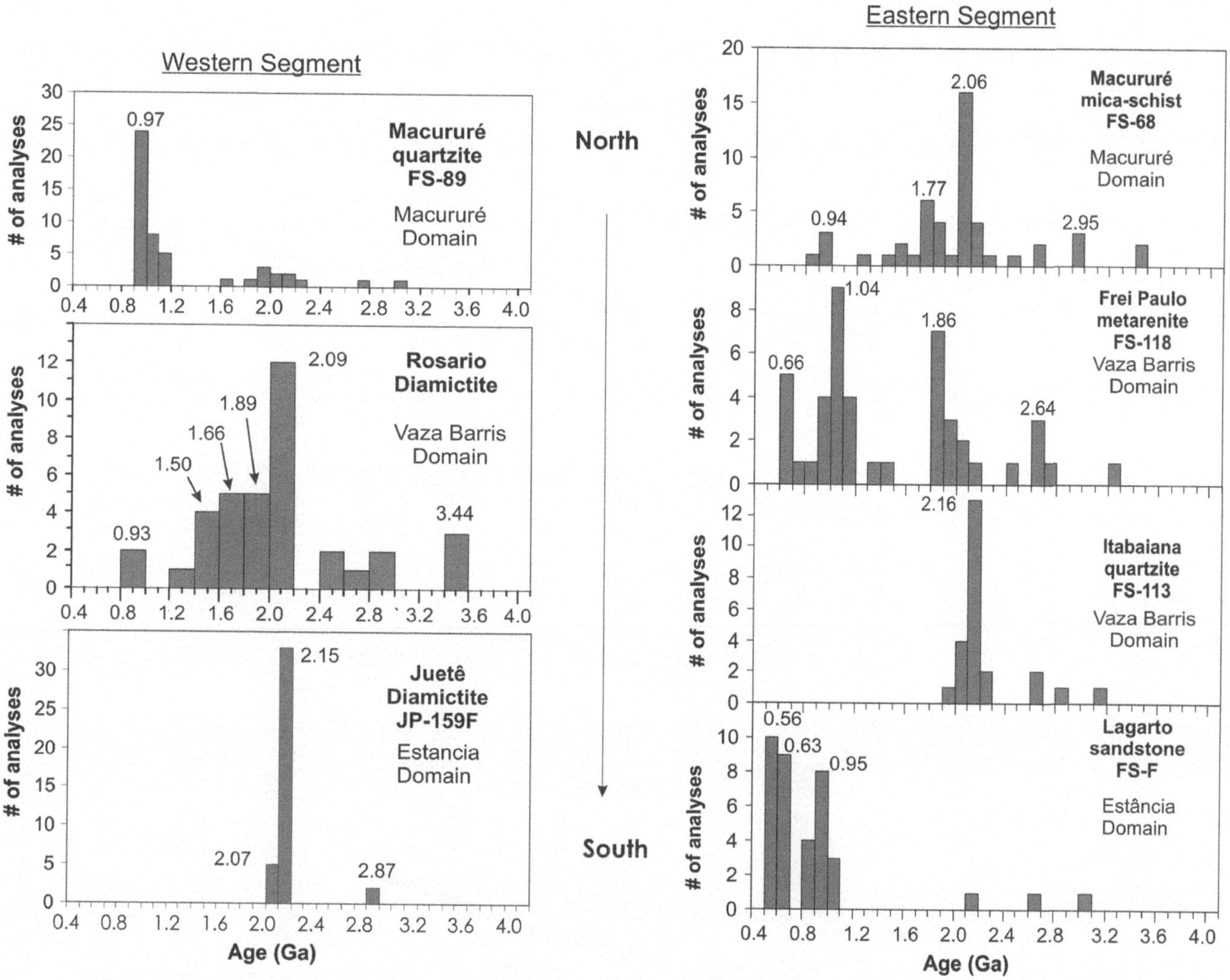

Fig. 13.8 SHRIMP U-Pb ages of detrital zircon grains from domains of the Sergipano belt (after Oliveira 2012; Oliveira et al. 2015b)

the Poço Redondo/Marancó domain; (iii) a passive margin probably developed on the southern edge of the Pernambuco-Alagoas block (basal quartzites of the Santa Cruz Formation in the Macururé domain). Another passive margin formed on the São Francisco Craton (basal clastic unit of the Vaza Barris domain—Itabaiana Formation). The absence of ophiolites suggests that ocean floor basalts, from an ocean that presumably separated the two opposing passive margins, were removed by subduction that later generated a continental magmatic arc between 630 and 620 Ma.

Deposition of sediments on the passive margin of the Pernambuco-Alagoas block began after ca. 900 Ma, i.e., the age of the youngest detrital zircons in sedimentary rocks of the Macururé domain and Marancó sub-domain. In the Canindé domain sedimentation probably started at about 715 Ma (U-Pb age of the A-type Garrote granite) and continued up to at least 625 Ma—the age of the youngest detrital zircon grains in the Novo Gosto-Mulungu unit. On the other hand, deposition of the Juetê and Itabaiana Formations on the SFC passive margin could have started any time after 1975 Ma (age of the youngest zircon grains in the Itabaiana Formation), or after 2150 Ma (age of the youngest zircon population in the Juetê Formation).

In the Canindé domain, rifting continued until approximately 640 Ma (Fig. 13.9d). This is suggested by emplacement of the bimodal igneous association of the Garrote A-type granite (715 Ma) and continental mafic volcanic rocks of the Novo Gosto-Mulungu unit, emplacement of the continental-type Canindé gabbroic complex (ca. 700 Ma), Gentileza microgabbros, quartz-monzodiorite (688 Ma), and rapakivi granite (684 Ma), as well as the Boa Esperança rapakivi granite (641 Ma). There is no conclusive evidence for the incipient ocean floor in the Canindé domain, although a few pillow-bearing amphibolites of the Novo Gosto-Mulungu unit resemble ocean-floor basalts (Oliveira et al. 2010a).

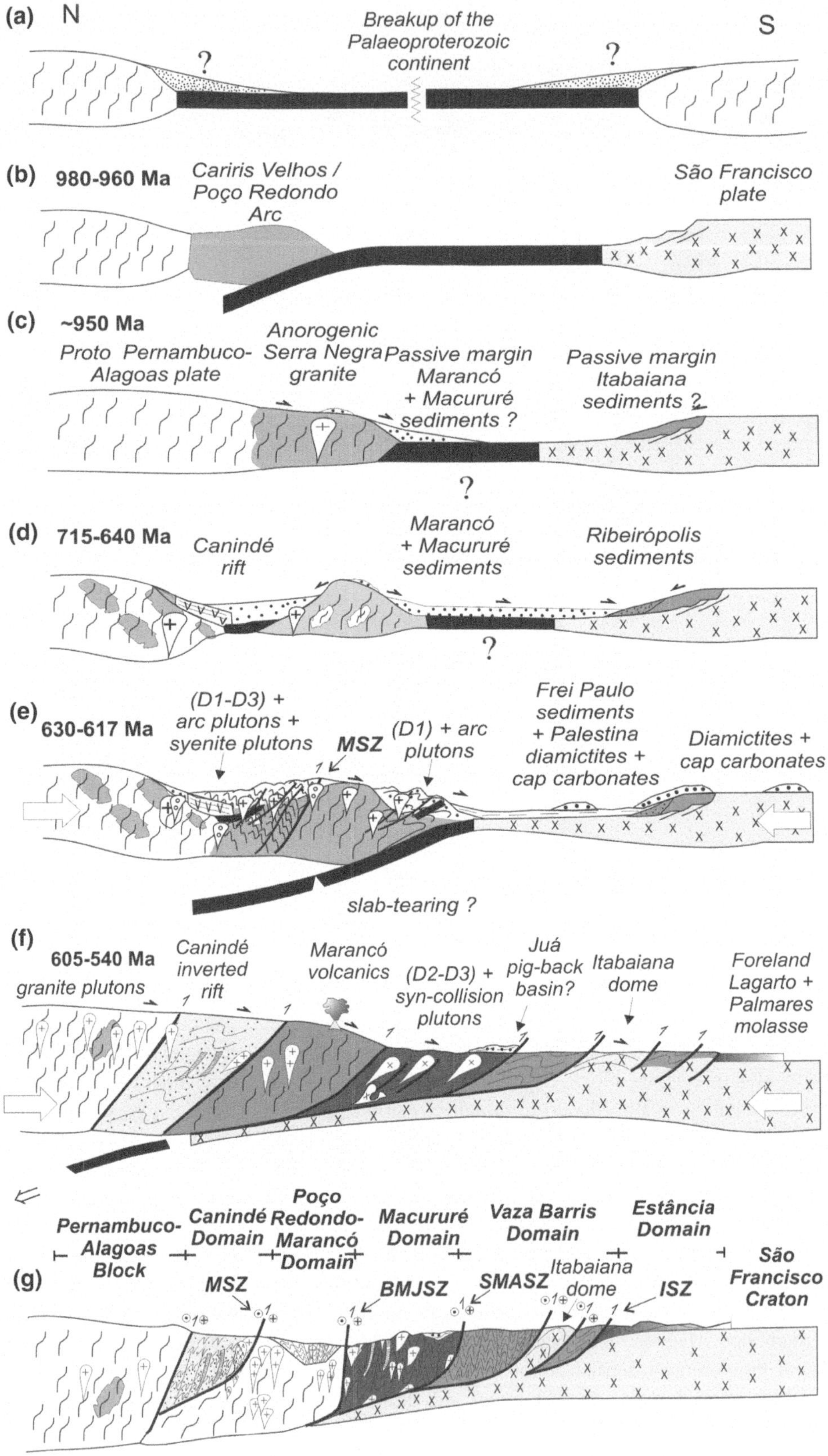

Fig. 13.9 Tectonic evolution of the Sergipano belt during the Neoproterozoic (after Oliveira et al. 2010a). Main shear zones—*MSZ* (Marancó shear zones), *BMJSZ* (Belo Monte/Jeremoabo shear zone), *SMASZ* (São Miguel do Aleixo shear zone), *ISZ* (Itaporanga shear zone)

Convergence of the Pernambuco-Alagoas block and the São Francisco craton led to deformation of shelf sediments, build up of a continental arc between 630 and 620 Ma (Fig. 13.9e) in the Macururé, Poço Redondo-Marancó and Canindé domains (621 Ma Lajedinho granite in the Canindé domain, 628–625 Ma granites in the Macururé domain, and ca. 623 Ma granite in the Poço Redondo-Marancó domain), and syn-collisional granite emplacement in the Macururé (590–570 Ma), Canindé and Poço Redondo-Marancó domains (leucogranite sheets of ca. 610 Ma) (Fig. 13.9e, f).

We infer that a small oceanic plate was subducted beneath the Poço Redondo-Marancó domain to explain the occurrence of 603 Ma-old, arc volcanic rocks in the Marancó sub-domain (Fig. 13.9f). Subsequent exhumation and erosion of the Pernambuco-Alagoas block and the latter three domains led to deposition of the uppermost clastic sediments in the Estância and Vaza Barris domains with 615–570 Ma-old detrital zircon grains, and deposition of the Juá sediments (piggy-back basin?) in the Macururé domain (Fig. 13.9f). At this time, the supracrustal rocks were thrust onto the continental margin of the São Francisco Craton in the south. Figure 13.9g shows the final domain configuration of the belt.

Acknowledgments The São Paulo State Research Foundation (FAPESP, Grant # 2002/03085-2) and the Brazilian National Research Council (CNPq, Grant # 305658/2015-8) provided research grants to EPO. This chapter is dedicated to all geologists who spent years of their professional life studying the Sergipano Belt, especially Gilles Allard, Fred Humphrey, Marinho Silva Filho, Ian Davison, Reginaldo Santos, and Luiz D'El-Rey Silva.

References

Basilici G, Luca PHV, Oliveira EP (2012) A depositional model for a wave-dominated open-coast tidal flat, based on analyses of the Cambrian–Ordovician Lagarto and Palmares formations, north-eastern Brazil. Sedimentology 59: 1613–1639.

Brito Neves BB, Sial AN, Albuquerque JPT (1977) Vergência centrífuga residual no sistema de dobramentos Sergipano. Revista Brasileira de Geociências 7: 102–114.

Brito Neves BB, Sial AN, Beurlen H (1978) O sistema de dobramentos sergipano – Análise do conhecimento. SBG, Núcleo Bahia. Boletim Especial 3: 369–398

Brito Neves BB, Van Schmus WR, Santos EJ, Campos Neto MC, Kozuch M (1995) O evento Cariris Velhos na província Borborema: Integração de dados, implicações e perspectivas. Revista Brasileira de Geociências 25: 279–296.

Brito MF, Silva Filho AF, Guimarães IP, Mariano G, Torres HHF (2006) Geoquímica isotópica de Sr e Nd do complexo granítico Sítios Novos, magmatismo neoproterozóico no sistema de dobramentos sergipano, Nordeste do Brasil. XLIII Congresso Brasileiro Geologia, Anais, ST08-P. 586, 312.

Bueno JF (2008) Geochemistry and chronology of collisional granites in the Sergipano Belt, Northeast of Brazil. Unpublished PhD thesis, University of Campinas, Brazil, 126 p. (in Portuguese)

Bueno JF, Oliveira EP, McNaughton N, Laux JH (2009) U–Pb dating of granites in the Neoproterozoic Sergipano Belt, NE-Brazil: Implications for the timing and duration of continental collision and extrusion tectonics in the Borborema Province. Gondwana Research 15: 86–97.

Carvalho MJ (2005) Tectonic evolution of the Marancó-Poço Redondo domain: Records of the Cariris Velhos and Brasiliano orogenesis in the Sergipano Belt, NE Brazil. Unpublished PhD thesis, University of Campinas, Brazil, 202 p. (in Portuguese)

Carvalho MJ, Oliveira EP, Dantas EL, McNaughton N (2005) Evolução tectônica do Domínio Marancó - Poço Redondo: registro das orogêneses Cariris Velhos e Brasiliana na margem norte da Faixa Sergipana. III Simpósio sobre o Cráton do São Francisco, Anais, 204–207.

D'el-Rey Silva LJH (1999) Basin infilling in the southern-central part of the Sergipano Belt (NE Brazil) and implications for the evolution of Pan-African/Brasiliano cratons and Neoproterozoic cover. Journal of South American Earth Sciences 12: 453–470.

D'el-Rey Silva LJH, McClay KR (1995) Stratigraphy of the southern part of the Sergipano Belt, NE Brazil: tectonic implications. Revista Brasileira de Geociências 25: 185–202.

Davison I, Santos RA (1989) Tectonic evolution of the Sergipano fold belt, NE Brazil, during the Brasiliano orogeny. Precambrian Research 45: 319–342.

Dominguez JML (1993) As coberturas do Cráton São Francisco: uma abordagem do ponto de vista da análise de bacias. In.: Dominguez J.M. L. and Misi A. (Eds.) O Cráton São Francisco, SBG. BA-SE, 137–159.

Huiqi L, McClay KR, Powell D (1990) Physical models of thrust wedges. In: McClay, K.R. (ed.), Thrust Tectonics. Chapman & Hall, pp. 71–81.

Humprey FL, Allard GO (1969) Geology of the Itabaiana Dome Area (Sergipe) and its bearing on the geology of the Propriá geosyncline: a newly recognized tectonic element in the Brazilian Shield. Petrobrás, CENPES, Rio de Janeiro, 104p.

Jardim de Sá EF, Moraes JAC, D'el-Rey Silva LJH (1986) Tectônica tangencial na Faixa Sergipana. Anais XXXIV Congresso Brasileiro de Geologia 3: 1246–1259.

Long LE, Castellana CH, Sial AN (2005) Age, origin and cooling history of the Coronel João Sá Pluton, Bahia, Brazil. Journal of Petrology 46: 255–273.

Menezes Filho NR, Santos RA, Souza JD (1988) Programa de Levantamentos Geológicos Básicos do Brasil, carta geológica, carta metalogenética/previsional, escala 1:100000, Folha SC.24-Z-A-II (Jeremoabo), Estados da Bahia. Brazilian Geological Survey (CPRM), Salvador, Brazil, 113p.

Nascimento RS (2005) The Canindé domain, Sergipano Belt, Northeast Brazil: A geochemical and isotopic study of a Neoproterozoic continental rift sequence. Unpublished PhD thesis, University of Campinas, Brazil, 159 p. (in Portuguese)

Nascimento RS, Oliveira EP, Carvalho MJ, McNaughton N (2005) Evolução Tectônica do Domínio Canindé, Faixa Sergipana, NE do Brasil. III Simpósio sobre o Cráton do São Francisco, Salvador, Bahia, Anais, 239–242.

Oliveira EP (2012) Geologia da Faixa Sergipana no Estado da Bahia. In: J. Barbosa; JF Mascarenhas, LC Correa Gomes, JML Dominguez, JS Souza (Org.). Geologia da Bahia. Pesquisa e Atualização. 1. ed. Salvador, Bahia: 2013. 2 vols. 1200p. Companhia Baiana de Pesquisa Mineral, 2012, v. 1: 179–190.

Oliveira EP, Tarney J (1990) Petrogenesis of the Canindé de São Francisco Complex: a major late Proterozoic gabbroic body in the Sergipe Fold Belt, northeastern Brazil. Journal of South American Earth Sciences 3: 125–140.

Oliveira EP, Tarney J (1995) Genesis of the copper-rich Caraiba norite-hypersthenite complex, Brazil. Mineralium Deposita 30: 351–373.

Oliveira EP, Carvalho MJ, Nascimento RS, Araújo MNC, Dantas D, Basilici G, Bueno JF, McNaughton N (2005) Evidence from detrital zircon geochronology and whole-rock Sm-Nd isotopes for off-craton provenance of clastic metasedimentary units of the Sergipano belt, NE Brazil. X Simpósio Nacional de Estudos Tectônicos, Curitiba, Boletim de Resumos Expandidos 308–311.

Oliveira EP, Toteu SF, Araújo MNC, Carvalho MJ, Nascimento RS, Bueno JF, McNaughton N, Basilici G (2006) Geologic correlation between the Neoproterozoic Sergipano belt (NE Brazil) and the Yaoundé schist belt (Cameroon, Africa). Journal of African Earth Sciences 44: 470–478.

Oliveira EP, Windley BF, Araújo MNC (2010a) The Neoproterozoic Sergipano orogenic belt, NE Brazil: a complete plate tectonic cycle in western Gondwana. Precambrian Research 181: 64–84.

Oliveira EP, McNaughton NJ, Armstrong R (2010b) MesoArchean to Paleoproterozoic growth of the northern segment of the Itabuna-Salvador-Curaçá Orogen, São Francisco Craton, Brazil. In: Kusky, T. M., Zhai, M.-G. & Xiao, W. (eds) The Evolving Continents: Understanding Processes of Continental Growth. Geological Society, London, Special Publications 338: 263–286.

Oliveira EP, Bueno JF, McNaughton NJ, Silva Filho AF, Nascimento RS, Donatti-Filho JP (2015a) Age, composition, and source of continental arc- and syn-collision granites of the Neoproterozoic Sergipano Belt, Southern Borborema Province, Brazil. Journal of South American Earth Sciences 58: 257–280.

Oliveira EP, McNaughton NJ, Windley BF, Carvalho MJ, Nascimento RS (2015b) Detrital zircon U–Pb geochronology and whole-rock Nd-isotope constraints on sediment provenance in the Neoproterozoic Sergipano orogen, Brazil: From early passive margins to late foreland basins. Tectonophysics 662: 183–194.

Oliveira RG (2008) Arcabouço Geofísico, Isostasia e Causas do Magmatismo Cenozóico da Província Borborema e de Sua Margem Continental (Nordeste do Brasil). Unpubl. PhD. thesis, Universidade Federal do Rio Grande do Norte, Natal, 411 p.

Oliveira RG, Medeiros WE (2012) Evidences of buried loads in the base of the crust of Borborema Plateau (NE Brazil) from Bouguer admittance estimates. Journal of South American Earth Sciences 37: 60–76.

Saes GS, Vilas Boas GS (1989) Depósitos de leques costeiros (fan deltas) e de plataforma marinha raso do Grupo Estância, Proterozóico Superior (Bahia e Sergipe). Revista Brasileira de Geociências 19: 343–349.

Santos EJ, Van Schmus WR, Kozuch M, Brito Neves B.B (2010) The Cariris Velhos tectonic event in Northeast Brazil. Journal of South American Earth Sciences 29: 61–76.

Santos EJ, Silva Filho MA (1975) Ensaio interpretativo sobre a evolução da Geossinclinal de Propriá, Nordeste do Brasil. Revista Mineração e Metalurgia 367: 3–22.

Santos RA, Martins AAM, Neves JP, Leal RA (1998) Geologia e recursos minerais do estado de Sergipe. Programa Levantamentos Geológicos Básicos do Brasil. Brazilian Geological Survey (CPRM), Salvador, Brazil. 156 p.

Seixas SRM, Moraes LC (2000) Geological map of Canindé area (unpublished simplified version). Brazilian Geological Survey (CPRM), Salvador, Brazil.

Silva Filho AF, Guimarães IP, Luna EB, Vila Verde VGR, Concentino L (2005) Características geológicas e geoquímicas do plutão Curituba; intrusão shoshonítica tardi a pós-colisional do Cinturão de Dobramentos Sergipano, Província Borborema. X Congresso Brasileiro de Geoquímica, Porto de Galinhas, Pernambuco, Brasil, Extended abstracts, CD 3 p.

Silva Filho MA (1976) A suíte ofiolítica da Geossinclinal de Propriá. XXIX Congresso Brasileiro Geologia, Anais 4: 51–58.

Silva Filho MA (1998) Arco Vulcânico de Canindé-Marancó e a Faixa Sul-Alagoana: seqüências orogênicas Mesoproterozóicas. XL Congresso Brasileiro Geologia, Anais 1: 16.

Silva Filho MA (2006) Lithogeochemistry and evolution of the Marancó Domain in the Sergipano System, Northeast Brazil. Unpublished PhD thesis, Federal University of Pernambuco, Brazil, 220 p. (in Portuguese)

Silva Filho MA, Bonfim LFC, Santos RA (1978) A geossinclinal sergipana: estrutura e evolução. 30 Congresso Brasileiro Geologia, Anais 2464–2477.

Trompette R (2000) Gondwana evolution; its assembly at around . ≈600 Ma. Comptes Rendus Académie Sciences de Paris 330: 305–315.

The Araçuaí Belt

14

Fernando F. Alkmim, Matheus Kuchenbecker, Humberto L.S. Reis, and Antônio C. Pedrosa-Soares

Abstract

The Araçuaí belts extends along the curved southeastern margin of the São Francisco craton between the Brazilian coast and Lat 21°S, where it merges with the Ribeira belt. It represents the external, basement-involved fold-thrust belt of the Araçuaí-West Congo confined orogen (AWCO), which formed due to the closure of the terminal branch of the Adamastor ocean during the amalgamation of West Gondwana in the Ediacaran and beginning of the Cambrian. Bounded to the east and southeast by the high grade and granitic core of the AWCO, the Araçuaí belt involves a basement assemblage older than 1.8 Ga, the 1.7–0.9 Ga rift to rift-sag successions of the Espinhaço Supergroup, the Tonian-Edicaran rift-passive margin Macaúbas Group, as well as the syn-orogenic Salinas Formation and crustal derived granitic intrusions. The Macaúbas Group, the type unit of the belt, contains a glaciomarine sequence made up of thick diamictites, sandstones and Rapitan-type banded iron formations. The units exposed along the belt were metamorphosed under greenschist to amphibolite facies conditions and affected by thrusts, reverse faults and cratonward verging folds, developed between 575 and 530 Ma. The Araçuaí orogenic front propagates into the craton interior and interacts with preexistent rift structures. This chapter describes the stratigraphic framework and overall structure of the Araçuaí belt, emphasizing the paleogeographic and tectonic significance of its sedimentary and volcanic assemblages.

Keywords

Araçuaí-West Congo orogen • Neoproterozoic • Espinhaço Supergroup • Macaúbas Group • Brasiliano event

14.1 Introduction

The Araçuaí belt, as defined by Almeida (1977), corresponds to the Brasiliano orogenic feature that fringes the São Francisco craton to the southeast. As the main fabric elements of Araçuaí belt, the author emphasizes the system of faults and folds that affect the basement gneisses and Proterozoic metasedimentary rocks (especially the Macaúbas Group, its type unit) along the southern Espinhaço range and the eastern portion of the Quadrilátero Ferrífero mining district in the Minas Gerais state (Fig. 14.1). As pointed out by Almeida (1977), the system of NS-trending structures of the belt in the Espinhaço range curves towards NE,

F.F. Alkmim (✉)
Departamento de Geologia, Escola de Minas, Universidade Federal de Ouro Preto, Morro Do Cruzeiro, Ouro Preto, MG 35.400.000, Brazil
e-mail: ffalkmim@gmail.com

M. Kuchenbecker · H.L.S. Reis
Laboratório de Estudos Tectônicos (LESTE)/Centro de Geociências, Instituto de Ciência e Tecnologia, Universidade Federal dos Vales do Jequitinhonha e Mucuri, Campus Juscelino Kubitscheck, Diamantina, MG 39.100.000, Brazil

M. Kuchenbecker · A.C. Pedrosa-Soares
Instituto de Geociências, Centro de Pesquisas Manoel Teixeira da Costa-CPMTC, Universidade Federal de Minas Gerais, Belo Horizonte, MG 31.270.901, Brazil

M. Heilbron et al. (eds.), *São Francisco Craton, Eastern Brazil*, Regional Geology Reviews, DOI 10.1007/978-3-319-01715-0_14

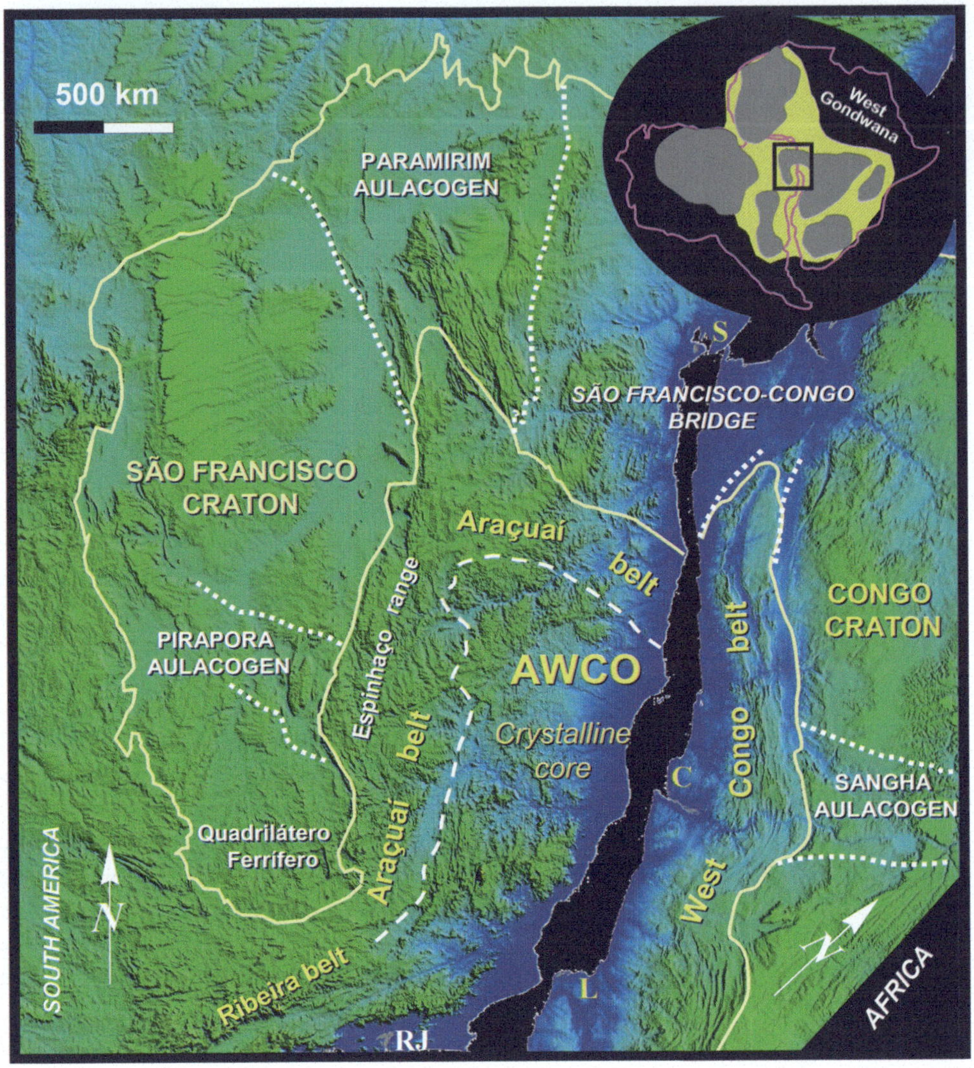

Fig. 14.1 Digital elevation model of southeastern Brazil and west Central Africa, juxtaposed according to the Gondwana best fit (De Wit et al. 1988). The image shows the topographic expression of the Araçuaí-West Congo orogen (AWCO) and the adjacent São Francisco and Congo cratons. The Araçuaí belt forms the external zone of the AWCO and interacts with the Pirapora and Paramirim aulacogens of the São Francisco craton. The crystalline core of AWCO is entirely located in the Brazilian territory. Cities: RJ Rio de Janeiro; V Vitória; S Salvador; L Luanda; C Cabinda

describes a large curve, continues further towards southwest, and reaches the Brazilian coast (Fig. 14.1).

Due to its mineral resources and quality of exposures, the majority of the studies performed in the Araçuaí belt, before and after Almeida's definition, focuses mainly the southern Espinhaço range and adjacent regions, which together with other portions of the São Francisco craton are considered classical areas of the Brazilian shield geology. Remarkably, many of these studies discuss the stratigraphic affinities between the Araçuaí belt and the adjacent cratonic domain.

The Araçuaí belt is presently viewed as one of the external domains of the Araçuaí-West Congo orogen (AWCO), the Ediacaran-Cambrian orogenic system developed between the São Francisco and the Congo cratonic regions during the assembly of West Gondwana (Fig. 14.1) (Pedrosa-Soares et al. 1992, 1998, 2001, 2008; Tack et al. 2001; Alkmim et al. 2006). The other external component of the AWCO corresponds to the West Congo belt of central West Africa, which straddles the coast of Angola, Democratic Republic of Congo, Republic of Congo and Gabon (Tack et al. 2001). In the reconstructions of Gondwana, the AWCO is bounded to the north (South American reference frame) by a crustal bridge that linked the São Francisco and Congo cratons (Porada 1989; Ledru et al. 1994; Silva et al. 2002a). In the south, the AWCO passes into the neighboring Ribeira belt (Heilbron et al. 2000, 2004, and this book) (Fig. 14.1).

As an external belt of the AWCO, the Araçuaí is bounded to east by the crystalline core of the orogen, which is made up essentially of high grade metamorphic rocks and Ediacaran to Cambrian granites. Amphibolite to granulite facies rocks of the AWCO's crystalline core are juxtaposed or thrust upon the greenschist to amphibolite grade rocks of the Araçuaí belt along the Abre Campo shear zone, between Lat 21° and 17°S (Haralyi and Hasui 1982; Cunningham et al. 1998; Peres et al. 2004; Silva et al. 2009) (Fig. 14.2). From Lat 17°S towards north and northeast, the boundary of the crystalline core is given by the first occurrence of syn- and post-collisional granitic bodies and granulite facies metasedimentary rocks (Fig. 14.2). Recent syntheses on the lithological assemblages and evolution of the crystalline core of the AWCO are provided by Gradim et al. (2014), Gonçalves et al. (2014, 2016), Richter et al. (2016), and Tedeschi et al. (2016).

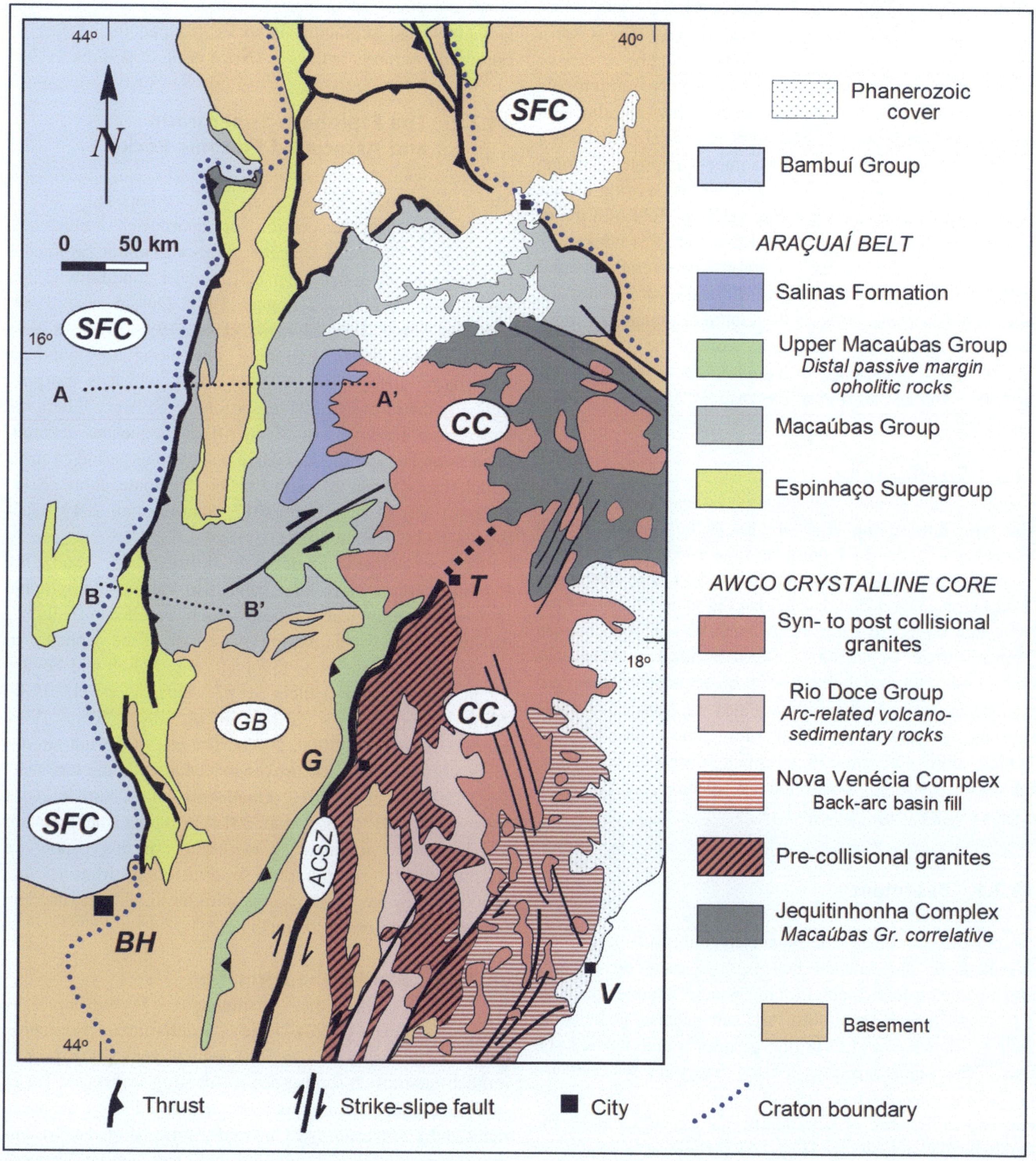

Fig. 14.2 Simplified geologic map of the Araçuaí belt and the high grade/granitic core of the AWCO. Cities: BH Belo Horizonte; G Governador Valadares; T Teófilo Otoni; V Vitória. SFC: São Francisco craton; GB: Guanhães block; ACSZ: Abre Campo shear zone; CC: Crystalline core of the orogen. Locations of the cross-sections shown on Fig. 14.9 are indicated by the lines AA′ and BB′ (modified after Pedrosa-Soares et al. 2001)

The southern limit of the Araçuaí belt is arbitrary. Around latitude 21°S, the NS-oriented fabric elements characteristic of the belt curve progressively towards southwest and merge with structures traditionally ascribed to the Brasiliano Ribeira belt (Pedrosa-Soares and Wiedemann-Leonardos 2000; Heilbron et al. 2000; Heilbron et al., this book). The western boundary to the São Francisco craton is marked by reverse or thrust faults, which juxtapose Meso and Neoproterozoic rocks involved in the belt with basement inliers or Ediacaran units of the São Francisco basin (see Reis et al., this book). Along the central segment of the southern Espinhaço range, these faults are replaced by a large-scale, W-verging monocline, which apparently represents the drape fold of a major blind thrust (Alkmim et al. 2007; Reis et al., this book).

This chapter focuses the stratigraphy and overall structure of the Araçuaí belt, emphasizing its relationships with the units and fabric elements of the cratonic domain. Components of the high grade and granitic core of the AWCO will be only briefly mentioned in the evolutionary synthesis we provide in the last section of the chapter.

14.2 Stratigraphy

The type unit of the Araçuaí belt is the Neoproterozoic Macaúbas Group, which dominates the exposures over most of the northern half of the belt (Fig. 14.2). The Macaúbas Group represents in South America the main fill unit of the basin that closed to form the Araçuaí-West Congo orogen (Pedrosa-Soares et al. 2001, 2008). Besides the Macaúbas Group, the Araçuaí belt also involves the Archean and Paleoproterozoic basement, the Paleo to Mesoproterozoic Espinhaço Supergroup, Statherian anorogenic plutons, Tonian anorogenic granites and mafic intrusions, as well as Ediacaran syn- to post-orogenic metasedimentary and igneous rocks (Figs. 14.2 and 14.3).

14.2.1 Basement

As an extension of the São Francisco craton substratum, the basement of the Araçuaí belt is also made up of rock assemblages older than 1.8 Ga, which include Archean TTGs and K-rich granitoids, Archean greenstone belts, as well as Rhyacian and Orosirian granitoids and supracrustal units. The main exposures of the basement assemblages occur in a relatively large area of the southern portion of the belt and in the core of large antiformal structures along the southern Espinhaço range (Fig. 14.2). These occurrences are currently referred to as the Mantiqueira, Guanhães, Gouveia and Porterinha complexes or blocks (Noce et al. 2007a, b). Smaller exposures correspond to thrust sheets and tectonic windows in various portions of the belt. In contrast with the above mentioned exposures, the basement occurrences in the crystalline core of the orogen (east of the Abre Campo shear zone) are made up of Paleoproterozoic juvenile granitoids and associated supracrustals rocks grouped in the Juiz de Fora and Pocrane complexes (Noce et al. 2007a, b).

14.2.2 The Espinhaço Supergroup and Associated Plutonic Rocks

The Espinhaço Supergroup is composed of a ca. 6000 m-thick succession of quartz-rich sandstones, pelites, conglomerates, volcanic rocks and minor carbonates metamorphosed under greenschist facies conditions (Pflug 1965, 1968; Scholl and Fogaça 1979; Dussin et al. 1984; Almeida-Abreu 1993; Uhlein et al. 1998; Martins-Neto 1998, 2000, 2009). Cropping out along the southern Espinhaço range (Figs. 14.1 and 14.4), which is morphologically the most expressive portion of the Araçuaí belt, the supergroup encompasses rift to rift-sag sequences accumulated in an intracontinental basin over a long period of time, which lasted from ca. 1.8–1.0 Ga (Chemale et al. 2012; Alkmim and Martins-Neto 2012; Santos et al. 2013; Guadagnin and Chemale 2015).

First investigated in detail by Pflug (1965, 1968), the Espinhaço strata were later subject to various stratigraphic and sedimentological studies, which resulted in slightly different subdivisions and interpretations of the depositional environments of the units involved (e.g., Scholl and Fogaça 1979; Garcia and Uhlein 1987; Dussin et al. 1984; Almeida-Abreu 1993; Uhlein et al. 1998; Silva 1998, Martins-Neto 1998, 2000, 2009). The use of detrital zircons in geochronological and provenance studies in the last years (e.g. Chemale et al. 2012; Guadagnin and Chemale 2015; Santos et al. 2013; Costa et al. 2014) provided a new view of the Espinhaço sedimentary evolution. In this paper, we adopted the lithostratigraphic subdivision by Martins-Neto (2000) with some modifications in order to take into account new geochronological data (Fig. 14.3).

14.2.2.1 Banderinha Formation

Incorporated in the Espinhaço Supergroup by Almeida-Abreu (1993), the Bandeirinha Formation (Fig. 14.3) consists of a ca. 200 m-thick package of pink to reddish quartz-rich sandstones, containing layers and lenses of coarse-grained conglomerates. Separated from the basement and younger units by unconformities (Fig. 14.5a), this unit represents fluvial, alluvial fan, and aeolian deposits (Silva 1998; Martins-Neto 1998), which occur in restrict fault bounded depocenters (Santos et al. 2013).

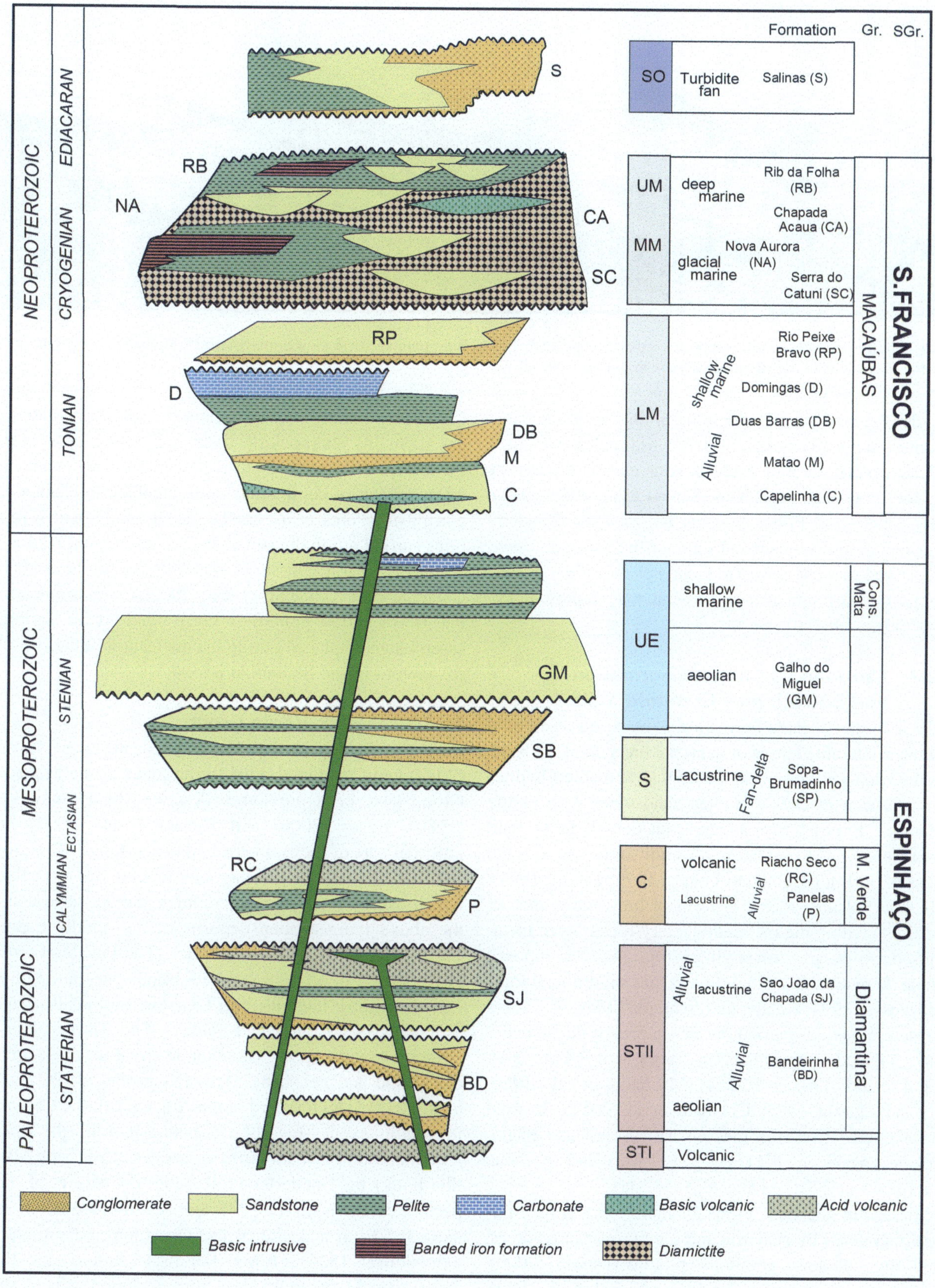

Fig. 14.3 Stratigraphic chart of the Araçuaí belt. Unconformity-bounded sequences: STI Statherian I; STII Statherian II; C Calymmian; S Stenian; UE Upper Espinhaço; LM Lower Macaúbas; MM Middle Macaúbas; UM Upper Macaúbas; SO Syn-orogenic. See text for explanation

Fig. 14.4 A typical landscape of southern Espinhaço range, showing an E-dipping succession of Sopa-Brumadinho quartzites overlain by thick beds of the aeolian Galho do Miguel quartzites in the background

The detrital zircon age spectrum of this unit indicates a maximum depositional age of 1.785 Ma and provenance from Paleoproterozoic and Archean sources (main peaks at 2155, 2468, 2664 and 2838 Ma) (Santos et al. 2013). Nevertheless, since gradational contacts between the Bandeirinha Formation and supracrustal units of the basement have been described in a few areas (Machado et al. 1989), at least part of its quartz-dominated succession may represent the record of a pre-1,8 Ga basin-fill.

14.2.2.2 São João da Chapada Formation and Borrachudos Granitoid Suite

The São João da Chapada Formation is made up of alluvial sandstones containing lenses of quartzite breccias at the base, layers of K-rich volcanic rocks (known as hematite-phyllite) and a succession of coarse-grained sandstones with pelitic intercalations. Exposed over an area considerably larger than the basal Bandeirinha Formation, this unit shows an average thickness of 300 m (Scholl and Fogaça 1979; Dussin 1994). The alkaline volcanic rocks, in form of intrusive bodies or intercalations within the metasedimentary rocks, were metamorphosed to grey-greenish phyllites (Fig. 14.5b), composed of sericite, hematite, chloritoid, quartz and subordinate tourmaline (Dussin 1994; Knauer and Schrank 1994).

Age determinations carried out in samples of the hematite phyllite yielded crystallization ages of 1715 ± 2, 1710 ± 12 and 1703 ± 16 Ma (Machado et al. 1989; Dussin and Dussin 1995; Chemale et al. 2012). Detrital zircons extracted from the metasandstones indicate Paleoproterozoic (peaks at 1711 and 2134 Ma) and Archean sources (peaks at 2701, 3151 and 3336 Ma).

The basement gneisses of the Guanhães block (Fig. 14.2) are locally covered by acid volcanics, and, over large areas, intruded by alkaline granitoid plutons (Dussin 2000), grouped in the Borrachudos Suite (Grossi-Sad et al. 1990). The acid volcanic rocks (rhyolites) were dated at 1770, 1719 and 1711 Ma (Brito Neves et al. 1979; Machado et al. 1989). Consisting mainly of leucogranites, the Borrachudos suite also contains charnockitic rocks and subordinate syenites. These rocks are intensively deformed and show microstructures compatible with amphibolite facies metamorphism (Peres et al. 2004). Geochronological investigations carried out in the suite have yielded crystallization ages around 1730 Ma (Dussin et al. 1994; Fernandes et al. 1994; Chemale et al. 1998; Silva et al. 2002b), which indicate that the Borrachudos granitoids correspond to an anorogenic assemblage emplaced during the nucleation of the Espinhaço rift system in the Statherian period.

14.2.2.3 Mato Verde Group

Recently re-defined and investigated in detail by Costa (2013) and Costa et al. (2014), the rocks of the Mato Verde Group have been previously described by Drumond et al. (1980), Egger (2006) and Knauer et al. (2007) in the northern part of the southern Espinhaço range. The group is subdivided into the Panelas and Riacho Seco formations, which overlie the basement complex and are separated from the above lying aeolian deposits of the middle Espinhaço Supergroup by an erosional surface. The Panelas Formation is composed of alluvial deposits (sandstones and conglomerates) grading laterally into lacustine and shallow marine sediments (sandstones and pelites). The Riacho Seco Formation is made up of sub-alkaline rhyolitic to dacitic volcaniclastic and epiclastic rocks (Knauer et al. 2007; Costa et al. 2014), which record in the Araçuaí belt the Calymmian magmatic event (Fig. 14.3) documented in the northern Espinhaço range (Paramirim aulacogen, Cruz and Alkmim, this book) by Danderfer et al. (2009). Costa et al. (2014) performed U-Pb LA-ICP-MS age determinations on zircons extracted from a dacite of the formation and estimated its extrusion at 1524 ± 6 Ma. For these authors, the Riacho Seco magmatism took place in association with a second rifting event experienced by the Espinhaço basin in the course of the Calymmian Period.

14.2.2.4 Sopa-Brumadinho Formation

Widely known for its diamond-bearing conglomerates, the Sopa-Brumadinho Formation (Pflug 1968) (Figs. 14.3 and 14.5c) comprises three coarsening upward cycles (20–50 m-thick), which often start with pelites, grade upward into fine-grained, poorly sorted sandstones (locally ferruginous), and end with very coarse-grained sandstones and conglomerates or breccias (Martins-Neto 2000). The whole sedimentary succession reaches a maximum thickness of 220 m. The conglomerates are in general clast-supported, containing pebbles and cobbles predominantly of quartzites, vein-quartz, iron-rich quartzites and subordinate fragments of acid volcanic rocks, imbedded in a sandy matrix (Pflug 1965; Scholl and Fogaça 1979; Martins-Neto 1996). Locally, the matrix consists of a green phyllite (Fig. 14.5c), which seems to be a metamorphosed tuffaceous pelite, rich in heavy minerals, such as fine-grained zircons (Chemale et al. 2012). This succession records a renewed rifting episode promoting the conditions for sediment deposition in braided fluvial systems, and lake fan-deltas (Martins-Neto 1996, 2000; Chemale et al. 2012).

Zircons extracted from the greenish tuffaceous matrix of the diamond bearing conglomerates were dated by Chemale et al. (2012) using the U-Pb LA-ICP-MS method. The ages obtained for the youngest grains constrain the deposition of this unit at ca. 1192 Ma, implying that a major hiatus (of ca. 500 Ma) exists between Sopa-Brumadinho Formation and the basal formations of the supergroup.

14.2.2.5 Galho do Miguel Formation

Deposits of a marine transgression onlapping the Sopa-Brumadinho sequence and older units mark the base of this formation, which is made up essentially of aeolian pure quartz arenites (Garcia and Uhlein 1987; Dussin and Dussin 1995; Uhlein et al. 1998; Martins-Neto 2000) (Fig. 14.5d). Reaching a maximum thickness of ca. 2000 m (Scholl and Fogaça 1979), and grading upward into shallow marine sandstones, the large-scale cross-bedded sandstones of this

Fig. 14.5 Outcrop and hand specimen expressions of the Espinhaço Supergroup rocks. **a** Unconformity separating the Bandeirinha and the São João da Chapada formations (view from NW) (Photo by M. Martins-Neto). **b** Statherian metavolcanic rocks (hematite phyllite) of the São João da Chapada Formation. **c** Diamond-bearing metaconglomerates of the Sopa-Brumadinho Formation. Large clasts of quartzite, quartz vein and hematite-rich metasandstones are imbedded in green phyllitic matrix. **d** Quartz-rich meta-sandstones of the Galho do Miguel Formation showing metric- to decametric-scale cross-stratifications

formation record a noticeable expansion of the Espinhaço basin, apparently associated with a thermal subsidence regime in the aftermath of the Sopa-Brumadinho rifting event. For Martins-Neto (2000), the Galho do Miguel Formation witnesses the transitional phase between the rift and the full stage sag, recorded respectively by the below and above lying Sopa-Brumadinho Formation and Conselheiro Mata Group.

The detrital zircon age spectrum obtained for the Galho do Miguel Formation differs from that of the Sopa Brumadinho for the absence of 1192 Ma grains. The youngest ages fall around 1581 Ma (Santos et al. 2015).

14.2.2.6 Conselheiro Mata Group

Subdivided into five formations (Dussin et al. 1984) and comprising a ca. 900 m-thick alternation of pelite and sandstone-dominated units, this group occurs in a synclinorium trough exposed along the westernmost portion of the southern Espinhaço range. Recording three marine transgressive-regressive cycles (Dupont 1995; Martins-Neto 2000, 2007), the group consists of offshore to shoreface deposits, partially influenced by tides and reworked by winds (Santos et al. 2015). The uppermost Rio Pardo Grande Formation also contains lenses and layers (Fig. 14.3) of stromatolite-rich dolostones representing a carbonate platform developed in the course of the youngest regressive event preserved in the group (Santos et al. 2015).

The detrital zircon age distributions are quite similar for the five formations of the group. However, a shift from a dominant 2000 Ma peak in the lower units to a 1900 Ma cluster in the upper formations can be observed in the age spectra obtained by Santos et al. (2015). The maximum and minimum depositional ages of the group, respectively of 1192 and 906 Ma, are constrained by the age of the Sopa Brumadinho Formation and by mafic dykes, which cut through the whole succession (Machado et al.1989; Chemale et al. 2012).

14.2.3 The Macaúbas Group, Tonian Granites and Mafic Intrusions

The Neoproterozoic Macaúbas Group represents an extensive and up to 10 km-thick sedimentary succession metamorphosed under greenschist to amphibolite facies conditions, which crops out along most of the northern Araçuaí belt (Figs. 14.2 and 14.3). Since the pioneer studies carried out by Moraes (1929) and Moraes and Guimarães (1931), the unit aroused the interest of geologists because of its glaciogenic rocks and their implications for the regional paleogeographic framework. The name "Macaúbas" was first applied only to the diamictite-bearing units exposed along the Araçuaí belt. Over the years, the concept of group has been expanded in order to incorporate non-glacial units (e.g., Karfunkel et al. 1985; Pedrosa-Soares et al. 1992; Noce et al. 1997; Martins et al. 2008).

Until recently, the Macaúbas Group was described as recording a single rift-passive margin basin fill sequence later involved in the Araçuaí-West Congo orogen (Pedrosa-Soares et al. 2011a; Babinski et al. 2012). New studies revealed, however, that the group encompasses at least three evolutionary stages, which are in part associated with distinct episodes of anorogenic magmatism (Kuchenbecker et al. 2015). Furthermore, two of them are separated by an expressive regional unconformity. Thus, considering the data presently available, the group can be subdivided into three major successions, which are: (i) a lower, pre-glacial sequence, comprising the Capelinha, Matão, Duas Barras, Domingas and Rio Peixe Bravo; (ii) a middle, glaciogenic sequence, including the diamictite-rich packages of the Serra do Catuni, Nova Aurora and Lower Chapada Acauã formations; and (iii) an upper, post-glacial succession, including the diamictite-free Upper Chapada Acauã and Ribeirão da Folha formations (Kuchenbecker et al. 2015) (Fig. 14.3).

14.2.3.1 Lower Macaúbas Group

The oldest basin-forming episode recorded by the Macaúbas Group is represented by a Tonian rifting succession accumulated between ca. 900 Ma and 750 Ma (Kuchenbecker et al. 2015). The onset of this event is marked by a bimodal magmatic event, which comprises: (i) the Pedro Lessa mafic dykes (930–905 Ma, Machado et al. 1989); (ii) the A-type granites of the Salto da Divisa Suite (c. 875 Ma, Silva et al. 2008; Menezes et al. 2012); and (iii) the Pedra Preta Amphibolite (Gradim 2012). These magmatic rocks may occur either as intrusions into the basement and the Espinhaço Supergroup or associated with the basal units of the group. Abortion of the extensional process was followed by an erosive period, recorded by a regional unconformity that separates the pre-glacial and glacial units of the group. Recognized already in the first regional studies (e.g. Karfunkel and Hoppe 1988; Uhlein et al. 1998, 1999; Martins-Neto et al. 2001), this unconformity was recently interpreted as a first-order sequence boundary, separating the two different basins-forming events (Kuchenbecker et al. 2015) (Fig. 14.3).

The Lower Macaúbas Group consists of continental to shallow marine deposits of the Capelinha, Matão, Duas Barras, Domingas, and Rio Peixe Bravo formations. U-Pb detrital zircon age determinations on these units indicate Archean to Neoproterozoic sources and a maximum depositional age compatible with the rift-related magmatism mentioned above (Babinski et al. 2012, Kuchenbecker et al. 2015). The Santo Onofre Group of the Paramirim aulacogen is considered a correlative of this portion of the group (Schobenhaus 1996; see Cruz and Alkmim, this volume).

Capelinha Formation
Formerly considered as a post-glacial unit, the Capelinha Formation was redefined as the basal portion of the Macaúbas Group by Castro (2014). The unit contains mica schists, quartz-schists and quartzites with high amount of garnet, staurolite and/or kyanite. Tholeiitic metabasalts occur interbedded with metasedimentary rocks of the lower portion of the formation, recording a syn-sedimentary magmatism dated at ca. 957 Ma. Furthermore, provenance studies recently carried out in these strata have yielded detrital zircon ages as young as 970 Ma (Castro 2014).

Matão Formation
The Matão Formation comprises breccias, conglomerates and sandstones deposited in fluvial to shallow marine conditions, during a transgressive episode (Karfunkel and Karfunkel 1976; Martins et al. 2008).

Duas Barras Formation
The Duas Barras Formation consists of sandstones and rare conglomerate, with variable contents of mica, feldspar, iron oxide and lithic fragments. It shows a maximum thickness of approximately 100 m and sedimentary structures suggesting fluvial to shallow marine environments (Martins 2006; Noce et al. 1997; Leite 2013).

Domingas Formation
The Domingas Formation occurs exclusively in the western portion of the Araçuaí Orogen, and comprises siltstones, mudstones and km-sized lenses of stromatolitic dolomite (Noce et al. 1997; Fraga et al. 2013), which record the most expressive shallow carbonate platform associated with the Macaúbas Group.

Rio Peixe Bravo Formation
The Rio Peixe Bravo Formation is an up to 700 m-thick succession of micaceous, ferruginous and/or feldspathic sandstone, pelites and rare clast-supported conglomerates (Viveiros et al. 1979; Grossi-Sad et al. 1997), which occur only along the eastern portion of the Porteirinha basement block.

14.2.3.2 Middle Macaúbas Group

After the erosive phase that affected the Tonian rift sequence, a renewed rifting event in the beginning of the Cryogenian period led to the accumulation of the glacial sediments characteristic of the group. The sedimentation during this phase is recorded by the diamictite-bearing Serra do Catuni, Nova Aurora and Lower Chapada Acauã formations, deposited in glaciomarine settings (Pedrosa-Soares et al. 2011a; Uhlein et al. 2007). The youngest detrital zircons found in these units show Tonian ages, reflecting the lack of Cryogenian volcanic centers in the source areas (Kuchenbecker et al. 2015). The Jequitaí Formation of the São Francisco basin is considered a glacio-terrestrial equivalent of this sequence (e.g. Pedrosa-Soares et al. 2011a; see Reis et al., this book).

Serra do Catuni Formation
The Serra do Catuni Formation comprises massive diamictite (Fig. 14.6a) with minor sandstone and pelite intercalations (Noce et al. 1997; Pedrosa-Soares et al. 2011a). Distinctive glacial sedimentary features include chaotic deposits, dropstones, boulders and striated flat-iron-shaped clasts (Karfunkel and Karfunkel 1976, 1977; Karfunkel and Hoppe 1988).

Nova Aurora Formation
The Nova Aurora represents a distal correlative of the Serra do Catuni Formation (Noce et al. 1997; Pedrosa-Soares et al. 2011a), and comprises diamictites, sandstones, and rare pelites. Thick layers of Rapitan-type diamictitic iron formation occur locally, representing an important mineral resource (Viveiros et al. 1979; Uhlein et al. 1999; Vilela et al. 2014).

Lower Chapada Acauã Formation
The Lower Chapada Acauã Formation consists of stratified diamictite, graded sandstones and pelites, composing a cyclic succession of debris flows and sandy to muddy turbidites (Pedrosa-Soares et al. 1992; Uhlein et al. 1999, 2007; Martins 2006).

14.2.3.3 Upper Macaúbas Group

As indicated by the occurrence of ophiolitic remnants in the central Araçuaí Orogen (Queiroga et al. 2007; Queiroga 2010) (Fig. 14.6b–d), the Macaúbas rift evolved into a passive margin complex, recorded by a sedimentary pile deposited in a inland sea basin, almost completely surrounded by the São Francisco-Congo paleocontinent. The transition to the passive margin took place after the end of the Cryogenian glacial event, as recorded by the post-glacial Upper Chapada Acauã and Ribeirão da Folha formations (Pedrosa-Soares et al. 2011a; Babinski et al. 2012; Kuchenbecker et al. 2015), which are the only units of the group containing Cryogenian detrital zircons (Peixoto et al. 2015; Kuchenbecker et al. 2015). The Jequitinhonha Complex, consisting of Al-rich paragneiss, quartzite, graphite gneiss and calc-silicate rocks exposed in the northern portion of the orogens' crystalline core, is currently viewed as correlative of the passive margin deposits of the Upper Macaúbas Group (Gonçalves-Dias et al. 2011).

Upper Chapada Acauã Formation
The Upper Chapada Acauã Formation (Fig. 14.3) comprises sandstone and pelite interpreted as post-glacial successions

Fig. 14.6 **a** Glaciogenic metadiamictites of the Serra do Catuni Formation, middle Macaúbas Group. Quartz, quartzite and carbonate pebbles and cobbles are dispersed in a grey matrix, composed of quartz, mica and carbonate. **b** Pillowed metabasalts (Rio Preto Member) that occur intercalated with passive margin deposits of the Upper Macaúbas Group. **c** Amphibolite cut by a small apophyse of plagiogranite representing the crustal section of the dismembered ophiolitic rocks of the Ribeirão da Folha Formation, Upper Macaúbas Group. **d** Component of the ultramafic section of the Ribeirão da Folha ophiolitic rocks containing cromitite pods

deposited in continental shelf settings during the passive margin stage (Pedrosa-Soares et al. 2011a; Noce et al. 1997; Martins-Neto et al. 2001).

Ribeirão da Folha Formation

The Upper Chapada Acauã Formation grades laterally into the Ribeirão da Folha Formation, which includes distal passive margin and ocean floor deposits (Pedrosa-Soares et al. 1992, 1998, 2011a, b; Queiroga et al. 2007). This formation is made up of fine-grained turbidites and chemical sediments such as chert, banded iron formation and limestone. Metamafic and meta-ultramafic rocks (Fig. 14.6c, d) with ocean-floor lithochemical signature occur closely associated with the Ribeirão da Folha Formation (Fig. 14.7) and represent tectonically dismembered ophiolite complexes (Pedrosa-Soares et al. 1992, 1998, 2001, 2008; Suita et al. 2004; Queiroga 2010; Peixoto et al. 2015). These complexes yielded Ediacaran magmatic crystallization U-Pb ages, suggesting oceanic spreading at least from ca. 600–660 Ma (Queiroga et al. 2007; Queiroga 2010).

14.2.4 The Salinas Formation

The Salinas Formation consists of a ca. 1600 m-thick succession of turbiditic wackes, sandstones, pelites and conglomerates metamorphosed under greenschist to amphibolite facies conditions (Fig. 14.7) (Lima et al. 2002; Pedrosa-Soares et al. 2008; Santos et al. 2009). Interpreted as a syn-orogenic assemblage, the Salinas turbidites occur in the internal portion of the belt (Fig. 14.2), covering or in tectonic contact with the distal passive deposits of the Upper Chapada Acauã and Ribeirão da Folha formations. The detrital zircon age spectra of the Salinas Formation

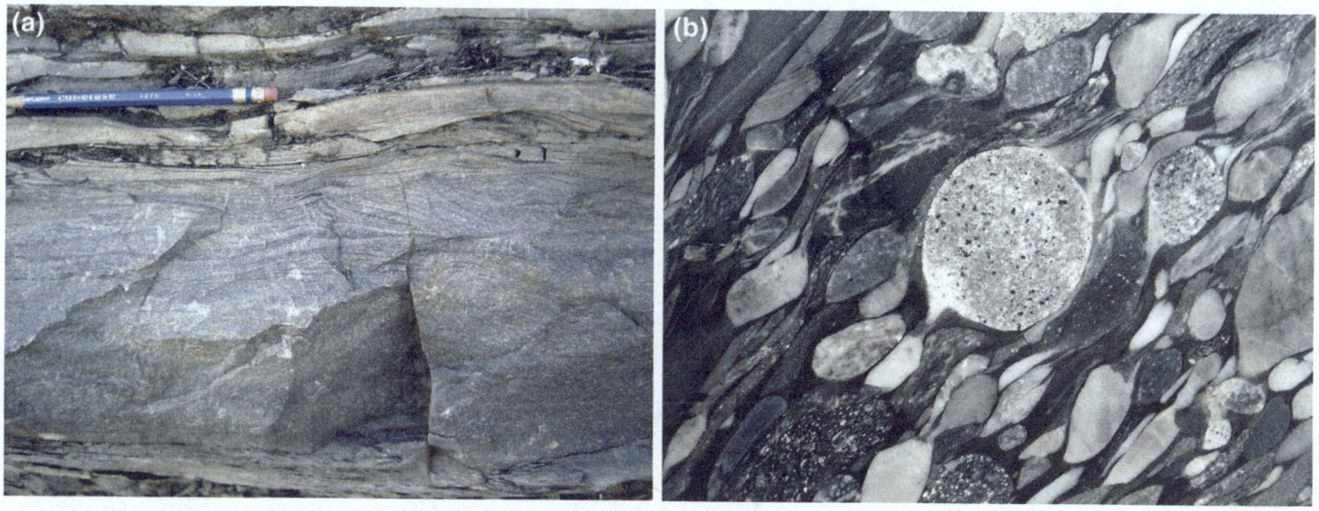

Fig. 14.7 **a** A turbidite bed of the Salinas Formation, composed of massive sandstones that grade upward into a fine-grained sandy layer exhibiting climbing ripples and capped by pelites. **b** Deformed conglomerate of the Salinas Formation containing abundant clasts of granites and acid volcanics

(Pedrosa-Soares et al. 2008; Kuchenbecker 2014) indicate that its main sources were the Rio Doce magmatic arc and Paleoproterozoic basement units. Developed between 630 and 580 Ma in the pre-collisional stage of the AWCO, the Rio Doce magmatic arc is materialized by granitoids of the G1 Supersuite and volcanic rocks of the Rio Doce Group (Pedrosa-Soares et al. 2011b; Gonçalves et al. 2014; Tedeschi et al. 2016), which occur in the crystalline core of the AWCO. As pointed out by Kuchenbecker (2014), the turbidites of the Salinas Formation accumulated between 548 and 520 Ma., i.e., coeval with the deposition of the Bambuí Group in the São Francisco basin (see Reis et al., this book).

14.2.5 Ediacaran Granites

Rocks of the Upper Macaúbas Group and Salinas Formations are cut by granites in a few localities. For instance, Salinas Schists are intruded by 544 Ma late-collisional and 520 Ma post-collisional granites in areas close the eastern boundary of the belt (Pedrosa-Soares et al. 2011b; Peixoto et al. 2015). The syn- to late collisional intrusions are grouped in the G2 Supersuite, which comprises S-type peraluminous granites dated between 590 and 540 Ma (Pedrosa-Soares et al. 2011b). The post-collisional granites characterize two distinct assemblages, the G4 and G5 Supersuites, which consist of 530–500 Ma leucogranites and 520–480 Ma high-K metaluminous I-type granitoids, respectively (Pedrosa-Soares et al. 2011b).

14.3 Tectonic Framework and Metamorphism

The Araçuaí belt is dominated by a system of folds, thrusts and reverse folds that verge towards the adjacent craton. From the transition to the Ribeira belt on the south, to its northeastern end in the continental margin, the following sectors and large-scale structures can be recognized in the belt (Fig. 14.8): (i) southern Espinhaço range fold-thrust belt; (ii) Rio Pardo salient; (iii) Itabepi shear zone and associated structures; (iv) interference zone with the Paramirim Aulacogen; (v) Dom Silvério shear zone and associated structures; (vi) Abre Campo shear zone; (vii) Guanhães block; (viii) São José da Safira schist belt; (ix) Chapada Acauã shear zone (CASZ); (x) Minas Novas transpressional corridor; (xi) the Salinas synclinorium (Alkmim et al. 2006, 2007).

The Araçuaí belt is connected to a foreland fold-thrust belt in the craton interior, i.e., in the São Francisco basin. Involving all Precambrian fill units of the basin, the foreland belt is also W-verging and thin-skinned along most of its extension (see Reis et al., this volume).

Rocks involved in the Araçuaí belt exhibit metamorphic parageneses characteristic of the greenschist and amphibolite facies. Greenschist assemblages predominate in the west, in the more external portions, whereas amphibolite facies rocks occur predominantly in the east and more internal domains. The Salinas synclinorium and adjacent areas bounded to the west by the CASZ (Figs. 14.8 and 14.9) make an exception to the mentioned facies distribution. Despite its location in the internal domain of the belt, part of the Salinas synclinorium shows mineral parageneses of greenschist facies. This fact seems to result from a post-collisional ESE-directed down-motion of the whole package bounded by the CASZ (see next section). Besides this, rocks of the Salinas Formation are also affected by a metamorphic aureole in areas close to the granitic intrusions of the crystalline core (Santos et al. 2009).

14.3.1 The Southern Espinhaço Range Fold-Thrust Belt

The roughly NS-trending southern Espinhaço range fold-thrust belt, as the largest and best exposed segment of the Araçuaí belt, extends for over 700 km along the eastern margin of the São Francisco craton (Fig. 14.8). Involving the basement, the Espinhaço Supergroup, the Macaúbas Group

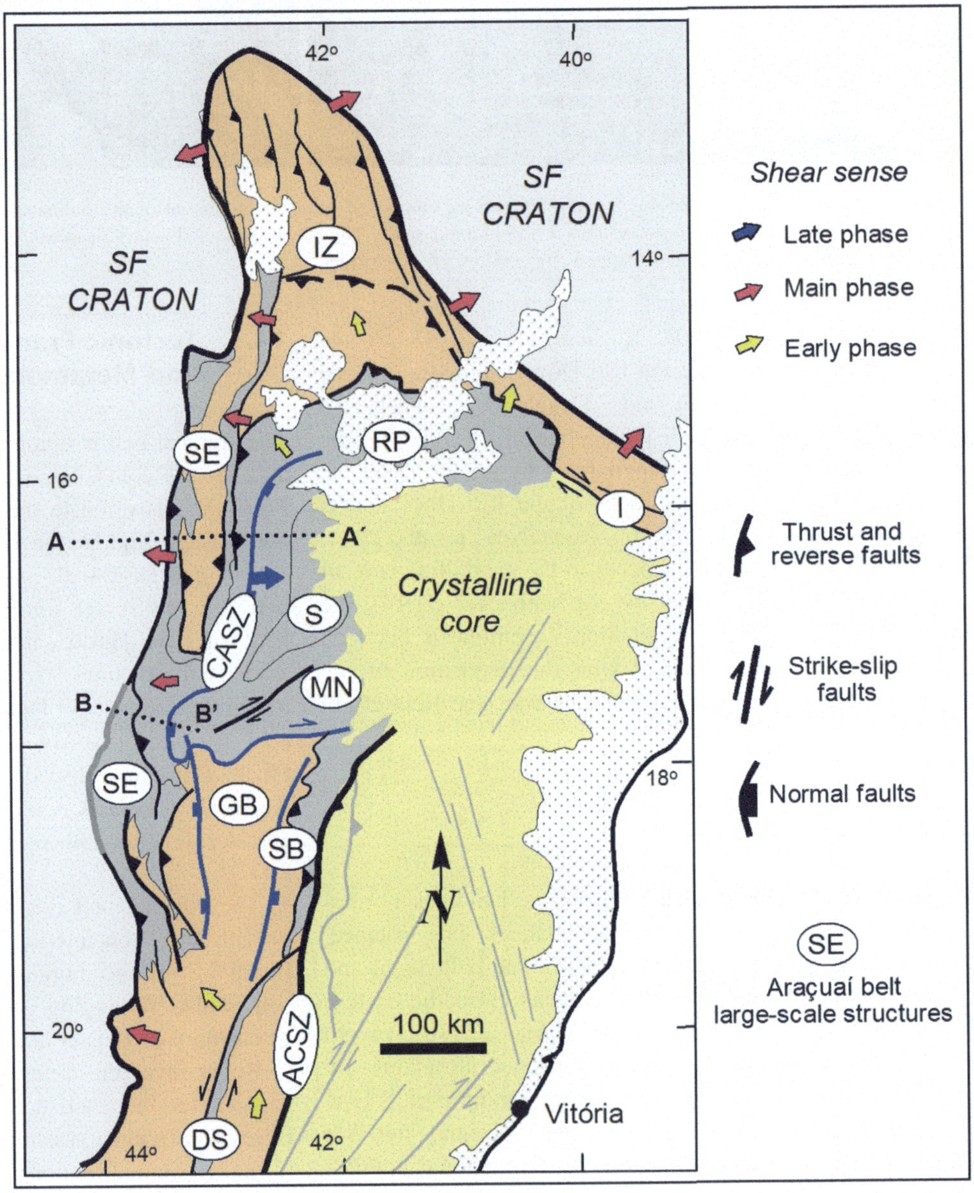

Fig. 14.8 Simplified tectonic map of the Araçuaí belt showing the main structures and the tectonic transport associated with three phases of its development. Large-scale structures: SE southern Espinhaço f-t-belt; RP Rio Pardo Salient; IZ Interference zone with the Paramirim aulacogen; I Itapebi shear zone; DS Dom Silvério shear zone; ACSZ Abre Campo shear zone; SB São Jose da Safira schist belt; GB Guanhães block; MN Minas Novas transpressional corridor; S Salinas synclinorium; CAZS Chapada Acauã shear zone. Location of the cross-sections of Fig. 14.9 are indicated by the lines AA′ and BB′ (modified from Alkmim et al. 2006)

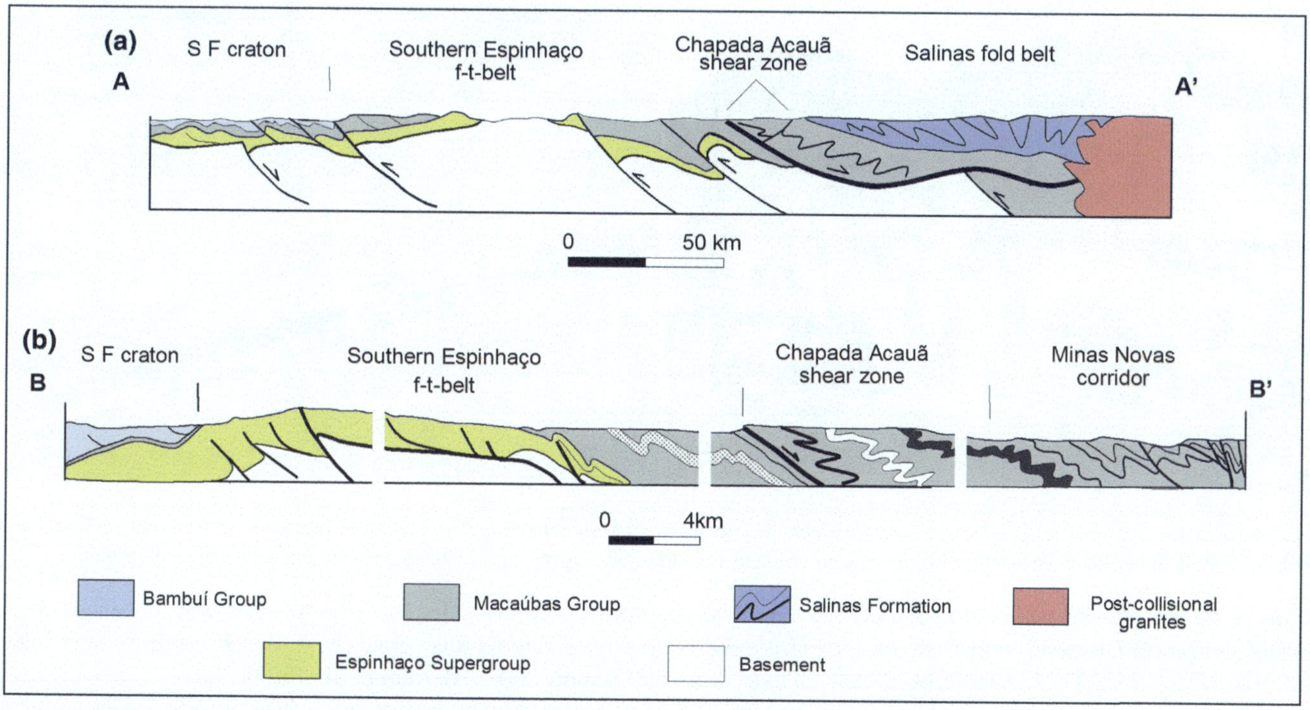

Fig. 14.9 Cross-sections of Araçuaí belt, emphasizing some of its large-scale structures. Locations of the sections are shown in the maps of Figs. 14.2 and 14.8 (modified from Alkmim et al. 2007)

and basic intrusives of the Pedro Lessa Suite, this belt is dominated by a system of folds, thrusts and reverse faults generated during two main deformational phases in response to an E-W contractional stress field (Dussin and Dussin 1995; Uhlein et al. 1986, 1998). An E-dipping foliation associated with a down-dip stretching lineation pervasively affects the units involved throughout the whole belt. Kinematic indicators associated with these and other larger scale structures attest a systematic W-directed (towards the craton) tectonic transport (Uhlein et al. 1986; Uhlein 1991; Dussin and Dussin 1995; Grossi-Sad et al. 1997) (Fig. 14.10a).

The f-t-belt exhibits the architecture of a deeply eroded basement-involved thrust system (Fig. 14.9). Stratigraphic and structural features documented in the region indicate that most of the major faults, especially reverse faults and thrusts that bound basement blocks, correspond to inverted normal faults (Marshak and Alkmim 1989; Alkmim et al. 1996).

The structural traces of the southern segment of the belt describe a relatively smooth salient in map-view, whose culmination is located in the region where the sedimentary succession involved shows its maximum thickness. This portion of the belt, marked by the absence of a frontal thrust, coincides with the occurrence at depth of the WWN-trending Pirapora aulacogen (Figs. 14.1 and 14.2), a partially inverted graben that cuts across the central portion of the adjacent craton (Alkmim and Cruz 2005; Reis et al., this book).

At its northern end, the external sector of the Serra do Espinhaço f-t-belt bends to NNW and continues in this direction as the western edge of the interaction zone with the Paramirim aulacogen (Cruz and Alkmim 2006) (Fig. 14.8). In the same region, the internal sector of the belt curves progressively towards northeast forming the Rio Pardo salient. At its southern end, the f-t-belt involves Archean and Paleoproterozoic basement units, as well as Statherian anorogenic plutons.

14.3.2 The Rio Pardo Salient and the Itapebi Shear Zone

Despite their map expressions and regional significance, the Rio Pardo salient, the Itapebi shear zone, and associated structures are the least studied segments of the Araçuaí belt. This results basically from the lack of good exposures, which are in both cases restricted to a relatively narrow ridge.

The Rio Pardo salient, the large antitaxial curve described by the Araçuaí belt along the craton southeastern margin (Fig. 14.8), involves essentially metadiamictites and schists of the Macaúbas Group. A ca. 200 m thick shallow-dipping mylonitic detachment developed along the basement-cover contact marks its leading edge. The dominant fabric ele-

Fig. 14.10 **a** Road cut exposing a W-directed thrust affecting metasandstones of the Sopa Brumadinho Formation in the southern Espinhaço range f-t-belt. **b** Small-scale asymmetric and W-verging folds in the turbiditic wackes of the Salinas Formation (Salinas synclinorium)

ments in the detachment zone and rock units above reflect an overall cratonward tectonic transport, i.e., towards NE, N and NW (Fig. 14.9). These structures, which include large isoclinal folds, ductile shear zones, and a variety of small-scale elements, are affected by younger family folds and strike-slip shear zones. Oriented preferentially in the NNW direction, the second generation of structures interferes with elements of the older set, thereby giving rise to the sinuous structural traces of the salient and local dome-and-basin fold interference patterns (Almeida et al. 1978; Cruz and Alkmim 2006).

The WNW-trending Itapebi shear zone (Fig. 14.8) extends for over 100 km in the northeastern segment of the belt, affecting the basement, Tonian anorogenic plutons and Macaúbas Group correlative units. Connected to the eastern end of the Rio Pardo salient, the Itapebi shear zone represents the central structure of a dextral transpressional system, which, together with NE- and SW-verging thrusts, forms a large-scale flower structure. Late-stage normal sinistral motions have been also detected along the system (Alkmim et al. 2006).

14.3.3 The Interference Zone with the Paramirim Aulacogen

The apex of the Rio Pardo salient (Fig. 14.8) is connected to the Paramirim deformation corridor (Alkmim et al. 1993; Cruz and Alkmim 2006), which cuts across the whole cratonic domain in the NNW direction. This deformation corridor corresponds to the zone of maximum inversion of the Paramirim aulacogen, a Proterozoic structure that extends over a substantial portion of the northern São Francisco craton (see Cruz and Alkmim, this book). The southern segment of the corridor corresponds to a basement cored uplift, considered for this reason to be part of Araçuaí belt, i.e., its interference zone with the Paramirim aulacogen (Alkmim 2004; Alkmim et al. 2007).

Bounded on both sides by reverse to oblique-slip faults of opposite vergences, the interference zone is characterized by three families of fabric elements, which are: (i) faults, folds, and a various small-scale structures that record an overall NNW- to N-directed tectonic transport; (ii) NNW-trending folds, reverse and thrust faults, also associated with a variety of small-scale structures nucleated under a ENE-WSW-oriented contractional stress-field; and (3) normal faults and normal reactivation of the pre-existent structures. The oldest set of structures reflects the northward propagation of the Araçuaí belt deformational front, whereas the dominant NNW-trending fabric elements were nucleated during the subsequent frontal inversion of the Paramirim aulacogen. Along the Rio Pardo salient apex, folds of this set overprint the older structures, thereby giving rise to a variety of interference patterns (Almeida et al. 1978; Cruz and Alkmim 2006).

14.3.4 Other Regional Structures of the Araçuaí Belt

The ca. 120 km long and NE-trending Dom Silvério shear zone (Endo 1997; Peres et al. 2004) is one of the most prominent morpho-structural elements of the internal domain of the belt (Fig. 14.8). This an up to 4 km-wide ductile-brittle strike-slip shear zone affects the basement gneisses, separating the terminal segment of the southern Espinhaço range fold-thrust belt on the west from a W-verging synclinorium to the east. Besides the basement, this synclinorium involves Upper Macaúbas Group correlative schists (Peres et al. 2004; Pedrosa-Soares et al. 2001; Gradim 2012). Exhibiting predominantly left-lateral motion, the Dom Silvério shear zone is cut by the Abre Campo shear zone that defines the eastern boundary of the Araçuaí belt (Peres et al. 2004).

The Guanhães block is a large basement high that probably acted as such during the deposition of the Espinhaço and Macaúbas successions. The São José da Safira schist belt that bounds the Guanhães block to east is an up to 36 km wide and 150 km long system of oblique-slip thrusts and folds that roots in the dextral strike-slip Abre Campo shear zone, the suture zone of the AWCO. Made up of Upper Macaúbas Group, Salinas Formation, tectonic slices of ophiolitic rocks, and syn-collisional granites, this belt is interpreted as a fragment of the accretionary prism of the AWCO, which became incorporated in its external belt (Peixoto et al. 2015).

The internal portion of the northern half of the Araçuaí belt underlain by the Upper Macaúbas Group and Salinas Formation encompasses three regional structures, namely, the Chapada Acauã shear zone (CAZS), the Salinas synclinorium and the Minas Novas transpressional corridor (Figs. 14.8 and 14.9). The ca. 35-wide CAZS extends along the northern boundary of the Guanhães block and eastern edge of the southern Espinhaço range fold-thrust belt. Differing remarkably from the adjacent structures, the CASZ is characterized by a variety of fabric elements that overprint the dominant W-verging set of structures characteristic of the belt and reflect an overall normal motion towards ESE (Gradim et al. 2005; Marshak et al. 2006) (Fig. 14.9). Among these elements, the most expressive are ESE-verging folds associated with a regionally pervasive WNW-dipping crenulation cleavage. Remarkably, the enveloping surfaces of these folds also dip ESE (Marshak et al. 2006, Alkmim et al. 2006). Down motion along the CASZ seems to be responsible for a decrease in the metamorphic grade in the hanging wall block, which includes the Minas Novas transpressional corridor and the Salinas synclinorium.

The Salinas synclinorium affects the Upper Macaúbas Group and the Salinas strata (Figs. 14.9 and 14.10b). Its NS-trending axial trace curves to the NE direction on both the northern and southern periclinal zones. Second and third order folds of the structure verge E and W (Uhlein 1991; Santos et al. 2009). The Minas Novas transpressional corridor corresponds to an NE-trending dextral strike-slip deformational zone, which also affects the Upper Macaúbas Group and the Salinas Formation (Pedrosa-Soares 1995; Alkmim et al. 2006).

14.4 Evolutionary Synthesis

As one of the external belts of the AWCO, the Araçuaí belt was generated through closure of the Macaúbas basin by the end of the Ediacaran Period, as the various plates that formed West Gondwana converged and collided (Pedrosa-Soares et al. 2001, 2008; Alkmim et al. 2006). In the pre-collisional scenario of the AWCO, the Macaúbas basin, representing one of the terminal branches of the Adamastor ocean, was probably an inland sea, located between a long peninsula, the present-day São Francisco craton, and a considerably larger landmass, represented by the Congo craton (Pedrosa-Soares et al. 2001, 2008; Alkmim et al. 2006) (Fig. 14.11a). As indicated by the occurrence of ophiolitic rocks along the internal zone of the Araçuaí belt (e.g., in the São José da Safira schist belt), a substantial portion of the Macaúbas basin must have been underlain by oceanic crust (Pedrosa-Soares et al. 1992, 1998, 2001; Queiroga et al. 2007; Peixoto et al. 2015).

Various models have been proposed for the development of the AWCO in the space between the São Francisco peninsula and Congo continent (e.g. Pedrosa-Soares et al. 1992; Trompette et al. 1992; Maurin 1993; Alkmim et al. 2006). According to the model by Alkmim et al. (2006), the Macaúbas basin closed in response to a clockwise rotation of the São Francisco peninsula in respect to the Congo landmass, a mechanism that resembles the operation of a nutcracker (Figs. 14.11 and 14.12).

The full development of the oceanic Macaúbas basin was preceded by a 1 Ga-long period, during which the São Francisco Congo crust underwent a protracted succession of rifting events, many of them associated with anorogenic magmatism and followed by thermal subsidence pulses (Pedrosa-Soares and Alkmim 2011). This history started with the Statherian rifting (Brito Neves et al. 1996), an event that extrapolates the Araçuaí belt region. It affected large areas of the SFC and its margins between 1780 and 1710 Ma, i.e., approximately 120 Ma after the Transmazonian orogeny (see Barbosa and Barbosa and Alkmim and Teixeira, this book).

The Statherian rifting is recorded in the Araçuaí belt by two pulses of felsic volcanism dated at 1770 and 1711 Ma, the emplacement of the Borrachudos anorogenic plutons around 1730 Ma, and the onset of the Espinhaço intracontinental rifting with the deposition of the Bandeirinha and São João da Chapada formations (Dussin et al. 1994, Chemale et al. 1998, 2012; Fernandes et al. 1994; Silva et al.

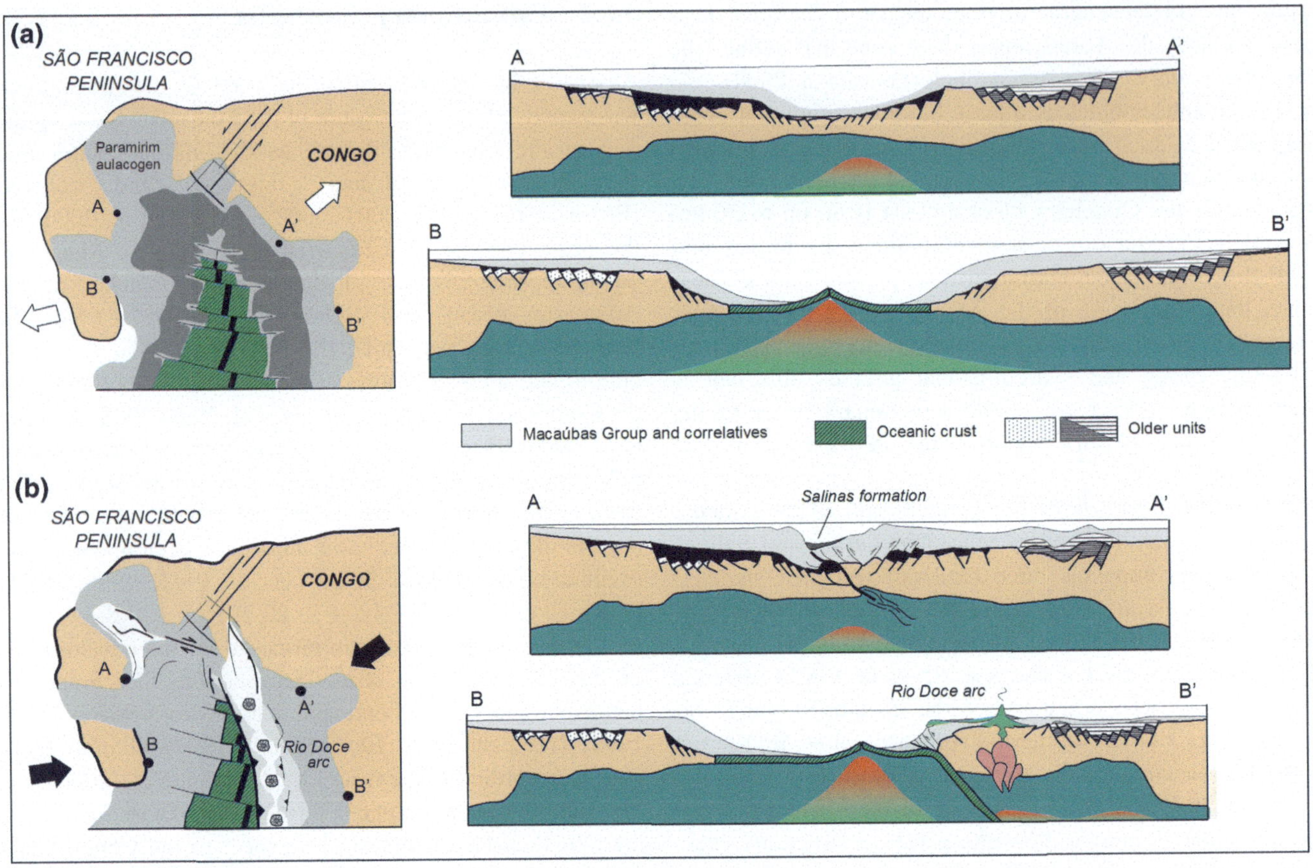

Fig. 14.11 Cartoon illustrating the pre-collisional scenario of the Araçuaí-West Congo orogen. **a** The Macaúbas basin as a large gulf located between the São Francisco peninsula and the Congo continent at ca. 680 Ma. Notice that the Macaúbas basin terminates against the crustal bridge connecting the peninsula to the continent and consequently was only partially floored by oceanic crust. **b** Initial closure of the Macaúbas basin. Subduction of the oceanic sector led to development of the Rio Doce magmatic arc along the Congolian margin between 630 and 580 Ma. In this stage, the ensialic portion of the basin already experienced contraction and inversion (modified from Alkmim et al. 2007)

2002b; Pedrosa-Soares and Alkmim 2011; Santos et al. 2015).

Two other rifting and magmatic events that took place during the Mesoproterozoic are recorded by the Espinhaço sequence: (i) the Calymmian 1524 Ma acid magmatism of the Richo Seco Formation (Costa et al. 2014); and (ii) the Early Stenian (1180 Ma) deposition of the diamond-bearing Sopa-Brumadinho Formation (Chemale et al. 2012; Santos et al. 2015). The accumulation of Sopa-Brumadinho Formations in grabens, superimposed on preexistent structures of the same nature, was followed by the generation of a large sag basin, in which the aeolian and shallow marine upper units of the Conselheiro Mata Group were deposited (Martins-Neto 2000; Santos et al. 2015).

The initiation of the Macaúbas basin in the Tonian period of the Neoproterozoic Era is marked in the Araçuaí belt region by the following events: (i) emplacement of the Pedro Lessa mafic dykes at 906 Ma (Machado et al. 1989); (ii) emplacement of the Salto da Divisa anorogenic granites at around 850 Ma (Silva et al. 2002b; Pedrosa-Soares and Alkmim 2011); (iii) deposition of the Lower Macaúbas Group rift-related units. In the West Congo belt (the African counterpart of the AWCO), the generation of the precursor basin was preceded and accompanied respectively by the intrusion of the anorogenic 999 Ma Noqui Granite (Tack et al. 2001), and the extrusion of the voluminous bimodal lavas of the Zadinian and Mayumbian groups between 920 and 910 Ma (Tack et al. 2001; Thiéblemont et al. 2011).

After the important erosive phase that affected the Tonian rift sequence, and before reaching the oceanic stage, the Macaúbas basin underwent a renewed extensional pulse, which probably took place in the course of the Sturtian global ice age that marks the beginning of the Cryogenian Period (Hoffman and Schrag 2002). The onset of the extension is marked by the intrusion of the anorogenic magmatic rocks of the South Bahia Alkaline Province exposed along the northern Araçuaí Orogen (730–675 Ma,

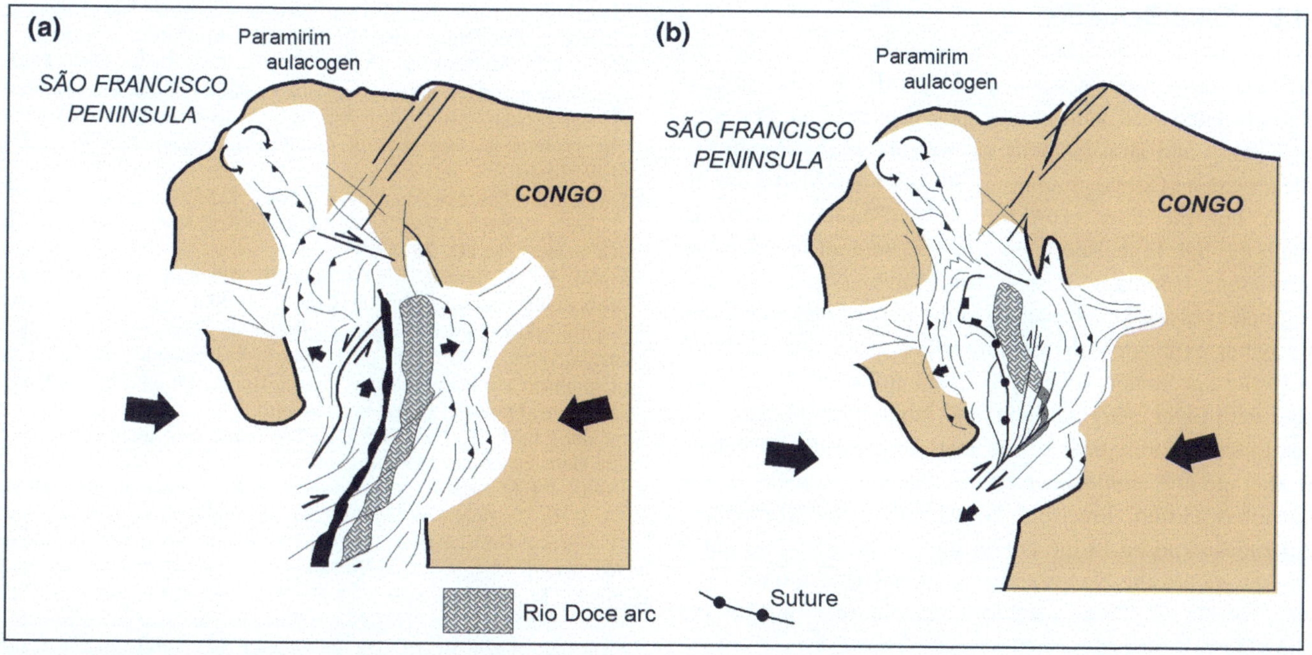

Fig. 14.12 Cartoon illustrating the closure of the Macaúbas basin and uplift of the AWCO. **a** The margins of the basin converged and collided at ca. 575 Ma. Consumption of the oceanic floor led to abortion of subduction and magmatism along the Rio Doce arc. **b** Continued convergence of the basin margins caused uplift and propagation of the orogenic fronts into the adjacent cratonic domains. The southern portion of AWCO experienced dextral transpressional deformation, as the São Francisco peninsula approached the Congo (modified from Alkmim et al. 2006, 2007)

Rosa et al. 2007), as well as the volcanic rocks associated with the Lower Diamictite (694 ± 4 Ma, Straathof 2011) and La Louila (713 ± 49 Ma, Thiéblemont et al. 2011) formations of the West Congo belt. The glacially influenced diamictites of the Middle Macaúbas Group witnesses the full glaciation of the São Francisco-Congo landmass, at that time probably individualized as major continental plate.

Taking into account the isotopic signature of the carbonates of the Macaúbas Group and their correlatives in the craton interior, Caxito et al. (2012) postulated that the Macaúbas glacial deposits more likely represent a manifestation of the Marinoan ice age that reaches low latitudes in the Early Ediacaran Period. Since the youngest detrital zircons of the Macaúbas glaciogenic deposits were dated at 880 Ma, the solution of this problem depends on other tools to better constrain the age of these rocks.

After the retreat of the ice cap, the expansion of the Macaúbas basin progressed, reaching its oceanic stage sometime before 660 Ma, the oldest age so far obtained from the ophiolithic rocks in the Araçuaí belt. Consumption of the oceanic segment of the basin led to the development of the Rio Doce magmatic arc (Fig. 14.11b), represented by the association of plutonic rocks of the G1 Supersuite with the volcanic units of the Rio Doce Group exposed in the crystalline core of the AWCO (Pedrosa-Soares et al. 2001, 2011; Gonçalves et al. 2014, 2016; Tedeschi et al. 2016). The oldest component of the Rio Doce arc so far dated in the orogen yielded a crystallization age of ca. 630 Ma. Closure of the Macaúbas basin must have started thus before 630 Ma and culminated with the abortion of the subduction system installed in its interior around 590 Ma, the age of the first generation of syn-collisonal granites, produced by melting of the basement and basin fill units (Pedrosa-Soares et al. 2011a, b; Gradim et al. 2014; Richter et al. 2016).

The propagation of the orogenic front from the internal parts of the AWCO towards the adjacent cratonic regions, with the consequent development of the Araçuaí belt, lasted from 575 Ma, the climax of the syn-kinematic metamorphism dated in the crystalline core, to ca. 530 Ma, the age of the oldest post-collisional granitic plutons. In fact, the uplift of the Araçuaí belt and the migration of the Brasiliano orogenic front into the São Francisco basin and Paramirim aulacogen must have occurred after or simultaneously to the deposition of the Três Marias Formation, the youngest Proterozoic unit exposed in the craton interior. Captured by the orogenic deformation, the Três Marias Formation contains detrital zircons dated at 558 Ma (Kuchenbecker et al. 2015; Rodrigues 2008). The post-collisional granites were emplaced as the AWCO underwent a phase of orogenic collapse (Pedrosa-Soares et al. 2011b), recorded by a variety of extensional structures, among them the large-scale Chapada Acauã shear zone (Marshak et al. 2006).

14.5 Final Remarks

The Paleo- to Mesoproterozoic rocks involved in the Araçuaí belt correlated well with units exposed in external zones of the Ribeira and Brasilia belts, as well as in the Paramirim aulacogen and Congo craton (e.g. Alkmim and Martins-Neto 2012). Together they record the ca. 700 Ma-long period, in which the São Francisco-Congo paleocontinent remained as a coherent landmass, despite successive extensional and magmatic pulses. Due basically to the lack of age constrains and better exposures, the evolution postulated for the Neoprotezoic succession is, however, not immediately comparable with other marginal belts and intracratonic basins. As mentioned before, some of the matters presently in debate are the age and significance of the glaciogenic rocks of the Macaúbas Group. Precisely, to which of the glaciogenic sequences exposed in the craton interior do they correlate? Do they record the Neoproterozoic global glaciations? How many of them? These questions remain unsolved at the moment.

Another matter that demands further investigations is age of the full development of the Araçuaí belt. Recent age determinations in Bambuí Group that fills the adjacent São Francisco basin in the craton interior led various authors (e.g. Warren et al. 2014; Paula-Santos et al. 2015) to argue in favor of maximum depositional ages as young as 540 Ma. If these assumptions are correct, and considering that the Bambuí Group is deformed along the craton margin, the Araçuaí orogenic front must have been active in the Cambrian, not ceasing in the Ediacaran as previously thought.

Acknowledgments This chapter synthesizes results obtained in various research projects. F.F. Alkmim and Pedrosa-Soares projects were supported by the CNPq (Conselho Nacional de Desenvolvimento Científico e Tecnológico) M. Kuchenbecker and H. Reis received support from CODEMIG-UFMG and CPRM-UFMG mapping programs. A review by U. Cordani greatly improved the original manuscript.

References

Alkmim, F. 2004. O que faz de um cráton um cráton? O Cráton do São Francisco e as revelações Almedianas ao delimita-lo. In: Mantesso-Neto et al. (eds) *Geologia do Continente Sul-Americano.Evolução da obra de Fernando Flávio Marques de Almeida*. Becca, São Paulo, p. 17–35.

Alkmim, F.F. & Cruz, S.C.P. 2005. Crátogenos, aulacógenos, orógenos e sua interação: O caso do Cráton do São Francisco-Congo e sistemas brasilianos/ pan-africanos adjacentes. Anais III Simpósio sobre o Cráton do São Francisco, SBG, Núcleo BA/SE, p. 185–187.

Alkmim, F.F., Martins-Neto, M.A., 2012. Proterozoic first-order sedimentary sequences of the São Francisco craton, eastern Brazil. Marine and Petroleum Geology, 33, 127–139.

Alkmim, F.F.; Brito Neves, B.B.; Castro Alves, J.A.1993. Arcabouço tectônico do Cráton do São Francisco - Uma Revisão. In: Dominguez, J.M.L. & Misi, A. (eds) O Cráton do São Francisco. Salvador, SBG/Núcleo BA/SE, SGM/BA. p. 45–62.

Alkmim, F.F., Marshak, S., Pedrosa-Soares, A.C., Peres, G.G., Cruz, S., Whittington, A. 2006. Kinematic evolution of the Araçuaí-West Congo orogen in Brazil and Africa: Nutcracker tectonics during the Neoproterozoic assembly of Gondwana. Precambrian Research, 149, 43–64.

Alkmim, F.F., Pedrosa-Soares, A.C., Noce, C.M., Cruz, S.C.P. 2007. Sobre a evolução tectônica do Orógeno Araçuaí-Congo Ocidental. Geonomos, 15 (1), 25–43.

Alkmim, F.F.; Chemale Jr., F.; Endo, I., 1996. A deformação das coberturas proterozóicas do Cráton do São Francisco e o seu significado tectônico. Revista da Escola de Minas, 49 (1): 22–38.

Almeida-Abreu, P.A., 1993. A Evolução Geodinâmica da Serra do Espinhaço Meridional, Minas Gerais, Brasil. (Ph. D. Thesis), Univ. Freiburg, Freiburg, Germany (150 pp.).

Almeida, F.F.M. 1977. O Cráton do São Francisco. Revista Brasileira de Geociências, 7, 285–295.

Almeida, F.F.M., Hasui, Y., Rodrigues, E.P., Yamamoto J.K., 1978. A faixa de dobramentos Araçuaí na região do Rio Pardo. In: Congresso Brasileiro de Geologia, 30, Recife. Sociedade Brasileira de Geologia, Anais v.1, 270–283.

Babinski, M., Pedrosa-Soares, A.C., Trindade, R.I.F., Martins, M., Noce, C.M., Liu, D., 2012. Neoproterozoic glacial deposits from the Araçuaí orogen, Brazil: age, provenance and correlations with the São Francisco Craton and West Congo belt. Gondwana Research, 21 (2–3), 451–465.

Brito Neves, B.B., Cordani, U.G., Kawashita, K., Delhal, J., 1979. A evolução geocronológica da Cordilheira do Espinhaço; dados novos e integração. Revista Brasileira de Geociências 9, 71–85.

Brito Neves, B.B., Sá, J.M., Nilson, A.A., Botelho, N.F., 1996. A tafrogênese estateriana nos blocos paleoproterozóicos da América do Sul e processos subseqüentes. Geonomos 3, 1–21.

Castro, M.P. 2014. Caracterização geológica da Formação Capelinha como uma Unidade Basal do Grupo Macaúbas emsua Área Tipo, Minas Gerais. Universidade Federal de Ouro Preto. MSc Thesis.

Caxito, F.A., Halverson, G.P., Uhlein, A., Stevensson, R., Dias, T.G., Uhlein, G.J. 2012. Marinoan glaciation in eastern Central Brazil. Precambrian Research, 203, 38–58.

Chemale-Jr F, Quade H and Van Schmus WR. 1998. Petrography, geochemistry and geochronology of the Borrachudo and Santa Bárbara metagranites, Quadrilátero Ferrífero, Brazil. Zbl Geol Paläont., 739–750.

Chemale Jr., F., Dussin, I.A., Alkmim, F.F., Martins, M.S., Queiroga, G., Armstrong, R., Santos, M.N., 2012. Unravelling a proterozoic basin history through detrital zircon geochronology: The case of the Espinhaço Supergroup, Minas Gerais, Brazil. Gondwana Research, 22, 200–206.

Costa, A.F.O. 2013. Estratigrafia e tectônica da borda oeste do Espinhaço Central no extremo norte da Faixa Araçuaí. MSc Dissertation, Departamento de Geologia/Escola de Minas, Universidade Federal de Ouro Preto, Ouro Preto, 170p.

Costa, A.F.O.; Danderfer, A., Lana, C. 2014. O registro do vulcanismo calimiano no Espinhaço Central (MG): Caracterização petrofaciológica, geoquímica e geocronológica. Geociências, 33(1), 119–135.

Cruz, S.C.P. & Alkmim, F.F. 2006 The tectonic interaction between the Paramirim Aulacogen and the Araçuaí belt, São Francisco Craton Region, Eastern Brazil. Anais da Academia Brasileira de Ciências, 78 (1), 151–174.

Cunningham, D., Alkmim, F.F., Marshak, S., 1998. A structural transect across the coastal mobile belt in the Brazilian Highlands (latitude 20°S): the roots of a Precambrian transpressional orogen. Precambrian Research 92, 251–275.

Danderfer, A., Waele, B.D., Pedreira, A.J., Nalini, H.A., 2009. New geochronological constraints on the geological evolution of

Espinhaço basin within the São Francisco Craton, Brazil, Precambrian Research 170, 116–128.
De Wit, M., Jeffery,M., Bergh, H., Nicolaysen, L., 1988. Geologic Map of Sectors of Gondwana. AAPG and University of Witwatersrand, Tulsa.
Drumond, J. B.V., Von Sperling, G, E.; Raposo, F.O. 1980. Projeto Porteirinha-Monte Azul. Belo Horizonte, DNPM-CPRM, 559p.
Dupont, H., 1995. O Grupo Conselheiro Mata no seu quadro paleogeográfico e estratigráfico. Simpósio de Geologia de Minas Gerais, Diamantina: Anais, Sociedade Brasileira de Geologia 13, pp. 9–10.
Dussin, I.A., Dussin, T.M., 1995. Supergrupo Espinhaço: Modelo de evolução geodinâmica. Geonomos 3, 19–26.
Dussin, I.A., Uhlein, A., Dossin, T.M., 1984. Geologia da Faixa Móvel Espinhaço em sua porção meridional, MG. Congresso Brasileiro de Geologia, 33, Rio de Janeiro: Anais, Sociedade Brasileira de Geologia 2, pp. 3118–3132.
Dussin, T.M., 1994. Associations plutono-volcaniques de l"EspinhaçoMeridional (SE-Brésil): un exemple d'évolution de la croûte protérozoïque. Ph.D.Thesis, Orleans,Universite d'Orleans, 177 pp.
Dussin, T.M., 2000. A tectônica extensional paleoproterozóica na borda sudeste do Cráton do São Francisco (SE, Brasil): Geoquímica e petrologia das meta-ígneas. Geonomos 8, 63–68.
Dussin TM, Rossi P, Dussin IA and Charvet J. 1994.The Borrachudos Suite, mesoproterozoic A-type granitic magmatism in the southeastrn São Francisco craton (SE Brazil). In: Annexe III, Evoluction structural de la region de l'Espinhaço Meridional, bordure sud-est du Craton São Francisco, Brésil. Doctor thesis, Univ d'Orléans, p. 1–35.
Egger, V. A. 2006. O Supergrupo do Espinhaço entre Serranópolis de Minas e Mato Verde (MG): Estratigrafia e implicações para o entendimento dos depósitos aluvionares de diamantes da região. MSc Dissertation, Instituto de Geociências, Universidade Federal de Minas Gerais, Belo Horizonte, 94p.
Endo, I., 1997. Regimes Tectônicos do Arqueano e Proterozóico o interior da placa SanFranciscana: Quadrilátero Ferrífero e áreas adjacentes, Minas Gerais. Doctoral Thesis, Instituto de Geociências, Universidade de São Paulo, São Paulo, 243 pp.
Fernandes MLS, Marciano VRPRO, Oliveira RC, Correia-Neves JM and Diláscio MV. 1994. Granitos Borrachudos: um exemplo de granitogênese anorogênica na porção central do Estado de Minas Gerais. Geonomos, 2, 23–29.
Fraga, L.M.S., Neves, S.C., Pires, G.L.P., Tibães, A.L., Uhlein, A. 2013. Estromatólitos colunares na base do Grupo Macaúbas, nordeste da Serra do Espinhaço (MG): paleontologia e ambiente de sedimentação. Geonomos, 21(1), 34–41.
Garcia, A.J.V., Uhlein, A., 1987. Sistemas deposicionais do Supergrupo Espinhaço na Região de Diamantina (MG). Simpósio sobre Sistemas Deposicionais no Pre-Cambriano, Ouro Preto, Atas, Sociedade Brasileira de Geologia, pp. 113–136.
Gonçalves, L., Farina, F., Lana, C., Pedrosa-Soares, A. C., Alkmim, F., Nalini, H.N., 2014. New U-Pb Ages and Lithochemical Attributes of the Ediacaran Rio Doce Magmatic Arc, Araçuaí Confined Orogen, Southeastern Brazil. Journal of South American Earth Sciences, 52, 1–20.
Gonçalves, L., Alkmim, F.F., Pedrosa-Soares, A.C., Dussin, I.A., Valeriano, C.M., Lana, C., Tedeschi, M. 2016. Granites of the intracontinental termination of a magmatic arc: an example from the Edicaran Araçuaí orogen, southeastern Brazil. Gondwana Research, 36, 439–458.
Gonçalves-Dias, T., Pedrosa-Soares, A.C., Dussin, I.A., Alkmim, F.F., Caxito, F.A., Silva, L.C., Noce, C.M., 2011. Idade máxima de sedimentação e proveniência do Complexo Jequitinhonha na área-tipo (Orógeno Araçuaí): primeiros dados U–Pb (LA-ICP-MS) de grãos detríticos de zircão. Geonomos 19 (2), 121–130.
Gradim,C., Roncato, J. Pedrosa-Soares, A.C., Cordani, U.G., Dussin, I. A., Alkmim, F.F., Queiroga, G., Jacobson, T., Silva, L.C., Babinski, M., 2014. The hot back-arc zone of the Araçuaí orogen, Eastern Brazil: from sedimentation to granite generation. Brazilian Journal of Geology, 44, 155–180.
Gradim, D.T., 2012. O Orógeno Araçuaí na Região de Viçosa, Sudeste de Minas Gerais. Universidade Federal de Minas Gerais. MSc. Thesis.
Gradim, R.J., Alkmim, F.F., Noce, Pedrosa-Soares, A.C., Babinski, M., Noce, C.M. 2005. Xistos verdes do Alto Araçuaí, Minas Gerais: Vulcanismo básico do rifte Neoproterozóico Macaúbas. Revista Brasileira de Geociências, 35(4), Suplemento, 59–69.
Grossi-Sad J.H., Chiodi F° C., Santos J.F., Magalhães, J.M.M. and Carelos P. 1990. Duas suítes graníticas do bordo sudeste do Cráton Sanfranciscano, em MinasGerais: petroquímica e potencial metalogenético. Congresso Brasileiro de Geologia, 36., Anais, Sociedade Brasileira de Geologia, v. 4, p. 1836–1848.
Grossi-Sad, J.H., Lobato, L.M., Pedrosa-Soares, A.C., Soares Filho, B. S., 1997. Projeto Espinhaço em CD-ROM (textos, mapas e anexos). COMIG, Belo Horizonte, 2693 pp.
Guadagnin, F. and Chemale Jr., F. 2015. Detrital zircon record of the Paleoproterozoic to Mesoproterozoic cratonic basins in the São Francisco Craton. Journal of South American Earth Sciences, 60, 104–116.
Haralyi, N.L.E., Hasui, Y., 1982. The gravimetric information and Archean-Proterozoic structural framework of eastern Brazil. Revista Brasileira de Geociências 112, 160–166.
Heilbron, M., Mohriak, W.U., Valeriano, C.M., Milani, E.J., Almeida, J., Tupinambá, M., 2000. From collision to extension: The roots of the southeastern continental margin of Brazil. In: Atlantic Rifts and Continental Margins, Geophysical Monograph 115, American Geophysical Union, 1– 32.
Heilbron, M., Pedrosa-Soares, A. C., Campos Neto, M. C., Silva, L. C., Trouw, R. & Janasi, V. A. 2004. Brasiliano orogens in Southeast and South Brazil. Journal of the Virtual Explorer, 17 (http://virtualexplorer.com.au/journal/2004/17/heilbron).
Hoffman P.F.and Schrag D.P., 2002. The snowball Earth hypotesis: testing the limits of global change. Terra Nova 14: 129–155.
Karfunkel, J. and Karfunkel, B. 1976. Estudos petro-faciológicos do Grupo Macaúbas na porção mediana da Serra do Espinhaço, Minas Gerais. In: Congresso Brasileiro de Geologia 29, Ouro Preto, Anais, 2. Sociedade Brasileira de Geologia, 179–188.
Karfunkel, J. And Karfunkel, B. 1977. Fazielle Entwicklung der mittleren Espinhaço-Zone mit besonderer Berucksichtigung des Tillit-Problems (Minas Gerais, Brasilien). Geologisches Jahrbuch, 24, 3–91.
Karfunkel, J.and Hoppe, A., 1988. Late Precambrian glaciation in central eastern Brazil: synthesis and model. Palaeogeogr. Palaeoclimatol. Palaeoecol. 65, 1–21.
Karfunkel, J., Pedrosa-Soares, A. C., Dossin, I. A. 1985. O Grupo Macaúbas em Minas Gerais: revisão dos conhecimentos. Boletim do Núcleo Minas Gerais – Sociedade Brasileira de Geologia, 5, 45–59.
Knauer, L.G., Schrank, A., 1994. A origem dos filitos hematíticos da Serra do Espinhaço Meridional, Minas Gerais. Geonomos 1, 33–38.
Knauer, L.G., Silva, L.L; Souza, F.B.B, Silva, L.R; Carmo, R.C. 2007. Folha Monte Azul - Programa Levantamentos Geológicos Básicos, folha SD.23-Z-D-II, escala 1:100.000. CPRM – Serviço Geológico do Brasil, BeloHorizonte,72p.
Kuchenbecker, M. 2014. Relações entre coberturas do Cráton do São Francisco e bacias situadas em orógenos marginais: O registro de datações U-Pb de grãos detríticos de zircão e suas impliações

geotectônicas. Universidade Federal de Minas Gerais, Instituto de Geociências, Unpl. PhD Thesis, 163p.

Kuchenbecker, M., Pedrosa-Soares, A.C., Babinski, M., Fanning, M. 2015. Detrital zircon age patterns and provenance assessment for pre-glacial to post-glacial successions of the Neoproterozoic Macaúbas Group, Araçuaí orogen, Brazil. Precambrian Research 266, 12–26.

Ledru, P. J., Johan, V., Milési, J.P., Tegycy, M., 1994. Markers of the last stage of the Paleoproterozoic collision: Evidence for a 2 Ga continent involving circum-South Atlantic provinces. Precambrian Research 69, 169–191.

Leite, M. M. 2013. Sistema deposicionais e estudos de proveniência sedimentar do Supergrupo Espinhaço e do Grupo Macaúbas na porção ocidental do Anticlinal de Itacambira (MG). Universidade Federal de Minas Gerais. MSc. Thesis.

Lima, S.A.A., Martins-Neto, M.A., Pedrosa-Soares, A.C., Cordani, U. G., Nutman, A., 2002. A Formação Salinas na área-tipo, NE de Minas Gerais: Uma proposta de revisão da estratigrafia da Faixa Araçuaí com base em evidências sedimentares, metamórficas e idades U-Pb SHRIMP. Revista Brasileira de Geociências 32, 491–500.

Machado, N., Schrank, A., Abreu, F.R., Knauer, L.G., Almeida-Abreu, P.A., 1989. Resultados preliminares da geocronologia U–Pb na Serra do Espinhaço Meridional. Bol. Núcl. Minas Gerais Soc. Bras. Geol. 10, 171–174.

Marshak, S., Alkmim, F.F., Whittington, A., Pedrosa-Soares, A.C. 2006 Extensional collapse in the Neoproterozoic Araçuaí orogen, eastern Brazil: A setting for reactivation of asymmetric crenulation cleavage. Journal of Structural Geology, 28, 129–147

Marshak, S., Alkmim, F.F., 1989. Proterozoic contraction/extension tectonics of the southern São Francisco region, Minas Gerais, Brazil. Tectonics 8, 555–571.

Martins, M.S. 2006. Geologia dos diamantes e carbonados aluvionares da bacia do Rio Macaúbas, MG. Universidade Federal de Minas Gerais, PhD Thesis.

Martins, M.S., Karfunkel, J., Noce, C.M., Babinski, M., Pedrosa-Soares, A.C., Sial, A.N., Liu, D., 2008. A Sequência Pré-Glacial do Grupo Macaúbas na área-tipo e o registro da abertura do rifte Araçuaí. Revista Brasileira de Geociências, 38, 761–772.

Martins-Neto, M.A., 1996. Lacustrine fan-deltaic sedimentation in a Proterozoic rift basin: the Sopa-Brumadinho Tectonosequence, southeastern Brazil. Sedimentary Geology, 106, 65–96.

Martins-Neto, M.A., 1998. O Supergrupo Espinhaço em Minas Gerais: Registro de uma Bacia Rifte-Sag do Paleo/Mesoproterozóico. Revista Brasileira de Geociências, 48, 151–168.

Martins-Neto, M.A., 2000. Tectonics and sedimentation in a paleo/mesoproterozoic rift-sag basin (Espinhaço basin, southeastern Brazil). Precambrian Research, 103, 147–173.

Martins-Neto, M.A., 2007. Proterozoic first-order sedimentary successions of the São Francisco Basin in eastern Brazil. Zeitschrift der Deutschen Gesellschaft für Geowissenschften 158, 31–44.

Martins-Neto, M.A., 2009. Sequence Stratigraphic framework of Proterozoic successions in eastern Brazil. Marine and Petroleum Geology 26, 163–176.

Martins-Neto, M.A., Pedrosa-Soares, A.C., Lima, S.A.A., 2001. Tectono-sedimentary evolution of sedimentary basins from Late Paleoproterozoic to Late Neoproterozoic in the São Francisco Craton and Araçuaí fold belt, eastern Brazil. Sedimentary Geology, 141–142, 343–370.

Maurin J.-C. 1993. La chaîne panafricaine ouest-congolienne: corrélation avec le domain est-brésilien et hypothese géodynamique. Bulletin de la Société Géologique de France, 164, 51–60.

Menezes, R.C.L., Conceição, H., Rosa, M.L.S., Macambira, M.J.B., Galarza, M.A., Rios, D.C., 2012. Geoquímica e geocronologia de granitos anorogenicos tonianos (c. 914–899 Ma) da Faixa Araçuaí no Sul do Estado da Bahia. Geonomos 20, 1–13.

Moraes, L. J. & Guimarães, D. 1931. The diamond-bearing region of Northern Minas Gerais, Brazil. Economic Geology, 26, 502–530.

Moraes, L. J. 1929. Geologia da região diamantina de Minas Gerais. In: Relato'rio Anual do Diretor 1928, Rio de Janeiro. Servic¸o Geológico e Mineralógico, 29–34.

Noce, C.M., Pedrosa-Soares, A.C., Grossi-Sad, J.H., Baars, F.J., Guimarães, M.V., Mourão, M.A.A., Oliveira, M.J.R. & Roque, N. C. 1997. Nova Subdivisão Estratigráfica Regional do Grupo Macaúbas na Faixa Araçuaí: O Registro de uma Bacia Neoproterozóica. Boletim do Núcleo Minas Gerais, Sociedade Brasileira de Geologia, 14, 29–31.

Noce, C.M., Pedrosa-Soaes, A.C., Silva, L.C., Alkmim, F.F. 2007a. O embasamento arqueano e paleoproterozoic do Orógeno Araçuaí. Geonomos, 15, 17–23.

Noce, C.M., Pedrosa-Soares, A.C., Silva, L.C., Armstrong, R. & Piuzana, D., 2007b. Evolution of polycyclic basement complexes in the Araçuaí orogeny, based on U–Pb SHRIMP data: Implications for Brazil – Africa links in Paloeproterozoic time. Precambrian Research, 159, 60–78.

Paula-Santos GM, Babinski M, Kuchenbecker M, Caetano-Filho S, Trindade RI, Pedrosa-Soares AC (2015). New evidence of na Ediacaran age for the Bambuí Group in Southern São Francisco craton (eastern Brazil) from zircon U-Pb data and isotope chemostratigraphy. Gondwana Research 28(2): 702–720.

Pedrosa-Soares, A.C., 1995. Potencial Aurífero do Vale do Araçuaí, MG: História da Exploração, Geologia e Controle Tectono-Metamórfico. Universidade de Brasília, Brazil, Ph.D. Thesis, 125 pp.

Pedrosa-Soares, A.C. and Alkmim, F.F. 2011. How many rifts preceded the development of the Araçuaí-West Congo Orogen? Geonomos, 19(2), 244–251.

Pedrosa-Soares, A.C., Noce, C.M., Vidal, P., Monteiro, R.L.B.P., Leonardos, O.H., 1992. Toward a new tectonic model for the Late Proterzoic Araçuaí (SE Brazil) - West Congolian (SW Africa) Belt. Journal of South American Earth Sciences 6, 33–47.

Pedrosa-Soares, A.C., Noce, C.M., Wiedemann, C.M. & Pinto, C. P. 2001. The Araçuaí–West Congo orogen in Brazil: An overview of a confined orogen formed during Gondwanland assembly. Precambrian Research, 110, 307–323.

Pedrosa-Soares, A.C., Vidal, P., Leonardos, O.H., Brito-Neves, B.B., 1998. Neoproterozoic oceanic remnants in eastern Brazil: Further evidence and refutation of an exclusively ensialic evolution for the Araçuaí-West Congo Orogen. Geology 26, 519–522.

Pedrosa-Soares, A.C., Wiedemann-Leonardos, C.M., 2000. Evolution of the Araçuaí belt and its connection to the Ribeira Belt, Eastern Brazil. In: Cordani, U.G., Milani, E.J., Thomaz Filho, A., Campos, D.A. (Eds.) Tectonic Evolution of South America, International Geological Congress, Rio de Jameiro, p. 265–285.

Pedrosa-Soares, A.C., Alkmim, F.F., Tack, L., Noce, C.M., Babinski, M., Silva, L.C., Martins-Neto, M., 2008. Similarities and differences between the Brazilian and African counterparts of the Neoproterozoic Araçuaí–West Congo Orogen. In: Pankhurst, J.R., Trouw, R.A.J., Brito Neves, B.B., De Wit, M.J. (Eds.), West Gondwana: Pre-Cenozoic Correlations across the South Atlantic Region. Geological Society, London, Special Publications, 294, pp. 153–172.

Pedrosa-Soares, A.C., Babinski, M., Noce, C. M., Martins, M., Queiroga, G., Vilela, F., 2011a. The Neoproterozoic Macaúbas Group (Araçuaí orogen, SE Brazil) with emphasis on the diamictite formations. In: Arnaud, E., Halverson, G.P., Shields-Zhou, G. (Org.). The Geological Record of Neoproterozoic Glaciations. Memoir of the Geological Society of London 36, 523–534.

Pedrosa-Soares, A.C., De Campos, C.P., Noce, C.M., Silva, L.C., Novo, T., Roncato, J., Medeiros, S., Castañeda, C., Queiroga, G., Dantas, E., Dussin, I., Alkmim F.F., 2011b. Late Neoproterozoic-Cambrian Granitic Magmatism in the Araçuaí Orogen (Brazil), the Eastern Brazilian Pegmatite Province and Related Mineral Resources. In: Sial, A.N., Bettencourt, J.S., De Campos, C.P., Ferreira, V.P. (Eds.), Granite-Related Ore Deposits. Geological Society, London, Special Publications, 350, pp. 25–51.

Peixoto, E.N., Pedrosa-Soares, A.C., Alkmim, F.F., Dussin, I.A., 2015. A suture-related accretionary wedge formed in the Neoproterozoic Araçuaí orogen (SE Brazil) during Western Gondwanaland assembly. Gondwana Res. 27, 878–896.

Peres, G.G., Alkmim, F.F., Jordt-Evangelista, H., 2004. The southern Araçuaí belt and the Dom Silvério Group: Geologic architecture and tectonic significance. Anais da Academia Brasileira de Ciências 76, 771–790.

Pflug, R. 1965. A Geologia da parte meridional da Serra do Espinhaço e zonas adjacentes, Minas Gerais. Boletim da Divisão de Geologia e Mineralogia do DNPM, 226, p. 1–51.

Pflug, R., 1968. Observações sobre a estratigrafia da Série Minas na região de Diamantina, Minas Gerais. Boletim da Divisão de Geologia e Mineralogia do Departamento Nacional de Produção Mineral: Notas Preliminares, 142 (20 pp.).

Porada, H., 1989. Pan-African rifting and orogenesis in southern to equatorial Africa and Eastern Brazil. Precambrian Research 44, 103–136.

Queiroga, G.N. 2010. Caracterização de restos de litosfera oceânica do Orógeno Araçuaí entre os paralelos 17o e 21o S. Universidade Federal de Minas Gerais. PhD Thesis.

Queiroga, G.N., Pedrosa-Soares, A.C., Noce, C.M., Alkmim, F.F., Pimentel, M.M., Dantas, E., Martins, M., Castaneda, C., Suita, M.T. F., Prichard, H., 2007. Age of the Ribeirão da Folha ophiolite, Aracuaí Orogen: the U–Pb zircon dating of a plagiogranite. Geonomos 15, 61–65.

Richter, F., Lan, C., Stevens, G., Buik, I., Pedrosa-Soares, A.C., Alkmim, F.F., Cutts, K. 2016. Sedimentation, metamorphism and granite generation in a back-arc region: Records from the Ediacaran Nova Venécia Complex (Araçuaí Orogen, Southeastern Brazil). Precambrian Research, 272, 78–100.

Rodrigues, J.B., 2008. Proveniência dos sedimentos dos grupos Canastra, Ibia, Vazante e Bambuí. Um estudo de zircões detriticos e idades modelo Sm-Nd. PhD thesis, Instituto de Geociências da Universidade de Brasília, Brasília, p. 129.

Rosa, M.L.S., Conceição, H., Macambira, M.J., Galarza, M.A., Cunha, M.P., Menezes, R.C.L., Marinho, M.M., Cruz-Filho, B.E., Rios, D. C., 2007. Neoproterozoic anorogenic magmatism in the Southern Bahia Alkaline Province of NE Brazil: U–Pb and Pb–Pb ages of the blue sodalite syenites. Lithos 97, 88–97.

Santos, R.F., Alkmim, F.F., Pedrosa-Soares, A.C., 2009. A Formação Salinas, Orógeno Araçuaí, MG: História deformacional e significado tectônico. Revista Brasileira de Geociências, 39(1), 81–100.

Santos, M.N., Chemale Jr., F., Dussin, I.A., Martins, M.S., Assis, T.A. R., Jelinek, A.R., Guadagnin, F., Armstrong, R., 2013. Sedimentological and paleoenvironmental constraints of the Statherian and Stenian Espinhaço Rift System, Brazil. Sedimentary Geology, 290, 47–59.

Santos M.N., Chemale Jr. F., Dussin I.A., Martins, M.S., Queiroga, G, Pinto, R.T.R., Santos, A. N., Armstrong, R. 2015 Provenance and paleogeographic reconstruction of a mesoproterozoic intracratonic sag basin (Upper Espinhaço Basin, Brazil). Sedimentary Geology, 318, 40–57.

Schobenhaus, C. 1996. As tafrogêneses superpostas Espinhaço e Santo Onofre, estado da Bahia: revisão e novas proposta. Revista Brasileira de Geociências 26(4), 265–276.

Scholl, W.U. and Fogaça, A.C. C. 1979. Estratigrafia da Serra do Espinhaço na região de Diamantina. Simpósio de Geologia de Minas Gerais, Diamantina: Anais, Sociedade Brasileira de Geologia, pp. 55–73.

Silva, L.C., Amstrong, R, Delgado I.M., Pimentel, M., Arcanjo, J.B., Melo, R.C., Teixeira, L.R., Jost, H., Cardoso Filho, J.M., Pereira, L. H.M. 2002a. Reavaliação da evolução geológica em terrenos pré-cambrianos brasileiros com base em novos dados U-Pb SHRIMP, Parte I: limite centro-oriental do Cráton do São Francisco na Bahia. Revista Brasileira de Geociências 32, 161–172.

Silva LC, Armstrong R, Noce CM, Carneiro MA,Pimentel M, Pedrosa-Soares AC, Leite CA,Vieiro VS, Silva MA, Paes VJC and Cardoso F°JM. 2002b. Reavaliação da evolução geológica emterrenos Pré-cambrianos brasileiros com base em novos dados U-Pb SHRIMP, parte II: Orógeno Araçuaí, Cinturão Mineiro e Cráton São Francisco Meridional. Revista Brasileira de Geociências, 32, 172–191.

Silva, L.C., Pedrosa-Soares, A.C., Teixeira, L.R., 2008. Tonian rift-related, A-type continental plutonism in the Araçuaí orogen, Eastern Brazil: new evidences for the breakup stage of the São Francisco–Congo Paleocontinent. Gondwana Research, 13, 527–537.

Silva, C.M.T., Alkmim, F.F., Pedrosa-Soares, A.C. 2009. Geometria e Evolução do feixe de zonas de cisalhamento Manhuaçu-Santa Margarida, Orogéno Araçuaí, MG. Revista da Escola de Minas, 62 (1), 23–34.

Silva, R.R., 1998. As bacias proterozóicas do Espinhaço e São Francisco em Minas Gerais: uma abordagem do ponto de vista da estratigrafia de sequências. Geonomos 6, 1–12.

Straathof, G.B. 2011. Neoproterozoic Low Latitude Glaciations: An African Perspective. University of Edinburgh, Scotland, UK. PhD Thesis.

Suita, M.T.F., Pedrosa-Soares, A.C., Leite, C., Nilson, A.A. & Prichard, H. 2004 Complexos Ofiolíticos do Brasil e a Metalogenia Comparada das Faixas Araçuaí e Brasília. *In*: E. Pereira, R. Castroviejo & Ortiz, F. (eds), *Complejos Ofiolíticos en Iberoamérica: guías de prospección para metales preciosos*. Ciencia y Tecnología para el Desarrollo-CYTED, Madrid, p. 101–132.

Tack, L., Wingate, M.T.D., Liégeois, J.P., Fernandez-Alonso, M., Deblond, A., 2001. Early Neoproterozoic magmatism (1000–910 Ma) of the Zadinian and Mayumbian Groups (Bas-Congo): onset of Rodinian rifting at the western edge of the Congo Craton. Precambrian Research, 110, 277–306.

Tedeschi, M., Novo T., Pedrosa-Soares, A.C., Dussin, I.A. Tassinari, C. G.T., Silva, L.C., Gonçalves, L., Alkmim, F.F., Lana C., Figueiredo, C., Dantas, E., Medeiros, S., Campos, C., Corrales, F., Heilbron, M. 2016. The Ediacaran Rio Doce magmatic arc revisited (Araçuaí-Ribeira orogenic system, SE Brazil). Journal of South American Earth Sciences, 68, 186–187.

Thiéblemont, D., Prian, J.P., Goujou, J.C., Boulingui, B., Ekogha, H., Kassadou, A.B., Simo-Ndounze, S., Walemba, A., Préat, A., Theunissen, K., Cocherie, A., Guerrot, C.,2011. Timing and characteristics of Neoproterozoic magmatism in SW-Gabon: first geochronogical and geochemical data on the West-Congolian orogen in Gabon (SYSMIN project, Gabon 2005–2009). In: 23 Colloquium of African Geology (CAG23). University of Johannesburg, Republic of South Africa, Abstracts.

Trompette R., Uhlein A., Egydio-Silva M.E., Karmann I. 1992. The Brasiliano São Francisco craton revisited (central Brazil). Journal of South American Earth Science, 6, 49–57.

Uhlein, A., 1991. Transição cráton-faixa dobrada: um exemplo do Cráton do São Francisco e da Faixa Araçuaí (ciclo Brasiliano) no Estado de Minas Gerais. Instituto de Geociências, Universidade de São Paulo, São Paulo, Doctoral Thesis, 295 pp.

Uhlein, A., R.R. Trompette, Egydio-Silva, M., 1998. Proterozoic rifting and closure, SE border of the São Francisco Craton, Brazil. Journal of South American Earth Sciences 11, 191–203.

Uhlein, A., Trompette, R., Alvarenga, C., 1999. Neoproterozoic glacial and gravitational sedimentation on a continental rifted margin: the Jequitaí-Macaúbas sequence (Minas Gerais, Brazil). J. S. Am. Earth Sci. 12, 435–451.

Uhlein, A., Trompette, R., Egydio-Silva, M. & Vauchez, A. 2007. A glaciação Sturtiana (~750 Ma), a estrutura do rifte Macaúbas-- Santo Onofre e a estratigrafia do Grupo Macaúbas, Faixa Araçuaí. Geonomos, 15, 45–60.

Uhlein, A., Trompette, R., Silva, M.E. 1986. A estruturação tectônica do Supergrupo Espinhaço na região de Diamantina, MG. Revista Brasileira de Geociências 16, 212–216.

Vilela, F.T., Pedrosa-Soares, A.C., Carvalho, M.T.N., Arimateia, R., Santos, E., Voll, E., 2014. Metalogênese da Faixa Arac¸uaí: O Distrito Ferrífero Nova Aurora (Grupo Macaúbas, Norte de Minas Gerais) no contexto dos recursos minerais do Orógeno Araçuaí. In: Silva, M.G., Rocha-Neto, M.B., Jost, H., Kuyumjian, R.M. (Eds.), Metalogênese das Províncias Tectônicas Brasileiras., 1st ed, pp. 415–430, CPRM (www.cprm.gov.br), Rio de Janeiro, Brazil.

Viveiros, J. F. M., Sá, E. L., Vilela, O. V., Santos, O. M., Moreira, J. M. P., Holder-Neto, F.&Vieira, V. S. 1979. Geologia dos vales dos rios Peixe Bravo e Alto Vacaria, norte de Minas Gerais. Boletim do Núcleo Minas Gerais-Sociedade Brasileira de Geologia, 1, 75–87.

Warren LV, Quaglio F, Riccomini C, Simões MG, Poiré DG, Strikis NM, Aneli LE, Strikis PC (2014) The puzzle assembled: Ediacaran guide fossil Cloudina reveals an old proto-Gondwana seaway. Geology 42(5): 391–394.

The Ribeira Belt

15

Monica Heilbron, André Ribeiro, Claudio Morisson Valeriano, Fábio V. Paciullo, Júlio Cesar H. Almeida, Rudolph Johannes A. Trouw, Miguel Tupinambá, and L.G. Eirado Silva

Abstract

The generation of the NW-verging Ribeira belt that fringes the São Francisco craton to the south resulted from collisional episodes dated at 620–605, 605–565, and 535–510 Ma. During these collisional events, micro-continents and magmatic arcs converged and accreted to a continental margin system formed along the previously amalgamated São Francisco-Paranapanema landmass. The São Francisco-Paranapanema collision at around 640–620 Ma led to the development of the Southern Brasilia belt. The Ribeira orogenic front overprinted the pre-existent southern end of east-verging Brasília belt, thereby creating a tectonically very complex interference zone. The Ribeira belt as whole comprises four tectono-stratigraphic terranes: the Occidental, Paraíba do Sul, Oriental, and Cabo Frio terranes. The Occidental terrane, representing the external sector of the belt, involves the reworked cratonic basement (Archean and Paleoproterozoic orthogneisses and orthogranulites), Mesoproterozoic intra-cratonic basins and a Neoproterozoic passive margin unit (Andrelândia Group). The uppermost unit of the Andrelândia Group is regarded to be deposited in an active margin setting and sourced by a magmatic arc installed in the Paranapanema plate. The Paraíba do Sul terrane, composed of an Archean-Paleoproterozoic basement, Neoproterozoic metasedimentary units, and a continental magmatic arc (the 640–595 Ma Serra da Bolívia complex), accreted to southeastern sector of the craton margin between 620 and 605 Ma. Afterwards, the juvenile to immature magmatic arc of the Oriental terrane (860–620 Ma Rio Negro and Serra da Prata complexes) collided to proto-Ribeira belt. The crustal thickening resulting from these collisions gave rise to widespread generation of I, S and hybrid granites that intrudes both basement and cover units in the most deformed sectors of the belt. The Cambrian (535–510 Ma) docking of the Cabo Frio terrane (an Angola craton fragment) reworked the previous accreted terranes, generating large scale folds and dextral transpresional shear zones that reached the SFC margin. Finally, a vigorous bimodal magmatic event associated with transtensional deformation episodes characterizes the orogenic collapse of the belt and marks its stabilization in the interior of the Gondwana supercontinent.

M. Heilbron (✉) · C.M. Valeriano · J.C.H. Almeida · M. Tupinambá · L.G. Eirado Silva
Tektos Research Group, Faculdade de Geologia, Universidade do Estado Rio de Janeiro (UERJ), Rua São Francisco Xavier, 524, Bloco A-4020, Rio de Janeiro, RJ 20.550-900, Brazil
e-mail: monica.heilbron@gmail.com

A. Ribeiro · F.V. Paciullo · R.J.A. Trouw
Depto de Geologia, Instituto de Geociências, Universidade Federal do Rio de Janeiro, Av. Athos Da Silveira 274-CCMN-Bloco G, Campus Ilha do Fundão-Cidade Universitária, Rio de Janeiro, 21949-000, Brazil

M. Heilbron et al. (eds.), *São Francisco Craton, Eastern Brazil*, Regional Geology Reviews, DOI 10.1007/978-3-319-01715-0_15

Keywords
Neoproterozoic • Magmatic arcs • Ribeira belt • Reworked Paleoproterozoic basement

15.1 Introduction and Tectonic Overview

The Ribeira and Araçuaí belts, bounding the São Francisco craton (SFC), respectively, to the east and south, integrate one of the largest Neoproterozoic Brasiliano orogenic systems of Brazil (Almeida et al. 1981; Cordani et al. 1973) (Fig. 15.1). The NE-trending Ribeira belt (Hasui et al. 1975; Trouw et al. 2000; Heilbron et al. 2000, 2004, 2008) runs parallel to the southeastern Brazilian coast between Lat. 20°S and Lat. 25°S (Fig. 15.1), and involves an Archean/Paleoproterozoic basement, Paleo- to Neoproterozoic metasedimentary units and Neoproterozoic granites. The northern boundary of the belt is arbitrarily traced along Lat. 21°S, where the NS-trending fabric elements and units of the Araçuaí belt bend toward NE. The NNW-verging structures of Ribeira belt overprint the southern tip of the Brasilia belt along the southern margin of the craton, thereby creating a structurally very complex interference zone (Trouw et al. 2013; Valeriano et al. 2008, this book; Heilbron et al. 2000; Campos Neto and Caby 1999).

In this chapter, we provide first a description of the stratigraphy and overall structure of Ribeira belt, focusing mainly its external portion and interference zone with the southern tip of the Brasília belt. In our descriptions (Heilbron et al. 2000, 2004, 2008), we adopted the concept of tectono-stratigraphic terranes sensu Howell (1989). Accordingly, the Ribeira belt is subdivided into four terranes, namely, the Occidental, Paraíba do Sul-Embu, Oriental and Cabo Frio terranes (Fig. 15.2). The external zone of the belt, which is continuous to the Araçuaí belt, corresponds to the Occidental terrane (Table 15.1). The final section of the chapter summarizes the tectonic evolution of belt (Fig. 15.2).

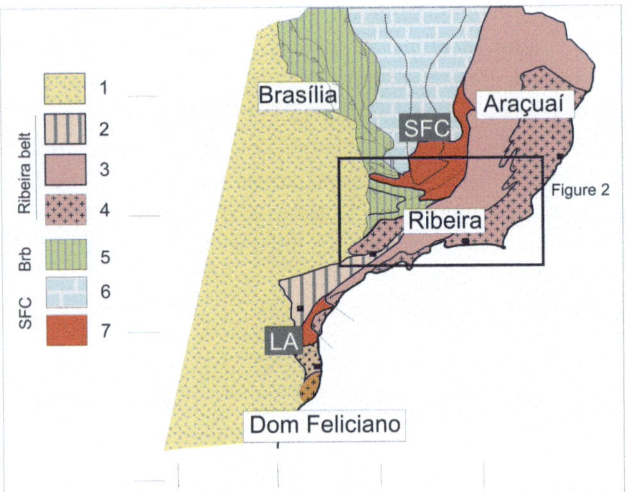

Fig. 15.1 The Ribeira belt and the southern tip of the Brasília belt. *1* Phanerozoic cover; *2–4* Ribeira belt units; *2* Apiaí terrane, *3* Reworked São Francisco margin, *4* Arc related and outboard terranes; *5* Brasília belt tectonic units; *6* Bambuí group; *7* exposed basement of the São Francisco (SF) and Luiz Alvez (LA) cratons

15.2 Stratigraphy

Heilbron et al. (2000, 2004, 2008) grouped the units involved in the Ribeira belt into seven major tectono-stratigraphic assemblages, as follows: (i) Archean and Paleoproterozoic basement rocks; (ii) intra-continental Mesoproterozoic basin fill successions; (iii) Neoproterozoic passive margin successions; (iv) Neoproterozoic arc-related orthogneisses and meta-gabbros; (v) Neoproterozoic active margin successions; (vi) syn-collisional granitic rocks; (vii) post-collisional collapse-related granites (Table 15.2). The main attributes of these assemblages in the various terranes of belt are described in the next sections.

The basement assemblages of the Ribeira belt record a long-term evolutionary history that extends from the Archean to the Upper Paleoproterozoic. Rocks units with SFC affinity were dismembered and reworked during the Neoproterozoic tectonic events, occuring in the Occidental and Paraíba do Sul terranes. The basement of the Oriental terrane is made up of Neoproterozoic arc-related granitoids and supracrustal rocks, whilst the Cabo Frio terrane in the easternmost portion is the belt (Fig. 15.3) is underlain by 1.9 Ga orthogneisses.

15.2.1 The Occidental Terrane

15.2.1.1 Basement

The basement of the Occidental terrane is made up mainly of Paleoproterozoic orthogneisses and orthogranulites with a minor contribution of basic rocks. Archean rocks are rare, occurring mainly as small inliers. The few mapped outcrops within the Ribeira belt consists of banded TTG orthogneisses with amphibolite layers. The presence of Archean components in the basement has also been detected as inheritance in the Paleoproterozoic units.

Table 15.1 Tectonic subdivisions of the Ribeira belt and the adjacent Araçuaí West-Congo orogen (AWCO)

	External zone	Internal zone	Accreted (exotic) terranes
Ribeira belt	Occidental terrane	Paraíba do Sul-Embú terrane, base of the Oriental terrane	Oriental and Cabo Frio terranes
AWCO	Araçuaí belt	Crystalline core	Not found

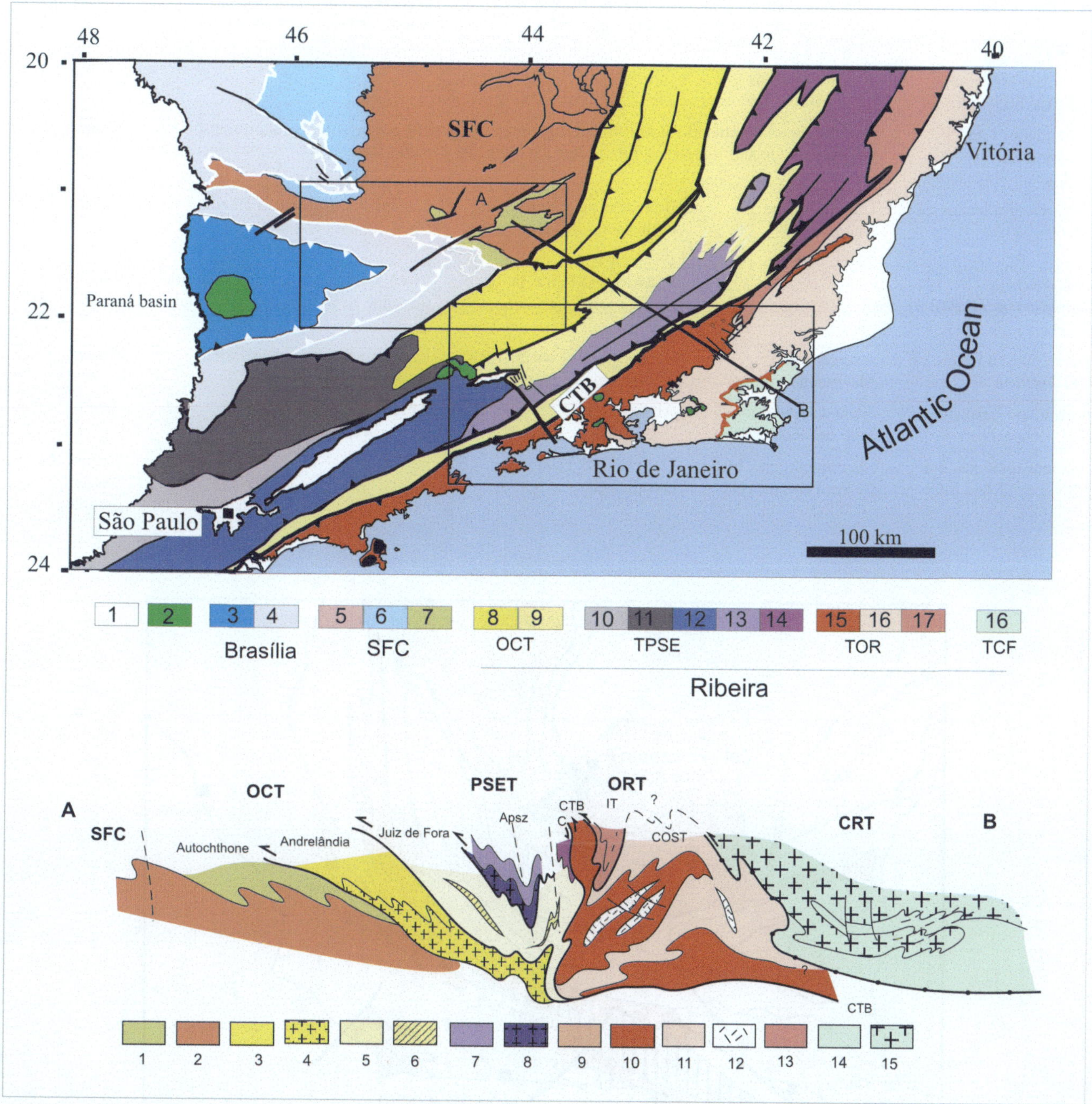

Fig. 15.2 Tectonic framework of the Ribeira belt and interference zone with the Brasília belt, compiled and modified from Trouw et al. (2000) and Heilbron et al. (2000, 2004), including a regional cross section. *1* Phanerozoic cover; *2* Upper Cretaceous alkaline plutons; *3* and *4* east-verging units of the Brasília Belt, including the Guaxupé nappe and lower nappes; *5–7* Units of the São Francisco craton: *5* Paleproterozoic to Archean basement, *6* Neoproterozoic cratonic cover, Bambuí Group, *7* Mesoproterozoic to Neoproterozoic metasediments of the autochthonous domain; *7–18* Terranes and structural domains of the Ribeira Belt: *8* Andrelândia and *9* Juiz de Fora domains of Occidental terrane, *10* Socorro Nappe; *11* Apiaí terrane; *12* Embú terrane; *13* Paraíba do Sul terrane, *14* Cambuci terrane; *15* Cryogenean to Ediacaran magmatic arc, *16* Neoproterozoic metasedimentary successions and *13* Tonian magmatic arc of the Oriental terrane; *12* Cabo Frio terrane.

Table 15.2 Tectono-stratigraphic units of the Ribeira belt (Heilbron et al. 2000, 2004, 2008)

Tectono-stratigraphic units	Terranes			
	Occidental	Paraíba do Sul-Embú	Oriental	Cabo Frio
Post-collision granites ca. 510–480 Ma	Absent	Very rare	Granites, minor diorites and gabbros, pegmatites	Only pegmatites
Syn-collision III granitoids ca. 535–510 Ma	Very rare as small plutons controlled by late shear zones. Slightly peraluminous leucogranitoids		Weakly foliated plutons, leucogranites and charnockites	Leucossomes
Syn-to late collision basins ca. 605–565 Ma	Transition basins			
Syn-collision I and II granitoids ca. 605–565 Ma	Metaluminous hornblende/biotite foliated granites, frequently porphyritic, peraluminous leucogranitoids and hybrid granitic rocks. Charnockitic granitoides			Absent
Active margin basins ca. 840–605 Ma	Unit A5 of the Andrelândia megasequence	Bom Jesus do Itapaboana basin (fore arc setting)	São Fidélis, Italva and Búzios basins (back arc setting)	Absent
Neoproterozoic magmatic arcs ca. 840–605 Ma		Socorro and Serra da Bolívia arcs (cordilleran seting)	Rio Negro and Serra da Prata arcs (juvenile to immature continental)	Absent
Passive margin basins ca. 1000–760 Ma	Andrelândia Megasequence	Paraíba do Sul basin	Lower units of the São Fidelis Group	Absent
Intracratonic basins ca. 1.7–1.3 Ga	Carandaí and São João del Rey basins	Absent	Absent	Absent
Basement rocks ca. 2.4–1.90 Ga ca. 2.8–2.6 Ga	Mantiqueira and Juiz de Fora complexes	Quirino and Taquaral complexes.	Absent	Região dos Lagos complex

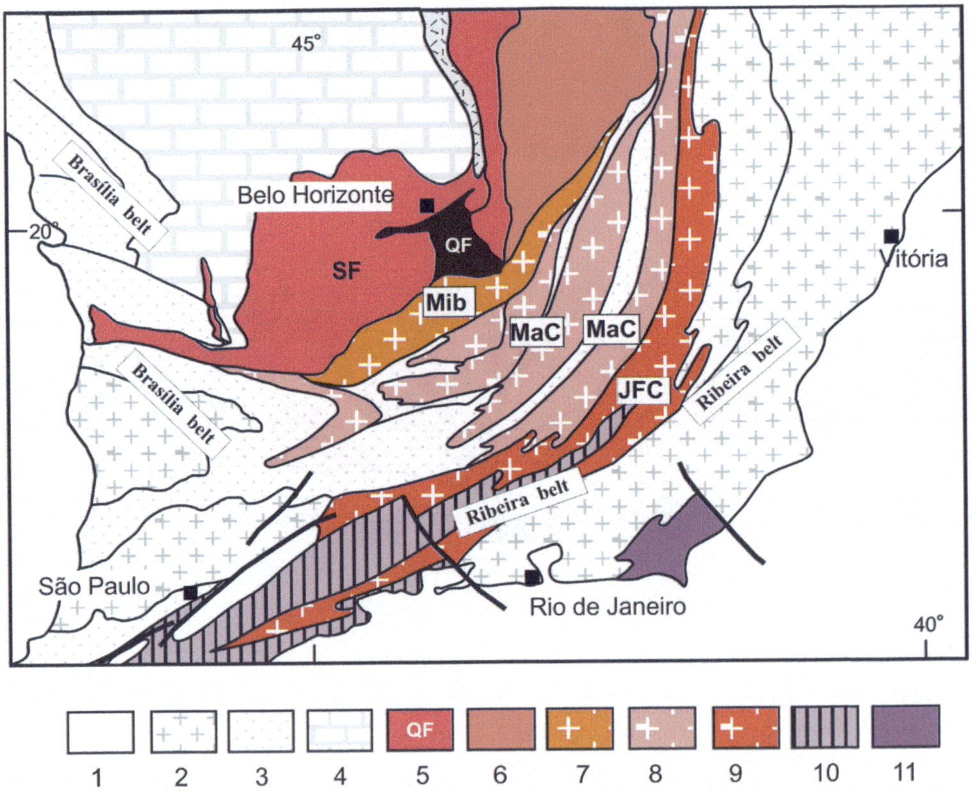

Fig. 15.3 Paleoproterozoic basement assemblages of the Ribeira belt (from Heilbron et al. 2013, reproduced with Elsevier permission). Legend: *1* Phanerozoic cover; *2–4* Neoproterozoic units of Brasília and Ribeira belts; *5* Exposed basement in the São Francisco craton; *6* The Archean Guanhães basement; *7–11* Paleoproterozoic basement units reworded in the Neoproterozoic; *7* Units of the Mineiro belt (Chap. 5 this book); *8* Mantiqueira complex; *9* Juiz de Fora complex; *10* Quirino Complex and other basement association of the Embú terrane; *11* Região dos Lagos Complex

The Paleoproterozoic basement of the Occidental terrane comprises the Mantiqueira and Juiz de Fora complexes. The Mantiqueira Complex (including other local units as the Amparo Complex, Ebert 1968), exposed in the areas close to craton boundary, consists of orthogneisses containing layers of amphibolites and meta-ultramafic lenses. The felsic orthogneisses correspond to calcalkaline quartz-diorites, tonalites, granodiorites, and porphyritic granites (Fig. 15.4). Geochemical and U–Pb geochronological data indicate that the Mantiqueira Complex was generated in a convergent margin setting, in which a 2.70–2.60 Ga Archean substratum received a substantial amount of magmatic additions in time interval of 2.22–2.05 Ma (Heilbron et al. 1998, 2010; Noce et al. 2007; Duarte et al. 2000, 2004). Despite the strong Neoproterozoic overprint, a metamorphic episode (related to the Transamazonian orogeny) dated at ca. 2.06–2.04 Ga was also detected in the Mantiqueira gneisses (Heilbron et al. 1998, 2010). The mafic components of the complex correspond to amphibolites that characterize a very heterogeneous group of tholeiitic rocks with MORB, IAT, and WPB signatures. The few available U–Pb age determinations point to crystallization ages around 2.1 Ga (Heilbron et al. 2010; Noce et al. 2007; Silva et al. 2002; Duarte et al. 2004).

The Juiz the Fora Complex occurs in the uppermost thrust sheet of the Occidental terrane, which is made up of granulite facies orthogneisses and metabasic rocks. The orthogranulites vary in composition from enderbites to charnockites (Fig. 15.4), which characterize calcalkaline and tholeiitic suites dated between 2.2 and 2.1 Ga. The geochemical and isotopic signatures of these suites attest their generation in intra-oceanic arc settings (Heilbron et al. 1998, 2010; Duarte et al. 1997; Noce et al. 2007).

Few Archean inliers were recently described within the occurrence area of the Juiz Fora Complex (Silva et al. 2002). Worth noting, Statherian (1.7–1.6 Ga) intraplate alkaline rocks intrude the Juiz Fora Complex in areas close to Minas Gerais-Rio de Janeiro boundary (Heilbron et al. 2010).

15.2.1.2 Mesoproterozoic Intra-continental Basin Fill Successions: The São João Del Rey Group

The São João del Rey Group (Figs. 15.5 and 15.6), consisting of a ca. 1500 m-thick succession of siliciclastic and subordinate carbonate rocks, occurs in the region close to the southern boundary of the craton (Ribeiro et al. 1990, 2013). Exhibiting very well preserved sedimentary structures, the rocks of this group are in markedly contrast with the remaining Neoproterozoic successions involved in the Ribeira belt, which are in general strongly deformed and metamorphosed.

The São João del Rey Group is subdivided into four formations (Ribeiro et al. 2013). The approximately 1000 m-thick Tiradentes Formation, at the base of the group, comprises four unconformity bounded sequences accumulated in shallow-marine to deltaic environments. The basal unit, also referred to as the Tiradentes sequence, comprises pebbly quartzites, quartz conglomerate, and quartzites with cross-beds and ripples, representing shoreface to foreshore deposits. Pure quartz-arenites accumulated in shallow marine sand bars, sourced by the previous unit, are the dominant lithofacies of the São José sequence. The shoreface, lagoon and tidal flat deposits of the Tejuco sequence consist of crossed-bedded quartzites and heterolithic facies (Figs. 15.5 and 15.6). The uppermost Lenheiro sequence comprises a swallowing upward succession of deltaic pelites and sandstones that record a significant change in the paleocurrent pattern of the Tiradentes Formation. The paleocurrent patterns of the lower units indicate NE-directed sediment input, whereas the deposition of Lenheiro deltaic sequence was associated with sediment transport towards NW.

The up to 30 m-thick Carandaí Formation (Leonardos 1940) occurs in form of lenses and beds of muddy diamictites containing angular to rounded clasts of basement gneiss, granitoid, phyllite and mafic rocks. Originally interpreted as glacial deposits (Leonardos 1940), these diamictites occur in close association with fault zones and are unconformably covered by the Barroso Limestone and Prados Pelites. According to Ribeiro et al. (2013), the Carandaí Diamictites more likely represent debris-flow deposits associated with normal fault scarps developed during a renewed rifting episode.

The 200 m-thick Barroso Formation, composed of gray limestones and subordinate pelites, characterizes a major onlap over the basement and previously described units. The Prados Formation, separated from the below and above lying units by unconformities, consists of an up to 500 m-thick package of marine pelites containing minor sandstone lenses.

Several NNW-trending mafic dykes intrude the São João del Rey Group. The dykes consist of apatite- and rutile-bearing, fine-grained basic rocks showing T_{DM} model ages between 1.7 and 1.4 Ga (Ribeiro et al. 2013).

Detrital zircons extracted from the basal portion of the São João Del Rey Group indicate Calymmian maximum depositional ages of 1540 Ma and derivation mainly from Paleoproterozoic sources (main age peaks at 1763 and 2123 Ma) for both the Tiradentes and Lenheiro strata. Detrital zircons from the Carandaí Diamictites characterize, on the other hand, a slightly younger deposition age of 1379 Ma and the same sources as the older units (Ribeiro et al. 2013).

According to Ribeiro et al. (2013), the São João Del Rey Group, as a correlative of the middle Espinhaço Supergroup

Fig. 15.4 Paleoproterozoic basement assemblages of the Ribeira belt: **a** banded tonalite to granodiorite gneisses with amphibolite layers, and **b** large amphibolite lenses and leucogneisses of the Mantiqueira complex; **c**, **d** enderbite and charnockite granulites of the Juiz de Fora Complex; **e** hornblende granodiorite orthogneisses of the Quirino complex, with mafic enclaves; **f** foliated variety of the Quirino Complex with a kinematic indication of top to NW; **g**, **h** orthogneisses and amphibolite layers and boudins of the Região dos Lagos complex

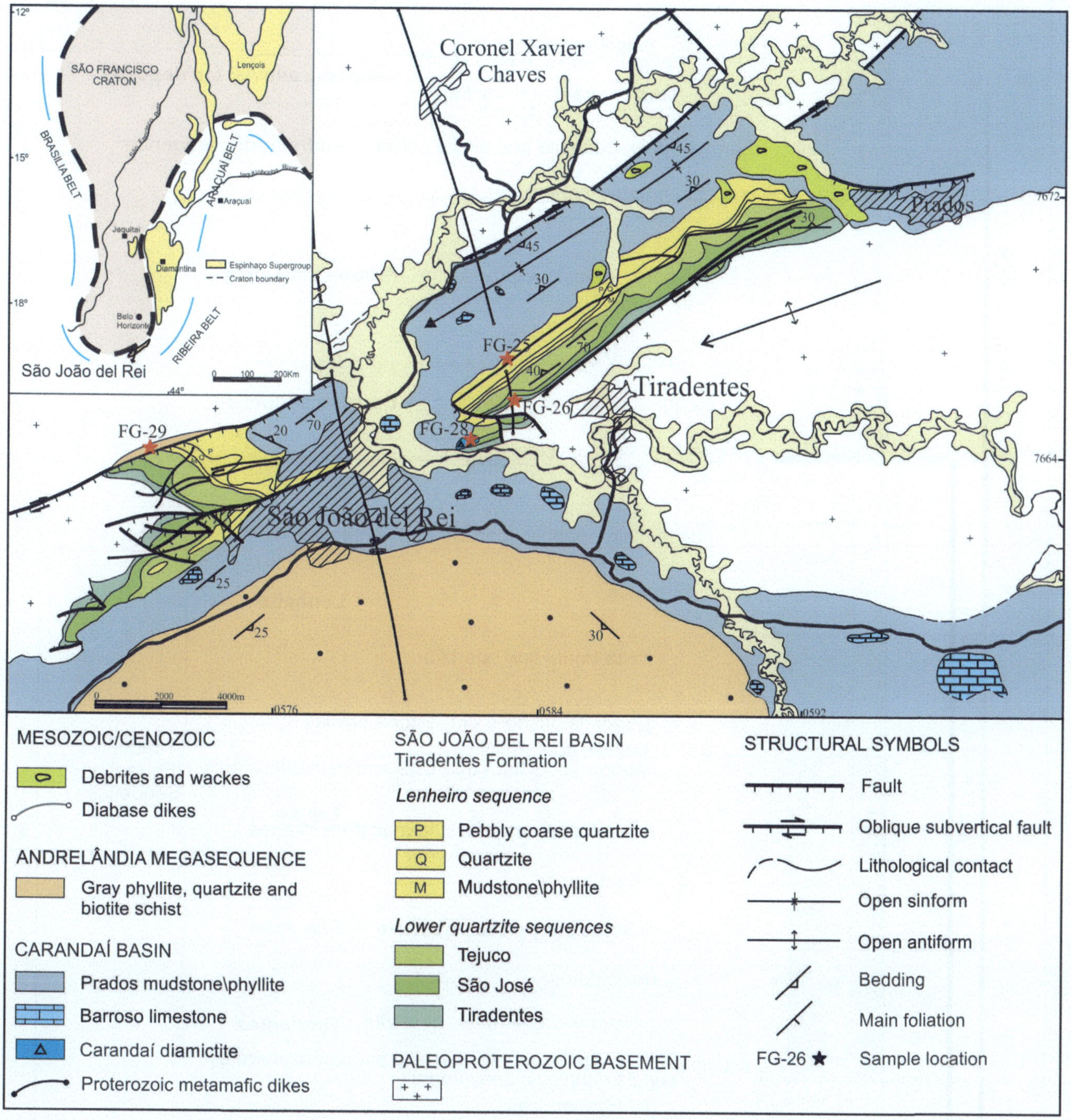

Fig. 15.5 Detailed geologic map of the São João del Rey and Carandaí basin successions (Ribeiro et al. 2013, reproduced with Elsevier permission)

exposed in the Paramirim aulacogen and Araçuaí belt (see Cruz and Alkmim, and Alkmim et al., this book), records rift and sag events experienced by the São Francisco landmass in the course of the Mesoproterozoic. As indicated by their lithofacies architecture and field relationships, the Tiradentes and Carandaí formations accumulated during two major rifting pulses that took place, respectively, in the Calymmian Period, and sometime after 1379 Ma, followed by periods of thermally driven subsidence of the basin system.

15.2.1.3 Neoproterozoic Passive to Active Margin Assemblages: The Andrelândia Megasequence

The Neoproterozoic Andrelândia megasequence (Ebert 1984; Paciullo 1997; Paciullo et al. 2000; Ribeiro et al. 2013)

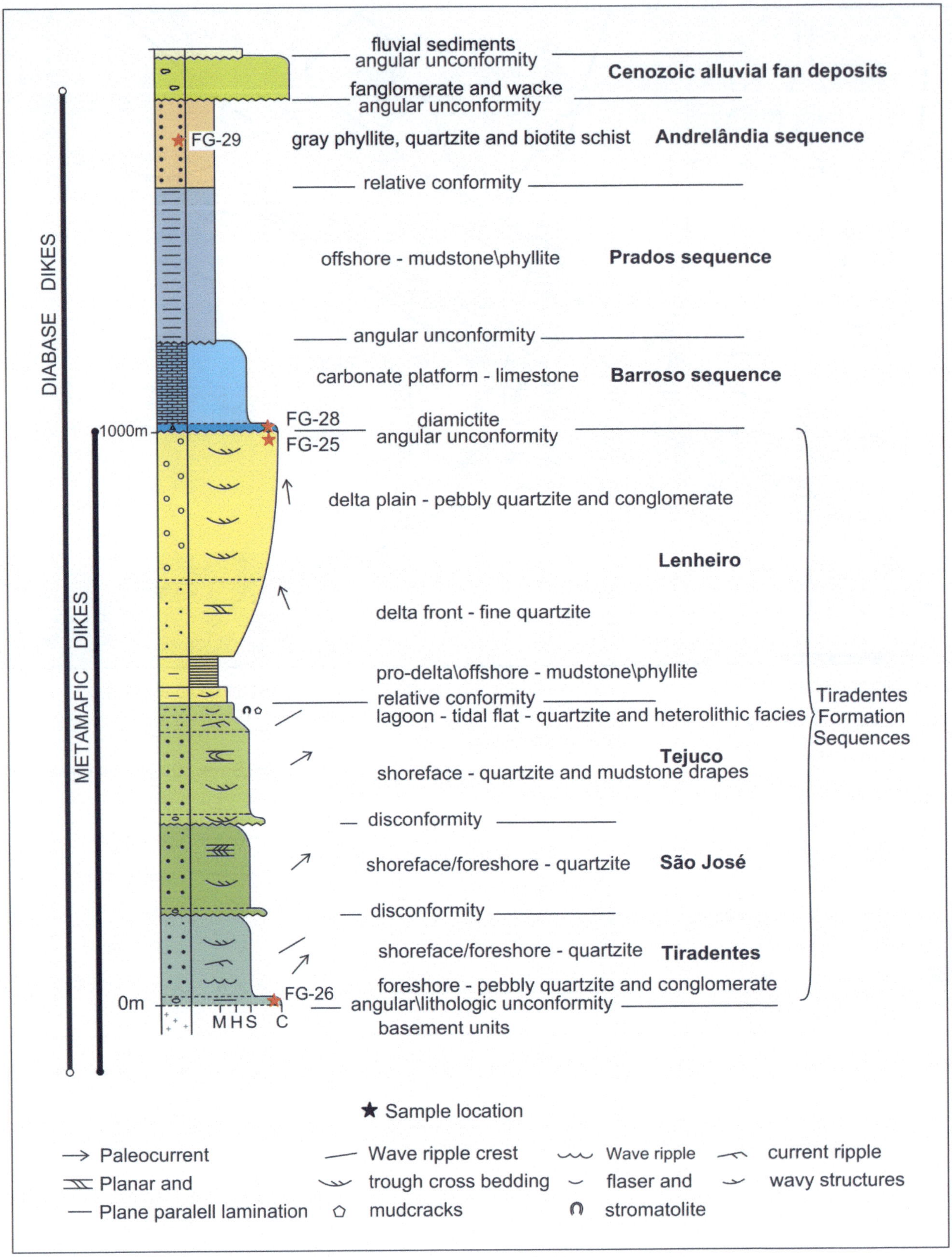

Fig. 15.6 Simplified stratigraphic column for the São João del Rey and Carandaí basins successions (modified from Ribeiro et al. 2013 and reproduced with Elsevier permission)

consists mainly of siliciclastic metasedimentary rocks that occur in the interference zone between the Ribeira and southern Brasília belt (Trouw et al. 2000, 2013; Campos Neto and Caby 2000), as well as within thrust sheets of eastern Occidental terrane. The sequence stratigraphy model discussed here is based on detailed mapping of lithofacies associations and their bounding unconformity surfaces (Figs. 15.7, 15.8 and 15.9).

The Andrelândia megasequence is further subdivided into two depositional sequences (Fig. 15.10). The basal sequence comprises four distinct lithofacies associations: banded arcosean meta-psamopelites (paragneisses) with pelitic layers and amphibolites (units A1 and A2); continuous layers of white mica quartzites (A3 unit); and an intrinsic intercalation of graphite bearing phyllites/schists with quartz-rich metasandstones (A4 unit) (Fig. 15.11).

The upper sequence, currently referred to as the Serra do Turvo sequence, comprises two lithofacies associations: plagioclase biotite schists/gneisses with quartzites and pelitic layers (A5 unit); pelitic schists and gneisses, with quartzites, amphibolites, calc-silicate and Mn-rich metacherts (A6 unit), which are especially abundant within the thrust sheets of the eastern portion of the Occidental terrane (Figs. 15.10 and 15.11).

Provenance studies based on detrital zircon age determinations revealed that the units A1, A2, A3, A4 and A6 of the Andrelândia megasequence derived from São Francisco basement assemblages, including the Paleo- and Mesoproterozoic intracratonic basin fill successions described above. The main sources of the megasequence are Rhyacian basement units; contributions from the Archean and Mesoproterozoic assemblages are subordinate (Machado and Gauthier 1996; Valladares et al. 2004, 2008; Valeriano et al. 2004; Belém et al. 2011; Santos 2011). The youngest detrital zircons and the data available for the basic rocks interlayered with the sequence constrain its depositional age to the Neoproterozoic, between 1.0 and 0.76 Ga. As an exception, the data available for the plagioclase-rich gneisses of A5 unit that occurs only in the interference zone with the southern Brasília belt indicates a provenance essentially from Neoproterozoic sources, and maximum depositional age o 620 Ma (Belém et al. 2011). These authors suggest provenance from the Guaxupé magmatic arc of the southern Brasilia belt and accumulation in a foreland basin setting.

Amphibolites occur as lenses or boudins in the A1, A2 and A6 units. Geochemical data indicate both intraplate and MORB signatures for these rocks, which might represent markers of extensional episodes within the Andrelândia Megasequence (Gonçalves and Figueiredo 1992 and Marins 2000). A single U–Pb age determination carried out on zircons extracted from on an amphibolite lens of the A6 unit yielded a crystallization age of 766 Ma. T_{DM} model ages obtained from distinct amphibolites of the megasequence fall in the 1.0–0.9 Ga interval (Heilbron et al. 2000; Paciullo et al. 2000).

Highly deformed granulite facies paragneisses that form the upper thrust sheets of the Occidental terrane are regarded as the distal deposits of the Andrelândia Megasequence. Two associations were mapped in this portion of the belt: biotite paragnaisses with intercalations of quartzites and calcsilicate rocks (distal A1 to A4 units) and (kyanite) sillimanite, garnet biotite gneisses (probably correlatives of the A6 units), with quartzites, Mn-rich rocks, amphibolites and calcsilicate rocks (Figs. 15.11, 15.12 and 15.13).

The Andrelândia megasequence is currently interpreted as the Neoproterozoic rift to passive margin succession developed along the southern margin of the São Francisco paleocontinente (Paciullo et al. 2000; Ribeiro et al. 1995). The sequence has been correlated to the Macaúbas Group of the Araçuaí belt and the Canastra and Vazante groups of the Brasília belt (see Alkmim et al. and Fuck et al., this book). The A5 unit, shed from southern Brasília belt magmatic arc, also has equivalent units in the Araçuaí and Brasília belts, namely, the Salinas Formation, and Araxá and Ibiá units, respectively.

15.2.2 The Guaxupé Nappe at the Southern Tip of the Brasília Belt

Rock assemblages of the southern tip of the Brasília belt are preserved along the border of the Paranapanema plate, which was thrust over and sutured to the São Francisco plate around 630 Ma, prior to the development of the Ribeira belt. These assemblages occur within an approximately 15 km-thick stack of nappes that characterize the Guaxupé terrane (Heilbron et al. 2004, Fig. 15.9). Three major units, representing different crustal levels, form the nappe stack: the Lower Granulite, the Intermediate Diatexite and the Upper Migmatite units (Campos Neto and Caby 2000).

The Lower Granulite unit comprises banded garnet-biotite-orthopyroxene granulites of enderbite to charnoenderbite composition with local intercalations of noritic to gabbroic gneisses. U–Pb ages in the range of 655–640 Ma (Basei et al. 1995; Ebert et al. 1996; Hackspacher et al. 2004) indicate that their magmatic protoliths were emplaced shortly before the granulite facies metamorphism. The Intermediate Diatexite unit is composed of meta-aluminous orthogneisses with discontinuous bodies of metasedimentary stromatic migmatites containing cordierite and spinel. The Upper Migmatite unit (Campos Neto and Basei 1983) correspond to a metasedimentary succession dominated by banded cordierite-sillimanite-garnet-biotite gneiss with stromatic leucosomes, in which the amount of melting decreases towards the upper portion of the nappe.

Fig. 15.7 Field aspects of the units of the São João del Rey and Carandaí basins: **a** couplets of rippled and cross-beds of the transgressive Tiradentes sequence; **b** thick cross beds separated by **c** herringbone-type cross-beds of the Tiradentes sequence; **d** mud cracks on top of wave ripples intertidal facies of the Tejuco sequence; **e** quartz lithic conglomerate of the deltaic Lenheiro sequence; **f** diamictite of the lower sequence of the Carandaí basin; **g** pelitic rocks of the Prados Fm (with *orange* and *pink* color) covering a paleokarst topography of the platformal limestones of the Barroso Fm (*grey* fresh rocks); **h** marine pelites containing minor sandstone lenses of the Prados Fm

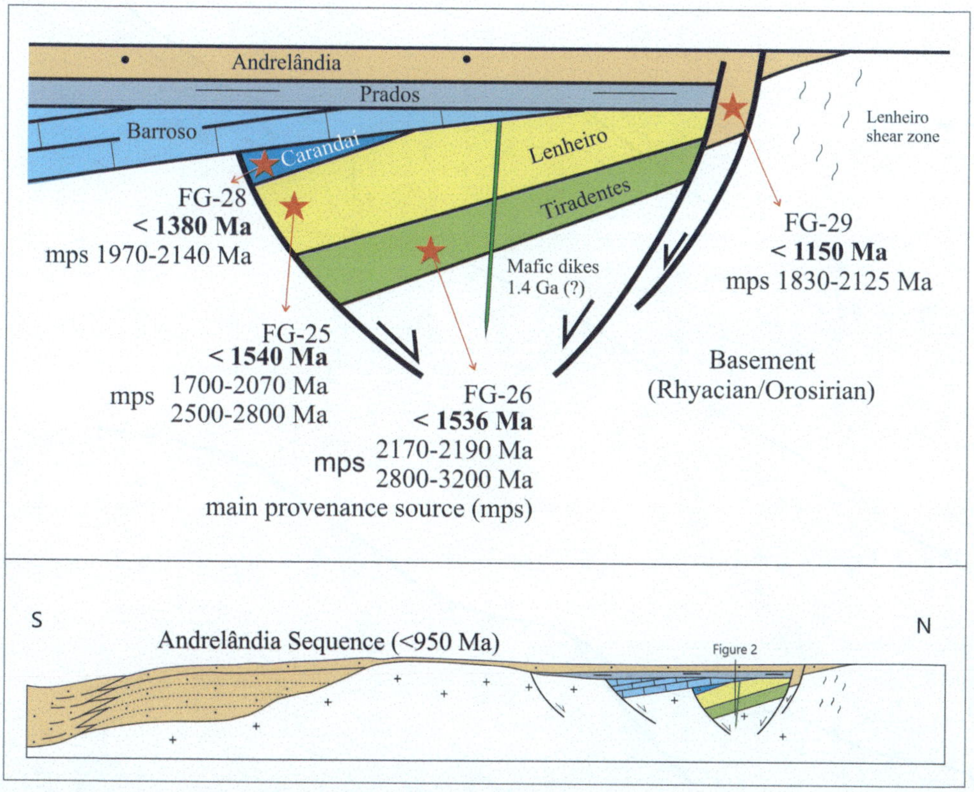

Fig. 15.8 Schematic section showing the spatial and chronological relationship between the Mesoproterozoic and Neoproterozoic basins fill units at the cratonic border. The main age peaks of the detrial zircons are indicated by the *numbers* (modified from Ribeiro et al. 2013, reproduced with Elsevier permission)

The Sm–Nd model ages available for these units vary between 1.4 and 1.7 Ga in the eastern part of the nappe stack in contrast with values higher than 2.0 Ga in the western part (Janasi et al. 2001, Janasi 2002). These ages are interpreted as results of a mixture between juvenile Neoproterozoic magmas with melts sourced by an older continental crust. Furthermore, they suggest that the frontal part of the Guaxupé nappes represents the thinned rim of the Paranapanema paleocontinent (Valeriano et al. 2008).

15.2.3 The Paraíba do Sul Terrane

As mentioned before, the Paraíba do Sul terrane is thrust over the Occidental terrane. It is made up of Paleoproterozoic hornblende bearing orthogneisses (Quirino basement complex), amphibolite facies paragneisses (Paraíba do Sul Group), and several Neoproterozoic granitoid bodies, including the arc-related rocks of the Serra da Bolívia complex (Heilbron et al. 2013) equivalent to the Rio Doce arc of the Araçuaí belt (Gonçalves et al. 2014, Alkmim et al. this book).

The Quirino Complex and correlative units are represented by hornblende and biotite-bearing orthogneisses with mafic and ultramafic enclaves (Fig. 15.4e, f). Geochemical data indicates two calcalkaline groups of medium and high-K affinities compatible with mature arc settings (Valladares et al. 2000, 2002). U–Pb geochronology and TDM model ages indicate crystallization ages between 2.3 and 2.1 Ga and Archean inheritance (Machado et al. 1996; Valladares et al. 2000, 2002; Ragatky et al. 1999).

The supracrustal assemblage of this terrane, the Paraíba do Sul Group, consists of sillimanite, garnet, biotite gneisses containing calc-silicate lenses and dolomitic marble intercalations (Eirado Silva et al. 2006) (Figs. 15.12, 15.13 and 15.14). Few available and ongoing data indicate Paleoproterozoic with minor Archean contributions as sources for the unit (Lobato et al. 2015). Metric lenses and boudins of amphibolites rocks with intraplate affinity were described by Marins (2000).

Around ca. 620 Ma or even before, the western margin of the Paraíba do Sul terrane was converted into an active setting with the development of a 620–595 Ma Serra da Bolívia cordilleran arc. Arc-related rocks, including calcalkaline orthogneisses and gabbroic rocks, intrude both the Paleoproterozoic basement and Neoproterozoic successions of the terrane (Figs. 15.11 and 15.14). Charnockitic members are also

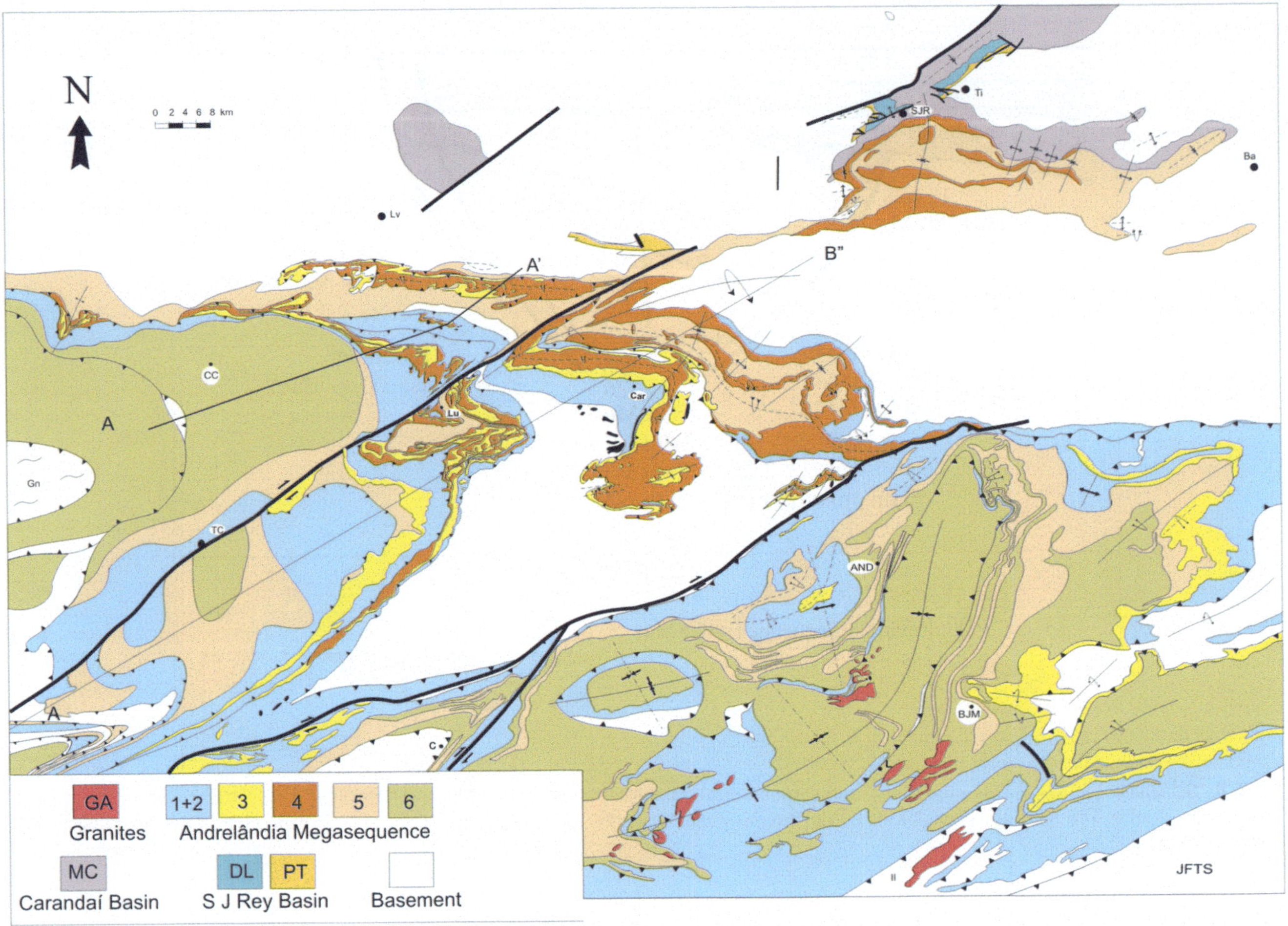

Fig. 15.9 Detailed geologic map of the Andrelândia Megasequence in the Ocidental terrane. Units A1 to A6 are discriminated on the map. See text for explanation

common in this assemblage. The Serra da Bolívia complex is a correlative of the Socorro and Rio Doce pre-collisional suites of the central Ribera and Araçuaí orogen, respectively (Heilbron et al. 2013; Pedrosa Soares et al. 2008).

15.2.4 The Oriental Terrane

The Oriental terrane docked onto the active margin represented by external portions of the Ribeira belt between 605 and 580 Ma. A distinctive characteristic of the Oriental alloctonous terrane is the absence of a Paleoproterozoic or older basement assemblage. It is made up of Tonian to Ediacaran intra-oceanic to immature continental arc-related rocks (Serra da Prata and Rio Negro complexes, Figs. 15.15 and 15.16) and Neoproterozoic high-grade metasedimentary successions (Italva and São Fidélis groups) (Heilbron et al. 2008).

The Serra da Prata Complex, consisting of hornblende and biotite- bearing tonalite to granodiorite orthogneisses with diorite lenses and leucogneisses, is the oldest and most primitive magmatic arc association documented in the Oriental terrane (Figs. 15.14 and 15.15). Geochronological data indicate crystallization ages of 860, 760, and 640–620 Ma. T_{DM} model ages of 1.1–0.9 Ga with ε_{nd} values in the range of +5 to +2 attest its juvenile nature (Peixoto and Heilbron 2010 and Peixoto et al. in press).

The Rio Negro Complex, exposed along the Serra do Mar, the most prominent mountain range of southeastern Brazil, comprises, besides gabbroic rocks, several associations of calc-alkaline and shoshonitic orthogneisses (Fig. 15.16). The orthogneisses are intrusive into the basal units of the granulite facies paragneisses of the São Fidélis Group, described below. U–Pb age data spreads within the interval of 790 and 605 Ma. Sm–Nd isotopic data suggests an initial stage characterized by juvenile tonalites and diorites of medium K calc-alkaline and tholeiitic series, followed by a progressive evolution towards to a mature stage, marked by granodiorites and granites of high K calc-alkaline series with cordilleran isotopic signatures (Cordani et al. 1973; Heilbron and Machado 2003; Tupinambá et al. 2000, 2012).

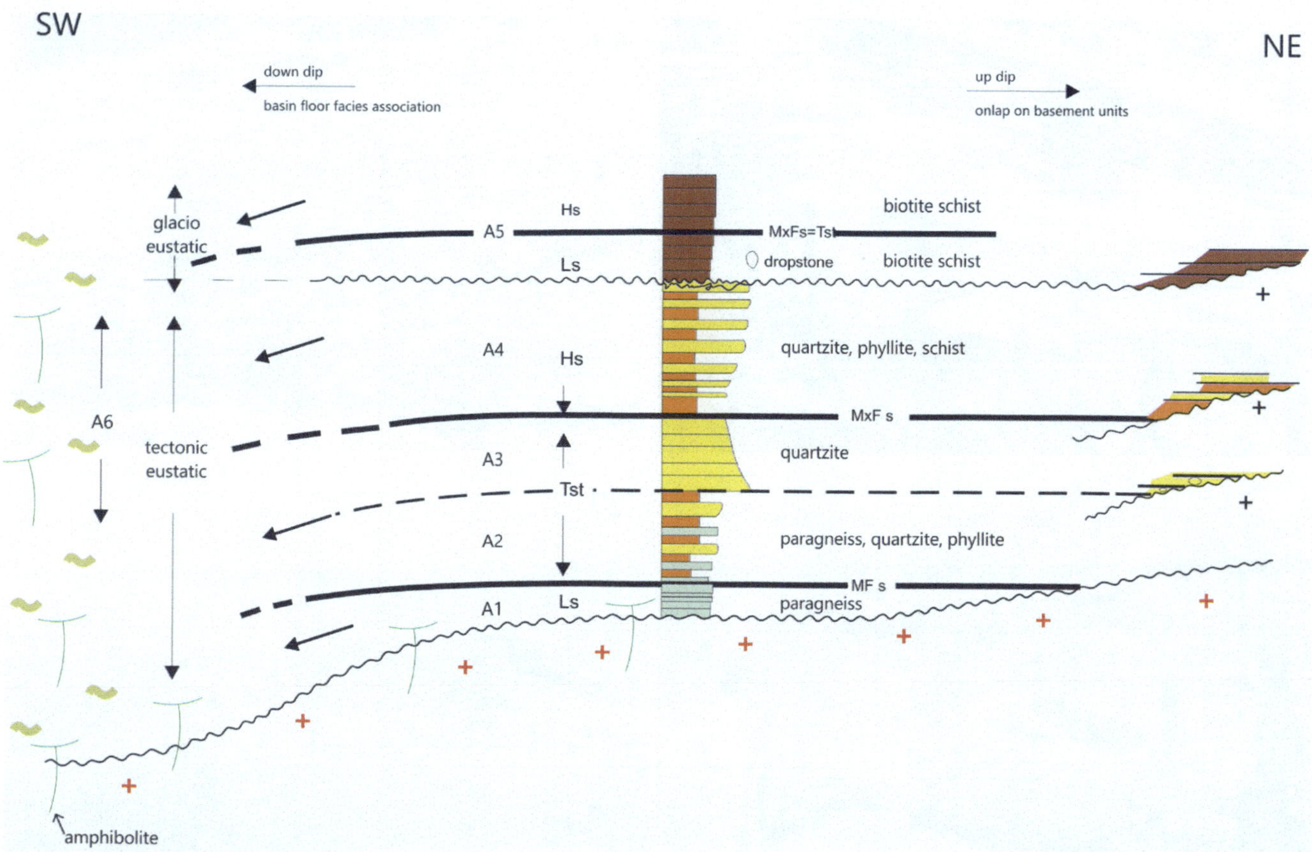

Fig. 15.10 Composed stratigraphic column for the units of the Andrelândia basin, as proposed by Paciullo et al. (2000)

Coeval with the development of the Rio da Prata and Rio Negrao arcs, supracrustal rocks of the Bom Jesus do Itabapoana, Italva and São Fidélis groups (see Table 15.3) were deposited in fore- and back-arc settings. Most of them are tectonically dismembered and represented by high-grade gneisses, with metamorphic parageneses ranging from the amphibolite to granulite facies. U–Pb geochronological data available for these units and interlayered amphibolites constrain their deposition ages to the Neoproterozoic. Besides contribution of Neoproterzoic arc-related rocks, Mesoproterozoic, Paleoproterozoic and Archean sources were also reported by Valladares et al. (2004) and Lobato et al. (2015), suggesting contribution from the African side (Congo craton). Table 15.3 summarizes the characteristics of these successions.

15.2.5 The Cabo Frio Terrane

The Cabo Frio terrane was docked to the system between 535 and 510 Ma (Schmitt et al. 2004, Fig. 15.2). This alloctonous terrane is composed of 1.9 Ga orthogneisses and amphibolites of the Região dos Lagos Complex and high-grade metasedimentary successions of the Búzios and Palmital groups (Heilbron et al. 1982; Table 15.2).

The Região dos Lagos orthogneisses are represented by calcalkaline granites and tonalites (Fonseca 1994; Viana 2003; Viana et al. 2008, Fig. 15.4g, h) containing diorite lenses and thick layers of amphibolites (Heilbron et al. 1982; Viana 2003; Viana et al. 2008). Available U–Pb age determinations point to crystallization ages of ca. 1.9 Ga and a Cambrian metamorphic overprint between 535 and 510 Ma (Zimbres et al. 1990; Schmitt et al. 2004). These data also suggest inheritance from a 2.09 to 1.95 Ga basement source (Zimbres et al. 1990; Schmitt et al. 2004).

The metasedimentary Búzios and Palmital sequences comprise (kyanite)-sillimanite garnet biotite gneisses with calc-silicate and garnet-amphibolite layers (Heilbron et al. 1982, 2000). The calc-silicate layers achieve decametric thickness and consist of diopside-rich banded gneisses, with hornblende, garnet and scapolite.

15.3 Structure and Metamorphism

The diachronic accretions along the southern São Francisco paleocontinent between 630 and 510 resulted in intense deformation and metamorphic overprint of the stretched craton margin and the overlying sedimentary successions that constitute the Occidental terrane. During the accretion

Fig. 15.11 Field aspects of the successions of the Andrelândia Megasequence: **a** Banded biotite gneisses of the A1 + A2 association; **b** basal meta-conglomerates and **c** green mica quartzites of the A3 Association; **d** quartzites and grey schist of the A4 Association, **e**, **f** Biotite schist of the A5 Association, with boulders of the basement orthogneisses (meta-wacke lenses, right)

process, the São Francisco paleocontinent acted as the lower plate, receiving the additions of the Guaxupé terrane on the southwest at the 630–605 Ma interval, as well as the Paraíba do Sul, Oriental and Cabo Frio terranes on the southeast, respectively, at 620–580, 605–595, and 535–510 Ma (Heilbron et al. 2008) (Figs. 15.1, 15.2 and 15.16). The suture zones between the various terranes are represented, respectively, by the basal thrust of the Guaxupé nappe, the Rio Preto-Mantiqueira thrust, the Central tectonic boundary (CTB), and the basal thrust of the Cabo Frio terrane, as shown on the map of Fig. 15.2.

Deformation and metamorphism increases from the SFC towards the Ribeira belt along both the southwest and southeast directions (Figs. 15.17 and 15.18). As expected,

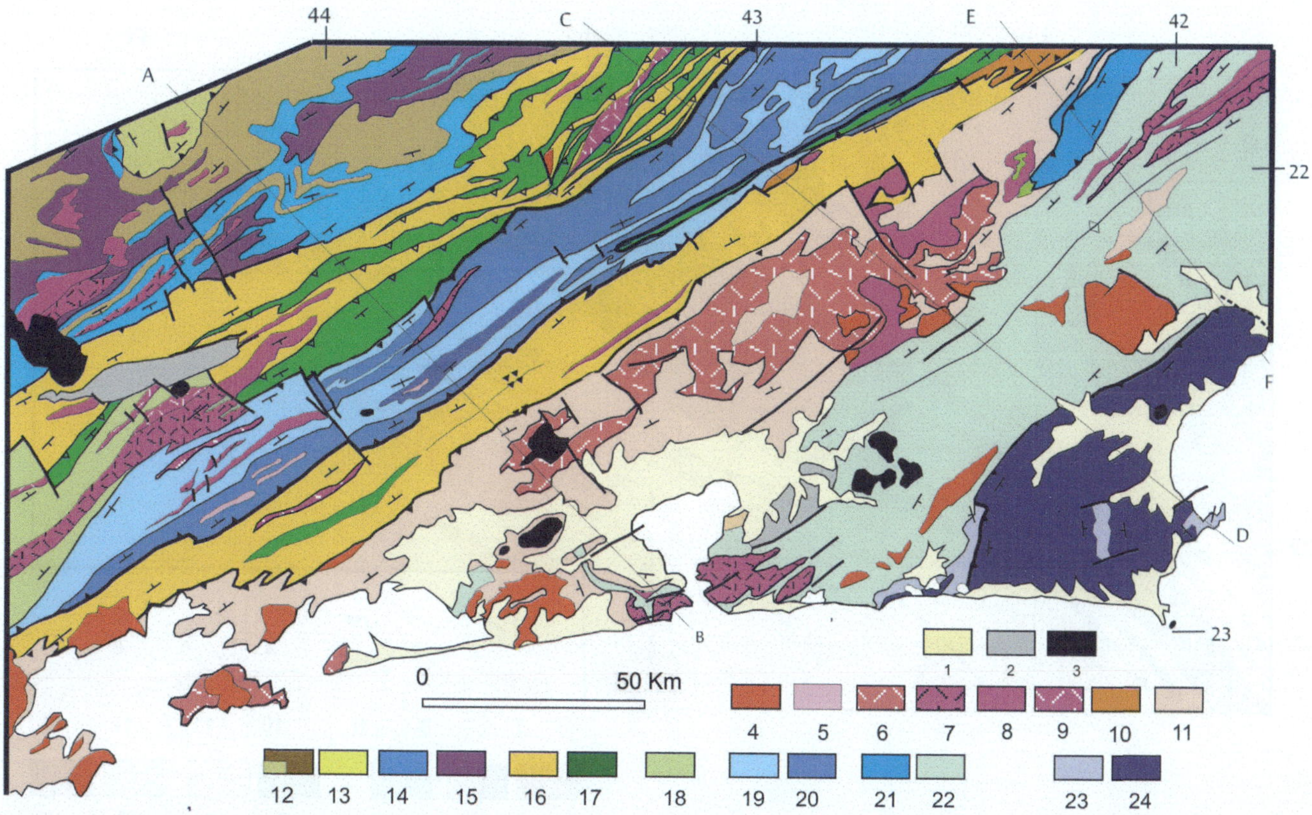

Fig. 15.12 Geological map of the central segment of the Ribeira belt, modified from Heilbron et al. (2000, 2004). *1* Quaternary cover; *2* Tertiary rift basins; *3* K-T alkaline rocks; (*4–9*): syn- to post-collisional granitoids; *4* post-collisional biotite granite (510–480 Ma, G5); *5* syn-D3 granitoids (535–520 Ma, G4); *6* late-collisional granites and charnockites (ca. 560 Ma, G3); *7* Porphyritic syn-collisional granites (590–560 Ma); *8* syn-collisional leucogranites and S-type to hybrid charnockites (ca. 580 Ma, G2); granitoids of undetermined ages (*9–10*): *9* Hornblende granite; *10* Anta and São Primo suites; *11* Rio Negro Magmatic Arc (790–620 Ma, G1-pre-collisional); Occidental Terrane (*12–17*): Andrelândia Megasequence (*12–14*): *12* A6–A5 Rio do Turvo Sequence (high P granulite facies); *13* Rio do Turvo Sequence; *14* A1 + A2 + A3 Carrancas Sequence; *15* Mantiqueira complex; *16* distal facies of the Andrelândia Megasequence in the Juiz de Fora Domain; *17* Juiz de Fora Complex; *18* Embú Complex; Paraíba do Sul Terrane (*19–20*): *19* Paraíba do Sul Group; *20* Quirino Complex; Oriental Terrane (*21–22* rich): *21* platformal carbonate Italva succession; *22* Costeiro succession; Cabo Frio Terrane (*23–24*): *23* Búzios and Palmital succession; *24* Região dos Lagos Complex

the metamorphic gradient is inverted on the lower plate represent by the Occidental terrane (Trouw et al. 2000, 2013; Campos Neto et al. 1999; Heilbron et al. 2000, 2004), and normal in the upper plate (Heilbron et al. 1995, 2000, 2008). Below, we present a summary of the main structural and metamorphic features of the Occidental terrane, followed by a brief discussion of the tectono-metamorphic picture of the remaining components of the belt.

15.3.1 Structures and Metamorphism Related to the Guaxupé Terrane Collision

The first collisional episode involving the southern margin of the craton led to the development of the southern branch of the Brasília belt. As described by Valeriano (this book), Trouw et al. (2000) and Campos Neto and Caby (1999, 2000, 2004), a thick stack of east-vergent nappes separated by transpressive sinistral shear zones dominate the structural picture southern Brasilia belt (Fig. 15.17). Within the thick-skinned nappes, the basement and the cover units are affected by large-scale recumbent folds. The metamorphic zoning, ranging from greenschist facies (chloritoid and garnet zones) to amphibolite and granulite facies (staurolite, kyanite, kyanite + K-feldspar zones), is inverted in the lower plate, represented by the craton margin. Detailed metamorphic maps of the region are provided by Ribeiro and Heilbron (1982) and Trouw et al. (2000, 2013).

The metamorphism in the lower nappes reflects relatively high pressures, typical for exhumed subduction zones. The upper nappes reach the granulite facies conditions with the development of kyanite-K-feldspar parageneses (Fig. 15.18). Pods and lenses of retro-eclogites were locally found (Trouw 1992; Trouw et al. 2000; Campos Neto and Caby 2000). In these rocks, the maximum pressures and temperatures were estimated at 13.5–15 kbar and 800–900 °C. High pressure

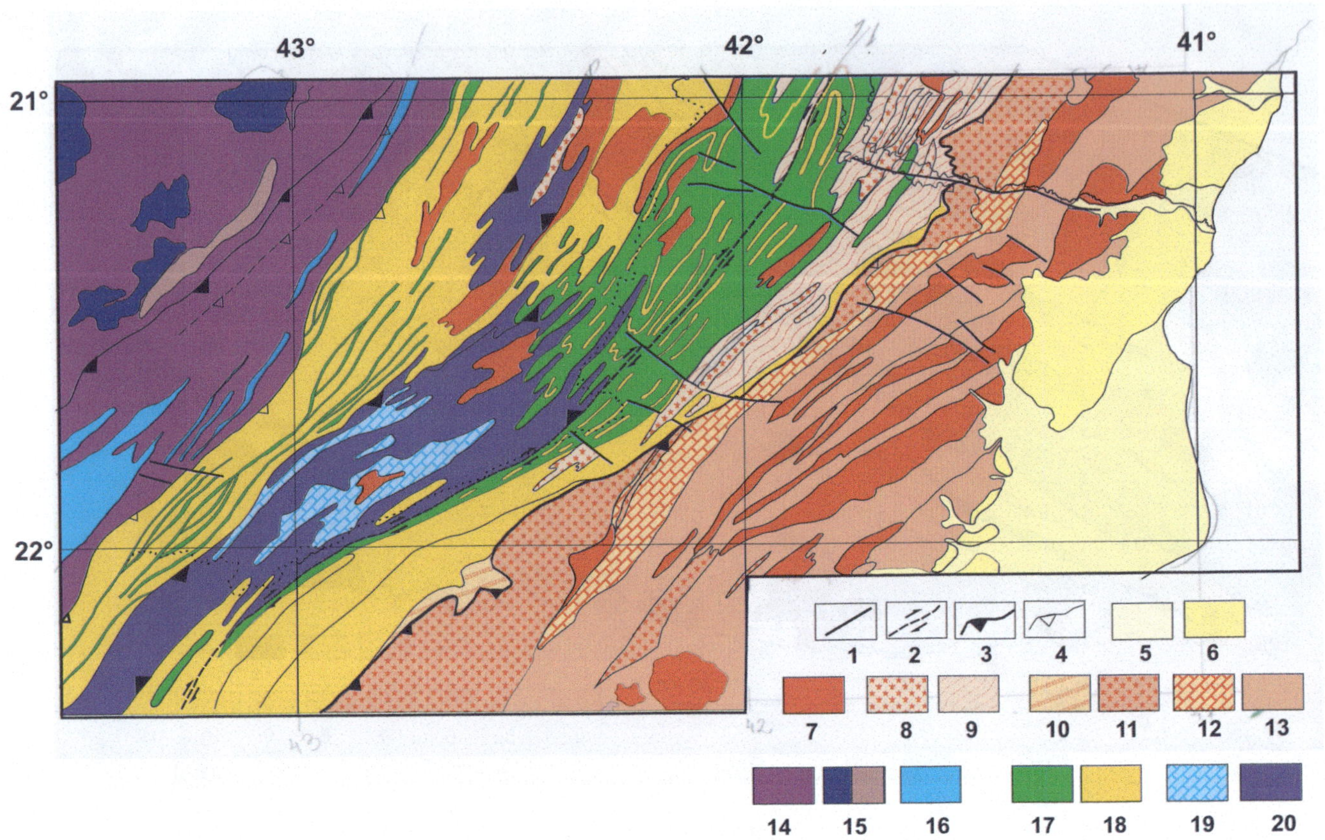

Fig. 15.13 Geological map of the northern segment of the Ribeira belt. Legend: *1* faults, *2* Além Paraíba shear zone, *3* main thrusts (suture-terrane boundary), *4* subordinated thrusts, *5* Cenozoic cover, *6* Neogene Barreiras formation *7* Syn to post tectonic Neoproterozoic-Ordovician granites (Brasiliano orogeny). Oriental terrane *8* Serra da Bolívia arc and associated *9* High grade metasediments, *10* Fore arc metasediments, *11* Rio Negro and Serra da Prata magmatic arcs, *12* carbonate rich intra/back-arc basin, *13* high grade metasediments (São Fidelis and group). Occidental terrane (reworked São Francisco margin): *14–15* orthogneisses and granitoid of the Mantiqueira complex, *16* Proximal Andrelândia passive margin, *17* orthogranulites of the Juiz de For a basement complex, *18* distal Andrelândia passive margin. Paraíba do Sul-Embú terrane: *19* Basement Quirino complex, *20* Carbonate rich Neoproterozoic margin (Paraíba do Sul group?)

conditions (14 kbar and 850 °C) were also attained in the Lower Granulite Unit of the Guaxupé nappe, indicating a tectonic environment compatible with the root zone of a magmatic arc (Del Lama et al. 2000; Campos Neto and Caby 2000; Freitas 2000; Garcia and Campos Neto 2003).

The metamorphic peak was probably attained at 640–630 Ma. Younger metamorphic ages between 610–605 Ma obtained in the rocks of the Guaxupé terrane were attributed to the exhumation of the nappe stack (Trouw and Pankhurst 1993; Campos Neto and Caby 2000; Trouw et al. 2000).

15.3.2 Structures and Metamorphism of Craton Margin Related to the Collision of the Paraíba do Sul and Oriental Terranes

In contrast with the structural style of the southern tip of the Brasília belt, the structures ($D_1 + D_2$) that resulted from the accretions to the southeastern margin of the craton correspond to NNW-verging crustal-scale thrust systems. Due to the oblique nature of the plate convergence, the deformations during these accretions were associated to a strong partition between thrust systems and dextral transpressive shear zones (Fig. 15.17c, d).

Based on structural style, the following tectonic subdivision is adopted for the Occidental terrane:

(a) Para-autochthonous to allochthonous domains, which include the pre-existent southern Brasilia belt;
(b) Andrelândia domain that represents the lower allochthonous unit;
(c) Juiz de Fora domain, representing the upper allochthonous unit.

Within the first two domains, asymmetric, open to tight folds and discrete thrust shear zones involving the basement assemblages are the main fabric elements (D1 + D2). The

Fig. 15.14 **a–e** Distal associations of the Andrelândia Megasequence, mylonitic banded biotite gneisses (**a**) with porphyroclasts of feldspar and sigma-objects of amphibolite (**b**); **c** Coarse quartzite layer with a metric calcsilicate layer; **d** Mn-rich layer; **e** symmetric calcsilicate boudins within the Além Paraíba shear zone. **f–g** Associations of the Paraíba do Sul group, with banded gneisses (**f**) and folded dolomitic marbles (**g**)

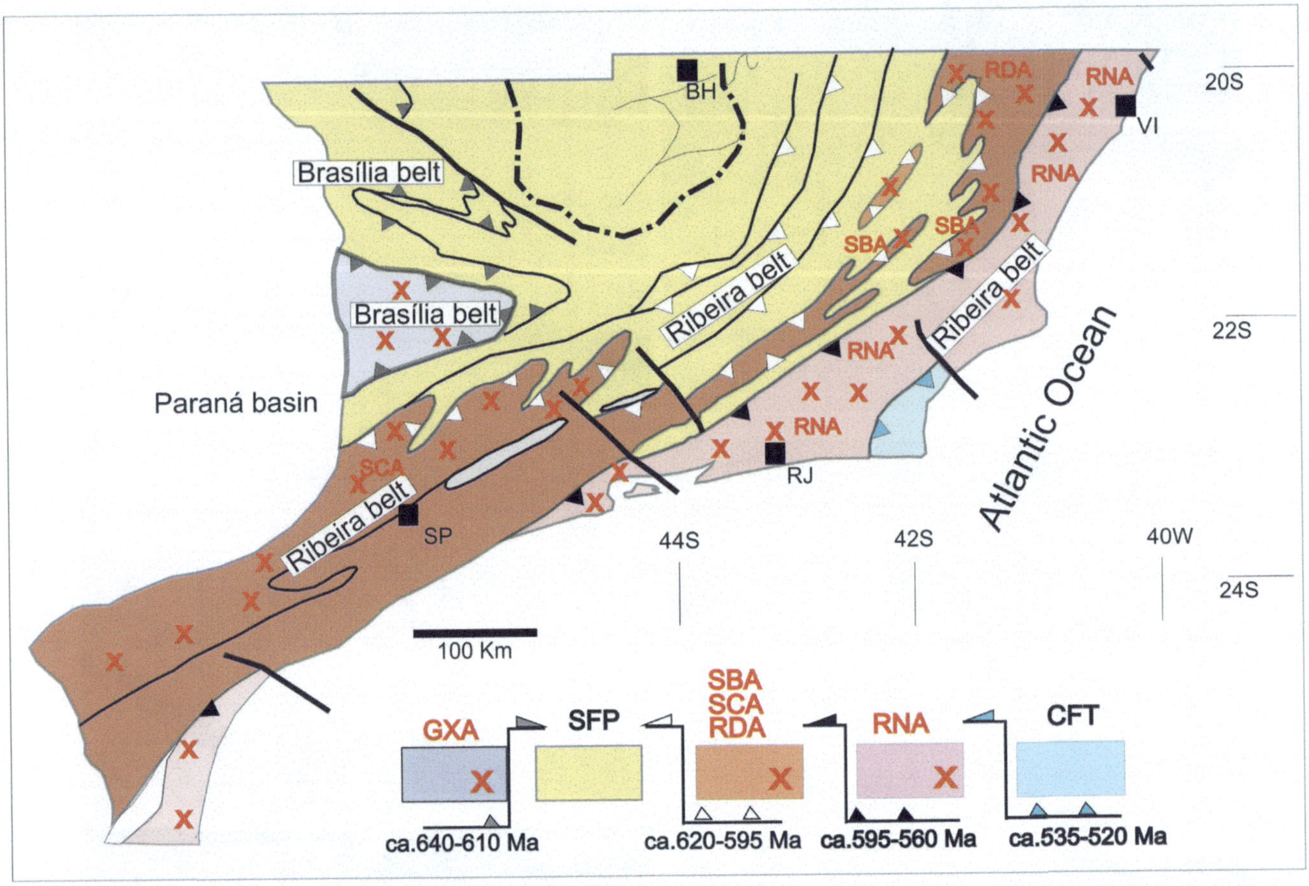

Fig. 15.15 Magmatic arcs around São Francisco Paleoplate (reproduced from Heilbron et al. 2013, with Elsevier permission). GXA: Guaxupé magmatic arc; SFP: São Francisco paleoplate; Cordilleran magmatic arcs: SBA Serra do Bolívia, RDA Rio Doce, and SCA Socorro; RNA: juvenile intra-oceanic to immature continental arcs (Serra da Prata and Rio Negro- arcs); CFT: Cabo Frio Terrane

basal contact of the Andrelândia domain is marked by a low to moderate angle mylonitic thrust zone exhibiting top to NW kinematic indicators. The deformation in high strain zones resulted in the generation of sheath folds, as proposed by Heilbron (1994).

The Juiz de Fora domain (the uppermost thrust unit of the Occidental Terrane) is a crustal-scale duplex system, in which the dextral transpressive component becomes progressively more important towards southeast. A pervasive mylonitic foliation associated with a well-developed stretching lineation and transposed tight to isoclinals folds are the most common features. Tectonic imbrication of basement and passive margin units often occur within the high strain zones (Figs. 15.2 and 15.17). At the basal portion of the thrust system, the faults dip 30°–45° SE. The transpressive steeply dipping shear zones predominate in the upper portion of system (Almeida et al. 1998; Heilbron et al. 2004).

The metamorphism in the authoctonous domains also show an inverted P-T gradient, varying from greenschist facies on the cratonic boundary to high-pressure granulite facies at the top units of the Andrelândia thrust system (Heilbron 1994, Heilbron et al. 1995; Trouw et al. 2013). Intermediate pressure granulite conditions (7 kbar and 700 °C) characterize the Juiz de Fora domain (Fig. 15.18). These conditions might however reflect only the decompression path of the system.

15.3.3 Structures and Metamorphism of the Remaining Terranes

The Paraíba do Sul and Oriental terranes are characterized by large-scale recumbent folds associated to a regionally penetrative foliation and a normal metamorphic gradient, typical for the upper plate domains of collisional systems (Fig. 15.18).

The structure picture the Paraíba do Sul terrane is dominated by a late stage (D3) regional syncline (Fig. 15.17). The main outcrop-scale structures are D2 tight folds and the

Fig. 15.16 Field aspects of the Magmatic arc granitoids of the Ribeira belt. **a** Foliated tonalitic orthogneisses with veins of leucogneiss, **b** less deformed tonalite with diorite enclaves of the Serra da Prat Arc (ca. 860–620 Ma), **c** foliated granitic leucogneiss and tonalite orthogneisses with leucosomes veins and **d** homogenous tonalite with diorite clots of the Rio Negro Arc (ca. 790–620 Ma)

Table 15.3 Tectono-stratigraphic units and lithology of the docked terranes of the Ribeira belt (Heilbron et al. 2000, 2004, 2008; Lobato et al., 2015)

Metasedimentary units of the docked terranes			
Unit	Lithology	Envisaged tectonic setting	Sedimentation age
Bom Jesus do Itabapoana	Psamo-pelitic gneisses, dolomitic marbles, amphibolites, calcsilicate	Fore arc basin to Serra da Bolívia (Rio doce) arc	ca. 630–595 Ma
Italva group (Euclidelândia unit)	Psamitc gneisses with marble lenses	Fore-arc basin to Serra da Prata	ca. 760–630 Ma
Italva group (São Joaquim and Macuco units)	Calcitic marbles, psamitc gneisses, amphibolites	Intra to proximal back-arc basin to Serra da Prata arc	ca. 860–760 Ma
São Fidelis group (upper unit)	Sillimanite garnet gneisses with quartzites and calc-silicate rocks	Back-arc basin to Serra da Prata/Rio negro arc	ca. 620–605 Ma
São Fidelis group (lower unit)	(Cordierite) kinzigitic gneisses with calcsilicate and marble lenses	Transition from passive margin to back arc setting to Rio negro arc	Not fixed yet

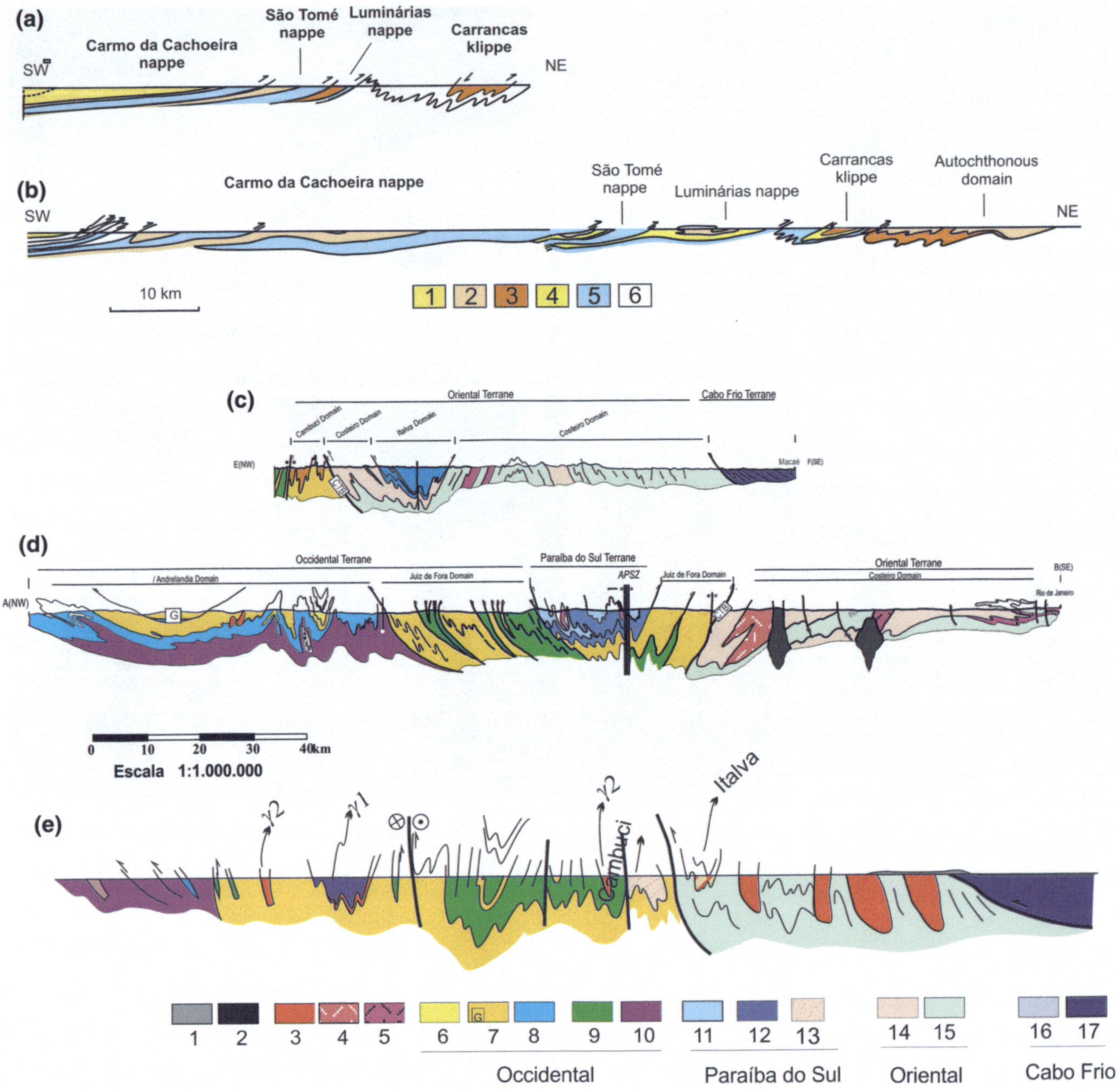

Fig. 15.17 Cross sections of the southern tip of the Brasília belt (**a**, **b**). *1* A6 unit, *2* A5 unit, *3* A4 unit, *4* A3 unit, *5* A1 + A2 units, *6* basement. Sections **c–e** present the structural style of the Ribeira belt Legend as Figs. 15.12 and 15.13

associated axial plane schistosity. The metamorphism varies from high amphibolite to granulite facies conditions. U–Pb data indicates that the main metamorphic episode took place between 615 and 585 Ma for the main metamorphic episode (Machado et al. 1996; Trouw et al. 2013).

The most conspicuous fabric elements of the Oriental terrane are large scale recumbent folds associated with a coarse schistosity (S_1) materialized by high temperature mineral assemblages. Two generations of open to tight upright folds and shear zones made up the structural style of the late deformations (D_3 and D_4 phases). Rocks of the pelitic succession of the São Fidélis Group were metamorphosed under medium to low pressure granulite facies conditions, whereas the Italva Group metasedimentary rocks display upper amphibolite facies parageneses. According to Kuhn et al. (2004) the peak metamorphic conditions of the São Fidelis Group are characterized by the assemblage Bt + Crd + Kfs + Pl + Grt + liq + Qtz, reflecting P-T conditions of 7

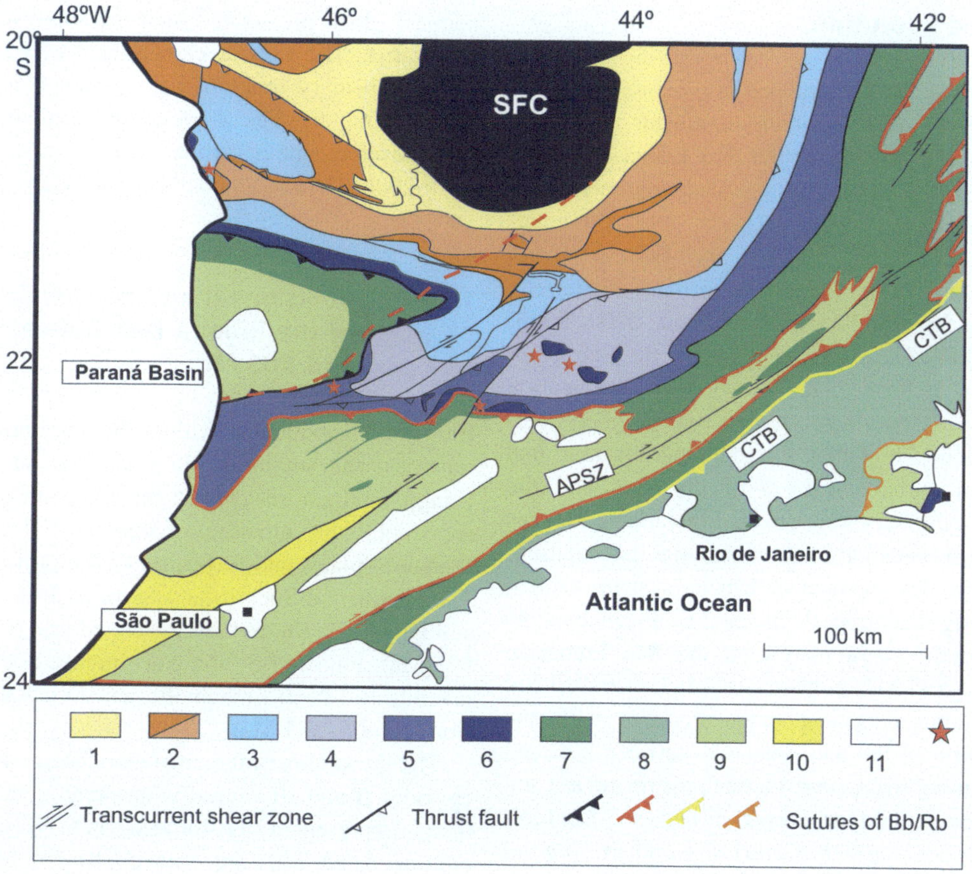

Fig. 15.18 Metamorphic map of the Ribeira beira belt and southern end of the Brasilia belt (reproduced from Trouw et al. 2013, with Elsevier permission and based on Heilbron et al. 1995, Valeriano et al. 2008). Note the superposition of Ribeira belt isogrades (E–W) over the previously NW-oriented Brasilia belt. *1* Greenschist facies without biotite; *2* greenschist facies with biotite and garnet; *3* amphibolite facies with garnet, staurolite and kyanite; *4* amphibolite facies with kyanite and sillimanite; *5* amphibolite/granulite facies with sillimanite and K-feldspar, ± orthopyroxene (without kyanite); *6* high pressure granulite facies kyanite and K-feldspar, +with/- sillimanite; *7* granulite facies with garnet, sillimanite, K-feldspar and orthopyroxene; *8* granulite facies with garnet, sillimanite and locally cordierite; *9* amphibolite facies with garnet and sillimanite; *10* low to medium pressure greenschist facies; *11* non metamorphosed younger rocks. *Red stars* indicate eclogitic rocks. Suture zones identified by *colors*: *black* ca. 640–630 Ma, *red* c. 620–65 Ma, *yellow* ca. 605–565 Ma and *orange* ca. 535–510 Ma. *Red dashed* line represent the Ribeira belt front. *CTB* Central tectonic boundary, *APSZ* Além Paraíba shear zone

kbar and 750–800 °C. Available U–Pb geochronological data constrains the syn-kinematic metamorphism of the Oriental in the 610–565 Ma interval (Machado et al. 1996; Heilbron and Machado 2003; Tupinambá et al. 2012).

Throughout both the Paraíba do Sul and Oriental terranes, the previously mentioned fabric elements are affected by upright to inclined folds and dextral transpressive shear zones related to the docking of the Cabo Frio Terrane in the Cambrian (the 530–510 Ma D_3 deformation phase). The most prominent folds of this generation are the Paraíba do Sul megasyncline and the Rio de Janeiro meganticline (Figs. 15.12a and 15.17). Examples of mylonitic dextral shear zones are the NE-trending Além Paraíba shear zone that extends from Rio de Janeiro to Minas Gerais states, and the Caxambú shear zone that offset the external nappes of the southern Brasília belt (Fig. 15.9). Syn-collision granites dated at 535–520 Ma constrain the timing of the D3 deformation.

The internal deformation the Cabo Frio terrane is very intense and resulted from three deformation phases. D_{1cf} + D_{2cf} are related to thrusting and the development of recumbent folds coeval with the metamorphic peak, which reaches high pressure granulite facies conditions (Heilbron et al. 1982; Schmitt et al. 2004). D_{3cf} gave rise to post- metamorphic peak recumbent folds that deforms the pre-existent structures. NW-trending brittle-ductile shear zones mark the latest deformation phase (D_{4cf}) in this terrane.

15.4 Tectonic Evolution

The development of the Ribeira belt is a consequence of diachronic collisions involving micro-continents, magmatic arcs and the southern margin of the São Francisco paleocontinent. These collisional episodes, collectively referred to as the Brasiliano tectonic collage or event, led to the amalgamation West Gondwana between the Ediacaran and Cambrian periods (2008; Campos-Neto and Figueiredo 1995; Campos and Caby 2000; Trouw et al. 2000, 2013; Schmitt et al. 2004). During these episodes, the São Francisco craton acted as lower plate, allowing the propagation of the orogenic front over a substantial portion of its margins. As a consequence of that, intracratonic and marginal basins underwent variable degrees of inversion and the craton margin assemblages were affected in variable degrees of intensity by syn-metamorphic contractional deformation.

The progressive development of Ribeira system resulted from the following diachronic collisions: (i) Paranapanema plate and the southwestern margin of the São Francisco between 640 and 595 Ma (collision I); (ii) Paraíba do Sul micro-continent and associated magmatic arcs with the southeastern margin of the São Francisco between 620 and 595 Ma (collision II); (iii) Oriental terrane magmatic arc and pre-existent system in the time interval of 605–565 Ma (collision III); and (iv) docking of the Cabo Frio (southern Angola) during the Cambrian period, between 535 and 510 Ma (collision IV) (Heilbron et al. 1982, 2000, 2008, 2013; Machado and Gauthier 1996; Trouw et al. 2000, 2013; Campos Neto and Caby 2000; Campos Neto 2000; Janasi et al. 2001; Cordani et al. 2002; Heilbron and Machado 2003; Valeriano et al. 2000, 2004; Silva et al. 2002, 2005; Schmitt et al. 2004). These stages are briefly discussed below.

15.4.1 Collision I: Development of the Southern End of the Brasília Belt

This episode is related to the closure of the Goianides Ocean along the western margin (present coordinates) of the SFC. The early stages of subduction within the Goianides Ocean led to the development of a magmatic arc (ca. 650–640 Ma) along the margin of the Paranapanema craton, presently hidden beneath the Paraná basin (Brito Neves et al. 1999; Campos Neto 2000). Remnants of this arc are represented by deep-seated batholiths (orthogneisses) of the Guaxupé terrane (Campos Neto 2000; Figs. 15.16 and 15.17).

The collision between the Paranapanema and São Francisco plates led to the formation of a pile of east-verging nappes, which propagated over the southwestern margin of the craton. The nappes carry slices of basement and the metasedimentary rocks of the Andrelândia passive margin (Fig. 15.16a, b). The amount of shortening associated to their development is estimated as at least 150 km (Trouw et al. 2000, 2013; Campos Neto and Caby 2004).

15.4.2 Collisions II and III: Development of the Ribeira Belt Between 620 and 565 Ma

E-directed subduction within the Adamastor Ocean that separated the southern São Francisco landmass from the Congo continent resulted in the accretion of two magmatic arcs onto the already amalgamated Paranapanema-São Francisco plates and formation of the proto-Ribeira system. The continental Socorro and Serra da Bolívia arcs (Socorro and Paraíba do Sul terranes) accreted first (Collision II, 620–595 Ma), followed by the convergence and collision of the juvenile Rio Negro magmatic arc (Oriental terrane) (Collision III, 605–565 Ma).

The thickening of the crust associated with the collisional episodes II and III caused widespread generation of crustal melts, represented by foliated and NE-trending granitic plutons, dated in the interval of 610–565 Ma.

15.4.3 Collision IV: Docking of the Cabo Frio Terrane Between 535 and 510 Ma

The docking of the Cabo Frio terrane, which probably represents a fragment of the Angola craton, caused the final closure of the Adamastor ocean and full development of the Ribeira system. The Cabo Frio terrane is now partially hidden beneath the offshore Campos and Santos basins of the Brazilian continental margin. Deformation and metamorphic overprint generated by this process is well documented in the basement (Região dos Lagos Complex) and metasedimentary units (Palmital-Búzios groups) of the terrane, envisaged as candidates for representing the accretionary prism of this tectonic episode.

The effects of the Cabo Frio terrane docking affect all previously accreted terranes, reaching the craton boundary. The associated deformation is highly partitioned between open to tight upright folds and an outstanding system of NE-trending dextral strike-slip shear zones, which channelized granitic melts and hydrothermal fluids, represented by late collision leucogranites (G4), pegmatites, and quartz veins.

15.4.4 Cambro-Ordovician Collapse of the Belt

The last tectonic episode document in the Ribeira belt is related to a transtensional deformation, during which a set of NW-trending normal shear zones nucleated. Some of the zones defined corridors of deformation that were reactivated during the episodes related to Gondwana break-up. Some of them reach the edge of the São Francisco craton.

Two generations of granites (G5 and G6) are coeval with this tectonic episode, sometimes occurring as minor dikes that follow the shear zones or as feeder dikes to the largest sill-like to laccolith-like bodies, with sharp contacts with the country rocks. Porphyritic to equigranular granites of this generation (Wiedemann et al. 2002) were dated between 510 and 480 Ma (Pedrosa-Soares and Wiedemann-Leonardos 2000, Machado et al. 1996; Heilbron and Machado 2003; Valeriano et al. 2011).

15.5 Final Remarks

The southern margin of the São Francisco craton hosts a complex and deeply eroded orogenic system, which records at least four terrane accretion episodes during time period of 640–510 Ma. Due of the geometry the craton margin as well as the polyphase and diachronic nature of the accretion processes, the Ribeira belt and its foreland are characterized by two remarkable features. First, they encompass the complex tectonic interference between the older E-verging nappe system of the Southern Brasília belt and the younger NW-verging folds and thrusts generated during closure of the Adamastor ocean. Second, the Ribeira itself records the superposition of fabric elements related to three distinct accretionary episodes, namely, the convergence and collision of the Paraíba do Sul, Oriental, and Cabo Frio terranes. Complex patterns of superposed fold generations and strong portioning of the deformation processes characterizes all domains of the belt.

Acknowledgments The authors thank the following funding agencies and companies of Brazil for the research support: CNPq, Faperj, CPRM and Petrobras. The chapter is a contribution to the IGCP 648 project. We also thank U. Cordani and F. Alkmim for the revisions that improve the manuscript.

References

Almeida F.F.M., Hasui Y., Neves B.B.B, Fuck, R.A, 1981. Brazilian structural provinces: an introduction. Earth-Science Reviews 17, 1–29

Almeida J.C.H, Tupinambá M., Heilbron, M., Trouw R., 1998. Geometric and Kinematic Analysis at the Central Tectonic Boundary of the Ribeira Belt, Southeastern Brazil. Congresso Brasileiro de Geologia, 1998, Belo Horizonte. Anais do Congresso Brasileiro de Geologia. Belo Horizonte: Sociedade Brasileira de Geologia, 40: 32

Basei MAS, Siga JrO, Sato K, Sproesser WM (1995) A metodologia Urânio-Chumbo na USP. Princípios metodológicos, aplicacões e resultados obtidos. Anais Academia Brasileira de Ciências 67: 221–237

Belém J, Pedrosa-Soares AC, Noce CM, Silva, LC, Armstrong R, Fleck A, Gradim Ct, Queiroga G N (2011) Bacia precursora versus bacias orogênicas: exemplos do Grupo Andrelândia com base em datações U-Pb (LA-ICP-MS) em zircão e análises litoquímicas. Geonomos, v. 19:224–243

Brito Neves, BB, Campos-Neto, MC, Fuck, RA (1999) From Rodinia to Western Gondwana: An approach to the Brasiliano Pan-African cycle and orogenic collage. Episodes, 22, 155–199

Campos Neto MC (2000) Orogenic systems from Southwestern Gondwana, an approach to Brasiliano-Pan African cycle and orogenic collage in south-eastern Brazil. In: Cordani UG, Milani EJ, Thomaz Filho A, Campos DA (ed), Tectonic Evolution of South America, SBG, Rio de Janeiro

Campos Neto MC, Basei MAS, Vlach SRF, CabyR, Szabó GAJ, Vasconcelos P (2004) Migração de orógenos e superposição de orogêneses: um esboço da colagem Brasiliana no sul do Cráton do São Francisco, SE -Brasil. Revista do Instituto de Geociências e USP. Geologia USP, Série Científica 4, 13–40

Campos Neto MC, Basei, MAS (1983). Evolução estrutural brasiliana do NE de São Paulo: Dobramentos superpostos e esboço estratigráfico e tectônico. In: Simpósio Regional de Geologia 4, 1: 61–78

Campos Neto MC, Caby R (1999) Neoproterozoic high-pressure metamorphism and tectonic constraint from nappe system south of the São Francisco craton, southeast Brazil. Precambrian Research 97:3–26

Campos Neto MC, Caby R (2000) Lower crust extrusion and terrane accretion in the Neoproterozoic nappes of southeast Brazil. Tectonics, 19, 669–687

Campos Neto MC, Caby R (2004) Lower crust extrusion and terrane accretion in the Neoproterozoic nappes of southeast Brazil. Tectonics, 19, 669–687

Campos Neto MC, Figueiredo MCH (1995) The Rio Doce Orogeny, South-eastern Brazil. J. South Am. Earth Sci. 8 (2),143–162

Cordani UG, Coutinho JMV, Nutman AP (2002) Geochronological constraints on the evolution of the Embú Complex, Sâo Paulo, Brazil. Journal of South American Earth Sciences, 14: 903–910

Cordani UG, Delhal J, Ledent D (1973) Orogeneses superposées dans le précambrien du Brésil Sud-Oriental (Etats de Rio de Janeiro et Minas Gerais). Revista Brasileira Geociências 3, 1–22

Del Lama EA, Zanardo A, Oliveira, MAF, Morales N (2000) Exhumation of high-pressure granulites of the Guaxupé Complex, Southeastern Brazil. Geological Journal (Chichester), Inglaterra, v. 35: 231–249

Duarte BP, Figueiredo MCH, Campos Neto M, HeilbronM (1997) Geochemistry of the granulite facies orthogneisses of the Juiz de Fora complex, central segment of Ribeira Belt, southeastern Brazil. Revista Brasileira de Geociências 27, 67–82

Duarte BP, Heilbron M, Campos Neto MC (2000). Granulite/charnockite from the Juiz de Fora Domain, central segment of the Brasiliano-Pan-African Ribeira belt.Revista Brasileira de Geociências 30, 358–362

Duarte BP, Valente S, Heilbron M, Campos Neto MC (2004) Petrogenesis of orthogneisses of the Mantiqueira Complex, Central Ribeira Belt, SE Brazil: an Archean to Paleoproterozoic basement unit reworked during the Pan-African Orogeny. Gondwana Research, 7, 437–450

Ebert, H. (1968). Ocorrência da fácies granulítica no sul de Minas Gerais e em áreas adjacentes, em dependência de sua estrutura orogênica: Hipótese sobre sua origem. Anais da Academia Brasileira de Ciências., 40:215–229

Ebert H (1984) Os Paraíbides entre São João del Rei, MG e Itapira, SP, e a bifurcação entre Paraíbiades e Araxaídes. São Paulo, SBG-SP, Publicação 12, pp. 72–103

Ebert HD, Chemale F, Babinski M, Artur AC, Van Schmus WR (1996) Tectonic setting and U/Pb zircon dating of the plutonic Socorro Complex in the transpressive Rio Paraı´ba do Sul Shear Belt,SE Brazil. Tectonics, 15, 688–699

Eirado Silva LG, Heilbron M, Almeida JCH (2006) Os terrenos tectônicos da Faixa Ribeira na Serra da Bocaina e na Baía da Ilha Grande, sudeste do Brasil. Revista Brasileira de Geociências, 36: 426–436

Fonseca, AC (1994) Esboço Geocronológico da região de Cabo Frio Rio de Janeiro. Phd Thesis. Universidade do Estado de São Paulo.

Freitas FC (2000) Geotermobarometria e evolução metamórfica das rochas granuíticas da região de Socorro-SP. Msc Dissertation, Universidade de São Paulo

Garcia MGM, Campos Neto MC (2003) Contrasting metamorphic conditions in the Neoproterozoic collision-realted nappes, south of São Francisco Craton, SE Brazil. Journal of South American Earth Sciences 15: 853–870

Gonçalves L, Farina, F, Lana C, Pedrosa-Soares AC, Alkmim, FF, Nalini HA (2014) New U-Pb Ages and Lithochemical Attributes of the Ediacaran Rio Doce Magmatic Arc, Araçuaí Confined Orogen, Southeastern Brazil. Journal of South American Earth Sciences, 52:1–20

Gonçalves ML, Figueiredo MCH (1992) Geoquímica dos anfibolitos de Santana do Garimbéu, MG: implicações sobre a evolução do Grupo Andrelândia. Geochimica Brasiliensis 6: 127–140

Hackspacher PC, Teixeira W, Dantas EL, Fetter A, Ebert HD, Trouw R, Vasconcelos, P (2004) Cooling and exumation of the final Brasiliano Orogeny at the southern Brasilia Belt and superposition with the Ribeira Belt: U/Pb and Ar/Ar methodologies. Journal of the Virtual Explorer (Online), Austrália, v. 17, n.2, p. 2

Hasui Y, Carneiro CDR, Coimbra AM (1975) The Ribeira Fold Belt. Revista Brasileira de Geociencias, 5, 257–267

Heilbron M (1994) Evolucao Tectono-Metamorfica da Secao Bom Jardim de Minas (Mg) -Barra do Pirai (Rj), Setor Central da Faixa Ribeira. Boletim Ig-Usp, Serie Cientifica, v. 25:128–130

Heilbron M, Duarte BP, Nogueira JR (1998) The Juiz de Fora Complex off The Central Ribeira Belt, Se Brazil: A Segment Of Paleoproterozoic Granulitic Crust Thust During The Panafrican Orogeny. Gondwana Research, v. 1, n.3:373–382

Heilbron M, Duarte BP, Valeriano CM, Simonetti A, Machado N, Nogueira, JR (2010) Evolution of reworked Paleoproterozoic basement rocks within the Ribeira belt (Neoproterozoic), SE-Brazil, based on U Pb geochronology: Implications for paleogeographic reconstructions of the São Francisco-Congo paleocontinent. Precambrian Research, v. 178:136–148

Heilbron M, Machado N (2003) Timing of terrane accretion in theNeoproterozoic–Eopaleozoic Ribeira Belt (SE Brazil). Precambrian Research 125:87–112

Heilbron M, Mohriak W, Valeriano CM, Milani E, Almeida JCH, Tupinambá M (2000) From collision to extension: the roots of the South-eastern continental margin of Brazil. In: Talwani, Mohriak (Eds.), Atlantic Rifts and Continental Margin. AGU Geophysical Monograph Series, vol. 115, 354 pp

Heilbron M, Pedrosa-Soares AC, Campos Neto M, Silva LC, Trouw RAJ, Janasi VC (2004) A Província Mantiqueira. In: Mantesso-Neto V, Bartorelli A, Carneiro CDR, Brito Neves BB (Eds.), O Desvendar de um Continente: A Moderna Geologia da América do Sul e o Legado da Obra de Fernando Flávio Marques de Almeida, XIII

Heilbron M, Simoes LSA, Alves RP Chrispim SJ (1982). Geologia do Cabo dos Buzios. Anais da Academia Brasileira de Ciencias,Rio de Janeiro, 54: 553–562

Heilbron M, Tupinambá M, Valeriano CM, Armstrong R, Eirado Silva LG, Melo RS, Simonetti A, Machado N (2013) The Serra da Bolívia complex: the record of a new neoproterozoic arc-related unit at ribeira belt. Precambrian Research, v. 50:1–35

Heilbron M, Valeriano CM, Tassinari CCG, Almeida JCH, Tupinamba M, Siga, O,Trouw R (2008) Correlation of Neoproterozoic terranes between the Ribeira Belt, SE Brazil and its African counterpart: comparative tectonic evolution andopen questions. Geological Society of London, Special Publication 294

Heilbron M, Valladares C, Valeriano C, Machado N (1995) A Orogenese Brasiliana No Segmento Central da Faixa Ribeira, Brasil. The Advance-Progress (Vidalia), v. 25, n.4: 32–50

Howell DG (1989) Tectonic of suspect terranes: mountain building and continental growth. Chapman & Hall, London.

Janasi VA (2002) Elemental and Sr-Nd isotope geochemistry of two Neoproterozoic mangerite suites in SE Brazil: implications for the origin of the mangerite-charnockite-granite series. Precambrian Research, v. 119: 301–327

Janasi VA, Leite RJ, Van Schmus WR (2001) U–Pb chronostratigraphy of the granitic magmatism in the Agudos Grandes Batholith (west of São Paulo, Brazil)—implications for the evolution of the Ribeira Belt. Journal of South American Earth Sciences, 14: 363–376

Kuhn, A, Stuwe, K, Trouw, RAJ (2004) Metamorphic evolution of the Ribeira Belt: evidence from outcrops in the Rio de Janeiro area, Brazil. Journal of Petrol. 43, 2303–2323

Leonardos, O.H. (1940) Geologia da bacia do Tocantins. B. Div. Fom. Prod. Min. Dep. Nac. Prod. Min., Rio de Janeiro, 41, 103–200

Lobato M, Heilbron M, Torós B, Ragatky D, Dantas, E (2015) Provenance of the Neoproterozoic High-grade Metasedimentary Rocks of the Arc-related Oriental Terrane of the Ribeira belt: Implications for Gondwana Amalgamation. Journal of South American Earth Sciences. 63: 260–278

Machado, N. & Gauthier, G. 1996. Determination of 207Pb/206Pb ages on zircon and monazite by laserablation ICPMS and application to a study of sedimentary provenance and metamorphism in southeastern Brazil. Geochimica et Cosmochimica Acta, 60,5063–5073.

Machado, N., Schrank, A., Noce, C. M. & Gauthier, G. 1996a. Ages of detrital zircon from Archean–Paleoproterozoic sequences: implications for greenstone belt setting and evolution of a Transamazonian foreland basin in Quadrilátero Ferrífero, southeast Brazil. Earth and Planetary Science Letters, 141, 259–276.

Marins GMS (2000) Petrologia dos Anfibolitos do Domínio Juiz de Fora e da Klippe Paraíba do Sul, Setor Central da Faixa Ribeira. Msc Dissertation, Universidade do Estado do Rio de Janeiro

Noce C M, Pedrosa-Soares AC, Silva LC, Alkmim FF (2007) O embasamento arqueano e paleoproterozóico do Orógeno Araçuaí. Geonomos, 15:17–23

Paciullo FVP (1997) A Sequência Deposicional Andrelândia, Instituto de Geociências, , Universidade Federal do Rio de Janeiro, Unpubl Doctoral Thesis

Paciullo FVP, Ribeiro A, Andreis RR., Trouw, RAJ (2000) The Andrelândia Basin, a Neoproterozoic intra-plate continental margin, southern Brasília Belt. Revista Brasileira de Geociências, 30: 200–202

Pedrosa-Soares AC, Alkmim FF, Tack L, Noce CM, Babinski M, Silva LC, Martins-Neto MA (2008) Similarities and differences between the Brazilian and African counterparts of the Neoproterozoic Araçuaí-West-Congo Orogen. Journal of the Geological Society of London, 294:.

Pedrosa-Soares AC, Vidal P, Leonardos OH, Brito-Neves BB (1998) Neoproterozoic Oceanic Remnants in Eastern Brazil: Further

Evidence and Refutation of an Exclusively Ensialic Evolution for the Araçuaí-West Congo Belt. Geology, 26:519–522
Pedrosa-Soares AC, Wiedemann-Leonardos CM (2000) Evolution of the Araçuaí Belt and its connection to the Ribeira Belt, Eastern Brazil. In: Cordani U, Milani E, Thomaz-Filho A, Campos, DA (Eds.), Tectonic Evolution of South America. Sociedade Brasileira de Geologia, São Paulo
Peixoto CA, Heilbron M (2010) Geologia da Klippe Italva na Regiâo entre Cantagalo e Itaocara, Nordeste do Estado do Rio de Janeiro. Geociências (UNESP. Impresso), 3:277–289
Ragatky D, Heilbron M, Tupinambá M, Sanros MG (1999) Sm/Nd data and provenance of metasediments of the central segment of Ribeira belt, Brazil Boletin Geologico y Minero, argentina, v. XXXIV, 437–440
Ribeiro A, Ávila CA, Teixeira W, Dussin IA, Nascimento D (2013) U-Pb LA-ICP-MS detrital zircon ages of the São João del Rei and Carandaí basins: New evidence of intermittent Proterozoic rifting in the São Francisco paleocontinent. Gondwana Research, v. 24:713–726
Ribeiro A, Trouw RAJ, Andreis RR, Paciullo FVP, Valença JG (1995) Evolução das bacias proterozóicas e o termo-tectonismo brasiliano na margem sul do cráton do São Francisco. Revista Brasileira de Geociências 25: 235–248
Ribeiro, A. & Heilbron, M., 1982. Estratigrafia e metamorfismo dos Grupos Carrancas e Andrelândia, sul de Minas Gerais. In : CONGR. BRAS. GEOL., 32, Salvador, 1982. *Anais...*, Salvador, SBG, vol.1, p. 177–186.
Ribeiro, A.; Paciullo, F. V. P.; Andreis, R. R.; Trouw, R. A. J. E Heilbron, M.; 1990. Evolução ; policíclica proterozóica no sul do Cráton do São Francisco : análise da região de São João del Rei e Andrelândia, MG. In: CONGR. BRAS. GEOL, 36. , Natal, 1990. *Anais...*, SBG, **6:** 2605–2614.
Santos, PS (2011) Geocronologia, área fonte e ambiente tectônico da Unidade Santo Antônio - Megassequencia Andrelândia, MG. 2011. Msc dissertation, Unversidade Federal does Rio de Janeiro
Schmitt RS, Trouw RAJ, Van Schmus WR, Pimentel MM (2004) Late amalgamation in the central part of Western Gondwana: new geochronological data and the characterization of a Cambrian collision orogeny in the Ribeira belt (SEBrazil). Precambrian Research 133: 29–61
Silva LC, Armstrong R, Noce, CM, Carneiro M,Pimentel M, Pedrosa-Soares, AC (2002) Leite, C.; Vieira, V.S.; Silva, M.; Paes, V.; Cardoso-Filho,J. Reavaliação da evolução geológica em terrenos pré-cambrianos brasileiros com base em novos dados U-Pb SHRIMP - Parte II: Orógeno Araçuaí, Cinturão Móvel Mineiro e Cráton São Francisco Meridional. Revista Brasileira de Geociências, 32: 513–528
Silva LC, Da Mcnaughton NJ, Armstrong R, Hartmann LA, Fletcher I (2005) The Neoproterozoic Mantiqueira Province and its African connections: a zircon-based U–Pb geochronological subdivision for the Brasiliano/Pan-African systems of orogens. Precambrian Research 166: 203–240
Trouw RAJ (1992) Evolução tectônica ao sul do Craton do São Francisco, baseado em análise metamórfica. In: 37° Congresso Brasileiro de Geologia, 1992, São Paulo. Boletim de resumos expandidos, v. 1: 327–329
Trouw RAJ, Heilbron M, Ribeiro A, Paciullo FVP, Valeriano CM, Almeida JCH, Tupinambá M, Andreis RR (2000) The central segment of the Ribeira belt. In: Cordani, U.G., Milani, E.J., Thomaz Filho, A., Campos, D.A. (Eds.), Tectonic Evolution of South America, 31th International Geological Congress, Rio de Janeiro, Brazil, pp. 287–310
Trouw RAJ, Pankhurst RJ (1993) Idades Radiometricas ao sul do Cráton de São Francisco: região da Folha Barbacena. In: 2 Simpósio do Cráton do São Francisco, 1993, Salvador. Anais do segundo Simpósio do Craton de São Francisco, p 260–262
Trouw RAJ, Peternel R, Ribeiro A, Heilbron M, Vinagre R, Duffles P, Trouw CC, Fontainha M, Kussama HH (2013). A New Interpretation for the Interference Zone between the southern Brasília belt and the central Ribeira belt, SE Brazil. Journal of South American Earth Sciences 48: 43–57
Tupinambá M, Heilbron M, Valeriano CM, Porto Jr R, Dios FRB, Machado N, Silva LGE, Almeida JCH (2012) Juvenile contribution of the Neoproterozoic Rio Negro Magmatic Arc (Ribeira Belt, Brazil): Implications for Western Gondwana amalgamation. Gondwana Research, 21: 422–438
Tupinambá M, Teixeira W, Heilbron M (2000) Neoproterozoic Western Gondwana assembly and subduction-related plutonism: the role of the Rio Negro Complex in the Ribeira Belt, South-eastern Brazil. Revista Brasileira de Geociencias, 30: 7–11
Valeriano CM (1995) Evolução Tectônica da Extremidade Meridional da Faixa Brasília, Região da Represa de Furnas, Sudoeste de Minas Gerais. Revista Brasileira de Geociências, 23(3): 335–336.
Valeriano CM, Machado N, Simonetti A,Valladares CS, Seer HJ, Simoes LS (2004) U–Pb Geochronology of the Southern Brası´lia Belt (SE Brazil): sedimentary provenance, Neoproterozoic orogeny and assembly of Western Gondwana. Precambrian Research, 130:27–55
Valeriano CM, Simões LSA, Teixeira W, Heilbron M (2000). Southern Brasilia belt (SE Brazil): tectonic discontinuities, K-Ar data and evolution during the Neoproterozoic Brasiliano orogeny.. Revista Brasileira de Geociências, 30 (1): 295–299
Valeriano CM, Pimentel MM, Heilbron M, Almeida JCH, Trouw RAJ (2008) Tectonic evolution of the Brasilia Belt, Central Brazil, and early assembly of Gondwana. Geological Society Special Publication, v. 294: 197–210
Valeriano CM, Tupinambá M, Simonetti A, Heilbron M, Almeida JCH, Eirado LG (2011) U-Pb LA-MC-ICPMS geochronology of Cambro-Ordovician post-collisional granites of the Ribeira belt, southeast Brazil: Terminal Brasiliano magmatism in central Gondwana supercontinent. Journal of South American Earth Sciences, 32:416–428
Valladares CS, Machado N, Heilbron M, Gauthier G. 2004. Ages of detrital zircon from siliciclastic successions of the Brasília Belt, southern border of the São Francisco craton, Brazil: implications for the evolution of Proterozoic basins. Gondwana Research, 7: 913–921.
Valladares CS, Machado N, Heilbron M,Duarte BP, Gauthier, G (2008) Sedimentary Provenance in the Central Ribeira belt based on Laser-Ablation ICPMS 207Pb/206Pb Zircon Ages. Gondwana Research, v. 13, p. 516/GR-00196–526
Valladares CS, Souza, SF, Ragatky D (2002) The Quirino Complex: a Transamazonian Magmatic Arc of the Central Segment of the Brasiliano/Pan-African Ribeira Belt, SE Brazil. Revista Universidade Rural. Série Ciências Exatas e da Terra, Rio de Janeiro, 21(1): 49–61
Valladares CS, Duarte BP, Heilbron M, Ragatky D (2000) Tectono-magmatic evolution of the western terrane and the Paraíba do Sul klippe of the Neoproterozoic Ribeira orogenic belt, southeastern Brazil. Revista Brasileira de Geociências, Brasil, 30 (1): 1–6
Viana SM (2003) Petrografia e Geoquímica dos ortognaisses do Complexo Região dos Lagos, Araruama-Cabo Frio (RJ). Master thesis, Universidade do Estado do Rio de Janeiro

Viana SM, Valladares CS, Tupinambá M, Simonetti A (2008) Orthogneisses of the Quirino Complex, Central Ribeira Belt, SE Brazil: U-Pb LA-ICPMS Zircon Ages and Lithogeochemical Studies. In: VI South American Symposium on Isotope Geology, 2008, San Carlos de Bariloche. In: Linares E, Cabareli NG, Do Campo MD, Ducós E, Panarello HO (Compilers) VI South American Symposium on Isotope Geology, Proceddings in CD-ROM. Buenos Aires: Ingeis, 2008. v. 1: 1–2.

Wiedemann CM, Medeiros SR, De Ludka IP, Mendes JC, Moura JC (2002) Architecture of late orogenic plutons in the Arac,uaı´-Ribeira fold belt, southeast Brazil. Gondwana Research, 5: 381–400

Zimbres E, Kawashita K, Van Schmus, WR (1990) Evidencias de um núcleo Transamazonico na Região de Cabo Frio, RJ e sua correlação com o cráton de Angola, Africa. Congresso Brasileiro de Geologia, 36, Sociedade Brasileira de Geologia. Natal, Brazil, 2735–2743.

Part V
Evolutionary Synthesis

The Paleomagnetic Record of the São Francisco-Congo Craton

16

Manoel S. D'Agrella-Filho and Umberto G. Cordani

Abstract

This chapter, based on paleomagnetic and geologic-geochronological evidence, discusses the position of the São Francisco craton and other South American and African cratonic blocks within paleo-continents, since the formation of Columbia supercontinent in the Paleoproterozoic up to the fragmentation of Pangea in the Mesozoic. In Paleoproterozoic times, between ca. 2.0 and 1.8 Ga, two large independent landmasses were formed. The first one involved several cratonic blocks that were leading to the formation of Laurentia. Later, Laurentia, proto-Amazonia, West Africa and Baltica amalgamated to form the nucleus of the supercontinent Columbia at about 1.78 Ga. The second landmass encompassed the São Francisco-Congo, Kalahari, Rio de la Plata and Borborema-Trans-Sahara, forming the Central African block. For the São Francisco-Congo and Kalahari cratons, two robust Paleoproterozoic poles are available. One is from the Jequié charnockites of Bahia (São Francisco Craton), and the other from the Limpopo high-grade metamorphics in South Africa (Kalahari Craton). They support the possible link between these two cratonic blocks at ca. 2.0 Ga. Columbia may have remained united until 1.25 Ga, when Baltica and Amazonia/West Africa broke apart. Their paleomagnetic record seems to indicate that both executed clockwise rotations, until they collided with Laurentia along the Grenville belt at ca. 1.0 Ga., culminating with the formation of Rodinia. For the Central African block, however, there are no reliable paleomagnetic poles available between 1.78 and 1.27 MA. Nevertheless, during this time interval, the geological-geochronological evidence indicates that no continental collisional episodes affected the São Francisco-Congo craton, where important intra-plate tectonic episodes occurred. Most probably, this large continental block drifted alone since the end of the Paleoproterozoic and did not take part of Columbia or Rodinia. At the end of the Mesoproterozoic, ca. 1100 MA, the robust Umkondo pole of the Kalahari craton, as part of the Central African block, and the equally robust Keweenawan pole of Laurentia at the center of Rodinia, indicated that these landmasses were very far apart. At that time a large oceanic realm, the Goiás-Pharusian Ocean, was indeed separating Amazonia-West Africa from the Central African block. This ocean closed by a continued subduction process that started at ca. 900 MA and ended in a collisional belt with Himalayan-type mountains at ca. 615 MA, as part of the few continental collisions which formed Gondwana. However, the age of the final convergence is still a matter of debate,

M.S. D'Agrella-Filho (✉)
Instituto de Astronomia, Geofísica e Ciências Atmosféricas, Universidade de São Paulo, Rua do Matão 1226, São Paulo, SP 05.508-090, Brazil
e-mail: dagrella@iag.usp.br

U.G. Cordani
Instituto de Geociências, Universidade de São Paulo, Rua do Lago 562, São Paulo, SP 05.508-080, Brazil

M. Heilbron et al. (eds.), *São Francisco Craton, Eastern Brazil*, Regional Geology Reviews, DOI 10.1007/978-3-319-01715-0_16

because paleomagnetic measurements for the Araras Group, which occurs within the Paraguay belt at the eastern margin of the Amazonian craton, would indicate that a large ocean was still in existence between it and São Francisco craton close to the Ediacaran/Cambrian boundary. Consensus about this matter awaits for further paleomagnetic data. Gondwana collided with Laurasia during the late Paleozoic, at about 300 Ma, originating Pangea, which not much later started splitting apart, near the Permian/Triassic boundary. As part of this present-time plate tectonic regime, the São Francisco Craton (in South America) started separation from the Congo craton (in Africa) in Jurassic times, giving rise of the present-day oceanic lithosphere of the Atlantic Ocean.

Keywords
São Francisco Craton • Paleomagnetism • Supercontinents

16.1 Introduction

The physical link of the São Francisco craton in South America with the Congo craton in Africa prior to the opening of the South Atlantic ocean, is well established since the works by Martin (1961), Hurley et al. (1967), Almeida and Black (1968), Smith and Hallam (1970) among many others. The main purpose of this chapter is to review the available evidence from paleomagnetic measurements of the wandering of the São Francisco-Congo craton since its amalgamation in the Paleoproterozoic until its breakup during the development of the South Atlantic Ocean in the Lower Cretaceous. In this way, we will try to interpret its possible paleogeographic positions in relation to other companion cratonic masses in a few windows of the geologic time. Our time-slices (Figs. 16.1, 16.2, 16.3, 16.4 and 16.5) were chosen in order to correspond roughly to the reconstructions of supercontinents, which are envisaged to have been in existence since about 2000 Ma, namely, Columbia (Nuna), Rodinia, Gondwana and Pangea.

The paleogeographic reconstructions proposed here are largely based on geological correlations. However, the relative positions of the continental masses for each of the selected time-slices are determined by the available paleomagnetic data.

In paleomagnetism, the model used for paleogeographic reconstructions is the Geocentric Axial Dipole (GAD) Field. The GAD establishes that the mean geomagnetic field over some thousands of years (>10,000 years) averages out its secular variation and it may be represented by the field of a dipole at the center of the Earth, and disposed along the Earth's rotation axis.

If the Earth's geomagnetic field, at least since Paleoproterozoic times, behaved as the GAD model, and if the rocks collected at several sampled sites of a geological unit fully registered the geomagnetic secular variation, the mean inclination (Im) of the magnetic field preserved in the rocks can be obtained and the paleolatitude (λ) of the investigated continental block may be calculated by the expression: $\mathrm{tg}(\lambda) = 2\,\mathrm{tg}(\mathrm{Im})$. Moreover, the mean declination (Dm) of the magnetic field, also preserved in the rocks, will give the rotation of the same continental block, i.e., its paleomeridian.

Due to symmetry of the GAD Field, paleolongitude cannot be constrained. Therefore, the position of a continental block based on a single paleomagnetic pole could occupy any position along the same geographic parallel. In addition, another ambiguity related with paleomagnetic poles is the uncertainty on polarity (south or north pole?), which implies that two positions of the continental block are possible, in the northern hemisphere or in the southern hemisphere, after rotation of 180° (see Buchan et al. 2000). Finally, there is always some uncertainty about the age of the primary magnetization. These are the main weaknesses of reconstructions based on paleomagnetic measurements only, and therefore paleomagnetism could only provide support to paleogeographic models based on geological evidence, such as lithostratigraphic, tectonic or geochronological correlations. The paleomagnetic poles we used for the scenarios prior to the agglutination of Columbia up to the formation of Pangea in the Phanerozoic are presented on Table 16.1.

16.2 The São Francisco Craton Paleomagnetic Data Set

The paleomagnetic dataset available for the São Francisco craton (SFC) is very limited and the reconstructions made in this work were only possible by making use of the assumption that the SFC was united with the Congo Craton at least since Paleoproterozoic times.

For the SFC we have only two paleomagnetic poles from presumed Paleoproterozoic rocks: (*i*) the Jequié pole, obtained for the high-grade metamorphic rocks from the Jequié Complex, whose magnetization age was interpreted to be of ca. 2.02 Ga (D'Agrella-Filho et al. 2011), and (*ii*) the Uauá pole, obtained for dykes from the Uauá mafic dyke swarm, situated

Table 16.1 Selected paleomagnetic poles used in reconstruction of Figs. 16.1, 16.2, 16.3 and 16.4

Continent/formation	Plat (°N)	Plong (°E)	A_{95} (°) dp/dm	Age (Ma)	Ref.
Laurentia					
1. Minto dykes (Superior craton)	30	183	13	1998 ± 2	1
2. Mackenzie dykes	04	190	05	1267 ± 2	2
3. Upper Bylot.	08	205	04	1204 ± 22	2
4. Abitibi dykes	43	208	14	1141 ± 1	2
5. Logan sills	49	220	04	1109 + 4 − 2	3
6. Halliburton intrusives	−33	143	06	980 ± 10	4
7. Tsezotene sills and dykes	2	138	5	779 ± 2	3
8. Wyoming dykes	13	131	4	785 ± 8; 782 ± 8	3
9. Mean of 7 and 8 (780 Ma)	08	135		785	5
10. Kwagunt Formation	18	166	7	742 ± 6	3
11. Long Range dykes	19	355	15/21	615	6
12. Cloud Mountain Basalt	−5	352	2/4	615	6
13. Garnish red sandstones	−5	328		607	7
14. Famine Back Cove basalt	1	342		607	7
15. Mean of 10, 11, 12 and 13 (610 Ma)	3	344		610	5
16. Skinner Cove Fm.	−16	338	7/11	550 ± 3	8
17. Tapeats sandstones	−5	338	3	508	4
Baltica					
18. Pudozhgora Intrusion	64	149	11/14	1984 ± 8	9
19. Hakefjorden intrusives	−05	069	04	916 ± 11	10
20. Mean 550 Ma	−42	116		550	11
21. Tornetrask Fm.	56	116	12/15	535	4
Amazonian craton					
22. Oyapok granitoids and volcanics	−28	346	14	2020 ± 4	12
23. Avanavero sills and dykes	−48	28	10	1789	13
24. Nova Floresta Fm.	25	165	06	1198 ± 3; 1201 ± 3	2
25. Fortuna Fm.	60	156	10	1149 ± 7	14
Kalahari					
26. Vredfort mean	23	042	10/16	2023 ± 4	15
27. Limpopo metamorphics	26	22	08	1950 − 1980?	16
28. Premier Mine Kinberlite	51	038	07	1200	17
29. Premier combined	41	055	16	1150	17
30. Unkondo combined	64	−36	04	1105 ± 2	17
31. Namaqua metamorphics	08	330	11	1000 ± 20	17
Congo/São Francisco					
32. Jequié metamorphic complex	−01	343	10	~2020	18
33. Post-Kibaran Intrusives	17	293	07	1250	17
34. Bahia coastal A-normal	−14	288	07	923 ± 3	10
35. Gagwe Lavas	29	103	13	790	17
36. Mbozi Mafic Complex	46	325	9	750	17

(continued)

Table 16.1 (continued)

Continent/formation	Plat (°N)	Plong (°E)	A_{95} (°) dp/dm	Age (Ma)	Ref.
Rio de la Plata					
37. Florida dykes	−78	162	10	1790 ± 5	19
Arabia-Nubia					
38. Sinyai metadolerites	−29	319	5	547 ± 4	17
Brasiliano belt					
39. Itabaiana dykes	35	315	7	525 ± 5	20

Plat, pole latitude; Plong, pole longitude; A95, dp, dm—Fisher's statistic parameters. References: 1. Mitchell et al. (2014); 2. Tohver et al. (2002); 3. Li et al. (2008); 4. Meert and Torsvik (2003); 5. This work; 6. Lubnina et al. (2014); 7. Piper (2010); 8. McCausland et al. (2011); 9. Lubnina et al. (2016); 10. Evans et al. (2016); 11. Klein et al. (2015); 12. Nomade et al. (2003); 13. Bispo-Santos et al (2014b); 14. D'Agrella-Filho et al. (2008); 15. Salminen et al. (2009); 16. Morgan (1985); 17. Tohver et al. (2006); 18. D'Agrella-Filho et al. (2011); 19. Teixeira et al. (2013); 20. Trindade et al. (2006)

in northern Bahia state (D'Agrella-Filho and Pacca 1998). These authors attributed an age between 1.90 and 2.00 Ga for the magnetization age recorded by the rocks, based on a few Rb-Sr and K-Ar ages obtained for the dykes. However, recent U-Pb dating on baddeleyite and zircon grains of some noritic and tholeiitic dykes from the Uauá region yielded precise Archean ages of about 2720 Ma (Oliveira et al. 2013).

A group of paleomagnetic poles were obtained for unmetamorphosed mafic dykes from Ilhéus, Olivença and Itaju do Colônia regions in Bahia State (D'Agrella-Filho et al. 1990), and also for the coeval unmetamorphosed Salvador mafic dykes (D'Agrella-Filho et al. 2004). Precise $^{40}Ar/^{39}Ar$ dating on biotite and plagioclase of two dykes from Olivença and Ilhéus yielded ages of 1.08 Ma and 1.01 Ga, respectively (Renne et al.1990), and a biotite from a granulite at the contact with a 30 m thick Salvador dyke yielded an age of 1.02 Ga (D'Agrella-Filho et al. 2004). These paleomagnetic poles were taken as defining an apparent polar wander (APW) path for the SFC between 1.08 and 1.01 Ga (D'Agrella-Filho et al. 2004). However, recent U-Pb dating on baddeleyite and zircon of several dykes from the same swarms yielded precise ages systematically around 924 Ma (Evans et al. 2016), putting serious doubts on the previous Ar-Ar data.

Rather less reliable Neoproterozoic paleomagnetic poles (with reliability factor $Q \leq 3$) were obtained for some dykes near Itabuna (Bahia state), with an age about 780 Ma, and for a few other dykes near Lavras-Pará de Minas (Minas Gerais state), whose age is yet poorly constrained (Tohver et al. 2006). Finally, two magnetic components were disclosed for the Neoproterozoic carbonates of the Bambuí Group in Minas Gerais (D'Agrella-Filho et al. 2000), and the Salitre Formation in Bahia (Trindade et al. 2004). These components were interpreted as due to remagnetization processes occurred at ca. 520 Ma, under the thermal influence of the the Brasiliano (Pan-African) orogeny.

16.3 São Francisco Craton in Pre-Columbia Times

Prior to the formation of the Columbia/Nuna supercontinent at about 1.8 Ga, any reconstruction must be considered very speculative because the cratonic masses that may have contributed to its formation, such as Laurentia, Baltica, Central Amazonia, West Africa, Congo/São Francisco, and Kalahari were yet in the process of assembling.

The better known amalgamation process is that of Laurentia. It was only assembled at ca. 1.85–1.80 Ga after several continental collisions (Mitchell et al. 2014; ST-Onge et al. 2009). According to Mitchell et al. (2014), the Archean Slave and Rae blocks collided at 1.97 Ga, Slave/Rae and Hearne blocks collided at 1.92 Ga, and this latter block collided with the Superior craton at 1.85 Ga. These authors, based on paleomagnetic poles from the Slave and Superior cratons for the interval 2.2 and 2.0 Ga, demonstrated that the mentioned blocks were separated by a very large ocean (Manikewan ocean) at 2.0 Ga. Using Mitchell et al. (2014)'s reconstruction, we attempted to draw a possible paleogeographic reconstruction at 2.0 Ga (Fig. 16.1). In this figure, the paleogeographic position of the Superior craton is established using the 1998 Ma pole determined for the Minto dykes (number 1 in Table 16.1) from this Craton. The relative positions of the Superior (Su) and Slave (S) cratons are the same proposed by Mitchell et al. (2014), and the Hae and Hearne blocks were tentatively plotted between both cratonic blocks. At that time, Baltica was not formed as well. Karelia and Kola Archean areas, originally separated from the Volgo-Uralia and Sarmatia cratonic blocks (Bogdanova et al. 2013), collided at about 1.9 Ga (Daly et al. 2006). In Fig. 16.1, Karelia position was constrained by the Pudozhgora Intrusion pole (number 18 in Table 16.1), and Kola craton is positioned close to Karelia. According to Daly et al. (2006), these two cratons were linked, forming the Archean

Kenorland supercontinent (Pesonen et al. 2003). After 2.5 Ga, with the break-up of this supercontinent, a Wilson cycle developed with separation of Karelia and Kola cratons. This was followed by the formation of an ocean and its later closure along the Lapland-Kola orogen which ended at ca. 1.9 Ga.

According to Bogdanova et al. (2013), the collision between Sarmatia and Volgo-Uralia occurred between 2.1 and 2.0 Ga, producing the Volgo-Sarmatia block. Also, around 2.0 Ga, Central Amazonia and West Africa were probably already formed and together, following the collision of Archean blocks along the 2.25–2.05 Ga Maroni-Itacaiunas and Birimian orogenic belts (Cordani and Teixeira 2007). In

Fig. 16.1 Reconstruction at 2.0 Ga partially based on paleomagnetic data (modified from D'Agrella-Filho et al. 2016). The Superior craton (Su) was constrained using the Minto dykes pole (number 1 in Table 16.1). The relative positions of the Superior (Su) and Slave (S) cratons are the same proposed by Mitchell et al. (2014). Rae (R) and Hearne (H) cratonic fragments were tentatively positioned between Superior and Slave. The Karelia (KAR) craton was constrained using the Pudozhgora Intrusion pole (number 18 in Table 16.1). The Kola (K) craton is positioned close, but apart from Karelia, since it collided with this craton only at ca. 1.9 Ga (Daly et al. 2006). Proto-Amazonia (PAm) was constrained using the Oyapok Granitoids pole (pole 22 in Table 16.1). It is suggested that a large land mass composed by Proto-Amazonia (PAm), West Africa (WA), Volgo-Uralia (V-U) and Sarmatia (SA) was already in existence at that time. The Central African block, as defined by Cordani et al. (2013a) and forming a coherent mass composed by the Congo/São Francisco (CSF), Kalahari (KAL), Rio de la Plata (RP) and Borborema/Trans-Sahara (BTS), was constrained using the Vredfort mean pole (pole 26 in Table 16.1). The link between Proto-Amazonia and West Africa was made according to Bispo-Santos et al. (2014b). Dashed lines indicate later borders of Laurentia, Baltica and Amazonian craton. See text for details

Fig. 16.1, we envisage that a large landmass was composed by Volgo-Sarmatia plus Central Amazonia and West Africa, amalgamated along Paleoproterozoic orogenic belts. This model follows the so-called SAMBA connection of Johansson (2009), who proposed that, at ~1.83 Ga, the nucleus of Columbia would include proto-Amazonia, West Africa and Baltica. The position of this landmass (Volgo-Sarmatia/Central Amazonia/West Africa) is constrained by the Oyapok granitoids pole (number 22 in Table 16.1) with an age of 2020 ± 4 Ma (see Nomade et al. 2003; Théveniaut et al. 2006 and D'Agrella-Filho et al. 2011). As shown on Fig. 16.1, subduction zones may have been active at the northern and western margins of Volgo-Sarmatia and Central Amazonia, respectively.

Finally, a large continental block in Fig. 16.1 was probably formed at about 2.0 Ga. It was called Central African Block by Cordani et al. (2013a), when describing the process of amalgamation of Gondwana. Only a few poles, whose ages are not well-constrained, are actually available (D'Agrella-Filho et al. 2011), and can be used to support a close location of the São-Francisco-Congo and Kalahari paleocontinents at the core of the Central African Block. The other units of the block, the Borborema-Transahara, Rio de La Plata and Paranapanema, are considered in this work to have been part of the same unit because they include important areas made up of terrains in which Paleoproterozoic ages predominate.

The best paleomagnetic pole for the São Francisco craton, with Paleoproterozoic age, was obtained for the Jequié charnockites (number 32 in Table 16.1), whose granulite metamorphism was imposed between 2.1 and 2.0 Ga (Silva et al. 2002). However, a slightly younger age of magnetization is suggested for the Jequié pole based on $^{40}Ar/^{39}Ar$ ages (D'Agrella-Filho et al. 2011). It plots close to the Limpopo metamorphics "A" pole (Morgan 1985, pole 27 in Table 16.1) and to the well-dated 2023 ± 4 Ma mean pole (number 26 in Table 16.1) determined for rocks of the Vredefort impact structure (Salminen et al. 2009)—both from the Kalahari craton—in a pre-drift Gondwana configuration (D'Agrella-Filho et al. 2011). The Limpopo belt is interpreted to have resulted from the collision of the Kaapvaal and Zimbabwe cratons at Paleoproterozoic times where high-grade metamorphism was imposed at ca. 2.03–2.01 Ga (e.g., Buick et al. 2006). These facts imply that the São Francisco-Congo and Kalahari cratons could have been united in the Paleoproterozoic.

The Central African block, which contains the São Francisco-Congo and the adjacent cratonic units, is constrained in Fig. 16.1 by the Jequié pole (number 32 in Table 16.1), whose age is around 2.02–2.00 Ga. As we will argue repeatedly in this work (Figs. 16.2, 16.3 and 16.4), this continental block will probably keep its individuality until the end of the Neoproterozoic.

16.4 São Francisco-Congo Out of Columbia?

According to Rogers and Santosh (2009), the Columbia (Nuna) supercontinent mostly assembled by about 1.90–1.85 Ga, as suggested by geologic correlations, age constraints and other lines of evidence. Mainly due to the scarcity of key palomagnetic poles, different paleogeographic scenarios for Columbia have been proposed (e.g., Zhao et al. 2002; Pesonen et al. 2012; Pisarevsky et al. 2014, and references therein). Moreover, some authors postulated that it was not fully formed until 1.6 Ga ago (Evans 2013; Pisarevsky et al. 2014; Pehrsson et al. 2016).

In Fig. 16.2 the relative positions of Laurentia, Baltica, proto-Amazonia and West Africa were constrained according to the model by Bispo-Santos et al. (2014a). Starting with the paleogeographic reconstruction for 2.0 Ga shown on Fig. 16.1, an oblique collision occurred between the proto-Amazonia/West-Africa/Volgo-Sarmatia continental block with Fennoscandian terranes along the NW part of Sarmatia between 1.83 and 1.80 Ga (Bogdanova et al. 2013). According to these authors, after that collision, Volgo/Sarmatia (together with Central Amazonia and West Africa in our reconstruction) performed a counterclockwise rotation, which activated older strike-slip faults, which accommodated mafic dyke swarms with ages between 1.79 and 1.75 Ga in the Ukrainian Shield, northwestern Sarmatia. At about the same time (1.79–1.78 Ga) the widespread Avanavero mafic intrusions occurred as dykes and sills in Venezuela, French Guyana and northern Brazil (Reis et al. 2013). After Columbia formation at about 1.78 Ga (Fig. 16.2, see also Bispo-Santos et al. 2014a) minor internal rotations occurred associated with 1.75 Ga mafic dyke intrusions in the Ukrainian Shield (Bogdanova et al. 2013).

Fig. 16.2 Paleogeography of Columbia at 1.78 Ga. Laurentia (LAU), Baltica (BA), Proto-Amazonia (PAm) and West Africa (WA) were plotted as in Bispo-Santos et al. (2014a). Siberia (SI), Mawson continent (MC—South Australia plus East Antarctica), North Australia (NAu), West Australia (WAu), North China (NC) and India (IN) were plotted as in Xu et al. (2014). In this interpretation, the Central African block (RP Rio de la Plata, KAL Kalahari, CSF São Francisco/Congo, and BTS Borborema/Trans-Sahara block), constrained by the 1790 Ma Florida mafic dykes (number 37 pole of Table 16.1), formed a large continental mass, which could be close to Columbia, but probably did not belong to this supercontinent

In our view, after 1.78–1.75 Ga, a great continental mass composed by proto-Amazonia, West Africa, Baltica, and Laurentia may have formed the nucleus of the Columbia supercontinent, which probably remained united until 1.26 Ga ago (see Evans 2013 and references therein for further discussion). Other cratonic blocks could be included in the paleogeography of Columbia (Fig. 16.2, see also Xu et al. 2014). Siberia is positioned at the present northern Artic coast of North America according to Li and Evans (2011). Proto-Australia (Evans and Mitchell 2011), together with North China and India, were located close to North America, according to Xu et al. (2014). In the paleogeography shown on Fig. 16.2, Laurentia, together with all the cratonic blocks described above was rotated 64.6° clockwise around the Euler pole at 16.8°N, 323.6°E (Bispo-Santos et al. 2014a).

The relation of other African and South American cratonic blocks with Columbia is still undefined. As already mentioned, the Proterozoic paleomagnetic database for the São Francisco-Congo craton, especially in the Paleo- and Mesoproterozoic, is very poor. In our Fig. 16.2, the Rio de la Plata and Borborema/Trans-Sahara blocks were united to the São Francisco-Congo/Kalahari continent, forming the Central African block of Cordani et al. (2013a). We tentatively constrained the position of this large landmass at 1.78 Ga using the paleomagnetic pole obtained for the Florida dykes (number 37 in Table 16.1) from the Rio de la Plata craton (Teixeira et al. 2013).

If other parts of South America and Africa, such as Rio de La Plata and Kalahari, were accreted to the Central African block, and also if this large block took part of the Columbia supercontinent is still very uncertain. Recently, Cederberg et al. (2016) published some U-Pb baddeleyite ages for the Pará de Minas mafic dyke swarm from southern São Francisco craton. Their geochronological study revealed three episodes of dyke intrusions. The oldest of them was dated at 1790 Ma. In our Fig. 16.2, a closer position of the São Francisco-Congo craton relative to Siberia and North China (as suggested by Cederberg et al. 2016) would be permitted by the data, due to the longitude indefinition in paleomagnetism. However, if the Central African block was already formed at that time, some cratonic blocks would be superposed with parts of Columbia at 1790 Ma.

16.5 From Columbia to Rodinia

Taking into account the discussion above, we have tentatively proposed that the Central African block, composed by Congo/São Francisco/Kalahari/Borborema/Trans-Sahara/Rio de la Plata (and probably other smaller cratonic blocks hidden below the Phanerozoic Paraná basin, such as the Paranapanema block), was not part of Columbia. As demonstrated by Cruz and Alkmim (this book) and Alkmim et al. (this book) the tectonic evolution of the São Francisco craton between the end of Paleoproterozoic—when Columbia was still assembling-, and during the whole Mesoproterzoic Era, is marked by a series of intra-plate events associated with the formation of rift basins. No late Paleoproterozoic or Mesoproterozoic collisional events have been documented in the SFC yet. In our view, this large cratonic unit behaved as an individual continental block up to the formation of Gondwana in Ediacaran/Cambrian times, not linked to the Columbia supercontinent.

As already mentioned, no reliable paleomagnetic pole is available for this continental block between 1780 and 1270 Ma, so that we cannot trace its drift from the end of the Paleoproterozoic through most of the Mesoproterozoic. Figure 16.3 shows how we tentatively envisage the paleogeographic reconstructions from 1270 Ma to 920 Ma, and the possible path of the Central African landmass.

At around 1270 Ma, the nucleus of Columbia (Laurentia/Baltica/Amazonia/West Africa) was probably still united (Salminen and Pesonen 2007; Evans 2013). Its position on Fig. 16.3a is constrained by the Mackenzie dykes paleomagnetic pole (number 2 in Table 16.1) from Laurentia (Irving et al. 1972). These igneous rocks are considered to be part of a giant radiating 1267 Ma dyke swarm (Hou et al. 2008). The same figure indicates the possible position of the Central African block, constrained by the post-Kibaran paleomagnetic pole (number 33 in Table 16.1) obtained for the Congo craton (Meert et al. 1994) at 1250 Ma. Later, at about 1200 Ma (Fig. 16.3b), Baltica and Amazonia/West Africa broke up and separated from Columbia, performing a clockwise rotation up to their final collision with Laurentia to form Rodinia (see Evans 2013; Johansson 2014).

Tentative paleogeographic reconstructions for 1200, 1150, 1100 and 1000 Ma are shown on Fig. 16.3b–e, based on the available paleomagnetic poles included in Table 16.1. They illustrate possible scenarios since the initial rupture of the nucleus of Columbia up to the final assembly of Rodinia. In our view, the Central African block did not take part of Columbia or Rodinia, as discussed in the next section.

As indicated on Fig. 16.3, a large subduction system was active along the Kalahari margin of the Central African block. Accretion along this system led to the development of the Namaqua-Natal orogenic belt by the end of the Mesoproterozoic. This belt was interpreted by De Beer and Meyer (1984) as an Andean-type orogen characterized by abundant arc-related calc-alkaline magmatism, low-P high-T metamorphism, and a major contribution of mantle derived melts. Two decades later, Eglington (2006) confirmed the accretionary character of the orogen, which is actually made up predominantly of juvenile crust with minor contribution from older continental sources. The ages of the dominant mafic to intermediate igneous units fall between 1.2 and 1.3 Ga, whereas the granitic intrusions define two main magmatic events dated at 1.15 and 1.08–1.03 Ga.

The positions of Kalahari and Laurentia on Fig. 16.3d (1100 Ma) were constrained, respectively, by the very well-dated Unkondo pole (combined pole, number 30 in Table 16.1) and Logan dykes pole (number 5 in Table 16.1) from the Keweenawan magmatic event. The coeval Unkondo and Keweenawan magmatic events led some authors (Hanson et al. 1998) to propose that they would be distinct parts of the same large igneous province (LIP), and that Kalahari would be connected to Laurentia. However, paleomagnetic poles of the same age from both magmatic events yielded paleolatitudes that differ for more than 30°, precluding a direct connection between them (Dalziel et al. 2000; Pisarevsky et al. 2003; Meert and Torsvik 2003). Moreover, the same geomagnetic polarity was obtained for coeval Unkondo and Keweenawan rocks, which eliminates polarity ambiguity (Hanson et al. 2004). So, this fact constrains the orientation of Kalahari relative to Laurentia as shown in Fig. 16.3d. In this figure,

Fig. 16.3 Paleogeographic reconstructions at 1250 Ma (**a**), 1200 Ma (**b**), 1150 Ma (**c**), 1100 Ma (**d**), 1000 Ma (**e**) and 920 Ma (**f**), based on paleomagnetic data (modified from D'Agrella-Filho et al. 2016). Laurentia (LAU) was constrained by poles 3, 4 and 5 (Table 16.1) for 1200 Ma (**b**), 1150 Ma (**c**) and 1100 Ma (**d**), respectively. Amazonia (AM) was constrained by poles 24 and 25 (Table 16.1) for 1200 and 1150 Ma, respectively. The Central African block, which includes the Paranapanema craton (PA) since at least 1250 Ma, was positioned out of Columbia and also out of Rodinia, and was constrained using the following poles (Table 16.1): 33 for 1250 Ma; 28 for 1200 Ma; 29 for 1150 Ma; 30 for 1100 Ma, 31 for 1000 Ma and 34 for 920 Ma. The sequence of diagrams illustrates our interpretation for the fragmentation of the nucleus of Columbia and the formation of Rodinia, as also suggested by other authors (e.g. Evans 2013 and references therein). The nucleus of Columbia, including Laurentia (LAU), Baltica (BA), Amazonia (AM) and West Africa (WA), remained united since 1780 Ma (Fig. 16.2) up to 1260–1250 Ma (Fig. 16.3a). Its position at 1250 Ma was constrained by pole 2 (MacEnzie dykes, Table 16.1). After that time (Fig. 16.3b, e), Baltica and Amazonia-West Africa executed a clockwise rotation and finally collided again with Laurentia in a different position to form Rodinia at ca. 1000 Ma (Li et al. 2008). Pole 6 in Table 16.1 was used to constrain this large continental mass at 1000 Ma (Fig. 16.3e) and pole 19 (Table 16.1) was used to constrain it at 920 Ma (Fig. 16.3f) ▶

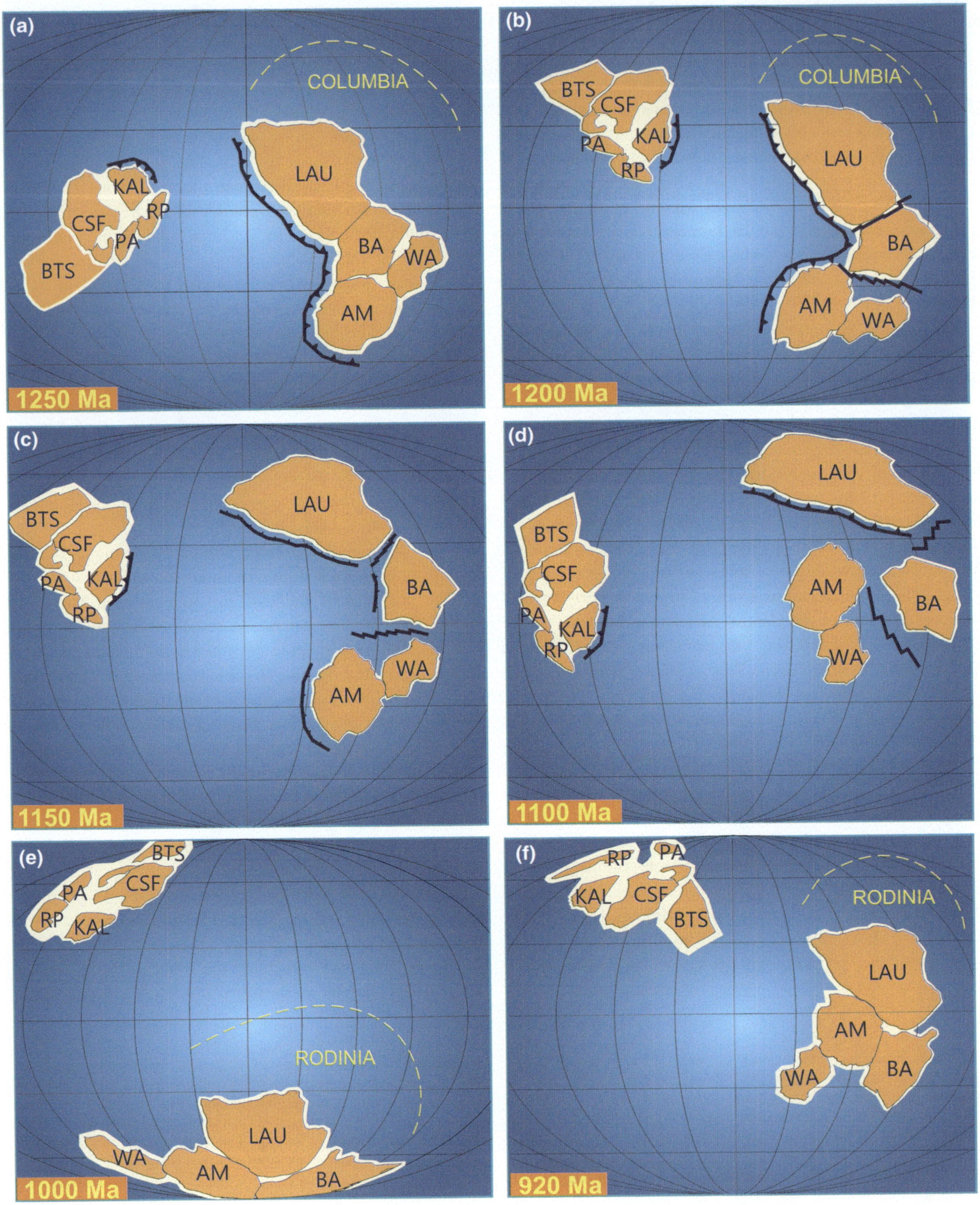
(a)
COLUMBIA
KAL
RP
CSF
PA
BTS
LAU
BA
WA
AM
1250 Ma
(b)
BTS
CSF
KAL
PA
RP
COLUMBIA
LAU
BA
AM
WA
1200 Ma
(c)
BTS
CSF
PA
KAL
RP
LAU
BA
WA
AM
1150 Ma
(d)
LAU
BTS
CSF
PA
KAL
RP
AM
BA
WA
1100 Ma
(e)
BTS
PA
CSF
RP
KAL
RODINIA
LAU
WA
AM
BA
1000 Ma
(f)
RP
PA
KAL
CSF
BTS
RODINIA
LAU
AM
BA
WA
920 Ma

Kalahari is plotted well far from Laurentia, although a closer approximation would also be permitted due to longitude indefinition in paleomagnetism.

At 1000 Ma (Fig. 16.3e), the position of the Central African block was constrained by the ca. 1000 Ma Namaqua pole (number 31 in Table 16.1) from the Kalahari craton (Renne et al. 1990). Laurentia plus Amazonia, West Africa and Baltica, appear in this figure as in the Rodinia reconstruction of Li et al. (2008), fixed by the 980 Ma Halliburton intrusions pole (number 6 in Table 16.1) (Buchan et al. 1983; Berger et al. 1979). Looking at the successive time-slices of Fig. 16.3, we note that, while Laurentia executed only a small (about 30° in latitude) movement to the northern hemisphere between 1250 Ma and 1100 Ma, the Central African block performed a clockwise rotation of almost 180° in the same time interval. In addition, between 1100 and 1000 Ma these continental blocks would proceed to opposite high latitudes.

Finally, at 920 Ma (Fig. 16.3f) the key paleomagnetic pole (number 34 in Table 16.1), obtained for the "normal" polarity mean directions disclosed for the Salvador, Olivença and Ilhéus dykes (Evans et al. 2016), was used to constrain the position of the Central African block. For this time, only Baltica has a reliable paleomagnetic pole (number 19 in Table 16.1), which was used to fix in Fig. 16.3f the Laurentia/Amazonia/West Africa/ Baltica continental mass, which would return to an equatorial position.

16.6 Was São Francisco-Congo Out of Rodinia?

Several attempts to correlate the São Francisco-Congo craton to the Mesoproterozoic Rodinia Supercontinent have been made (see Evans 2013 and references therein). First, D'Agrella-Filho et al. (1990) and Renne et al. (1990), based on the paleomagnetic poles and $^{40}Ar/^{39}Ar$ ages at that time available for the Ilhéus, Olivença and Itaju do Colônia mafic dykes (northern SFC, see Girardi et al. this book) suggested an APW path for this cratonic unit between 1.08 and 1.01 Ga. A comparison of the obtained APW path with those traced for Laurentia and Baltica for the same time interval, led D'Agrella-Filho et al. (1998) and Weil et al. (1998) to propose a Rodinian paleogeography, in which the São Francisco-Congo (together with Kalahari) faced Laurentia along the southern Llano Grenvillian orogenic belt at ca. 1000 Ma. In this reconstruction the Amazonian craton also faced Laurentia along the Sunsás and Grenville belts, in a position close to the São Francisco-Congo craton. A similar reconstruction was proposed by Li et al. (2008), based on the same paleomagnetic poles. An alternative paleomagnetically-based reconstruction for Rodinia was later suggested by Evans (2009), in which the São Francisco-Congo was linked to northern Laurentia.

Based on geological and geochronologial evidence, Kröner and Cordani (2003), Cordani et al. (2003) and Pisarevsky et al. (2003) suggested that the São Francisco-Congo and Kalahari cratons did not take part of Rodinia. The main reason was that a large ocean existed between Amazonia and Congo-São Francisco at about 940 Ma ago, a fact supported by juvenile intraoceanic magmatism recorded all along the present-day central Brazil (Pimentel et al. 1999). D'Agrella-Filho et al. (2004) used the same reasoning to re-interpret the paleomagnetic and the age data for the Salvador mafic dykes. A plateau ^{40}Ar–^{39}Ar age (biotite) of 1021 ± 8 Ma was obtained for a granulite at the contact of one of the Salvador dykes, helping to establish the APW path of the Congo-São Francisco craton between 1.08 and 1.01 Ga. These authors proposed then three different configurations for Rodinia, all of them considering the São Francisco-Congo craton (and Kalahari) located quite far from the core of Rodinia, an interpretation also adopted by Tohver et al. (2006).

However, as already mentioned in Sect. 16.2, new U-Pb age determinations on baddeleyite and zircon from the Salvador, Ilhéus and Olivença dykes yielded systematically younger ages, close to 920 Ma, which were interpreted as the time of the intrusions (Evans et al. 2016), casting doubts on the previous Ar-Ar datings. New paleomagnetic results obtained on the same dykes by these authors replicate the previous results published by D'Agrella-Filho et al. (1990, 2004). From the old and new paleomagnetic data, Evans et al. (2016) calculated new means using site mean directions for both, "normal" and "reverse" polarities: Dec = 86.5°, Inc = −72.3° (N = 33, α_{95} = 4.1°, K = 38), and Dec = 299.9°, Inc. = 62.1°, (N = 13, α_{95} = 5.9°, K = 51), which yielded the following paleomagnetic poles: 288.3°E, 14.3°S (A_{95} = 6.8°) and 282.0°E, 10.2°N, respectively (A_{95} = 8.5°). Unfortunately, all the U-Pb 920 Ma ages are from dykes with "normal" polarity, so that no U-Pb age is presently available for the "reverse" directions that could constrain the age of the related pole described above. Therefore, the pole number 34 in Table 16.1 is the only key paleomagnetic pole dated at 920 Ma for the São Francisco-Congo craton that can be used to constrain the paleogeography of this unit in respect to the Rodinia supercontinent.

For the age interval of 935–910 Ma, Baltica is the only cratonic block for which reliable paleomagntic poles are presently available (Pisarevsky and Bylund 2006). Using these poles and the 920 Ma key paleomagnetic pole for the São Francisco-Congo craton described above, Evans et al. (2016) tested two configurations proposed for Rodinia by Li et al. (2008) and Evans (2009). None of them pass the test.

Other possible configurations of the São Francisco-Congo relative to Baltica/Laurentia are also possible, such as those proposed by Cawood and Pisarevsky (2006). For example, São Francisco-Congo could be linked to the western or southwestern Laurentia (with North America in its present position), or to the present northeastern or southeastern Baltica (see Evans et al. 2016). Following Kröner and Cordani (2003), Cordani et al. (2003), Pisarevsky et al. (2003), D'Agrella-Filho et al. (2004) and Tohver et al. (2006), we suggest that São Francisco-Congo was not part of Rodinia at 920 Ma ago, as shown on Fig. 16.3f.

16.7 São Francisco-Congo in Gondwana and Pangea

The dispersal of Rodinia and the wandering followed by collision of its fragments gave rise to the Gondwana continent, which was only completely amalgamated in Cambrian times (Meert and Van der Voo 1997). However, the time when Amazonia plus West Africa broke-up from Rodinia, as well as the time they finally collided with the Central African block are yet in dispute.

On geological-tectonic grounds, Ganade de Araujo et al. (2014) demonstrated that a large oceanic realm, the Goias-Pharusian ocean, once separated Amazonia-West-Africa from the Central African block. Closure of this ocean involved a continued subduction process from about 900 up to 600 Ma, giving rise to Himalaya-type mountains. The related suture, exhumed at around 615 Ma, is more than 2500 km long and marked by eclogitic rocks formed at depths of about 130 km.

Trindade et al. (2006), interpreting paleomagnetic data from the Araras Group of the Paraguay belt in central Brazil, concluded that a large Ediacaran ocean (named by them as the Clymene ocean) separated Amazonia from the São Francisco-Congo craton. According to these authors closure of the Clymene took place only during Cambrian (550–520 Ma). However, the directions disclosed for the Araras Group, although registering both polarities along a sedimentary profile (Trindade et al. 2003), are close to the present geomagnetic field, which puts some doubt about their primary origin (Pisarevsky et al. 2008). More recently, Tohver et al. (2010) proposed that folding, trusting and remagnetization of carbonates occurring along the oroclinal inflection of the Paraguay belt dated at 528 ± 36 Ma caused the coherent change in declination observed in the remanent magnetization disclosed for the rocks collected in the northern and southern segments of the orocline, reflecting thus the closure of the Clymene ocean at Cambrian times.

These disputing hypotheses gave rise of an interesting debate (see discussions in Cordani et al. 2013b, 2014 and Tohver and Trindade 2014). The main argument of Cordani et al. (2013b), although recognizing the existence of a major seaway like the Clymene in South America, is that it should have been not an ocean, but a large epicontinental sea, formed over the pre-existent continental crust of West Gondwana. Indeed, many authors advocate a final collision between Amazonia plus West Africa against the Central African block at around 650–600 Ma, after the closure of the great Goiás-Pharusian ocean along the Transbrasiliano-Kandi megashear (e.g., Trompette 1994; Cordani et al. 2000, 2013b; Ganade de Araujo et al. 2014; Cordani et al., this book, among many others).

Figure 16.4 shows a possible scenario for the formation of Gondwana considering the hypothesis that at 650–600 Ma the large Goias-Pharusian ocean that once separated São Francisco-Congo- and Amazonia was already closed. At 790 Ma and 750 Ma (Fig. 16.4a, b), Central Africa and Laurentia/Amazonia/West Africa/Baltica (and probably other cratonic blocks that formed Rodinia, see Li et al. 2008) behaved as independent entities. At 610 Ma, given that West Gondwana was already formed, the paleogeographic reconstruction shown on Fig. 16.4c was constrained by the mean of ca. 610 Ma paleomagnetic poles from Laurentia (number 15 in Table 16.1). After 600 Ma, Baltica and Laurentia broke-up from West Gondwana forming the Iapetus Ocean as shown on Fig. 16.4d and Fig. 16.4e, for at 550 Ma and

Fig. 16.4 Paleogeographic reconstructions at 790 Ma (**a**), 750 Ma (**b**), 610 Ma (**c**), 550 Ma (**d**) and 525 Ma (**e**), based on paleomagnetic data. The sequence of diagrams shows the formation of Gondwana, with the closure of the Goiás-Pharusian and Mozambique oceans, and the separation of Laurentia (LAU) and Baltica (BA), with the opening of the Iapetus Ocean. For 790 Ma, the Gagwe lavas pole (number 35 in Table 16.1) from the Congo craton, and the mean pole (number 9 in Table 16.1) calculated for Laurentia (785 Ma) were used for the reconstruction. For 750 Ma, the reconstruction was made using the Mbozi Complex pole (number 36 in Table 16.1) from the Congo craton and Kwagunt Formation pole (number 10 in Table 16.1) from Laurentia. Between 750 Ma and 610 Ma there is a lack of paleomagnetic control for the different cratonic fragments. A mean pole (number 15 in Table 16.1) was used for Laurentia to constrain the block formed by Laurentia/Amazonia/West Africa/Baltica at 610 Ma. Moreover, we also suppose that West Gondwana was practically formed at that time. For 550 Ma, the Sinyai metadolerites pole (number 38 in Table 16.1) from Arabia-Nubia (A-N) block was used to constrain West Gondwana position, the Skinner Cove Formation pole (number 16 in Table 16.1) was used for Laurentia and a mean pole (number 20 in Table 16.1) was used for Baltica. At this age, the Iapetus Ocean is wide open and the Gondwana supercontinent is practically formed, with the amalgamation of India (IN), Madagascar (M), Antarctica (AN) and Australia (AU), and the Central African block occupying its nucleus. Finally, for 525 Ma, the Itabaiana dykes pole (number 39 in Table 16.1) was used to constrain West Gondwana, Tapeats Formation pole (number 17 in Table 16.1) for Laurentia and Tornetrask Formation pole (number 21 in Table 16.1) for Baltica

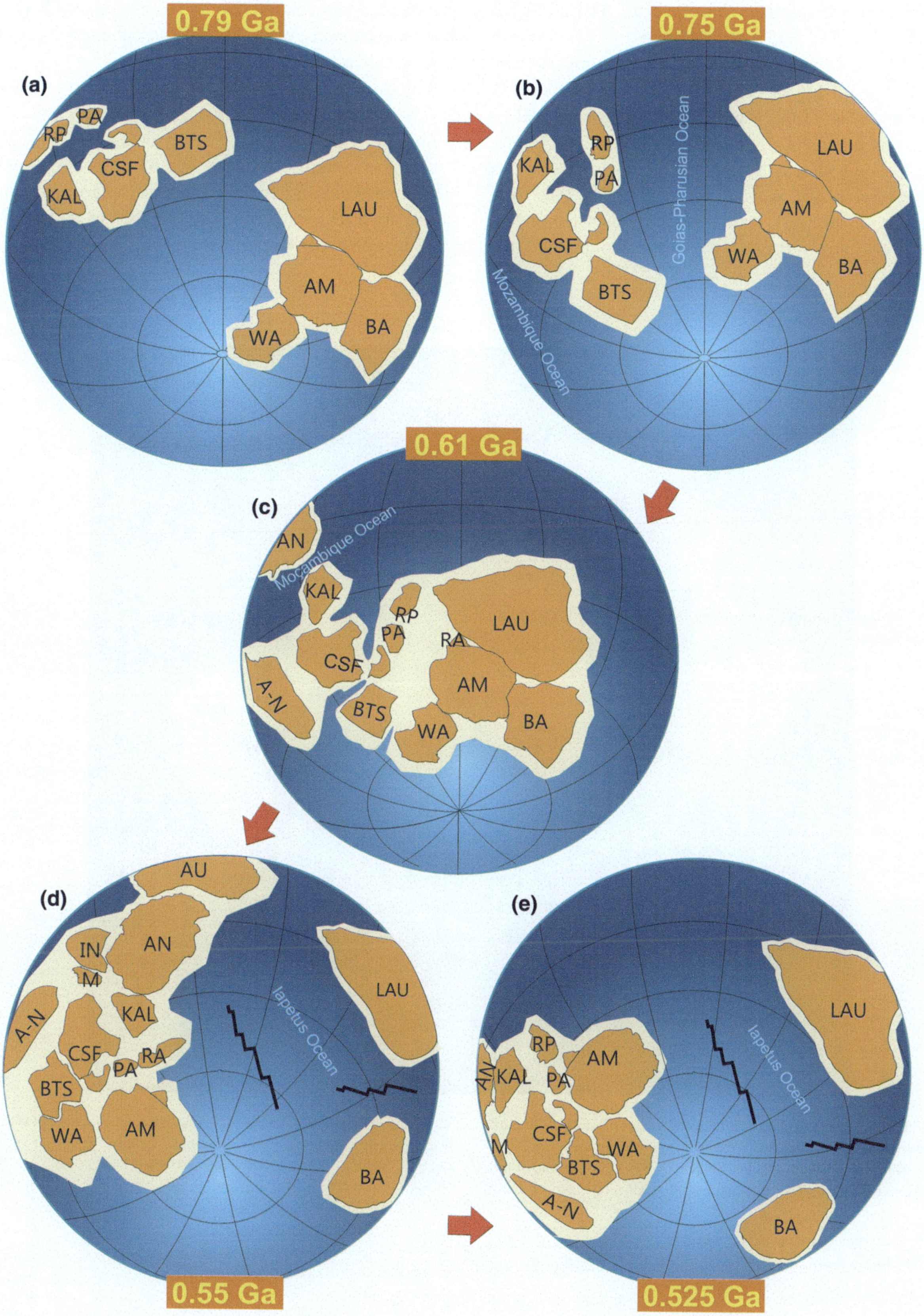
0.79 Ga
(a)
PA
RP
BTS
CSF
KAL
LAU
AM
WA
BA
0.75 Ga
(b)
RP
PA
KAL
CSF
BTS
Goias-Pharusian Ocean
Mozambique Ocean
LAU
AM
WA
BA
0.61 Ga
(c)
AN
Moçambique Ocean
KAL
RP
PA
RA
LAU
CSF
A-N
BTS
AM
WA
BA
AU
(d)
IN
AN
M
KAL
A-N
CSF
RA
PA
BTS
WA
AM
Iapetus Ocean
LAU
BA
0.55 Ga
(e)
RP
AM
AN
KAL
PA
CSF
WA
M
BTS
A-N
Iapetus Ocean
LAU
BA
0.525 Ga

525 Ma, respectively (see also Cawood et al. 2001 and Klein et al. 2015).

In any case, new paleomagnetic data for the 900–600 Ma interval is required for all Gondwana cratonic blocks. Figure 16.5 shows one of the many configurations adopted for the Gondwana continent (Schmitt et al. 2008). The Central African block, as shown by this figure, includes the São Francisco-Congo craton in its nucleus. In the late Paleozoic, Gondwana collided with Laurentia and other continental fragments forming Pangea, a supercontinent that encompassed all continental masses existing on Earth at that time (see Domeier et al. 2012 for a comprehensive discussion on this matter).

The evolution of the São Francisco-Congo craton during the Phanerozoic is well known. It remained as part of Gondwana during the formation of Pangea at about 300–280 Ma. During the rupture of Pangea, the São Francisco craton yet incorporated in South America and the Congo craton in Africa started to drift away from one another at ca. 130 Ma, as sea-floor spreading began in the young South Atlantic.

Fig. 16.5 Gondwana configuration adapted with some modifications from Schmitt et al. (2008). WA—West Africa; AM—Amazonia; AQ—Arequipa; BTS—Borborema-Trans-Sahara; CSF—Congo/São Francisco; KAL—Kalahari; PA—Paranapanema; RP—Rio de la Plata; M—Madagascar; IN—India; AN—Antarctica; WAu—West Australia; Nau—North Australia; GW—Gawler Craton. EAO—East African Orogen; WGO—West Gondwana Orogen

16.8 Conclusions

Based on geologic-geochronological and paleomagnetic evidence, this chapter discusses the trajectory of the São Francisco craton and other cratonic blocks from South America and Africa from ca. 2.1 Ga in the pre-Columbia Paleoproterozoic up to at ca. 300 Ma, as they become incorporated into Pangea. Our main conclusions are the following:

1. At 2.0 Ga at least two large landmasses with independent drifts are envisaged. The first involved proto-Amazonia, West Africa and Volgo-Sarmatia, which collided with the Kola-Karelia to form the nucleus of Columbia at ca. 1.78 Ga. The second, referred to as Central African block (Cordani et al. 2013a), encompassed São Francisco, Congo, Kalahari, Rio de la Plata and Borborema-Trans-Sahara.
2. The Central African block most likely did not take part of Columbia or Rodinia, drifting alone from 2.0 Ga until the formation of Gondwana at the Ediacaran/Cambrian boundary.
3. At 2.0 Ga, amalgamation of Laurentia was in progress. A large ocean (Manikewan Ocean) separated several cratonic blocks, such as the Superior, Rae, Hearne, Slave, and parts of Greenland and Baltica, which became completely agglutinated only at ca. 1.85–1.80 Ga ago.
4. The nucleus of Columbia, formed at ca. 1.78 Ga by the proto-Amazonia, West Africa, Baltica and Laurentia, remained united up to 1.25 Ga. After that, proto-Amazonia/West Africa and Baltica broke up and experienced a clockwise rotation until docking again with Laurentia along the Grenville belt and becoming incorporated into Rodinia at ca. 1000 Ma. During this time, the Central African block underwent a clockwise rotation, drifting from low to high northern latitudes.
5. The timing of the final assembly of West Gondwana is matter of debate. Some models propose a collision of Amazonian and São Francisco during closure of the Clymene ocean at the Edicaran/Cambrian boundary (550–530 Ma) (e.g., Tohver et al. 2010). Other authors (e.g., Ganade de Araujo et al. 2014) postulate an early collision of these cratonic blocks, which would occur during closure of the Goiás-Pharusian ocean at ca. 650 Ma. After that, Laurentia and Baltica broke apart, opening the Iapetus Ocean. At the same time other continental blocks (Antarctica, Australia and India) collided with the West African block to form Gondwana.
6. Gondwana remained as an integrated continental mass for more than 300 Ma. It collided with Laurasia at about 300–280 Ma, forming the Pangea, which started to split apart at around 180 Ma. With the opening of the South Atlantic at about 130 Ma, the São Francisco craton (in South America) and the Congo Craton (in Africa) drifted away from each other reaching their present positions.

Acknowledgments We thank Fernando F. Alkmim for its helpful comments on the text.

References

Almeida, F.F.M., Black, R., 1968. Geological comparison of northeastern South America and western Africa. Anais da Academia Brasileira de Ciências, Rio de Janeiro, Brazil, 40 (supplement), 317–319.

Berger, G.W., York, D., Dunlop, D.J., 1979. Calibration of Grenvillian paleopoles by ^{40}Ar-^{39}Ar dating. Nature 277, 46–48.

Bispo-Santos, F., D'Agrella-Filho, M.S., Janikian, L., Reis, N.J., Reis, M.A.A.A. Trindade, R.I.F., 2014a. Towards Columbia: Paleomagnetism of 1980–1960 Ma Surumu volcanic rocks, Northern Amazonian Craton. Precambrian Research 244, 123–138.

Bispo-Santos, F., D'Agrella-Filho, M.S., Trindade, R.I.F., Janikian, L., Reis, N.J., 2014b. Was there SAMBA in Columbia? Paleomagnetic evidence from 1790 Ma Avanavero mafic sills (Northern Amazonian craton). Precambrian Research 244, 139–155.

Bogdanova, S.V., Gintov, O.B., Kurlovich, D.M., Lubnina, N.V., Nilsson, M.K.M., Orlyuk, M.I.O., Pashkevich, I.K., Shumlyanskyy, L.V., Starostenko, V.I., 2013. Late Palaeoproterozoic mafic dyking in the Ukranian Shield of Volgo-Sarmatia caused by rotation during the assembly of supercontinent Columbia (Nuna). Lithos 174, 196–216.

Buchan, K.L., Fahrig, W.F., Freda, G.N., Frith, R.A., 1983. Paleomagnetism of the Lac St-Jean anorthosite and related rocks, Grenville Province, Quebec. Canadian Journal of Earth Sciences 20, 246–258.

Buchan, K.L., Mertanen, S., Park, R.G., Pesonen, L.J., Elming, S.-A, Abrahamsen, N. and Bylund, G., 2000. Comparing the drift of Laurentia and Baltica in the Proterozoic: the importance of key palaeomagnetic poles. Tectonophysics 319, 167–198.

Buick, I.S., Hermann, J., Williams, I.S., Gibson, R.L., Rubatto, D., 2006. A SHRIMP U-Pb and LA-ICP-MS trace element study of the petrogenesis of garnet-cordierite-orthoamphibole gneisses from the Central Zone of the Limpopo Belt, South Africa. Lithos 88 150–172.

Cawood, P.A., Pisarevsky, S. A., 2006. Was Baltica right-way-up or upside-down in the Neoproterozoic? Journal of the Geological Society, London. 163, 753–759.

Cawood, P.A., McCausland, P.J.A., Dunning, G.R., 2001. Opening Iapetus: Constraints from the Laurentian margin in Newfoundland. Geological Society of American Bulletin 113, 443–453.

Cederberg, J., Söderlund, U., Oliveira, E.P., Ernst, R.E., Pisarevsky, S. A., 2016. U-Pb baddeleyite dating of the Proterozoic Pará de Minas dyke swarm in the São Francisco craton (Brazil) – implications for the tectonic correlation with the Siberian, Congo and North China cratons. GFF 138, 219–240.

Cordani, U.G., Teixeira, W., 2007. Proterozoic accretionary belts in the Amazonian Craton. Geological Society of American, Memoir 200, 297–320.

Cordani, U.G., D'Agrella-Filho, M.S., Brito-Neves, B.B., Trindade, R. I.F., 2003. Tearing up Rodinia: the Neoproterozoic paleogeography of South American fragments. Terra Nova 15, 350–359.

Cordani, U.G., Sato, K., Teixeira, W., Tassinari, C.C.G., Basei, M.A. S., 2000. Crustal evolution of the South American platform. In: Cordani, U.G., Milani, E.J., Thomaz-Filho, A., Campos, D.A. (eds.) Tectonic Evolution of South America, Rio de Janeiro, p 19–40.

Cordani, U.G., Pimentel, M.M., Ganade de Araújo, C.E.G., Fuck, R.A., 2013a. The significance of the Transbrasiliano-Kandy tectonic corridor for the amalgamation of West Gondwana. Brazilian Journal of Geology 43, 583–597.

Cordani, U.G., Pimentel, M.M., Ganade de Araújo, C.E., Basei, M.A. S., Fuck, R.A., Girardi, V.A.V., 2013b. Was there an Ediacaran Clymene Ocean in central South America? American Journal of Science 313, 517–539.

Cordani, U.G., Pimentel, M.M., Ganadede de Araújo, C.E.G., Basei, M.A.S., Fuck, R.A., Girardi, V.A.V., 2014. Reply to the comment by Tohver E, Trindade RIF, on was there a Clymene ocean in central South America? 313, 517–539. American Journalof Science 314(3), 814–819.

D'Agrella-Filho, M.S., Pacca, I.G., 1998. Paleomagnetism of Paleopreterozoic mafic dyke swarm from the Uauá region, northeastern São Francisco Craton, Brazil: tectonic implications. Journal of South American Earth Sciences 11, 23–33.

D'Agrella-Filho, M.S., Bispo-Santos, F., Trindade, R.I.F., Antonio, P. Y.J., 2016. Paleomagnetism of the Amazonian Craton and its role in paleocontinents. Brazilian Journal of Geology 46(2), 275–299. doi: 10.1590/2317-4889201520160055.

D'Agrella-Filho, M.S., Pacca, I.G., Renne, P.R., Onstott, T.C., Teixeira, W., 1990. Paleomagnetism of Middle Proterozoic (1.01 to 1.08 Ga) mafic dykes in southeastern Bahia State-São Francisco Craton, Brazil. Earth Planetary Science Letters 101, 332–348.

D'Agrella-Filho, M.S., Babinski, M., Trindade, R.I.F., Van Schmus, W.R., Ernesto, E., 2000. Simultaneous remagnetization and U-Pb isotope resetting in Neoproterozoic carbonates of the São Francisco Craton, Brazil. Precambrian Research, 99, 179–196.

D'Agrella-Filho, M.S., Pacca, I.G., Trindade, R.I.F., Teixeira, W., Raposo, M.I.B., Onstott, T.C., 2004. Paleomagnetism and $^{40}Ar/^{39}Ar$ ages of mafic dykes from Salvador (Brazil): new constraints on the São Francisco Craton APW path between 1080 and 1010 Ma. Precambrian Research 132, 55–77.

D'Agrella-Filho, M.S., Trindade, R.I.F., Siqueira, R., Ponte-Neto, C.F., Pacca, I.G., 1998. Paleomagnetic constraints on the Rodinia Supercontinent: Implications for its Neoproterozoic break-up and the Formation of Gondwana. International Geology Review 40, 171–188.

D'Agrella-Filho, M.S., Tohver, E., Santos, J.O.S., Elming, S-A., Trindade, R.I.F., Pacca, I.I.G., Geraldes, M.C., 2008. Direct dating of paleomagnetic results from Precambrian sediments in the Amazon craton: Evidence for Grenvillian emplacement of exotic crust in SE Appalachians of North America. Earth and Planetary Science Letters 267, 188–199.

D'Agrella-Filho, M.S., Trindade, R.I.F., Tohver, E., Janikian, L., Teixeira, W., Hall, C., 2011. Paleomagnetism and ^{40}Ar-^{39}Ar geochronology of the high-grade metamorphic rocks of the Jequié block, São Francisco Craton: Atlantica, Ur and beyond. Precambrian Research 185, 183–201.

Daly, J.S., Balagansky, V.V., Timmerman, M.J., Whitehouse, M.J., 2006. The Lapland-Kola orogen: Palaeoproterozoic collision and accretion of the northern Fennoscandian lithosphere. In: Gee, D.G., Stephenson, R.A., (eds.) European Lithosphere Dynamics. Geological Society, London, Memoirs 32, 579–598.

Dalziel, I.W.D., Mosher, S., Gahagan, L.M., 2000. Laurentia-Kalahari collision and the assembly of Rodinia. Journal of Geology 108, 499–513.

De Beer, J.H., Meyer, R., 1984. Geophysical characteristics of the Namaqua-Natal Belt and its boundaries, South Africa. Journal of Geodynamics 1, 473–494.

Domeier, M., Van der Voo, R., Torsvik, T.H., 2012. Paleomagnetism and Pangea: The road to reconciliation. Tectonophysics 514–517, 14–43.

Eglington, B.M., 2006. Evolution of the Namaqua-Natal Belt, southern Africa – A geochronological and isotope geochemical review. Journal of African Earth Science 46, 93–111.

Evans, D.A.D., 2009. The palaeomagnetically viable, long-lived and all inclusive Rodinia supercontinent reconstruction. In: Murphy, J.B., Keppie, J.D., Hynes, A.J. (Eds.), Ancient Orogens and Modern Analogues. Geological Society, London, Special Publication 327, 371–404.

Evans, D.A.D., 2013. Reconstructing pre-Pangean supercontinents. GSA Bulletim 125, 1735–1751.

Evans, D. A. D., Mitchell, R. N., 2011. Assembly and breakup of the core of Paleoproterozoic–Mesoproterozoic supercontinent Nuna. Geology 39, 443–446.

Evans, D.A.D., Trindade, R.I.F., Catelani, E.L., D'Agrella-Filho, M.S., Heaman, L.M., Oliveira, E.P., Söderlund, U., Ernst, R.E., Smirnov, A.V., Salminen, J.M., 2016. Return to Rodinia? Moderate to high paleolatitude of the São Francisco/Congo craton at 920 Ma. In: Li, Z.X., Evans, D.A.D., Murphy, J.B. (eds.), Supercontinent Cycles Through Earth History. Geological Society, London, Special Publications, 424, 167–190. doi:10.1144/SP424.1.

Ganade de Araújo, C.E.G., Rubatto, D., Hermann, J., Cordani, U.G., Caby, R., Basei, M.A.S., 2014. Ediacaran 2.500-km-long synchronous deep continental subduction in the West Gondwana orogen. Nature Communications 5:5198, 1–8. Doi: 10.1038/ncomms6198.

Hanson, R.E., Martin, M.W., Bowring, S.A., Munyanyiwa, H., 1998. U-Pb zircon age for the Umkondo dolerites, eastern Zimbabwe: 1.1 Ga large igneous province in southern Africa-East Antarctica and possible Rodinia correlations. Geology 26, 1143–1146.

Hanson, R.E., Crowley, J.L., Bowring, S.A., Ramezani, J., Gose, W.A., Dalziel, W.D., Pancake, J.A., Seidel, E.K., Blenkinsop, T.G., Mukwakwami, J., 2004. Coeval large-scale magmatism in the Kalahari and Laurentian Cratons during Rodinia Assembly. Science 304, 1126–1129.

Hou, G., Santosh, M., Qian, X., Lister, G.S., Li, J., 2008. Tectonic constraints on 1.3 ~ 1.2 Ga final breakup of Columbia supercontinent from a giant radiating dyke swarm. Gondwana Research 14, 561–566.

Hurley, P.M., Almeida de, F.F.M., Melcher, G.C., Cordani, U.G., Rand, J.R., Kawashita, K., Vandoros, P., Pinson, W.H., Fairbairn Jr., H.W., 1967. Test of continental drift by comparison of radiometric ages. A pre-drift reconstruction shows matching geologic age provinces in West Africa and Northern Brazil. Science 157, 495–500.

Irving, E., Park, J.K.., McGlynn, J.C., 1972. Paleomagnetism of Et-Then Group and MacEnzie diabase in Great Slave Lake area. Canadian Journal of Earth Science 9, 744–767.

Johansson, A., 2009. Baltica, Amazonia and the SAMBA connection – 1000 million years of neighbourhood during the Proterozoic? Precambrian Research 175, 221–234.

Johansson, A., 2014. From Rodinia to Gondwana with the 'SAMBA' model – a distant view from Baltica towards Amazonia and beyond. Precambrian Research 244, 226–235.

Klein, R., Salminen, J., Mertanen, S., 2015. Baltica during the Ediacaran and Cambrian: A paleomagnetic study of Hailuoto sediments in Finland. Precambrian Research 267, 94–105.

Kröner, A., Cordani, U.G., 2003. African and South American cratons were not part of the Rodinia supercontinent: evidence from field relationships and geochronology. Tectonophysics 375, 325–352.

Li, Z-X., Evans, D.A.D., 2011. Late Neoproterozoic 40° intraplate rotation within Australia allows for a tighter-fitting and longer-lasting Rodinia. Geology 39, 39–42.

Li, Z.X., Bogdanova, S.V., Collins, A.S., Davidson, A., De Waele, B., Ernst, R.E., Fitzsimons, I.C.W., Fuck, R.A., Gladkochub, D.P., Jacobs, J., Karlstrom, K.E., Lu, S., Natapov, L.M., Pease, V., Pisarevsky, S.A., Thrane, K. and Vernikovsky, V., 2008. Assembly, configuration, and break-up history of Rodinia: A synthesis. Precambrian Research 160, 179–210.

Lubnina, N.V., Pisarevsky, S.A., Puchkov, V.N., Kozlov, V.I., Sergeeva, N.D., 2014. New paleomagnetic data from Late Neoproterozoic sedimentary successions in Southern Urals, Russia: implications for the Late Neoproterozoic paleogeography of the Iapetan realm. International Journal of Earth Science (Geol. Rundsch) 103, 1317–1334.

Lubnina. N.V., Stepanova, A.V., Ernst, R.E., Nilsson, M., Söderlund, U., 2016. New U-Pb baddeleyite age, and AMS and paleomagnetic data for dolerites in the Lake Onega belonging to the 1.98–1.95 Ga regional Pechenga-Onega Large Igneous province. GFF 138(1), 54–78. doi: 10.1080/11035897.2015.1129549.

McCausland, P.J.A., Hankard, F., Van der Voo, R., Hall, C.M., 2011. Ediacaran paleogeography of Laurentia: Paleomagnetism and ^{40}Ar-^{39}Ar geochronology of the 583 Ma Baie des Moutons syenite, Quebec. Precambrian Research 187, 58–78.

Martin, H., 1961. The hypothesis of continental drift in the light of recente advances of geological knowledge in Brazil and in South West Africa. Geological Society of South Africa, vol. LXIV, annex. 1–47.

Meert, J.G., Torsvik, T.H., 2003. The making and unmaking of a supercontinent: Rodinia revisited. Tectonophysics 375, 261–288.

Meert, J.G., Van der Voo, R., 1997. The assembly of Gondwana (800–550 Ma). Journal of Geodynamics 23, 223–235.

Meert, J.G., Hargraves, R.B., Van der Voo, R., Hall, C.M., Halliday, A. N., 1994. Paleomagnetic and $^{40}Ar/^{39}Ar$ studies of Late Kibaran Intrusives in Burundi, East Africa: Implications for Late Proterozoic Supercontinents. Journal of Geology 102, 621–637.

Mitchell, R.N., Bleeker, W., Van Breemen, O., Lacheminant, T.N., Peng, P., Nilsson, M.K.M., Evans, D.A.D., 2014. Plate tectonics before 2.0 Ga: Evidence from Paleomagnetism of cratons within Supercontinent Nuna. American Journal of Sciences 314, 878–894.

Morgan, G.E., 1985. The paleomagnetism and cooling hystory of metamorphic and igneous rocks from the Limpopo Mobile Belt, Southern Africa. Geological Society of America, Bulletim 96, 663–675.

Nomade, S., Chen, Y., Pouclet, A., Féraud, G., Théveniaut, H., Daouda, B.Y., Vidal, M., Rigolet, C., 2003. The Guiana and the West African Shield Palaeoproterozoic grouping: new palaeomagnetic data for French Guiana and the Ivory Coast. Geophysical Journal International 154, 677–694.

Oliveira, E.P., Silveira, E.M., Söderlund, U., Ernst, R.E., 2013. U-Pb ages and geochemistry of mafic dyke swarms from the Uauá Block, São Francisco Craton, Brazil: LIPs remnants relevant for Late Archaean break-up of a supercraton. Lithos 174, 308–322.

Pehrsson, S.J., Eglington, B.M., Evans, D.A.D., Huston, D., Reddy, S., 2016. Metallogeny and its link to orogenic style during the Nuna supercontinent cycle. In: Li, Z.X., Evans, D.A.D., Murphy, J.B., Supercontinent Cycles Through Earth History, Geological Society, London, Special Publications, 424, 83–94. doi: 10.1144/SP424.5.

Pesonen, L.J., Elming, S.-Å., Mertanen, S., Pisarevsky, S., D'Agrella--Filho, M.S., Meert, J.G., Schmidt, P.W., Abrahamsen, N. and Bylund, G., 2003. Palaeomagnetic configuration of continents during the Proterozoic. Tectonophysics 375, 289–324.

Pesonen, L.J., Mertanen, S., Veikkolainen, T., 2012. Paleo-Mesoproterozoic Supercontinents – A paleomagnetic view. Geophysica 47, 5–47.

Pimentel, M.M., Fuck, R.A., Botelho, N.F., 1999. Granites and the geodynamic history of the neoproterozoic Brasília belt, Central Brazil: a review. Lithos 46, 463–483.

Piper, J.D.A., 2010. Palaeopangea in Meso-Neoproterozoic times: The palaeomagnetic evidence and implications to continental integrity, supercontinent form and Eocambrian break-up. Journal of Geodynamics 50, 191–223.

Pisarevsky, S.A., Bylund, G., 2006. Palaeomagnetism of 935 Ma mafic dykes in southern Sweden and implications for the Sveconorwegian Loop. Geophysical Journal International 166, 1095–1104.

Pisarevsky, S.A., Wingate, M.T.D., MCA Powell, C., Johnson, S., Evans, D.A.D., 2003. Models of Rodinia assembly and fragmentation. In: Yoshida, M., Windley, B.F., Dasgupta, S. (Eds.), Proterozoic East Gondwana: Supercontinent Assembly and Breakup. Geological Society, London, Special Publications 206, 35–55.

Pisarevsky, S.A., Murphy, J.B., Cawood, P.A., Collins, A.S., 2008. Late Neoproterozoic and Early Cambrian palaeogeography: models and problems. In: Pankhurst, R.J., Trouw, R.A.J., Brito Neves, B. B., De Wit, M.J. (eds.) West Gondwana: Pre-Cenozoic Correlations Across the South Atlantic Region. Geological Society, London, Special Publications 294, 9–31.

Pisarevsky, S.A., Elming, S.-Å., Pesonen, L.J., Li, Z.-X., 2014. Mesoproterozoic paleogeography: Supercontinent and beyond. Precambrian Research 244, 207–225.

Reis, N.J., Teixeira, W., Hamilton, M.A., Bispo-Santos, F., Almeida, M.E., D'Agrella-Filho, M.S., 2013. The Avanavero mafic magmatism, a late Paleoproterozoic LIP in the Guiana Shield, Amazonian Craton: U-Pb TIMS baddeleyite, geochemical and paleomagnetic evidence. Lithos 174, 175–195.

Renne, P.R., Onstott, T.C., D'Agrella-Filho, M.S., Pacca, I.G., Teixeira, W., 1990. $^{40}Ar/^{39}Ar$ dating of 1.0–1.1 Ga magnetizations from the São Francisco and Kalahari Cratons: tectonic implications for Pan-African and Brasiliano mobile belts. Earth and Planetary Science Letters 101, 349–366.

Rogers, J.J.W., Santosh, M., 2009. Tectonics and surface effects of the supercontinent Columbia. Gondwana Research 15, 373–380.

Salminen, J., Pesonen, L.J., 2007. Paleomagnetic and rock magnetic study of the Mesoproterozoic sill, Valaanisland, Russian Karelia. Precambrian Research 159, 212–230.

Salminen, J., Pesonen, L.J., Reimold, W.U., Donadini, F., Gibson, R. L., 2009. Paleomagnetic and rock magnetic study of the Vredfort impact structure and the Johannesburg Dome, Kaapvaal Craton, South Africa - Implications for the apparent polar wander path of the Kaapvaal Craton during the Mesoproterozoic. Precambrian Research 168, 167–184.

Schmitt R.S., Trouw R.A.J., Van Schmus W.R., Passchier C.W. 2008. Cambrian orogeny in the Ribeira belt (SE Brazil) and correlation within West Gondwana: ties that bind underwater. In: Pankhurst R. J., Trouw R.A.J., Brito Neves B.B., de Wit M.J. (eds.), West Gondwana Pre-Cenozoic Correlations Across the South Atlantic Region. The Geological Society, London, Special Publication 294, 279–296.

Silva, L.C. da, Armstrong, R., Delgado, I.M., Pimentel, M., Arcanjo, J. B., Melo, R.C. de, Teixeira, L.R., Jost, H., Cardoso Filho, J.M., Pereira, L.H.M., 2002. Reavaliação da evolução geológica em terrenos Pré-Cambrianos Brasileiros com base em novos dados U-Pb SHRIMP, Parte I: Limite centro-oriental do Craton do São Francisco na Bahia. Revista Brasileira de Geociências 32, 161–172.

Smith, A.G., Hallam, A., 1970. The fit of the southern continents. Nature 225, 139.

ST-Onge, M.R., Van Gool, J.A.M., Gerde, A.A., Scott, D.J., 2009. Correlation of Archaean and Palaeoproterozoic units between Canada and western Greenland constraining the pre-collisional upper plate accretionary history of the Trans-Hudson orogen. In: Cawood, P.A., Kröner, A. (eds.) Earth Accretionary Systems in Space and Time. The Geological Society, London, Special Puvblications 318, 193–235.

Teixeira, W., D'Agrella-Filho, M.S., Hamilton, M.A., Ernst, R.E., Girardi, V.V., Mazzucchelli, M., Bettencourt, J.S., 2013. U-Pb baddeleyite ages and paleomagnetism of 1.79 and 1.50 Ga tholeiitic dyke swarms, and position of the Rio de La Plata Craton within the Columbia Supercontinent. Lithos 174, 157–174.

Théveniaut, H., Delor, C., Lafon, J.M., Monié, P., Rossi, P., Lahondère, D., 2006. Paleoproterozoic (2155–1970 Ma) evolution of the Guiana Shield (Transamazonian event) in the light of new paleomagnetic data from French Guiana. Precambrian Research 150, 221–256.

Tohver, E., Trindade, R.I.F., 2014. Comment on was there an ediacaran Clymene ocean in central South America by Cordani UG and others. American Journal of Science 314(3), 805–813.

Tohver, E., van der Pluijm, B.A., Van der Voo, R., Rizzotto, G., Scandolara, J.E., 2002. Paleogeography of the Amazon craton at 1.2 Ga: early Grenvillian collision with the Llano segment of Laurentia. Earth and Planetary Science Letters 199, 185–200.

Tohver, E., D'Agrella-Filho, M.S., Trindade, I.F., 2006. Paleomagnetic record of Africa and South America for the 1200–500 Ma interval, and evaluation of Rodinia and Gondwana assemblies. Precambrian Research 147, 193–222.

Tohver, E., Trindade, R.I.F., Solum, J.G., Hall, C.M., Riccomini, C., Nogueira, A.C., 2010. Closing the Clymene ocean and bending a Brasiliano belt: Evidence for the Cambrian formation of Gondwana, southeast Amazon craton. Geology 38, 267–270.

Trindade, R.I.F., Font, E., D'Agrella-Filho, M.S., Nogueira, A.C.R., Riccomini, C. 2003. Amazonia at low-latitude by the end of the ~600 Ma Puga glaciation. Terra Nova 15, 441–446.

Trindade, R.I.F., D'Agrella-Filho, M.S., Babinski, M., Font, E., Brito-Neves, B.B., 2004. Paleomagnetism and geochronology of the Bebedouro cap carbonate: evidence for continental-scale Cambrian remagnetization in the São Francisco Craton, Brazil. Precambrian Research, 128, 83–103.

Trindade, R.I.F., D'Agrella-Filho, M.S., Epof, I., Brito-Neves, B.B., 2006. Paleomagnetism of the Early Cambrian Itabaiana mafic dykes, NE Brazil, and implications for the final assembly of Gondwana and its proximity to Laurentia. Earth and Planetary Science Letters 244, 361–377.

Trompette, R., 1994. Geology of Western Gondwana (2000–500 Ma). Pan-African-Brasiliano Aggregation of South America and Africa. A.A. Balkema, Rotterdam, Brookfield, p 366.

Weil, A.B., Van der Voo, R., Niocaill, C.M., Meert, J.G., 1998. The Proterozoic supercontinent Rodinia: paleomagnetically derived reconstructions for 1100 to 800 Ma. Earth and Planetary Science Letters 154, 13–24.

Xu, H., Yang, Z., Peng, P., Meert, J.G., Zhu, R., 2014. Paleo-position of the North China craton within the Supercontinent Columbia: Constraints from new paleomagnetic results. Precambrian Research 255, 276–293.

Zhao, G., Cawood, P.A., Wilde, S.A., Sun, M., 2002. Review of global 2.1–1.8 Ga orogens: implications for a pre-Rodinia supercontinent. Earth-Science Reviews 59, 125–162.

Tectonic Genealogy of a Miniature Continent

17

Monica Heilbron, Umberto G. Cordani, Fernando F. Alkmim, and Humberto L.S. Reis

Abstract

The São Francisco craton (SFC), although small in size, exhibits elements of some of the Earth principal evolutionary phases. Repeated cycles of granite-greenstone terrain formation recorded in the craton basement attest the high-heat tectonic regime that characterize the Archean Earth. Like in many other places of the world, these terrains amalgamated to form a coherent and stable continental mass in the Late Archean. In the course of the Paleoproterozoic, subduction-driven accretionary orogens, which incorporated island arcs and continental terranes, were added to the Archean continental nuclei. Typical of the Proterozoic plate tectonic regime, these events took place around the important 2.1 Ga age peak of juvenile crustal production. The collage of several blocks concurred to form the São Francisco paleocontinent, which always united to the Congo landmass, experienced a series of intraplate processes, such as rifting and associated bimodal magmatism during the following 1300 Ma-long period. Very likely, the São Francisco-Congo did not take part of the Columbia and Rodinia supercontinents. Later, in Ediacaran and Cambrian times, it was involved in the collage that resulted in the formation of the Gondwana Supercontinent. Driven by slab-pull subduction and collisional processes, typical of modern-type plate tectonics, the Rio Preto, Riacho do Pontal, Sergipano, Araçuaí-Ribeira and Brasilia belts surrounded and shaped the present configuration of the SFC. Finally, the SFC was separated from the Congo craton following the Brazil–Africa continental drift and the formation of the South Atlantic Ocean. This chapter explores the significance of rock assemblages and tectonic features exhibited by the miniature São Francisco continent in terms of Earth global processes.

Keywords

São Francisco craton • Archean • Paleoproterozoic • Neoproterozoic • Brasiliano event

M. Heilbron
Tektos Research Group, Faculdade de Geologia, Universidade do Estado Rio de Janeiro (UERJ), Rua São Francisco Xavier, 524, Bloco A-4020, Rio de Janeiro, RJ 20550-900, Brazil

U.G. Cordani (✉)
Instituto de Geociências, Universidade de São Paulo, Rua do Lago 562, São Paulo, SP 05508-080, Brazil
e-mail: ucordani@usp.br

F.F. Alkmim
Departamento de Geologia, Escola de Minas, Universidade Federal de Ouro Preto, Morro do Cruzeiro s/n, Ouro Preto, MG 35400-000, Brazil

H.L.S. Reis
Laboratório de Estudos Tectônicos (LESTE)/Núcleo de Geociências e Instituto de Ciência e Tecnologia, Universidade Federal dos Vales do Jequitinhonha e Mucuri, Campus Juscelino Kubitscheck, 39100-000 Diamantina, MG, Brazil

M. Heilbron et al. (eds.), *São Francisco Craton, Eastern Brazil*, Regional Geology Reviews, DOI 10.1007/978-3-319-01715-0_17

17.1 Introduction

Planet Earth started hot and was losing heat through time. Indeed, the Earth's mantle temperature decreased by several hundred degrees Celsius since Archean times. This is the main factor that governs the geodynamic evolution of the planet (Condie and Kroener 2008; Brown 2008).

In the Archean, the high rate of heat production by radioactive decay induced mantle convection and growth of new oceanic lithosphere. The consumption of it produced TTG calc-alkaline granitoids (Tarney et al. 1979; Condie and Kroener 2008), which intruded the crust. They were later deformed and metamorphosed into the continental gneisses and granulites which appear in association with greenstone belts. These features are made up of supracrustal metamorphic rocks (metavolcanics and metasedimentary), formed within oceanic domains, in tectonic environments such as island arcs or spreading centres. Such granite-greenstone terrains are widespread within the Archean domains. They are typical of high-heat regime, formed basically by mantle-derived granitoids and by metamorphic rocks which may include UHT granulites. Plate tectonics came early into the Earth's geodynamics evolution, but it did not appear as a single event in a given time. Returns of oceanic lithosphere into the mantle may have started to cool enough in the early Archean, to make possible some localized subduction. Gradually, as the planet continued to cool, more and more slabs could have been able to descend at steep angles. Perhaps, the first documented subduction zone occurred within the 3.8 Ga Isua greenstone belt (Komiya et al. 2002), and other examples were registered, such as the Barberton greenstone belt in South Africa (Moyen et al. 2006). By the Late Archean, steep-mode subduction was a quite important tectonic regime in the planet.

After the Archean, plate tectonics became the main mechanism by which the Earth loses its internal heat. Stern (2005) initiated an interesting controversy, arguing that subduction tectonics only started in the Neoproterozoic. The main arguments were the lack of some clear indicators for that tectonic regime, such as blueschists, ophiolites and UHP metamorphic rocks, prior to 1.0 Ga. Different tectonic regimes, or different proto-plate tectonic models, may have been in place. On the contrary, Cawood and Pisarewsky (2006) and Condie and Kroener (2008), although recognizing the lack of the specific rock indicators mentioned by Stern (2005), claimed that there is enough evidence for subduction-driven tectonic processes in the entire Proterozoic. Brown (2008), looking at the metamorphic record in geologic time, suggested a dramatic change in the thermal environments of crustal metamorphism at the Archean/Proterozoic boundary, leading to a "Proterozoic plate tectonics regime", characterized by the subduction of oceanic lithosphere and continental lithospheric stability, when cratons formed the nuclei of major continental masses. Condie (2000), suggested that subduction could be episodic, which explain the important peak in crustal production in the Paleoproterozoic, at about 2.0 Ga.

At about 1.0 Ga, after a long time of continued heat loss, the Earth cooled sufficiently to allow the oceanic lithosphere to become negative buoyant in a relatively short time, and the new low T and high P tectonic regime was characterized by thermal gradients that allowed the transformation of basalts into eclogites in the descending slabs. Such negative buoyancy of dense oceanic lithosphere led to the "slab-pull type" of driving force for the slabs that slide away from mid-ocean ridges. According to Brown (2008), UHP metamorphic terrains, bearing coesite or diamond, started to appear in the Earth, indicating subduction of continental crust to depths >100 km, followed by rapid exhumation. In his view, this marks the transition towards the new tectonic regime, related to "modern-type subduction zones" and "subduction-to-collision" orogenic belts, widespread in the Phanerozoic. The oldest eclogites so far described (Caby 1994) were dated by robust U-Pb zircon ages of ca. 615 Ma (Ganade de Araujo et al. 2014).

The aim of this chapter is to provide an integrated synthesis on the geologic history and tectonic evolution of the São Francisco miniature continent (see Heilbron et al., Chap. 1). It is surprizing to see how, in a small cratonic fragment such as the São Francisco miniature continent, we can find a clear evidence for some of the relevant phases of the Earth tectonic evolution. In the next sections, compiling information presented in the preceding chapters of this volume, we will present and discuss the tectonic events recorded in the São Francisco craton and its margins, exploring their significance in terms of the global tectonic processes that governed the Earth's dynamics.

17.2 High-Heat Tectonic Regime and the Formation of Granite-Greenstone Terrains

Archean rocks are widespread in both the northern (Bahia State) and southern (Minas Gerais State) segments of the São Francisco craton (SFC), forming a series of granite-greenstone terrains of different age. The oldest ages of ca. 3.6 Ga are found in some TTG granitoids that crop out in the Gavião block of central Bahia (see Teixeira et al. Chap. 3, and Barbosa and Barbosa, Chap. 4) (Fig. 17.1). Sm–Nd model ages of some of these rocks reach ca. 3.9 Ga. Together with one single detrital zircon of ca. 4.1 Ga, found in a supracrustal rock from a younger greenstone belt sequence, these ages are indicative of the existence of even older

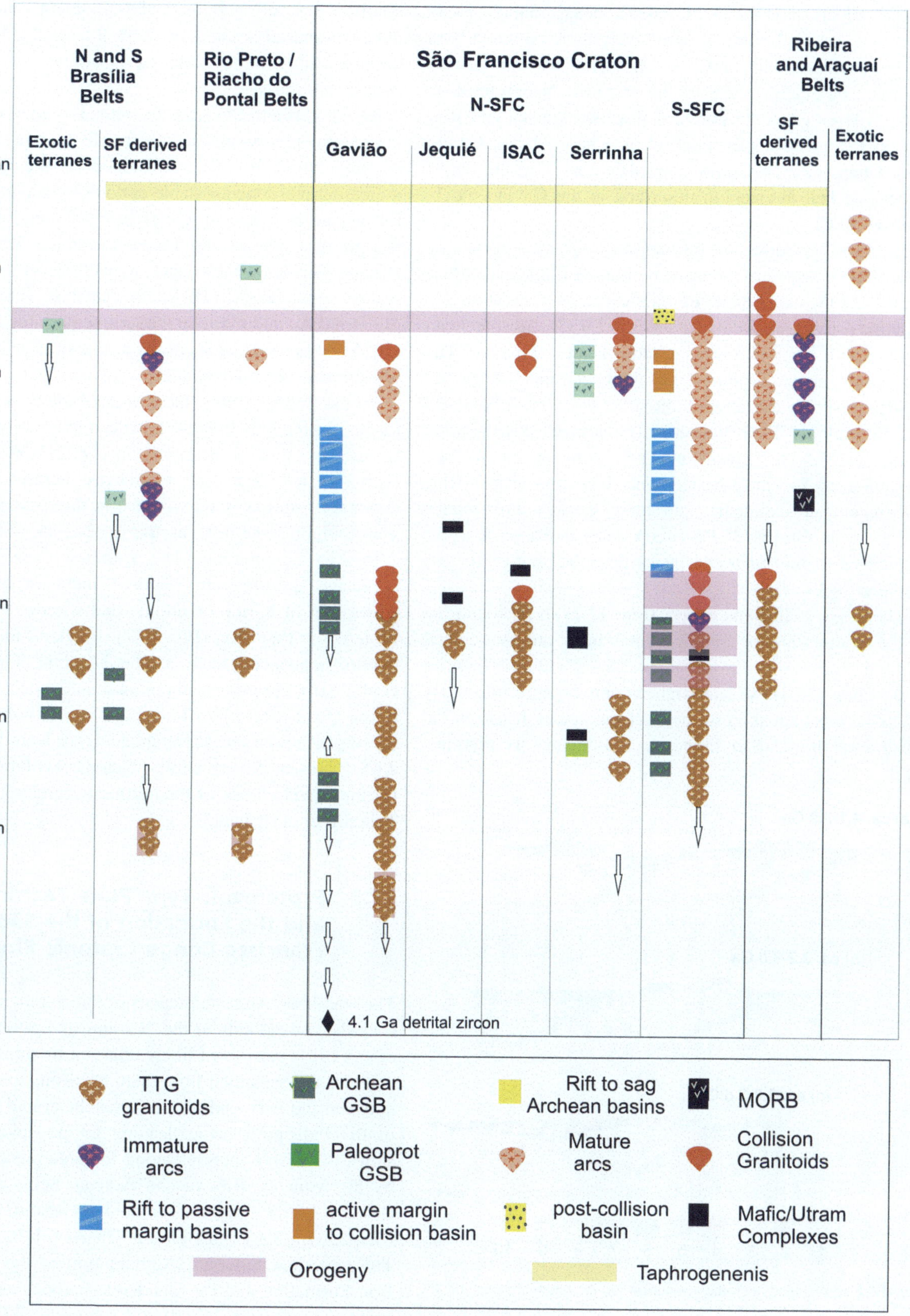

Fig. 17.1 Archean to Paleoproterozoic schematic tectono-stratigraphic chart

crustal precursors (Fig. 17.1). Paleoarchean granitoid rocks of ca. 3.4–3.2 Ga became progressively more common both in the southern and northern segments, as well within reworked basement inliers of the marginal Neoproterozoic belts, as reported in the northern Brasília, Araçuaí and Riacho do Pontal belts (see Fuck et al. Chap. 11 and Caxito et al. Chap. 12). This time interval is coeval of the oldest greenstone belt association described in the Gavião block, northern SFC.

Renewed episodes of formation of granite-greenstone terrains are detected in the same regions of Bahia and Minas Gerais, and the main pulses of magmatic rock generation due to plume activity were dated at ca. 3.2–2.9 Ga (Mesoarchean time) and 2.8–2.6 Ga (Neoarchean time). The tectono-magmatic evolution of these terrains developed through juvenile accretion/differentiation events characterized by multiple TTG plutonism (Lana et al. 2013; Farina et al. 2015). The greenstone belt associations include komatiitic rocks, meta-basalts and meta-andesites, BIFs, other chemical sediments, and lithic wackes. Provenance studies made through U-Pb zircon dates indicated the predominance of the main age peaks described above for the granitoid rocks.

The oldest collisional events (Fig. 17.2) so far identified in the craton interior and marginal belts are the crustal forming processes occurring at ca. 2.8–2.7 Ga (see Teixeira et al., Chap. 3). These late Archean convergence ultimately led to the amalgamation of the Archean nuclei that form the substratum of the São Francisco craton and its margins.

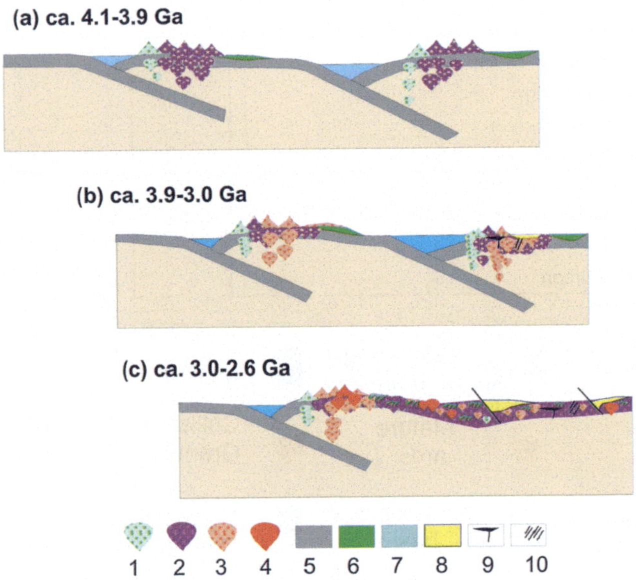

Fig. 17.2 Envisaged tectonic evolution of Archean terranes of the SFC. *1* IAT basic rocks, *2* TTG granitoids, *3* calc-alkaline tonalites and granodiorites *4* K-rich granitoids, *5* oceanic crust, *6* greenstone belts, *7* sag to passive margin basins, *8* rift basins, *9* mafic to ultramafic complexes, *10* basic dikes

Anorthosites, as well as mafic-ultramafic complexes and dykes emplaced at ca. 3.16, 2.65–2.63 and 2.58 Ga, testify the growing and the increased stability of the Archean continental mass.

In the northern SFC, the Jacobina Group, representing an intracontinental basin filled essentially by mature sandstones and conglomerates, was developed on the Gavião block, indicating that a major stable cratonic-type area was already formed in the Archean. Quartzites and conglomerates from this group displayed only Paleoarchean (ca. 3.5–3.2) detrital zircons. Based on these ages, as well as on the crustal signatures obtained from Hf isotopes analyses and the presence of detrital Au and pyrite in the sedimentary rocks, Teles et al. (2015) compared the Jacobina succession to the Witwatersrand basin of South Africa.

In the southern SFC, the Rio das Velhas greenstone belt, located in the Quadrilátero Ferrífero mineral province, is so far the best studied of the craton (Teixeira et al., Chap. 3). Farina et al. (2015 and references therein) described its tectonic evolution, subdividing it in three events marked by granitoid emplacement at the 2.92–2.85, 2.80–2.76, and 2.76–2.68 Ga intervals. The youngest event is related to pervasive deformation, regional metamorphism and the generation of K-rick granitoids that witness the Archean—Proterozoic transition (Fig. 17.2). Foreland to intramontane mature sequences, such as the Maquiné Group (Moreira et al. 2016), closed this important continental growing event in the Quadrilátero Ferrífero. Late to post-collisional *strictu* sensu granites and basic intraplate dykes were intruded at ca. 2.65–2.56 Ga. A very similar tectonic evolution is reported for the Gavião block of the northern segment of the SFC by Teixeira et al. (Chap. 3).

17.3 Proterozoic-Type Plate Tectonics and the Formation of the São Francisco-Congo Cratonic Block

The stabilized Archean nucleus of the craton, a region which roughly corresponds to the "Paramirim craton" of Almeida et al. (1981), was most likely adjacent to similar continental masses, and separated from them by rifting events sometime between the very end of the Neoarchean and the beginning of the Paleoproterozoic, leading to the development of a series of passive margins along the area presently occupied by the Mineiro, Ribeira and Araçuaí belts, as well as the Eastern Bahia orogenic domain (Alkmim and Teixeira, Chap. 5, Barbosa and Barbosa, Chap. 4, Ledru et al. 2004). The particular Siderian sedimentation of the Cauê Banded Iron Formation and the Gandarela carbonates in the Minas basin (Quadrilátero Ferrífero, southern SFC) took place around 2.42 Ga (Fig. 17.3) in the course of the Earth's *Great Oxygenation Event*. Peculiar passive margin units are

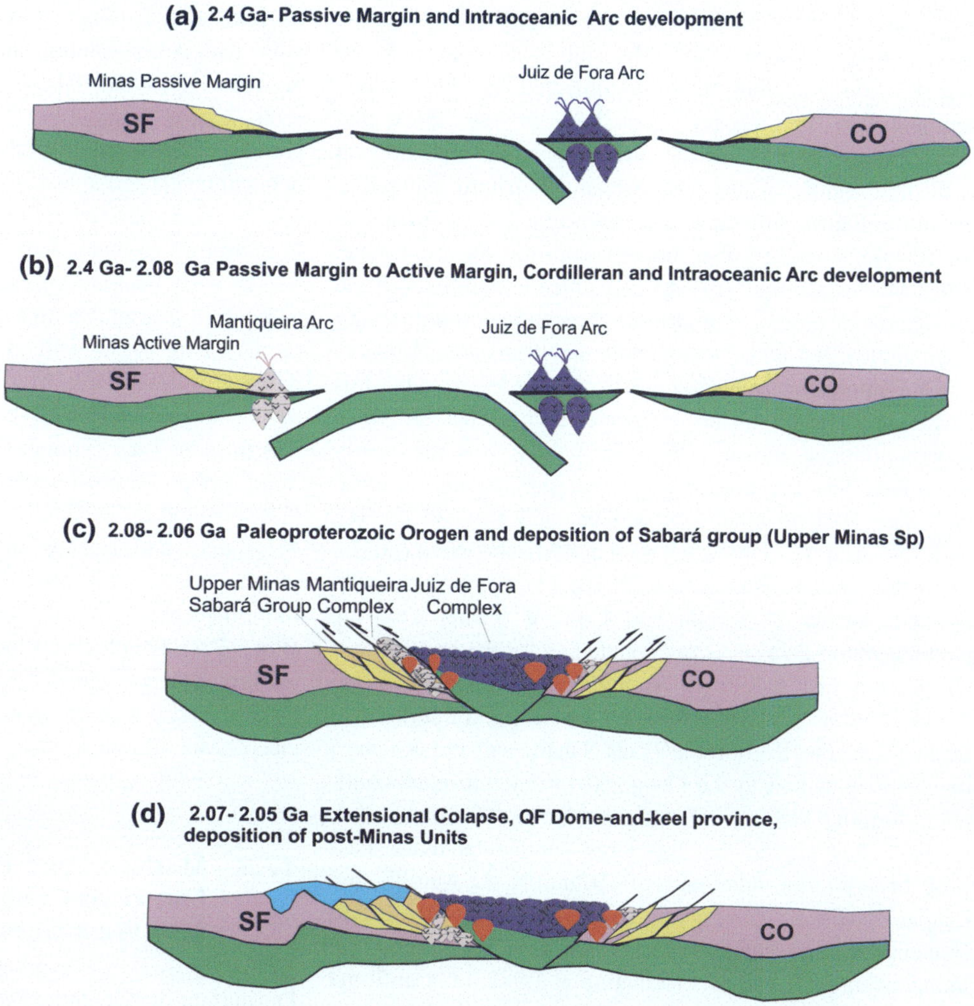

Fig. 17.3 Envisaged tectonic evolution of Paleoproterozoic terranes of the SFC. *1* Post-Minas units, *2* collisional granites, *3* cordilleran arc, *4* juvenile intraoceanic arc, *5* syn-collision Sabará Group (upper Minas Supergroup), *6* passive margin lower Minas Supergroup, *7* oceanic crust, *8* Archean continental crust, *9* lithospheric mantle. SF São Francisco; CO Congo.

also recorded in the Eastern Bahia belt (Barbosa and Barbosa, Chap. 4), such as Lake Superior-type BIFs, dolomitic marbles and manganese-rich schists. In various aspects, these specific metasedimentary successions could be correlative to the basal Francevillian A and B formations, exposed in the northwestern portion of the Congo craton in Gabon (Ledru et al. 1994; Feybesse et al. 1998).

Some greenstone belt-like volcano-sedimentary sequences and associated juvenile granitoids of Rhyacian age, detected within the Mineiro, Serrinha and Brasilia belts (see Fig. 17.1 and Chaps. 4 and 5) record the development of back-arc to intra-arc basins in oceanic realms. Despite their restricted expression in the geologic scenario, these assemblages witness the development of subduction systems along the margins of the Archean cratonic nuclei in the Early Paleoproterozoic. The rock record of the craton and its margins relative to the subsequent time interval of 2.40–2.10 Ga is limited to some granitic intrusions and volcano-sedimentary sequences, in agreement with the global magmatic quietude pointed out by Condie et al. (2009) for this period. A Siderian age of ca. 2.4 Ga was obtained from a metamafic unit within a reworked basement inlier of the Ribeira belt, with a MORB-type signature, probably marking the distal oceanic portion of the Minas basin (Heilbron et al. 2010).

Within the Minas Supergroup, after a long depositional hiatus that followed the initial Siderian rift-to-passive margin phase (Alkmim and Teixeira, Chap. 5), the syn-orogenic assemblage of the Sabará Group was deposited at ca. 2125 Ma (Fig. 17.3). The deposition of this sequence, coeval to the C and D formations of the Francevillian Group in Gabon, marks the onset of the continental collision between the São Francisco and Congo landmasses. On the Brazilian side, the generation of several magmatic arcs and microcontinents of Rhyacian age, such as the Mantiqueira and Quirino cordilleran arcs, the Juiz de Fora intraoceanic arc in the basement of the Ribeira belt, and the Jequié and Serrinha blocks within the Eastern Bahia belt, was described by Noce et al. (2007), Heilbron et al. (2010), Teixeira et al. (2014), and Barbosa and Sabaté (2004). The eastern margin of the São Francisco Archean nucleus, already converted into a back-arc domain, started to be fed with sediments sourced by the approaching collisional front. The climax of this process was reached around 2080 Ma, the age peak of the syn-collision metamorphism documented in the various segments of the orogen preserved in the intra and extra-cratonic sectors. In the aftermath of the collision, at least the region of the Quadrilátero Ferrífero entered a gravity collapse, subsiding and collecting the sediments of the Itacolomi Group, also derived from the orogenic zone. Thus, after receiving recycled and juvenile additions, these amalgamated pieces of cratonic lithosphere formed a very large continental mass, tectonically stabilized at ca.1.9 Ga.

One interesting question arises at this point. Did this Rhyacian\early Orosirian orogenic edifice belong to a supercontinental mosaic? In other words, was the newly assembled São Francisco-Congo incorporated into Columbia? This was postulated by Rogers and Santosh (2004), Brito-Neves et al. (1999), Alkmim and Martins-Neto (2012), Zhao et al. (2004), Nance et al. (2013). On the other hand, as pointed out by D'Agrella and Cordani (Chap. 16), the paleomagnetic data set is not conclusive for the inclusion of all the cratonic blocks in a single supercontinent at ca. 1.9 Ga. In their Fig. 16.2, these authors suggest that the Rio de la Plata, the Borborema/Trans-Sahara and the Kalahari blocks were united to the Congo-São Francisco paleocontinent, forming the Central African block of Cordani et al. (2013). They argued that this large continental mass most probably did not take part of Columbia. D'Agrella and Cordani (Chap. 16) also suggested that the São Francisco-Congo paleocontinent may have experienced an independent drift since its formation at 1.9–2.0 Ga until the formation of West Gondwana in Ediacaran time.

17.4 1300 Ma of Intraplate Events

Looking at the Earth tectonic evolution in geological time, from ca. 1900 to ca. 600 Ma, Columbia was amalgamated and later dispersed in many fragments that eventually reassembled into Rodinia (e.g. Nance et al. 2013), and Rodinia was also disrupted not long after its formation. During this very long period of 1300 Ma, no clear evidence of collisional tectonism affecting the SFC was found. On the contrary, plenty of within-plate episodes such as rifting and anorogenic magmatism are registered in the SFC, as pointed out by Cruz and Alkmim and Reis et al. (Chaps. 6 and 7). Mafic dike swarms within the craton also testify the extensional tectonic events (see Girardi et al., Chap. 8). This is somewhat endorsed by many of the Rodinia's paleomagnetic reconstructions (e.g., Tohver et al. 2006; Nance et al. 2013), which show the São Francisco-Congo as an isolated landmass.

A large system of Statherian rifts nucleated in the central part of the SFC and its margins, receiving successive sequences of continental sediments accumulated in association with A-type plutons and bi-modal volcanic rocks dated between 1.78 and 1.71 Ga (Cruz and Alkmim, Alkmim et al., Chaps. 6 and 14). The intraplate tectonic events that led to the development of the various Espinhaço Supergroup depocenters lasted at least until the end of the Tonian Period, when several passive margin basins were formed in the São Francisco-Congo landmass (Fig. 17.4).

The early Neoproterozoic extensional tectonism, from rift to passive margin, included mafic sills and dikes (Girardi et al., Chap. 8), as well as bimodal magmatic associations dated at 1.0–0.80 Ga. Many of these units record onset of rifting events, followed by full development of passive margins in Late Tonian-Cryogenian times. The passive margin phase is documented in all Brasiliano orogenic belts that surround the SFC, such as the Brasilia, Ribeira, Araçuaí, Sergipano, and Riacho do Pontal belts. Some lithostratigraphic units of this stage exhibit evidence of glacial origin, like the presence of diamictites and cap carbonates (Fig. 17.4), which could likely be attributed to the Cryogenian global ice age. The best example is the Macaubas basin-cycle, documented in the Araçuaí belt and São Francisco basin (Babinski et al. 2012 and references therein).

17.5 Modern-Type Plate Tectonics: São Francisco-Congo Incorporated into West Gondwana

During almost the entire Neoproterozoic, the oceanic basins around the São Francisco-Congo paleocontinent underwent a continued process of slab-pull subduction, that eventually led to the amalgamation of West Gondwana. For the São Francisco craton, the consumption of three oceanic realms is fundamental: the very large Goiás-Pharusian Ocean, the more restricted Adamastor Ocean and the minor Canindé oceanic domain (Fig. 17.5 and Chaps. 10–15). Magmatic arcs were registered and characterized within these oceanic domains, since the early Tonian, especially widespread in the Brasilia marginal belt and within the Ribeira-Araçuaí system (see Fuck et al., Chap. 11, Alkmim et al., Chap. 14, Heilbron et al., Chap. 15). The primitive arcs display in common juvenile signatures suggestive for intra-oceanic settings. Later, within all the marginal belts, the development of cordilleran arc successions is described, pointing to successive collisions and the progressive closing of the intervening oceanic spaces. In the Ediacaran, at ca. 600 Ma, several continental masses converged to form West Gondwana, and one of these is the SFC, still attached to the companion Congo craton. The advanced interaction of all continental masses resulted in the uplift of the Brasiliano/PanAfrican orogenic systems and configuration of the craton boundaries (Almeida 1977). The

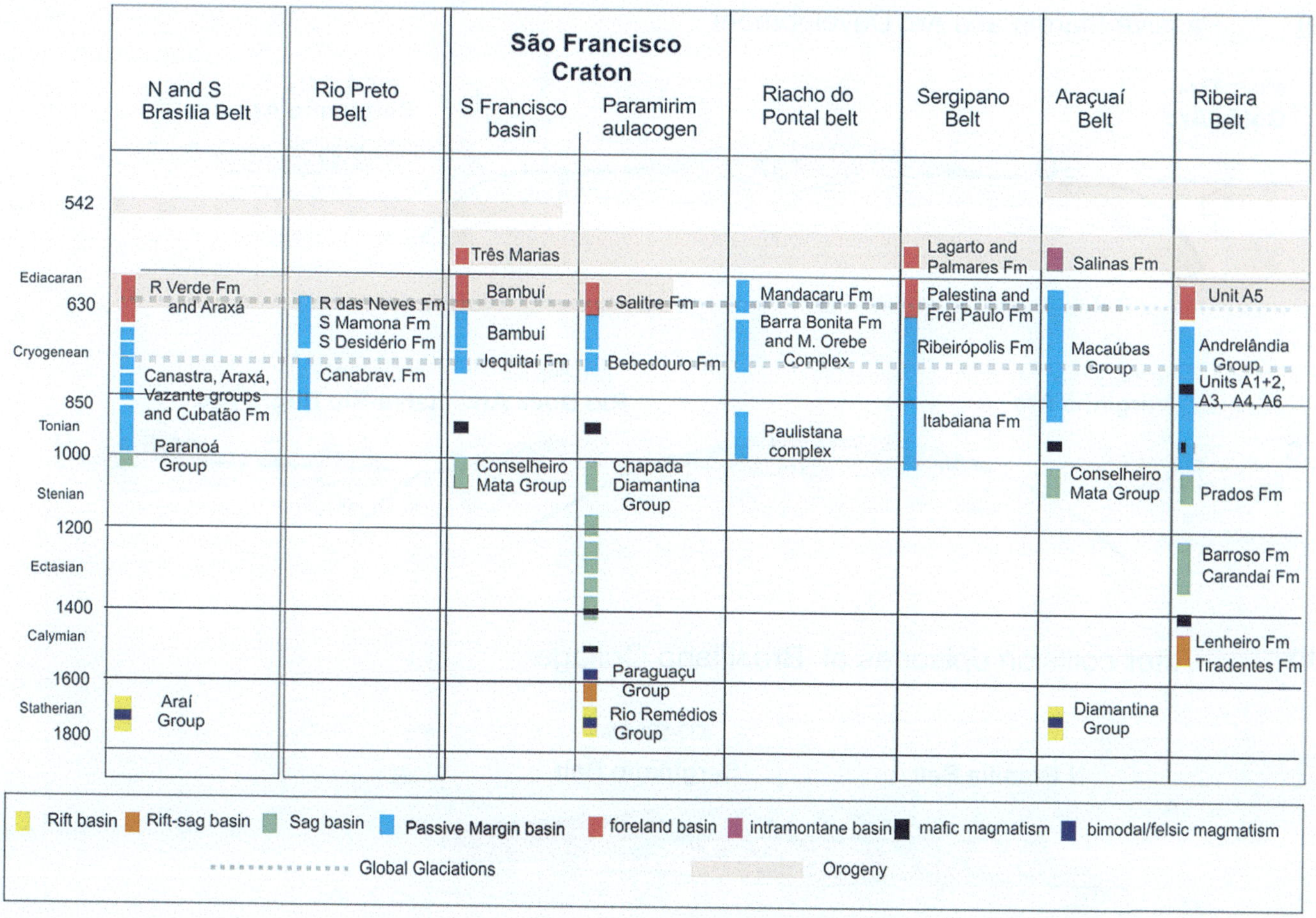

Fig. 17.4 Mesoproterozoic to Neoproterozoic schematic tectono-stratigraphic chart

oceanic realms, Goiás-Pharusian, Adamastor and Canindé were consumed by subduction-to-collision tectonic episodes.

The Goiás-Pharusian Ocean was closed after a long duration phase that started with the formation of a series of subduction-driven intraoceanic island arcs and was ended with the generation of the West Gondwana orogen (Ganade de Araujo et al. 2014), which is represented by the region affected by the Transbrasiliano-Kandi mega-shear zone. The final continental collision brought together, at one side the Amazonian and the West African cratonic masses, and at the other side the São Francisco-Congo, the Rio de La Plata and the Saharan counterparts. In the case of the SFC, as a result of this collision, the Brasilia orogenic belt was developed (see Valeriano, Chap. 10, and Fuck et al., Chap. 11).

On the Adamastor Ocean, the initial rifting of a region located on the São Francisco-Congo paleocontinent occurred in the late Tonian, at about 900 Ma. However, the rifting did not succeed in separating the São Francisco and the Congo cratonic elements, which remained attached through the Bahia-Gabon cratonic bridge. The main portion of the Adamastor occupied an extended oceanic region that later would become the site of the amalgamation of smaller cratonic masses and massifs, such as the Rio de La Plata, Paranapanema, Luiz Alves and Kalahari. With the closure of one of the terminal segments of the Adamastor, the Macaúbas basin, the Araçuaí and Ribeira orogenic belts were developed (see Alkmim et al., Chap. 14 and Heilbron et al., Chap. 15).

The Canindé restricted oceanic domain was closed by a continental collision that produced the tectonic structures of the Sergipano, Riacho do Pontal and Rio Preto belts (Caxito et al., Chap. 12, and Oliveira et al., Chap. 13). It was the result of the convergence of the SFC with the proto-Borborema Province, which must have initially acted as a coherent mass. During this episode parts of the Borborema were thrust onto the northern part of the craton along a zone of ca. 1000 km.

From the above discussion, we may infer that the Brasiliano belts surrounding the SFC are members of a tectonic collage, which concluded the convergence of a few cratonic elements of different size by a series of continental collisions, in the Ediacaran period. All the Brasiliano belts comprise tectonic-stratigraphic sections including some rift to passive margin-type deposits, with some oceanic crust

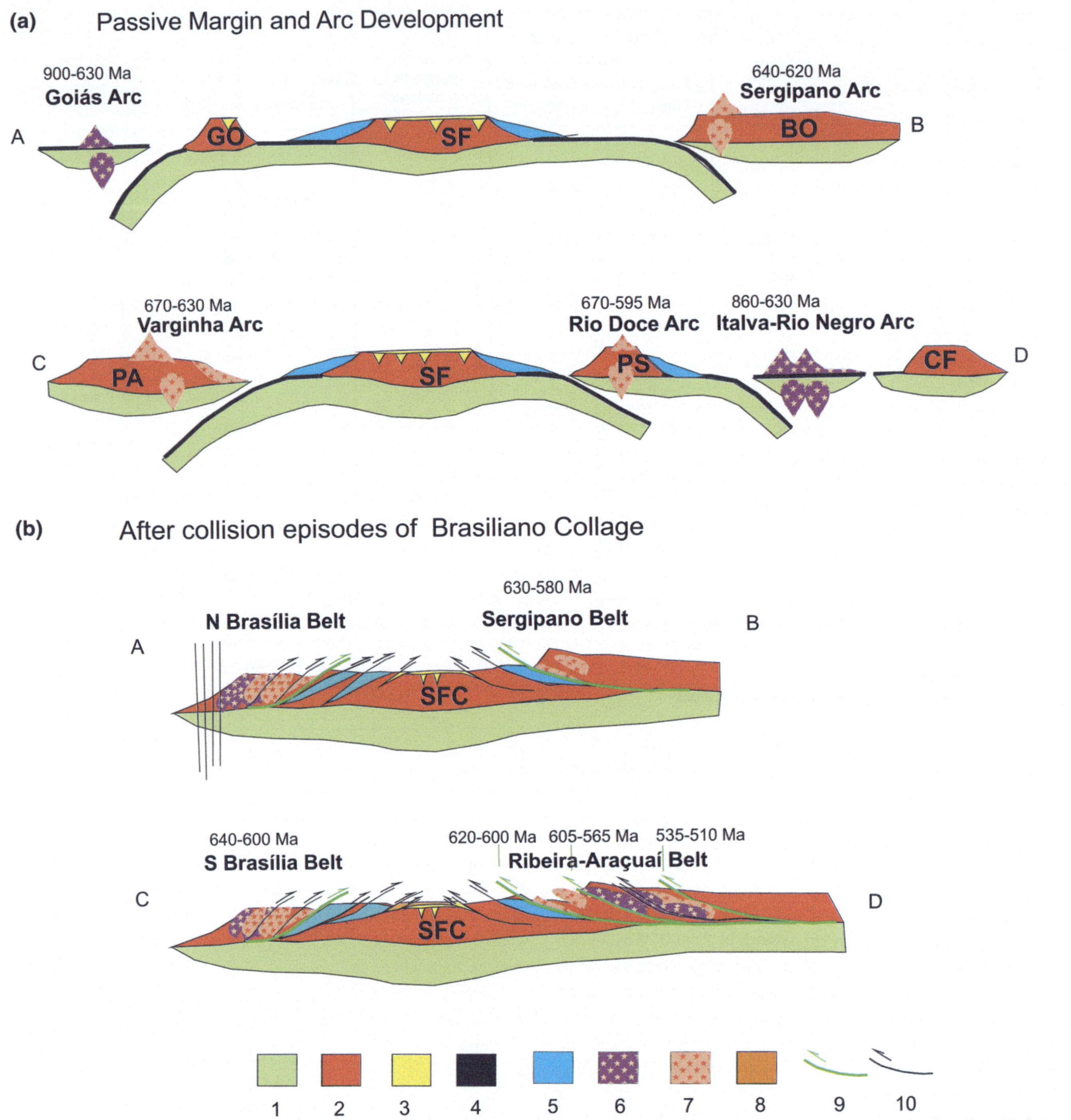

Fig. 17.5 Envisaged Neoproterozoic tectonic evolution the SFC and outboard terranes. *1* Lihospheric mantle, *2* continental crust, *3* oceanic crust, *4* rift to sag intracratonic basins, *5* passive margin basin, *6* juvenile arc and related basins, *7* cordilleran arc and related basins, *8* foreland to successor basins, *9* main sutures, *10* subordinated thrusts GO-Goiást, SF-São Francisco, BO-Borborema, PS-Paraíba do Sul, CF-Cabo Frio

fragments and turbiditic successions linked to coeval magmatic arcs, formed usually between ca. 900 and ca. 500 Ma. Figure 17.5 presents a few schematic profiles illustrating the Neoproterozoic evolution of the belts around the SFC. The Brasiliano fold and thrust belts now expose rocks assemblages metamorphosed under low-to-medium grade, but areas of high-grade metamorphism also occur in their internal portions. Tectonically reworked basement inliers are quite common, and those of Paleoproterozoic age are very frequent (Chaps. 10–15 for a detailed review).

To the west of the SFC, the collapse of the Goiás-Pharusian Ocean and the convergence of the Goiás magmatic arc affected the passive margin of the craton. The collisional phases occurred between 650 and 560 Ma. They involved exotic allochthonous terranes like the Central Goiás Massif and produced a thrust-and-fold belt comprising a series of eastern vergence nappes that were emplaced over the craton (Fuck et al., Chap. 11). Moreover, in the final stages of the West Gondwana assembly, in the northern Brasilia belt, a second contractional deformation phase was produced along the preexistent southern Brasilia belt and the foreland sequences on the craton interior became involved in the deformation.

To the north of the SFC, the most important collisional episode took place between ca. 605–565 Ma and was responsible for the generation of the Sergipano, Riacho do Pontal and Rio Preto belts, with the consumption of the Canindé oceanic domain, finally shaping the present boundaries of the SFC (Caxito et al., Chap. 12 and Oliveira et al., Chap. 13). The deformational front invaded the border of the São Francisco continent, involving the previous Mesoproterozoic intracratonic basin in the deformation.

To the south of the CSF, the earliest Ediacaran orogenic event represents the Paranapanema-São Francico/Congo collision, responsible for the uplift of the southern Brasília metamorphic belt around 630 Ma (see Valeriano, Chap. 10). After this event, a major transgression affected the São Francisco peninsula, which started to behave as a downwarp basin, where the mixed carbonatic-siliciclastic sediments of the Bambuí 1st-order sequence started to accumulate after 610 Ma (Fig. 17.4).

At the eastern and southeastern corners of the SFC, within the Araçuaí and Ribeira belts, the tectono-magmatic evolution was similar, and pre, syn and late-collisional granitoid rocks were formed in continental magmatic arcs, in successive pulses between 630 and 530 Ma (Chaps. 14 and 15). The Araçuaí orogenic front propagated towards the SFC thereby affecting the Bambuí strata and causing partial inversion of the Pirapora aulacogen within the foreland (Reis et al., Chap. 8). Within the Ribeira belt, cordilleran arc systems developed at the margins of Paleoproterozoic microcontinents and collided to the E-SE margin of the craton. Moreover, the subducted passive margin sediments of the Andrelândia basin underwent HP metamorphism and intense deformation that resulted on the nappe stacking at the Ribeira belt. Finally, at the easternmost Ribeira belt, a Cambrian collisional episode was responsible for the docking of the Cabo Frio terrane, and its deformation front reached the previously amalgamated terranes (Heilbron et al., Chap. 15).

The Neoproterozoic marginal belts of the São Francisco craton were developed within the "modern-type slab-pull subduction-to-collision tectonics regime" proposed by Brown (2008). A good indication is the presence of HP metamorphic terrains in some of the belts, like the Ribeira or the southern Brasilia (Chaps. 10 and 15 of Valeriano and Heilbron et al.). In the latter, coesite inclusions within zircon and garnet from a kyanite-garnet metamorphic UHP rock were described by Parkinson et al. (2001). Retro-eclogites and HP metamorphic rocks were described by a few authors for the Brasília belt and its interference zone with the Ribeira belt (Campos Neto and Caby 2004; Heilbron et al. 2008; Trouw et al. 2013 and Chaps. 14 and 15).

After the termination of the tectonic pulses within the marginal belts, the tectonically stable interior of the SFC subsided in response to the orogenic loads developed along its margins. It received sediments shed from the newly uplifted areas around and was caught in some places by the compressional deformation fronts (see Reis et al., Chap. 5). From its residence in West Gondwana and Pangea, the São Francisco basin only preserves the Santa Fé glaciogenic sediments, which accumulated when the supercontinent wandered along polar latitudes.

17.6 Breakup of the São Francisco-Congo Cratonic Bridge

The breakup of West Gondwana in the Lower Cretaceous, as one of the main outcomes of the current plate tectonics regime, resulted in the opening of the South Atlantic Ocean and separation of the São Francisco and Congo cratons (Porada 1989; Pedrosa-Soares et al. 2001; Gordon et al. Chap. 9). The main segments of the large rift system developed in the initial phase of this process evolved to the present-day passive and transform margins of South America and Africa. Other branches of the system, such as the Abaeté graben in the São Francisco basin, failed and became important depocenters in the interior of both continents (Reis et al., Chap. 7). On the other side, important intraplate stress rearrangements in the course of the South Atlantic event caused the uplift of the Alto Paranaíba arch, whose ascension most likely started prior to the Cretaceous, being successively reactivated afterwards.

Later, in the Upper Cretaceous, the Alto Paranaíba arch experienced a new reactivation, associated with the extensive intraplate magmatic episode that culminated with the intrusion of NW-trending dykes, alkaline plutons, and the extrusion of the kamafugitic lavas and pyroclastics of the Mata da Corda Group. This magmatic episode may have occurred in response to the action of a mantle plume beneath the craton border in the Late Cretaceous (see Chap. 7).

In the last 100 Ma, South America and Africa drifted away and the Atlantic Ocean reached its largest wideness. The São Francisco and Congo cratons, that were good companions and drifted together at the surface of the Earth for about 2.0 Ga, are now more than 5000 km apart. Most likely, separation will be forever.

17.7 Conclusions

In this chapter a brief description of the tectonic evolution of the miniature São Francisco continent, for a few time-intervals, was made. Although small in size, we have seen that this tectonic unit exhibits many features of some of the principal phases of the Earth global geodynamics.

(1) For the Archean, in the northern and southern segments of the craton, granite-greenstone terrains are found in a few different places. They comprise TTG-type granitoids of different ages, formed from mantle plumes, deformed and associated to greenstone belts. Like in many other Archean terrains of the world, they amalgamated to form a coherent and stable continental mass in the late Archean.

(2) For the ca. 2.0 Ga time slice of the Paleoproterozoic, a few relevant subduction-driven accretionary belts, including juvenile island arcs and allochthonous terranes, within the Mineiro and Eastern Bahia orogens, were added to the Archean continental crust. They are typical of the Proterozoic plate tectonic regime of Brown (2008) and were formed near the important peak in crustal production indicated by Condie (2000).

(3) In the following long period of more than 1 billion years, orogenic activity was not identified in the São Francisco craton. It was united to the Congo craton, within the Central African block, and possibly it did not take part of the Columbia and Rodinia supercontinents. During this time slice it was affected by a series of important intraplate processes, such as anorogenic magmatism and the formation of aulacogenic-type basins.

(4) Later, in Ediacaran to Cambrian times, the São Francisco craton, within the African block, took part in the formation of the Gondwana Supercontinent by means of the collage of practically coeval subduction-to-collision orogenic belts, as typical examples of the slab-pull subduction processes typical of the modern-type plate tectonics regime. The Rio Preto, Riacho do Pontal, Sergipano, Araçuaí-Ribeira and Brasilia belts surrounded completely the SFC.

(5) Finally, with the disruption of Pangea supercontinent in Meso-Cenozoic time, the São Francisco craton was finally separated from its counterpart, the Congo craton, by means of the formation of the South Atlantic Ocean, as a typical element of the present-day plate tectonics regime.

Acknowledgments The authors wish to acknowledge several Brazilian institutions, such as CNPq, FAPERJ, FAPESP, FAPEMIG, Petrobrás and others, for funding our research in the last three decades and also to thank Springer that brought to us such a pleasure and challenging task.

References

Alkmim, F.F. and Martins-Neto, M.A., 2012. Proterozoic first-order sedimentary sequences of the São Francisco Craton, eastern Brazil. Mar. Pet. Geol. 33, 127–139.

Alkmim, F.F. and Marshak, S., 1998. Transamazonian Orogeny in the Southern São Francisco Craton Region, Minas Gerais, Brazil: evidence for Paleoproterozoic collision and collapse in the Quadrelátero Ferrífero. Precambrian Research, 90, 29–58

Almeida FFM (1977) O Cráton do São Francisco. Revista Brasileira de Geociências 7(4): 349–364.

Almeida, F.F.M. (1981) O cráton do Paramirim e suas relações com o do São Francisco. Simpósio sobre o Cráton do São Francisco e suas Faixas Marginais. Salvador, 1981. Anais da SBG-Núcleo Bahia, p.1-10.

Babinski, M., Pedrosa-Soares, A.C., Trindade, R.I.F., Martins, M., Noce, C.M., Liu, D., 2012. Neoproterozoic glacial deposits from the Araçuaí orogen, Brazil: Age, provenance and correlations with the São Francisco Craton and West Congo belt. Gondwana Research 21: 451–456

Barbosa, J.S.F., Sabaté, P., 2004. Archean and Paleoproterozoic crust of the São Francisco Craton, Bahia, Brazil: geodynamic features. Precambrian Research, 133:1–27.

Brito-Neves B.B. de, Campos-Neto M.C. da, Fuck R.A. (1999) From Rodinia to Western Gondwana: An approach to the Brasiliano-Pan African Cycle and orogenic collage. Episodes 22(3): 155–166.

Brown, M. 2008. Characteristic thermal regimes of plate tectonics and their metamorphic imprint throughout Early history: When did Earth first adopt a plate tectonics mode of behaviour?. In: When Did Plate Tectonics begin on Planet Earth? GSA Special Paper 440, 97–128.

Caby, R.1994. Precambrian Coesite from northern Mali: First record and implications for plate tectonics in the Trans-Sahara segment of the Pan-African belt. European Journal of Mineralogy 6: 235–244.

Campos Neto, M.C & Caby, R. 2004. Lower crust extrusion and terrane accretion in the Neoproterozoic nappes of southeast Brazil. Tectonics, 19, 669–687

Cawood, P.A. & Pisarewsky, A. 2006. Precambrian Plate Tectonics: Criteria and Evidence. GSA Today 16#7:4–11.

Condie, K. C. 2000. Episodic continental growth model- Afterthoughts and extensions. Tectonophysics 322: 153–162.

Condie, K.C. & Kroener, A. 2008. When Did Plate Tectonics begin: Evidence from the geological record. In: When Did Plate Tectonics begin on Planet Earth? GSA Special Paper 440, 281–294.

Condie, K.C., O'Neill, C., Aster, R.C. 2009. Evidence and implications for a widespread magmatic shutdown for 250 My on Earth. Earth and Planetary Science Letters 282: 294–298.

Cordani U.G., Pimentel M.M., Araújo C.E.G., Fuck R.A. 2013. The Significance of the Transbrasiliano-Kandi Tectonic Corridor for the Amalgamation of West Gondwana. Brazilian Journal of Geology 43 (3), 583–597

Farina, F., Albert, C., Martínez-Dopico, C. Aguilar Gil, C., Moreira, H., Hippertt, J.P. Cutts, K., Alkmim, F.F., Lana, C. 2015. The Archean-Paleoproterozoic evolution of the Quadrilátero Ferrífero (Brasil)- Current models and open questions. Journal of South American Earth Sciences 68: 4–21.

Feybesse J.L, Johan V, Triboulet C, Guerrot C, Mayaga-Mikolo F, Bouchot V, Eko N'dong J. 1998. The West Central African belt: a model of 2.5–2.0 Ga accretion and two-phase orogenic evolution. Precambrian Research, 87, 161–216

Ganade de Araujo, C.E., Rubatto, D., Hermann, J., Cordani, U., Caby, R., Basei, M.A.S. 2014. Ediacaran 2,500-km-long synchronous deep continental subduction in the West Gondwana Orogen. Nature Communications: 5:5198:1–7

Heilbron M, Valeriano C.M., Tassinari C.C.G., Almeida J.C.H., Tupinamba M, Siga, O., Trouw R. 2008. Correlation of Neoproterozoic terranes between the Ribeira Belt, SE Brazil and its African counterpart: comparative tectonic evolution and open questions. Geological Society of London, Special Publication 294

Heilbron M, Duarte B.P., Valeriano C.M., Simonetti A, Machado N, Nogueira, J.R. (2010) Evolution of reworked Paleoproterozoic basement rocks within the Ribeira belt (Neoproterozoic), SE-Brazil, based on U-Pb geochronology: Implications for paleogeographic reconstructions of the São Francisco-Congo paleocontinent. Precambrian Research, v. 178:136–148

Komiya, T., Hayahi, M., Yurimoto, H. 2002. Intermediate-P-type Archean metamorphism of the Isua supracrustal belt: Implications for secular change of geothermal gradients at subduction zones and for Archean plate tectonics. American Journal of Science 302: 806–826.

Lana, C., Alkmim, F.F., Armstrong, R., Scholz, R., Romano, R., Nalini Jr., H.A., 2013. The ancestry and magmatic evolution of Archaean TTG rocks of the Quadrilátero Ferrífero province, southeast Brazil. Precambrian Research 231: 157–173.

Ledru, P.J., Johan, V., Milési, J. P., Teguy, M. 1994. Markers of the last stage of the Paleoproterozoic collision: Evidence for 2 Ga continent involving circum-South Atlantic provinces. Precambrian Research, 69 :169–191.

Moreira, H., Lana, C. & Nalini, H.A.Jr., 2016. The detrital zircon record of an Archaean convergent basin in the Southern São Francisco Craton, Brazil. Precambrian Research, 272, 84–99.

Moyen, J., Stevens, G., Kisters, A. 2006. Record of mid-Archean subduction from metamorphism in the Barbeton terrain, South Africa. Nature 442:559–562

Nance, R.D. & Murphy, J. B. 2013. Origins of supercontinent cycle. Geoscience Frontiers 4(4): 439–448.

Nance R.D., Murphy J.B., Santosh M. 2013. The supercontinent cycle: A retrospective essay: Gondwana Research, 25: 4–29.

Noce, C; Pedrosa-Soares, A.C.; Silva, L.C; Armstrong, R; Piuzana, D 2007. Evolution of polycyclic basement complexes in the Araçuaí Orogen, based on U-Pb SHRIMP data: Implications for Brazil Africa links in Paleoproterozoic time?. Precambrian Research 159: 60–78, 2007.

Rogers, J.J.W. & Santosh, M. 2004. Continents and supercontinents. Oxford Press, New York, 289p.

Parkinson, C.D., Motoki, A., Onishi, C.T., Maruyama, S. 2001. Ultrahigh-pressure pyrope-kyanite granulites and associated eclogites in Neoproterozoic nappes of southeast Brazil. UHPM Workshop, Waseda University, Japan. Extended Abstract, 87-90.

Rogers, J.J.W. & Santosh, M., 2009. Tectonics and surface effects of the supercontinent Columbia. Gondwana Research 15, 373–380.

Stern, R.J. 2005. Evidence from ophiolites, blueschists and ultrahigh metamorphic terranes that the modern episode of subduction tectonics began in Neoproterozoic time. Geology 33:557–560.

Tarney, J., Weaver, B., Drury, S.A. 1979. Geochemistry of Archean trondhjemitic and tonalitic gneisses from Scotland and Greenland: In Barker, F. ed, Trondhjemites, dacites and related rocks. Elsevier: 275–299.

Teixeira W, Ávila, C.A., Dussin, I.A., Corrêa Neto, A.V., Bongiolo, E.M., Santos, J.O.S., Barbosa, N. 2014. Zircon U-Pb-Hf, Nd-Sr constraints and geochemistry of the Resende Costa orthogneiss and coeval rocks: new clues for a juvenile accretion episode (2.36–2.33 Ga) in the Mineiro belt and its role to the long-lived Minas accretionary orogeny. Precambrian Research, 256: 148–169

Teles, G., Chemale Jr., F., Oliveira, C.G. 2015. Paleoarchean record of the detrital pyrite-bearing Jacobina Au–U deposits, Bahia, Brazil. Precambrian Research 256 (2015) 289–313

Tohver, E., D'Agrella-Filho, M.S., Trindade, I.F., 2006. Paleomagnetic record of Africa and South America for the 1200–500 Ma interval, and evaluation of Rodinia and Gondwana assemblies. Precambrian Research 147, 193-222.

Trouw R.A.J., Peternel R., Ribeiro, A., Heilbron, M., Vinagre, R., Duffles, P., Trouw, C.C., Fontainha, M., Kussama, H.H. 2013. A New Interpretation for the Interference Zone between the southern Brasília belt and the central Ribeira belt, SE Brazil. Journal of South American Earth Sciences 48: 43–57

Zhao, G. C., Sun, M., Wilde, S.A. 2004. A paleo-Mesoproterozoic supercontinent: assembly, growth and breakup. Earth Science Reviews 67:91–123.

The manufacturer's authorised representative in the EU is Springer Nature Customer Service Centre GmbH, Europaplatz 3, 69115 Heidelberg, Germany. If you have any concerns regarding our products, please contact ProductSafety@springernature.com

Printed and bound by CPI Group (UK) Ltd, Croydon, CR0 4YY
05/10/2026
02228163-0003